Springer Series in Chemica

MW00760958

Volume 111

Series editors

Albert W. Castleman, University Park, USA
J. Peter Toennies, Göttingen, Germany
Kaoru Yamanouchi, Tokyo, Japan
Wolfgang Zinth, München, Germany

The purpose of this series is to provide comprehensive up-to-date monographs in both well established disciplines and emerging research areas within the broad fields of chemical physics and physical chemistry. The books deal with both fundamental science and applications, and may have either a theoretical or an experimental emphasis. They are aimed primarily at researchers and graduate students in chemical physics and related fields.

More information about this series at http://www.springer.com/series/676

Wolfgang Becker
Editor

Advanced Time-Correlated Single Photon Counting Applications

 Springer

Editor
Wolfgang Becker
Becker & Hickl GmbH
Berlin
Germany

ISSN 0172-6218
Springer Series in Chemical Physics
ISBN 978-3-319-35842-0   ISBN 978-3-319-14929-5   (eBook)
DOI 10.1007/978-3-319-14929-5

Springer Cham Heidelberg New York Dordrecht London

Printed on acid-free paper

Springer International Publishing AG Switzerland is part of Springer Science+Business Media (www.springer.com)

Preface

Time-correlated single photon counting (TCSPC) was born 54 years ago, in 1960, when Bollinger and Thomas published a paper on the 'Measurement of Time Dependence of Scintillation Intensity by a Delayed Coincidence Method'. Desmond O'Oonnor and David Phillips published their book, *Time-Correlated Single Photon Counting*, 31 years ago. *Advanced Time-Correlated Single Photon Counting Techniques* appeared 10 years ago.

Can a measurement technique survive for more than half a century? Yes, if it is constantly developing. And there has been a lot of development in TCSPC: The electronics have become orders of magnitude faster, and high-repetition-rate lasers and reversed start-stop have resulted in dramatically shorter acquisition times. Faster detectors are now available and fast excitation sources are more affordable.

A big step forward was the introduction of multidimensional TCSPC in 1993. Optical signals could now be observed not only as a function of the time after an excitation pulse but simultaneously as multidimensional functions of wavelength, polarization, experiment time, spatial parameters, or any other physical quantity describing the momentary state of the measurement object. This paved the way for another breakthrough, the combination of TCSPC with microscopy. Fluorescence lifetime measurements could now be performed in biological tissue, in single cells, and even in single biomolecules. The fluorescence lifetime, formerly considered simply the time a molecule stays in the excited state and possibly undergoes photo-induced reactions, became an indicator of the molecular environment of the molecule.

Techniques based on FRET and other conformation-dependent effects have been developed to investigate the function of biological systems. The combination of spatial resolution on the microscopic and submicroscopic scale with molecular information is one of the big advances of the past several decades, recognized by the award of the 2014 Nobel prize to Eric Betzig, Stefan W. Hell, and William E. Moerner.

Microscopy has created another level of multidimensionality. Histograms of fluorescence parameters can be built up over a large number of pixels in a fluorescence lifetime image or over a large number of molecules that have diffused through a femtoliter observation volume. These data reveal subpopulations of biomolecules, subpopulations of molecules of different conformations, and transitions between different conformations.

High sensitivity, high time resolution, and the ability of TCSPC to obtain multiple biomedically relevant parameters from a single measurement have led to a variety of clinical applications. These range from fluorescence lifetime measurements on the spatial scale of single cells, to imaging of millimeter- and centimeter-size areas, to diffuse optical imaging techniques that reveal biochemical information from deep inside living tissue.

The complexity of these applications has, however, also created a problem: Understanding advanced TCSPC requires complex, almost magical, thinking, and a solid understanding not only of the TCSPC technique but also of modern microscopy techniques and other optical methods. There is a similar situation in J.K. Rowling's Harry Potter stories (I recommend them as supplementary literature): There are people who can do magic (witches and wizards) and normal people (Muggles) who cannot understand or even see magic when it happens. But there is hope: Hermione Granger was Muggle-born and became the best of her year at Hogwarts.

This book should be considered a continuation of W. Becker, *Advanced Time-Correlated Single Photon Counting Techniques*, Springer (2005). It is an attempt to spread the ideas of advanced time-resolved optical techniques more widely in the scientific community. It contains both chapters about the basics of multidimensional TCSPC and about applications in biology and medicine. The chapters are written by an originator of the technique and by successful users. Our hope is that this combination helps potential users better understand the technique, its various implementations, and encourages its adoption in their own research.

Berlin   Wolfgang Becker

Contents

Klaus Suhling, Liisa M. Hirvonen, James A. Levitt,
Pei-Hua Chung, Carolyn Tregido, Alix le Marois,
Dmitri A. Rusakov, Kaiyu Zheng, Simon Ameer-Beg,
Simon Poland, Simon Coelho and Richard Dimble

Contributors

Simon Ameer-Beg Randall Division of Cell and Molecular Biophysics, King's College London, London, UK

Jillian Bartusek Department of Chemistry and Biochemistry, College of Pharmacy, University of Minnesota Duluth, Duluth, MN, USA

Wolfgang Becker Becker & Hickl GmbH, Berlin, Germany

Osman Bilsel Department of Biochemistry and Molecular Pharmacology, University of Massachusetts Medical School, Worcester, MA, USA

Paul S. Blank Eunice Kennedy Shriver National Institute of Child Health and Human Development, Bethesda, MD, USA

Michael Börsch Single-Molecule Microscopy Group, Jena University Hospital, Friedrich Schiller University Jena, Jena, Germany

Gianluca Boso Dipartimento di Elettronica, Informazione e Bioingegneria, Politecnico di Milano, Milan, Italy

Kathryn G. Christopher University of Virginia, Charlottesville, VA, USA

Pei-Hua Chung Department of Physics, King's College London, London, UK

Simao Coelho Randall Division of Cell and Molecular Biophysics, King's College London, London, UK

Davide Contini Dipartimento di Fisica, Politecnico di Milano, Milan, Italy

Rinaldo Cubeddu Dipartimento di Fisica, Politecnico di Milano, Milan, Italy

Alberto Dalla Mora Dipartimento di Fisica, Politecnico di Milano, Milan, Italy

Richard N. Day Departments of Medicine and Cell Biology, University of Virginia, Charlottesville, VA, USA

Hugues de Rocquigny UMR 7213 CNRS, Laboratoire de Biophotonique et Pharmacologie, Faculté de Pharmacie, Illkirch, France

Laura Di Sieno Dipartimento di Fisica, Politecnico di Milano, Milan, Italy

Pascal Didier UMR 7213 CNRS, Laboratoire de Biophotonique et Pharmacologie, Faculté de Pharmacie, Illkirch, France

Richard Dimble Department of Cancer, Research, Division of Cancer Studies, New Hunt's, House, Guy's Campus, King's College London, London, UK

Ruslan I. Dmitriev School of Biochemistry and Cell Biology, University College Cork, Cork, Ireland

Christoph Fahlke Institute of Complex Systems 4 (ICS-4, Cellular Biophysics), Forschungszentrum Jülich, Jülich, Germany

Arne Franzen Institute of Complex Systems 4 (ICS-4, Cellular Biophysics), Forschungszentrum Jülich, Jülich, Germany

Samuel Frere Department of Physiology and Pharmacology, Sackler School of Medicine, Tel Aviv University, Tel Aviv, Israel

Katja Fuchs Physikalisch-Technische Bundesanstalt (PTB), Berlin, Germany

Thomas Gensch Institute of Complex Systems 4 (ICS-4, Cellular Biophysics), Forschungszentrum Jülich, Jülich, Germany

Anna Gerega Polish Academy of Sciences, Institute of Biocybernetics and Biomedical Engineering, Warsaw, Poland

Martin Hammer Klinik für Augenheilkunde, Friedrich-Schiller-Universität Jena, Jena, Germany

Isha N. Haridass Therapeutics Research Centre, Princess Alexandra Hospital, Woolloongabba, QLD, Australia

Ahmed A. Heikal Department of Chemistry and Biochemistry, College of Pharmacy, University of Minnesota Duluth, Duluth, MN, USA

Liisa M. Hirvonen Department of Physics, King's College London, London, UK

James Jenkins School of Biochemistry and Cell Biology, University College Cork, Cork, Ireland

Michal Kacprzak Polish Academy of Sciences, Institute of Biocybernetics and Biomedical Engineering, Warsaw, Poland

Sagar V. Kathuria Department of Biochemistry and Molecular Pharmacology, University of Massachusetts Medical School, Worcester, MA, USA

Karsten König Department of Biophotonics and Laser Technology, Saarland University, Saarbruecken, Germany

Peter Kovermann Institute of Complex Systems 4 (ICS-4, Cellular Biophysics), Forschungszentrum Jülich, Jülich, Germany

Alix le Marois Department of Physics, King's College London, London, UK

James A. Levitt Department of Physics, King's College London, London, UK

Adam Liebert Polish Academy of Sciences, Institute of Biocybernetics and Biomedical Engineering, Warsaw, Poland

Rainer Macdonald Physikalisch-Technische Bundesanstalt (PTB), Berlin, Germany

Roman Maniewski Polish Academy of Sciences, Institute of Biocybernetics and Biomedical Engineering, Warsaw, Poland

Alzbeta Marcek Chorvatova Department of Biophotonics, International Laser Centre, Bratislava, Slovakia

Nirmal Mazumder University of Virginia, Charlottesville, VA, USA

Mikhail Mazurenka Physikalisch-Technische Bundesanstalt (PTB), Berlin, Germany

Yves Mély UMR 7213 CNRS, Laboratoire de Biophotonique et Pharmacologie, Faculté de Pharmacie, Illkirch, France

Daniel Milej Polish Academy of Sciences, Institute of Biocybernetics and Biomedical Engineering, Warsaw, Poland

Tuan A. Nguyen Division of Intramural Clinical and Biological Research, National Institutes of Health, National Institute of Alcohol Abuse and Alcoholism, Bethesda, MD, USA

Dmitri B. Papkovsky Laboratory of Biophysics and Bioanalysis, School of Biochemistry and Cell Biology, University College Cork, Cork, Ireland

Michael Pastore Division of Health Sciences, University of South Australia, School of Pharmacy and Medical Science, Adelaide, SA, Australia

Ammasi Periasamy W.M. Keck Center for Cellular Imaging, University of Virginia, Charlottesville, VA, USA

Antonio Pifferi Dipartimento di Fisica, Politecnico di Milano, Milan, Italy

Simon Poland Randall Division of Cell and Molecular Biophysics, King's College London, London, UK

Giovanna Quarto Dipartimento di Fisica, Politecnico di Milano, Milan, Italy

Ludovic Richert UMR 7213 CNRS, Laboratoire de Biophotonique et Pharmacologie, Faculté de Pharmacie, Illkirch, France

Michael S. Roberts Therapeutics Research Centre, Princess Alexandra Hospital, Woolloongabba, QLD, Australia

Dmitri A. Rusakov Laboratory of Synaptic Imaging, Department of Clinical and Experimental, Epilepsy, University College London, London, UK

Washington Y. Sanchez Therapeutics Research Centre, Princess Alexandra Hospital, Woolloongabba, QLD, Australia

Piotr Sawosz Polish Academy of Sciences, Institute of Biocybernetics and Bio-medical Engineering, Warsaw, Poland

Dietrich Schweitzer Klinik für Augenheilkunde, Friedrich-Schiller-Universität Jena, Jena, Germany

Amy T. Shah Department of Biomedical Engineering, Vanderbilt University, Nashville, TN, USA

Joe T. Sharick Department of Biomedical Engineering, Vanderbilt University, Nashville, TN, USA

Vladislav Shcheslavskiy Becker & Hickl GmbH, Berlin, Germany

Melissa C. Skala Department of Biomedical Engineering, Vanderbilt University, Nashville, TN, USA

Inna Slutsky Department of Physiology and Pharmacology, Sackler School of Medicine, Tel Aviv University, Tel Aviv, Israel

Lorenzo Spinelli Dipartimento di Fisica, Politecnico di Milano, Milan, Italy

Hauke Studier Becker & Hickl GmbH, Berlin, Germany

Klaus Suhling Department of Physics, Experimental Biophysics and Nanotech-nology, King's College London, London, UK

Yuansheng Sun University of Virginia, Biology, Charlottesville, VA, USA

Paola Taroni Dipartimento di Fisica, Politecnico di Milano, Milan, Italy

Randi Timerman Department of Chemistry and Biochemistry, College of Phar-macy, University of Minnesota Duluth, Duluth, MN, USA

Alessandro Torricelli Dipartimento di Fisica, Politecnico di Milano, Milan, Italy

Alberto Tosi Dipartimento di Elettronica, Informazione e Bioingegneria, Politecnico di Milano, Milan, Italy

Carolyn Tregido Department of Physics, King's College London, London, UK

Verena Untiet Institute of Complex Systems 4 (ICS-4, Cellular Biophysics), Forschungszentrum Jülich, Jülich, Germany

B. Wieb van der Meer Western Kentucky University, Department of Physics and Astronomy, Kentucky, USA

Steven S. Vogel Division of Intramural Clinical and Biological Research, National Institutes of Health, National Institute of Alcohol Abuse and Alcoholism, Bethesda, MD, USA

Heidrun Wabnitz Physikalisch-Technische Bundesanstalt (PTB), Berlin, Germany

Alex J. Walsh Department of Biomedical Engineering, Vanderbilt University, Nashville, TN, USA

Dhanushka Wickramasinghe Department of Chemistry and Biochemistry, University of Minnesota Duluth, Duluth, MN, USA

Kaiyu Zheng Laboratory of Synaptic Imaging, Department of Clinical and Experimental, Epilepsy, University College London, London, UK

Chapter 1
Introduction to Multi-dimensional TCSPC

Wolfgang Becker

Abstract Classic time-correlated single photon counting (TCSPC) detects single photons of a periodic optical signal, determines the times of the photons relative to a reference pulse, and builds up the waveform of the signal from the detection times. The technique achieves extremely high time resolution and near-ideal detection efficiency. The modern implementation of TCSPC is multi-dimensional. For each photon not only the time in the signal period is determined but also other parameters, such as the wavelength of the photons, the time from the start of the experiment, the time after a stimulation of the sample, the time within the period of an additional modulation of the excitation light source, spatial coordinates within an image area, or other parameters which can either vary randomly or are actively be modulated in the external experiment setup. The recording process builds up a photon distribution over these parameters. The result can be interpreted as a (usually large) number of optical waveforms for different combination of the parameters. The advantage of multi-dimensional TCSPC is that the recording process does not suppress any photons, and that it works even when the parameters vary faster than the photon detection rate. Typical multi-dimensional TCSPC implementations are multi-wavelength recording, recording at different excitation wavelengths, time-series recording, combined fluorescence and phosphorescence decay recording, fluorescence lifetime imaging, and combinations of these techniques. Modern TCSPC also delivers parameter-tagged data of the individual photons. These data can be used to build up fluorescence correlation and cross-correlation spectra (FCS and FCCS), to record fluorescence data from single molecules, or to record time-traces of photon bursts originating from single molecules diffusing through a small detection volume. These data are used to derive multi-dimensional histograms of the changes in the fluorescence signature of a single molecules over time or over a large number of different molecules passing the detection volume. The chapter describes the technical principles of the various multi-dimensional TCSPC configurations and gives examples of typical applications.

W. Becker (✉)
Becker & Hickl GmbH, Berlin, Germany
e-mail: becker@becker-hickl.com

© Springer International Publishing Switzerland 2015
W. Becker (ed.), *Advanced Time-Correlated Single Photon Counting Applications*,
Springer Series in Chemical Physics 111, DOI 10.1007/978-3-319-14929-5_1

1

1.1 Time-Resolved Optical Detection at Low Light Intensity

When light propagates through matter it can interact with the molecules in various ways. In the simplest case, the photons of the incident light may be reflected, absorbed, or scatted. When photons are absorbed the molecules within a sample enter an excited state from which they may return by emitting a photon of longer wavelength. At high intensities several photons may interact with the molecules simultaneously, resulting in nonlinear effects like multiphoton excitation, or second second-harmonic generation. These interactions change the intensity, the spectral properties, and the temporal properties of the light. Light transmitted trough, scattered in, or otherwise emitted from a sample thus carries information on molecular parameters. In other words, optical techniques can be used to probe molecular parameters inside a sample.

Often the effects involved in the conversion of the light are extremely weak, or the concentration of the molecules involved in the conversion processes is low. Moreover, the light intensity tolerated by the sample may be limited. As a result, the intensity of light signals to be detected may be very low. The situation is further complicated when signals are recorded at high time resolution. Optical signals have to be considered a stream of photons. At a given photon rate, the signal-to-noise ratio decreases with the square root of the detection bandwidth. A typical situation is shown in Fig. 1.1. An optical signal was detected by a photomultiplier tube at bandwidth of approximately 350 MHz, or a time resolution of 1 ns. The detected photon rate was approximately 10^7 photons/s. This is much higher than typically obtained in molecular imaging experiments in live sciences. As can be seen from Fig. 1.1, the signal is a random trace of extremely short pulses. The pulses represent the detection of single photons of the light arriving at the detector.

Now consider the detection of a fast optical waveform, such as a fluorescence decay excited by short laser pulses. The fluorescence decay time is on the order of a

Fig. 1.1 Optical signal detected at a time resolution of 1 ns. Detected photon rate 10^7 photons/s

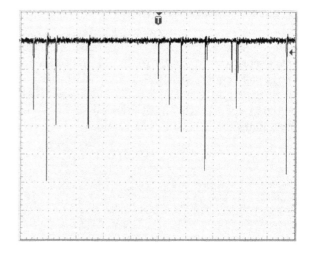

few nanoseconds. At first glance, it seems impossible to retrieve the signal shape from the signal shown in Fig. 1.1. Nevertheless, there is a solution to the problem: Repeat the experiment, in this case the excitation of the fluorescence decay, for a large number of times, and average the signal detected within a given period of time after the excitation pulses. There is a variety of signal averaging techniques that can be used to recover signals hidden in noise. However, a weak optical signal is special in that it is not a superposition of the signal itself and a noise background. Instead, it is a sequence of individual photon detection events. The signal waveform can therefore more efficiently be reconstructed by determining the arrival times of the photons, and counting them in several time bins according to their times after the excitation pulses.

1.2 Principle of Time-Correlated Single Photon Counting

Time-correlated single photon counting, or TCSPC, is based on the detection of single photons of a periodic light signal, the measurement of the detection times, and the reconstruction of the waveform from the individual time measurements [19, 117]. Technically, TCSPC was derived from the 'Delayed Coincidence' method used in nuclear physics to determine the lifetime of unstable nuclei. The earliest publication on the use of the method to detect the shape of light pulses dates back to 1961 [34], the first applications to spectroscopy of excited molecules were published in the 1970s [50, 96, 97, 133, 151].

TCSPC makes use of the fact that for low-level, high-repetition rate signals the light intensity is usually low enough that the probability to detect more than one photon in one signal period is negligible. The situation is illustrated in Fig. 1.2.

Fluorescence of a sample is excited by a laser of 80 MHz pulse repetition rate (a). The expected fluorescence waveform is (b). However, the detector signal, (as measured by an oscilloscope) has no similarity with the expected fluorescence

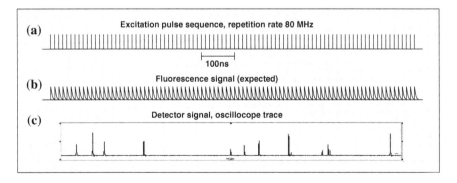

Fig. 1.2 Detector signal for fluorescence detection at a pulse repetition rate of 80 MHz. Average photon rate $10^7 \, s^{-1}$

waveform. Instead, it is a sequence of extremely narrow pulses randomly spread over the time axis (c). A signal like this often looks confusing to users not familiar with photon counting. Of course, there is a simple explanation to the odd signal shape: The pulses represent single photons of the light signal arriving at the detector. The shape of the pulses has nothing to do with the waveform of the light signal. It is the response of the detector to the detection of a single photon. Please note that the photon detection rate of (c) was about 10^7 s^{-1}. This is on the order of the maximum permissible detection rate of most photon counting detectors. Even at a detection rate this high, the detector signal is far from being a continuous waveform.

There are two conclusions from the signal shape in Fig. 1.2c. First, the waveform of the optical signal is not the detector signal. Instead, it is the distribution of the detector pulses over the time in the excitation pulse periods. Second, the detection of a photon within a particular excitation pulse period is a relatively unlikely event. The detection of two or more photons is even more unlikely. Therefore, only the first photon within a particular pulse period has to be considered. The build-up of the photon distribution over the pulse period then becomes a relatively straightforward process. The principle is shown in Fig. 1.3.

When a photon is detected, the arrival time of the corresponding detector pulse in the signal period is measured. The event is transferred into a digital memory by adding a '1' in a memory location proportional to the measured detection time (Fig. 1.3a). The same is done for a second photon (Fig. 1.3b). The process is continued until a large number of photons has been detected and accumulated in the memory (Fig. 1.3c). The result is the distribution of the photons over the time after the excitation pulses. The distribution represents the 'waveform' of the optical pulse (Please note that there is actually no such waveform, only a distribution of the photon probability, see Fig. 1.2). The principle is similar to the 'delayed coincidence' method used in nuclear physics to record the decay times of unstable nuclei. TCSPC has, in fact, been derived form the techniques used there [34, 110].

Although the principle of TCSPC looks complicated at first glance, it has a number of intriguing features. The first one is that the time resolution is better than the width of the single-electron response (SER) of the detector. The explanation is given in Fig. 1.4. The times of the photons are derived from the arrival times of the detector pulses. These times can be measured at an accuracy much better than the width of the pulses. Thus, the 'instrument response function', or IRF, of a TCSPC system is essentially given by the transit time dispersion (or transit time spread, TTS) of the photon pulses in the detector. The TTS can be more than 10 times shorter than the single-photon response of the detector. The IRF is therefore much narrower than the detector response. This is a considerable advantage over techniques based on direct analog recording of the detector signal [19, 117].

The second advantage is that there is no loss of photons in the recording process. Provided the timing electronics is fast enough every photon seen by the detector arrives in the photon distribution. TCSPC therefore reaches a near-ideal counting efficiency [83]. In particular, the efficiency is much higher than for recording techniques that shift a time-gate over the optical waveform [19, 120].

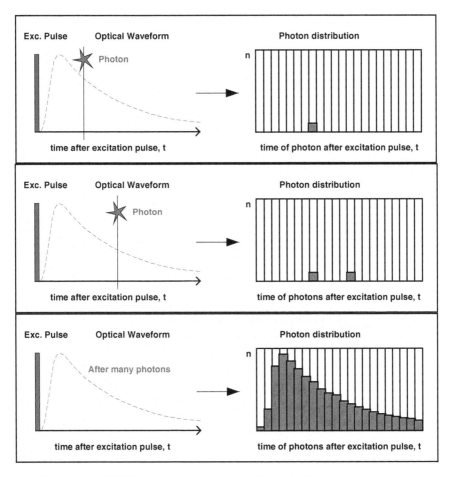

Fig. 1.3 Principle of TCSPC: the recording process builds up the distribution of the photons over the time after the excitation pulses

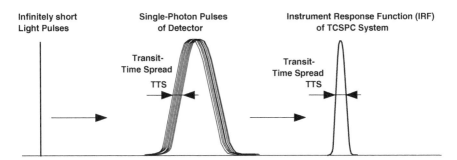

Fig. 1.4 Response of detector to infinitely short light pulses and instrument response of TCSPC system

There is also a third advantage that has been noticed only recently. If there is variation in the optical signal waveform over the time the data are acquired TCSPC records an *average waveform*. This can be important in biomedical applications where the optical waveforms may vary due to metabolic processes. A technique based on temporal scanning, i.e. by shifting a time-gate over the signal, would record a distorted waveform in these cases.

As a serious drawback of TCSPC it is often stated that the technique can record only one photon per signal period. If the light intensity is high a possible second or third photon in the same excitation pulse period with the first one is lost. The result is a distortion of the recorded signal waveform. Pile-up was indeed a serious problem in early TCPC experiments working at pulse repetition rates on the order of 10 kHz. It is far less a problem in modern implementations which work at pulse repetition rates on the order of 50–80 MHz. A derivation of the systematic error in the detected fluorescence lifetime has been given in [19]. For detection rates not exceeding 50 % of the excitation rate the measured intensity-weighted lifetime, τ_{meani}, of a single-exponential decay of the true lifetime τ can be approximated by

$$\tau_{meani} \approx \tau \left(1 - P/4\right)$$

P is the average number of photons per excitation period. The ratio of the recorded lifetime, τ_{meani}, and the true lifetime, τ, as a function of the average number of photons, P, is shown in Fig. 1.5. The lifetime error induced by pile-up is surprisingly small: Even for an average detection rate of 20 % of the excitation rate ($P = 0.2$) the lifetime error is no larger than 5 %. For excitation with a Ti:Sa laser (80 MHz) $P = 0.2$ would correspond to a detection rate of 16 MHz. This is much more than the detection rates in typical TCSPC experiments. With state-of-the-art laser sources and TCSPC electronics the count rate—and thus the available acquisition speed—is rather limited by the photostability of the samples than by pile-up effects [22]. This is especially the case in fluorescence lifetime microscopy where the fluorescence comes from a pico- or femtoliter sample volume.

The real limitation of the classic TCSPC principle is that it is *intrinsically one-dimensional*. That means, it records only the *waveform* of the signal. To record, for instance, the fluorescence decay function for a range of wavelengths, or over the

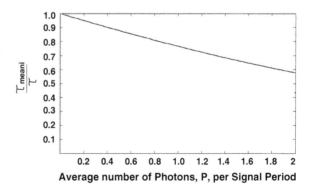

Fig. 1.5 Ratio of measured intensity-weighted lifetime, τ_{meani} and the true lifetime, τ, as a function of the number of photons, P, detected per signal period

pixels of an image area, the decay functions had to be recorded sequentially for a series of wavelengths or for the individual pixels of the image.

In applications were only a single optical waveform has to be recorded, or the experiment allows a series of waveforms to be recorded sequentially the classic principle is still used to a large extend. The typical application is recording of fluorescence decay curves. Examples are shown in Fig. 1.6. Figure 1.6, left, shows fluorescence decay curves of quinine sulphate quenched by Cl⁻ ions. The fluorescence lifetime changes with the quencher concentration [92]. Figure 1.6, right shows decay curves of DODCI (3,3′diethyloxadicarbocyanine iodide) for different wavelength. The data were obtained by scanning the detection wavelength by a monochromator and recording the decay curves sequentially.

The classic TCSPC principle is often used to record fluorescence anisotropy decay times. A sample is excited by linearly polarised light, and the fluorescence decay function measured in two directions of polarisation. An example is shown in Fig. 1.7. Fluorescein in a water-glycerol solution was excited at 474 nm. The blue curve shows the fluorescence decay measured at a polarisation parallel, $I_{par}(t)$, the red curve the fluorescence decay at a polarisation perpendicular to the excitation, $I_{perp}(t)$. The black curve shows the anisotropy, $R(t)$, (arbitrary units) calculated by

$$R(t) = I_{par}(t) - I_{perp}(t)/I_{tot}(t)$$

$I_{tot}(t)$ is the total intensity. For measurement in a 90° parallel-beam configuration it is $I_{tot}(t) = I_{par}(t) + 2\,I_{perp}(t)$ [92, 144], for excitation and detection from the full half space it approaches $I_{tot}(t) = I_{par}(t) + I_{perp}(t)$ [22]. Please see also Chaps. 3 and 12.

The problem of anisotropy-decay measurement is that noise and systematic errors in the difference $I_{par}(t) - I_{perp}(t)$ are much larger than in the signals themselves. Therefore, high-accuracy data are required which can only be obtained by TCSPC. Please see Chap. 12 for details.

The advantage of fluorescence-anisotropy experiments is that the anisotropy decay time bears information on the size of the fluorophore molecule, possible binding to other molecules, and on the viscosity of the molecular environment. It is therefore used to investigate the configuration of fluorescence-labelled bio-molecules. Such

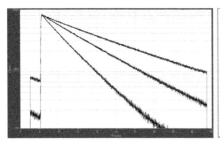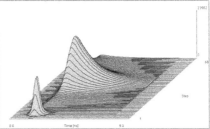

Fig. 1.6 Fluorescence decay measurement by TCSPC. *Left* Fluorescence decay curves of quinine sulphate for different quencher concentration. *Right* Fluorescence decay curves of DODCI recorded for different wavelengths

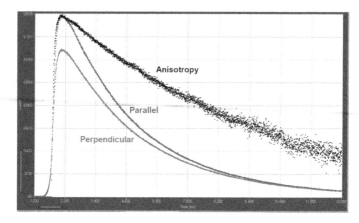

Fig. 1.7 Fluorescence anisotropy measurement. The fluorescence of a sample was measured at polarisation angles of 0° and 90° to the polarisation of the excitation; the anisotropy was calculated by $R(t) = I_{par}(t) - I_{perp}(t)/I_{tot}(t)$

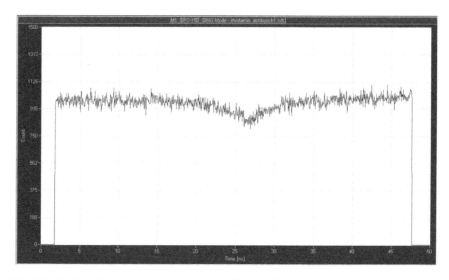

Fig. 1.8 Anti-bunching curve recorded at a diluted solution of fluorescein in a confocal microscope. About 7 molecules are in the focus at a time

experiments and their interpretation can be real detective stories [93, 94, 114, 115, 132, 142].

Another application of the classic principle is recording of anti-bunching effects. A single optical emitter, such as a single molecule or a quantum dot, can only emit a single photon within its fluorescence decay time. The experiment is run in the classic 'Hanbury-Brown-Twiss' setup [72]: The light is split in two detection channels, and the detection events in one detector are used as a start, the ones in the other as a stop signal for TCSPC. The resulting curve has a dip at a time at which the events coincide. Two examples are shown in Fig. 1.8.

Until today, some of these basic applications are performed by TCSPC setups based on the classic NIM (Nuclear Instrumentation Module) architecture [117]. Of course, modern TCSPC devices (see below) have the traditional recording modes implemented as well. Even in simple applications as the ones shown above the modern devices have advantages over NIM-based instruments: They work at higher pulse repetition rates and higher count rates, achieve shorter acquisition times, and are fully computer controlled. The real advantage of the modern implementation is, however, that the recording process is multi-dimensional. As a result, experiments can be performed which are entirely out of the reach of the classic design.

1.3 Multi-dimensional TCSPC

1.3.1 Principle of Multi-dimensional Recording

Development of multi-dimensional TCSPC goes back to the realisation that TCSPC records a photon distribution. In case of classic TCSPC, the distribution is over only one parameter, which is the time after an excitation pulse. However, if, by any means, additional parameters can be associated to the individual photons, the photon distribution can be made multi-dimensional. Typical parameters are the wavelength of the photons, the time from the start of an experiment, the distance along a one-dimensional scan, or the spatial coordinates within an image area. The difference between classic TCSPC and multi-dimensional TCSPC is illustrated in Fig. 1.9.

Depending on which and how many additional parameters are used, different photon distributions are obtained. Three possibilities are shown in Fig. 1.10. Wavelength-resolved fluorescence decay data (shown left) are obtained by using the wavelength as additional recording parameter. Dynamic fluorescence decay data (middle) are obtained by using the time from a stimulation event. Fluorescence lifetime imaging (right) uses the spatial coordinates, x, y, within an image as additional parameters.

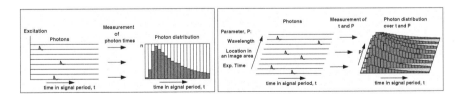

Fig. 1.9 *Left* Principle of classic TCSPC. The result is a one-dimensional photon distribution over the time in the signal period. *Right* Multi-dimensional TCSPC. The result is a multi-dimensional distribution over the time in the signal period and one or several additional parameters determined for the individual photons

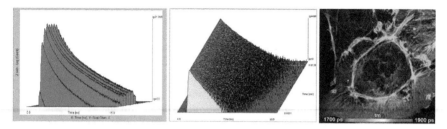

Fig. 1.10 Examples of photon distributions recorded by multidimensional TCSPC. *Left to right* Wavelength resolved fluorescence decay, dynamic change of fluorescence decay after a stimulation of the sample, fluorescence lifetime image obtained by fast confocal scanning

It is sometimes believed that the same results an also be obtained by classic TCSPC if only the parameter of interest would be varied and a sequence of waveforms be recorded. This is correct, but there is an important difference. The classic procedure would have to acquire a full waveform until the value of the parameter is changed, and then start a new recording. That means, any change in any of the additional photon parameters had to be slow, and predictable.

Multi-dimensional TCSPC is free of this limitation. Any parameter of the photon distribution can vary at any speed. It can even vary randomly. Every photon is just put into its place in the photon distribution according to its time after the excitation and the value of the additional parameters in the moment of the photon detection. As the parameters keep varying, the photon distribution is just accumulated until enough photons have been collected for any particular combination of parameters. In other words, the rate at which the parameters are varying is decoupled from the rate at which the photons are recorded.

Typical examples are the results shown in Fig. 1.10: The wavelength of the photons in Fig. 1.10 was determined by a multi-wavelength detector. It varies randomly, and is different for every photon. The variation in the fluorescence decay functions in Fig. 1.10, middle, is induced by repetitive stimulation of the sample. The data were accumulated over a large number of stimulation periods. Thus, a high signal-to noise ratio was obtained despite of the fact that the variation in the decay curves was recorded at a resolution of 100 μs/curve. The lifetime image on the right was recorded by scanning the sample with a pixel dwell time of one microsecond. At an average photon rate on the order of 10^6 s^{-1} one would expect that the result would contain no more than a few photons per pixel. Fluorescence decay data acquired this way would be useless. However, because the photons were accumulated over a large number of frames of the scan, an excellent accuracy of the decay time is obtained.

Despite of its large potential the idea of multidimensional TCSPC has spread only slowly. The introduction was hampered by the fact that the principles behind TCSPC in general and multi-dimensional TCSPC in particular were not commonly understood. The first patent dates back to 1998 [8], a paper at least partially related to multidimensional TCSPC was published in 1991 [9], and two other patents followed in 1993 [10, 11]. The first multidimensional-TCSPC devices on the market were the SPC-300 and SPC-330 boards of Becker & Hickl. These boards had memory space for up to 128 waveforms. The SPC-430 boards introduced in 1994 already had space for up to 2048 waveforms, and had fast time-series recording and parameter-tag functions for single-molecule spectroscopy [12, 121]. The first TCSPC module with imaging capability was the SPC-535 introduced in 1995. The final breakthrough came in 2001 with the introduction of TCSPC FLIM into fluorescence laser scanning microscopy [13–16].

1.3.2 Architecture of TCSPC for Multidimensional Data Acquisition

The general architecture of a TCSPC device with multi-dimensional data acquisition is shown in Fig. 1.11, left.

The core of the TCSPC device is the time measurement block. This block receives the single-photon pulses from a detector, and the timing reference pulses from a pulsed light source. The time is measured from a start pulse to a stop pulse. Depending on the principle used in the timing electronics the time between start and stop can be measured at an accuracy from about 20 ps (rms) down less than 3 ps (rms). Please see [19] for technical details.

In early TCSPC applications the start pulse came from the light source, the stop from the detector. When high-repetition rate lasers were introduced as light sources start and stop has been reversed: The time measurement is started with the photon, and stopped with a reference pulse from the light source. This 'reversed start-stop' principle has the advantage that the timing electronics need not work at the high repetition rate of the light source but only at the much lower photon detection rate. Please see [19] or [22] for technical details.

The time measurement block delivers the photon time, t, relative to the stop pulse as a digital data word. The data word addresses the time-channel in a selected waveform data block in the memory. The photon is added in the addressed memory location. As more and more photons are detected the waveform of the optical signal builds up.

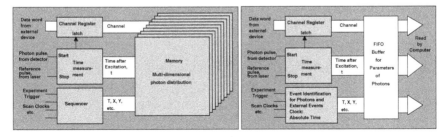

Fig. 1.11 Multi-dimensional TCSPC architecture. *Left* Photon distribution built up in memory of TCSPC device. *Right* Photon distributions built up in computer

Up to this point the function of the TCSPC device is identical with classic TCSPC. The difference is, however, that the memory has space for a large number of waveform data blocks, and that these blocks can be addressed via additional building blocks of the TCSPC device. The address part delivered by these blocks carries additional information related to the individual photons. Photon parameters available directly as digital data words are fed into the TCSPC device via the 'Channel' register. Examples are the wavelength channel in which a photon was detected by a multi-wavelength detector (see Sect. 1.4.1), or the number of one of several multiplexed lasers or laser wavelengths (see Sect. 1.4.2). The external data word is read into an the 'Channel' register in the moment when the corresponding photon is detected.

Additional parts of the address can be generated by a 'Sequencer' block. The sequencer generates an address word either by counting synchronisation pulses from an external experiment control device (e.g. an optical scanner), or by counting pulses from an internal clock.

Figure 1.11, left, the photon distributions are built up by on-board logics in the memory of the TCSPC device. The advantage of this principle is that the acquisition runs virtually without interaction of the measurement control software.

A second architecture is shown in Fig. 1.11, right. Here, the parameters of the individual photons are directly transferred into the system computer. To buffer the information for periods when the computer is not able to read the data a FIFO (first-in-first-out) buffer is inserted in the data path. The computer builds up the photon distribution by software or directly stores the data of the individual photons for further processing.

When data of individual photons are generated there is no need to use a sequencer to direct photons into different waveform data blocks of the memory. The function of the sequencer is therefore performed by 'Event Identification' logics. This block adds an absolute time (from the start of the recording) to the data words of the individual photons. Moreover, it puts external events, such as trigger pulses from an experiment or synchronisation pulses from a scanner, into the data stream and marks them with an absolute time and an identifier.

The advantage of this architecture is that more memory size is available to build up the photon distributions, and that more complex data operations can be performed. Because the principle in Fig. 1.11, right, uses a stream of data for photons marked with detection times and other parameter it is also called 'time tag' or 'parameter-tag' mode.

1.4 Multidimensional TCSPC Implementations

The system architecture shown in Fig. 1.11 can be used for a wide variety of recording tasks. Depending on the information fed into the TCSPC module photons can be recorded versus the time after the excitation, detection wavelength, excitation wavelength, distance along a line scan, coordinates in a scan area, excitation and detection positions multiplexed by a fibre switch, time from a stimulation of the sample, or time within the period of an additional modulation of the excitation laser. Photon distributions can be built up over almost any combination of these parameters. The photon data can be processed internally by the architecture shown in Fig. 1.11, left, or externally by the architecture shown in Fig. 1.11, right. A comprehensive description of all combinations would be out of the framework of this book. The sections below therefore concentrate on frequently used implementations.

1.4.1 Multi-detector and Multi-wavelength TCSPC

Multidimensional TCSPC is able to detect several optical signals simultaneously by several detectors. Although the principle can be used to record any signals originating from the same excitation source the main application is multi-wavelength TCSPC. The principle of multi-wavelength TCSPC is illustrated in Fig. 1.12. The light is dispersed spectrally, and the spectrum is projected on an array of detectors. The detector electronics determines the detector channel at which a particular photon has arrived. The channel number thus represents the wavelength of the photon. It is used as a second dimension of the photon distribution. The result is a photon distribution over time and wavelength.

To understand why multi-wavelength detection works, please remember that the average number of photons detected per signal period is far less than one, see Fig. 1.3. It is therefore unlikely that several photons per signal period will be detected. Now consider an array of detectors over which the same photons flux is dispersed spectrally. Because it is unlikely that the complete array detects several photons per period it is also unlikely that several detectors of the array will detect a photon in one signal period. This is the basic idea behind multi-detector TCSPC. Although several detectors are *active simultaneously they are unlikely to deliver a photon pulse in the same signal period*. Therefore, only single detection events in one of the detector channels have to be considered. Consequently, the times of

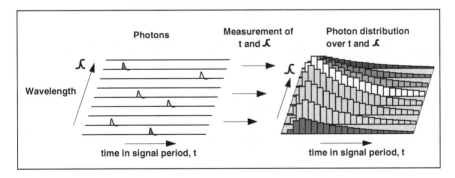

Fig. 1.12 Multi-wavelength TCSPC determines the wavelengths of the individual photons and builds up a distribution over the arrival time in the signal period and the wavelength

the photons detected in all detector channels can be measured by a single time-measurement block [11, 19, 22].

The technical principle is illustrated in Fig. 1.13. The photons of four detectors are combined into a common timing pulse line. Simultaneously, a 'Channel' signal is generated that indicates in which of the detectors a particular photon was detected. The combined photon pulses are sent through the normal time measurement procedure of the TCSPC device. The detector number is used as a 'channel' signal for multi-dimensional TCSPC. Considered from the point of view of multi-dimensional TCSPC the TCSPC device builds up a photon distribution over the times of the photons after the excitation and the detector channel number. Because the individual photons are routed into different memory blocks according to the detector in which they were detected the technique has been called 'routing', and the device that generates the channel number 'router'.

Routing has already been used in classic NIM-based TCSPC setups [27, 140]. Each of the detectors had its own constant-fraction discriminator (CFD). The CFD output pulses were combined into one common TAC stop signal, and simultaneously used to control one or two higher address bits of the multichannel analyser. Because separate CFDs were used for the detectors, the number of detector channels was limited. The modern implementation uses a single CFD for all detector channels, as shown in Fig. 1.13. The 'router' combines the single-photon pulses into one common timing pulse line, and generates a channel signal that indicates at which of the detectors the current photon arrived, see Fig. 1.13.

Routing is often used in time-resolved anisotropy decay measurements. An advantage in this application is that the photons for parallel and perpendicular polarisation are processed in the same timing electronics and are thus recorded at exactly the same time scale. In other applications with no more than four channels routing systems are more and more replaced with fully parallel TCSPC systems [21, 76]. Routing is used, however, to further extend parallel TCSPC systems and to obtain large detection channel numbers with single TCSPC modules. Typical applications are diffuse optical imaging by near-infrared spectroscopy (NIRS) and multi-wavelength TCSPC. NIRS systems with up to four TCSPC channels and 32 detectors (8 for each TCSPC channel) are used [48, 49]. Multi-wavelength TCSPC

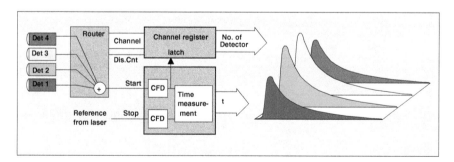

Fig. 1.13 TCSPC multidetector operation, several individual detectors. By the 'channel signal' from the router, the photons of the individual detectors are routed into separate memory blocks

with 16 spectral channels is used in tissue spectrometers (see below) and in fluorescence lifetime imaging systems (see Sect. 1.4.5.2).

The architecture of a multi-wavelength TCSPC system is shown in Fig. 1.14.

A spectrum of the light to be recorded is projected on the photocathode of a multi-anode PMT. The routing electronics and the multi-anode PMT are integrated in a common housing. The combination of the photon pulses of all channels is achieved by deriving the pulses from the last dynode of the PMT [19, 22].

The high counting efficiency and the elimination of any wavelength scanning makes multi-wavelength TCSPC attractive to biological and biomedical applications. A typical setup of a multi-wavelength tissue lifetime spectrometer is shown in Fig. 1.15. It consists of a picosecond diode laser, a fibre probe, a polychromator, a 16-anode PMT with routing electronics, and a TCSPC module. [19, 22]. Multi-spectral fluorescence decay data of human skin obtained this way are shown in Fig. 1.15, right.

A portable instrument of this type has been described by De Beule et al. [51]. It uses two multiplexed ps lasers of 355 and 440 nm wavelength. Recording of the multiplexed signals was obtained by the technique described in Sect. 1.4.2. The results have 32 decay-curve channels for different combinations of excitation and detection wavelength.

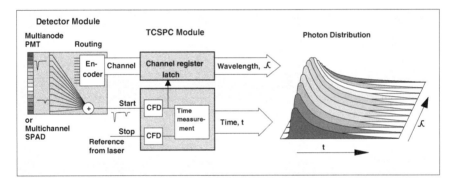

Fig. 1.14 Multi-wavelength TCSPC

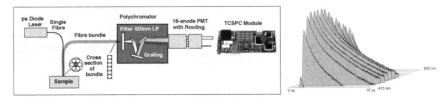

Fig. 1.15 *Left* Optical setup for single-point autofluorescence measurement. *Right* Multi-wavelength fluorescence decay data of human skin. Time scale 0–15 ns, wavelength scale 410–600 nm, intensity scale logarithmic from 500 to 30,000 counts/channel

Other non-scanning multi-wavelength TCSPC applications use a micro-spectrometric setup with a microscope and a confocal detection beam path to obtain fluorescence from a specific location within a cell [40].

Most multi-wavelength applications target at the measurement of the metabolic state of cells and tissue. Information is obtained from the fluorescence decay functions of NADH and FAD, see Chaps. 13, 14 and 15. Extracting accurate decay parameters is, however, difficult because both the excitation and the emission spectra of NADH and FAD overlap. Spectrally resolved detection of decay parameters therefore helps separate the individual decay components [40, 41, 46, 153]. An overview on multi-wavelength detection of NADH and FAD has been given by Chorvat and Chorvatova [42]. Using a micro-spectrometry setup, Chorvat et al. found changes in the decay signature during rejection of transplanted hearts in human patients [43]. The effect of Ouabain on the metabolic oxidative state in living cardiomyocytes was shown in [44], the effect of oxidative stress in [45].

De Beule et al. used a tissue spectrometer setup and demonstrated the applicability of NADH and FAD multi-wavelength lifetime imaging to cancer detection and diagnosis [51]. Using a similar optical setup Coda et al. found significant differences in the fluorescence lifetime of normal and neoplastic tissue [47].

Multi-wavelength detection becomes even more attractive when its is combined with fluorescence lifetime imaging microscopy [15, 19, 20, 22]. Please see Sect. 1.4.5, and Chap. 2 of this book.

1.4.2 Excitation Wavelength Multiplexing

The routing capability of TCSPC can be used to multiplex several light signals and record them quasi-simultaneously [9, 19]. The most common application of multiplexed TCSPC is recording of fluorescence signals excited by several multiplexed lasers. Each laser is turned periodically for a short period of time, and a routing signal is generated indicating which of the lasers is active at a given time. The routing signal is used to identify the photons excited by different lasers, and to store them in different waveform blocks of a common photon distribution. The principle is shown in Fig. 1.16.

Excitation wavelength multiplexing is easy with picosecond diode lasers. Diode lasers can be turned on and off electronically at a speed in the ns range. Multiplexing can also be implemented elegantly by using a super-continuum laser with an acousto-optical filter (AOTF). The transition from one filter wavelength to another takes only a few microseconds. All that is needed to multiplex several lasers or filter wavelengths is a simple controller that generates the on/off selection signal to the diode lasers or to the AOTF, and the routing signal to the TCSPC device. Please see Chap. 18.

Laser wavelength multiplexing can be combined with operation of several detectors at different detection wavelength, with multi-wavelength detection [51], and with fluorescence lifetime imaging [19, 22], see Sect. 1.4.5.3.

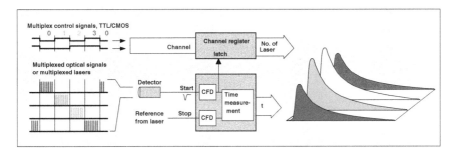

Fig. 1.16 Multiplexed TCSPC operation. Several signals are actively multiplexed into the detector, or several lasers are periodically switched. The destination in the TCSPC memory is controlled by a multiplexing signal at the 'channel' input

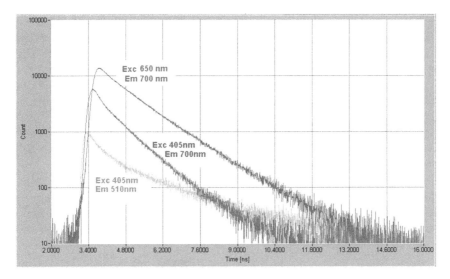

Fig. 1.17 Laser wavelength multiplexing. *Left* Fluorescence decay of a plant leaf, multiplexed excitation at 405 and 650 nm. Detection by two detectors at 510 and 700 nm

An example is shown in Fig. 1.17. A leaf was excited by two multiplexed diode lasers of 405 and 650 nm wavelength. Fluorescence was detected by two detectors at 510 and 700 nm and recorded in one TCSPC module via a router. The result are three fluorescence decay curves for different combination of excitation and emission wavelength: 405/510, 405/700, and 650/700 nm. The forth combination (650/510 nm, not shown) does not contain photons because the detection wavelength is shorter than the excitation wavelength.

It should be noted here that multiplexing of lasers can also be obtained by 'pulse-interleaved excitation', or PIE, see Sect. 1.4.5.3. In that case the lasers are multiplexed pulse by pulse. The multiplexing technique shown in Fig. 1.16 has, however, a few advantages. The most important one is that it is free of crosstalk.

The tail of the fluorescence decay excited by one laser does not show up in the decay curve excited by the next one. Moreover, there is no mutual influence of the signals via pile-up or counting-loss effects [22].

1.4.3 Spatial Multiplexing

The multiplexing technique shown in Fig. 1.16 is also used to multiplex spatial excitation or detection locations at a sample. This principle is commonly used in diffuse optical imaging of the human brain, see Fig. 1.18. A number of source fibres and detection fibres is attached to the head. The injection of the laser(s) into the source fibres is controlled by a fibre switch. The controller of the fibre switch switches though subsequent source positions and sends a status signal to the TCSPC module(s) indicating the current source position. This signal is used as a multiplexing (or routing) signal in the TCSPC modules(s) to build up separate time-of-flight distributions or fluorescence decay curves for the different source positions. The multiplexing of the source position can be combined with multiplexing of several lasers. Because diffuse-optical imaging works at high count rates the signals from the different detection positions are usually detected by individual detectors and recorded by separate TCSPC modules.

To increase the number of detector positions the setup can be extended by routers. Up to 32 detectors have been used which were connected to four TCSPC channels via four routers [48, 49].

The setup is mainly used for dynamic brain imaging, i.e. for recording relative changes in the concentration of oxy- and deoxyhemoglobin. In this case, the TCSPC systems records a time series of data, see section below. The fibre switch period is on the order of 10 ms, and the time per step of the time series is between 50 ms and a few seconds. Please see Chap. 17 of this book.

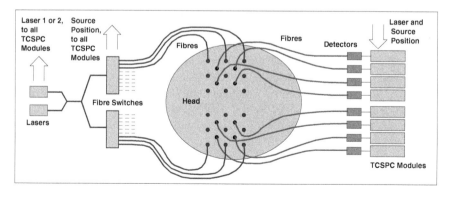

Fig. 1.18 Architecture of the 8-channel parallel DOT system described in [76]

1.4.4 Time-Series Recording

When a time-series of TCSPC data is to be recorded the first idea of classic-TCSPC users is normally to run a record-and-save procedure. The measurement would be run for a defined acquisition time, the data be read from the TCSPC device, and saved into a data file by the associated system computer. This procedure would be continued until the desired number of time steps have been performed. A procedure like this has the charm that the number of time steps is virtually unlimited. However, it has also disadvantages. An obvious one is that the speed of the sequence is limited by readout and save times. Another one is that the signal-to noise ratio of the subsequent recordings depends on the speed of the sequence. It is therefore difficult, if not impossible, to record transient changes in the recorded waveforms at millisecond or microsecond resolution.

The limitations of the record-and-save procedure can be avoided by multi-dimensional recording. The principle is shown in Fig. 1.19. The memory of the TCSPC device has space for a large number of waveforms. The number of the waveform block into which the photons are recorded is controlled by the sequencer logics of the TCSPC device. The recording starts with a user command or with an external trigger. The sequencer starts recording into the first waveform block, and then switches through the blocks in regular intervals of time.

At first glance, the principle shown in Fig. 1.19 may look like a record-and-save procedure. It is, however, different in that it records subsequent waveforms into *one single photon distribution*. There is no time needed to read and save the data. More importantly, the complete distribution can be accumulated: The experiment, e.g. the induction of a change in the fluorescence decay of a sample would be repeated, the run of the sequencer through the waveform blocks triggered by the stimulation, and the photons accumulated into a distribution according to their times after the excitation pulse and their times after the stimulation. This way, the available speed of the sequence gets entirely decoupled from the photon detection rate. It does no longer depend on the time of the individual steps of the sequence but only on the

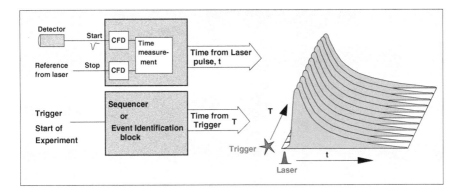

Fig. 1.19 Time-series recording

total acquisition time. The procedure even works if the sequence is so fast that only one (or less than one) photon is recorded per step. The recording would just be continued until all time channels of all waveform blocks have been filled with a reasonable amount of photons.

An example is shown in Fig. 1.20. A change in the lifetime of chlorophyll in a plant [69] was initiated by turning on the excitation light. The recording procedure steps through the curves at a rate of 50 µs/curve. Within this time, about 50 photons are recorded. To obtain a reasonable signal-to-noise ratio the procedure was repeated and the photons accumulated. The result has a high signal-to-noise ratio, despite of the short time per curve.

The procedure shown above does, of course, not require that the data in each step of the time series are only a single waveform. The data can be multi-dimensional themselves. The steps of the sequence can contain multi-wavelength data, data recorded with laser multiplexing, or combinations of both. Time-series recording can even be combined with imaging, see Sect. 1.4.5.4, and Chap. 2, Sect. 2.4.4.

Time-Series With Memory Swapping

Modern TCSPC modules have space for hundreds if not thousands of waveforms in their internal memories. A time series recorded by the principle shown in Fig. 1.19 can have a correspondingly large number of steps. However, there are applications which require to record a time-series with step times in the millisecond range over a period of time that can reach hours. Such data can be recorded by a memory swapping procedure.

The principle is shown in Fig. 1.21. The memory of the TCSPC device is split in two independent memory banks. When the measurement is started the sequencer starts to record in the first waveform block of the first bank. After a defined acquisition time it switches to the next waveform block until all blocks of the

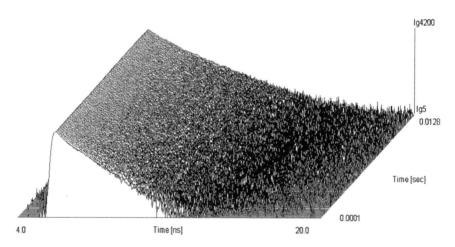

Fig. 1.20 Photochemical transient of chlorophyll in a live plant. Time per curve 100 µs, 10,000 on/off cycles were accumulated. Time-series starts from the front

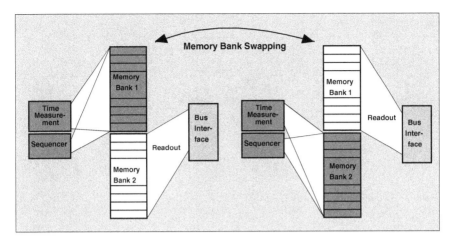

Fig. 1.21 Unlimited sequential recording by memory bank swapping. When one memory bank is full, the sequencer swaps the banks. While the sequencer writes into one bank, the other bank is read by the computer

memory bank are filled with data. Then the memory banks are swapped, and the recording continues in the other memory bank. During the time the second bank is filled with photons the system computer reads the data from the first bank and saves them into a file. This way, a virtually unlimited sequence of waveforms can be recorded without time gaps between the recordings. Also here, it is not required that a single waveform data block contains only a single waveform. The data in the subsequent blocks can be multi-wavelength data, laser multiplexing data, combinations of such data, or even FLIM data.

The memory swapping technique (or 'continuous flow mode') was originally developed for DNA analysis and single-molecule detection in a capillary gel electrophoresis setup [12]. It has been used in conjunction with TCSPC FLIM to record fast time series of chlorophyll transients [79]. It is commonly used for dynamic brain imaging by diffuse optical tomography (DOT) techniques [76, 105–108, 111, 122], see Chaps. 17 and 18 of this book. The memory swapping technique can be used with a trigger signal that starts either the recording of each bank or the recording of each data block within the current bank. It is also possible to run triggered accumulation within one memory bank, and, after a defined number of accumulations, pass to the next bank. With these options, the Continuous Flow mode provides fast and efficient recording procedures for a large number of complex experiments.

1.4.5 Fluorescence Lifetime Imaging (FLIM)

The general architecture of a TCSPC FLIM system is shown in Fig. 1.22. A laser scanning microscope (or another scanning device) scans the sample with a focused

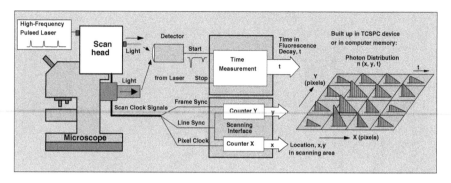

Fig. 1.22 Multidimensional TCSPC architecture for FLIM

beam of a high-repetition-rate pulsed laser. Depending on the laser used, the fluorescence in the sample can either be excited by one-photon or by multiphoton excitation. The TCSPC detector is attached either to a confocal or non-descanned port of the scanning microscope, see Chap. 2, Sect. 2.2. For every detected photon the detector sends an electrical pulse into the TCSPC module. From this pulse, the TCSPC module determines the time, t, of the photon within the laser pulse sequence (i.e. in the fluorescence decay). Moreover, the TCSPC module receives scan clock signals (pixel, line, and frame clock) from the scan controller of the microscope. A system of counters (the scanning interface) counts the pixels within each line and the lines within each frame. The counter outputs deliver the position of the laser beam, x and y in the scan area.

The information, t, x, y, is used to address a memory in which the detection events are accumulated. Thus, in the memory the distribution of the photon density over x, y, and t builds up. The result is a data array representing the pixel array of the scan, with every pixel containing a large number of time channels with photon numbers for consecutive times after the excitation pulse. In other words, the result is an image that contains a fluorescence decay curve in each pixel [13, 14, 17, 19, 22].

The procedure described above does not require that the laser beam stays at the same position until enough photons have been acquired in the decay curve of the corresponding pixel. It is only necessary that the *total pixel time*, over a large number of subsequent frames, is large enough to record a reasonable number of photons per pixel. In other words, the signal-to-noise ratio depends on the total acquisition time, not on the san rate. Thus, TCSPC FLIM works even at the highest scan rates available in laser scanning microscopes. At pixel rates used in practice, the recording process is more or less random: A photon is just stored in a memory location according to its time in the fluorescence decay and the position of the laser spot in the moment of detection.

In early TCSPC FLIM systems the photon distribution was built up directly in the hardware of the TCSPC module (see Fig. 1.11, left). The advantage of hardware accumulation is that the acquisition runs independently of the computer. It can therefore be used at extremely high photon rates and scan speeds. The disadvantage is that the available memory space is limited. A typical application of hardware-accumulation is the 'Preview' mode of FLIM systems where images must be recordable up to almost any count rate and displayed at high image rate, but with a relatively moderate number of pixels and time channels.

The way to record larger images is to use software accumulation, i.e. the configuration shown in Fig. 1.11, right. The TCSPC device transfers the data of the individual photons and the clock pulses to the computer, and the photon distribution is built up by software. The required hardware architecture and suitable acquisition modes were available even in early TCSPC modules [22, 121]. However, the bus transfer speed and the processing capabilities of the PCs of the 1990s were insufficient for this kind of operation. At count rates above 100 kHz and pixel rates on the order of 1 MHz, as they are routinely used in FLIM, the computer was not able to read the data from the SPC module and simultaneously process them to build up a photon distribution.

The situation changed with the introduction of fast multi-core PCs. The tasks of data transfer and data processing could now be shared between the CPU cores. Count rate in the parameter-tag mode was no longer a problem. Modern TCSPC FLIM system therefore use almost exclusively software accumulation. Typical images sizes with 32-bit operating systems were 512×512 pixels, with 256 time channels per pixel [22]. With the introduction of Windows 7, 64 bit, and corresponding 64-bit data acquisition software, the maximum image size increased to 2048×2048 pixels $\times 256$ time channels and more [139]. Pixel numbers this large are enough to record images of the maximum field of view even of the best microscope lenses. 64-bit FLIM applications are therefore rarely limited by memory size.

FLIM results are normally displayed as pseudo-colour images. The brightness represents the number of photons per pixel. The colour can be assigned to any parameter of the decay profile: The lifetime of a single-exponential approximation of the decay, the average lifetime of a multi-exponential decay, the lifetime or amplitude of a decay component, or the ratio of such parameters [22]. An example is shown in Fig. 1.23. The colour shows the amplitude-weighted mean lifetime of a double-exponential decay. The image format is 512×512 pixels, every pixel contains 256 time channels. Decay curves in two selected pixels are shown on the right. Similar curves are contained in any pixel of the image.

FLIM systems of the architecture shown in Fig. 1.22 are used for a wide range of applications. An overview is given in Chap. 3 of this book. The applications can essentially be divided into three classes: Measurement of parameters of the local molecular environment of the fluorophores (Chaps. 4 and 5), protein interaction experiments by FRET (Förster Energy Transfer, Chaps. 7 and 8), and imaging of metabolic parameters via the fluorescence lifetimes of endogenous fluorophores (Chaps. 13, 14, and 15). Examples for the three classes of applications are shown in Figs. 1.24 and 1.25.

Measurements of molecular environment parameters are based on changes in the fluorescence lifetime of a fluorophore with its molecular environment [92]. A commonly know effect is fluorescence quenching: The fluorescence lifetime is

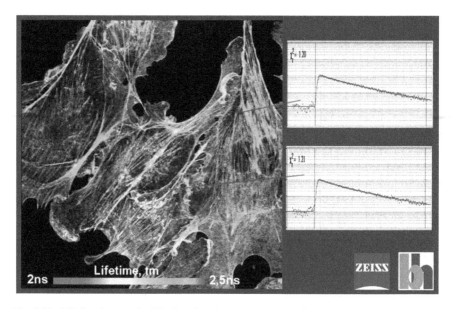

Fig. 1.23 Lifetime image of a BPAE cell, stained with Alexa 488. FLIM data 512 × 512 pixels, 256 time channels per pixel. Fluorescence decay shown for two selected pixels. Zeiss LSM 710 Intune system with Becker & Hickl Simple-Tau 150 FLIM system

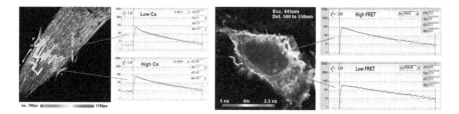

Fig. 1.24 *Left* Barley root tip, stained with *Oregon green*. Courtesy of Feifei Wang, Zhonghua Chen and Anya Salih, University of Confocal Bioimaging Facility, University of Western Sydney, Australia. Leica SP5 MP with bh SPC-150 FLIM module. *Right* Cell containing a expressing a GFP fusion protein (donor) and Cy3-labelled antibody (acceptor). Donor image, amplitude-weighted lifetime of double-exponential decay model. Zeiss LSM 710 microscope with Becker & Hickl Simple-Tau 152 FLIM system

proportional to the reciprocal concentration of the quencher, and can thus be used as a probe function for the concentration of the quencher [65, 68, 77]. A fluorophore may also have two forms of different fluorescence quantum efficiency and thus different fluorescence lifetimes. The concentration ratio of the forms, and thus the average fluorescence lifetime, may depend on the environment. Examples are Ca^{2+} sensors [88, 90, 92, 112] and pH sensors [73, 130]. There are also effects of local viscosity [89], local refractive index [143], proximity to nanoparticles and metal

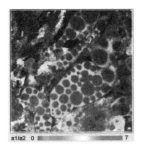
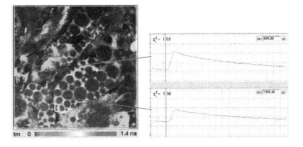

Fig. 1.25 FLIM of human salivary gland stem cells. *Left* Image of amplitude ratio, a1/a2, of fast and slow decay component. *Middle* Image of amplitude-weighted lifetime, tm. *Right* Decay curves in indicated pixels of image. Data courtesy of Aisada Uchugonova and Karsten König, Saarland University, Saarbrücken

surfaces [66, 109, 113, 124], and aggregation of the fluorophore [80]. Please see also [24, 26, 125], and Chap. 3 of this book.

The most frequent FLIM application is probably Förster resonance energy transfer (FRET), see Chaps. 7 and 8 of this book. FRET experiments use the fact that energy can be transferred from a donor to an acceptor molecule if the donor emission spectrum overlaps the acceptor excitation spectrum [63, 64]. A condition for FRET to occur is that the distance between donor and acceptor is on the order of a few nm or less. FRET is therefore used as an indicator for protein interaction. The advantage of FLIM-based FRET experiments is that the FRET efficiency is obtained from a single fluorescence-lifetime image of the donor. The problems of steady-state FRET measurements, such as directly excited acceptor fluorescence, extension of the donor emission in the acceptor channel, and unknown concentration ratio, are therefore avoided. Moreover, multi-exponential decay analysis can be used to separate the interacting donor fraction from the non-interacting one [17, 22]. There are hundreds of FRET papers based on TCSPC FLIM. An overview on the FLIM FRET applications and the corresponding literature has been given in [19, 22]. Please see also Chaps. 3, 7, and 8.

The third class of FLIM applications is the extraction of metabolic parameters from the lifetimes of endogenous fluorophores. Most autofluorescence FLIM applications use the fact that that the fluorescence lifetimes of NADH and FAD depend on the binding to proteins [91, 119]. Bound and unbound fractions can therefore be distinguised by double-exponential decay analysis and characterised by the amplitudes and lifetimes of the decay components. The ratio of bound and unbound NADH depends on the metabolic state [29, 40, 42, 56, 67, 137, 138]. FLIM data of NADH and FAD are therefore used to detect precancerous and cancerous alterations [32, 78, 95, 128, 137, 138]. It has also been shown that the fluorescence decay parameters of the NADH fluorescence change with maturation of the cells, during apoptosis and necrosis, and in response to treatment with cancer drugs [67, 86, 131, 146–148]. Autofluorescence FLIM is closely related to clinical applications [59, 84, 85, 134–136]. Please see Chaps. 13, 14 and 15 of this book.

1.4.5.1 FLIM in Parallel TCSPC Channels

A TCSPC FLIM system with a single detector detecting in a single wavelength interval can be used for a wide range of applications, see examples above. The reason that FLIM at a single wavelength works so well is that the fluorescence lifetime is inherently 'ratiometric'. The fluorescence lifetime can be considered an intensity ratio in two time intervals of the decay curve. The result thus does not depend on the concentration of the fluorophore.

The situation changes if several fluorophores with different lifetimes and different emission spectra are involved, or if fluorescence anisotropy decay data are to be recorded. Images then have to be recorded simultaneously in different wavelength intervals, or under 0° and 90° angles of polarisation. Early TCSPC FLIM systems used routing to record these signals. Modern systems more and more use fully parallel TCSPC channels.

The advantage of the parallel architecture is that the maximum count rate is higher, and that the channels are fully independent. Even if one channel saturates or a detector shuts down by overload the other channel(s) may still record correct data. Systems with two channels are standard [6, 7, 22], systems with four channels are easily possible [18], and systems with 8 parallel SPC-150 channels have been demonstrated [21, 22]. FLIM data obtained by this system are shown in Fig. 1.26.

1.4.5.2 Multi-wavelength FLIM

Multi-wavelength FLIM (also called Multi-Spectral FLIM or spectral lifetime imaging, SLIM) uses a combination of the basic FLIM technique shown in Fig. 1.22 and the multi-wavelength detection technique shown in Fig. 1.14.

The architecture of multi-wavelength FLIM is shown in Fig. 1.27. A spectrum of the fluorescence light is spread over an array of detector channels. For every photon, the time in the laser pulse period, the wavelength-channel number in the detector array, and the position, x, and y, of the laser spot in the scan area are determined. These pieces of information are used to build up a photon distribution

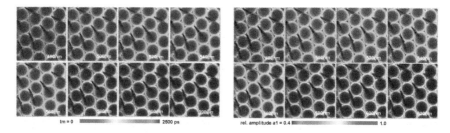

Fig. 1.26 Eight-channel parallel FLIM of a drosophila eye. Autofluorescence. Excitation at 407 nm, detection wavelength from 480 to 620 nm. *Left* Mean Lifetime of double-exponential decay. *Right* Relative amplitude of fast decay component. bh DCS-120 confocal scanning system

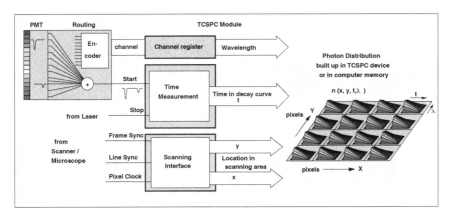

Fig. 1.27 Principle of multi-wavelength TCSPC FLIM

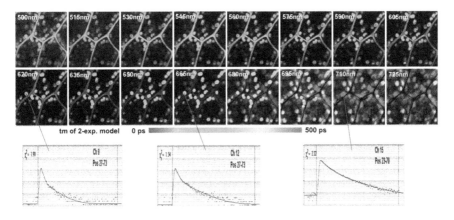

Fig. 1.28 *Top* Multi-wavelength FLIM of plant tissue, 16 wavelength intervals, 128 × 128 pixels, 256 time channels. Amplitude-weighted lifetime of double-exponential fit, normalised intensity. *Bottom* Decay curves in selected pixels of 620, 656, and 710 nm images. Two-photon excitation at 850 nm, detection from 500 to 725 nm. Zeiss LSM 710, bh Simple-Tau 150 FLIM system [7]

over the arrival times of the photons in the fluorescence decay, the wavelength, and the coordinates of the image [15, 19, 20, 22].

As for single-wavelength FLIM, the result of the recording process is an array of pixels. However, the pixels of multi-wavelength FLIM contain several decay curves for different wavelength. Each decay curve contains a large number of time channels; the time channels contain photon numbers for consecutive times after the excitation pulse. Multi-wavelength FLIM data can also be considered a set of lifetime images for different wavelength. A result of a multi-wavelength FLIM measurement is shown in Fig. 1.28.

Multi-wavelength FLIM requires a large amount of memory: For the data shown in Fig. 1.28 memory space must be provided for 16 images, each with

128×128 pixels, and 265 time channels per pixel. Multi-wavelength FLIM can therefore performed at reasonable pixel numbers only by building up the photon distribution in the computer memory. A significant progress has been made with the introduction of 64 bit software. In the 64-bit environment, data with 16 wavelength channels can be resolved into images of 512×512 pixels, and 256 time channels [139]. This is about the maximum resolution achieved for a single FLIM image in a 32 bit environment. Examples are shown in Chap. 2, Figs. 2.25 and 2.26, and in Chap. 3, Fig. 3.17.

Multi-wavelength FLIM is predominately used in applications where fluorescence signals of several fluorophores are present, and the emission of the fluorophores cannot cleanly separated by filters, or there are just too many fluorophores to get them separated by filters and dichroic beamsplitters.

Multi-wavelength FLIM was first demonstrated in 2002 by Becker et al. for recording decay data in the complete donor-acceptor wavelength range of a FRET experiment [20]. Bird et al. demonstrated the technique for lifetime imaging of stained kidney tissue samples [28]. Rück et al. used of multi-wavelength FLIM for monitoring the conversion of photosensitisers for PDT and the generation of photoproducts [126, 127]. FRET measurements by multi-wavelength FLIM were described in [30, 31, 76].

The majority of multi-wavelength FLIM applications are in autofluorescence imaging, especially imaging of the coenzymes NADH and FAD [42, 98, 128, 150]. Chorvatova and Chorvat worked out spectral unmixing techniques based on multi-wavelength FLIM data and used them to determine metabolic parameters in cardiomyocytes [40–46] and investigate their response to drugs and stress conditions.

Li et al. used multi-wavelength FLIM for NADH imaging. They found different bound/unbound ratios (represented by the a_1/a_2 ratio of the amplitudes of the lifetime components) in the cytosol and in the mitochondria, and changes induced by variable concentration of deoxyglucose [98]. They also found changes in the bound/unbound ratio when cells were exposed to sub-lethal concentrations of cadmium [99, 150].

Using two-photon excitation and multi-wavelength FLIM Li et al. were able to detect and characterise two-photon excited fluorescence from haemoglobin [154]. Multi-wavelength FLIM was used to discriminate the haemoglobin fluorescence from fluorescence of other endogenous fluorophores. This way, haemoglobin fluorescence was used for label-free imaging of microvasculature in live tissue [103, 104].

A multiphoton multi-colour excitation system with a Titanium Sapphire laser, super-continuum-generation in a photonic crystal fibre and a Becker & Hickl PML-16/SPC-150 multi-wavelength FLIM system has been developed by Li et al. [100]. The system excites tryptophane and NADH simultaneously and separates the fluorescence of both compounds spectrally. The authors found that the ratio of NADH and Tryptophane fluorescence is a sensitive indicator of cell metabolism. The same instrument was used to record Tryptophane, and NADH lifetime images in combination with SHG images in different depth of epithelial tissue [101] and for investigation of squamous intraepithelial neoplasia [81]. Simultaneous recording of tryptophane, SHG, NADH intensity images by the system has been described in [141]. Other applications are autofluorescence lifetime imaging of leukocytes [152], multi-modal label-free imaging of zebra fish [102], and skeletal muscle tissue [141].

Multi-wavelength FLIM in combination with multiphoton tomography of human skin [84, 85, 125] was used by Dimitrov et al. [55]. The authors found changes in the fluorescence spectra and shorter fluorescence lifetime in malignant melanoma compared to normal skin.

1.4.5.3 FLIM with Excitation Wavelength Multiplexing

FLIM can be combined with excitation wavelength multiplexing. The general principle of multiplexed TCSPC is shown in Fig. 1.16. The extension of the principle to FLIM is shown in Fig. 1.29. Excitation at different wavelength is achieved by multiplexing (on/off switching) of several lasers, or by switching the wavelength of the acousto-optical filter (AOTF) of a super-continuum laser. A multiplexing signal that indicates which laser (or laser wavelength) is active is fed into the routing input of the TCSPC module. The signal represents the excitation wavelength. The TCSPC module is running the normal FLIM acquisition process: It builds up a photon distribution over the coordinates of the scan area, the photon times, and the excitation wavelength. The result is a data set that contains images for the individual excitation wavelengths. (It can also be interpreted as a single image that has several decay curves for different excitation wavelengths in its pixels.)

Different combinations of excitation and emission wavelengths can be obtained by using several detectors and a router. Modern implementations normally use several parallel TCSPC modules.

To avoid interference of the multiplexing frequency with the pixel, line, or frame frequency of the scanner multiplexing is normally synchronised with the pixels, the lines, or the frames of the scan. Details of the optical system are described in Chap. 2, Sect. 2.2.7. An typical result is shown in Fig. 1.30.

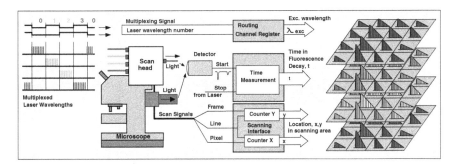

Fig. 1.29 FLIM with laser wavelength multiplexing

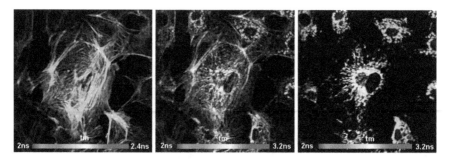

Fig. 1.30 FLIM by wavelength multiplexing, Supercontinuum laser with AOTF, two parallel SPC-150 TCSPC modules. *Left to right* Excitation 500 nm emission 525 ± 50 nm, excitation 500 nm emission 620 ± 30 nm, excitation 580 nm emission 620 ± 30 nm. 256 × 256 pixels, 256 time channels, frame-by-frame multiplexing

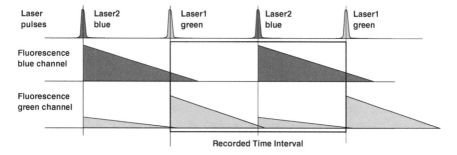

Fig. 1.31 Principle of laser wavelength multiplexing by pulse-interleaved excitation

A similar effect can be obtained by 'Pulse-Interleaved Excitation', or 'PIE'. PIE does not use the laser wavelength as a dimension of the photon distribution. Instead, the pulse trains of several (usually two) lasers are interleaved pulse by pulse, and the fluorescence decay functions excited by both lasers are recorded into a single waveform, see Fig. 1.31. The recorded waveform contains the fluorescence decay excited by one laser followed by the decay excited by the other. The data of the two excitation channels are selected by time-gating in the data analysis. Data for different combinations of excitation and emission wavelengths are recorded by using several parallel TCSPC channels or a single TCSPC channel with a router. A typical result is shown in Fig. 1.32.

The advantage of PIE in FLIM applications is that it is also applicable to lasers that cannot be on-off modulated at high speed. The disadvantage is that PIE is not entirely free of crosstalk. The tails of the fluorescence decay excited by one laser may extend into the decay excited by the other. Moreover, there is intensity crosstalk due to counting loss, and mutual influence by detector afterpulsing. Please see [22] for a comparison of PIE and laser multiplexing.

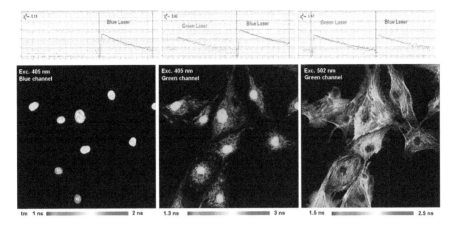

Fig. 1.32 FLIM by PIE. Zeiss LSM 710 Intune laser (*green laser*) and 405 nm ps diode laser (*blue laser*). Images recorded by two parallel SPC-150 TCSPC modules. *Left* Blue laser, blue detection channel. *Middle* Blue laser, green detection channel. *Right* Green laser, green detection channel. *Top* Decay curves in selected pixels. *Bottom* FLIM images, analysed by SPCImage, amplitude-weighted lifetime of double-exponential decay fit, time-gated intensity

1.4.5.4 FLIM Time-Series Recording

A time-series of FLIM images can be recorded by a simple record-and-save procedure. A FLIM image is recorded for a defined acquisition time or for a number of frames of the scan, and saved to a file. The procedure is repeated until the desired number of images—or steps of the time series—have been acquired [19]. The procedure is simple and has the advantage that it does not require more memory space for the photon distribution than a single FLIM recording. The disadvantage is that the save operation takes time: At least one frame of the scan is lost during the time the data are saved. A continuous sequence can be recorded by dual-memory recording [19, 22, 79]. Memory blocks are provided for two photon distributions, and the FLIM system records into one block while the data from the other one are saved. However, the time per step cannot be made faster than the time needed for saving, and the rate of the sequence is limited by the decrease of the signal-to-noise ratio with decreasing acquisition time per step of the sequence. The fastest image rate that can be achieved is 1–2 images per second, see Fig. 1.33.

A more efficient way of time-series FLIM has been provided by 64-bit data acquisition software [139]. In the 64-bit environment, the available amount of memory is so large that a large number of reasonable-size FLIM images can be recorded into one and the same photon distribution. The technique is derived from spatial 'Mosaic' or 'Tile' imaging. A FLIM mosaic is a data array (either one- or two-dimensional) which has space for a large number of FLIM data sets. The recording process starts to record FLIM data in the first mosaic element. After a defined number of frames of the scan it switches to the next element. Thus, all

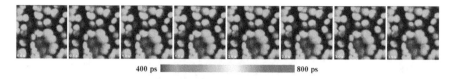

400 ps ▓▓▓▓▓▓▓▓▓▓▓▓▓▓▓▓▓▓▓▓▓▓▓▓ 800 ps

Fig. 1.33 Time series recorded at a speed of 2 images per second. Chloroplasts in a moss leaf. Dual-memory recording, images 128 × 128 pixels

elements of the mosaic are filled with data one after another. The time per element is determined by the frame time, and by the number of frames per element.

Mosaic time series recording has two advantages over the conventional record-and-save procedure. First, the transition from one mosaic element to the next occurs instantaneously. There are no time gaps between the steps of the sequence. Therefore, very fast time series can be recorded. Second, mosaic time series data can be accumulated: A lifetime change in a sample would be stimulated periodically, and the start of the mosaic recording be triggered by the stimulation. With every new stimulation the recording procedure runs through all elements of the mosaic, and accumulates the photons. Accumulation allows data to be recorded without the need of trading photon number and lifetime accuracy against the speed of the time series: The signal-to noise-ratio depends on the total acquisition time, not on the speed of the sequence. Please see Chap. 2, Sect. 2.4, Mosaic FLIM.

1.4.6 FLITS

By the technique described above a FLIM time series can be recorded down to the frame time of a fast optical scanner. Faster effects can be resolved by a combination of TCSPC and line scanning. The technique has been named 'FLITS', fluorescence *lifetime-transient* scanning [22, 23, 25]. FLITS is based on building up a photon distribution over the distance along the line of the scan, the experiment time after a stimulation of the sample, and the arrival times of the photons after the excitation pulses. The principle is shown in Fig. 1.34.

FLITS uses the same recording procedure as FLIM. Similar as for FLIM, the result is an array of pixels, each of which contains a fluorescence decay curve in form of photon numbers in subsequent time channels. The recoding is synchronised with the scan by pixel clock pulses (which indicate the transition to the next pixel of the line) and line clock pulses, which indicate the transition from the end of the line back to the start. The procedure differs from FLIM in that the frame clock pulses does not come from the scanner but from an external event that stimulates a fluorescence-lifetime change in the sample. The TCSPC module thus records a photon distribution which has the Y coordinate of FLIM replaced with a coordinate that represents how often the scanner scanned along the line since the stimulation. In other words, the Y coordinate is the time after the stimulation given in multiples of the line time.

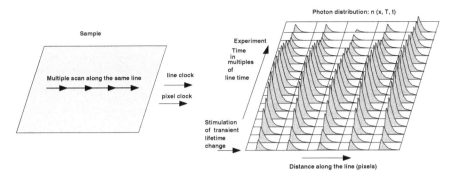

Fig. 1.34 Principle of FLITS. A sample is scanned along a line. The TCSPC system builds up a photon distribution over the distance along the line, the time after a stimulation of the sample, and the times of the photons after the excitation pulses

As long as the stimulation occurs only once the recording process is simple: The sequencer of the TCSPC module [19, 22] starts to run with the stimulation, and puts the photons in consecutive experiment-time channels along the T axis.

A FLITS recording is shown in Fig. 1.35, left. It shows the non-photochemical lifetime transient of chlorophyll [69] recorded after the start of illumination. The horizontal axis is the distance along the line scanned, the vertical axis (bottom to top) is the experiment time, T, in this case the time from the start of the illumination. Decay curves for a selected pixel within the line are shown for T = 0.5, 7.5, and 13.4 s.

The result shown in Fig. 1.35, left, is identical with a time-series of line scans. However, there is an important difference: The data are still in the memory (either in the on-board memory of the TCSPC module or in the computer) when the sequence is completed. Thus, the recording process can be made repetitive: The sample would be stimulated periodically, and the start of T triggered by the stimulation. The recording then runs along the T axis periodically, and the photons are accumulated into one and the same photon distribution.

With repetitive stimulation of the sample, it is no longer necessary that each T step acquires enough photons to obtain a complete decay curve in each pixel and T channel. No matter when and from where a photon arrives, it is assigned to the right location in x, the right experiment time, T, and to the right arrival time, t, after the laser pulse. As in the case of FLIM, a desired signal-to-noise-ratio is obtained by simply running the recording process for a sufficiently long acquisition time. Obviously, the resolution in T is limited by the period of the line scan only, which is about 1 ms for the commonly used galvanometer scanners. For resonance scanners and polygon scanners (see Chap. 2) the line time can be made even shorter.

An example of an accumulated FLITS recording is shown in Fig. 1.35 right. It shows the photochemical transient [69] of the chlorophyll in a plant. The result was obtained by turning on and off the laser periodically, and recording the photons by

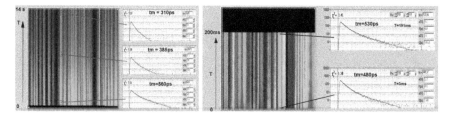

Fig. 1.35 FLITs recordings of chlorophyll transients. *Left* Non-photochemical transient, vertical time scale 0–14 s, decay curves shown for a selected pixel at T = 0.5, 7.5, and 13.4 s after turn-on of the laser. *Right* Photochemical transient. Vertical time scale 0–200 ms, decay curves shown for a selected pixel at T = 3 and 191 ms into the laser-on phases. The amplitude-weighted lifetime increases from 480 to 530 ps

triggered accumulation. The T scale is from 0 to 200 ms. Decay curves are shown for a selected pixel within the line for T = 3 and 191 ms into the laser-on phases. The amplitude-weighted lifetime increases from 480 to 530 ps.

The application of FLITS to Ca^{2+} imaging in neurons is described in Chap. 5 of this book.

1.4.7 Phosphorescence Lifetime Imaging (PLIM)

Phosphorescence occurs when an excited molecule transits from the first excited singlet state, S1, into the first triplet state, T1, and returns from there to the ground state by emitting a photon [92]. Both the S1-T1 transition and the T1-S0 transition are 'forbidden' processes. The transition rates are therefore much smaller than for the S1-S0 transition. That means that phosphorescence is a slow process, with lifetimes on the order of microseconds or even milliseconds. The usual approach to phosphorescence recording is to reduce the excitation pulse rate and extend the time scale of decay recording to microseconds or milliseconds. Except for fluorophores of high intersystem crossing rate, the results obtained this way are disappointing.

There are several reasons why simply decreasing the repetition rate does not work well. Both are related to the low S1-T1 transition rate. The excitation energy injected into the molecules is preferentially dissipated by the S1-S0 transition, and not deposited in the T1 state. The T1 population therefore remains low, and so does the phosphorescence intensity. Simply increasing the peak power of the excitation pulse meets technical constraints, or causes nonlinear effects in the sample. The second problem is that high excitation peak power also causes high peak intensity of fluorescence. The fluorescence pulse not only leads to the detection of one or several photons per excitation pulse and thus violates the rules of TCSPC detection. It can become so strong that it causes temporary overload in the detector.

These problems can be solved by exciting phosphorescence with laser pulses longer than the fluorescence lifetime. The long pulse width increases the population of the triplet state without increasing the peak fluorescence intensity. Similarly, a group of laser pulses instead of a single pulse can be used. The principle is shown in Fig. 1.36.

A high-frequency pulsed laser is modulated on/off at a period several times longer than the phosphorescence lifetime. The laser pulses within the 'on' phase of the modulation period excite fluorescence and build up phosphorescence. The 'off' phase contains only phosphorescence. The advantage of this excitation principle is not only that it can be used for multi-photon excitation but also that it can be used *to record fluorescence and phosphorescence decay data simultaneously*. This is achieved by assigning two times to every photon, one of which is the time in the laser pulse period, t, the other the time in the laser modulation period, T. Two separate photon distributions are built up, one over t, the other over T. The distribution over t is the fluorescence decay, the distribution over T the phosphorescence decay. The principle can be used both for single-curve recording and for lifetime imaging [22].

The architecture of a combined FLIM and PLIM system is shown in Fig. 1.37. The TCSPC module works in the parameter-tag mode. The times of the photons, t,

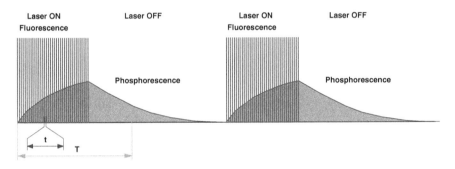

Fig. 1.36 Excitation principle for combined FLIM/PLIM

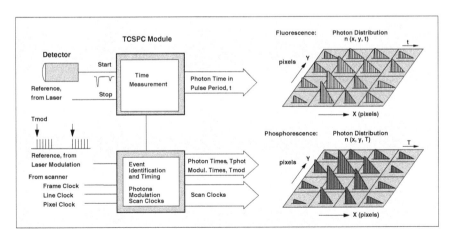

Fig. 1.37 Combined FLIM/PLIM system

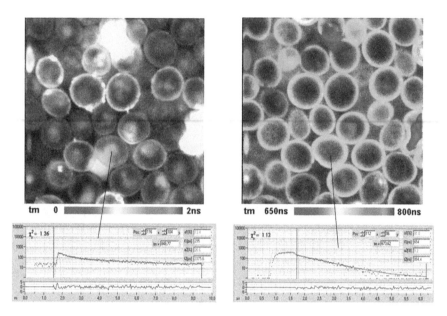

Fig. 1.38 Combined FLIM/PLIM recording. Yeast cells stained with a ruthenium dye. FLIM (*left*) shows autofluorescence, PLIM (*right*) phosphorescence of the Ruthenium dye

within the laser pulse period are measured by the normal time-measurement block. The 'Event Identification' block (see Fig. 1.11 right) determines absolute times, T_{phot}, of the photons. Moreover, it receives a reference signal from the laser modulation and clock pulses from the scanner. From these signals, it determines absolute times for the modulation reference, T_{mod}, and for the scan clocks, T_{pixel}, T_{line}, and T_{frame}. The values are put in a data stream together with the photon data. From these data, the computer software determines the locations, x, y, of the individual photons in the scan area, and the times of the photons within the modulation period, $T = T_{phot} - T_{mod}$. The software builds up two photon distributions, one over the scan coordinates and the times, t, and another over the scan coordinates and the times, T. The first one is a fluorescence lifetime image during the on-phase of the laser modulation period, the second one a phosphorescence lifetime image.

An example of a FLIM/PLIM measurement is shown in Fig. 1.38. The figure shows yeast cells stained with tris (2,2′-bipyridyl) dichlororuthenium (II) hexahydrate. The FLIM image is shown Fig. 1.38, left. It is dominated by autofluorescence of the cells. PLIM is shown in Fig. 1.38, right. The phosphorescence comes from the ruthenium dye. Decay curves in selected pixels are shown at the bottom. The time scale of FLIM is 0–10 ns, the time scale of PLIM 0–6.5 μs.

PLIM applications aim at suppression of autofluorescence [4, 5], and, more importantly, oxygen-concentration sensing. Phosphorescence from almost any phosphorescing compound is strongly quenched by oxygen. The phosphorescence lifetime can therefore be used to determine oxygen concentrations [53, 54, 129]. Combined FLIM/PLIM is especially promising because it is able to record metabolic information via FLIM while monitoring the oxygen concentration via PLIM. Please see Chap. 6 of this book.

1.4.8 Multi-dimensional Recording by Modulation of Experiment Parameters

The lifetime imaging techniques described above scan the spatial location of the fluorescence excitation and detection and record the shape of the optical signal as a function of the scan coordinates. There are other TCSPC applications that aim on recording optical signals as functions of parameters which are not necessarily spatial ones. The task can be solved in a similar way as FLIM: Parameters in the sample or in the experiment setup are modulated, the TCSPC process is synchronised with the parameter modulation via scan clock pulses, and a photon distribution is recorded over the parameters and the time in the optical signal. A typical experiment of this class is used in plasma physics for the investigation of barrier discharges.

Barrier discharges occur if the electric field between two isolator-coated electrodes exceeds a critical value, see Fig. 1.39. The visible discharge phenomenon consists of a large number of micro-discharges with nanosecond duration and kHz frequency.

Barrier discharges are technically highly relevant: They are a degradation mechanism of insulators in electrical systems, and they are used in a wide variety of plasma applications, e.g. for cleanup of exhaust gases. The investigation of the discharges is difficult: They occur at random times, the duration of the light pulses is on the order of a few nanoseconds, and the intensity is low. Ideally, it would be desirable to record the shapes of the optical pulses as a function of the gap voltage and the distance along the gap for selectable wavelength. Exactly this task can be solved by multi-dimensional TCSPC. The principal setup is shown in Fig. 1.40.

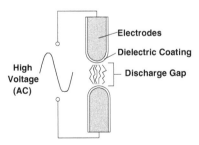

Fig. 1.39 Barrier discharge between dielectrically coated electrodes

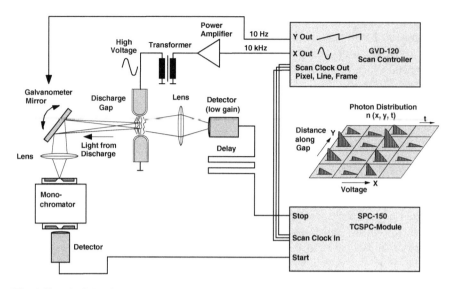

Fig. 1.40 Principle of the setup for barrier discharge measurements

One output of a digital signal synthesizer (in the present case a bh GVD-120 scan controller) delivers a sinusoidal output voltage. The voltage is amplified, transformed to a voltage in the 10 kV range, and applied across the discharge gap. A second output of the synthesizer (the Y output of the scan controller) drives a galvanometer mirror. The galvanometer mirror periodically scans the spot from which the light is detected along the discharge gap. A monochromator selects the desired detection wavelength. Single photons of the selected signal are detected by a PMT module. The photon pulses are connected to the start input of an SPC-150 TCSPC module.

The stop input for the TCSPC module comes from a second PMT. This PMT detects light from the entire discharge gap in a wide spectral range. For every discharge, it detects about 1000 photons. Operated at a relatively low gain, it delivers a timing reference for the TCSPC module. The recording process is synchronised with the operation of the scan controller via scan clock pulses. Consequently, the TCSPC module records a photon distribution over the phase of the gap voltage (X), the distance along the discharge gap (Y), and the times of the photons in the optical pulses generated by the discharges.

Figure 1.41, shows the waveform of the optical pulses as a function of the distance over the gap integrated over the positive and the negative half-waves of the gap voltage. It can be seen that the optical pulse varies along the gap. It also differs between the positive and the negative half-wave of the gap voltage.

Figure 1.42 shows the intensity over the time in the optical pulse and the distance along the gap for four different phases in the gap voltage, Fig. 1.43 the intensity over the time in the pulse and the phase in the gap voltage for four subsequent distance intervals along the gap. The images reveal a complex

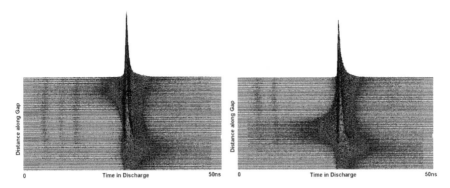

Fig. 1.41 Waveform of optical pulses for different distance along the gap for positive (*left*) and negative (*right*) half-wave of the gap voltage. Data courtesy of Ronny Brandenburg, INP Greifswald, Germany

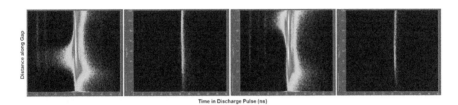

Fig. 1.42 Intensity over time in discharge pulse and distance along the gap for four subsequent phase intervals (*quarter waves*) in the gap voltage. Horizontal scale 0–50 ns. Data courtesy of Ronny Brandenburg, INP Greifswald, Germany

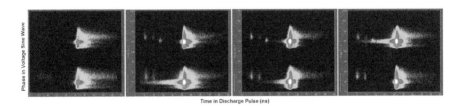

Fig. 1.43 Intensity over time in the discharge pulse and phase in the gap voltage for four subsequent distance intervals along the gap. Horizontal scale 0–50 ns. Data courtesy of Ronny Brandenburg, INP Greifswald, Germany

behaviour of the light pulses. The shape varies both with the location in the gap and with the gap voltage at which the discharges occur. There are also pre-pulses that occur some 10 ns before the main pulse.

Similar as the other multi-dimensional TCSPC applications, the principle shown above has no inherent limitation of the rate at which an experiment parameter is modulated. This is very important for barrier-discharge measurements: The frequency of the gap voltage must be in the 10 kHz range, and matched to the resonance frequency of the transformer-gap combination. Sequential recording of the voltage-dependence—e.g. by keeping the voltage constant until enough discharges have occurred and enough photons been recorded—is therefore not an option.

Barrier-discharge systems exist in several variants. Some are using sub-sets of the principle shown above, e.g. point measurements at selected location in the gap, or slow scanning along the gap distance. However, all are using the phase of the gap voltage as a parameter of a multi-dimensional TCSPC process. Please see [37, 38, 71, 74, 75, 82, 87, 145] for technical details and for results.

1.5 Using Parameter-Tagged Single-Photon Data

The techniques described above transfer parameter-tagged single-photon data into the system computer and build up photon distributions by software. Once a photon has been put into the photon distribution the information associated to it is no longer needed and, normally, discarded.

Parameter-tagged photon data may, however, be used to build up other results than multi-dimensional photon distributions. When the data are recorded it may even not be clear how exactly they are to be processed. User-interaction during the data processing may be required, or the processing may be so time-consuming that it cannot be performed online. In these cases the single photon data may be saved for later off-line processing. The general structure of parameter-tagged single-photon data and the use of such data in a few typical applications will be described below.

1.5.1 Structure of Parameter-Tagged Single-Photon Data

Parameter-tagged single-photon data contain information about the individual photons detected by the TCSPC device [19, 22]. An example of the structure of such data is shown in Fig. 1.44.

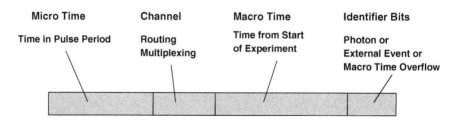

Fig. 1.44 Structure of parameter-tagged data

Each photon is tagged with its time after the excitation pulse (the 'micro time'), the time from the start of the recording (the 'macro time'), and the data word from the 'channel' register of the TCSPC device. External events, such as transitions of scan clock signals, are also put in the stream of parameter-tagged data. Several identifier bits mark a particular entry either as a photon or as an external event. Of course, the entries of external events do not contain valid micro time information. The micro time bits of external events may therefore be used to transfer other information.

The macro-time clock has normally a resolution of on the order of few 10 ns. However, an experiment can be run over a time of seconds, minutes, or even hours. Because the number of macro time bits in the photon data words is limited the macro time will overflow in regular intervals during the recording. To allow the data processing software to account for the overflows these are marked in the data. Macro time overflows can either be attached to the photon data words or be put in the data stream the same way as external events.

1.5.2 Calculation of FCS from Parameter-Tagged Data

Fluorescence Correlation Spectroscopy is based on recording fluorescence from a limited number of fluorescing molecules in a small sample volume, and correlating intensity fluctuations caused by the motion of the molecules [123]. The data processing procedure for FCS is usually described by the correlation functions

$$G(\tau) = \lim_{T \to \infty} \frac{1}{2T} \int\limits_{-T}^{+T} I(t)I(t + \tau)dt \quad G_{12}(\tau) = \lim_{T \to \infty} \frac{1}{2T} \int\limits_{-T}^{+T} I_1(t)I_2(t + \tau)dt$$

where $G(\tau)$ is the autocorrelation of a single signal, $I(t)$, and $G_{12}(\tau)$ the cross-correlation function of two signals, $I_1(t)$ and $I_2(t)$.

For photon counts, N, in consecutive, discrete time channels $G(\tau)$ and $G_{12}(\tau)$ can be obtained by calculating

$$G(\tau) = \sum N(t) \cdot N(t + \tau) \quad \text{and} \quad G_{12}(\tau) = \sum N_1(t) \cdot N_2(t + \tau)$$

The general behaviour of the auto- and cross-correlation functions is illustrated in Fig. 1.45. For a randomly fluctuating signal, $I(t)$, the autocorrelation function $G(\tau)$ delivers high values only if the intensity values, I, at a given time, t, and at a later time, $t + \tau$ are correlated. Uncorrelated fluctuations of I cancel over the integration time of the experiment. Similarly, $G_{12}(\tau)$ delivers high values if the fluctuations in both signals, $I_1(t)$ and $I_2(t + \tau)$, correlate with each other. The drop of $G(\tau)$ and $G_{12}(\tau)$ over the shift time, τ, shows over which length of time the fluctuations are correlated.

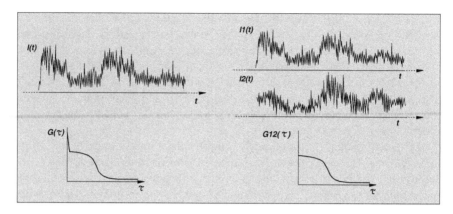

Fig. 1.45 General behaviour of the autocorrelation function, $G(\tau)$, of a signal I and the cross-correlation function, $G_{12}(\tau)$, of the signals I_1 and I_2

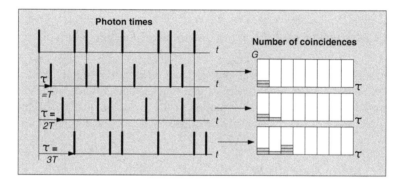

Fig. 1.46 Calculation of the autocorrelation function from TCSPC time-tag data

The equations shown above are based on the assumption that the intensity waveforms are analog signals. However, TCSPC data are not a continuous wave-form as those shown in Fig. 1.45. Instead, the data describe a random sequence of individual detection events. In fact, the events can be (and usually are) so rare that there is less than one photon within the range of τ over which the intensities are to be correlated. The equations shown above are therefore inappropriate to calculate correlation functions from parameter-tagged photon data recorded by TCSPC.

The general correlation procedure for parameter-tagged photon data is illustrated in Fig. 1.46. In typical TCSPC data, the time-channel width, T, is the period of the macro-time clock. It is shorter than the dead time of the detector/photon counter combination. Therefore only one photon can be recorded in a particular macro time period. Consequently, $N(t)$ and $N(t + \tau)$ can only be 0 or 1. The calculation of the autocorrelation function therefore becomes a simple shift, compare, and histo-gramming procedure. The times of the individual photons are subsequently shifted by one macro-time clock period, T, and compared with the original detection times. The coincidences found between the shifted and the unshifted data are transferred into a histogram of the number of coincidences, G, versus the shift time, τ. The obtained $G(\tau)$ is the (un-normalised) autocorrelation function.

The cross-correlation function between two signals is obtained by a similar procedure. However, the photon times of one detector channel are shifted versus the photon times of the other channel.

To obtain intensity-independent correlation curves the results of the algorithm shown in Fig. 1.46 need to be normalised. Normalisation can be considered the ratio of the number of coincidences found in the recorded signal and the number of coincidences expected for an uncorrelated signal of the same count rate. The normalised autocorrelation and cross-correlation functions are

$$G_n(\tau) = G(\tau) \, \frac{n_T}{N_P^2}$$

with n_T = total number of time intervals, Np = total number of photons, and

$$G_{12n}(\tau) = G_{12}(\tau) \, \frac{n_t}{N_{P1}N_{P2}}$$

with n_T = total number of time intervals, N_{p1} = total number of photons in signal 1, N_{p2} = total number of photons in signal 2.

The procedure illustrated in Fig. 1.46 yields $G(\tau)$ in equidistant τ channels. The width of the τ channels is equal to the macro time clock period of the SPC module, T. The algorithm is known as the 'Linear-Tau' algorithm. Unfortunately, for long correlation times the algorithm results in an extremely large number of τ channels in $G(\tau)$, and in intolerably long calculation times.

Therefore, the algorithm is usually modified by applying binning steps to the photon data during the correlation procedure. The procedure is illustrated in Fig. 1.47.

The procedure starts with a number of shift-and-compare steps as shown in Fig. 1.46. After a number of steps the remaining photon data are binned. Then the correlation procedure is continued on the binned data. The new time bins can contain several photons. Therefore, the photon numbers in the original and the shifted data have to be multiplied. After a number of shift steps the data are binned again, and the procedure is continued. The procedure is called 'Multi-Tau' algorithm. Despite of the binning operations, it yields shorter calculation times than the Linear-Tau algorithm. It also delivers the logarithmic τ axis commonly used for FCS curves.

The correlation procedures shown above can be used to calculate FCS curves on-line. In this case, photons are constantly recorded, and the data transferred into the system computer. In certain intervals, usually a few seconds, the software calculates a correlation function (or several correlation functions if several detectors are active) on the data that have arrived within this time. The coincidences for the

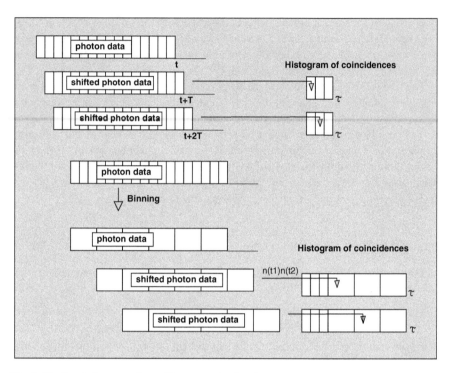

Fig. 1.47 Correlation procedure with progressive binning

subsequent intervals are summed up in a common histogram, see Figs. 1.46, and 1.47, right. Of course, the procedure requires that the recording-time intervals be longer than the maximum time, τ, over which the photons are correlated. Moreover, photons recorded in one interval must still be available in the next one to correlate late photons in one interval with early photons in the next.

An example of FCS calculation from parameter-tagged data is shown in Fig. 1.48. The data were recorded in a confocal microscope [6] from a diluted solution of fluorescein. On average, about 0.4 molecules were in the detection volume at a time. The average count rate was about 5000 photons/s.

Figure 1.48, top, shows an intensity trace of the recorded signals at time-bin width of 10 μs, and over a time interval of 10 ms. Due to the low detection rate there is no more than 1 photon per time bin, and there are about 50 photons over a period of time of 10 ms.

The autocorrelation function is shown at the bottom of Fig. 1.48. Note that the τ scale of the FCS curve is the same as the length of the time intervals of the intensity trace on the left. However, the FCS curve was calculated over the complete acquisition time of the measurement, in this case 120 s. It shows a clear correlation curve of the intensity fluctuations caused by diffusion of the molecules. That a correlation curve is obtained from the data shown in Fig. 1.48, top, may surprise at first glance. It is a result of the fact that the data have been correlated over a period of 120 s, not

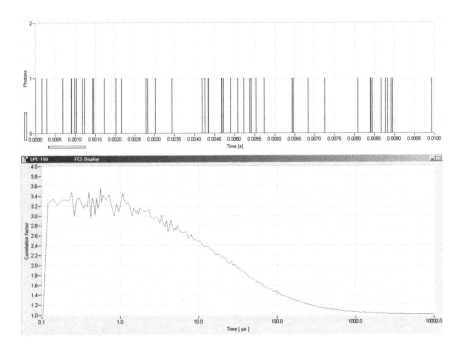

Fig. 1.48 *Top* Intensity trace, time bin width 10 μs, displayed time interval 10 ms. *Bottom* Autocorrelation curve from the same signal correlation time from 100 ns to 10 ms. Calculated over acquisition time of 120 s

only over the time intervals shown in the intensity trace in Fig. 1.48, top. This is another example that the signal-to-noise ratio in single-photon counting experiments depends on the total acquisition time, not on the time over which the photons are correlated.

Most FCS applications aim at the estimation of the size of labelled biomolecules, on the detection of interaction between biomolecules, and on the determination of the number of fluorescent molecules attached to a special biomolecule. The size of the molecules can be derived from the diffusion time, i.e. the point of the auto-correlation function at which drops to 50 % of its maximum. More accurately, the diffusion time is obtained by fitting the autocorrelation function with an appropriate model. Interaction of molecules is derived from the cross-correlation functions. Different biomolecules are labelled with fluorophores of different absorption and emission spectra. The cross-correlation (FCCS) function of the two signals indicates whether the constructs diffuse together or independently.

In terms of TCSPC, data acquisition FCS and FCCS are very simple applications. They even do not need the micro-times of the photons, and do not need a pulsed excitation source. FCS and FCCS are, however, anything but simple in respect to the optical system and the detectors: Excellent optical efficiency, low chromatic and spherical aberration, excellent alignment of the confocal optics, and excellent detection efficiency are necessary to obtain good FCS results.

The basic optical setup for a dual-colour FCS experiment is shown in Fig. 1.49. Two diode lasers are used to excite fluorescence in the sample. The sample contains two fluorophores, Atto 488 and Atto 647N, each of them excited by one of the lasers. The fluorescence is detected by two detectors through different filters. The detection (macro) times of the photons are recorded by two TCSPC channels. The photons of the channels are auto-correlated and cross-correlated by the instrument software. For the setup shown in Fig. 1.49 a bh DCS-120 confocal FLIM system was used. The principle can, however, be implemented in almost any confocal laser scanning microscope.

Auto- and cross-correlation functions of free Atto 488 and Atto 647N fluorophores are shown in Fig. 1.50, left. There is autocorrelation for the signals from both fluorophores (Red and blue curves) but no cross-correlation between the

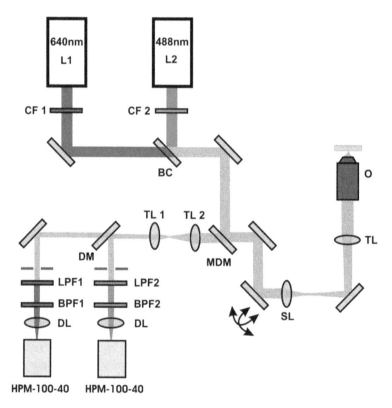

Fig. 1.49 Dual-colour FCCS setup based on BDL-SMN lasers and DSC-120 confocal FLIM system. *L1*, *L2* BDL-SMN picosecond/CW diode laser, operated in CW mode. *BC* Beam combiner. *MDM* Dual-wavelength dichroic mirror, 488/647 nm. *SL* Scan lens of DCS-120 scan head. *TL* tube lens of microscope. *O* Microscope objective lens. *TL1*, *TL2* Lenses, forming a telscope projecting the beam into comfocal pinholes. *DM* Dichroic mirror, splits the light of the two fluorophores. *LPF1*, *LPF2* Long-pass filters. *BPF1* and *BPF2* band-pass filters. *DL* Lens centering the light on the detectors

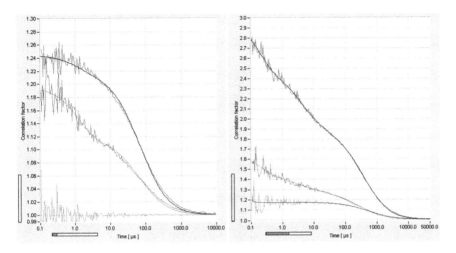

Fig. 1.50 *Left* Atto 488/7 nM Atto 647N, molecules not linked. *Right* Alexa 488 and Cy5 partially linked by double-stranded DNA. *Blue* and *Red* Autocorrelation. *Green* Cross-correlation

signals (green). Curves for a 40-base pair double-stranded DNA labelled with Alexa 488 and Cy5 are shown in Fig. 1.50, right. There is a significant cross-correlation (green curve) showing that DNA strands labelled with different fluorophores are partially linked to each other.

A third application of FCS—determination of the number of fluorophore molecules attached to a biomolecule—is used to obtain information on the structure and conformation of the molecules. Special subunits of the biomolecule are labelled by fluorescent antibodies or—more reliably—by expressing fluorescent proteins. From the amplitude of the autocorrelation function and the number of counts the brightness of the labelled biomolecule can be obtained. For a given fluorophore, this is proportional to the number of fluorophores attached to it. Provided the labelling or expression is complete it represents the number of subunits under investigation. The technique has been used to determine the structure of CaMKII by combined fluorescence anisotropy, and fluorescence correlation techniques [115].

The correlation function is also used for perfusion measurements in diffuse optical imaging. The technique is based on correlating intensity fluctuations of photons scattered in the blood flowing in the tissue [39]. The basic setup is shown in Fig. 1.51, left. A laser injects light into the tissue via a fibre. The light at a different spot is collected by another fibre and detected by a photon counting detector. An autocorrelation function of the intensity is calculated from the absolute times of the photons. To obtain a useful correlation function it is required that the laser have a coherence length longer than the average path length in the tissue, and that the fibres have diameters on the order of 20 μm or less. A correlation function recorded at a human forearm is shown in Fig. 1.51, right.

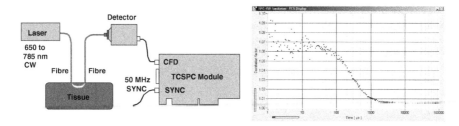

Fig. 1.51 Diffuse optical correlation. *Left* Basic experiment setup. *Right* Correlation curve recorded at a human forearm, bh SPC-150 TCSPC module

1.5.2.1 Including the Micro-Time of TCSPC in FCS

The calculation of FCS by the algorithms shown above are based exclusively on the macro times of the photons. FCS and Cross-FCS can therefore also be obtained by excitation with CW lasers. If the fluorescence is excited by pulsed lasers the micro time of the photons can be used in several ways.

Simultaneous Recording of Fluorescence Decay Functions

Obviously, fluorescence decay curves can be obtained in parallel with the FCS measurement. All that is necessary is to build up a photon distribution over the micro times of the photons. The fluorescence decay is helpful to identify problems, such as contamination by autofluorescence or Raman scattering. Fluorescence decay times can also be used to improve FCS analysis of correlation curves with several components.

Time Gating

FCS signals are sometimes contaminated by Raman scattering. Raman light shows up as a sharp peak at the beginning of the fluorescence decay. The peak can be gated off, either by setting a time gate in the TCSPC hardware [22], or by excluding photons with macro times outside a defined time window (Fig. 1.52). Time gating can also be necessary when FCS is recorded with pulse-interleaved excitation (PIE).

Filtered FCS

Micro times can be used to separate FCS signals of fluorophores of different fluorescence decay times. In that case, the photons are weighted with a coefficient derived from the micro time via a filter function [33, 62]. The technique can also be

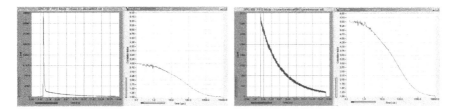

Fig. 1.52 Gated FCS, detected light contains Raman scattering. *Left* Fluorescence decay with Raman peak, FCS curve calculated from entire signal. *Right* Raman peak gated out, FCS calculated inside selected time window. Note the increased FCS amplitude. Zeiss Intune system, Becker & Hickl SPC-150 TCSPC FLIM Module with HPM-100-40 hybrid detectors

used to reject detector background signals, Raman scattering, or unwanted background fluorescence.

Global Fit of FCS and Decay Data

Another way to improve the separation of several diffusion components in FCS analysis has been developed by Anthony and Berland [2]. Assume there are several fluorescing species of similar or nearly similar excitation and emission spectra. These species diffuse at different speed. In principle, the two diffusion components can be separated by fitting the FCS data with a model that contains two diffusion components. The problem is that, unless the diffusion times are very different, the components can be separated only at limited accuracy. However, if the two species have different fluorescence lifetimes the relative amplitudes can be obtained via double-exponential decay analysis. The amplitudes can then be used in the FCS analysis to obtain a more reliable fit. An even more robust fit is obtained if all parameters, i.e. the fluorescence lifetimes, the component amplitudes, and the diffusion times of the components are determined in a global fit [2, 3].

FCS Down to the Picosecond Time Scale

By including the micro times, photons can be correlated down to the picosecond time scale. The principles has been introduced by Felekyan et al. in 2005 [60]. They used two synchronised TCSPC modules in the setup shown in Fig. 1.53, left. Fluorescence in a sample is excited by a CW laser in a femtoliter sample volume. The fluorescence light is split into two detectors, the signals of which are recorded by two TCSPC modules. The stop pulses for the time measurement come from an external 80 MHz clock generator. The internal macro time clock is synchronised to the signal at the stop inputs of the TCSPC modules [19]. The TCSPC modules work in the parameter-tag-mode; the single-photon data are constantly read by the system computer. By using both the micro times and the macro times FCS data were

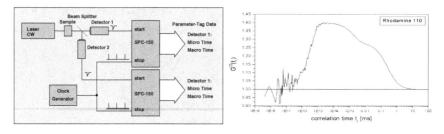

Fig. 1.53 Picosecond fluorescence correlation with two synchronised TCSPC modules

obtained at a resolution of about 20 ps. The algorithm is described in [60], and included in a single-molecule analysis software package of the University of Düsseldorf [61].

At first glance, it may look surprising that this method works: The TCSPC modules used in the setup determine the micro times by advanced TAC/ADC principles. This delivers high resolution micro times at excellent stability. However, the time scales of the micro time and the macro time are not strictly comparable, and there may be minor overlaps or gaps in the two time scales. It can also happen that photons with arrival times too close to a stop pulse are not recorded. However, all these effects are correlated to the signal from the external clock oscillator. They are *not* correlated to the photons. The correlation of the photon data therefore delivers correct results.

Applications of picosecond correlation to dye-exchange dynamics of R123 in micellar solutions and supra-molecular cyclodextrin-pyronin complexes are described in [116, 1]. Combined with FRET measurement, the method has been used to study conformational changes in a yellow chameleon Ca^+ sensor [36]. It has also been used to study complexes of adamantane and cyclodextrin [70], and micellar exchange dynamics [35].

Picosecond correlation is easy with TCSPC devices using direct time-to-digital conversion by TDC chips, especially if the chips contain several channels operated by the same internal clock. A picosecond correlation curve recorded by a Becker & Hickl DPC-230 photon correlator is shown in Fig. 1.54.

1.5.3 Single-Molecule Burst Analysis

Consider a solution of fluorescent molecules, excited by a focused laser beam through a microscope lens, with the emitted photons being detected through a confocal pinhole that transmits light only from a volume of diffraction limited size. When the concentration of fluorescent molecules is small enough only one molecule will be in the detection volume at a time. As the molecule diffuses through the excitation/detection volume it emits photons. Thus, the detection signal consist of bursts of photons caused by individual molecules. The photon bursts can be

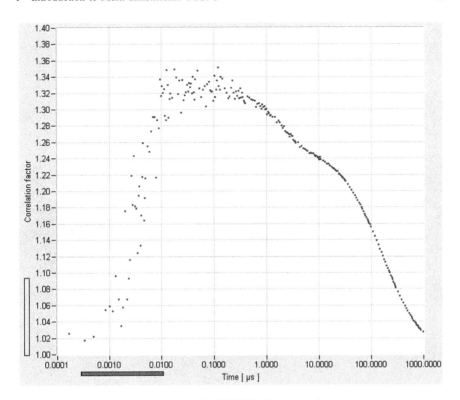

Fig. 1.54 Picosecond photon correlation by DPC-230 photon correlator

recorded in the parameter-tag mode of a TCSPC device. Photon bursts from an Alexa 488 Dextrane construct are shown in Fig. 1.55.

The idea behind single-molecule burst analysis is to determine as many as possible fluorescence parameters for the individual bursts and build up multi-dimensional histograms of the frequency of the bursts versus the parameter values. A demonstration of the technique has been given by Widengren et al. [149]. FRET pairs were constructed by linking different fluorophores to a deoxyoligonucleotide. A mixture of these constructs was investigated in a confocal optical setup. The photons were detected simultaneously under 0° and 90° polarisation and in two wavelength intervals, and recorded by TCSPC in the parameter-tag mode. Individual bursts were identified in the parameter-tag data. Within the bursts, the fluorescence intensities and the fluorescence lifetime in both wavelength channels and the fluorescence anisotropy were determined. From these data, two-dimensional histograms of the burst frequency were built up versus the intensity ratio in the two wavelength channels, the fluorescence lifetime and the fluorescence anisotropy. Donor-only constructs, donor-acceptor constructs, and donor-acceptor constructs of different donor-acceptor distance could be clearly identified in the histograms, see Fig. 1.56.

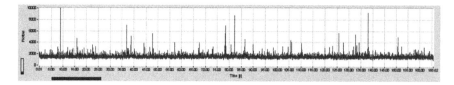

Fig. 1.55 Photon bursts from single molecules in an intensity trace calculated from parameter-tag data

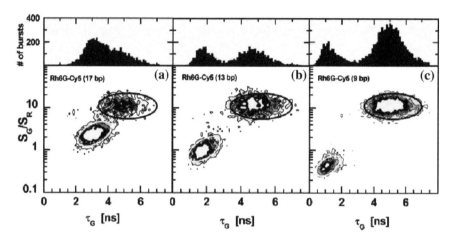

Fig. 1.56 Frequency of photon bursts in a histogram of the donor/acceptor intensity ratio versus the donor lifetime. The *black ellipses* mark donor-only events, the *yellow ellipses* donor-acceptor events. *Left to right* Donor-acceptor distance 17, 13, and 9 base pairs. With permission, from [149], by American Chemical Society

A frequent application of burst analysis is the detection of conformational changes in biomolecules by single-molecule FRET techniques. Please see Chap. 9 of this book.

The technique has been used to detect conformational changes in the sarcoen-doplasmic reticulum of calcium ATPase (SERCA). Two constructs were used. One was had Cerulean at the N-terminus and an EYFP intra-sequence tag in the nucleotide-binding domain (CY-CERCA, Fig. 1.57a). The other was labelled with GFP in the N-domain and a TagRFP in the N-terminus (Fig. 1.57b). The molecules were freely diffusing through the laser focus of a Leica SP5 two-photon micro-scope. Fluorescence was detected simultaneously in two emission wavelength intervals corresponding to the emission spectra of Cerulean and GFP. The photons were recorded by a bh SPC-830 TCSPC module via a router. Parameter-tag data were recorded and analysed as shown in Fig. 1.57c–f.

Figure 1.57c shows the intensity bursts caused by the transition of the molecules through the laser focus. The bursts were identified in the data stream and a lifetime

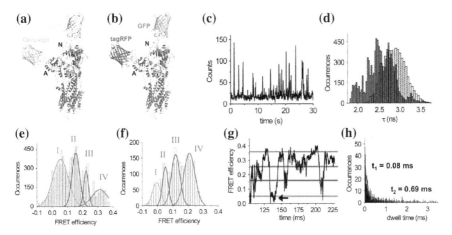

Fig. 1.57 Single-molecule FRET of SERCA. **a** CY-SERCA comprises a Cerulean FRET donor fused to the N-terminus of SERCA (*grey*) in the actuator, or A-domain, and an intrasequence FP inserted in the nucleotide-binding, or N-domain. **b** RG-SERCA is an analogous construct composed of a TagRFP acceptor fused to the N-terminus in the A-domain and GFP (*green*) donor inserted in the N-domain. **c** Time-trace of fluorescence bursts from single molecules. **d** Histogram of the fluorescence lifetimes obtained from RG-SERCA (*red*) or control GFP-SERCA (*black*). **e** Histogram of the FRET efficiency calculated from the lifetime measurements shown in panel **d**. A Gaussian fit reveals subpopulations consistent with four discrete conformations of SERCA. **f** CY-SERCA also exhibited four discrete RET states. The reduced FRET efficiency of these states (compared to panel **e**) was consistent with a larger Förster radius. **g** RG-SERCA single-molecule FRET time trajectory. The *horizontal lines* indicate States *I–IV* (*orange* through *green*) identified by Gaussian analysis of the distributions of the FRET efficiency. **h** Analysis of the dwell time for SERCA-sampling States *I–IV* revealed a biphasic distribution of dwell times characterized by fast (t_1) and slow (t_2) kinetics. With permission, from [118]

analysis performed in the individual bursts. A histogram of the lifetimes obtained over a large number of bursts is shown in Fig. 1.57d. Changes in the conformation of the molecules result in different FRET efficiency and thus in different fluorescence lifetimes. The lifetimes were therefore translated into FRET efficiencies. Histograms of the FRET efficiencies are shown in Fig. 1.57e, f. Four different FRET states are visible in the figures. The FRET efficiencies were extracted by fitting the distributions with four gaussian distributions. Figure 1.57g shows the fluctuations of the FRET efficiency within a single burst. Also here, four different FRET efficiencies can be identified. Figure 1.57h shows a distribution of the duration the molecules spent in each FRET state. The distribution was fitted with a bi-exponential model. This revealed two characteristic times, t1 and t2, the system stayed in the individual FRET states.

The experiments were extended to SERCA-expressing cultured rabbit cardio myocytes. Also here, four discrete structural states were found. The relative populations of these states oscillated with electrical pacing. Low-FRET states were most populated in the low-Ca phase (diastole), high-FRET states correlated with high Ca (systole).

Burst recording and molecule FRET analysis has been used to detect conformational changes [52, 57] also in F_0F_1-ATP synthase and even determine the step size of its molecular rotor [58], see Chap. 9. EGFP was fused to the C terminus of subunit α, and Alexa 568 attached to residue E2C at one of the c subunits, see Fig. 1.58, left. FRET occurs between the EGFP and the Alexa 568. The entire C ring (blue) rotates against the upper part of the structure. This causes a change in the distance between the EGFP and the Alexa 568 and, consequently, a change in the FRET efficiency. The FRET efficiency was determined by recording fluorescence in a red (acceptor, Alexa) and a green (donor (EGFP)) detection channel and calculating ratios of the count numbers. Bursts from constructs containing both the donor and the acceptor were selected by wavelength-multiplexed excitation and analysing the count numbers in both channels. Intensity traces and FRET distances for two selected bursts are shown in Fig. 1.58, right.

There are two different models of the rotation of the C ring against the upper part of the structure. One model predicts a step size of the rotation of 120°, the other a step size of 36°. To decide between the two models the distances of subsequent FRET states were identified in the data. The distances for subsequent states were determined, and the density plot of distances shown in Fig. 1.59 was built up. It shows the frequency of data pairs for subsequent FRET states (one distance versus the subsequent one). The result is compatible with the prediction for 36° step size (white line) but not with 120° step size (black line). Please see Chap. 9 for more details.

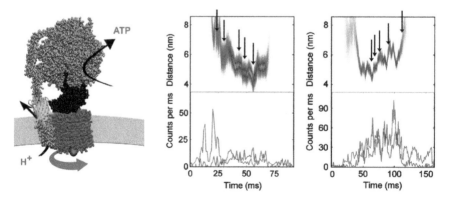

Fig. 1.58 *Left* Model of *E. coli* F0F1-ATP synthase labelled with EGFP (donor, *green*, fused to the C terminus of subunit α, *orange*) and Alexa 568 (acceptor, *red*, at residue E2C at one of the c subunits (*blue*). *Right* Counts in *green* (donor) and *red* (acceptor) detection channel (*bottom*) and donor-acceptor distances derived from FRET intensities for two selected photon bursts. The *arrows* mark steps in the FRET distance. With permission, from [58]

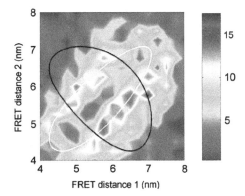

Fig. 1.59 FRET transition density plot. Distance pairs for subsequent FRET states, distance 1 and distance 2. The result is compatible with the prediction for 36° step size (*white line*) but not with 120° (*black line*). With permission, from [58]

1.6 Summary

Multi-dimensional TCSPC builds up photon distributions over various parameters of the photons. These can be inherent parameters of the photons, such as wavelength, polarisation, or the location they are coming from, parameters describing the state of the sample in the moment of the photon detection, or parameters actively varied in the experiment setup. The advantage over classic approaches is that no photons are suppressed in the recording process, and that the speed of the parameter variation gets de-coupled from the photon detection rate. The signal-to-noise rate of the results depends only on the total acquisition time, not on the time scale at which individual parameters are resolved. Typical applications of multi-dimensional TCSPC are multi-wavelength recording of fluorescence decay data, diffuse optical imaging experiments, multi-spectral lifetime imaging of biological tissue, recording of physiological changes in biological objects, fluorescence lifetime imaging (FLIM), or combinations of these techniques like multi-wavelength FLIM, excitation-wavelength multiplexed FLIM, time-series FLIM, recording of transient changes in the fluorescence of a sample along a line (FLITS) or across a selected area, and simultaneous fluorescence and phosphorescence lifetime imaging (FLIM/PLIM). Moreover, data of the individual photons can be stored and used to build up FCS and FCCS data or multi-dimensional histograms of single-molecule parameters over a longer period of time or a large number of molecules. There are possible TCSPC applications which have not finally been explored yet: TCSPC could be combined with electro-physiology, with protein folding experiments, temperature-jump techniques, and experiments that record changes in the metabolic state of cells and tissue.

References

1. W. Al-Soufi, B. Reija, M. Novo, Suren Felekyan, R. Kühnemuth, C.A.M. Seidel, Fluorescence correlation spectroscopy, a tool to investigate supramolecular dynamics: inclusion complexes of pyronines with cyclodextrin. J. Am. Chem. Soc. **127**, 8775–8784 (2005)
2. N. Anthony, K. Berland, Global Analysis in Fluorescence Correlation Spectroscopy and Fluorescence Lifetime Microscopy, in *Methods in Enzymology 518, Fluorescence Fluctuation Spectroscopy (FFS), Part A*, ed. by S.Y. Tetin (Academic Press, San Diego, 2013)
3. N.R. Anthony, K.M. Berland, Global analysis enhances resolution and sensitivity in fluorescence fluctuation measurements. PLoS ONE **9**, e90456, www.plosone.ord
4. E. Baggaley, M.R. Gill, N.H. Green, D. Turton, I.V. Sazanovich, S.W. Botchway, C. Smythe, J.W. Haycock, J.A. Weinstein, J.A. Thomas, Dinuclear Ruthenium(II) complexes as two-photon, time-resolved emission microscopy probes for cellular DNA. Angew. Chem. Int. Ed. Engl. **53**, 3367–3371 (2014)
5. E. Baggaley, S.W. Botchway, J.W. Haycock, H. Morris, I.V. Sazanovich, J.A.G. Williams, J.A. Weinstein, Long-lived metal complexes open up microsecond lifetime imaging microscopy under multiphoton excitation: from FLIM to PLIM and beyond. Chem. Sci. **5**, 879–886 (2014)
6. Becker & Hickl GmbH, DCS-120 Confocal Scanning FLIM Systems, user handbook, edition 2012, www.becker-hickl.com
7. Becker & Hickl GmbH, Modular FLIM systems for Zeiss LSM 510 and LSM 710 family laser scanning microscopes. User handbook, 5th edition. www.becker-hickl.com
8. W. Becker, H. Stiel, Verfahren zur mehrdimensionalen zeitaufgelösten Messung von Lichtsignalen durch Photonenzählung, Patent WP 282 518, G01 J/327 7903 (1998)
9. W. Becker, H. Stiel, E. Klose, Flexible Instrument for time-correlated single photon counting. Rev. Sci. Instrum. **62**, 2991–2996 (1991)
10. W. Becker, Verfahren und Vorrichtung zur zeitkorrelierten Einzelphotonenzählung mit hoher Registrierrate, Patent DE 43 39 784 (1993)
11. W. Becker, Verfahren und Vorrichtung zur Messung von Lichtsignalen mit zeitlicher und räumlicher Auflösung, Patent DE 43 39 787 (1993)
12. W. Becker, H. Hickl, C. Zander, K.H. Drexhage, M. Sauer, S. Siebert, J. Wolfrum, Time-resolved detection and identification of single analyte molecules in microcapillaries by time-correlated single photon counting. Rev. Sci. Instrum. **70**, 1835–1841 (1999)
13. W. Becker, K. Benndorf, A. Bergmann, C. Biskup, K. König, U. Tirlapur, T. Zimmer, FRET measurements by TCSPC laser scanning microscopy. Proc. SPIE **4431**, 94–98 (2001)
14. W. Becker, A. Bergmann, K. König, U. Tirlapur, Picosecond fluorescence lifetime microscopy by TCSPC imaging. Proc. SPIE **4262**, 414–419 (2001)
15. W. Becker, A. Bergmann, C. Biskup, T. Zimmer, N. Klöcker, K. Benndorf, Multi-wavelength TCSPC lifetime imaging. Proc. SPIE **4620**, 79–84 (2002)
16. W. Becker, A. Bergmann, G. Weiss, Lifetime Imaging with the Zeiss LSM-510. Proc. SPIE **4620**, 30–35 (2002)
17. W. Becker, A. Bergmann, M.A. Hink, K. König, K. Benndorf, C. Biskup, Fluorescence lifetime imaging by time-correlated single photon counting. Micr. Res. Techn. **63**, 58–66 (2004)
18. W. Becker, A. Bergmann, G. Biscotti, K. Koenig, I. Riemann, L. Kelbauskas, C. Biskup, High-speed FLIM data acquisition by time-correlated single photon counting. Proc. SPIE **5323**, 27–35 (2004)
19. W. Becker, *Advanced Time-Correlated Single-Photon Counting Techniques* (Springer, Berlin, 2005)
20. W. Becker, A. Bergmann, C. Biskup, Multi-spectral fluorescence lifetime imaging by TCSPC. Micr. Res. Tech. **70**, 403–409 (2007)

21. W. Becker, B. Su, A. Bergmann, Fast-acquisition multispectral FLIM by parallel TCSPC. Proc. SPIE **7183**, 718305 (2009)
22. W. Becker, The bh TCSPC handbook. 6th edition. Becker & Hickl GmbH (2015), www.becker-hickl.com, Printed copies available from Becker & Hickl GmbH
23. W. Becker, B. Su, A. Bergmann, Spatially resolved recording of transient fluorescence lifetime effects by line-scanning TCSPC. Proc. SPIE **8226**, 82260C-1–82260C-6 (2012)
24. W. Becker, Fluorescence lifetime imaging—techniques and applications. J. Microsc. **247**, 119–136 (2012)
25. W. Becker, V. Shcheslavkiy, S. Frere, I. Slutsky, Spatially resolved recording of transient fluorescence-lifetime effects by line-scanning TCSPC. Microsc. Res. Techn. **77**, 216–224 (2014)
26. M.Y. Berezin, S. Achilefu, Fluorescence lifetime measurement and biological imaging. Chem. Rev. **110**, 2641–2684 (2010)
27. D.J.S. Birch, A.S. Holmes, R.E. Imhof, B.Z. Nadolski, K. Suhling, Multiplexed array fluorometry. J. Phys. E. Sci Instrum. **21**, 415 (1988)
28. D.K. Bird, K.W. Eliceiri, C.-H. Fan, J.G. White, Simultaneous two-photon spectral and lifetime fluorescence microscopy. Appl. Opt. **43**, 5173–5182 (2004)
29. D.K. Bird, L. Yan, K.M. Vrotsos, K.E. Eliceiri, E.M. Vaughan, Metabolic mapping of MCF10A human breast cells via multiphoton fluorescence lifetime imaging of coenzyme NADH. Cancer Res. **65**, 8766–8773 (2005)
30. C. Biskup, T. Zimmer, L. Kelbauskas, B. Hoffmann, N. Klöcker, W. Becker, A. Bergmann, K. Benndorf, Multi-dimensional fluorescence lifetime and FRET measurements. Micr. Res. Tech. **70**, 403–409 (2007)
31. C. Biskup, B. Hoffmann, K. Benndorf, A. Rueck, Spectrally Resolved Lifetime Imaging Microscopy, in *FLIM Microscopy in Biology and Medicine*, ed. by A. Periasamy, R.M. Clegg (CRC Press, Boca Raton, 2009)
32. T.S. Blacker, Z.F. Mann, J.E. Gale, M. Ziegler, A.J. Bain, G. Szabadkai, M.R. Duchen, Separating NADH and NADPH fluorescence in live cells and tissues using FLIM. Nat. Commun. **5**, 3936-1–3936-6 (2014)
33. M. Böhmer, M. Wahl, H.-J. Rahn, R. Erdmann, J. Enderlein, Time-resolved fluorescence correlation spectroscopy. Chem. Phys. Lett. **353**, 439–445 (2002)
34. L.M. Bollinger, G.E. Thomas, Measurement of the time dependence of scintillation intensity by a delayed coincidence method. Rev. Sci. Instrum. **32**, 1044–1050 (1961)
35. J. Bordello, M. Novo, W. Al-Soufi, Exchange dynamics of a neutral hydrophobic dye in micellar solutions studied by fluorescence correlation spectroscopy. J. Colloid Interface Sci. **345**, 369–376 (2010)
36. J.W. Borst, S.P. Laptenok, A.H. Westphal, R. Kühnemuth, H. Hornen, N.V. Visser, S. Kalinin, J. Aker, A. van Hoek, C.A.M. Seidel, A.J.W.G. Visser, Structural changes of yellow cameleon domains observed by quantitative FRET analysis and polarized fluorescence correlation spectroscopy. Biophys. J. **95**, 5399–5411 (2008)
37. R. Brandenburg, H.-E. Wagner A. M. Morozov, K. V. Kozlov, Axial and radial development of microdischarges of barrier discharges in N_2/O_2 mixtures at atmospheric pressure. J. Phys. D: Appl. Phys. **38**, 1649–1657 (2005)
38. R. Brandenburg, H. Grosch, T. Hoder, K.-D. Weltmann, Phase resolved cross-correlation spectroscopy on surface barrier discharges in air at atmospheric pressure. Eur. Phys. J. Appl. Phys. **55**, 13813-p1–13813-p6 (2011)
39. C. Cheung, J.P. Culver, K. Takahashi, J.H. Greenberg, A.G. Yodh, In vivo cerebrovascular measurement combining diffuse near-infrared absorption and correlation spectroscopies. Phys. Med. Biol. **46**(8), 2053–2065 (2001)
40. D. Chorvat, A. Chorvatova, Spectrally resolved time-correlated single photon counting: a novel approach for characterization of endogenous fluorescence in isolated cardiac myocytes. Eur. Biophys. J. **36**, 73–83 (2006)

41. D. Chorvat, A. Mateasik, J. Kirchnerova, A. Chorvatova, Application of spectral unmixing in multi-wavelength time-resolved spectroscopy. Proc. SPIE **6771**, 677105-1–677105-12 (2007)

42. D. Chorvat, A. Chorvatova, Multi-wavelength fluorescence lifetime spectroscopy: a new approach to the study of endogenous fluorescence in living cells and tissues. Laser Phys. Lett. **6**, 175–193 (2009)

43. D. Chorvat Jr., A. Mateasik, Y.g Cheng, N.y Poirier, J. Miro, N.S. Dahdah, A. Chorvatova, Rejection of transplanted hearts in patients evaluated by the component analysis of multi-wavelength NAD(P)H fluorescence lifetime spectroscopy. J. Biophotonics **3**, 646–652 (2010)

44. A. Chorvatova, F. Elzwiei, A. Mateasik, D. Chorvat, Effect of ouabain on metabolic oxidative state in living cardiomyocytes evaluated by time-resolved spectroscopy of endogenous NAAD(P)H fluorescence. J. Biomed. Opt. **17**(10), 101505-1–101505-7 (2012)

45. A. Chorvatova, S. Aneba, A. Mateasik, D. Chorvat Jr., B. Comte, Time-resolved fluorescence spectroscopy investigation of the effect of 4-hydroxynonenal on endogenous NAD(P)H in living cardiac myocytes. J. Biomed. Opt. **18**(6), 067009-1–067009-11 (2013)

46. A. Chorvatova, A. Mateasik, D. Chorvat Jr, Spectral decomposition of NAD(P)H fluorescence components recorded by multi-wavelength fluorescence lifetime spectroscopy in living cardiac cells. Laser Phys. Lett. **10**, 125703-1–125703-10 (2013)

47. S. Coda, A.J. Thompson,1,5 G.T. Kennedy, K.L. Roche, L. Ayaru, D.S. Bansi, G.W. Stamp, A.V. Thillainayagam, P.M.W. French, C. Dunsby, Fluorescence lifetime spectroscopy of tissue autofluorescence in normal and diseased colon measured ex vivo using a fiber-optic probe. Biomed. Opt. Expr. **5**, 515–538 (2014)

48. D. Contini, A. Torricelli, A. Pifferi, L. Spinelli, F. Paglia, R. Cubeddu, Multi-channel time-resolved system for functional near infrared spectroscopy. Opt. Express **14**, 5418–5432 (2006)

49. R.J. Cooper, E. Magee, N. Everdell, S. Magazov, M. Varela, D. Airantzis, A.P. Gibson, J.C. Hebden, MONSTIR II: a 32-channel, multispectral, time-resolved optical tomography system for neonatal brain imaging. Rev. Sci. Instrum. **85**(5), 0531052014 (2014)

50. S. Cova, M. Bertolaccini, C. Bussolati, The measurement of luminescence waveforms by single-photon techniques. Phys. Stat. Sol. **18**, 11–61 (1973)

51. P.A.A. De Beule, C. Dunsby, N.P. Galletly, G.W. Stamp, A.C. Chu, U. Anand, P. Anand, C.D. Benham A. Naylor, P.M.W. French, A hyperspectral fluorescence lifetime probe for skin cancer diagnosis. Rev. Sci. Instrum. **78**, 123101 (2007)

52. M. Diez, B. Zimmermann, M. Börsch, M. König, E. Schweinberger, S. Steigmiller, R. Reuter, S. Fe-lekyan, V. Kudryavtchev, C.A.M. Seidel, P. Gräber, Proton-powered subunit rotation in single membrane-bound F0F1-ATP synthase. Nat. Struct. Mol. Biol. **11** (2), 135–141 (2004)

53. R.I. Dmitriev, A.V. Zhdanov, Y.M. Nolan, D.B. Papkovsky, Imaging of neurosphere oxygenation with phosphorescent probes. Biomaterials **34**, 9307–9317 (2013)

54. R.I. Dmitriev, A.V. Kondrashina, K. Koren, I. Klimant, A.V. Zhdanov, J.M.P. Pakan, K.W. McDermott, D.B. Papkovsky, Small molecule phosphorescent probes for O_2 imaging in 3D tissue models. Biomater. Sci. **2**, 853–866 (2014)

55. E. Dimitrow, I. Riemann, A. Ehlers, M.J. Koehler, J. Norgauer, P. Elsner, K. König, M. Kaatz, Spectral fluorescence lifetime detection and selective melanin imaging by multiphoton laser tomography for melanoma diagnosis. Exp. Dermatol. **18**, 509–515 (2009)

56. K. Drozdowicz-Tomsia, A.G. Anwer, M.A. Cahill, K.N. Madlum, A.M. Maki, M.S. Baker, E.M. Goldys, Multiphoton fluorescence lifetime imaging microscopy reveals free-to-bound NADH ratio changes associated with metabolic inhibition. J. Biomed. Opt. **19**, 08601 (2014)

57. M. Düser, Y. Bi, N. Zarrabi, S.D. Dunn, M. Börsch, The proton-translocating a subunit of F0F1-ATP synthase is allocated asymmetrically to the peripheral stalk. J. Biol. Chem. **48**, 33602–33610 (2008)

58. M. Düser, N.d Zarrabi, D.J. Cipriano, S. Ernst, G.D. Glick, S.D. Dunn, M. Börsch, 36° step size of proton-driven c-ring rotation in FoF1-ATP synthase. EMBO J. **28**, 2689–2696 (2009)

59. C. Dysli, G. Quellec, M Abegg, M.N. Menke, U. Wolf-Schnurrbusch, J. Kowal, J. Blatz, O. La Schiazza, A.B. Leichtle, S. Wolf, M.S. Zinkernagel, Quantitative analysis of fluorescence lifetime measurements of the macula using the fluorescence lifetime imaging ophthalmoscope in healthy subjects. IOVS **55**, 2107–2113 (2014)

60. S. Felekyan, R. Kühnemuth, V. Kudryavtsev, C. Sandhagen, W. Becker, C.A.M. Seidel, Full correlation from picoseconds to seconds by time-resolved and time-correlated single photon detection. Rev. Sci. Instrum. **76**, 083104 (2005)

61. S. Felekyan, Software package for multiparameter fluorescence spectroscopy, full correlation and multiparameter imaging. www.mpc.uni-duesseldorf.de/seidel/software.htm

62. S. Felekyan, S. Kalinin, A. Valeri, C.A.M. Seidel, Filtered FCS and species cross correlation function. Proc. SPIE **7183**, 71830D (2009)

63. Th. Förster, Zwischenmolekulare Energiewanderung und Fluoreszenz, Ann. Phys. (Serie 6) **2**, 55–75 (1948)

64. Th. Förster, Energy migration and fluorescence. Translated by Klaus Suhling. J. Biomed. Opt. **17**, 011002-1–011002-10

65. K. Funk, A. Woitecki, C. Franjic-Würtz, Th Gensch, F. Möhrlein, S. Frings, Modulation of chloride homeostasis by inflammatory mediators in dorsal ganglion neurons. Mol. Pain **4**, 32 (2008)

66. C.D. Geddes, H. Cao, I. Gryczynski, J. Fang, J.R. Lakowicz, Metal-enhanced fluorescence (MEF) due to silver colloids on a planar surface: Potential applications of indocyanine green to in vivo imaging. J. Phys. Chem. A **107**, 3443–3449 (2003)

67. V. Ghukassian, F.-J. Kao, Monitoring cellular metabolism with fluorescence lifetime of reduced nicotinamide adenine dinucleotide. J. Phys. Chem. C **113**, 11532–11540 (2009)

68. D. Gilbert, C. Franjic-Würtz, K. Funk, T. Gensch, S. Frings, F. Möhrlen, Differential maturation of chloride homeostasis in primary afferent neurons of the somatosensory system. Int. J. Devl. Neurosci. **25**, 479–489 (2007)

69. R. Govindjee, Sixty-three years since Kautsky: chlorophyll α fluorescence, Aust. J. Plant Physiol. **22**, 131–160 (1995)

70. D. Granadero, J. Bordello, M.J. Perez-Alvite, M. Novo, W. Al-Soufi, Host-guest complexation studied by fluorescence correlation spectroscopy: adamantane-cyclodextrin inclusion. Int. J. Mol. Sci. **11**, 173–188 (2010)

71. H. Grosch, T. Hoder, K.-D. Weltmann, R. Brandenburg, Spatio-temporal development of microdischarges in a surface barrier discharge arrangement in air at atmospheric pressure. Eur. Phys. J. D **60**, 547–553 (2010)

72. R. Hanbury-Brown, R.Q. Twiss, Nature **177**, 27–29 (1956)

73. K.M. Hanson, M.J. Behne, N.P. Barry, T.M. Mauro, E. Gratton, Two-photon fluorescence imaging of the skin stratum corneum pH gradient. Biophys. J. **83**, 1682–1690 (2002)

74. T. Hoder, R. Brandenburg, R. Basner1, K.-D. Weltmann, K.V. Kozlov, H.-E. Wagner, A comparative study of three different types of barrier discharges in air at atmospheric pressure by cross-correlation spectroscopy. J. Phys. D: Appl. Phys. **43**, 124009-1–124009-8 (2010)

75. T. Hoder, M. Cernak, J. Paillol, D. Loffhagen, R. Brandenburg, High-resolution measurements of the electric field at the streamer arrival to the cathode: A unification of the streamer-initiated gas-breakdown mechanism. Phys. Rev. E **86**, 055401-1–055401-5 (2012)

76. M. Kacprzak, A. Liebert, P. Sawosz, N. Zolek, R. Maniewski, Time-resolved optical imager for assessment of cerebral oxygenation. J. Biomed. Opt. **12**, 034019-1–034019-14 (2007)

77. H. Kaneko, I. Putzier, S. Frings, U.B. Kaupp, Th Gensch, Chloride accumulation in mammalian olfactory sensory neurons. J. Neurosci. **24**(36), 7931–7938 (2004)

78. S.R. Kantelhardt, J. Leppert, J. Krajewski, N. Petkus, E. Reusche, V. M. Tronnier, G. Hüttmann, A. Giese, Imaging of brain and brain tumor specimens by time-resolved multiphoton excitation microscopy ex vivo. Neuro-Onkology **9**, 103–112 (2007)

79. V. Katsoulidou, A. Bergmann, W. Becker, How fast can TCSPC FLIM be made? Proc. SPIE **6771**, 67710B-1–67710B-7 (2007)

80. L. Kelbauskas, W. Dietel, Internalization of aggregated photosensitizers by tumor cells: Subcellular time-resolved fluorescence spectroscopy on derivatives of pyropheophorbide-a ethers and chlorin e6 under femtosecond one- and two-photon excitation. Photochem. Photobiol. **76**, 686–694 (2002)

81. S. Khoon Teh, W. Zheng, S. Li, D. Li, Y. Zeng, Y. Yang, J. Y. Qu, Multimodal nonlinear optical microscopy improves the accuracy of early diagnosis of squamous intraepithelial neoplasia. J. Biomed. Opt. **18**(3), 036001-1–036001-11 (2013)

82. P. Kloc, H.-E. Wagner, D. Trunec, Z. Navratil, G. Fedosov, An investigation of dieclectric barrier dis-charge in Ar and Ar/NH$_3$ mixture using cross-correlation spectroscopy. J. Phys. D Appl. Phys. **43**, 34514–345205 (2010)

83. M. Köllner, J. Wolfrum, How many photons are necessary for fluorescence-lifetime measurements? Phys. Chem. Lett. **200**, 199–204 (1992)

84. K. König, Clinical multiphoton tomography. J. Biophoton. **1**, 13–23 (2008)

85. K. Koenig, A. Uchugonova, in *Multiphoton Fluorescence Lifetime Imaging at the Dawn of Clinical Application*, ed by A. Periasamy, R.M. Clegg, FLIM Microscopy in Biology and Medicine (CRC Press, Boca Raton, 2009)

86. K. König, A. Uchugonova, E. Gorjup, Multiphoton fluorescence lifetime imaging of 3D-stem cell spheroids during differentiation. Microsc. Res. Techn. **74**, 9–17 (2011)

87. K.V. Kozlov, R. Brandenburg, H.-E. Wagner, A.M. Morozov, P. Michel, Investigation of the filamentary and diffuse mode of barrier discharges in N2/O2 mixtures at atmospheric pressure by cross-correlation spectroscopy. J. Phys. D Appl. Phys. **38**, 518–529 (2005)

88. K.V. Kuchibhotla, C.R. Lattarulo, B. Hyman, B. J. Bacskai, Synchronous hyperactivity and intercellular calcium waves in astrocytes in Alzheimer mice. Science **323**, 1211–1215 (2009)

89. M.K. Kuimova, G. Yahioglu, J.A. Levitt, K. Suhling, Molecular rotor measures viscosity of live cells via fluorescence lifetime imaging. J. Am. Chem. Soc. **130**, 6672–6673 (2008)

90. J.R. Lakowicz, H. Szmacinski, M.L. Johnson, Calcium imaging using fluorescence lifetimes an long-wavelength probes. J. Fluoresc. **2**, 47–62 (1992)

91. J.R. Lakowicz, H. Szmacinski, K. Nowaczyk, M.L. Johnson, Fluorescence lifetime imaging of free and protein-bound NADH. PNAS **89**, 1271–1275 (1992)

92. J.R. Lakowicz, *Principles of Fluorescence Spectroscopy*, 3rd edn. (Springer, Berlin, 2006)

93. G.S. Lakshmikanth, G. Krishnamoorthy, Solvent-exposed tryptophans probe the dynamics at protein surfaces. Biophys. J. **77**, 1100–1106 (1999)

94. G.S. Lakshmikanth, K. Sridevi, G. Krishnamoorthy, J.B. Udgaonkar, Structure is lost incrementally during the unfolding of barstar. Nat. Struct. Biol. **8**, 799–804 (2001)

95. J. Leppert, J. Krajewski, S.R. Kantelhardt, S. Schlaffer, N. Petkus, E. Reusche, G. Hüttmann, A. Giese, Multiphoton excitation of autofluorescence for microscopy of glioma tissue. Neurosurgery **58**, 759–767 (2006)

96. B. Leskovar, C.C. Lo, Photon counting system for subnanosecond fluorescence lifetime measurements. Rev. Sci. Instrum. **47**, 1113–1121 (1976)

97. C. Lewis, W.R. Ware, The measurement of short-lived fluorescence decay using the single photon counting method. Rev. Sci. Instrum. **44**, 107–114 (1973)

98. D. Li, W. Zheng, J.Y. Qu, Time-resolved spectroscopic imaging reveals the fundamentals of cellular NADH fluorescence. Opt. Lett. **33**, 2365–2367 (2008)

99. D. Li, M.S. Yang, W. Zheng, J.Y. Qu, Study of cadmium-induced cytotoxicity using two-photon excitation endogenous fluorescence microscopy. J. Biomed. Opt. **14**(5), 054028-1–054028-8 (2009)

100. D. Li, W. Zheng, .J.Y. Qu, Two-photon autofluorescence microscopy of multicolor excitation. Opt. Lett. **34**, 202–204 (2009)

101. D. Li, W. Zheng, .J.Y. Qu, Imaging of epithelial tissue in vivo based on excitation of multiple endogenous nonlinear optical signals. Letter **34**, 2853–2855 (2009)

102. D. Li, W. Zheng, Y. Zeng, and J. Y. Qu, In vivo and simultaneous multimodal imaging: Integrated multiplex coherent anti-Stokes Raman scattering and two-photon microscopy. Appl. Phys. Lett **97**, 223702-1–223702-3 (2010)

103. D. Li, W. Zheng, Y. Zeng, Y. Luo, J.Y. Qu, Two-photon excited hemoglobin fluorescence provides contrast mechanism for label-free imaging of microvasculature in vivo. Opt. Lett. **36**, 834–836 (2011)
104. D. Li, W. Zheng, W. Zhang, S. Khoon Teh, Y. Zeng, Y. Luo, J.Y. Qu, Time-resolved detection enables standard two-photon fluorescence microscopy for in vivo label-free imaging of microvasculature in tissue. Opt. Lett. **36**, 2638–2640 (2011)
105. A. Liebert, H. Wabnitz, M. Möller, A. Walter, R. Macdonald, H. Rinneberg, H. Obrig, I. Steinbrink, Time-Resolved Diffuse NIR-Reflectance Topography of the Adult Head During Motor Stimulation, in *OSA Biomedical Optics Topical Meetings on CD ROM* (The Optical Society of America, Washington, DC, WF34 2004)
106. A. Liebert, H. Wabnitz, J. Steinbrink, H. Obrig, M. Möller, R. Macdonald, A. Villringer, H. Rinneberg, Time-resolved multidistance near-infrared spectroscopy at the human head: Intra- and extracerebral absorption changes from moments of distribution of times of flight of photons. Appl. Opt. **43**, 3037–3047 (2004)
107. A. Liebert, H. Wabnitz, J. Steinbrink, M. Möller, R. Macdonald, H. Rinneberg, A. Villringer, H. Obrig, Bed-side assessment of cerebral perfusion in stroke patients based on optical monitoring of a dye bolus by time-resolved diffuse reflectance. NeuroImage **24**, 426–435 (2005)
108. A. Liebert, P. Sawosz, D. Milej, M. Kacprzak, W. Weigl, M. Botwicz, J. Maczewska, K. Fronczewska, E. Mayzner-Zawadzka, L. Krolicki, R. Maniewski, Assessment of inflow and washout of indocyanine green in the adult human brain by monitoring of diffuse reflectance at large source-detector separation. J. Biomed. Opt. **16**(4), 046011-1–046011-7 (2011)
109. J. Malicka, I. Gryczynski, C.D. Geddes, J.R. Lakowicz, Metal-enhanced emission from indocyanine green: a new approach to in vivo imaging. J. Biomed. Opt. **8**, 472–478 (2003)
110. W. Meiling, F. Stary, *Nanosecond Pulse Techniques* (Akademie-Verlag, Berlin, 1963)
111. D. Milej, A. Gerega, N. Zolek, W. Weigl, M. Kacprzak, P. Sawosz, J. Maczewska, K. Fronczewska, E. Mayzner-Zawadzka, L. Krolicki, R. Maniewski, A. Liebert, Time-resolved detection of fluorescent light during inflow of ICG to the brain—a methodological study. Phys. Med. Biol. **57**, 6725–6742 (2012)
112. A. Minta, J.P.Y. Kao, R.Y. Tsien, Fluorescent indicators for cytosolic calcium based on rhodamine and fluorescein chromophores. J. Biol. Chem. **264**, 8171–8178 (1989)
113. H.S. Muddana, T.T. Morgan, J.H. Adair, P.J. Butler, Photophysics of Cy3-encapsulated calcium phosphate nanoparticles. Nano Lett. **9**(4), 1556–1559 (2009)
114. S. Mukhopadhyay, P.K. Nayak, J.B. Udgaonkar, G. Krishnamoorthy, Characterization of the formation of amyloid Protofibrils from Barstar by mapping residue-specific fluorescence dynamics. J. Mol. Biol. **358**, 935–942 (2006)
115. T.A. Nguyen, P. Sarkar, J.V. Veetil, S.V. Koushik, S.S. Vogel, Fluorescence polarization and fluctuation analysis monitors subunit proximity, stoichiometry, and protein complex hydrodynamics. PLoS ONE **7**, e38209-1–e38209-13 (2012)
116. M. Novo, S. Felekyan, C.A.M. Seidel, W. Al-Soufi, Dye-exchange dynamics in micellar solutions studied by fluorescence correlation spectroscopy. J. Phys. Chem. B **111**, 3614–3624 (2007)
117. D.V. O'Connor, D. Phillips, *Time-Correlated Single Photon Counting* (Academic Press, London, 1984)
118. S. Pallikkuth, D.J. Blackwell, Z. Hu, Z. Hou, D.T. Zieman, B. Svensson, D.D. Thomas, S.L. Robia, Phosphorylated phospholamban stabilizes a compact conformation of the cardiac calcium-ATPase. Biophys. J. **105**, 1812–1821 (2013)
119. R.J. Paul, H. Schneckenburger, Oxygen concentration and the oxidation-reduction state of yeast: determination of free/bound NADH and flavins by time-resolved spectroscopy. Naturwissenschaften **83**, 32–35 (1996)
120. J.P. Philip, K. Carlsson, Theoretical investigation of the signal-to-noise ratio in fluorescence lifetime imaging. J. Opt. Soc. Am. **A20**, 368–379 (2003)

121. M. Prummer, C. Hübner, B. Sick, B. Hecht, A. Renn, U.P. Wild, Single-molecule identification by spectrally and time-resolved fluorescence detection. Anal. Chem. **72**, 433–447 (2000)
122. R. Re, D. Contini, M. Caffini, R. Cubeddu, L. Spinelli, A. Torricelli, A compact time-resolved system for near infrared spectroscopy based on wavelength space multiplexing. Rev. Sci. Instrum. **81**, 113101 (2010)
123. R. Rigler, E.S. Elson (eds.), *Fluorescence Correlation Spectroscopy* (Springer, Berlin, 2001)
124. T. Ritman-Meer, N.I. Cade, D. Richards, Spatial imaging of modifications to fluorescence lifetime and intensity by individual Ag nanoparticles. Appl. Phys. Lett. **91**, 123122 (2007)
125. M.S. Roberts, Y. Dancik, T.W. Prow, C.A. Thorling, L. Li, J.E. Grice, T.A. Robertson, K. König, W. Becker, Non-invasive imaging of skin physiology and percutaneous penetration using fluorescence spectral and lifetime imaging with multiphoton and confocal microscopy. Eur. J. Pharm. Biopharm. **77**, 469–488 (2011)
126. A. Rück, F. Dolp, C. Hülshoff, C. Hauser, C. Scalfi-Happ, Fluorescence lifetime imaging in PDT. An overview. Med. Laser Appl. **20**, 125–129 (2005)
127. A. Rück, Ch. Hülshoff, I. Kinzler, W. Becker, R. Steiner, SLIM: a new method for molecular imaging. Micr. Res. Tech. **70**, 403–409 (2007)
128. A. Rück, C. Hauser, S. Mosch, S. Kalinina, Spectrally resolved fluorescence lifetime imaging to investigate cell metabolism in malignant and nonmalignant oral mucosa cells. J. Biomed. Opt. **19**(9), 096005-1–096005-9 (2014)
129. S. Sakadžic, E. Roussakis, M.A. Yaseen, E.T. Mandeville, V.J. Srinivasan1, K. Arai, S. Ruvinskaya, A. Devor, E.H. Lo, S.A. Vinogradov, D.A. Boas, Two-photon high-resolution measurement of partial pressure of oxygen in cerebral vasculature and tissue. Nat. Methods **7**(9), 755–759 (2010)
130. R. Sanders, A. Draaijer, H.C. Gerritsen, P.M. Houpt, Y.K. Levine, Quantitative pH Imaging in cells using confocal fluorescence lifetime imaging microscopy. Anal. Biochem. **227**, 302–308 (1995)
131. W.Y. Sanchez, T.W. Prow, W.H. Sanchez, J.E. Grice, M.S. Roberts, Analysis of the metaboloic deterioration of ex-vivo skin, from ischemic necrosis, through the imaging of intracellular NAD(P)H by multiphoton tomography and fluorescence lifetime imaging microscopy (MPT-FLIM). J. Biomed. Opt. 09567RR in press (2010)
132. A.M. Saxena, J.B. Udgaonkar, G. Krishnamoorthy, Characterization of intra-molecule distances ans site-specific dynamics in chemically unfolded barstar: Evidence for denaturant-dependent non-random structure. J. Mol. Biol. **359**, 174–189 (2006)
133. R. Schuyler, I. Isenberg, A monophoton fluorometer with energy discrimination. Rev. Sci. Instrum. **42**, 813–817 (1971)
134. D. Schweitzer, A. Kolb, M. Hammer, E. Thamm, Basic investigations for 2-dimensional time-resolved fluorescence measurements at the fundus. Int. Ophthalmol. **23**, 399–404 (2001)
135. D. Schweitzer, S. Schenke, M. Hammer, F. Schweitzer, S. Jentsch, E. Birckner, W. Becker, Towards metabolic mapping of the human retina. Micr. Res. Tech. **70**, 403–409 (2007)
136. D. Schweitzer, Metabolic Mapping, in *Medical Retina, Essential in Opthalmology*, ed. by F. G. Holz, R.F. Spaide (Springer, New York, 2010)
137. M.C. Skala, K.M. Riching, D.K. Bird, A. Dendron-Fitzpatrick, J. Eickhoff, K.W. Eliceiri, P J. Keely, N. Ramanujam, In vivo multiphoton fluorescence lifetime imaging of protein-bound and free nicotinamide adenine dinucleotide in normal and precancerous epithelia. J. Biomed. Opt. **12**, 02401-1–02401-10 (2007)
138. M.C. Skala, K.M. Riching, A. Gendron-Fitzpatrick, J. Eickhoff, K.W. Eliceiri, J.G. White, N. Ramanujam, In vivo multiphoton microscopy of NADH and FAD redox states, fluorescence lifetimes, and cellular morphology in precancerous epithelia. PNAS **104**, 19494–19499 (2007)
139. H. Studier, W. Becker, Megapixel FLIM. Proc. SPIE **8948**, 89481K (2014)
140. K. Suhling, D. McLoskey, D.J.S. Birch, Multiplexed single-photon counting. II. The statistical theory of time-correlated measurements. Rev. Sci. Instrum. **67**, 2230–2246 (1996)

141. Q. Sun, Y. Li, S. He, C. Situ, Z. Wu, J.Y. Qu, Label-free multimodal nonlinear optical microscopy reveals fundamental insights of skeletal muscle development. Biomed. Opt. Expr. **5**, 158–166 (2013)
142. C. Thaler, S.V. Koushik, H.L. Puhl, P.S. Blank, S.S. Vogel, Structural rearrangement of CaMKIIα catalytic domains encodes activation. PNAS **106**, 6369–6374 (2009) doi:10.1073/pnas.0901913106
143. Tregido, J.A. Levitt, K. Suhling, Effect of refractive index on the fluorescence lifetime of green fluorescent protein. J. Biomed. Opt. **13**(3), 031218-1–031218-8 (2008)
144. S.S. Vogel, C. Thaler, P.S. Blank, S.V. Koushik, Time-Resolved Fluorescence Anisotropy, ed by A. Periasamy, R.M. Clegg, FLIM Microscopy in Biology and Medicine (CRC Press, Taylor & Francis, Boca Raton, 2010)
145. H.E. Wagner, R. Brandenburg, K.V. Kozlov, Progress in the visualisation of filamentary gas discharges, part1: milestones and diagnostics of dielectric-barrier discharges by cross-correlation. J. Adv. Oxid. Technol. **7**, 11–19 (2004)
146. A.J. Walsh, R.S. Cook, H.C. Manning, D.J. Hicks, A. Lafontant, C.L. Arteaga, M.C. Skala, Optical metabolic imaging identifies glycolytic levels, subtypes, and early-treatment response in breast cancer. Cancer Res. **73**, 6164–6174 (2013)
147. A.J. Walsh, R.S. Cook, M.E. Sanders, L. Aurisicchio, G. Ciliberto, C.L. Arteaga, M.C. Skala, Quantitative optical imaging of primary tumor organoid metabolism predicts drug response in breast cancer. Cancer Res. **74**, OF1-OF11 (2014)
148. H.-W. Wang, V. Ghukassyan, C.T. Chen, Y.H. Wei, H.W. Guo, J.S. Yu, F.J. Kao, Differentiation of apoptosis from necrosis by dynamic changes of reduced nicotinamide adenine dinucleotide fluorescence lifetime in live cells. J. Biomed. Opt. **13**(5), 054011-1–054011-9 (2008)
149. J. Widengren, V. Kudryavtsev, M. Antonik, S. Berger. M. Gerken, C.A.M. Seidel, Single-molecule detection and identification of multiple species by multiparameter fluorescence detection. Anal. Chem. **78**, 2039–2050 (2006)
150. M.S. Yang, D. Li, T, Lin, J.J. Zheng, W. Zheng, J.Y. Qu, Increase in intracellular free/bound NAD[P]H as a cause of Cd-induced oxidative stress in the HepG$_2$ cells. Toxicology **247**, 6–10 (2008)
151. J. Yguerabide, Nanosecond fluorescence spectroscopy of macromolecules. Meth. Enzymol. **26**, 498–578 (1972)
152. Y. Zeng, B. Yan, Q. Sun, S. Khoon Teh, W. Zhang, Z. Wen, Jianan Y. Qu, Label-free in vivo imaging of human leukocytes using two-photon excited endogenous fluorescence. J. Biomed. Opt. **18**(4), 040103-1–040103-3 (2013)
153. W. Zheng, D. Li, J.Y. Qu, Monitoring changes of cellular metabolism and microviscosity in vitro based on time-resolved endogenous fluorescence and its anisotropy decay dynamics. J. Biomed. Opt. **15**(3), 037013-1–037013-11 (2010)
154. W. Zheng, D. Li, Y. Zeng, Y. Luo, J.Y. Qu, Two-photon excited hemoglobin fluorescence. Biomed. Opt. Expr. **2**, 71–79 (2011)

Chapter 2
TCSPC FLIM with Different Optical Scanning Techniques

Wolfgang Becker, Vladislav Shcheslavskiy and Hauke Studier

Abstract Scanning a sample with a focused laser beam and detecting light from the illuminated spot delivers images of superior quality. The images are largely free of lateral scattering, and data can be obtained from a defined plane inside the sample. Moreover, nonlinear optical techniques and optical near-field techniques like multiphoton microscopy, STED or NSOM can be applied to record images from deep sample layers or at a spatial resolution below the diffraction limit. Multi-dimensional TCSPC combines favourably with these techniques. The result is a FLIM technique with excellent spatial resolution, excellent image contrast, optical sectioning capability, near-ideal photon efficiency, excellent time resolution, and resolution of multi-exponential decay profiles in the individual pixels of the image. Moreover, it can be used to record multi-wavelength FLIM data, FLIM data for several excitation wavelengths, spatial mosaics of FLIM data, FLIM Z stacks, and temporally resolved FLIM data of physiological changes on the millisecond time scale. This chapter describes the implementation of multi-dimensional TCSPC in the optical systems of confocal and multiphoton laser scanning microscopes, the extension of the wavelength range into the near infrared, the combination of TCSPC with galvanometer scanners, piezo scanners and polygon scanners, the combination with endoscopic systems and with optical systems for imaging of millimeter and centimeter-size objects, and the use of TCSPC FLIM in near-field optical scanning (NSOM) and stimulated emission-depletion (STED) microscopy systems.

W. Becker (✉) · V. Shcheslavskiy · H. Studier
Becker & Hickl GmbH, Berlin, Germany
e-mail: becker@becker-hickl.de

V. Shcheslavskiy
e-mail: vis@becker-hickl.de

H. Studier
e-mail: studier@becker-hickl.de

© Springer International Publishing Switzerland 2015
W. Becker (ed.), *Advanced Time-Correlated Single Photon Counting Applications*,
Springer Series in Chemical Physics 111, DOI 10.1007/978-3-319-14929-5_2

2.1 Introduction

Already in the 50th of the last century it was discovered that scanning improves the image contrast. Consider a focused beam of light scanning over the sample, a single-point detector measuring the light intensity returned from the excited spot, and some kind of image reconstruction device, such as a CRT screen, that builds up an image from the intensity values at each point of the scan. A device like that records scattered light and out-of-focus blur *only from the scanned pixel* but not from the other pixels of the sample. The contrast is thus better than for wide-field illumination and detection by an image sensor: In this case, the image sensor in every pixel detects scattered light from *all other pixels* of the illuminated area. The difference between scanning and wide-field imaging can be dramatic, see Fig. 2.1.

Scanning alone, however, does not solve the general problem of microscopy of thick samples: The image contains light not only from the focal plane but from the entire depth of the sample. The images therefore quickly lose contrast with increasing thickness of the sample. The problem was solved with the invention of the confocal microscope by Minsky [65, 66]. Minsky combined scanning with detection through a confocal pinhole. Light from outside the focal plane is not focused into the pinhole and thus substantially suppressed. Although Minski used confocal imaging for transmission microscopy the principle can be applied to fluorescence microscopy as well.

After Minsky, confocal microscopy was 'forgotten and then reinvented about once every decade', as James Pawley wrote in the introduction of the second edition of the Handbook of Biological Confocal Microscopy [69].

The breakthrough came with the introduction of fluorescence techniques in biological microscopy. Fluorescence techniques not only record images of the spatial structure of a sample but also deliver information about biochemical parameters on

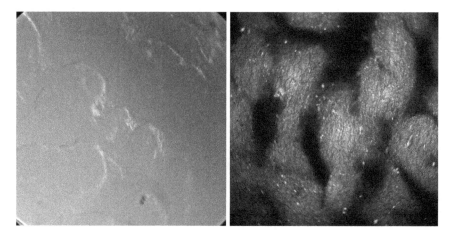

Fig. 2.1 Images obtained from a pig skin in the same microscope. *Left* Wide-field image. *Right* Scan image. Focal plane about 20 μm below the tissue surface, image size 200 μm × 200 μm

the molecular scale. While the spatial structure of a sample can still be retrieved from a low-contrast image molecular information cannot. Molecular imaging relies on the correct measurement of fluorescence intensities and fluorescence spectra in exactly defined locations of the sample. Contamination of the signals by out-of-focus light is the last thing one would like to have. Suddenly, there was a huge market for laser scanning microscopes, and a 'virtual explosion in the field of biological confocal microscopy' [26, 69, 81]. The next step came with the invention and introduction of multiphoton microscopy [12, 82]. Different than confocal detection, which avoids out-of-focus *detection*, multiphoton excitation avoids out-of-focus *excitation*.

The interest in the measurement of molecular parameters naturally also created an interest in fluorescence lifetime detection. Different than the fluorescence intensity, the *fluorescence lifetime* or, more exactly, the *fluorescence decay function*, depends on the molecular environment of a fluorophore but not on its concentration. Using the fluorescence decay functions, molecular interactions can thus be determined unimpaired by the variable, and usually unknown concentration of the fluorophores [12, 16, 76]. See also Chaps. 3, 4 and 7 of this book.

The optical sectioning capability of confocal and multiphoton laser scanning microscopes is especially important in FLIM applications: The fluorescence decay functions recorded from biological objects are usually multi-exponential. The decay components have to be separated to obtain quantitative FRET results, distinguish different proteins, different metabolic states, or to derive biochemical information from autofluorescence. Separating several decay components is generally difficult and requires a large number of photons in the pixels of the image [49]. The situation becomes virtually hopeless if the pixels contain not only fluorescence of different fluorophore populations but also fluorescence from different sample layers. Suppression of out-of-focus signals is therefore a precondition for obtaining high-accuracy lifetime images.

It is a lucky coincidence, or perhaps a result of the intuition of the designers, that the most accurate and sensitive technique of fluorescence decay measurement, time-correlated single photon counting, is almost ideally compatible with laser scanning techniques. Multi-dimensional TCSPC is able to record images with a fully resolved decay curve in each pixel of an image. The results are independent of the scan rate of the scanning system: The signal-to-noise ratio depends only on the total acquisition time, not on the scan rate. TCSPC delivers an excellent time resolution: The IRF—the temporal point-spread function—is short and clean, and virtually free of background. Thus, the combination of laser scanning techniques and multi-dimensional TCSPC combines near-ideal resolution in space with near-ideal resolution in time.

This chapter describes the implementation of multi-dimensional TCSPC in confocal and multiphoton laser scanning microscopes, in imaging systems for macroscopic objects and endoscope imaging systems, and in stimulated emission depletion (STED) and near-field optical scanning (NSOM) microscopy.

2.2 Laser Scanning Microscopes

2.2.1 Confocal Microscopes

2.2.1.1 Optical Principle

The basis of confocal detection is a pinhole in the upper focal plane of the microscope, see Fig. 2.2, left. The principle was invented by Minsky in 1957 [65, 66]. Consider a point source located on the optical axis in the focal plane. The light from this point (shown red) will be focused into a diffraction-limited spot in the upper focal plane of the microscope. A pinhole placed on the optical axis in this plane will transmit the light from the point source in the lower focal plane. Light coming from points above or below the focal plane is focused into a plane below or above the pinhole. As a result, only a small fraction of this light passes the pinhole. The pinhole thus suppresses light from sample planes other than the focal plane.

A system as the one shown in Fig. 2.2, left, does, of course, not deliver an image of the sample. To obtain a fluorescence image of the sample both the spot excited by the laser and the point from which the light is detected must be scanned over the sample. Minsky moved the sample by a electromagnetically driven resonance device. Modern laser scanning microscopes use beam scanning by galvanometer mirrors. The principle is shown in Fig. 2.2, right.

The collimated laser beam is deflected by the galvanometer mirrors. It then passes the 'scan lens'. The scan lens performs two tasks. It focuses the laser into the upper focal plane, which is conjugate with the focal plane of the microscope

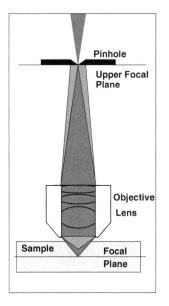
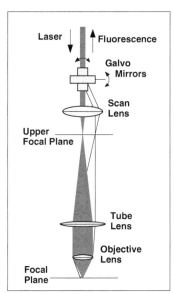

Fig. 2.2 *Left* Suppression of out-of-focus light by a confocal pinhole. *Right* Principle of scanning

objective lens in the sample. Thus, the angular motion of the galvanometer mirrors scans a focused spot of laser light over the focal plane in the sample. The second task performed by the scan lens is to project the galvanometer-mirror axis into the principal plane of the microscope lens. The motion of the laser beam at the back aperture of the microscope lens is therefore mainly an angular one. This avoids obstruction of the scan field by the small diameter of the microscope lens.

Fluorescence light emitted in the focal plane is returned via the same beam path. After being reflected at the galvanometer mirrors the fluorescence light forms a stationary collimated beam. In other words, the fluorescence beam is 'descanned'. It can therefore be focused into a stationary pinhole.

A comparison of a wide-field image, an ordinary (non-confocal) scan image, and a confocal image is shown in Fig. 2.3. The increase in image contrast can clearly be seen: The scan image is free of lateral scattering, the confocal image is also free of out-of-focus blur.

A schematic drawing of the scan head of a confocal laser scanning microscope is shown in Fig. 2.4. The collimated laser beam enters the scanner at the left. It is reflected down into the microscope beam path by a dichroic mirror. It is further reflected by the galvanometer mirrors. The scan lens projects the axis of the scan mirrors on the microscope lens. The microscope lens focuses the laser beam into the sample.

The fluorescence light from the excited spot is collected by the microscope lens, and projected up to the galvanometer mirrors. It is de-scanned by the galvanometer mirrors, separated from the excitation light by a dichroic mirror, and projected into the pinhole. As described above, only light from the focal plane passes the pinhole. Light from outside the focal plane is suppressed. The light that has passed the pinhole goes through one or several filters and enters the detector.

It should be noted that an optical system of the type shown in Fig. 2.4 achieves an out-of-focus suppression effect even without a pinhole. Light from far outside the focal plane is de-collimated so strongly that is not projected on the detector. Some early confocal microscopes therefore had no pinhole, but, nevertheless, achieved reasonably good images. Also the surprisingly high quality of the image in Fig. 2.3, middle, may be in part due to this effect.

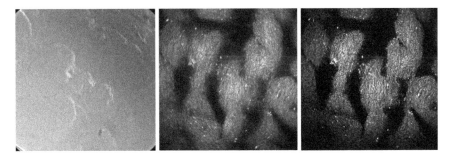

Fig. 2.3 *Left* to *right* Wide-field image, non-confocal scan image, confocal image

Fig. 2.4 Principle of the scan head of a confocal microscope

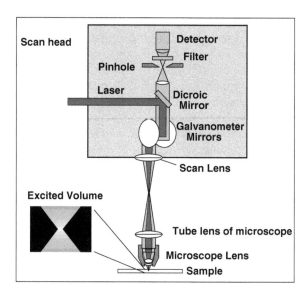

The optical design of a confocal scanner is anything but trivial. Ideally, the pinhole has a diameter on the order of the diameter of the Airy disc in the upper focal plane. Keeping the system aligned for different scan angles, scan amplitudes, microscope lenses, wavelengths, and main dichroic beamsplitters requires extraordinary mechanical reproducibility and optical and mechanical perfection.

2.2.1.2 FLIM with Confocal Microscopes

TCSPC FLIM is based on excitation of the sample with a high-frequency pulsed laser, detecting single photons of the fluorescence light, and determining the times of the photons after the laser pulses and the spatial coordinates of the excited spot in the sample in the moment of the photon detection. The spatial coordinates are determined by transferring scan synchronisation signals from the scanner into the TCSPC module: A 'frame' pulse indicates the beginning of a new frame, a 'line' pulse the beginning of a new line, and a 'pixel' pulse the transition to the next pixel within a line. The spatial coordinates of the photons are determined by counting these pulses. The process is described in detail in Chap. 1, Sect. 1.4.5.

Adding a FLIM system to a confocal microscope may look simple at first glance: A pulsed laser, typically a ps diode laser, and a fast photon counting detector would be added to the system, and the detector output, a synchronisation signal from the laser, and the scan clock pulses from the microscope scan controller be connected to the TCSPC module.

In practice the task can be much more difficult. To be compatible with FLIM the microscope must have one or several unused input ports for pulsed lasers and at least one output port to a FLIM detector. A simple 'confocal' usually does not have these ports. Even if a port for the laser and a port for the detector are available the scan head must have the laser beam combiners and the main dichroic beam splitter suitable for the wavelength of the FLIM laser. These problems have severely hampered the introduction of TCSPC FLIM into confocal microscopy. For these reasons, TCSPC FLIM was introduced into multiphoton microscopy several years earlier than in confocal microscopy [7].

Some confocal microscopes, such as the Zeiss LSM 710/780 family, are available with pulsed one-photon excitation sources (ps diode lasers and a tuneable fibre laser) and with a confocal output port, see Fig. 2.5, left. These systems are fully compatible with FLIM [6]. A FLIM image recorded with the Zeiss LSM 710 and a bh SPC-150 TCSPC FLIM module is shown in Fig. 2.5, right.

Compatibility problems are, of course, avoided in confocal scanning systems that are primarily designed for FLIM. An example is the Becker & Hickl DCS-120 system [5]. Picosecond diode lasers or super-continuum lasers are fibre-coupled to the scan head. The scanner can thus be used with various excitation sources. Both the scanner and the lasers are fully controlled by the FLIM system: The lasers can be multiplexed synchronously with the pixels, lines, or frames of the scan for dual-wavelength excitation [11], see Chap. 1, Sect. 1.4.5.3, or on-off modulated synchronously with the pixels for phosphorescence lifetime imaging [9], see Chap. 1, Sect. 1.4.7, and Chap. 6 of this book. The detectors are attached to the back of the scan head and can easily be replaced. The system can thus be operated with various detectors: GaAsP hybrid detectors [10] are used for the visible range, GaAs hybrid detectors for the NIR, bi-alkali hybrid detectors or MCP PMTs for ultra-high resolution FLIM, and GaAsP multi-wavelength detectors for spectrally resolved FLIM [11].

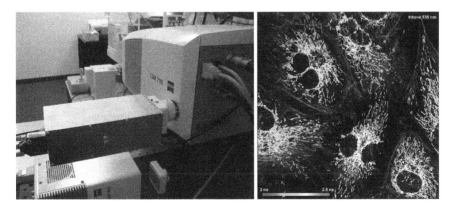

Fig. 2.5 *Left* Confocal output from the scan head of the Zeiss LSM 710/780 microscope family. *Right* Lifetime image of BPAE cells stained with Alexa 488 and Mito Tracker *red*. Excitation by InTune (tuneable) laser. Becker & Hickl HPM-100-40 GaAsP detector and SPC-150 TCSPC FLIM module. 1024 × 1024 pixels, 63× NA = 1.4 oil immersion lens, pinhole 1 AU, excitation wavelength 535 nm, laser power 1 %, detection from 550 to 700 nm

2.2.2 Multi-photon Microscopes

2.2.2.1 Optical Principle

Multi-photon microscopes use a picosecond or femtosecond Ti:Sa laser to excite the sample by two-photon absorption. Simultaneous absorption of two (or more) photons has been theoretically predicted by Göppert-Mayer [40]. It has been suggested for laser scanning microscopy by Wilson and Sheppard in 1984 [82] and practically introduced by Denk and Strickler [27] in 1990. The efficiency of two-photon excitation increases with the square of the excitation power density. With a femtosecond titanium-sapphire laser the excitation process becomes so efficient that no more than a few milliwatt of laser power in the focal plane are required to obtain clear fluorescence images with endogenous [21, 25, 50, 51] and exogenous fluorophores [19, 83]. 'A few milliwatt' may sound a lot of power compared to the μW power levels used for one-photon excitation. It should, however, be noted that only the absorbed power is essential for possible damage effects, not the incident power. The power absorbed by the fluorophores is, however, no higher than for one-photon excitation.

Two-photon excitation has a number of advantages over one-photon excitation.

First, the absorption—both one-photon and two-photon—of the sample at the NIR wavelength of the Ti:Sa laser is very low. Also the scattering coefficient is lower than at visible or ultraviolet wavelengths. Two-photon excitation therefore penetrates deeper into tissue than one-photon excitation, see Fig. 2.6, left and middle. Two-photon excitation can be used to excite fluorescence in tissue layers as deep as 100 μm, in some cases even 1 mm [31, 51, 74, 77].

Second, noticeable two-photon excitation is obtained only in the focus (Fig. 2.6, middle). Thus, two-photon excitation is a second way to obtain optical sectioning, depth resolution and suppression of out-of-focus fluorescence. Different than one-photon excitation with confocal detection, which avoids out-of focus *detection*, two-photon excitation avoids out-of-focus *excitation*.

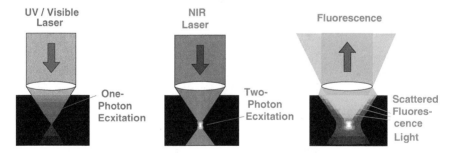

Fig. 2.6 *Left* One-photon excitation. The effective excitation power decreases rapidly with increasing depth. *Middle* Two-photon excitation. The NIR laser penetrates deeply into the sample. *Right* The fluorescence from a deep focus is scattered on the way out of the sample. It leaves the back aperture of the microscope lens in a wide cone

Third, absorption and possible sample damage happens only within a thin layer in and around the focal plane of the microscope. Two-photon excitation is therefore benign to live cells and tissues [33, 34, 50–53]. It can even be used in clinical applications [54, 55], see Chap. 15.

Non-descanned Detection

The fact that the fluorescence signal comes only from the focus leads to a another advantage of two-photon excitation: A two-photon microscope is able to detect photons which are scattered on the way out of the sample. Scattered photons leave the back aperture of the microscope lens in a wide cone of light (Fig. 2.6, right). In a one-photon system with a descanned confocal beam path these photons would be lost. They can neither be fed through the narrow beam path of the scanner, nor can they be focused into a pinhole. In a two-photon microscope, however, there is no out-of focus excitation and, consequently, no need to use a pinhole. Two-photon microscopes therefore divert the fluorescence from the excitation beam directly behind the microscope lens and project it on a large-area detector, see Fig. 2.7. The principle is called non-descanned (or 'direct') detection (NDD). A two-photon microscope with non-descanned detection not only excites fluorescence in deep sample layers, it also detects these photons efficiently [4, 31, 68].

The NDD beamsplitter is usually placed in the filter carousel of the microscope. A lens in the NDD beam path projects a de-magnified image of the microscope lens (not of the sample!) on the detector. A laser blocking filter suppresses laser light scattered in the sample and in the optical system, a second filter selects the detection wavelength range.

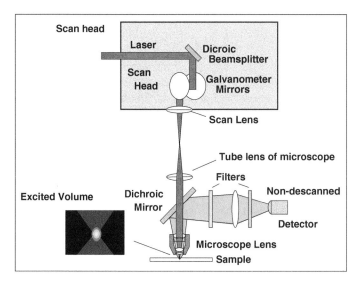

Fig. 2.7 Two-photon excitation with non-descanned detection (NDD)

A practical advantage of NDD in combination with multiphoton excitation is that the scanner is far less complicated than for confocal detection. There is no need for exactly cancelling the beam motion for different wavelengths, scan fields, or objective lenses. It is also not necessary to focus the fluorescence in a diffraction-limited size pinhole, and keep the alignment stable when filters or dichroics are replaced. All that is required for multiphoton NDD imaging is that a laser beam of the correct diameter is cleanly transferred into the back aperture of the microscope lens, and angle-scanned. The simplifications in the scanner optics at least partially compensate for the high price of a femtosecond laser. Two-photon microscopes can therefore be built by attaching the laser and an optical scanner to a conventional microscope [29] or by upgrading a one-photon system with a Ti:Sapphire laser [30, 60].

Two-Photon Excitation with Descanned Detection

For sake of completeness, it should be mentioned that two-photon excitation is also being used with descanned detection. The advantage over non-descanned detection is that the pinhole rejects daylight and optical reflections. Confocal detection may also help suppress spurious fluorescence light if there is a highly fluorescence layer close to the image plane. The disadvantage is that fluorescence photons scattered on the way out of the sample are not detected. Descanned detection therefore does not exploit the deep-tissue imaging capability of multiphoton excitation.

2.2.2.2 FLIM with Multiphoton Microscopes

The titanium-sapphire laser of multi-photon microscopes is a near-ideal excitation source for FLIM: The pulse width is so short that it does not broaden the TCSPC IRF, and the wavelength is tuneable over a wide range. Fluorophores with excitation maxima shorter than 400 nm can easily be reached by two-photon or three-photon excitation. As a drawback it is often stated that the repetition rate (typically around 80 MHz) is too high for the fluorescence lifetimes of a number of fluorophores. In practice, this is rarely a problem: Residual fluorescence from the previous pulses can be taken into account in the data analysis by an 'incomplete decay' model [5, 6, 11], and increased background by pile-up of afterpulses can be avoided by using hybrid detectors [10].

Most multiphoton microscopes have non-descanned detector ports to which the FLIM detector(s) can be attached. NDD ports of the Zeiss LSM 710/780 NLO family are shown in Fig. 2.8.

An example of two-photon NDD FLIM is shown in Fig. 2.9. The images show an unstained pig skin sample exited by two-photon excitation at 800 nm. The left image shows the wavelength channel below 480 nm. This channel contains both fluorescence from endogenous fluorophores and SHG signals. The SHG fraction of the signal has been extracted from the FLIM data and displayed by colour. The right image is from the channel >480 nm. It contains only fluorescence, the colour corresponds to the amplitude-weighted mean lifetime of the multi-exponential decay functions.

Fig. 2.8 NDD detection at the LSM 710. *Left* Upright microscope. *Middle* Inverted microscope. *Right* NDD beamsplitter cube for two FLIM detectors. The 90° output is shown with the detector de-attached

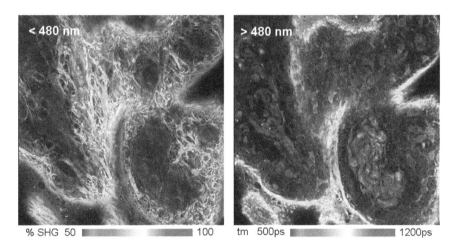

Fig. 2.9 Two-photon FLIM of pig skin. LSM 710 NLO, excitation 800 nm, HPM-100-40 hybrid detector, NDD. *Left* Wavelength channel <480 nm, colour shows the percentage of SHG in the recorded signal. *Right* Wavelength channel >480 nm, colour shows amplitude-weighted mean lifetime. Analysis by Becker & Hickl SPCImage software [11]

2.2.3 Image Size in FLIM

FLIM for laser scanning microscopy was introduced with the SPC-730 modules of Becker & Hickl in 1998. The FLIM data were built up by the on-board processing logics in the memory of the TCSPC module. The size of the FLIM images was therefore limited by the size of the on-board memory. Typical images had 128 × 128 pixels, and 256 time channels per pixel. The image size increased by a factor of four with the introduction of the SPC-830 modules.

After the introduction of multi-core CPUs larger images could be recorded by using software accumulation of the data, see Chap. 1, Sect. 1.3.2, Fig. 1.11. Limited by the amount of memory addressable by Windows 32 bit, the typical image size

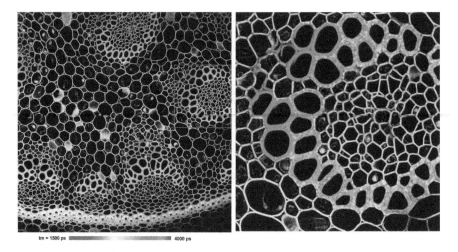

tm = 1500 ps ████████████   ████████████ 4000 ps

Fig. 2.10 FLIM recorded with 2048 × 2048 pixels and 256 time channels per pixel. *Left* Entire image. *Right* Digital zoom into indicated area of left image. bh DCS-120 confocal scanning FLIM system, Zeiss Axio Observer microscope, ×20 NA = 0.5 air lens

was 512 × 512 pixels, 256 time channels. This is more than enough to obtain diffraction-limited microscopy images of cells and tissue samples at a time resolution (time-channel width) on the order of a few 10 ps.

However, there are FLIM applications that make it desirable to record FLIM data of larger size. Examples are imaging of a large number of cells in the same field of view, tissue imaging, or FLIM measurements over additional parameters, such as wavelength, depth in the sample, or time after a stimulation. Such applications require an amount of memory space which is only available in a 64-bit environment. A typical PC running Windows 64 bit is able to use about 16 GB of memory, as compared to 4 GB for Windows 32 bit. With 16 GB of memory, FLIM data can be recorded at a resolution of 2048 × 2048 pixels and 256 time channels, or 1024 × 1024 pixels and 1024 time channels. Several such data sets can even be recorded simultaneously if the FLIM system has several TCSPC channels.

An image of a convallaria sample recorded with 2048 × 2048 pixels is shown in Fig. 2.10. The image on the left is the entire FLIM image, the image on the right is a digital zoom into the area marked in the in the image shown left.

2.2.4 Signal-to-Noise Ratio and Acquisition Time of FLIM

There is probably no TCSPC parameter that is discussed more controversially than the acquisition time of FLIM. Therefore, this section gives an overview on the signal-to-noise ratio of fluorescence lifetime measurement, on the acquisition time of FLIM, and on the influence of the imaging conditions on these parameters.

Signal-to-Noise Ratio of Fluorescence Lifetime Measurement

From a single-exponential fluorescence decay recorded under ideal conditions the fluorescence lifetime can theoretically be obtained with relative standard deviation, or signal-to-noise ratio, SNR_τ, of

$$SNR_\tau^2 = N^{1/2}$$

with N = number of recorded photons [3, 39, 49]. In other words, the fluorescence lifetime can be obtained at the same accuracy as the intensity. Measurement under ideal condition means the decay function is recorded with an instrument response function that is short compared to the decay time, into a large number of time channels, within a time interval several times longer than the decay time, and with negligible background of environment light, detector dark counts or afterpulsing from previous excitation periods [49]. Carefully conducted TCSPC and TCSPC FLIM measurements come very close to this limit.

Acquisition Time

The equation given above can be used to estimate the number of photons and the acquisition time needed to record a fluorescence lifetime image. For a given signal-to-noise ratio expected from the measurement, a number of photons, N, of

$$N = SNR_\tau^2$$

in one pixel is required. For the entire lifetime image, this number of photons has to be multiplied with the number of pixels in the image, i.e.

$$N = SNR_\tau^2 \cdot p_x \cdot p_y$$

with p_x, p_y being the number of pixels in x and y. The time to acquire these photons is

$$T = SNR_\tau^2 \cdot p_x \cdot p_y / R$$

with R being the photon rate in photons per second. Figure 2.11, left, shows the acquisition time as a function of the desired signal-to-noise ratio for different image size. A photon rate of 10^6/s was assumed. A count rate on this order can be obtained from reasonably bright samples in good optical systems with hybrid detectors. Higher count rates are, in principle, possible by TCSPC FLIM (see Chap. 1, Sect. 1. 2) but often cause photobleaching and photo-induced intensity and lifetime changes in the samples.

As can be seen from Fig. 2.11, left, an image of 512×512 pixels with a signal-to-noise ratio of 10 (a relative standard deviation of 10 %) is obtained within less than 30 s. For smaller pixel numbers the acquisition time is correspondingly shorter: A 128×128 pixel image of the same accuracy is obtained in 1.6 s. However, an acquisition time of almost 1 h would be needed to record a lifetime image of 512×512 pixels with a signal-to-noise ratio of 100 (a relative standard deviation of 1 %).

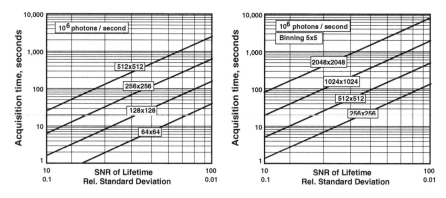

Fig. 2.11 *Left* Acquisition time versus desired signal-to-noise ratio of the lifetime for different image size at a photon rate of 10^6 s^{-1}. *Right* Acquisition time for 5 × 5 pixel binning of the lifetime data

Surprisingly, the acquisition time of FLIM for a given accuracy is often *shorter* than the times shown in Fig. 2.11, left. The explanation is that the lifetime analysis software uses, or can use, binning of the fluorescence decay data, see [11]. That means the decay information for every pixel is taken not only from a single pixel but also from pixels around it. This works well especially for large images: These images are often oversampled. That means there are several pixels inside the area of the Airy disc of the microscope lens. Lifetime data of pixels within the radius of the Airy disc are, of course, highly correlated. Binning them for lifetime analysis is therefore justified. With the definition of the binning factor, n, used in SPCImage time-domain FLIM analysis software [11], the effective number of pixels, pxl_{eff}, of an image with $p_x \cdot p_y$ pixels is

$$pxl_{eff} = p_x \cdot p_y / (2n + 1)^2$$

and the acquisition time becomes

$$T = SNR_\tau^2 \cdot p_x \cdot p_y / R(2n + 1)$$

The acquisition time for a binning parameter of 2 (a 5 × 5 pixel area with the current pixel in the centre) is shown in Fig. 2.11, right. For a 512 × 512 pixel image an accuracy of 10 % is reached in 5.2 s, an accuracy of 1 % in 500 s. Even images with 1024 × 1024 and 2048 × 2048 pixels and moderate accuracy are within reach. FLIM images recorded within 5 s acquisition time are shown in Fig. 2.12.

Multi-exponential Decay Functions

The required number of photons increases if multi-exponential decay functions are to be resolved into their decay components. In [49] the number of photons required to resolve a double-exponential decay was estimated to be N = 400,000. A number of photon per pixel this high is, of course, entirely beyond the capabilities of a typical biological sample [49] is therefore often used as an argument that double-

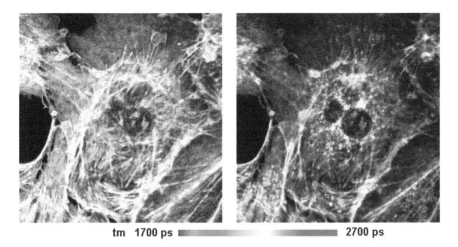

tm 1700 ps ▬▬▬▬▬▬ ▬▬▬▬▬▬ **2700 ps**

Fig. 2.12 FLIM of a BPAE sample stained with Alexa 488 and Mito Tracker *red*. Acquisition time 5 s, 256 × 256 pixels, SPCImage binning parameter 2. Average count rate 10^6 s^{-1}. Two-channel FLIM, 485–560 nm (*left*), 560–650 nm (*right*)

exponential analysis of FLIM data is impossible. Fortunately, the prospects of separating two lifetime components improve dramatically with the ratio of the two lifetimes and with the amplitude factor of the short lifetime component. The lifetime components assumed in [49] were 10 % of 2 ns and 90 % of 4 ns. This is an extremely unfavourable situation which indeed requires an exceedingly high number of photons. Fortunately, the decay profiles encountered in FRET and autofluorescence measurements have a much more favourable composition. Usually the lifetime components are separated by a factor of 5–10, and the amplitude of the fast component is 50–90 %. Under such conditions double or even triple exponential analysis is feasible on no more than a few 1000 photons per pixel, and very satisfactory results are obtained from 10,000 photon per pixel.

Where Are the Photons Coming from?

A parameter that is often ignored when different FLIM techniques are compared is from where in the sample the photon are detected, see Fig. 2.13. Different FLIM techniques use different optical systems. Confocal or multiphoton FLIM detects only from a thin layer around the focal plane of the microscope, wide-field FLIM from the entire depth of the sample. As a result, wide-field FLIM detects much more photons per pixel and second than confocal of multiphoton FLIM. That means the photon rate, R, becomes higher, and the acquisition time correspondingly shorter. However, this comes at a price: The additional photons come from the wrong volume elements inside the sample. On this background, some moderate binning in TCSPC FLIM appears acceptable: At least, the photons in the individual pixels are coming from the correct image plane, and from a clearly defined area around the current pixel.

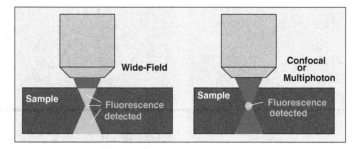

Fig. 2.13 Wide-field imaging detects photons from the entire depth of the sample, confocal and multiphoton imaging only from the focal plane

Why Does Lifetime Imaging Need More Photons Than Intensity Imaging?

The fact that, for the same number of photons, FLIM achieves the same signal-to-noise ratio as intensity imaging is in apparent contradiction with common experience: From the same sample imaged under similar excitation conditions, intensity images are obtained in a far shorter acquisition time than FLIM images. What is wrong?

The mistake is to implicitly assume that 'intensity imaging' (in the sense of spatial imaging) and FLIM are recording the same physical properties of the sample. This is wrong: Intensity imaging aims at resolving the spatial structure of the sample. To do so, it uses the fluorescence intensity of a fluorophore attached to the internal sample structures. In other words, intensity imaging records the concentration variation of a fluorophore in the sample. The relative concentration in stained and unstained structures of the sample can vary by a factor 100 and more. Structural information can therefore be obtained with no more than a few photons per pixel.

FLIM aims at resolving changes in the fluorescence lifetime, i.e. in the quantum efficiency of the fluorophore. Relative changes in these parameters are much smaller than the concentration changes. Lifetime changes are usually in the 1–10 % range. Consequently, a higher signal-to-noise ratio, and, because of the SNR^2 dependence, a far higher number of photons is needed. A higher number of photons cannot be obtained by simply increasing the excitation power.

Sample-Limited Count Rates

The difference described above are further enhanced by photobleaching. Moderate photobleaching has no effect on the spatial structure of a sample. Intensity measurements (in the sense of spatial imaging) can therefore be performed at relatively high excitation power.

It may be expected that photobleaching also has no effect on FLIM: FLIM is independent of the fluorophore concentration. Some loss in fluorophore concentration therefore should not have an impact on the recorded lifetimes. Unfortunately, this is usually not the case. Biological samples may contain several fluorophores, and, especially those used for FLIM, fluorophores in different states of interaction with the molecular environment. Different fluorophores and different fluorophore fractions bleach at different rate. As a result, unpredictable changes in the composition of the fluorophores and thus in the fluorescence decay parameters can occur. Moreover, photobleaching generates radicals, and these have a destructive effect on the proteins. Photobleaching may therefore not only change the fluorophores but also the molecular structure of the specimen itself. This is a general problem for all molecular-imaging experiments, not only for those using FLIM.

Another feature of molecular-imaging experiments is that the fluorophores are usually linked to highly specific sites of selected proteins. That means the effective fluorophore concentration is low. For a given emission rate, the molecules perform more excitation-emission cycles, and, consequently, photobleach faster. Any attempt to increase the emission from the sample (to reduce the acquisition time) by increasing the excitation power makes the situation worse. The only way to keep the acquisition time short and photobleaching artefacts low is to use high photon collection efficiency (high NA objective lenses and good detectors). Under such 'sample-limited' conditions, TCSPC FLIM delivers a better lifetime accuracy in a given acquisition time (and vice versa) than any other FLIM technique with a comparable optical system. It also delivers better results than techniques based on intensity imaging, as has been confirmed for FRET experiments [70].

2.2.5 Nonlinear Scanning

Some laser scanning microscopes, such as the Leica SP5 and SP8, use a sinusoidal scan in x direction. The advantage of a sinusoidal scan is that the scanner can be operated at a higher line frequency. A positive side effect is that an intensity drop towards the edge of the field is compensated by longer pixel integration time in these areas.

It is sometimes believed that TCSPC FLIM delivers distorted images when combined with sinusoidal scanning. This is not correct. The reason is that the microscopes compensate for the nonlinearity of the scan by a non-equidistant pixel clock. The pixel clock periods are shorter in the centre and longer at the outer parts of the line, see Fig. 2.14, left. The distance along the line then becomes a linear function over the number of pixel clocks. A FLIM system that synchronises the recording along the lines with the pixel clock of the microscope thus records an undistorted image, see Fig. 2.14, right.

2.2.6 Tuneable Excitation

Most confocal microscopes work with a number of discrete laser wavelengths. For use with a wide range of fluorophores it is, of course, desirable to tune the excitation wavelength exactly to the absorption maximum of the fluorophore used. Tuneable excitation is, in principle, available in multiphoton microscopes with titanium-

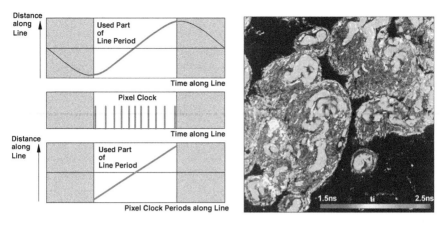

Fig. 2.14 *Left* Sinusoidal scan and linearization by non-equidistant pixel clock. *Right* Autofluorescence FLIM image of a pig skin sample. Leica SP5 MP, two-photon excitation at 800 nm, Leica hybrid detectors, bh SPC-830 TCSPC FLIM module running at external pixel clock

sapphire lasers. However, the price of such systems is high, the tuning speed is low, and the tuning range is not sufficient to cover the excitation spectra of all dyes used in fluorescence microscopy.

Scanning microscopy with tuneable excitation has a long history. The first confocal scanning microscope developed by Minsky [65, 66] in 1957 used wide-band illumination by a conventional lamp and a 50 % wideband beamsplitter. Minsky's microscope was, in principle, a tuneable system: By inserting different filters, the microscope could be used at any desired wavelength. Unfortunately, Minsky did not use fluorescence for image formation. The reason was certainly not the detector—photomultipliers for low-intensity detection were routinely used at this time. Presumably, the benefits of using fluorescence had just not been discovered yet. In 1962, Freed and Engle [35] described a scanning microscope that used an arc lamp as a light source. The wavelength of the illumination light was selected from the lamp spectrum by a monochromator. Also this system was described for reflected light only. Fluorescence was not mentioned although the instrument must have been able to record fluorescence images by simply inserting the right filter in the detection path.

The first tuneable excitation scanning fluorescence microscope was described by Van der Oord et al. [78, 79]. They used synchrotron radiation for excitation. The excitation wavelength was selected from the wideband continuum by a filter in front of the input of the laser scanning microscope. A similar instrument was described in [67]. The instrument described in [78] was even used for FLIM by parallel-gated photon counting.

With the introduction of light sources based on super-continuum generation in photonic crystal fibres a bright tuneable light source became available for laser scanning microscopy [61, 62]. By an acousto-optical tuneable filter (AOTF) any

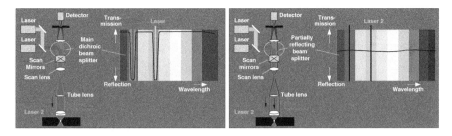

Fig. 2.15 Confocal scanner with dichroic beamsplitter (*left*) and wideband beamsplitter (*right*)

wavelength from less than 400 nm to more than 1000 nm can be selected from the super-continuum.

However, if the tuneability of the light source is to be practically used there is an optical problem. The laser scanning microscope uses a dichroic beamsplitter to reflect the excitation light down to the microscope lens, and transmit the emission light to the detectors, see Fig. 2.15, left. This beamsplitter can only be made only for a single, or a limited number of excitation wavelengths. If the microscope has to be used for a wide range of excitation wavelengths the beamsplitter must be replaced for every wavelength. It is possible to use several dichroic mirrors on a wheel, but this has other disadvantages: To maintain confocal alignment for different beamsplitters extraordinary mechanical stability is required. Moreover, switching dichroics by rotating a wheel takes time. This makes fast wavelength multiplexing (see below) impossible.

The easiest way to avoid these problems is to go back to Minsky's original design and use a partially reflective mirror, see Fig. 2.15, right.

At first glance, a wideband beamsplitter may appear a very poor design: A considerable part of the emission and the excitation light would be lost. A somewhat closer look, however, shows that Minsky's approach is not so poor after all: The vast majority of TCSPC FLIM experiments (probably also of other LSM applications) is performed at less than 10 % of the available laser power. Under these conditions, a loss in excitation power at the beamsplitter can easily be compensated by increasing the laser power at the input of the scanner. The loss in the detection beam path can be kept low with a beamsplitter with a reflection/transmission ratio of more than 50/50. An 80/20 beamsplitter—with 80 % transmission in the fluorescence path—loses only 20 % of the light. A loss of 20 % of the fluorescence photons results only in a 10 % decrease in lifetime accuracy, see Sect. 2.2.4 of this chapter.

Wideband beamsplitters are available in most commercial laser scanning microscopes. Adding tuneable excitation to a confocal laser scanning system thus does not require more than a suitable laser input port. An example is shown in Fig. 2.16. The figure shows autofluorescence FLIM of a pig skin sample excited by a supercontinuum laser at 450 nm (left) and 532 nm (right). A 60/40 transmission/reflection beamsplitter was used in the scanner.

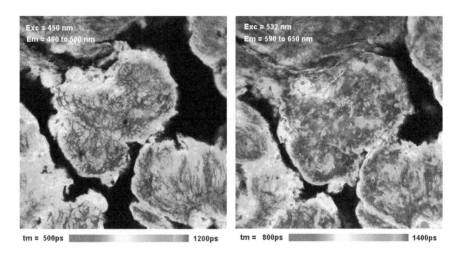

<figure>
tm = 500ps [gradient bar] 1200ps   tm = 800ps [gradient bar] 1400ps
</figure>

Fig. 2.16 Tuneable excitation with supercontinuum laser and 60/40 wideband beamsplitter. Excitation at 450 and 532 nm, detection from 460 to 500 nm and 590 to 650 nm. Pig skin, autofluorescence

2.2.7 Excitation Wavelength Multiplexing

TCSPC FLIM is able to work with several multiplexed excitation wavelengths. To avoid interference with the scanning the wavelength must be switched synchronously with the frames, lines, or pixels of the scan. The current wavelength is indicated to the TCSPC module by one or more bits of the routing signal [8, 11]. The TCSPC module then records the photons excited by different laser wavelengths into separate photon distributions. The principle is described in detail in Chap. 1, Sect. 1.4.5.3.

The multiplexing and the routing signals can be generated directly by the scan controller [5, 11] or by external electronics [11], see also Chap. 18. Wavelength switching on the laser side can be achieved by controlling several picosecond diode lasers via their on/off inputs or by selecting different wavelengths from the spectrum of a super-continuum laser via its AOTF, see also Chap. 18. In both cases the switching time is on the order of a few microseconds or faster.

On the optical side, excitation wavelength multiplexing faces the same problems as tuneable excitation: Two or three fixed wavelengths can still be coupled into the beam path by a multi-band dichroic. However, if more wavelengths are multiplexed, or if the wavelengths must be freely selectable the only practicable solution is a wideband beam splitter. Another requirement is, of course, that the excitation wavelengths must not overlap with the transmission bands of the filters in the detection beam path. Examples of wavelength multiplexing in FLIM microscopy are shown in Chap. 1, Fig. 1.30. Excitation wavelength multiplexing in a scanning system for diffuse optical imaging is shown in Chap. 18, Fig. 18.12.

Excitation wavelength multiplexing is a promising technique to obtain better separation of signals from endogenous fluorophores. It could also become important for clinical applications: It has recently been shown that excitation and fluorescence decay detection at two wavelength combinations yields increased discrimination between healthy tissue and cancer tissue [71].

2.3 FLIM with Near-Infrared Dyes

Near-infrared dyes are used as contrast agents and as fluorophores in diffuse optical imaging applications [20, 47, 57, 59]. Diffuse optical imaging (or diffuse optical tomography, DOT) reconstructs internal structures and scattering and absorption parameters of tissue from diffusely transmitted or reflected photons, see Chaps. 17–19 of this book. By using wavelengths between 650 and 900 nm information about the tissue constitution can be obtained up to a depths of several centimetres. For these applications it is important to have information about binding of the dyes to proteins, DNA, collagen, and other cell constituents available. It is also important to know whether the dyes change their fluorescence lifetimes—and thus quantum efficiencies—on binding, and whether these lifetime changes depend on the binding targets [16, 17]. Possible lifetime changes may interfere with the reconstruction of the tissue structure and tissue parameters from time-resolved data, but may also be exploitable to gain additional biological information [38, 18, 84]. Moreover, the penetration depth in the NIR—even for a confocal system—can be expected to be noticeably higher in the NIR than for excitation and detection in the visible range of the optical spectrum.

2.3.1 Optical System

At first glance, FLIM at NIR wavelengths should neither be a problem on the detection nor on the excitation side: FLIM detection in the 700–900 nm range can be achieved by hybrid detectors or PMTs with GaAs cathodes, or with single-photon avalanche photodiodes. TCSPC FLIM electronics works equally well with all these detectors [8, 11]. Picosecond diode lasers for the red and near-infrared range are available with 640, 685, and 785 nm, the Intune laser of the Zeiss LSM 710/780/880 microscopes can be tuned up to 645 nm, and super-continuum lasers with acousto-optical filters deliver any wavelength from the visible range to more than 1000 nm. Multiphoton microscopes have a Ti:Sapphire laser that works from about 700 m to 1000 nm. The wavelengths available from these light sources are compatible with the (one-photon) excitation spectra of a wide variety of NIR dyes.

It is often believed that NIR FLIM requires nothing than an NIR sensitive detector and an NIR laser of appropriate wavelength. Unfortunately, this is not correct. The problem is in the main dichroic beamsplitter optics of the laser

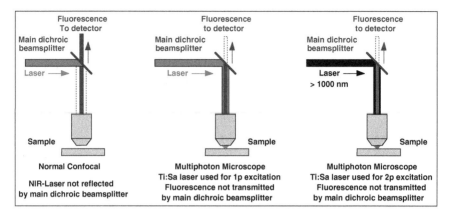

Fig. 2.17 Beamsplitter configurations in laser scanning microscopes. *Left* Confocal, the NIR laser is not reflected to the sample. *Middle* Ti:Sa laser of multiphoton microscope used for one-photon excitation in the NIR, the fluorescence is not transmitted to the detector. *Right* Two-photon excitation, the fluorescence is not transmitted to the detector

scanning microscope, see Fig. 2.17. A combination of a 'normal confocal' with a near-infrared laser is shown in Fig. 2.17, left. It does not work because the main dichroic beamsplitter does not reflect the laser down to the sample. It cannot do so because it is designed to transmit the fluorescence signal from the visible range all the way up to the NIR. The attempt to use the Ti:Sa laser of a multiphoton microscope as a conventional (one-photon) excitation source is shown in Fig. 2.17, middle. Also this configuration does not work: The main dichroic beamsplitter is designed to transmit fluorescence at the short-wavelength side of the laser. However, the one-photon excited NIR fluorescence is on the long-wavelength side. The beamsplitter therefore does not transmit the fluorescence to the detector. The third configuration, shown in Fig. 2.17, right, tries to excite the fluorescence by two-photon excitation. Also this configuration does not work: The beamsplitter is designed to reflect the laser from about 700 nm all the way up to >1000 nm. Consequently, it does not transmit the fluorescence.

A possible way to solve these problem is to use a wideband beamsplitter. This is a semi-transparent mirror that splits the light into two components independently of the wavelength. Many microscopes have an '80/20' (wideband) beamsplitter in their beamsplitter wheels. With With this beamsplitter the confocal setup works both with an NIR laser and with the titanium-sapphire laser (used for one-photon excitation), see Fig. 2.18, left and middle. A disadvantage is that the wideband beamsplitter loses light. However, this is less problematic than commonly believed, see Sect. 2.2.6.

The configuration in Fig. 2.18, right, uses two-photon excitation and non-descanned detection. In that case, the fluorescence light is split off by a dichroic mirror in the filter carousel directly behind the microscope lens. Of course, the mirror normally inserted in this place cannot be used—it is designed to transmit the laser from 700 nm up to >1000 nm. It thus does not reflect the fluorescence. However,

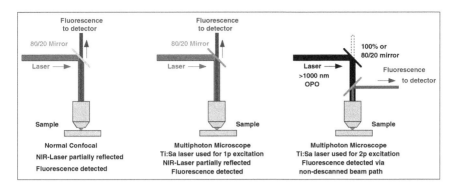

Fig. 2.18 Solutions to NIR FLIM. *Left* Confocal system with 80/20 wideband beamsplitter. *Middle* Multiphoton system, but Ti:Sa laser used for 1-photon excitation, 80/20 wideband beamsplitter. *Right* Multiphoton system, multiphoton excitation by OPO, non-descanned detection

replacing the dichroic mirror in the filter carousel is easy. The problem of the configuration of Fig. 2.18, right, is that a femtosecond laser of long wavelength is needed to excite NIR dyes efficiently [84]. This is usually an OPO. A two-photon NIR-detection system with two-photon excitation is therefore not a cheap solution.

FLIM images recorded in these three configurations will be presented below.

2.3.2 Confocal System, One-Photon Excitation with ps Diode Laser

Figure 2.19 shows a FLIM image recorded by a Becker & Hickl DCS-120WB confocal FLIM system with picosecond diode laser excitation and a wideband beamsplitter [5]. The photons were detected by a HPM-100-50 GaAs hybrid detector [11]. As a test sample we used pig skin stained with indocyanine (ICG) green. The skin samples were immersed for 30 min in a 30 μm ICG solution and then immediately used for imaging.

ICG has an absorption maximum around 780 nm. At concentrations higher than about 50 μm/l ICG forms aggregates which have an absorption maximum at 690 nm. The fluorescence is emitted around 820 nm [28]. The small stokes shift between the absorption and the emission makes it difficult to suppress scattered laser light and wideband spectral background of the diode laser. Fortunately, ICG has sufficient absorption down to 630 nm. We therefore used a 640 nm BDL-SMC ps diode laser for the experiments. This provided convenient spacing between the excitation and emission wavelengths, and solved any filter leakage problems.

The image shown in Fig. 2.19, left, is from a focal plane about 20 μm below the top of the skin. There are surprisingly large changes in the fluorescence lifetime throughout the sample. Two fluorescence decay functions from selected spots with the corresponding decay parameters are shown in Fig. 2.19, right.

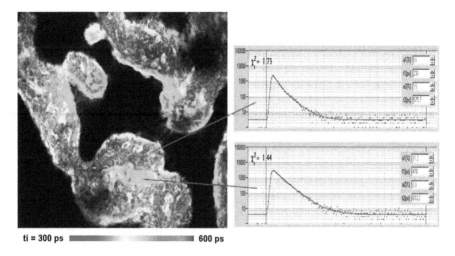

ti = 300 ps ■■■■■■■■■■■■■■■■■■■■■■■■■ 600 ps

Fig. 2.19 Pig skin sample stained with ICG. *Left* Lifetime image about 20 µm from the top of the skin. *Right* Decay curves in different areas of the image. Fit with double-exponential model, intensity-weighted lifetime. DCS-120 WB system, HPM-100-50 detector. FLIM data format 512 × 512 pixels, 256 time channels. Data analysis with Becker & Hickl SPCImage [5, 11]

It is known that ICG has a lifetime of about 200 ps in water, and about 600 ps when bound to serum albumin. In the short-lifetime regions (yellow-orange in Fig. 2.19, left) the lifetimes of the decay components are 231 and 576 ps. This is marginally compatible with a mixture of bound and unbound ICG. In the long-lifetime areas (blue-green in the images) the fit delivers decay components with lifetimes of 470 and 833 ps. This is clearly incompatible with a simple mixture of bound and unbound ICG. It may be an effect of aggregation. However, in that case it should rather be observed at the edge of the tissue where the staining concentration is high.

Whatever the reasons of the lifetime changes are, the results show that the assumption of an essentially invariable fluorescence lifetime (and thus quantum efficiency) of ICG in biological systems is not correct. The variability may have an impact on the reconstruction of tissue parameters in diffuse optical imaging experiments.

An image of a pig skin sample stained with DTTCC (3,3′-diethylthiatricarbo-cyanine) is shown in Fig. 2.20, right. The incubation time was 16 h. There is not only excellent contrast in the image, but also different fluorescence lifetime in different tissue structures. What the mechanism of the lifetime variation is, and to which tissue parameters the lifetime reacts must be verified by further experiments.

Fig. 2.20 Pig skin samples stained with 3,3′-diethylthiatricarbocyanine. bh DCS-120 WB confocal FLIM system, excitation 650 nm, detection 68–900 nm, HPM-100-50 detector

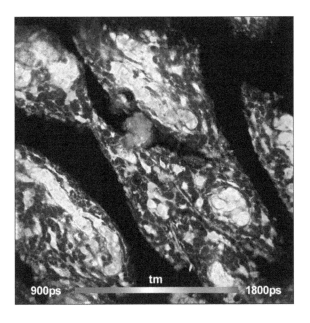

2.3.3 Confocal System, One-Photon Excitation by Ti:Sa Laser

For the images shown in Fig. 2.21 we used a Zeiss LSM 710 NLO multiphoton microscope. The Ti:Sa laser was used for one-photon excitation at a wavelength of 780 nm. The FLIM data were recorded by a Becker & Hickl Simple-Tau 150 FLIM system [11]; the photons were detected by a Becker & Hickl HPM-100-50 hybrid detector [11]. The detector was attached to a BiG type confocal output port of the LSM 710 scan head [6], see Fig. 2.5. The excitation wavelength was 780 nm. The 80/20 beamsplitter of the LSM 710 scan head was used. Scattered laser light was suppressed by an 800 nm long-pass filter in front of the HPM-100-50 detector. Despite of the 80 % loss in the 80/20 beamsplitter the excitation power in the sample was too high even when the power was turned down to <1 % by the AOM of the microscope. We therefore inserted an additional attenuator in the laser beam path.

FLIM images of a pig skin sample stained with DTTCC (3,3′-diethylthiatri-carbocyanine) are shown in Fig. 2.21. Also in these images, large lifetime differences are seen. However, the behaviour is not the same as in Fig. 2.20, right. We believe that this is a result of the incubation time. The sample in Fig. 2.20 was incubated for 12 h, the sample in Fig. 2.21 for only 20 min. Therefore the dye has not fully diffused into the tissue, and not entirely bound to the tissue constituents. Slow binding has also been observes for other carbocyanines [42] and thus must be taken into consideration also for DTTCC.

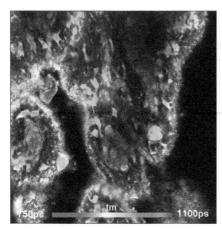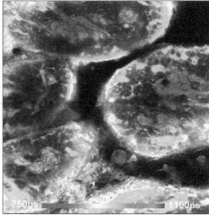

Fig. 2.21 Pig skin samples stained with DTTCC. Zeiss LSM 710 NLO, Ti:Sa laser used for one-photon excitation. Excitation wavelength 780 nm, detection wavelength 800–900 nm. HPM-100-50 hybrid detector, Simple-Tau 152 FLIM system. Lateral size of the images 212 × 212 μm, depth about 30 μm from surface. Note the high contrast of the images

2.3.4 Two-Photon FLIM with OPO Excitation

The images shown below were recorded by a Zeiss LSM 710 NLO multiphoton microscope with OPO excitation. A HPM-100-50 hybrid detector was attached to the non-descanned output of the microscope. The data were recorded by a Simple-Tau 150 FLIM system.

Figure 2.22 shows an image obtained from pig skin stained with methylene blue. Methylene Blue is a biomedically interesting compound. It has anti-viral and anti-bacterial effects, and it has been evaluated as a drug against malaria [73]. It has been applied to induce apoptosis in cancer cells [80], and to treat psoriatic skin lesions by photodynamic therapy [72]. The use of methylene blue to treat Alzheimer disease has been under clinical trial [63].

Methylene Blue has an absorption band from 550 to 690 nm, with a maximum at 660 nm. Fluorescence is emitted from 650 to 750 nm, with a maximum at 680 nm. It can thus be expected that two-photon excitation works well in the range from 1000 to 1300 nm.

A FLIM image is shown in Fig. 2.22, left. The sample was excited at 1200 nm, the fluorescence was detected from 680 to 780 nm. Excitation was surprisingly efficient. No more than 4 % of the available OPO power were needed to obtain a count rate on the order of 1 MHz. Decay curves from characteristic spots are shown in Fig. 2.22, right. The decay profiles are multi-exponential but can be reasonably fitted by a double-exponential decay model. The amplitude-weighted lifetime in different parts of the tissue varies by almost a factor of two.

To prove that the OPO system is able to excite also dyes with longer absorption wavelength we recorded FLIM Images of a similar sample stained with ICG.

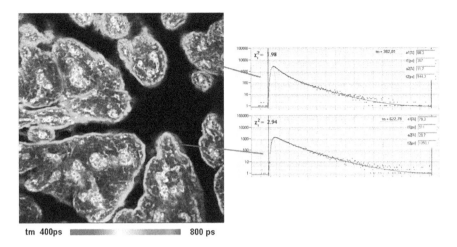

tm 400ps ▮▮▮▮▮▮▮▮▮▮▮ 800 ps

Fig. 2.22 Pig skin stained with methylene Blue. *Left* Lifetime image, double-exponential decay model, amplitude-weighted lifetime. Two-photon excitation at 1200 nm, 512 × 512 pixels, 256 time channels. *Right* Decay curves in characteristic spots of the image. Data analysis by Becker & Hickl SPCImage [11]

The fluorescence was excited at 1200 nm, fluorescence was detected from 780 to 850 nm. Results are shown in Fig. 2.23, a FLIM Z stack recorded at the same sample is shown in Fig. 2.28. As for the Methylene Blue, about 3 % of the available excitation power were sufficient to obtain a count rate on the order of 1 MHz averaged over the entire image.

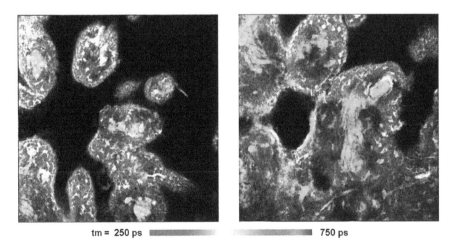

tm = 250 ps ▮▮▮▮▮▮▮▮▮▮ 750 ps

Fig. 2.23 Pig skin stained with Indocyanin Green. Two-photon excitation at 1200 nm, detection from 780 to 850 nm. Amplitude-weighted lifetime of double-exponential decay. Depth from top of tissue 10 μm (*left*) and 40 μm (*right*). 512 × 512 pixels, 256 time channels

The range of the lifetime and the variability is in agreement with FLIM data obtained by one-photon excitation, see [13]. Again, the results show that the common assumption of an essentially invariable fluorescence lifetime (and thus quantum efficiency) of ICG in biological systems is not correct.

2.4 Mosaic FLIM

2.4.1 X–Y Mosaic

To record images larger than the maximum field diameter of the microscope lens laser scanning microscopes use a 'tile imaging' function. The microscope scans an image at one position of the sample, then offsets the sample by the size of the scan area, and scans a new image. The process is repeated, and the images of the individual frames are stitched together. This way, a sample area larger than the field of view of the microscope lens can be imaged.

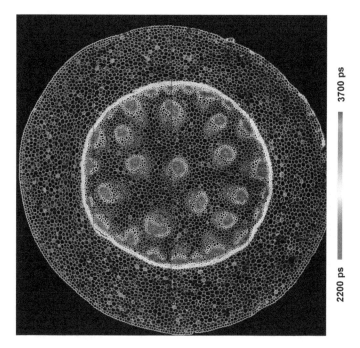

Fig. 2.24 Mosaic FLIM of a Convallaria sample. The mosaic has 4 × 4 elements, each element has 512 × 512 pixels, each pixel has 256 time channels. Zeiss LSM 710 Intune system with bh Simple-Tau 150 FLIM system

The large memory size available in the 64 bit environment allows tile imaging to be combined with FLIM. In the memory of the FLIM system a two-dimensional mosaic of FLIM data blocks is defined. The number and arrangement of the data blocks is the same as for the tiles in the imaging procedure of the microscope. Moreover, for each element of the FLIM mosaic a number of frames is defined which is identical with the number of frames per tile defined in the microscope. When the microscope starts the scan the FLIM system starts to record an image in the first mosaic element. It continues to do so until the defined number of frames has been completed, and then switches to the next mosaic element. This way, the FLIM system switches through all mosaic elements, filling all of them with data of the corresponding tiles. The procedure ends when the microscope has completed the last frame of the last tile.

The FLIM data structure can thus be considered one large FLIM image that contains all tiles in the correct arrangement. An example of a Mosaic FLIM is shown in Fig. 2.24. The image was recorded by a Zeiss LSM 710 Intune (tuneable excitation) system in combination with a bh Simple-Tau 150 FLIM system. The mosaic has 4 × 4 elements, each element has 512 × 512 pixels with 256 time channels. The complete mosaic has thus 2048 × 2048 pixels, each pixel holding 256 time channels. The sample area covered by the mosaic is 2.5 mm × 2.5 mm.

2.4.2 Multi-wavelength FLIM Mosaic

From the TCSPC point of view a spatial mosaic is a photon distribution over the times of the photons in the laser pulse period, the scan coordinates, and the coordinates of the x–y displacement of the tile scan. The idea of mosaic FLIM can therefore be applied also for other recording tasks with complex photon distributions.

Figure 2.25 shows a multi-wavelength FLIM recording of a Convallaria sample acquired in the Mosaic mode of an SPC-150 TCSPC FLIM module. The sample was scanned by a bh DCS-120 confocal scanning system with a bh MW-FLIM GaAsP 16-channel detector. The data were recorded at a resolution of 512 × 512 pixels, and 256 time channels and 16 wavelength channels per pixel. For technical principles, see Chap. 1, Sect. 1.4.5.2.

Figure 2.26 demonstrates the true spatial resolution of the data. The image from one wavelength channel, 565 nm, of the data shown Fig. 2.25 is displayed at larger scale and with individually adjusted lifetime range. The spatial resolution is comparable with what previously could be reached for FLIM at a single wavelength. Nevertheless, the decay data are recorded at a temporal resolution of 256 time channels. A decay curve for a selected pixel of the image is shown in Fig. 2.27.

Multi-wavelength mosaic FLIM essentially uses the same recording procedure and contains the same data as traditional multi-wavelength FLIM, see Chap. 1, Sect. 1.4.5.2. However, the data presentation is different. Traditional multi-

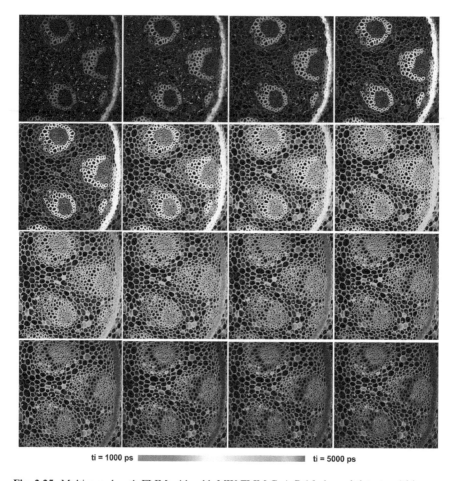

ti = 1000 ps ▬▬▬▬▬▬▬▬▬▬▬▬▬▬▬▬▬▬▬ ti = 5000 ps

Fig. 2.25 Multi-wavelength FLIM with a bh MW-FLIM GaAsP 16-channel detector. 16 images with 512 × 512 pixels and 256 time channels were recorded simultaneously. Wavelength from *upper left* to *lower right*, 490 to 690 nm, 12.5 nm per image. DCS-120 confocal scanner, Zeiss Axio Observer microscope, ×20 *NA* = 0.5 air lens

wavelength FLIM presents the data as a number of images for different wavelength, mosaic FLIM presents the data as one large image containing sub-images of different wavelength. The advantage is that data analysis can process the whole mosaic in one single run. This makes it easier to use analysis with global parameters for the whole array.

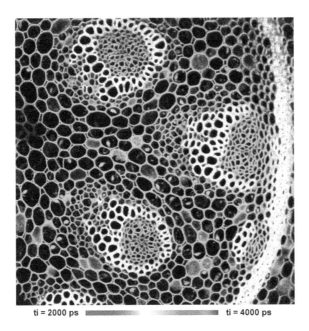

ti = 2000 ps ▬▬▬▬▬▬▬▬ ▬▬▬▬▬▬▬▬ ti = 4000 ps

Fig. 2.26 Single image from the array shown in Fig. 2.25, displayed in larger scale and with individually adjusted lifetime range. Wavelength channel 565 nm. The images size is 512 × 512 pixels, with 256 time channels per pixel

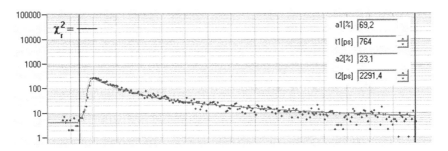

Fig. 2.27 Decay curves at selected pixel position in the image shown above. *Blue dots* Photon numbers in the time channels. *Red curve* Fit with a double-exponential model

2.4.3 Z-Stack Recording by Mosaic FLIM

The mosaic procedure can be used to record z-stacks of FLIM images. The number of mosaic elements is made identical with the number of z planes the microscope scans. The number of frames per mosaic element is selected according to the number of frames per z plane. Compared with Z-stack recording by a record-and-save procedure [6, 11] mosaic FLIM has the advantage that the synchronisation between the microscope and the FLIM system is easier, and no time has to be

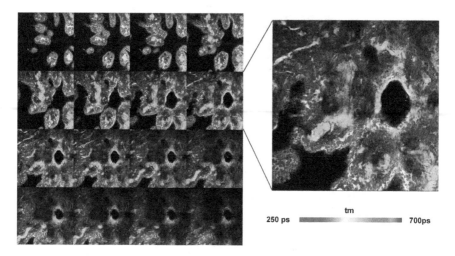

Fig. 2.28 Z-stack recorded by Mosaic FLIM. Pig skin stained with Indocyanin Green. Zeiss LSM 7 MP OPO system, 16 planes from 0 to 60 μm from top of sample, each plane 512 × 512 pixels, 256 time channels. Amplitude-weighted lifetime of double-exponential decay

reserved for the saving actions. Moreover, the data of all z planes are recorded in one and the same photon distribution. The data analysis can therefore be run for the data of all z planes together. As for multi-wavelength mosaic, this makes it easier to maintain exactly the same fit conditions and model parameters for all z planes, and to fit the decay data in the entire array with global model parameters. An example of a mosaic Z stack is shown in Fig. 2.28.

2.4.4 Temporal Mosaic FLIM

The usual way to record time series by FLIM is to record a sequence of images and save the subsequent images in consecutive files [8, 11]. Mosaic FLIM provides a second way to record a time series. In that case, every element of the mosaic represents one step of the series. Each element can be a single frame or a defined number of frames of the scan.

An example is shown in Fig. 2.29. The sample was a fresh moss leaf, the microscope a bh DCS-120 confocal FLIM system [5]. The time runs from lower left to upper right. The excitation light causes a non-photochemical transient in the chlorophyll fluorescence, resulting in a decrease in the fluorescence lifetime over the time of exposure [43].

Mosaic time series recording has two advantages over the conventional record-and-save procedure. First, the transition from one mosaic element to the next occurs instantaneously. There are no time gaps between the steps of the sequence. Therefore, very fast time series can be recorded. Second, mosaic time series data

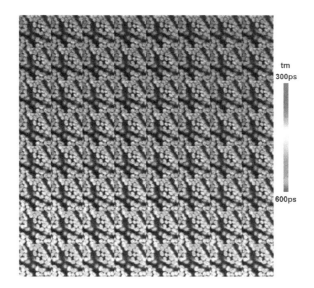

tm
300ps

600ps

Fig. 2.29 Time series recorded by mosaic imaging. Non-photochemical transients in chloroplasts of a moss leaf. 64 mosaic elements for consecutive times after turn-on of excitation light. Acquisition time per element 1 s, total time of sequence 64 s, image size of each element 128 × 128 pixels, 256 time channels. Time runs from *lower left* to *upper right*. bh DCS-120 confocal FLIM system with SPC-150 TCSPC modules, SPCImage data analysis software

can be accumulated: A sample would be repeatedly stimulated by an external event, and the start of the mosaic recording be triggered by the stimulation. With every new stimulation the recording procedure runs through all elements of the mosaic, and accumulates the photons. Accumulation allows data to be recorded without the need of trading photon number (and thus lifetime accuracy) against the speed of the time series. Consequently, the time per step (or mosaic element) is only limited by the minimum frame time of the scanner. For pixel numbers on the order of 128 × 128 and small scan areas galvanometer scanners reach frame times on the order of 50 ms. Resonance scanner are even faster. This brings the time-series resolution into the range where even lifetime changes induced by Ca^{2+} transients in neurons can be recorded. Previously, this was only possible by FLITS, i.e. by recording the decay functions within the pixels of a line scan [14], see Chap. 1, Sect. 1.4.6, and Chap. 5.

Figure 2.30 shows a mosaic FLIM recording of the Ca^{2+} transient in cultured neurons after stimulation with an electrical signal. Oregon Green was used as a Calcium sensor. A Zeiss LSM 7 MP multiphoton microscope was used to scan the sample. The scan format was 64 × 64 pixels. With 64 × 64 pixels and a zoom factor of 5, the LSM 7 MP reaches a frame time of 38 ms. The FLIM data were recorded by a bh SPC-150 module. A FLIM mosaic size of 8 × 8 elements was defined. The stimulation period was 3 s. 150 ms before every stimulation a recording through the entire 64-element mosaic was started. With the frame time of 38 ms, the acquisition thus runs through the entire mosaic in 2.43 s. 100 such acquisition runs were

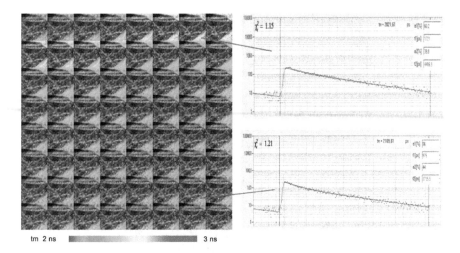

Fig. 2.30 Temporal mosaic FLIM of the Ca^{2+} transient in cultured neurons after stimulation with an electrical signal. The time per mosaic element is 38 ms, the entire mosaic covers 2.43 s. Experiment time runs from *upper left* to *lower right*. Photons were accumulated over 100 stimulation periods. Recorded by Zeiss LSM 7 MP and bh SPC-150 TCSPC module. Data courtesy of Inna Slutsky and Samuel Frere, Tel Aviv University, Sackler Faculty of Medicine

accumulated. The result shows clearly the increase of the fluorescence lifetime of the Ca^{2+} sensor in the mosaic elements 4–6, and a return to the resting state over the next 10–15 mosaic elements (380–570 ms).

The element time of 38 ms in Fig. 2.30 is about the fastest which can be achieved by a normal galvanometer scanner. Faster scan times can probably achieved by resonance scanners. Another way to higher resolution is line scanning, see Chap. 1, Sect. 1.4.6 and Chap. 5, Sect. 5.2.

2.5 Macroscopic FLIM

Fluorescence Lifetime Imaging (FLIM) is often associated with Fluorescence Lifetime Imaging Microscopy. Consequently, it is often believed that FLIM, especially TCSPC FLIM, can only be obtained from microscopic objects. This is, of course, incorrect. TCSPC FLIM can be combined with any optical technique that scans a sample with a single focused beam of light.

Centimetre-size objects can be scanned by placing them directly in the image plane of a confocal scan head. The optical principle for a macroscopic scanning system with two excitation and two detection channels is shown in Fig. 2.31, left.

The image plane of the scan lens is brought in coincidence with the sample surface. As the galvanometer mirrors change the beam angle the laser focus scans across the sample. Fluorescence light excited in the sample is collimated by the scan lens, de-scanned by the galvanometer mirrors, and separated from the excitation

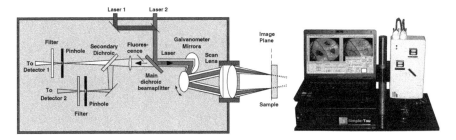

Fig. 2.31 *Left* Principle of the DCS-120 scanner for imaging macroscopic objects (simplified). *Right* DCS-120 MACRO system

light by the main dichroic beamsplitter. As usual, the fluorescence beam is further split into two spectral or polarisation components, and focused into pinholes. Light passing the pinholes is sent to the detectors.

For the results presented here we used the Becker & Hickl DCS-120 MACRO confocal FLIM system. Both Becker & Hickl picosecond diode lasers and super-continuum lasers with acousto-optical filters were used as excitation sources. See [5] for details. A photo of the system is shown in Fig. 2.31, right.

2.5.1 Examples

Figure 2.32 shows a leaf with a fungus infection. The fluorescence lifetime in the infected areas is considerably longer than in the non-infected areas, see fluorescence decay curves on the right.

Figure 2.33 shows that good images are obtained even from weakly fluorescent samples. The image on the left was obtained from the feather of a songbird, the image on the right from a *coccinella* beetle. Both objects delivered an average count rate of about 100,000 counts per second when excited with the ps diode laser.

2.5.2 Image Area and Resolution

The maximum diameter of the image area in the primary image plane of the scanner is about 15 mm. Smaller areas can be scanned by using the 'Zoom' function of the scanner, see [5]. The size of the laser spot in the image plane is about 15 μm. The resolution in the detection path is even higher because the numerical aperture in the focus of the scan lens is larger than for the laser beam. That means full-size images could, in principle, be reasonably scanned with at least 1024 × 1024 pixels. With 64-bit data acquisition software, the number of time channels per pixel at a spatial resolution of 1024 × 1024 pixels can be increased up to 1024 [75], see also Sect. 2.2.3.

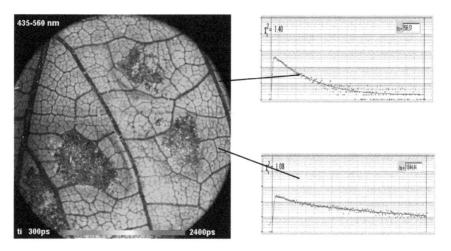

Fig. 2.32 FLIM in the primary image plane of a DCS-120 scanner. *Left* Leaf with a fungus infection. ps diode laser excitation, 405 nm, scan format 512 × 512 pixels. *Right* Decay functions of healthy and infected areas. Analysis by Becker & Hickl SPCImage lifetime analysis software [5], pixel binning 3 × 3

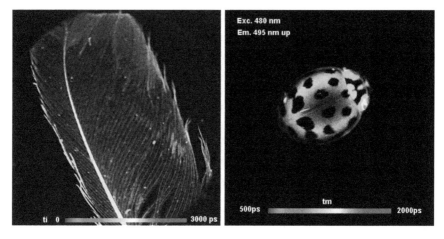

Fig. 2.33 *Left* Feather of a songbird. Excitation by 405 nm picosecond diode laser. *Right* *Coccinella* beetle, excited by super-continuum laser with acousto-optical filter, wavelength 480 nm. Scan format 512 × 512 pixels

2.5.3 Collection Efficiency and Required Laser Power

The numerical aperture of the detection beam path is given by the beam path diameter (about 3 mm) and the focal length of the scan lens (40 mm). The collection efficiency is thus considerably lower than in combination with a microscope. However, macroscopic imaging can use much higher laser power: The power is distributed over a large area so that photobleaching is not a problem. High laser power can therefore be used to compensate for low collection efficiency. The image shown in Fig. 2.32 was obtained with ps diode laser excitation at about 20 μW in the sample plane. This is about 4 % of the available laser power. The image shown in Fig. 2.33, right, was obtained by super-continuum laser excitation. The power in the sample plane was about 100 μW. Only 5 % of the available power was used.

2.5.4 Acquisition Times

The acquisition time depends on the number of pixels in the image and the requirements to the accuracy of the lifetimes. The image shown in Fig. 2.32 was recorded at a spatial resolution of 512 × 512 pixels and contains about 250 photons per pixel. It was analysed with a spatial binning of 3 × 3 pixels. Under these condition the lifetime accuracy is about 5 % [15, 49]. The images were recorded within an acquisition time of about one minute. For images recorded at 128 × 128 pixels the same lifetime accuracy is obtained in less than 5 s.

2.6 Scanning Through Periscopes

The principle used above can be extended to endoscopy by scanning the beam over the input face of a fibre bundle endoscope, see Fig. 2.34. The excitation spot at the input face of the bundle is transferred to the distal bundle face. The distal face is either brought in direct contact with the sample, or imaged into deeper sample layers by a transfer lens. The fluorescence excited in the sample is transferred back to the input face, and back through the confocal scanner to the detectors. The setup has optical sectioning capability: Light from a point not in contact with the distal bundle face (or, if a transfer lens is used, not conjugate with it) does not form a sharp spot on the fibre bundle input face. It is thus not focused into the pinholes.

An instrument of this type has been described by Kennedy et al. [46]. Excitation pulses of 488 nm wavelength were generated by a frequency-doubled Ti:Sapphire

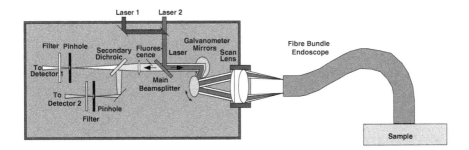

Fig. 2.34 Combination of a confocal scanner with a fibre bundle endoscope

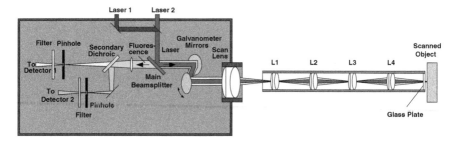

Fig. 2.35 Scanning through a rigid endoscope (not to scale). *L2* and *L4* are transfer lenses, *L1* and *L3* are field lenses

laser. The excitation beam was scanned over the input face of a fibre bundle consisting of 30,000 single-mode fibres. By terminating the distal fibre end with a miniature objective lens images at a resolution of 1.4 µm were obtained.

A significant problem of the setup shown in Fig. 2.34 is fluorescence from the fibre bundle. With the commonly used imaging fibre bundles we found it virtually impossible to record FLIM from biologically relevant samples at excitation wavelengths shorter than 450 nm. That means that the most interesting endogenous fluorophore, NADH, cannot be excited.

Good results were obtained from a similar setup using a rigid periscope. The principle is shown in Fig. 2.35. L2 transfers the image plane of the scan lens into the plane of L3, L4 transfers the image in the plane of L3 into the plane of the object. L1 and L3 are field lenses: L1 forms an image of the scan lens on L3, L3 form an image of L2 on L4. The lenses thus avoid that the field of view is obstructed by the transfer lenses, L2 and L4. The periscope is sealed with a glass plane at the distal end. The measured object is brought into direct contact with the glass plate. This minimises motion of the sample surface relative to the endoscope.

For recording the images shown below we used a bh DCS-120 scan head with an SPC-152 dual-channel FLIM system. Excitation was performed by bh BDL-SMC diode lasers of 405 and 473 nm wavelength. The lenses in the periscope were achromatic doublets of 45 mm focal length and 8 mm diameter. The endoscope had a length of 38 cm and an outer diameter of 9 mm. The field of view has a maximum diameter of 6 mm.

Figure 2.36 shows a lifetime image of a plant leaf. The scanned area is 5 × 5 mm, the image format is 512 × 512 pixels, 256 time channels per pixel. The excitation wavelength was 473 nm, the detection wavelength from 560 to 690 nm. Figure 2.36 shows that the optical resolution is not impaired by the endoscope optics.

Figure 2.37 shows lifetime images of a basal cell papilloma. No staining was used, the emission comes exclusively from endogenous fluorophores. The excitation wavelength was 405 nm. The left image is from a detection wavelength interval of 490–560 nm, the right image from an interval of 560–690 nm. An area of 5 × 5 mm was scanned, the image format was 512 × 512 pixel, 256 time channels. The total acquisition time was 10 s. 10 frames were accumulated within this time. Although some blurring is caused by motion during the acquisition time the

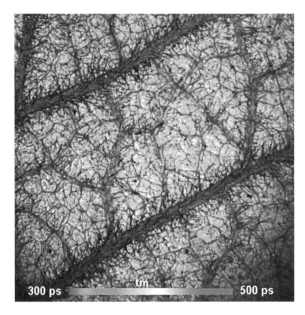

Fig. 2.36 Lifetime image of a leaf scanned through the periscope. 512 × 512 pixels, 256 time channels. Amplitude-weighted lifetime of double-exponential decay. Excitation 473 nm, detection 560–690 nm

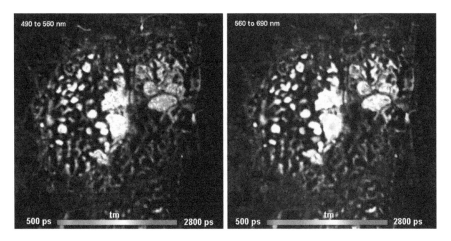

Fig. 2.37 Human basal cell papilloma, scanned through the endoscope. image size 5 × 5 mm, scanned with 512 × 512 pixels. Becker & Hickl DCS-120 confocal scanner and Simple-Tau 152 TCSPC FLIM system. Amplitude-weighted lifetime of double-exponential decay fit. Excitation wavelength 405 nm, detection wavelength 490–560 nm (*left*) and 560–690 nm (*right*). Excitation power 50 µW, acquisition time 10 s

anatomical structure is still reasonably resolved. The images were recorded at an excitation power of 50 µW, measured in the sample plane. Note that this power was distributed over an area of 25 mm², corresponding to an excitation power of 2 µW per square millimeter. This is far below the permissible exposure for human tissue. It is even within the range considered safe to the human retina, see Chap. 16 of this book.

2.7 Polygon Scanners

In vivo imaging systems often use fast-rotating polygon scanners to obtain imaging at near-video rate. If such systems have or can provide an input for a picosecond diode laser or a super-continuum laser and an output to a FLIM detector they can be used for TCSPC FLIM. The principle of a polygon scanner and its connection to a bh FLIM system are shown in Fig. 2.38.

Polygon scanners can achieve pixel rates on the order of 10 MHz, i.e. the pixel time is on the order of 100 ns. As long as the scanner provides correct scan clock pulses such pixel rates are no basic problem for multi-dimensional TCSPC. A problem can occur when a software-based imaging mode (see Chap. 1, Sect. 1.3.2) is used with external pixel clock: The pixel clock rate can be so high that the transfer of pixel clocks alone saturates the computer bus. The problem can easily be solved by using hardware-acquisition of the images (see Chap. 1, Sect. 1.3.2) or by recording with internal pixel clock [11]. Internal pixel clock does not cause the same inconveniencies as for galvanometer scanners because the polygon scanner runs at constant speed. A pixel clock is therefore not needed, and, in some instances, also not provided by polygon scanners.

bh FLIM has been used in combination with the 'Vivascope' system of Lucid, Rochester (see also Chap. 15). A lifetime image recorded with the Vivascope and a bh SPC-150 TCSPC FLIM modules is shown in Fig. 2.39. Despite of the high scan rate data of excellent quality are obtained, and no synchronisation problems, such as missing lines or shifted pixels have been found. Resolution and contrast are excellent, and the decay data are clean, see Fig. 2.40.

See also Chap. 15 for application of the system to FLIM of skin biopsies.

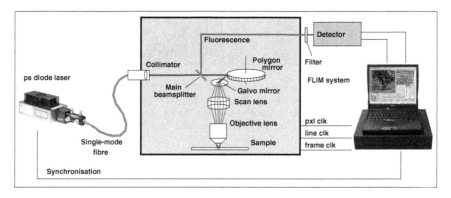

Fig. 2.38 FLIM with a polygon scanner

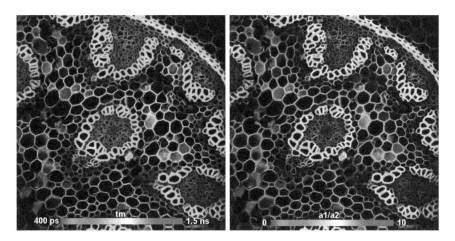

Fig. 2.39 Convallaria sample, 512 × 512 pixels, 256 time channels. Double-exponential analysis. *Left* Amplitude-weighted lifetime of decay components. *Right* Ratio of amplitudes of double-exponential decay. Becker & Hickl BDL-SMC 473 nm laser and Simple-Tau 150 connected to Lucid Vivascope

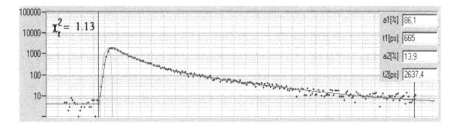

Fig. 2.40 Fluorescence decay in a selected spot of Fig. 2.39. No optical reflections are visible

2.8 Scanning with Piezo Stages

2.8.1 Optical System

The scanning systems described above use beam scanning to move the laser spot over the sample. However, scanning can also be performed by moving the sample. The optical principle of a confocal system with sample scanning is shown in Fig. 2.41.

The system shown on the left injects the laser beam into the microscope via the port for the fluorescence lamp. The beamsplitter and the emission filter are inserted in a standard microscope filter cube in the filter carousel of the microscope. The detector is attached to another port of the microscope. A pinhole in the upper image plane of the microscope suppresses out-of-focus light. The setup is easy to implement but has a few disadvantages: The lamp port is blocked by the excitation beam path, and the laser beam diameter must be increased by a beam expander.

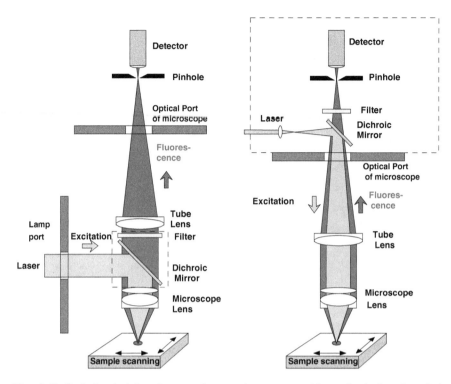

Fig. 2.41 Optical principle of a sample scanning system with confocal detection. *Left* Beamsplitter and emission filter in microscope beamsplitter cube, pinhole and detector attached to a port of the microscope. *Right* Beamsplitter filter, pinhole and detector in compact optical assembly attached to a port of the microscope

The most significant problem is alignment. The laser must be focused into exactly the same spot in the sample the pinhole is looking at. Alignment critically depends on the angle of the dichroic mirror in the filter cube. Microscopy filter cubes are not designed for this level of precision.

The system on the right has the laser input, the dichroic beamsplitter, the emission filter, the pinhole, and the detector placed in a compact optical assembly. The entire assembly is attached to an optical port of the microscope. The alignment of the laser beam with respect to the pinhole is therefore more stable.

2.8.2 Synchronisation of the Scanner with the TCSPC Module

Piezo scanning can be combined with multi-dimensional TCSPC the same way as beam scanning: The recording is synchronised with the scanning by pixel clock,

line clock, and frame clock pulses, see Chap. 1, Sect. 1.4.5. The only difference is in
the scan rate. Stage scanning is typically 100 times slower than beam scanning. As
a consequence, piezo scan data are often acquired in a single frame whereas beam-
scanning data are normally accumulated over many frames. In any case, the
recording process is fully integrated in the hardware and software of state-of-the art
TCSPC devices and does not require any user-specific software.

One way to combine a piezo scanner with TCSPC FLIM is to run the piezo
controller software in parallel with the TCSPC FLIM software. The piezo controllers can usually be programmed to generate the clock synchronisation pulses
required for TCSPC.

Another way of integrating a scan stage in a TCSPC system is via a bh GVD-120
scan controller card. The GVD-120 card is part of the bh DCS-120 confocal
scanning system. It has analog outputs which are normally used to drive galvanometer scanners. The card can, however, be configured to run the slow scan ramps
required for piezo stages. The control of the card is fully integrated in the bh SPCM
data acquisition software. This way, the piezo stage gets fully integrated into the
SPCM instrument software [5, 11]. Different pixel numbers, scan formats, and scan
speeds can be selected via the SPCM software. The scanner can be parked in a
selected spot for FCS recording or single-molecule spectroscopy, see Chap. 9. The
scan controller also controls two ps diode lasers. The lasers can be multiplexed
synchronously with the pixels, lines, or frames of the scan, or be modulated for
simultaneous FLIM/PLIM operation [5, 11], see Chap. 1, Sect. 1.4.7.

Figure 2.42 shows a typical lifetime image recorded in the setup of Fig. 2.41,
right. A Nikon TE 2000 microscope with an 63×, NA = 1.3 oil immersion lens was
used. A PI P733.3DD piezo stage with analog inputs was driven by a bh GVD-120

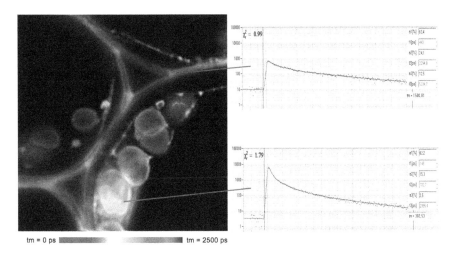

tm = 0 ps   tm = 2500 ps

Fig. 2.42 FLIM image obtained piezo-stage scanning. PI P733.3DD piezo stage controlled by bh
GVD-120 scan controller. Recording by bh SPC-150 TCSPC FLIM module. Image size
30 μm × 30 μm, data format 512 × 512 pixels, 256 time channels. Single scan, scan speed 5 lines
per second

scan controller card. A bh BDL-488 SMN (488 nm) laser was used for excitation. The fluorescence was detected by a bh HPM-100-40 GaAsP hybrid detector [10], and recorded by a bh SPC-150 TCSPC FLIM module. The scanning and the recording was controlled by standard bh SPCM software. The data FLIM data were analysed by bh SPCImage software.

The sample was a convallaria preparation as it is commonly used in microscope demonstrations. The scanner was run at a line time of 3.33 s (including line fly-back). The scan format was 512 × 512 pixels, resulting in a scan time of 1700 s for the entire image.

The advantage of stage scanning is that there are only a few optical elements in the beam path, and there are no moving elements. The excitation and detection beams remain on the optical axis for all locations in the image. This way, aberrations and transmission losses are kept at a minimum. However, these advantages come at a price: The scanning is painfully slow, acquisition speed is limited by the scanner, a fast preview function is not available, focusing into the desired image plane is a nightmare, and deliberately selecting an image area of interest is difficult. Moreover, the scanning exerts inert ion forces to the sample. Imaging of live cells in a cell culture can be a problem because the cells are shaken around by the scanning action.

Stage scanning is traditionally used in STED microscopy (see Sect. 2.10) and in NSOM microscopy, see Sect. 2.9. There is actually no technical reason to use scan stages for STED: A galvanometer scanner achieves the required accuracy. In NSOM, however, there is no way around the piezo stage: The required resolution is on the order of a nanometer, and the sample has to be moved with respect to the tip.

2.9 NSOM FLIM

The scanning near-field optical microscope (SNOM or NSOM) combines the principles of the atomic force microscope (AFM) and the laser scanning microscope [32]. The AFM system scans a sharp tip over the sample, and keeps it at a distance comparable to the diameter of a single molecule. Optical near-field excitation is obtained either by using a tapered optical fibre as a tip (see Fig. 2.43, left) or by using a metallic tip and focusing the laser on it. The evanescent field at the tip is then used to probe the sample structure (Fig. 2.43, right). In both cases the fluorescence photons are collected through the microscope objective lens. Scanning is performed by a piezo-driven scan stage. Scanning must be slow enough to give the AFM system enough time to track the topography of the sample surface.

Because NSOM uses scanning it can easily be combined with TCSPC FLIM. Because the image is obtained in a single slow scan the fluorescence decay curves are often recorded and read out on a pixel-by-pixel basis. However, a better way to record NSOM FLIM is by the usual imaging procedure of multi-dimensional TCSPC, see Chap. 1. The advantage of using multidimensional TCSPC is that no user-specific programming is necessary, that readout times during the acquisition

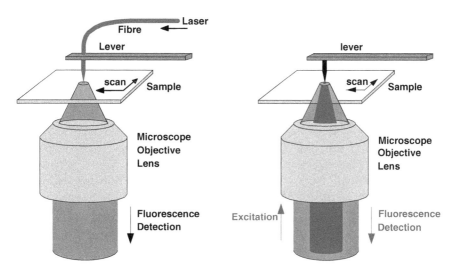

Fig. 2.43 Optical near-field microscope. *Left* Excitation by tapered fibre probe. *Right* Excitation by tip enhancement

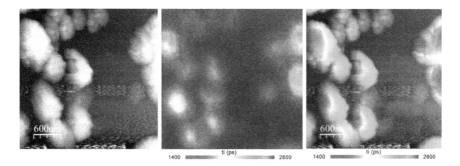

Fig. 2.44 NSOM FLIM of nano-islands on a semiconductor surface. *Left* Topography from AFM system. *Middle* FLIM. *Right* Overlay of AFM topography with lifetime. Amplitude-weighted lifetime of double-exponential decay. Nanonics AFM/NSOM system with bh SPC-150 FLIM system

are avoided, and that the data can be processed by standard TCSPC-FLIM analysis software. An example of NSOM FLIM is shown in Fig. 2.44.

The optical near-field microscope reaches an optical resolution on the order of 50–100 nm, i.e. about 2–4 times better than a confocal or two-photon laser scanning microscope. However, there is also topography data from the atomic-force part of the system. Atomic-force data have a resolution on the order of a few nanometers. Sample topography can thus be combined with spectroscopy data.

Early applications of TCSPC to SNOM have focused mainly on using the lifetime information as an additional contrast mechanism [45, 56, 64]. Near-field FLIM Images of good quality were presented by Cadby et al. [23]. The authors used a Veeco Aurora SNOM system upgraded with an R3809U MCP PMT and an SPC-830 TCSPC module. The system was used to investigate the nano-structure of blends of fluorescent and non-fluorescent polymers [23, 24] and to microcapillary cavities containing a fluorescent dye [1].

Gokus et al. used tip-enhanced near-field excitation to identify single carbon nanotubes and measure their fluorescence lifetime. A titanium-sapphire laser was used to excite the sample at 760 nm. The fluorescence was detected by an MPD PDM 50 SPAD, and the decay curves were recorded by a bh SPC-140 TCSPC module. The instrument response width was 27 ps (!), so that fluorescence lifetimes of 4–36 ps could reliably be recorded [41].

Gaiduk et al. used the SNOM principle for combined optical and force spectroscopy of single labelled DNA molecules [36, 37]. When a DNA molecule is captured between the tip and the surface simultaneous force and fluorescence data are acquired. By moving the tip up and down complete distance cycles were recorded.

2.10 STED FLIM

Stimulated-Emission-Depletion (STED) microscopy exploits the nonlinearity of stimulated emission to obtain optical super-resolution [44, 48]. The fluorescence excited by the excitation beam in a scanning microscope is depleted by stimulated emission induced by a second (STED) laser beam. The wavefront of the STED beam is manipulated to obtain a diffraction pattern that either has doughnut shape laying in the x–y plane or a dumbbell shape oriented along the z axis. Fluorescence remains un-depleted in the centre part of the doughnut or the dumbbell. Because stimulated emission is highly nonlinear the un-depleted volume can be made considerably smaller than the point-spread function of the excitation beam. Images are obtained by sample scanning or beam scanning.

At first glance, it may appear impossible to combine STED with fluorescence lifetime imaging, because the decay curves are distorted by stimulated emission. Nevertheless, Auksorius et al. have shown that FLIM in combination with STED is possible [2]. The authors used a femtosecond Ti:sapphire laser to generate the excitation beam and the STED beam. The excitation beam was obtained by sending a part of the Ti:Sapphire laser emission through a micro-structured optical fibre. The fibre generated a super-continuum from which the excitation wavelength was selected by a filter. For the STED beam the Ti:Sapphire pulses were broadened to 300 ps by 100 m of single-mode fibre. The broad pulses prevent multiphoton excitation by the STED beam. FLIM was performed the usual way by recording the photons in the Scan Sync In mode of a bh SPC-730 TCSPC module [11]. Figure 2.45 shows a group of fluorescent beads imaged by confocal detection (with the STED beam turned off) and by STED (with the STED beam turned on). Fluorescence decay curves for confocal detection and STED detection are shown in Fig. 2.45, right.

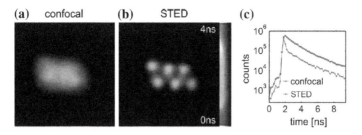

Fig. 2.45 Confocal (**a**), STED lifetime images (**b**), and confocal and STED decay curves (**c**). From [2]

The depletion of the fluorescence by the STED beam clearly shows up in the STED decay curves. The peak at the beginning is caused by the un-depleted fluorescence. A few 100 ps later the fluorescence in the outer parts of the Airy disc is depleted by the STED beam. At later times the fluorescence decays with the same decay time as in the confocal case. The peak of un-depleted fluorescence in the STED data can be gated off either directly in the TCSPC process or removed from the intensity data during FLIM data analysis. This way, TCSPC detection can be used to increase the contrast of STED images.

Bückers et al. used a super-continuum laser to generate two pairs of excitation and STED beams [22]. The optical principle is shown in Fig. 2.46, left. The pulses of the both pairs were offset in time by 40 ns. Both excitation/STED beams are used simultaneously to excite and deplete different fluorophores. The fluorescence signals were separated from the excitation and STED beams by a dichroic beam splitter, DC, split spectrally by a second dichroic beam splitter, and projected into confocal pinholes, PH. Light passing the pinholes was detected by Perkin Elmer SPC-AQR detectors. Scanning was achieved by moving the sample by a piezo scan stage. The photon pulses from the detectors were processed by a bh SPC-730 TCSPC FLIM module via a router. The TCSPC module records separate lifetime images for the two detection wavelengths. Moreover, each pixel of each lifetime image contains two subsequent decay curves for the two excitation pulse trains (see Chap. 1, Sect. 1.4.5.3, Fig. 1.32). From the recorded data, images for fluorophores of different lifetimes and excitation/emission wavelengths were extracted.

A result is shown in Fig. 2.46, right. A confocal image (with the STED beams turned off) is shown on the left, a STED image on the right. The sample was a cell in which lamin was stained with KK 114, tubulin with ATTO 647N, and clathrin with ATTO 590. The fluorescence signals from KK 114 and ATTO 647N, were segregated via their fluorescence lifetime, ATTO 590 via its fluorescence wavelength. From the intensities of the separated signals images showing the location of the cells constituents were constructed.

Lesoine et al. [58] used the same basic optical setup but with a single excitation beam and STED beam. The photons were detected by an HPM-100-40 hybrid detector [10] and the image recorded by an SPC-830 FLIM module. The authors detected lifetime changes of Alexa Flour 594 phalloidin, see Fig. 2.47. They

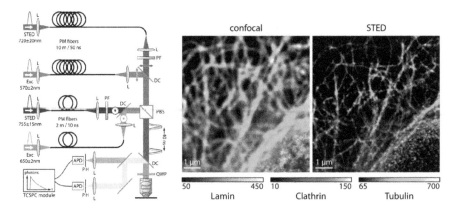

Fig. 2.46 Dual-Wavelength STED system with super-continuum laser. *Left* Optical principle. *Right* Confocal image and STED image. Combined spectral and lifetime separation. Lamin and tubulin were stained with KK 114 and ATTO 647N, respectively, which were segregated by lifetime analysis. The ATTO 590 fluorescence (clathrin) was spectrally separated from those of the other dyes. From [22]

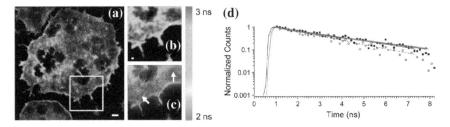

Fig. 2.47 *Left* Fluorescence lifetime images of a cell with Alexa Fluor 594-phalloidin labelled actin. Overview confocal image (**a**), confocal image (**b**) and STED images (**c**) of the $10 \times 10 \ \mu m^2$ region marked in (**a**). *Right* **d** Fluorescence decay curves from two regions (denoted by *white arrows*) from the image (**c**). Fit with single exponentials of 2.9-ns decay time (*solid square*) and 2.2-ns decay time (*open circle*). *Blue curve* Fluorescence decay obtained from the Alexa Fluor 594-phalloidin conjugate in the medium used to prepare the cells. From [58]

attribute these changes to electron transfer between the Alexa dye as an electron acceptor and tryptophan as an electron donor.

STED imaging is currently a bit outshone by super-resolution techniques like PALM or STORM. These techniques are based on switchable fluorophores. The fluorophores are optically controlled so that only a small fraction of molecules is fluorescing at a time. These molecules are localised at high precision, and the image reconstructed from the positions of the individual molecules. However, these techniques are difficult if not impossible to be combined with FLIM: They require simultaneous detection of the whole field by a wide-field detector of high pixel number. There is no such detector that acquires fluorescence decay curves simultaneously in many time channels and without the need of temporal scanning.

Moreover, PLM and STORM are restricted to special fluorophores. Whether the lifetimes of these fluorophores report biologically interesting parameters is not known. STED FLIM is currently the best technique to combine optical super-resolution with fluorescence lifetime detection.

2.11 Summary

The combination of multi-dimensional TCSPC with laser scanning techniques has amazing capabilities: Lifetime images containing one or several fluorescence decay curves in the individual pixels are obtained at extremely high contrast, high spatial resolution, high photon efficiency and high time resolution. Applications range from imaging of macroscopic objects and imaging through endoscopes over laser scanning microscopy to super-resolution techniques like NSOM and STED. Combinations of multi-dimensional features of laser scanning microscopy and of advanced TCSPC result in techniques like multi-spectral FLIM, FLIM at multiple excitation wavelengths, spatial mosaic FLIM, Z stack FLIM, and combined FLIM/PLIM. Spatial mosaic FLIM and FLITS are able to record changes in the fluorescence behaviour of a sample down to the millisecond time scale. No other FLIM technique offers a similar variety of features. We therefore believe that there is still unused and, probably, unknown potential of TCSPC FLIM in a wide variety of applications.

References

1. A.M. Adawi, A. Cadby, L.G. Connolly, W.-C. Hung, R. Dean, A. Tahraoui, A.M. Fox, A.G. Cullis, D. Sanvitto, M.S. Skolnick, D.G. Lidzey, Spontaneous emission control in micropillar cavities containing a fluorescent molecular dye. Adv. Mater. **18**, 742–747 (2006)
2. E. Auksorius, B.R. Boruah, C. Dunsby, P.M.P. Lanigan, G. Kennedy, M.A.A. Neil, P.M.W. French, Stimulated emission depletion microscopy with a supercontinuum source and fluorescence lifetime imaging. Opt. Lett. **33**, 113–115 (2008)
3. R.M. Ballew, J.N. Demas, An error analysis of the rapid lifetime determination method for the evaluation of single exponential decays. Anal. Chem. **61**, 30 (1989)
4. E. Beaupaire, M. Oheim, J. Mertz, Ultra-deep two-photon fluorescence excitation in turbid media. Opt. Commun. **188**, 25–29 (2001)
5. Becker & Hickl GmbH, *DCS-120 Confocal Scanning FLIM Systems, User Handbook*, edition 2012. www.becker-hickl.com
6. Becker & Hickl GmbH, *Modular FLIM Systems for Zeiss LSM 510 and LSM 710 Family Laser Scanning Microscopes, User Handbook*, 5th edn (Becker & Hickl GmbH, 2012). www.becker-hickl.com
7. W. Becker, K. Benndorf, A. Bergmann, C. Biskup, K. König, U. Tirlapur, T. Zimmer, FRET measurements by TCSPC laser scanning microscopy. Proc. SPIE **4431**, 94–98 (2001)
8. W. Becker, *Advanced time-correlated single-photon counting techniques* (Springer, Berlin, 2005)

9. W. Becker, B. Su, A. Bergmann, K. Weisshart, O. Holub, Simultaneous fluorescence and phosphorescence lifetime imaging. Proc. SPIE **7903**, 790320 (2011)

10. W. Becker, B. Su, K. Weisshart, O. Holub, FLIM and FCS detection in laser-scanning microscopes: increased efficiency by GaAsP hybrid detectors. Microsc. Res. Technol. **74**, 804–811 (2011)

11. W. Becker, *The bh TCSPC Handbook*, 6th edn (Becker & Hickl GmbH, 2015). www.becker-hickl.com

12. W. Becker, Fluorescence lifetime imaging—techniques and applications. J. Microsc. **247**, 119–136 (2012)

13. W. Becker, V. Shcheslavskiy, Fluorescence lifetime imaging with near-infrared dyes. Proc. SPIE **8588**, 85880R (2013)

14. W. Becker, V. Shcheslavkiy, S. Frere, I. Slutsky, Spatially resolved recording of transient fluorescence-lifetime effects by line-scanning TCSPC. Microsc. Res. Technol. **77**, 216–224 (2014)

15. W. Becker, Fluorescence lifetime imaging techniques: time-correlated single photon counting, in *Fluorescence Lifetime Spectroscopy and Imaging*, ed. by L. Marcu, P.W.M. French, D.S. Elson (CRC Press, Boca Raton, 2015)

16. M.Y. Berezin, H. Lee, W. Akers, S. Achilefu, Near infrared dyes as lifetime solvatochromic probes for micropolarity measurements of biological systems. Biophys. J. **93**, 2892–2899 (2007)

17. M.Y. Berezin, W.J. Akers, K. Guo, G.M. Fischer, E. Daltrozzo, A. Zumbusch, S. Achilefu, Long lifetime molecular probes based on near-infrared pyrrolopyrrole cyanine fluorophores for in vivo imaging. Biophys. J. **97**, L22–L24 (2009)

18. M.Y. Berezin, S. Achilefu, Fluorescence lifetime measurement and biological imaging. Chem. Rev. **110**, 2641–2684 (2010)

19. F. Bestvater, E. Spiess, G. Strobrawa, M. Hacker, T. Feurer, T. Porwol, U. Berchner-Pfannschmidt, C. Wotzlaw, H. Acker, Two-photon fluorescence absorption and emission spectra of dyes relevant for cell imaging. J. Microsc. **208**(Pt. 2), 108–115 (2002)

20. M. Brambilla, L. Spinelli, A. Pifferi, A. Torricelli, R. Cubeddu, Time-resolved scanning system for double reflectance and transmittance fluorescence imaging of diffusive media. Rev. Sci. Instrum. **79**, 013103-1 to -9 (2008)

21. H.-G. Breunig, H. Studier, K. König, Multiphoton excitation characteristics of cellular fluorophores of human skin in vivo. Opt. Expr. **18**(8), 7857–7871 (2010)

22. J. Bückers, D. Wildanger, G. Vicidomini, L. Kastrup, S.W. Hell, Simultaneous multi-lifetime multi-colour STED imaging for colocalization anlysis. Opt. Expr. **19**, 3130–3143 (2011)

23. A. Cadby, R. Dean, A.M. Fox, R.A.L. Jones, D.G. Lidzey, Mapping the fluorescence decay lifetime of a conjugated polymer in a phase-separated blend using a scanning near-field optical microscope. Nano Lett. **5**, 2232–2237 (2005)

24. A.J. Cadby, R. Dean, C. Elliott, R.A.L. Jones, A.M. Fox, D.G. Lidzey, Imaging the fluorescence decay lifetime of a conjugated-polymer blend by using a scanning near-field optical microscope. Adv. Mater. **19**, 107–111 (2007)

25. A. Chorvatova, D. Chorvat, Tissue fluorophores and their spectroscopic characteristics, in *Fluorescence lifetime spectroscopy and imaging*, ed. by L. Marcu, P.W.M. French, D.S. Elson (CRC Press, Boca Raton, 2015)

26. G. Cox, *Optical Imaging Techniques in Cell Biology* (Taylor & Francis, Boca Raton, 2007)

27. W. Denk, J.H. Strickler, W.W.W. Webb, Two-photon laser scanning fluorescence microscopy. Science **248**, 73–76 (1990)

28. T. Desmettre, J.M. Devoisselle, S. Mordon, Fluorescence properties and metabolic features of indocyanine green (ICG) as related to angiography. Surv. Ophthalmol. **45**, 15–27 (2000)

29. A. Diaspro, in *Building a Two-Photon Microscope Using a Laser Scanning Confocal Architecture*, ed. by A. Periasamy. Methods in Cellular Imaging (Oxford University Press, New York, 2001), pp. 162–179

30. A. Diaspro, M. Corosu, P. Ramoino, M. Robello, Adapting a compact confocal microscope system to a two-photon excitation fluorescence Imaging architecture. Microsc. Res. Technol. **47**, 196–205 (1999)
31. A. Diaspro (ed.), *Confocal and Two-Photon Microscopy: Foundations, Applications and Advances* (Wiley-Liss, New York, 2001)
32. R.C. Dunn, Near-field scanning optical microscopy. Chem. Rev. **99**, 2891–2927 (1999)
33. F. Fischer, B. Volkmer, S. Puschmann, R. Greinert, W. Breitbart, J. Kiefer, R. Wepf, Risk estimation of skin damage due to ultrashort pulsed, focused near-infrared laser irradiation at 800 nm. J. Biomed. Opt. **13**(4), 041320 (2008)
34. F. Fischer, B. Volkmer, S. Puschmann, R. Greinert, W. Breitbart, J. Kiefer, R. Wepf, Assessing the risk of skin damage due to femtosecond laser irradiation. J. Biophoton. **1**, 470–477 (2008)
35. J.J. Freed, J.L. Engle, Development of the vibrating-mirror flying spot microscope for untraviolet spctrophotometry. Ann. N.Y. Acad. Sci. **97**, 412–488 (1962)
36. A. Gaiduk, R. Kühnemuth, S. Felekyan, M. Antonik, W. Becker, V. Kudryavtsev, M. Koenig, C. Sandhagen, C.A.M. Seidel, Time-resolved photon counting allows for new temporal and spacial insights imnto the nanoworld. Proc. SPIE **6372**, 637203-1 to -13 (2006)
37. A. Gaiduk, R. Kühnemuth, S. Felekyan, M. Antonik, W. Becker, V. Kudryavtsev, C. Sandhagen, C.A.M. Seidel, Fluorescence detection with high time resolution: from optical microscopy to simultaneous force and fluorescence spectroscopy. Microsc. Res. Technol. **70**, 403–409 (2007)
38. I. Gannot, I. Ron, F. Hekmat, V. Chernomordik, A. Ganjbakhche, Functional optical detection based on pH dependent fluorescence lifetime. Lasers Surg. Med. **35**, 342–348 (2004)
39. H.C. Gerritsen, M.A.H. Asselbergs, A.V. Agronskaia, W.G.J.H.M. van Sark, Fluorescence lifetime imaging in scanning microscopes: acquisition speed, photon economy and lifetime resolution. J. Microsc. **206**, 218–224 (2002)
40. M. Göppert-Mayer, Über elementarakte mit zwei quantensprüngen. Ann. Phys. **9**, 273–294 (1931)
41. T. Gokus, A. Hartschuh, H. Harutyunyan, M. Allegrini, F. Hennrich, M. Kappes, A.A. Green, M.C. Hersam, P.T. Araujo, A. Jorio, Exiton dynamics in individual carbon nanotubes at room temperature. Appl. Phys. Lett. **92**, 153116 (2008)
42. H.H. Gorris, S.M. Saleh, D.B. Groegel, S. Ernst, K. Reiner, H. Mustroph, O.S. Wolfbeis, Long-wavelength absorbing and fluorescent chameleon labels for proteins, peptides, and amines. Bioconjug Chem. **22**, 1433–1437 (2011)
43. Govindjee, Sixty-three years since kautsky: chlorophyll α fluorescence. Aust. J. Plant Physiol. **22**, 131–160 (1995)
44. S.W. Hell, J. Wichmann, Breaking the diffraction resolution limit by stimulated emission: stimulated-emission-depletion fluorescence microscopy. Opt. Lett. **19**, 780–782 (1994)
45. D. Hu, M. Micic, N. Klymyshyn, Y.D. Suh, H.P. Lu, Correlated topographic and spectroscopic imaging beyond diffraction limit by atomic force microscopy metallic tip-enhanced near-field fluorescence lifetime microscopy. Rev. Sci. Instrum. **74**, 3347–3355 (2003)
46. G.T. Kennedy, H.B. Manning, D.S. Elson, M.A.A. Neil, G.W. Stamp, B. Viellerobe, F. Lacombe, C. Dunsby, P.M.W. French, A fluorescence lifetime imaging scanning confocal endomicroscope. J. Biophoton. **3**, 103–107 (2010)
47. D. Kepshire, N. Mincu, M. Hutchins, J. Gruber, H. Dehghani, J. Hypnarowski, F. Leblond, M. Khayat, B.W. Pogue, A microcomputed tomography guided fluorescence tomography system for small animal molecular imaging. Rev. Sci. Instrum. **80**, 043701-1 to -10 (2009)
48. T.A. Klar, S.W. Hell, Subdiffraction resolution in far-field fluorescence microscopy. Opt. Lett. **24**, 954–956 (1999)
49. M. Köllner, J. Wolfrum, How many photons are necessary for fluorescence-lifetime measurements? Phys. Chem. Lett. **200**, 199–204 (1992)
50. K. König, P.T.C. So, W.W. Mantulin, B.J. Tromberg, E. Gratton, Two-Photon excited lifetime imaging of autofluorescence in cells during UVA and NIR photostress. J. Microsc. **183**, 197–204 (1996)

51. K. König, Multiphoton microscopy in life sciences. J. Microsc. **200**, 83–104 (2000)
52. K. König, in *Cellular Response to Laser Radiation in Fluorescence Microscopes*, ed. by A. Periasamy. Methods in Cellular Imaging (Oxford University Press, New York, 2001), pp. 236–254
53. K. König, in *Multiphoton-induced Cell Damage*, ed. by B.R. Masters, P.T.C. So. Handbook of Biomedical Nonlinear Optical Microscopy (Oxford University Press, New York, 2008)
54. K. Koenig, Clinical multiphoton tomography. J. Biophoton. **1**, 13–23 (2008)
55. K. Koenig, A. Uchugonova, in *Multiphoton Fluorescence Lifetime Imaging at the Dawn of Clinical Application*, ed. by A. Periasamy, R.M. Clegg. FLIM Microscopy in Biology and Medicine (CRC Press, Boca Raton, 2009)
56. E.S. Kwak, T.J. Kang, A.A. Vanden Bout, Fluorescence lifetime imaging with near-field scanning optical microscopy. Anal. Chem. **73**, 3257–3262 (2001)
57. E. Lapointe, J. Pichette, Y. Berube-Lauziere, A multi-view time-domain non-contact diffuse optical tomography scanner with dual wavelength detection for intrinsic and fluorescence small animal imaging. Rev. Sci. Instrum. **83**, 063703-1 to -14 (2012)
58. M.D. Lesoine, S. Bose, J.W. Petrich, E.A. Smith, Supercontinuum stimulated emission depletion fluorescence lifetime imaging. J. Phys. Chem. B **116**, 7821–7826 (2012)
59. A. Liebert, H. Wabnitz, H. Obrig, R. Erdmann, M. Möller, R. Macdonald, H. Rinneberg, A. Villringer, J. Steinbrink, Non-invasive detection of fluorescence from exogenous chromophores in the adult human brain. NeuroImage **31**, 600–608 (2006)
60. J.J. Mancuso, A.M. Larson, T.G. Wensel, P. Saggau, Multiphoton adaptation of a commercial low-cost confocal microscope for live tissue imaging. J. Biomed. Opt. **14**(3), 034048 (2009)
61. G. McConnel, Confocal laser scanning fluorescence microscopy with a visible continuum source. Op. Expr. **13**, 2844–2850 (2004)
62. G. McConnel, J.M. Girkin, S.M. Ameer-Beg, P.R. Barber, B. Vojnovic, T. Ng, A. Banerjee, T. F. Watson, R.J. Cook, Time-correlated single-photon counting fluorescence lifetime confocal imaging of decayed and sound dental structures with a white-light supercontinuum source. J. Microsc. **225**, 126–136 (2007)
63. D.X. Medina, A. Caccamo, S. Oddo, Methylene blue reduces Aβ levels and rescues early cognitive deficit by increasing proteasome activity. Brain Pathol. **21**, 140–149 (2011)
64. M. Micic, D. Hu, Y.D. Suh, G. Newton, M. Romine, H.P. Lu, Correlated atomic force microscopy and fluorescence lifetime imaging of live bacterial cells. Colloids Surf., B **34**, 205–212 (2004)
65. M. Minsky, US Patent 3013467, 1957
66. M. Minsky, Memoir on inventing the confocal microscope. Scanning **10**, 128–138 (1988)
67. I.H. Munro, G.R. Jones, M. Tobin, D.A. Shaw, Y. Levine, H. Gerritsen, K. van der Oord, F. Rommerts, Confocal imaging using synchrotron radiation. J. Electron Spectrosc. Relat. Phenom. **80**, 343–347 (1996)
68. M. Oheim, E. Beaurepaire, E. Chaigneau, J. Mertz, S. Charpak, Two-photon microscopy in brain tissue: parameters influencing the imaging depth. J. Neurosci. Methods **111**, 29–37 (2001)
69. J. Pawley (ed.), *Handbook of Biological Confocal Microscopy*, 3rd edn (Springer Science and Business Media LLC, New York, 2006)
70. S. Pelet, M.J.R. Previte, P.T.C. So, Comparing the quantification of Förster resonance energy transfer measurement accuracies based on intensity, spectral, and lifetime imaging. J. Biomed. Opt. **11**, 034017-1 to -11 (2006)
71. L. Pires, M.S. Nogueira, S. Pratavieira, L.T. Moriyama, C. Kurachi, Time-resolved fluorescence lifetime for cutaneous melanoma detection. Biomed. Opt. Expr. **5**, 3080–3089 (2014)
72. M. Salah, N. Samy, M. Fadel, Methylene blue mediated photodynamic therapy for resistant plaque psoriasis. J. Drags Dermatol. **8**, 42–49 (2009)
73. R.H. Schirmer, B. Coulibaly, A. Stich, M. Scheiwein, H. Merkle, J. Eubel, K. Becker, H. Becher, O. Müller, T. Zich, W. Schiek, B. Kouyate, Methylene blue as an antimalarial agent. Redox Rep. **8**, 272–275 (2003)

74. P.T.C. So, K.H. Kim, L. Hsu, P. Kaplan, T. Hacewicz, C.Y. Dong, U. Greuter, N. Schlumpf, C. Buehler, Two-photon microscopy of tissues, in *Handbook of Biomedical Fluorescence*, ed. by M.-A. Mycek, B.W. Pogue (Marcel Dekker, Basel, 2003), pp. 181–208

75. H. Studier, W. Becker, Megapixel FLIM. Proc. SPIE **8948**, 89481K (2014)

76. K. Suhling, P.M.W. French, D. Phillips, Time-resolved fluorescence microscopy. Photochem. Photobiol. Sci. **4**, 13–22 (2005)

77. P. Theer, M.T. Hasan, W. Denk, Multi-photon imaging using a Ti:sapphire regenerative amplifier. Proc. SPIE **5139**, 1–6 (2003)

78. C.J.R. Van Der Oord, H.C. Gerritsen, F.F.G. Rommerts, D.A. Shaw, I.H. Munro, Y.K. Levine, Micro-volume time-resolved fluorescence spectroscopy using a confocal synchrotron radiation microscope. Appl. Spectrosc. **49**, 1469–1473 (1995)

79. C.J.R. Van der Oord, J.R. Jones, D.A. Shaw, I.H. Munro, Y.K. Levine, H.C. Gerritsen, High-resolution confocal microscopy using synchrotron radiation. J. Microsc. **182**, 217–224 (1996)

80. G.T. Wondrak, NQQO1-activated phenothiazinium redox cyclers for the targeted bioreductive induction of cancer cell apaptosis. Free Radic. Biol. Med. **15**, 178–190 (2007)

81. J.G. White, W.B. Amos, M. Fordham, An evaluation of confocal versus conventional imaging of biological structures by fluorescence light microscopy. J. Cell Biol. **105**, 41–48 (1987)

82. T. Wilson, C. Sheppard, *Theory and Practice of Scanning Optical Microscopy* (Academic Press, London, 1984)

83. C. Xu, W.W. Webb, Measurement of two-photon excitation cross sections of molecular fluorophores with data from 690 to 1050 nm. J. Opt. Soc. Am. B **13**, 481–491 (1996)

84. S. Yazdanfar, C. Joo, C. Zhan, M.Y. Berezin, W.J. Akers, S. Achilefu, Multiphoton microscopy with near infrared contrast agents. J. Biomed. Opt. **15**(3), 030505-1 to -3 (2010)

Chapter 3
Fluorescence Lifetime Imaging (FLIM): Basic Concepts and Recent Applications

Klaus Suhling, Liisa M. Hirvonen, James A. Levitt, Pei-Hua Chung, Carolyn Tregido, Alix le Marois, Dmitri A. Rusakov, Kaiyu Zheng, Simon Ameer-Beg, Simon Poland, Simon Coelho and Richard Dimble

Abstract Fluorescence lifetime imaging (FLIM) is a key fluorescence microscopy technique to map the environment and interaction of fluorescent probes. It can report on photo physical events that are difficult or impossible to observe by fluorescence intensity imaging, because FLIM is independent of the local fluoro-

An erratum of this chapter can be found under DOI 10.1007/978-3-319-14929-5_21

K. Suhling (✉)
Department of Physics, King's College London, London WC2R 2LS, UK
e-mail: klaus.suhling@kcl.ac.uk

L.M. Hirvonen · J.A. Levitt · P.-H. Chung · C. Tregido · A. le Marois
Department of Physics, King's College London, London WC2R 2LS, UK
e-mail: liisa.2.hirvonen@kcl.ac.uk

J.A. Levitt
e-mail: james.levitt@kcl.ac.uk

P.-H. Chung
e-mail: m9314001@gmail.com

C. Tregido
e-mail: carolyn@tregidgo.com

A. le Marois
e-mail: alix.le_marois@kcl.ac.uk

D.A. Rusakov · K. Zheng
Institute of Neurology, University College London, London WC1N 3BG, UK
e-mail: d.rusakov@ucl.ac.uk

K. Zheng
e-mail: k.zheng@ucl.ac.uk

S. Ameer-Beg · S. Poland · S. Coelho
Randall Division of Cell and Molecular, Biophysics, King's College London,
London SE1 1UL, UK

R. Dimble
Department of Cancer, Research, Division of Cancer Studies, New Hunt's, House,
Guy's Campus, King's College London, London SE1 1UL, UK

© Springer International Publishing Switzerland 2015
W. Becker (ed.), *Advanced Time-Correlated Single Photon Counting Applications*,
Springer Series in Chemical Physics 111, DOI 10.1007/978-3-319-14929-5_3

119

phore concentration and excitation intensity. A FLIM application relevant for biology concerns the identification of FRET to study protein interactions and conformational changes, and FLIM can also be used to image viscosity, temperature, pH, refractive index and ion and oxygen concentrations, all at the cellular or sub-cellular level, as well as autofluorescence. The basic principles and some recent advances in the application of FLIM, FLIM instrumentation and molecular probe development will be discussed.

3.1 Introduction

Much of our knowledge of biological processes at the cellular and sub-cellular level comes from the microscope's ability to directly visualize them: optical imaging is compatible with living specimens, as light is non-ionizing, non-destructive and minimally invasive. Fluorescence microscopy in particular combines advantages of single-molecule sensitivity, molecular specificity, sub-micron resolution and real-time data collection from live cells with negligible cytotoxicity. This allows not only the study of the structure of the sample, but also the observation of dynamics and function [2, 452].

Among the various fluorescence microscopy methods, Fluorescence Lifetime Imaging (FLIM) has emerged as a key technique to image the environment and interaction of specific probes in living cells [17, 33, 52, 452]. There are several technological implementations of FLIM, but they all can report on photo physical events that are difficult or impossible to observe by fluorescence intensity imaging: The fluorescence lifetime provides an absolute measurement which is independent of the fluorophore concentration and, compared to fluorescence intensity, less susceptible to artefacts arising from scattered light, photobleaching, non-uniform illumination of the sample, light path length, or excitation intensity variations.

FLIM has long been used to detect Förster resonance energy transfer (FRET) for the identification protein interactions or conformational changes of proteins in the life and biomedical sciences [69, 100, 120, 200, 319, 324, 440]. However, applications in diverse areas such as forensic science [42], combustion research [103, 298], luminescence lifetime mapping in diamond [245, 246], microfluidic systems [29, 31, 109, 153, 259, 331, 338], art conservation [81, 82], remote sensing [114, 250, 251], lipid order problems in physical chemistry [410], the spatial distribution of dopants in perovskite oxides [339] and temperature sensing [28, 31, 153, 272] have also been reported. FLIM has been carried out from the UV [243] to the near infrared [23]. It is not surprising that fluorescence lifetime-based imaging is widely used in the biomedical sciences, and that this trend shows no signs of abating.

The observation of fluorescence and the use of microscopy stretches back many hundreds of years [174, 175], as illustrated in Fig. 3.1. However, the understanding of fluorescence-related phenomena and the creation of an appropriate theoretical framework to quantitatively interpret and predict fluorescence and to design a

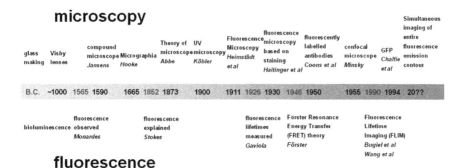

Fig. 3.1 A brief history of microscopy and fluorescence. The development of glass making was essential for making lenses, which in turn was essential for the development of the microscope. It allowed the discovery and detailed study of the cell, the basic and universal building block of all living organisms. The combination of microscopes with lasers, powerful computers, sensitive detectors and cameras, and the genetically encoded green fluorescent protein (GFP) and its variants have led to revolutionary advances in the life sciences over the last two decades

fluorescence microscope only occurred 100–150 years ago.[1] Over the last 10 or 20 years the field has advanced rapidly and enormously [452], mainly due to the combination of lasers and beam scanning [2], powerful computers and data analysis software, sensitive detectors and cameras [59, 106, 163, 194, 277], and genetic engineering [75, 361, 362]—the latter effort being recognized by the award of the Nobel Prize in Chemistry in 2008 to Osamu Shimomura, Martin Chalfie and Roger Tsien "for the discovery and development of the green fluorescent protein, GFP". A year later, in 2009, half of the Nobel Prize in Physics was given to Willard Boyle and George Smith "for the invention of an imaging semiconductor circuit—the CCD sensor" in 1969 [370]—a device which has also played a significant role in advancing fluorescence microscopy. The sensitivity of fluorescence detection allows single molecules to be detected, and fluorescence imaging and localization well below the spatial resolution limit given by classical optical diffraction can be achieved with super-resolution techniques—a currently fast moving and rapidly expanding field [85, 137, 178, 184], culminating in the award of the 2014 Nobel Prize in Chemistry to Eric Betzig, W.E. Moerner and Stefan Hell for "for the development of super-resolved fluorescence microscopy".

While the idea of nanosecond time-resolved fluorescence measurements of samples under a microscope dates back to the 1950s [433], the emergence of FLIM as a technique for mapping fluorescence lifetimes only began in 1989. In this year, the first reports were published describing a fluorescence imaging technique where the contrast in the image is provided by the fluorescence lifetime [58, 447].

Since then, the power of FLIM has increased dramatically with the extension to spectrally resolved FLIM, polarization-resolved FLIM and rapid acquisition with

---

[1] The Nature milestone website contains a wealth of information on the history and impact of optical microscopy: http://www.nature.com/milestones/milelight/index.html.

single photon sensitivity. FLIM has also been combined with other techniques, such as fluorescence correlation spectroscopy (FCS) [21, 56, 297], scanning near-field optical microscopy (SNOM) [228], atomic force microscopy (AFM) [279], selective plane illumination microscopy, [159] fluorescence recovery after photobleaching (FRAP) [239, 337, 436], total internal reflection (TIRF) microscopy [57, 92], Stimulated Emission Depletion (STED) microscopy [5, 237, 251], coherent anti-Stokes Raman scattering (CARS) [369] and tomography [271].

The increasing popularity is facilitated by commercial availability of key enabling technology: FLIM add-on units for conventional microscopes, for wide-field, confocal and multiphoton excitation microscopy, including data analysis software, are available from a number of specialist companies.

3.2 Fluorescence

Fluorescence as a phenomenon has been observed for hundreds or even thousands of years, but the understanding and explanation of it took a long time, especially its distinction from incandescence, iridescence and scattered light [423] see Fig. 3.1. In 1852, George Stokes, building on previous work by Robert Boyle, Isaac Newton, David Brewster, John Herschel and others, explained in a 100 page long publication that the emitted light was of a longer wavelength than the absorbed light [378]—an effect now known as the Stokes' shift. Above all, Stokes coined the term fluorescence [378, 379]. Despite this breakthrough, some confusion remained, but it eventually faded away like fluorescence itself [261]. After some theoretical considerations regarding fluorescence lifetimes [377], the first reports on measuring nanosecond fluorescence lifetimes experimentally appeared in the mid-1920s [141].

Upon absorption of a photon, one of the weakly bound electrons of the fluorescent molecule—a fluorophore—is promoted to a higher energy level. The fluorophore is then said to be in an excited state, A^*. This state is metastable, and therefore the fluorophore will return to its stable ground state, A. It can do so either radiatively by emitting a fluorescence photon $h\nu$,

$$A^* \rightarrow A + h\nu \tag{3.1}$$

or non-radiatively, for example, by dissipating the excited state energy as heat [149, 230, 350, 424].

$$A^* \rightarrow A + heat \tag{3.2}$$

The depopulation of the excited state depends on the de-excitation pathways available. Fluorescence is the radiative deactivation of the lowest vibrational energy level of the first electronically excited singlet state, S_1, back to the electronic ground state, S_0. The singlet states are the energy levels that can be populated by the weakly bound electron without a spin flip. The absorption and emission processes

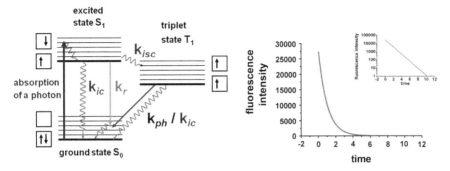

Fig. 3.2 *Left* A schematic energy level diagram, named after Aleksander Jablonski, of a fluorescent molecule, depicting the molecular singlet and triplet electronic energy levels, each with vibrational energy levels as well as excitation and de-excitation pathways. Excitation of the fluorophore into an excited state by the absorption of a photon promotes a weakly bound electron to a higher energy level. If the spin of the electron (indicated by the *arrow*) remains the same, the fluorophore is in a singlet energy state (S_1), if the spin flips, it is in a triplet state (T_1). k_{ic} is the rate constant for internal conversion, k_{isc} the rate constant for intersystem crossing to the triplet state, k_r is the fluorescence rate constant and k_{ph} is the phosphorescence rate constant. *Right* A schematic fluorescence decay, where the fluorescence intensity decays over time according to an exponential decay law. The *inset* is a semi-logarithmic plot of the same fluorescence decay, which, convenient for easy visual inspection, appears as a *straight line*

are illustrated by an energy level diagram named after Aleksander Jablonski [197], as shown in Fig. 3.2, left.

The fluorescence lifetime, τ, is the average time a fluorophore remains in the electronically excited state S_1 after excitation. τ is defined as the inverse of the sum of the rate parameters for all excited state depopulation processes:

$$\tau = \frac{1}{k_r + k_{nr}} \tag{3.3}$$

where k_r is the radiative rate constant. The non-radiative rate constant k_{nr} is the sum of the rate constant for internal conversion, k_{ic}, and the rate constant for intersystem crossing to the triplet state, k_{isc}, so that $k_{nr} = k_{ic} + k_{isc}$. The fluorescence emission always occurs from the lowest vibrational level of S_1, a rule known as Kasha's rule, [207] indicating that the fluorophore has no memory of its excitation pathway, e.g. one and two-photon excitation yields the same fluorescence spectrum, quantum yield and lifetime.

$\tau_0 = k_r^{-1}$ is the natural or radiative lifetime which is related to the fluorescence lifetime τ via the fluorescence quantum yield ϕ:

$$\phi = \frac{\tau}{\tau_0} = \frac{k_r}{k_r + k_{nr}} = \frac{1}{1 + \frac{k_{nr}}{k_r}} \tag{3.4}$$

The fluorescence quantum yield can be thought of as the ratio of the number of fluorescence photons emitted to the number of photons absorbed (regardless of their energy) and is always less than or equal to one, $0 \leq \phi \leq 1$. And since $\phi \tau_0 = \tau$, τ_0 can be thought of as the longest possible lifetime of the fluorophore, i.e. when $k_{nr} = 0$.

Both the fluorescence lifetime and the fluorescence quantum yield are key spectroscopic parameters, the measurement of which allows the explicit calculation of the radiative rate constant k_r and the non-radiative rate constant k_{nr}.

The radiative rate constant k_r is related to the absorption and fluorescence spectra and is a function of the refractive index of the medium surrounding the fluorophore:

$$k_r = \frac{1}{\tau_0} = \frac{8000\pi c_0}{N_A} \frac{n_{fl}^3}{n_{abs}} \frac{g_{ex}}{g_{er}} \langle \tilde{v}^{-3} \rangle^{-1} \int \frac{\varepsilon(\tilde{v})}{\tilde{v}} d\tilde{v} \tag{3.5}$$

where c_0 is the speed of light in vacuum, N_A Avogadro's number, n_{fl} the mean refractive index over the emission spectrum, n_{abs} the mean refractive index over the absorption spectrum, g_{ex} and g_{gr} are the so-called multiplicities (related to the spin of the electron of the excited state, g = 1 for a singlet, g = 3 for a triplet state) for the excited and the ground state, respectively, ε is the extinction coefficient and $\tilde{v}$ the wavenumber. $\langle \tilde{v}^{-3} \rangle^{-1}$ is the reciprocal of the mean value of $\tilde{v}^{-3}$ in the fluorescence emission spectrum, independent of the fluorescence intensity or quantum yield and given by

$$\langle \tilde{v}^{-3} \rangle^{-1} = \frac{\int F(\tilde{v}) d\tilde{v}}{\int F(\tilde{v}) \tilde{v}^{-3} d\tilde{v}} \tag{3.6}$$

where F is the fluorescence emission spectrum. This equation is known as the Strickler-Berg equation [381], and for $n_{fl} = n_{abs}$ and $g_{ex} = g_{gr} = 1$ often quoted thus:

$$k_r = 2.88 \times 10^{-9} n^2 \frac{\int I(\tilde{v}) d\tilde{v}}{\int I(\tilde{v}) \tilde{v}^{-3} d\tilde{v}} \int \frac{\varepsilon(\tilde{v})}{\tilde{v}} d\tilde{v} \tag{3.7}$$

Essentially, the Strickler-Berg equation is a version of the Einstein coefficients for absorption and spontaneous and stimulated emission [104, 105] but adapted for molecules with broad absorption and emission spectra, rather than atomic line spectra. Different models exist for the refractive index dependence of k_r [182, 364], and a more detailed treatment taking into account the transition dipole moment, an intrinsic property of the molecule, has been devised by Toptygin et al. [413]. Toptygin has also written a detailed review of the subject [412].

The time-dependence of the depopulation of the excited state—the decay of the excited state—can be explained as follows. After excitation, N fluorophores will populate the excited state S_1 (see Fig. 3.2left). In a time interval dt, the number of excited fluorophores dN returning to the ground state S_0 is given by the following rate equation:

$$dN = (k_r + k_{nr})N(t)dt \tag{3.8}$$

where t is the time.

Integration of (3.8), and taking into account that the fluorescence intensity $F(t)$ is proportional to the number of excited fluorophores $N(t)$ yields

$$F(t) = F_0 e^{-t/\tau} \tag{3.9}$$

where F_0 represents the fluorescence intensity at $t = 0$, and τ is the fluorescence lifetime as defined in (3.3). The decay of the fluorescence intensity thus follows an exponential decay law [196], schematically shown in Fig. 3.2, right. τ is the time it takes for the fluorescence intensity to decay from its peak value to $e^{-1} \approx 37\%$ of its peak value. This applies both to repeatedly excited single molecules—where the fluorescence lifetime represents a measure of the emission probability after a certain time—and to the fluorescence decay of an ensemble of fluorophores after a single excitation. Note that on a logarithmic fluorescence intensity scale (y-axis), a mono-exponential decay conveniently appears as a straight line, as shown in the inset of Fig. 3.2, right. Plotting decay data this way thus aids simple and rapid visual inspection of the fluorescence decay behaviour.

3.3 Fluorescence Probes

Some minerals fluoresce, and naturally occurring fluorescent dyes have been known for a long time [230, 424]. The first synthetic dye was mauve, synthesized by William Perkin in Manchester in 1856 [469]. It had a low quantum yield, but shortly afterwards, in 1871, the much brighter dye fluorescein was synthesized by Adolf von Baeyer. He was awarded the Nobel Prize in Chemistry in 1905, "in recognition of his services in the advancement of organic chemistry and the chemical industry, through his work on organic dyes and hydroaromatic compounds". This work was closely linked to colour chemistry, i.e. the research into dyes for staining fabrics and other materials [469]. Often, these dyes were not fluorescent, but they did absorb light, and were of major interest for the textile industry—not only in the West, but also in China for staining silk, for example.

Today, fluorescence sensing and microscopy can be performed by labelling a sample with fluorescent dyes, fluorescent proteins, quantum dots [335] or other nanoparticles [158, 190, 191], including nano diamonds [118, 227, 285, 295] and nano-ruby [102], as reviewed recently [375], as well as imaging intrinsically fluorescent molecules naturally occurring within the sample—autofluorescence [110, 419]. This is endogenous fluorescence from, for example, tryptophan, melanin, keratin, elastin, lipofuscin, nicotinamide adenine dinucleotide (NADH) or flavin adenine dinucleotide (FAD) or, in the case of plants, chlorophyll. In addition to fluorescence dyes, quantum dots and other nanoparticles have also recently found favour in cell imaging applications due to their high fluorescence quantum yield, low photobleaching susceptibility and narrow, size-dependent emission spectra which can be excited with a single wavelength [156, 157, 276, 335]. Widely-used probes in biology are organic fluorophores and genetically encoded fluorescent proteins [75, 361, 362]. Imaging autofluorescence is also increasingly employed as discussed in detail in Chaps. 13–16 of this book.

The ratio of radiative and non-radiative decay rates of a fluorophore (Fig. 3.2)—and thus its fluorescence quantum yield and its fluorescence lifetime—change with its local molecular environment. There are several mechanisms of lifetime changes: In the simplest case the lifetime is influenced by the refractive index of the surrounding medium according to (3.5). The fluorophores may also undergo conformational changes or changes in their electronic structure. These changes may depend on the polarity of the environment, or on the binding to proteins and lipids. Changes in the electronic structure may also be induced when the fluorophore gets protonated or de-protonated, or when metal ions bind to it. Substantial lifetime changes can also be induced by collisional quenching and by electron transfer. A mechanism especially important to biological research is Förster resonance energy transfer (FRET), see Chaps. 7 and 8 of this book, and Sect. 3.5.1 of this chapter.

The dependence of the fluorescence lifetime on the local molecular environment of a fluorophore is the basis for a wide range of molecular sensing applications. The most common ones are protein interaction measurements by FRET, measurements of the concentration of biologically relevant ions, and measurements of pH, local viscosity, polarity, and temperature. Fluorescence lifetime techniques are also used for determination of the concentration of biologically relevant compounds, like glucose. Especially promising is the measurement of metabolic activity via the fluorescence decay parameters of NADH and FAD. The most common lifetime applications and the fluorescence sensors used are summarised in Table 3.1.

3.4 FLIM Implementations

Ideally, a FLIM technique should record data at high spatial resolution and high time resolution. It should also achieve the best possible accuracy of the detected decay data for a given number of photons emitted by the sample, it should resolve multi-exponential decay profiles, and it should be able to record several wavelength channels simultaneously. It should do all this in the minimum possible acquisition time. It is also important that the decay data are recorded from a defined focal plane to avoid decay data with out-of-focus components.

FLIM can be based on various optical and electronic principles, see Fig. 3.3. Optical image acquisition can be based on wide-field (camera) techniques or on point-scanning, Fig. 3.3, top. Point scanning techniques can be further divided into beam scanning and sample scanning techniques [2]. Electronic acquisition of the fluorescence decay data can be performed in the frequency domain, or in the time domain [422], Fig. 3.3, middle: Frequency-domain techniques measure the phase and the degree of modulation of the fluorescence signal in comparison to the excitation. Time-domain techniques directly measure the waveform of the fluorescence signal. Time-domain techniques can be further separated into techniques which scan a time gate over the decay functions, simultaneously recording intensity

Table 3.1 Some features and parameters that can be sensed with fluorescence lifetime techniques, some dyes and references (not exhaustive)

Feature to be sensed	Fluorescence parameter yielding information	Fluorophore	References	Remarks
Inter- or intramolecular interactions				
Hetero—FRET	Lifetime, spectrum, polarization	Many, provided there is spectral overlap, fluorescent proteins for polarization	[1, 100 , 120, 200, 265, 266, 319, 324, 389]	Conformational changes and interaction with other molecules
Homo—FRET	Polarization	Many, provided they have a small Stokes shift necessary for spectral overlap	[10, 11, 337, 406, 432, 434, 458]	The only[a] way to detect homo-FRET is by polarization (when the fluorescent lifetimes of donor and acceptor are the same, see discussion in [391]). For this approach fluorescent proteins are best, due to their large rotational correlation time
Physical parameters				
Viscosity	Polarization	Many, provided the rotational correlation time is no longer than 10 times the fluorescence lifetime	[53, 78, 393]	
Viscosity	Lifetime of fluorescent molecular rotors, intensity (ratiometric, using a viscosity-independent reference dye)	E.g. BODIPY-C_{12}, DCVJ, CCVJ, Thioflavin T, DASMPI, Cy dyes	[14, 145, 164, 168, 186, 193, 222, 223, 241, 253, 302, 303, 316, 341, 441, 442, 467]	
Polarity	Lifetime, spectrum	Nile red, laurdan, prodan, di-4-ANEPPDHQ	[150, 310, 311]	
Temperature	Lifetime	Kiton red, rhodamine B, temperature-sensitive polymers	[28, 31, 153, 272, 308]	The polymers are quenched by water at low temperatures, but at higher temperatures, they shrink, releasing water molecules, increasing their fluorescence quantum yield and lifetime [308]

(continued)

Table 3.1 (continued)

Feature to be sensed	Fluorescence parameter yielding information	Fluorophore	References	Remarks
Refractive index	Lifetime	GFP, quantum dots, nanodiamonds, fluorophores which are not sensitive to anything else	[51, 258, 326, 394, 409, 418, 427, 453]	Default effect due to Strickler-Berg equation
Solute/fluorophore concentration				
pH	Lifetime	BCECF	[26, 65, 172, 181, 248, 264, 306, 347]	
O_2	Lifetime	Complexes of ruthenium, iridium, osmium, rhenium	[144, 185, 189, 444]	Hundreds of nanoseconds, microseconds lifetime
Ca^{2+}	Intensity, lifetime	Calmodulin sensing GFP, Quin-2, Calcium Green and derivatives	[179, 219, 235, 284, 348, 449, 450]	
Cl^-	Lifetime	6-Methoxy-quinolyl acetoethyl ester (MQAE)	[133, 147, 206, 229]	
Cu^{2+}	Lifetime	Oregon Green or GFP-FRET sensor	[187, 188, 270]	
Na^+	Lifetime	CFP-YFP FRET, sodium green	[49, 229, 401]	
K^+	Lifetime	CFP-YFP FRET	[48]	
Mg^{2+}	Lifetime	Mag-quin-2, magnesium green	[399]	
PO_4^{3-} (phosphate)	Lifetime		[312]	
NO (nitric oxide)	Lifetime		[464]	Not yet employed in imaging

(continued)

Table 3.1 (continued)

Feature to be sensed	Fluorescence parameter yielding information	Fluorophore	References	Remarks
Extrinsic fluorophore concentration	Lifetime		[45, 46, 134, 135]	Evaluation of fluorescence intensity (proportional to quantum yield and probe concentration) and lifetime (proportional to quantum yield only, (3.3) and (3.4)) allows calculation of relative probe concentration
Intrinsic fluorophore concentration	Lifetime	NADH, FAD	[86, 146, 212, 234, 356, 357, 368, 459, 462]	Bound/unbound ratio
Glucose	Lifetime	Ruthenium-malachite green FRET, glucose/ galactose binding protein (GBP) labelled Badan	[353, 354, 411]	

[a] In the special case of the Cerulean Fluorescent Protein, the fluorescence lifetime has been reported to change due to homo-FRET, but the reason remains unclear [214]. The yellow fluorescent protein Venus did not show this effect

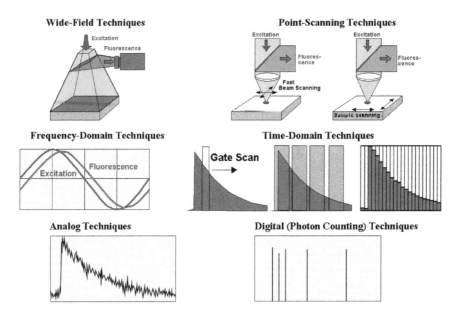

Fig. 3.3 Different optical and electronic principles used for recording FLIM data. *Top* Optical image acquisition. *Middle* Electronic principles of detecting decay functions or decay times. *Bottom* The detected signals may be considered continuous waveforms or single-photon detection events

data into a small number of time windows, or simultaneously recording the data into a large number of time channels. Finally, the detected signals can be considered to be analogue (continuous) waveforms or detection events of single photons of the fluorescence light, Fig. 3.3, bottom.

Almost all combinations are in use, leading to a confusing variety of different FLIM techniques. Which FLIM technique is best depends partly on the application, but has to take into account the specific conditions of biological microscopy like sample-limited emission rates, the need (or not) for lifetime accuracy, imaging conditions such as pixel numbers, sample size, excitation power, laser repetition rate, fluorophore concentration, thickness of the layer where the signal originates, pinhole size, etc.

In addition, it is also very important to note that the performance of an instrument not only depends on its general operating principle, but also on the technical details of its particular implementation. One example is the idea that frequency-domain and time-domain techniques are equivalent because they are connected via the Fourier transform [176, 422]. This may be mathematically correct [422] but not technically: There are time-domain techniques that record a waveform simultaneously into a large number of time channels, but there is no frequency-domain technique that simultaneously records into a large number of frequency channels. We will therefore not discuss the technical parameters of the different principles here but only give a non-exhaustive short introduction into the commonly used technical combinations. For a more detailed discussion please see [18, 116, 143, 155, 321, 387].

3.4.1 Wide-Field Imaging with a Gated Camera

The entire image area is illuminated by a pulsed laser, and the fluorescence light is detected by a gated image intensifier [98, 358, 445, 446] (Fig. 3.3 upper left and middle right). An image intensifier is a detector that consists of a photocathode, a stack of microchannel plates (MCPs) and a phosphor screen for the read-out of the signal. The phosphor screen is imaged by a camera, and the image intensifier acts as a shutter—it can be turned on and off within picoseconds, thus providing a gate for the light to pass through. The gate is shifted sequentially over the decay, and a series of images for different times after the excitation recorded. Directly gated CCD cameras have also been developed [282, 283], but their time resolution is much lower than of gated image intensified cameras and they are more suited to imaging long lifetime probes.

Gated camera techniques can achieve very short acquisition times: All pixels of the image are recorded simultaneously. Because there are usually more pixels than temporal steps of the gate scan the technique is faster than a technique based on spatial scanning and simultaneous recording in time. The advantage can, however, only be exploited if the sample is able to emit high fluorescence intensities without photobleaching or photodamage, which may not necessarily be the case in many FLIM applications. The photon efficiency of the technique, i.e. the number of photons to be emitted by the sample for a given accuracy [143, 211, 321] is low because most of the photons are rejected by the gating. This problem has been mitigated by recording in only a few time gates simultaneously [111]. This increases the efficiency but significantly impairs the ability to record the parameters of multi-exponential decay profiles. However, in biomedical applications the information is often contained just in these parameters, see Chaps. 8, 13 and 14. Moreover, the camera technique suffers from the general shortcomings of wide-field imaging: scattered light is detected from all pixels, not only from a single one, and there is no inherent depth resolution, see Chap. 2, Figs. 2.1 and 2.3. Depth resolution can be obtained by structured illumination techniques [80] but this further reduces the efficiency. Other solutions to the depth-resolution problem are multi-beam multi-photon excitation scanning approaches [30, 380], or light-sheet techniques [159].

3.4.2 Streak Cameras

Streak cameras can be used with point scanning to record images at a large number of wavelengths simultaneously [49]. In that case, a spectrum of the fluorescence light is projected on the photocathode in perpendicular to the streak direction. A streak camera can also been used to record a whole line of the scan simultaneously [49, 216–218, 252, 281, 329]. The line can either be scanned by a single laser spot, or the whole line can be illuminated simultaneously. The streak camera delivers a streak of the signal waveform along the fluorescence decay time axis. The signal is recorded in a large number of time channels with a resolution up to a few ps per time channel. If the streak camera can work at the full repetition rate of the laser the photon efficiency is high. A problem can be the nonlinearity of the time axis and the absolute stability of the time scale. Streak cameras are definitely a high end solution for FLIM. They are, however, not often used, probably because of the high price.

3.4.3 Wide-Field Imaging with Modulated Camera

The principle is similar to the gated camera. However, the image intensifier is not gated but modulated with a frequency similar to the modulation of the excitation light [136, 232, 287, 296, 317, 447]. By changing the phase of the intensifier modulation, images for different phases are recorded, from which the fluorescence lifetime is derived. The modulation of both the light source and the detector need not necessarily be sinusoidal. Also pulsed excitation or rectangular modulation of the intensifier are possible, for example to avoid aliasing [108, 428] or other artefacts [171, 429]. The photon efficiency critically depends on these parameters, and on the modulation depth in the intensifier [321]. The technique has a limited dynamic range [355] and the signal intensity has to be high enough so its modulation is practical. The scattered light and depth resolution caveats apply, as stated above in the context of time-gated wide-field FLIM, as do the approaches to tackle them.

3.4.4 Wide-Field TCSPC

Wide-field Time-Correlated Single Photon Counting (TCSPC) FLIM with picosecond time resolution can be performed with MCP-based photon counting image intensifiers [275, 277]. In this case, the image intensifier is operated with a saturated gain producing a distinct pulse height distribution for individual photon events, capable of detecting single photons across the field of view [388, 390]. Instead of using a phosphor as in gated or modulated image intensifiers, an electronic read-out is employed. Various architectures exist, such as crossed delay line anodes, wedge and strip anodes or quadrant anodes. Quadrant anode [113, 320, 373, 435, 436] and delay line anode [274, 278] detectors have successfully been combined with picosecond timing and fluorescence microscopy. They typically detect one photon per pulse for the whole field of view, but advanced read-out architectures allowing multi-photon hits to be detected have been designed [198]. In addition, microsecond time-resolution wide-field TCSPC with a Complementary Metal-oxide Semiconductor (CMOS) camera and a photon counting image intensifier with a phosphor screen readout has also been demonstrated, and applied to imaging europium [360] and ruthenium [183] containing, long lifetime compounds in living cells. Microwatt excitation powers are sufficient for this approach, and due to the parallel readout, acquisition times are in the order of seconds [183]. The advantage of wide-field TCSPC is that position-sensitive TCSPC can be performed, no scanning is needed, and the advantages of TCSPC such as single photon sensitivity—important in view of the limited photon budget available from fluorophores [463]—clearly defined Poissonian statistics, excellent signal-to-noise ratio, a wide dynamic range and easy visualization of fluorescence decays are retained. This is important for microscopy techniques that are difficult or impossible to implement with scanning, e.g. TIRF and lightsheet microscopy.

3.4.5 Point Scanning with Detection in the Frequency Domain

The sample is scanned with a focused beam of light, the excitation is either pulsed or modulated, and the fluorescence is detected by a modulated detector. Normally, a heterodyne principle is used: the detector is modulated with a slightly different frequency than the excitation. The result is that the phase shift and the modulation degree are transferred to low frequency where it can be detected by digital signal processing techniques. The photon efficiency depends on the temporal shape of the excitation signal, the temporal shape of the detector modulation, and, critically, on the depth of the detector modulation [50, 64, 131]. The technique usually does not aim to explicitly detect lifetimes or lifetime components of the fluorescence decay. Instead, it delivers a 'phasor' which can be considered a signature of the fluorophore or fluorophore combination in the particular pixels [94].

3.4.6 Point Scanning with Analog Detection

The sample is scanned with a pulsed laser beam, and the detector signal is recorded with a fast digitizer [70]. The technique circumvents the pile-up problem of TCSPC —should it occur—and is thus able to record with fast acquisition times. The efficiency of the technique decreases rapidly at low intensities, when the photon rate drops below one photon per excitation pulse. The time-channel width is on the order of 150 ps per time channel, which may not be enough to reliably resolve multi-exponential decay functions with fast components, or the Instrumental Response Function (IRF).

3.4.7 Point Scanning with Gated Photon Counting

The technique scans the sample with a pulsed laser beam, detects single photons of the fluorescence, and counts the photons in a small number of pre-set time intervals [60]. The fluorescence lifetime is derived from the photon numbers in the time intervals. The technique delivers excellent photon efficiency [143], and works up to very high count rates. If the sample is able to deliver high intensities, then short acquisition times can be obtained. Multi-exponential analysis is, in principle, possible by using more than two time intervals. However, there is a possible pitfall: If the decay curve does not start exactly at the beginning of the first interval, or if the laser pulse shape or the temporal detector response are not known or variable (they cannot be measured with the same technique) the determination of amplitudes and lifetimes of fast decay components becomes unreliable.

3.4.8 Point Scanning with Multi-dimensional TCSPC

The technique is based on building up photon arrival time distributions over various parameters of the detected photons [15, 41, 305]. The technique is described in detail in Chap. 1, the combination with various optical scanning techniques in Chap. 2. The principle of a TCSPC FLIM system is shown in Fig. 3.4. The sample is scanned with a high-repetition rate laser and single photons of the fluorescence signal are detected. For each photon, the time in the laser period, and the position of the laser beam in the scan area are determined. The recording process builds up a photon distribution over these parameters [15]. The result can be interpreted as an array of pixels, each containing a full fluorescence decay curve. When a photon is detected it is put into a time channel according to its time in the decay function, and

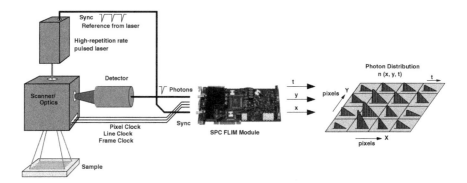

Fig. 3.4 Principle of TCSPC FLIM. The sample is scanned at a high rate. For every photon detected, the TCSPC module determines the time, t, in the fluorescence decay, and the position of the laser beam, x, y, in the scan area. A photon distribution over x, y, t is built up from the times and the positions

a pixel according to the position of the laser beam at the moment of its detection. In other words, the sample is probed by randomly emitted photons. The digital nature of photon detection (yes/no) is the underlying reason for the strength of the technique [15, 41, 305].

The acquisition is continued over a sufficiently large number of frames of the scan, until the desired signal-to-noise ratio is obtained. The photon efficiency comes very close to the theoretical value, i.e. the standard deviation of the lifetime is the reciprocal square root of the photon number [211].

TCSPC FLIM records the decay data into a large number of time channels. As a rule of thumb, the fastest decay components present in the signal should be sampled with about 5 time channels to avoid aliasing artefacts. Fast decay components (e.g. of Flavin Adenine Dinucleotide (FAD)) have lifetimes on the order of 200 ps. The time channel width should therefore be on the order of 50 ps or less, and 250 time channels are required to cover a recording-time interval of 10 ns. This is no problem for modern TCSPC FLIM electronics.

A frequent concern against TCSPC FLIM is that, in principle, pile-up errors can occur. This effect is due to the electronics recording no more than one photon per excitation pulse [305]. If there are more photons these are lost, and a distorted waveform is recorded. Pile-up was a serious problem in early TCSPC experiments with low-repetition rate light sources. Therefore, various schemes have been devised to avoid pile up. In practice the best solution is to design the experiment in such a way that pile-up does not occur or its effect on the decay data is negligible. The best way to do so is to use high-repetition rate excitation [305]. It has been theoretically shown [15] and experimentally verified [16] that count rates of 10 % of the laser repetition rate can be used without introducing more than 2.5 % lifetime error [4], see Chap. 1, Sect. 1.2. For an 80 MHz Ti: Sapphire laser the count rate would be 8 MHz. Acquisition times as short as 0.5 s per image (128 × 128 pixels) have been obtained under such conditions [208].

Pile-up errors should not be confused with counting loss. Counting loss means that photons are lost in the dead time of the internal processing logics of the TCSPC module. The loss increases with increasing count rate, and thus causes a nonlinearity in the intensity scale [392]. Because the intensity information is not normally used in FLIM experiments the nonlinearity can normally be ignored. The nonlinearity problem has recently been solved by introducing an additional counter channel that bypasses the timing electronics [386]. The intensity information is derived from the counter channel, the decay information from the TCSPC timing electronics.

FLIM based on multi-dimensional TCSPC can be extended to record simultaneously in several (currently up to 16) wavelength intervals [15, 19] (Chap. 1, Sect. 1. 4.5.2), to record dynamic changes in the fluorescence along a one-dimensional scan [22] (Chap. 1, Sect. 1.4.6 and Chap. 5), to record at several laser wavelengths multiplexed at high rate [16] (Chap. 1, Sect. 1.4.5.3) or to simultaneously record phosphorescence and fluorescence lifetime images [24] (Chap. 1, Sect. 1.4.7 and Chap. 6). Because TCSPC is based on single-photon timing it can also record Fluorescence Correlation Spectroscopy (FCS) Curves (Chap. 1, Sect. 1.5.2).

TCSPC FLIM has obtained a considerable push with the introduction of computers with 64-bit operating systems. The large memory space available in the 64-bit environment made it possible to record more complex photon distributions. Arrays of images can be recorded for different scan areas, subsequent times after a stimulation of the sample, or for different depth of the focal plane [386]. The entire data are recorded in one large multi-dimensional photon distribution. This not only avoids that time is wasted for saving operations, it makes it also easier to analyse the data with global fitting routines.

TCSPC FLIM has also benefited from the introduction of hybrid detectors [25, 273]. The advantage of these devices over photomultiplier and MCP detectors, or single photon avalanche diodes [194] is a clean temporal response (IRF), extremely high sensitivity (quantum efficiency almost 50 % with GaAsP cathodes), and an absence of afterpulsing. The absence of this signal-induced afterpulsing is not only an advantage for FCS applications, it also leads to a much lower counting background than for conventional detectors: it has long been known that a large part of the background signal in conventional photomultipliers and single-photon avalanche photodiodes (SPADs) comes from afterpulsing [79, 288]. With this part being absent, the hybrid detectors deliver a much lower background, although their thermal background count rate is not particularly low [273].

3.4.9 Excitation Sources

Tunable mode-locked solid state lasers such as Ti:sapphire lasers provide picosecond or femtosecond pulses over a wider tuning range ($\approx$680–1080 nm), are user-friendly and commercially available as turn-key systems. They can have an average power up to several watts, a fixed repetition rate of about 80 MHz which corresponds to 12.5 ns between pulses (roundtrip time of a pulse in the laser cavity) and are often used as excitation sources for TCSPC FLIM, in particular for two-photon excitation FLIM, but also frequency doubled for single photon excitation FLIM ($\approx$340–540 nm). The repetition rate can be reduced by pulse pickers or cavity dumpers, which employ acousto-optical devices to select only a specified fraction of the pulses in the pulse train, or by long cavity lasers [238]. Alternatively, the incomplete decay of the fluorophore can be taken into account by suitable fit models [344]. Small and inexpensive low average power ($\sim$1 mW) picosecond diode lasers at fixed wavelengths with variable repetition rates have also been employed for TCSPC FLIM [215, 223, 343, 418], and their variable repetition rate is particularly suited to measuring long fluorescence decays, e.g. those of quantum dots [455]. Another way to record long-lifetime decays is by modulating a laser of high repetition rate with a pulse period in the microsecond range. TCSPC FLIM is then used to determine photon times both within the laser pulse period and within the modulation period. These times are used to build up fluorescence and phosphorescence lifetime images (FLIM and PLIM) simultaneously [24], see Chap. 1, Sect. 1.4.7 and Chap. 6.

An innovative development is the use of a photonic crystal fibre as a tunable super-continuum excitation source for FLIM [101]. Ti:sapphire laser pulses at 790 nm were coupled into a 30 cm long, 2 μm diameter, micro-structured photonic crystal fibre to produce a continuum of pulses from 435 to 1150 nm. Appropriate spectral selection allowed the excitation of GFP and autofluorescence in confocal TCSPC and wide-field time-gated FLIM. The ease and simplicity with which the tunability is achieved over such a large range is a distinct advantage of this approach. Synchrotron radiation which has similar spectral broadband features has also been used for confocal microscopy (as far back as 1995 [431]), but this remains a specialist application. The optical pulse of these super-continuum sources is around 10 ps [205], they are now commercially available and have been used for FLIM [115, 154, 205, 269]. Several excitation wavelengths can be multiplexed and used simultaneously with such a source, see Chap. 1, Sect. 1.4.5.3.

3.4.10 FLIM Implementation

FLIM techniques continue to be improved, and the relative merits of the various FLIM implementations are summarized in [387, 389]. Some microscopy techniques such as TIRF, supercritical angle fluorescence or selective plane illumination (light sheet) are difficult or impossible to implement with scanning, and image acquisition has to be performed in wide-field mode with a camera. In combination with FLIM, this has until recently meant that gated or frequency-domain camera-based FLIM had to be used, but wide-field TCSPC methods have been improved to take advantage of the high signal-to-noise ratio available by using this type of FLIM [275, 277].

3.5 FLIM Applications in the Life Sciences

3.5.1 Förster Resonance Energy Transfer (FRET)

One of the most widespread applications of FLIM in cell biology is the identification of Förster resonance energy transfer (FRET), see Table 3.1. FRET is a bimolecular fluorescence quenching process where the excited state energy of a donor fluorophore is non-radiatively transferred to a ground state acceptor molecule. The mechanism is schematically illustrated in Fig. 3.5.

The phenomenon is based on a dipole–dipole coupling process and was quantitatively correctly described by Förster in 1946 [127]. FRET only occurs if the donor and acceptor fluorophores are within close proximity (typically <10 nm), and the emission spectrum of the donor and the absorption spectrum of the acceptor overlap, as indicated in Fig. 3.5b. In addition, the transition dipole moments of the donor and acceptor must not be perpendicular—otherwise the transfer efficiency is zero, irrespective of the donor-acceptor distance or the spectral overlap. Finally, the

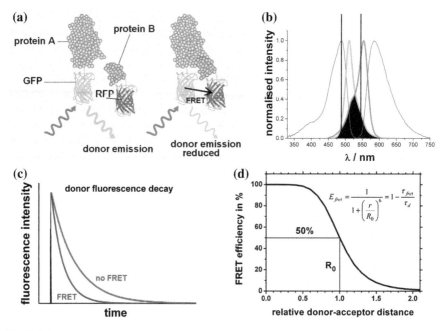

Fig. 3.5 Förster resonance energy transfer (FRET). **a** FRET schematic illustrating the use of this photophysical phenomenon to elucidate protein interaction between protein A, labelled with GFP, and protein B, labelled with RFP. **b** The spectral overlap between the GFP donor emission spectrum (*green*) and the RFP acceptor absorption spectrum (*orange*) is indicated in *black* ("resonance"). FLIM to identify FRET can be performed by measuring the fluorescence decay of the donor in the spectral window indicated by the *black vertical bars* over the donor emission spectrum. Close proximity of donor and acceptor and favourable orientation of their transition dipole moments is also required for FRET to occur. The excited donor transfers its energy to the acceptor, whereupon the donor returns to the ground state, and the acceptor finds itself in the excited state. Note that no photons are emitted in FRET, it is a non-radiative transfer of excited state energy from the donor to the acceptor. **c** FRET effect on donor fluorescence decay. FRET is a quenching process, i.e. offers an additional non-radiative decay pathway in (3.3) and thus shortens the donor fluorescence decay. **d** The distance dependence of FRET. The FRET efficiency varies in proportion to r^{-6} where r is the distance between donor and acceptor, idealized as point dipoles

multiplicity must be preserved by the transitions, singlet-triplet transitions are forbidden as they require a spin flip [350]. (In this context, note that the important singlet oxygen generation in photodynamic therapy [307], or as one of the photobleaching processes, by energy transfer from the fluorophore's triplet state occurs via Dexter-type electron exchange which does not need to conserve multiplicities [93].) The critical transfer distance, R_0, where FRET and fluorescence emission are equally likely, can be calculated[2] from the spectral overlap integral J_F [96, 99] according to

[2] *Free Photochem-CAD software to calculate R_0 from donor and acceptor spectra can be downloaded from* http://photochemcad.com [96, 99].

$$R_0^6 = \frac{9000\kappa^2 \phi \ln 10}{128\pi^5 N_A n^4} J_F \tag{3.10}$$

with κ^2 the orientation factor and

$$J_F = \int_0^\infty \frac{F(\tilde{v})\varepsilon(\tilde{v})}{\tilde{v}^4} d\tilde{v} = \int_0^\infty F(\lambda)\varepsilon(\lambda)\lambda^4 d\lambda \tag{3.11}$$

which only yields a non-zero value in the region of spectral overlap, shown in black in Fig. 3.5b. If the fluorophore's Brownian rotation is slow compared to the energy transfer rate, then "static averaging" of κ^2 in 3 dimensions yields 0.476, whereas when the donor and acceptor can have all orientations during the transfer time "dynamic averaging" yields $\kappa^2 = 2/3$ [88, 97, 107, 425, 437].

The FRET efficiency, E, varies with the inverse 6th power of the distance between donor and acceptor, and is usually negligible beyond 10 nm, as shown in Fig. 3.5d. FRET can therefore be used as a "spectroscopic ruler" to probe inter- and intra-molecular distances on the scale of the dimensions of the proteins themselves [97, 384, 385]. This is a significant advantage over co-localisation studies with two fluorophores which is limited by the optical resolution of light microscopy (approximately 200 nm laterally, 500 nm axially [178]—although for single molecule co-localisation this resolution limit is somewhat relaxed). Thus if one type of protein is labelled with a donor and another type of protein is labelled with an acceptor, the detection of FRET yields proximity information well below the optical resolution limit that can be achieved by co-localisation imaging of the two fluorophores, and is interpreted as the interaction of the two proteins.

An example is shown in Fig. 3.6. CFP-labelled growth factor receptor–bound protein 2 (CFP-Grb2) interacts with RFP-tagged wild-type SH2 domain–containing protein tyrosine phosphatase 2 (RFP-[WT]Shp2) and after stimulation with 20 ng/ml fibroblast growth factor 9 (FGF9). After 15 min stimulation with FGF9, protein interaction is induced as indicated by the reduced average fluorescence lifetime of the CFP donor [1].

In addition, FRET is also frequently used to study conformational changes within a protein [195], or cleavage of a protein [173], or as a sensor, e.g. for Cu^{2+} ions [187], or for Ca^{2+} ions [284]. The chameleon Ca^{2+} sensor, for example, consists of cyan fluorescent protein (CFP) and yellow fluorescent protein (YFP) and induces FRET through a conformational change upon binding of four Ca^{2+} ions, whereas in the green fluorescent protein (GFP)-based Cu^{2+} sensor, the Cu^{2+} ion itself acts as the acceptor due to its absorption in the red. The glycosylation state of cell surface glycoproteins has also been imaged with two-photon excitation FLIM, using a dye-labelled antibody fragment and FRET [27].

FRET, as a fluorescence quenching process, reduces the quantum yield and the fluorescence lifetime of the donor according to (3.3) and (3.4). If the acceptor is fluorescent—which is not a necessary requirement for FRET to occur—FRET can lead to sensitized acceptor emission. To identify and quantify FRET in biological

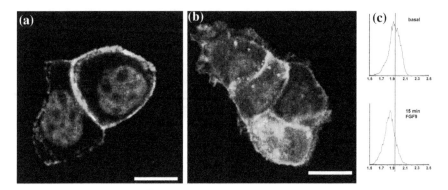

Fig. 3.6 An example of FRET between CFP-donor-labelled proteins, and RFP-acceptor-labelled proteins, reproduced from [1]. Interaction of CFP-labelled growth factor receptor–bound protein 2 (CFP-Grb2) with RFP-tagged wild-type SH2 domain–containing protein tyrosine phosphatase 2 (RFP-^{WT}Shp2) at basal and after stimulation with 20 ng/ml fibroblast growth factor 9 (FGF9). **a** No interaction at basal. **b** Upon 15 min stimulation with FGF9, protein interaction is induced as indicated by the reduced average fluorescence lifetime of the CFP donor. **c** Average fluorescence lifetime histogram of CFP donor at basal (no interaction) and 15 min FGF9 stimulation, shifted to shorter lifetimes, consistent with FRET (interaction). The scale bar is 20 μm

applications, the fluorescence decay of the donor can be measured in the absence and presence of the acceptor, see Fig. 3.5c. The advantage of time-resolved over intensity-based measurements is the ability to directly distinguish between effects due to FRET or probe concentration. For example, a low donor fluorescence intensity can be caused by either a low donor concentration or efficient quenching by FRET—but only in the latter case is the fluorescence decay shortened. Indeed, FLIM is the most reliable method to identify FRET [315, 405]. It only requires measurement of the donor fluorescence and allows the separation of energy transfer efficiency and FRET population, independent of local concentration and stoichiometry of donor and acceptor. If the stoichiometry is not known, i.e. the sample contains both interacting and non-interacting donors, then a bi-exponential donor fluorescence decay would result. The non-interacting donors do not undergo FRET, and thus emit fluorescence with the lifetime of the unquenched donor. The donors undergoing FRET exhibit a shortened fluorescence decay. The ratio of the pre-exponential factors, or amplitudes, of the bi-exponential decay represents the ratio of interacting donors undergoing FRET to those not interacting [122, 366]. In practice, however, note that the complex photophysics of fluorescent proteins means they have multi-exponential decays even before undergoing FRET [84, 177, 195, 394, 420]. Moreover, donors not undergoing FRET photobleach faster than those undergoing FRET. This is due to the extra de-excitation pathway via FRET, which means the donors participating in FRET spend less time in the excited state (i.e. have a shorter lifetime). The photobleaching is usually associated with the triplet state, and the probability of populating it is reduced when the fluorescence lifetime is short. Thus, the photobleaching rate of the donor is slowed down in FRET. This effect may not only have to be taken into account for quantitative FRET analysis, but can also (and has been) exploited to study FRET, as donor photo-bleaching FRET [202].

Accurate determinations of molecular separation are rarely quoted in the literature, due to uncertainty in the real value of R_0. However, the principal goal is usually the detection of FRET to infer proximity of donor and acceptor and thus interaction of the proteins they are tagged to, or conformational changes, rather than obtaining precise molecular separation.

While single point FRET-studies on cells were performed through a microscope well before the development of FLIM [119], imaging FRET can map interactions between proteins, lipids, DNA and RNA, as well as conformational changes or cleavage of a protein in a two-dimensional, position-sensitive manner, so that the FRET signal provides the contrast in the image [200]. The high-resolution and optical sectioning capabilities of confocal or two-photon excitation scanning FLIM allows FRET to be mapped with great detail, and protein interactions to be located accurately within different cell organelles, such as the nucleus, the cytoplasm or the membrane.

Although FLIM of FRET now often involves fluorescent proteins, this technique was already performed before their availability. For example intracellular fusion of endosomes, or the dimerisation of epidermal growth factor (EGF) or the role of the protein kinase C (PKC) family of proteins in cellular signal transduction was studied with FLIM of FRET, as reviewed previously [389]. Today, fluorescent proteins can be used for genetically encoding a fluorescent label [75, 361, 362]. Due to their large size, the donor and acceptor fluorescent proteins cannot come closer than 2.5–3 nm and thus their fluorophores do not undergo collisional quenching. However, this also limits their FRET efficiencies, they cannot therefore reach the maximum 100 % level, see Fig. 3.5d. The excitation and emission spectra of the green fluorescent protein (GFP) and its derivatives span the entire visible range [362], but the photophysics of the fluorescence proteins is complex [468]. The widely used mutant enhanced GFP (F64L, S65T), for example, has at least two emitting states [84, 177, 394, 420]. Nonetheless, FLIM of GFP and its spectral variants [362] with average fluorescence lifetimes in the 2–3 ns region, has proved extremely valuable to the fluorescence microscopy community.

3.5.2 Mapping Viscosity by Lifetime of Molecular Rotors

Fluorescent molecular rotors are fluorophores whose fluorescence quantum yield ϕ and fluorescence lifetime τ are functions of the viscosity η of their environment [166, 167, 220, 221, 421] where

$$\phi = z\eta^x \tag{3.12}$$

and

$$\tau = z' \eta^x \tag{3.13}$$

according to a model proposed by Förster and Hoffmann, with x = 2/3 [128] or later in a more general form with $0 < x \leq 1$ by Loutfy [254]. z and z' are constants, and $\phi \ll 1$.

A key characteristic of a fluorescent molecular rotor is that, in the excited state, it can rotate one segment of its structure around a single bond and thus form a twisted state. It is this intramolecular rotation which depends strongly on the viscosity of the environment, so that the radiative de-excitation pathway of fluorescent molecular rotors competes with radiationless decay by intramolecular twisting in the excited state. This twisting motion is slowed in viscous media. Thus, the fluorescence lifetime and fluorescence quantum yield of fluorescent molecular rotors are high in viscous microenvironments, and low in non-viscous microenvironments. Fluorescent molecular rotors have been used to measure the microviscosity in polymers [254], sol-gels [193, 333], microbubbles [186], micelles [236], ionic liquids [68, 161, 256, 314], blood plasma [168], liposomes [226, 302], membranes of bacillus spores [253], and biological structures such as tubulin [225] and living cells [14, 165, 204, 223, 241, 257, 316, 441]. The viscosity measurement can be accomplished either by ratiometric spectral measurements with a rigid reference fluorophore whose fluorescence quantum yield and lifetime are independent of viscosity [145, 164, 222, 302, 316, 441, 442] or by fluorescence lifetime measurements [14, 223, 241, 316]. Switchable fluorescent molecular rotors than can be activated or locked have also been designed [467].

FLIM of fluorescent molecular rotors has been employed to image viscosity in living cells, [14, 223, 241, 316] microbubbles [186] and the membranes of bacillus spores [253]. A double logarithmic plot of τ of the fluorescent molecular rotor in various solvents versus the viscosity of the solvent should yield a straight line with a gradient of x, according to (3.13) [128]. Such a plot serves as a calibration graph, which allows conversion of the fluorescence lifetime into viscosity, as shown in Fig. 3.7c [166, 167, 223, 241, 255]. A typical example of FLIM of bodipy-C_{12} in lipid droplets is shown in Fig. 3.7a, with the corresponding fluorescence lifetime histogram with contributions from short lifetimes around 1.6 ns in lipid droplets, and longer lifetimes around 1.8 ns in other locations in the cell in Fig. 3.7b.

The big advantage of time-resolved measurements of fluorescent molecular rotors is that the fluorescence lifetime is independent of the fluorophore concentration [145, 164, 222, 302, 316, 441, 442]. Thus, FLIM intrinsically separates concentration and viscosity effects. There is no need to conjugate the fluorescent molecular rotors to other viscosity-independent fluorophores in order to account for variations in dye concentrations as in ratiometric intensity imaging [167]. Moreover, fluorescence lifetime measurements can detect heterogeneous viscosity environments via multi-exponential fluorescence decays, potentially within a single pixel, and the lifetime calibration does not depend on the spectral sensitivity of the detection system. Furthermore, FLIM of fluorescent molecular rotors allows not

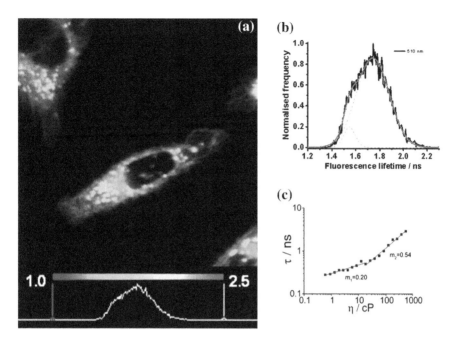

Fig. 3.7 HeLa cells stained with fluorescent molecular rotor bodipy-C_{12}. **a** FLIM image, with a lifetime range of 1 ns (*blue*) to 2.5 ns (*red*), indicating viscosity. **b** The fluorescence lifetime histogram, with contributions from short lifetimes around 1.6 ns in lipid droplets, and longer lifetimes around 1.8 ns in other locations in the cell. **c** A calibration plot of fluorescence lifetime τ in different methanol/glycerol mixtures versus the viscosity η of these mixtures. m is the gradient of a *straight line* fit according to (3.13), and allows the conversion of lifetime into viscosity. In this case, the gradient is 0.20 in the low viscosity region below 10 cp, and 0.54 above 10 cp

only mapping the viscosity in living cells, but also monitoring dynamic cellular processes in real time. Possible techniques are conventional time-series FLIM (Chap. 1, Sect. 1.4.5.4) temporal mosaic FLIM (Chap. 2, Sect. 2.4.4), or Fluorescence Lifetime-Transient Scanning (FLITS) (Chap. 1, Sect. 1.4.6 and Chap. 5). Moreover, compared to viscosity imaging with a viscosity-independent reference fluorophore FLIM-based measurement frees the spectral region occupied by the reference fluorophore for simultaneous mapping of other parameters, e.g. polarity. Thus FLIM of suitable fluorescent molecular rotors represents a major advance in terms of straight-forward calibration and rapid, real-time and ultra-sensitive viscosity mapping [220].

3.5.3 Temperature Mapping

One of the latest advances in the use of FLIM is to use it in combination with special temperature-sensitive polymers to map the temperature in living cells. While

FLIM of rhodamine B in methanol was used to map the temperature in a glass microchip from 10 to about 95 °C with a ±3 °C accuracy [31], and FLIM of Kiton red, a water-soluble rhodamine B derivative, was used to map thermal and solution transport processes in a microfluidic T-mixer [272], these dyes have a limited sensitivity to temperature. They may cover a large dynamic range from 10 to 100 °C, but they are not very sensitive to temperature variations around 37 °C.

Novel temperature-sensitive polymers, fluorescent polymeric thermometers, have been designed that are not very sensitive to temperature over a wide dynamic range, but rather display a large fluorescence lifetime variation near 37 °C [308]. At low temperatures, a thermo-responsive polymer assumes an extended configuration, where a water-sensitive unit can be quenched by water molecules in its vicinity. At higher temperatures, hydration is weakened and the structure shrinks, releasing water molecules and thus increasing its fluorescence quantum yield and lifetime. These sensitive fluorescent polymeric thermometers have been used in combination with TCSPC-based FLIM to map the temperature in living cells to a fraction of a degree, as shown in Fig. 3.8. The resulting temperature maps illustrated thermogenesis in the mitochondria, showed that the temperature of the nucleus is about 1 °C higher than that of the cytoplasm, and that these observations depend on the cell cycle [308].

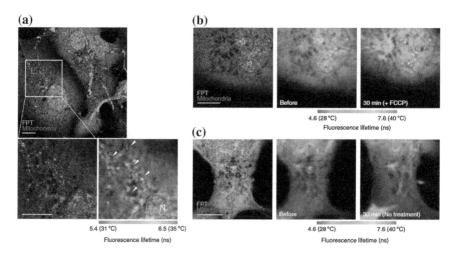

Fig. 3.8 FLIM to map the temperature in living COS7 cells using a temperature-sensitive polymer probe. **a** Confocal fluorescence images of the probe (*green*) and MitoTracker Deep Red FM (*red*; *upper* and *left lower*) and fluorescence lifetime image of the probe (*right lower*). The region of interest, as shown in the *square* in the *upper panel*, is enlarged in the *lower panels* (N is the nucleus). The *arrowheads* point to local heat production, i.e. thermogenesis, near the mitochondria. **b** The temperature increases near the mitochondria after the inhibition of ATP synthesis. Confocal fluorescence images of the probe (*green*) and MitoTracker Deep Red FM (*red*; *left*) and fluorescence lifetime images of the probe (*middle* and *right*). **c** Control experiment for (**b**). No significant change in the probe's fluorescence lifetime was detected without a chemical stimulus. In (**a–c**), the temperature of the medium was maintained at 30 °C. The *scale bar* represents 10 μm. Reproduced from [308]

3.5.4 Mapping of the Refractive Index

The fluorescence decay time of GFP is a function of the refractive index of its environment [394, 417, 418]. A similar observation has been made for CFP and YFP [51]. The reason for this is that the radiative rate constant, k_r in (3.5) and (3.7) is a function of the refractive index, n [412]. This effect, expressed empirically as a n^2-dependence of the radiative rate constant in the Strickler-Berg formula, has been predicted theoretically and demonstrated experimentally for fluorescent dyes, lanthanides, quantum dots and nanodiamonds over the years, by varying the refractive index by solvent composition (see Fig. 3.9a, b) or pressure, including supersonic jet-spectroscopy in vacuum [394]. In the particular case of GFP, the non-radiative rate constant seems to be insensitive of the environment, as the GFP fluorophore is tightly bound inside its barrel, protected from solvent effects, oxygen quenching and other diffusion-controlled collisional quenching effects—influences fluorescent dyes in solution are generally subjected to. The range over which the GFP decay senses the refractive index can be large, in the order of the wavelength of the light,

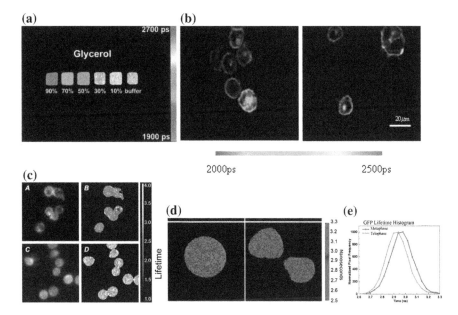

Fig. 3.9 The GFP fluorescence decay is a function of the refractive index of its environment, and can thus be used to sense it. **a** FLIM of GFP in mixtures of buffer and glycerol, from [394]. **b** FLIM of GFP-tagged MHC proteins in fixed cells in buffer and glycerol, from [418]. **c** FLIM of GFP-tagged membrane proteins (*A*—fluorescence intensity, *B*—fluorescence lifetime) and cytosolic GFP-tagged proteins (*C*—fluorescence intensity, *D*—fluorescence lifetime), from [427]. The higher refractive index of the cell membrane decreases the average GFP fluorescence lifetime of the GFP-labelled constructs. **d** GFP in cells at different stages of the cell cycle with (**e**) the corresponding average GFP lifetime histogram, from [326]. The average GFP lifetime has been proposed to be sensitive to the protein concentration during the cell cycle

depending on the experimental details [418]. It plays a role in TIRF FLIM, since GFP in close proximity to a glass water interface has a lower average decay time than far away from the interface [418], and has been used to study GFP infiltration into the nanochannels of mesoporous silica particles [258].

In combination with FLIM, this effect has been exploited to show that GFP-tagged proteins have a faster decay in the cell membrane compared to the cyto-plasm, owing to the membrane's higher refractive index [427], as shown in Fig. 3.9c. In another study, the fluorescence decays of cytoplasmic GFP and also of tdTomato, a red fluorescent protein, were mapped during mitosis, showing that the average GFP and tdTomato lifetimes remained constant during mitosis but rapidly shortened at the final stage of cell division [326], as shown in Fig. 3.9d with the corresponding average GFP lifetime histograms in Fig. 3.9e. The interpretation of this observation put forward was that the concentration of proteins—which have a higher refractive index than the cytoplasm—in the cell changes during the cell cycle. Furthermore, it has been found that fixation of cells also decreases the average lifetime of fluorescent proteins [139, 201], and that this can be related to the refractive index of the mounting solutions [201]. Reports that the average GFP fluorescence lifetime of maltreated cells changes may be related to this effect too [293]. Moreover, using flow cytometry and the GFP fluorescence lifetime as the cytometric parameter, it has been shown that the GFP fluorescence lifetime can be correlated to changes in the subcellular localization of GFP-LC3 fusion protein to the autophagosome during autophagy [148].

3.5.5 Metal-Modified Fluorescence

While fluorescence lifetime changes due to the effect the refractive index has on the radiative rate constant k_r are modest [412], metal-induced fluorescence lifetime modifications can be much stronger [129]. In the presence of a metal, the excited-state molecular dipole can couple with surface plasmons, i.e. collective electron oscillations in the metal, creating additional radiative k_r^* and non-radiative decay channels k_{nr}^* [13, 129, 231].

In such a case, (3.3) for the fluorescence lifetime has to be modified and the metal-enhanced fluorescence lifetime τ is then given by

$$\tau = \frac{1}{k_r^* + k_r + k_{nr} + k_{nr}^*} \tag{3.14}$$

with the corresponding modified (3.4) for the metal enhanced quantum yield ϕ

$$\phi = \frac{k_r + k_r^*}{k_r^* + k_r + k_{nr} + k_{nr}^*} \tag{3.15}$$

The additional de-activation pathways are strongly dependent on the separation between the emitting fluorophore and the metal; hence, (3.14) and (3.15) predict that as k_r^* increases near a metal surface, the fluorescence quantum yield increases while the fluorescence lifetime decreases. However, within 5–10 nm of the metal, the additional non-radiative channel k_{nr}^* dominates, leading to a strong quenching of the fluorescence, reducing the quantum yield as well.

This metal-modified fluorescence effect was exploited to study a multilayered polyelectrolyte film incorporating aluminium tetrasulfonated phthalocyanine (Al-PcTS), a dye also used as a photosensitiser, and gold nanoparticles. The authors found that fluorescence enhancement can be tuned by the number of polyelectrolyte layers separating AlPcTS and the gold nanoparticles [403].

Moreover, FLIM of metal-enhanced fluorescence can provide increased axial specificity in fluorescence microscopy. After demonstration of the fluorescence enhancement effect with fluorescently labelled beads on a gold film, a calibration system that closely mimics a cell imaging geometry, Cade et al. studied mammary adenocarcinoma cells expressing GFP-labelled membrane proteins grown on a 30 nm gold film [61]. A result is shown in Fig. 3.10.The FLIM image shows a significantly reduced GFP lifetime in the membrane near the gold film, but the GFP lifetime is unmodified in parts of the cell further above the gold film. Thus, the GFP fluorescence lifetime yields information about the proximity of the GFP to the gold film within the confocal volume without resorting to techniques such as TIRF, SNOM, [209] or 4Pi microscopy [178, 184]. This was then exploited to study receptor internalization, i.e. protein redistribution, during receptor-mediated

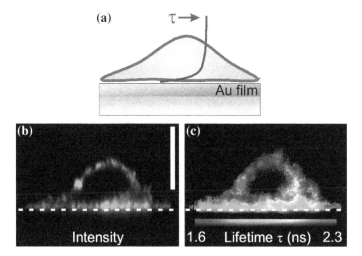

Fig. 3.10 Metal-modified fluorescence. **a** The fluorescence lifetime is a function of the distance of the fluorophore from the gold surface. **b** Confocal xz-cross-section of mammary adenocarcinoma cell (MTLn3E) with GFP-labelled CXCR4 in its cell membrane. The *vertical scale bar* is 5 μm. **c** FLIM image of (**b**), showing a shortened GFP fluorescence lifetime in the vicinity of the gold surface. Reproduced from [61]

endocytosis [61], a technique which has recently been improved by using a bespoke plasmonic nanostructure-coated glass substrate [62]. A similar approach was used to obtain axial distances in tilted microtubules up to 100 nm above a metal surface [35], and the recent 3-dimensional reconstruction of the position of the basal cell membrane of various cell types stained with a CellMask Deep Red plasma membrane dye above a gold film [71].

3.5.6 Mapping of Glucose

Among the reporters for fluorescence-based glucose sensing, the glucose/galactose binding protein (GBP) undergoes a conformational change upon glucose binding [322]. This can either be detected with FRET, or by labelling with an environmentally sensitive fluorophore such as Badan near its glucose binding site. The latter path was chosen, and it was found that glucose binding resulted in a large increase of fluorescence quantum yield and lifetime [354]. Nickel-nitrilotriacetic acid agarose beads with bound GBP–Badan were imaged by FLIM, and addition of glucose resulted in a Badan lifetime shift from 2.2 ns at zero glucose to around 2.7 ns in a 100 mM saturated glucose solution [353], as shown in Fig. 3.11.

The authors point out that the fluorescence lifetime is a particularly useful parameter to perform glucose sensing, since it is relatively independent of light scattering in tissue, signal amplitude fluctuations and fluorophore concentration. The fluorescence lifetime is thus a good alternative to electro-chemistry or glucose

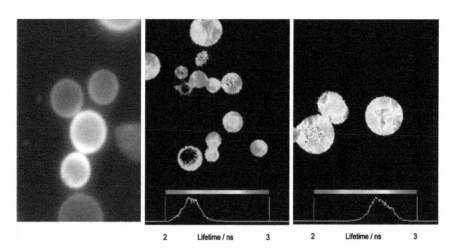

Fig. 3.11 Nickel-nitrilotriacetic acid agarose beads with bound glucose/galactose binding protein labelled with the fluorophore Badan. *Left* Epi-fluorescence intensity image (λ_{ex} = 400 nm and λ_{em} = 550 nm). Average fluorescence lifetime images of beads in 10 mM PBS: *Middle* Zero glucose. *Right* 100 mM saturated glucose. The average Badan lifetime shifts from 2.2 ns at zero glucose to around 2.7 ns in 100 mM glucose solution. Reproduced from [353]

oxidase methods, which have limited accuracy and impaired responses in vivo, possibly due to interfering electroactive substances in the tissues, coating of the sensor by protein and cells and changes in blood flow that alter oxygen access [353].

3.5.7 Mapping of Ion Concentrations

Ions play a major role in living cells and organisms, and mapping and measuring ion concentrations and to dynamically observe changes and fluctuations is of great interest to cell biologists and physiologists [47]. The most important ions are Ca^{2+}, Na^+, K^+, and Cl^-, and a number of different strategies employing fluorescence-based ion-sensing exist. Mapping ion concentrations via the fluorescence lifetime with FLIM in principle offers the advantage of being independent of the fluorophore concentration. FLIM is also unaffected by variations of illumination intensity, or photobleaching—provided the probes do not aggregate, and the photoproducts do not fluoresce. However, in practice, multi-exponential decays and undesirable photoproducts may hamper applications of some probes [235].

For example, instead of using intensity-based imaging of ratiometric probes, the fluorescence lifetime of the Ca^{2+} sensor Quin-2 [233, 235] has been used to image Ca^{2+} concentration in cells. Quin-2, excited at 340 nm, unfortunately forms a photoproduct with a different Ca^{2+} affinity [235], but calcium crimson [179], Calcium Green [66], Fluo-3 [348] and Fluo-4 do not suffer from this problem. However, the lifetime change of Calcium Green upon Ca^{2+} binding is not as large as in the case of Quin-2. Fluo-3 and Fluo-4 do not show noticeable lifetime changes at all. The reason is that these dyes are virtually non-fluorescent when not bound to Ca^{2+}. As a result, only fluorescence from the Ca^{2+} bound form is observed, and there is no change in the average lifetime. Oregon Green BAPTA 1-AM is another Ca^{2+} dye which changes its fluorescence lifetime when the stimulation induces a change in the Ca^{2+} ion concentration. It is this thus suitable for FLIM-based Ca^{2+} measurement [22], see Chap. 5 of this book.

The photophysical properties of Ca^{2+} sensing dyes have been evaluated and their calcium sensing mechanisms classified into static quenching mechanisms (the presence of a non-fluorescent unbound and a fluorescent Ca^{2+}-bound form), the presence of an unbound form and a Ca^{2+}-bound form of different lifetimes, and collisional quenching (calcium green-1, Oregon Green BAPTA-1, magnesium green, Oregon Green BAPTA5 N) [449]. Even an effect of the Ca^{2+} concentration on the extinction coefficient of the dyes (e.g. fluo-3, calcein, fura-2) has been considered. This evaluation was performed with a view of selecting appropriate Ca^{2+} sensing dyes for two-photon excitation TCSPC-based FLIM in brain slices (where Oregon Green BAPTA-1 was chosen) [450]. A powerful application of two-photon excitation FLIM is the imaging of astrocytic calcium homeostasis in living mouse models of Alzheimer's disease, again using the calcium dye Oregon Green BAPTA-1 [219]. Studying calcium levels under different conditions, the authors found that although neurotoxicity is observed near amyloid-β deposits, a more general astrocyte-based network response to pathology also exists.

Ca^{2+} levels in neurons can change on a time scale in the millisecond range. Recording such changes with TCSPC FLIM has long been considered impossible. The changes can, however, be clearly resolved by FLITS, see Chap. 5, and recorded at least qualitatively by temporal mosaic FLIM, see Chap. 2, Sect. 2.4.4.

The fluorescence lifetime of the Cl^- sensing dye N-(ethoxycarbonylmethyl)-6-methoxy-quinolinium bromide (MQAE) has been used to probe Cl^- concentrations in cockroach salivary acinar cells [229], in mammalian olfactory sensory neurons [206] and in nociceptors, neurons for pain and temperature, in the dorsal root ganglion [133, 147]. The dye's sensitivity to Cl^- is due to collisional quenching which obeys Stern-Volmer kinetics [377], see Fig. 3.12a. The ion concentration can thus be mapped with FLIM. An example is shown in Fig. 3.12c.

A Cu^{2+} sensor based on FRET between GFP as the donor and Cu^{2+} has been reported [187] and employed for mapping Cu^{2+} ion uptake and release in plant cells via FLIM [188]. An Oregon Green-labelled apocarbonic anhydrase II Cu^{2+} sensor has also recently been reported, where the Oregon Green fluorescence is quenched by FRET upon binding of Cu^{2+} to the apocarbonic anhydrase II [270]. Furthermore, FRET-based K^+ and Na^+ probes are important for hypertension measurements in blood, and fluorescence lifetime measurements for this purpose have been reported [400], albeit without imaging. A range of Mg^{2+} lifetime probes has also been tested, but again without imaging [399]. Phosphate ions have also been mapped in living cells using FLIM [312], and nitric oxide lifetime probes have been reported, but without use in FLIM as yet [464].

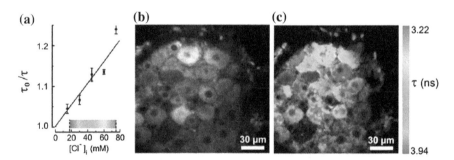

Fig. 3.12 FLIM of chloride ions in somatosensory neurons. **a** A Stern-Volmer calibration plot of relative fluorescence lifetime of chloride ion sensor MQAE versus chloride ion concentration. It follows a linear relationship, and allows the conversion of lifetime into chloride ion concentration. **b** MQAE fluorescence intensity image of the dorsal root ganglion obtained by two-photon excitation, showing an intense signal from the cytosol and a weak signals from the nuclei. **c** Two-photon excitation FLIM image of the same dorsal root ganglion. The fluorescence lifetime, τ, is colour-coded with *warmer colours* representing higher chloride ion concentrations. Reproduced from [147]

Fig. 3.13 FLIM of the pH-
sensitive dye BCECF in skin.
BCECF is present in a
protonated and un-protonated
(BCECF) form. The average
fluorescence lifetimes of each
form are known, and the
fluorescence ratio depends on
pH. Courtesy of Theodora
Mauro, University of San
Francisco, reproduced from
[17]

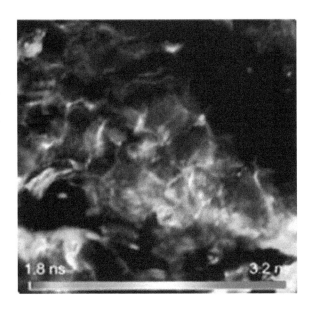

3.5.8 FLIM to Map the pH

Other examples of FLIM are mapping the pH—or H^+ ion concentration—in single
cells [65, 181, 248, 347] and skin [26, 172], see Fig. 3.13. Here, the pH sensor 2,7-
bis-(2-carboxyethyl)-5-(and-6) carboxyfluorescein (BCECF) was used to image pH
in the skin stratum corneum. BCECF is present in a protonated and un-protonated
(BCECF$^-$) form, the fluorescence ratio of which depends on pH. The average
fluorescence lifetimes of each form are known from a calibration (2.75 ns for
protonated BCECF and 3.90 ns BCECF$^-$ in buffer solution) [172]. Two-photon
excitation FLIM was used to non-destructively obtain pH maps at various depths,
which is difficult to achieve by non-optical methods. Moreover, as the authors point
out, intensity-based fluorescence imaging of the pH probe could not have been used
for their study as the observation of a variation in fluorescence intensity could be
ascribed to either a change in pH or a variation of the local probe concentration
[26, 172].

FLIM of GFP excited at 405 nm where the extinction coefficient is very low and
the neutral fluorophore is predominantly excited, has been reported to be pH sensitive. The average lifetime increases as the pH increases, and this has been applied
to measure the pH between 4.5 and 7.5 in HeLa cells [294]. The same team has
repeated this feat without GFP, using autofluorescence of cells, namely the nicotinamide adenine dinucleotide (NADH) fluorescence lifetime, upon excitation at
370 nm. The authors found that the NADH lifetime decreases as pH increases
[306]. Mercaptopropionic acid-capped quantum dots have also recently been
employed as pH sensors in cells, and imaged with FLIM [309]. The advantage here

is that the long fluorescence lifetimes of the quantum dots, from around 9 ns at pH 6 to about 15 ns at pH 8, make it easy to discriminate their signal from short-lived cellular autofluorescence, as the authors point out.

3.5.9 Lifetime Imaging to Map Oxygen

Metal-ligand complexes of ruthenium, or other metal ions such as iridium, osmium or rhenium can be used as optical oxygen sensors [444]. The transition metal complexes can undergo metal–ligand charge transfer to form an excited triplet state, and the collisional quenching of oxygen and the sensor reduces the luminescence intensity and lifetime [117, 121, 144, 185]. TCSPC-based FLIM has been used to detect pericellular oxygen concentrations around isolated viable chondrocytes seeded in three-dimensional agarose gel, revealing a subpopulation of cells with spatial oxygen gradients [185]. Furthermore, FLIM of a long-lived ruthenium-based oxygen sensor with an unquenched decay time of 760 ns has been used to map oxygen concentrations in macrophages [144]. Lifetime measurements are particularly advantageous, since intensity based fluorescence imaging of oxygen in cells would require a calibration of the intensity of the probe unquenched by oxygen as well as knowing its concentration in the cell. This is not practically possible. Temporal focussing for two-photon wide-field excitation with a frequency-domain FLIM system has recently been reported to image ruthenium lifetimes in cells [72]. This approach allows rapid optical sectioning with wide-field excitation and camera detection. Line-scanning multi-photon excitation FLIM where each excitation point along the line is intensity modulated by a unique frequency generated by a spatial light modulator has also been used to image microsecond ruthenium lifetimes in the brain vasculature in living mice [189].

A combined fluorescence and phosphorescence lifetime imaging (FLIM/PLIM) technique based on TCSPC imaging is described in Chap. 1, Sect. 1.4.7. Applications are described in Chap. 6. The technique is based on scanning the sample with a high-frequency pulsed laser that is modulated synchronously with the pixel scanning [24]. The advantage of the technique is that it simultaneously delivers FLIM and PLIM images, and that it can be used with two-photon excitation [17]. Please see Chap. 1, Sect. 1.4.7 and Chap. 6.

3.5.10 FLIM of Autofluorescence of Tissue, Eyes and Teeth

FLIM of autofluorescence has recently expanded rapidly [33, 76, 263]. The advantage of this approach is that no specific labelling is needed, as the fluorescence signal is provided by endogenous fluorophores such as NADH or FAD [146]. The fluorescence lifetimes provide a readout of the metabolic state of the samples under investigation, see Chaps. 13–16. Over 20 years ago it was demonstrated that

FLIM can map free and protein-bound NADH [234], and is has been known for even longer that the redox state of mitochondria can be monitored by NADH fluorescence, as reviewed by the discoverer of this effect, Britton Chance [268]. Breast cancer cells have been studied by FLIM of NADH [44, 83, 328, 402, 459], and this approach has been extended to include FAD [367, 368]. Autofluorescence of cardiac myocytes has been studied with FLIM [73, 74], and it has been shown that FLIM of autofluorescence can distinguish necrosis from apoptosis [443], which may be relevant for cancer treatment optimization by adapting drug dosage for maximal apoptotic response and to limit inflammatory reaction and increase the given therapy efficiency. Employing phasor analysis of the FLIM data, bacteria and mammalian cells have also recently been studied with this method in real time [382, 383, 414].

FLIM of autofluorescence has potential as a label-free clinical diagnostic tool for in vivo optical biopsies, in particular for skin [89, 138, 313, 345, 346, 359], see Chap. 15. An example is shown in Fig. 3.14.

In diagnostic applications it is important to have both morphological information and fluorescence lifetime information available. FLIM of skin is therefore often combined with Z-stack imaging. Z stacks can be recorded by TCSPC FLIM by the traditional record-and-save procedure, or by mosaic FLIM, see Chap. 2, Sect. 2.4.3. Examples is are shown in Figs. 3.15 and 3.16. Images in two wavelength intervals, 380–480 and 480–600 nm, were recorded simultaneously by a dual-channel

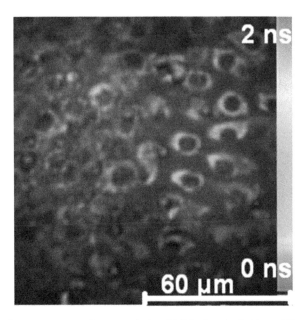

Fig. 3.14 Autofluorescence two-photon excitation FLIM of excised skin, at 75 μm depth, with basal cell carcinoma. The excitation wavelength of 760 nm excites NAD(P)H and melanin in the cytosol. The nucleus does not show any autofluorescence. Reproduced from [138]

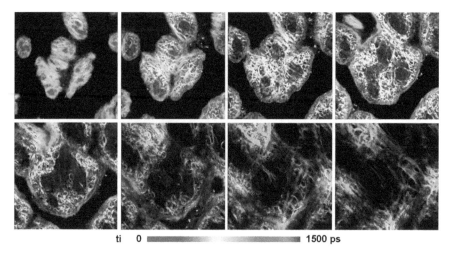

ti   0 ▬▬▬▬▬▬▬▬▬▬▬▬▬▬▬▬▬ 1500 ps

Fig. 3.15 FLIM Z stack of pig skin, detection wavelength 380–480 nm. Two-photon excitation at 800 nm. Zeiss LSM 710 NLO multiphoton microscope with Becker & Hickl Simple-Tau 152 dual-channel FLIM system. Image depth from 5 to 40 μm, acquisition time per z-plane 25 s. Intensity-weighted lifetime of triple-exponential fit

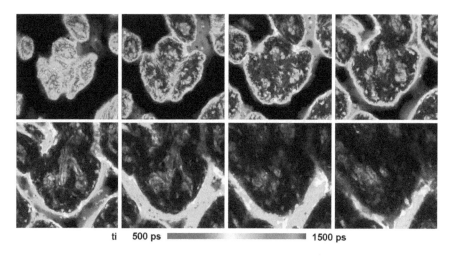

ti   500 ps ▬▬▬▬▬▬▬▬▬▬▬▬▬▬▬ 1500 ps

Fig. 3.16 FLIM z-stack of pig skin, detection wavelength 480–600 nm. Two-photon excitation at 800 nm. Zeiss LSM 710 NLO multiphoton microscope with Becker & Hickl Simple-Tau 152 dual-channel FLIM system. Image depth from 5 to 40 μm, acquisition time per z-plane 25 s. Intensity-weighted lifetime of triple-exponential fit

TCSPC FLIM system. The sample was excited by two-photon excitation at 800 nm. A z-stack of lifetime images was recorded with a vertical step width of 5 μm, and an acquisition time of 25 s per step. The image from 380 to 480 nm interval contains both SHG from collagen and fluorescence, mainly from NADH. The 480–600 nm contains only fluorescence, most likely from FAD.

3-Dimensional optical biopsies do not require any removal of tissue samples, or any other mechanical or chemical treatment. They provide information on morphology and metabolism at a subcellular level, and it has been shown that FLIM of skin autofluorescence can distinguish basal cell carcinoma from the surrounding skin [138] and benign melanocytic nevi from malignant melanocytic lesions [3]. A combination of FLIM of autofluorescence with coherent anti-Stokes Raman Spectroscopy (CARS) could add information about chemical vibrational fingerprint and also lipid and water content to the optical biopsy [55]. Apart from early detection of skin diseases, these approaches could also be used to monitor the progression of wound healing and the effect of cosmetics on the skin [249].

Another application with direct clinical relevance is autofluorescence FLIM of the eye. The autofluorescence decay of the retina is multi-exponential, and a scatter plot of short versus long autofluorescence lifetimes appears to be different for healthy retinas, and retinas at the onset of age-related macular degeneration (AMD) [356, 357]. This approach may offer the opportunity for early detection and diagnosis of this debilitating eye disease. Please see Chap. 16 of this book.

Moreover, the autofluorescence of teeth has also been studied with FLIM [212, 269, 365], and efforts are underway to use FLIM, possibly combined with endoscopy, for clinical diagnostics [132, 334] and brain tumour image-guided surgery [395].

Plant autofluorescence, i.e. FLIM of chlorophyll in algae has recently been used to study cadmium toxicity. After careful calibration of the chlorophyll fluorescence under different excitation conditions, it was found that cadmium exposures appear to lengthen the average chlorophyll fluorescence decay, possibly due to disruption of the electron transport system in photosynthesis [460]. The authors point out that the characteristics of the chlorophyll fluorescence decay could serve as a non-invasive indicator of cadmium toxicity in algae. Autofluorescence of eucalyptus leaves have also been studied with FLIM, revealing a unique spatial organization of cavity metabolites whereby the non-volatile component forms a layer between the secretory cells lining the lumen and the essential oil [180].

Finally, the supra-molecular organization of DNA has been probed with FLIM [290–292, 430], amyloid beta plaques, relevant for e.g. Alzheimer's disease, have been investigated with FLIM [8, 34, 205] and even hematoxylin and eosin staining, a standard technique in histology, has been subjected to FLIM in a quest for more information than from the hematoxylin and eosin intensity images alone [83, 160].

3.5.11 Simultaneous FLIM of Fluorophores with Different Emission Spectra

Modern FLIM systems, in particular TCSPC-based ones, usually have two to four parallel detection and recording channels to simultaneously record images in several wavelength intervals [16]. Signals from several fluorophores can thus be observed simultaneously. However, to obtain true spectral information, e.g. to unmix the signals of fluorophores with overlapping emission spectra, a larger number of spectral channels is required.

Spectrally resolved FLIM can be obtained by streak camera systems or by multi-dimensional TCSPC. Multi-dimensional TCSPC uses the wavelength of the photons as an additional dimension of the recording process of TCSPC FLIM [15, 19], see Chap. 1, Sect. 1.4.5.2. A typical application of spectrally resolved FLIM is autofluorescence imaging. The excitation and emission spectra of the endogenous fluorophores are broad, and usually overlapping. Spectrally resolved FLIM helps to separate the signals from different fluorophores and fluorophore fractions. Please see Chap. 13. A multi-wavelength image of pig skin is shown in Fig. 3.17.

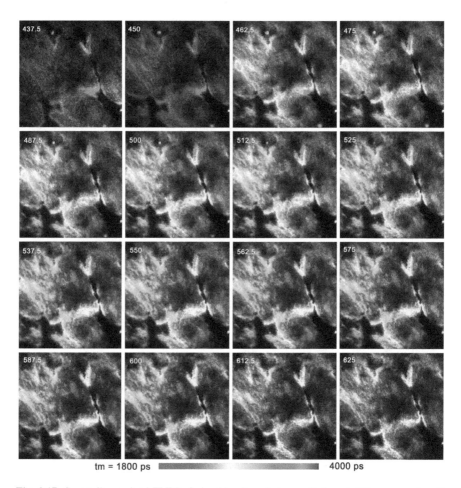

tm = 1800 ps   4000 ps

Fig. 3.17 Spectrally resolved FLIM of pig skin. One-photon excitation at 405 nm, wavelength channel width 12.5 nm, centre wavelength from 437.5 nm (*upper left*) to 625 nm (*lower right*), indicated in images. Multi-wavelength TCSPC FLIM, Becker & Hickl DCS-120 system with 16-channel GaAsP detector. Images 256 × 256 pixels, 256 time channels. Amplitude-weighted lifetime of double-exponential decay, normalised intensity

Spectrally resolved FLIM is advantageous also in FRET studies, where the donor fluorescence lifetime can be monitored in one spectral channel, and the acceptor fluorescence in another. A shortening of the average fluorescence lifetime of the donor CFP due to FRET to the acceptor YFP, both linked by a short amino acid chain, was accompanied by an initial rise of the YFP fluorescence lifetime in the acceptor channel (acceptor in-growth) due to sensitized emission [20]. Other spectrally-resolved FLIM applications concern studies where the fluorescence lifetime of fluorophores emitting in different spectral regions is monitored simultaneously [64], including single molecule studies [408]. The spectral resolution in these cases is really a spectral separation, namely between the two spectral regions of fluorescence emission. However, true spectrally-resolved FLIM with 10 nm bandwidth over a wide spectral range has been reported, both in the frequency domain [169] and the time-domain (e.g. using a 16 anode photomultiplier) [20, 43, 90] or a streak camera [252, 329], allowing sophisticated analysis of multiple fluorophores sensing multiple biophysical parameters, FRET and possibly multiple donor-acceptor FRET pairs [154].

3.5.12 Polarization-Resolved FLIM

3.5.12.1 Rotational Depolarisation

In order to maximize the information available from a limited fluorescence photon budget [463], it is advantageous to record multiple fluorescence parameters—such as lifetime, spectrum and polarization—in a single imaging experiment [242]. Fluorescence is polarized due to the existence of a transition dipole moment in the fluorophore, hence the electric dipole characteristics of the emission. Polarization-resolved fluorescence measurements have been performed since the 1920s [127], and the use of fluorescence anisotropy in imaging and for single molecule work has been reviewed recently [67, 152, 199, 415, 457]. When using fluorescence as a probe, polarization-resolved measurements can yield information on the properties of a sample that cannot be extracted by intensity and lifetime methods alone [39, 325]. Please see Chap. 12 of this book.

In a polarization-resolved fluorescence microscopy experiment, a fluorescently-labelled sample is excited using linearly polarized light, and the time-resolved fluorescence intensity is measured at polarizations parallel and perpendicular to that of the exciting light [230, 424, 438]. The fluorescence decay parallel to the polarization of the excitation, $F_\parallel$, is given by

$$F_\parallel(t) = \frac{1}{3} F_0 \exp\left(-\frac{t}{\tau}\right) \cdot \left[1 + 2r_0 \exp\left(-\frac{t}{\theta}\right)\right] \qquad (3.16)$$

and the fluorescence decay perpendicular to the polarization of the excitation, $F_\perp$, is

$$F_\perp(t) = \frac{1}{3} F_0 \exp\left(-\frac{t}{\tau}\right) \cdot \left[1 - r_0 \exp\left(-\frac{t}{\theta}\right)\right] \tag{3.17}$$

where r_0 is the initial anisotropy, and θ the rotational correlation time. The difference between the parallel and perpendicular fluorescence signal is due to depolarization of the fluorescence. The fluorescence anisotropy r is then defined as

$$r(t) = \frac{F_{//}(t) - GF_\perp(t)}{F_{//}(t) + zGF_\perp(t)} \tag{3.18}$$

where $F_{\parallel}(t)$ and $F_\perp(t)$ are the fluorescence intensity decays parallel and perpendicular to the polarization of the exciting light. The denominator of 3.18 describes the total intensity emitted in the full space around the sample. If the intensity is excited and/or detected within a limited solid angle, the detection efficiency is different for molecules of different orientation, see Chap. 12, Fig. 12.2. This is taken into account by the factor z. The value of z depends on the NA of the microscope objective, where $1 \leq z \leq 2$ ($z \approx 1$ for a high NA objective, $z = 2$ for a collimated beam) [6, 123–126, 162, 213, 456]. Although a rigorous treatment of the effect of high NA objectives to "see around" the fluorophore and therefore collect all three emission components F_x, F_y, F_z leads to a slightly more complex description than (3.18) [7, 152, 162, 213], this empirical approach is attractive due to its simplicity and similarity with that of a collimated beam and has worked well in our laboratory and others [15]. The empirical constant z is a function of the NA of the microscope objective and is chosen such that (i) a time-resolved fluorescence anisotropy decay starts at the correct initial anisotropy r_0, (as determined by spectroscopic measurements using collimated excitation light) and (ii) the total fluorescence intensity decay $F_{\parallel}(t) + z F_\perp(t)$ is the same as a decay collected using collimated beams with magic angle detection such that polarization contributions are removed [15]. The denominator is proportional to the total fluorescence emission, and G accounts for differences in the transmission and detection efficiencies of the imaging system at parallel and perpendicular polarization. If necessary, an appropriate background has to be subtracted [393]. Due to the nature of the photoselection for absorption and emission transition dipoles, multiphoton excitation provides a greater dynamic range for anisotropy measurements than single photon excitation [40].

The depolarization of the fluorescence, i.e. the decay of the anisotropy r as a function of time, can either be due to the rotational diffusion of the fluorophore in its excited state before emission of a fluorescence photon, or due to energy-migration or homo-FRET.

The rotational diffusion of the fluorophore in its excited state before emission of a fluorescence photon depends on its volume, and the viscosity and temperature of its environment. For a spherical molecule, r(t) decays as a single exponential and is related to the rotational correlation time θ according to:

$$r(t) = (r_0 - r_\infty)e^{-t/\theta} + r_\infty \tag{3.19}$$

where r_0 is the initial anisotropy (the maximum value is 0.4 for single photon excitation) and r_∞ accounts for a restricted rotational mobility. $r_\infty = 0$ for freely rotating fluorophores, e.g. in isotropic, homogeneous solution. For a spherical molecule in an isotropic medium, θ is directly proportional to the viscosity η of the solvent and the hydrodynamic volume V of the rotating molecule:

$$\theta = \frac{\eta V}{kT} \tag{3.20}$$

where k is the Boltzmann constant and T the absolute temperature. Therefore, if the volume of the fluorophore is known, the rotational correlation time can report on the viscosity of the fluorophore's immediate environment. Alternatively, as the rotational diffusion can be slowed down by binding or sped up by cleavage, θ can yield information about the size of the tumbling unit. In addition, evidence of a hindered rotation of the fluorophore due to geometrical restrictions, e.g. in the cell membrane, can be gleaned from r_∞.

Note that the steady-state anisotropy is calculated from the fluorescence intensities, i.e. integrated fluorescence decays, from (3.16) and (3.17) and obeys the Perrin equation

$$r = \frac{r_0 - r_\infty}{1 + \frac{\tau}{\theta}} + r_\infty \tag{3.21}$$

where τ is the fluorescence lifetime, defined in (3.3) [247]. Whilst the steady-state anisotropy r is relatively easy to measure, and in particular to image [199] it may not be unambiguous to interpret in the absence of time-resolved measurements: r depends on the three parameters τ, θ and r_∞, and their respective contribution cannot be disentangled from steady state measurements alone.

Steady-state fluorescence anisotropy imaging has, for example, been used to study viscosity or enzyme activity in cells [37, 38, 95, 130, 151, 240, 244, 397], DNA digestion [63] or to identify FRET between fluorescent proteins [262, 265, 266, 336, 374]. However, it is difficult to obtain information about a hindered rotational mobility as indicated by a non-zero r_∞, and time-resolved measurements are needed to determine this parameter.

Time-resolved fluorescence anisotropy has been used on cells for single point measurements [210, 372, 397, 416] and for mapping solvent interactions in microfluidic devices [29], as well as the viscosity in the cell cytoplasm [78, 247, 393], as shown in Fig. 3.18, and membrane [53].

In the brain, the speed with which neurotransmitters diffuse in the interstitial space contributes critically to the shaping of elementary signals transferred by neural circuits [342]. Indeed, experimental alterations of extracellular medium viscosity could reveal a clear impact of the interstitial diffusion rate on neural signal formation, both inside and outside the synaptic cleft [280, 301, 318, 323, 351, 352].

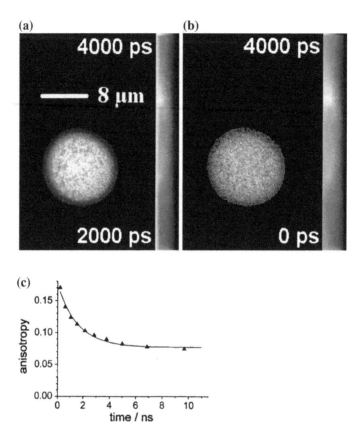

Fig. 3.18 a Fluorescence lifetime and **b** rotational correlation time map of the fluorescein derivative carboxyfluorescein diacetate succinimidyl ester (CFSE) in a B cell. **c** The fluorescence anisotropy decay curve averaged over the entire cell. A mono-exponential fit yields 1.48 ± 0.15 ns at 20 °C, which, assuming an effective hydrodynamic radius of 0.62 nm for CFSE, corresponds to an average cytoplasmic viscosity of 6.0 cp. Reproduced from [393]

Furthermore, it has been suggested that medium micro-viscosity could influence rapid movements of protein domains during ion channel opening: in the squid giant axon, a 30–40 % increase in the local viscosity slows down the gating time of sodium channels by more than two-fold [224]. Similarly, rapid intracellular diffusion of molecular messengers in the protein-crowded microenvironment of small cellular compartments sets the rates of diffusion-limited cellular signalling cascades throughout the central nervous system. Cytosolic mobility and protein crowding have been demonstrated to play an important role in controlling the intracellular spread of molecular signals generated by synaptic signal exchange [36, 289, 349, 396]. In the context of neural coding mechanisms, it would seem reasonable to suggest that understanding the mobility of small signalling molecules in the microenvironment of functional connections in the brain bears as much importance as deciphering their rapid reaction kinetics per se.

Measurements of bulk extracellular diffusion in the brain have a long history. An important advance came with the point-source iontophoresis technique [299], which has been used extensively in various brain areas (reviewed in [398]). It was subsequently complemented by imaging methods which analyze profiles of fluorescence indicators ejected from a point source [192, 300, 351, 454, 466] also employing quantum dots as a diffusing probe [407]. Recent developments in the spot-imaging of extracellular fluorescent probes using microfibre optics have improved spatial resolution of such methods to just a few microns [260, 461]. However, these approaches deal with the apparent diffusion speed which incorporates steric hindrance, or tissue tortuosity, arising from geometric obstacles such as cell walls and membranes of cellular organelles. Molecular mobility on the scale of local biochemical reactions, i.e. within the range of several nanometers, remains poorly understood.

One way of measuring translational diffusion coefficients is FCS [112]. This technique relies on the fluctuation of the fluorescence signal from fluorophores diffusing through the focal spot of the microscope objective, and requires a low fluorophore concentration [21]. FCS can be obtained from TCSPC data as shown in Chap. 1, Sect. 1.5.2. Perhaps the most well-established experimental approach to gauge intra-cytosolic diffusion has been fluorescence recovery after photobleaching, or FRAP (recently reviewed in [91, 439]). Combining FRAP and real-time imaging of photo-activated molecular probes has been highly instrumental in unveiling spatiotemporal aspects of molecular reactions in small dendritic compartments of neurons in situ [36, 87, 289, 349, 396]. A sufficiently fast image acquisition rate can be obtained by TCSPC FLIM via conventional time-series recording or by temporal mosaic FLIM, see Chap. 1, Sect. 1.4.5.4 and Chap. 2, Sect. 2.4.4. The spatial resolution of this method could be as good as the diffraction-limited resolution in the optical imaging system. Even at this resolution level, however, estimated diffusion will incorporate the effect of macromolecular obstacles, intracellular organelles and membrane geometry features, potentially masking the speed at which small molecules shuttle within nanoscopic cellular compartments. Time-resolved fluorescence anisotropy imaging (TR-FAIM) [78, 393] is ideally suited to enable diffusion monitoring at the molecular scale, or in other words to gauge quasi-instantaneous molecular mobility [465].

In the case of severely restricted or even absence of rotational diffusion, polarization-resolved measurements can be used to elucidate the orientation of fluorophores, e.g. in the membrane [32, 332], muscle fibres [54] or in DNA [286]. In these cases, neither the depolarization due to Brownian rotational motion nor homo-FRET is measured, but rather the angle between the electric vector of the light exciting the sample, and the transition dipole moment of the static fluorophore, thus yielding its orientation.

3.5.12.2 Homo-FRET

FRET can occur if the absorption spectrum of the acceptor overlaps with the emission spectrum of the donor, the fluorophores are in close proximity and their orientation is favourable (i.e. orientation factor $\kappa^2 \neq 0$ [88, 97, 425], see (3.10)), as extensively discussed in reference [200]. These conditions can apply to fluorophores with a small Stokes shift and hence lead to the donor and acceptor being the same type of fluorophore. Thus resonance energy transfer between the same type of fluorophore can take place, known as energy migration or homo-FRET. This phenomenon depolarizes the fluorescence emission (see Chap. 12, Sect. 12.7.3) [127]. It has been exploited in single-point measurements and imaging, e.g. to monitor the proximity of iso forms of the glycosyl phosphatidylinositol (GPI)-anchored folate receptor bound to a fluorescent analogue of folic acid to study lipid rafts [363, 432], to monitor actin polymerisation [434], or to image the aggregation of protein α-synuclein, relevant for Parkinson's disease [426].

Time-resolved fluorescence anisotropy measurements to identify homo-FRET have be carried out to study conformational changes in G-protein coupled receptors [376], dimerisation [140], and quantification of protein cluster sizes [9–11, 458]. It has also been used to show that a neuronal iso form of Venus-tagged calcium-calmodulin dependent protein kinase-II alpha (CaMKIIa) holoenzyme forms catalytic domain pairs, and that glutamate receptor activation in neurons triggered an increase in anisotropy consistent with a structural transition from a paired to unpaired conformation [297, 406]. Moreover, time-resolved fluorescence anisotropy measurements have been employed to study the homodimerization of amyloid precursor protein at the plasma membrane, relevant for Alzheimer's disease [92]. In these cases, it is advantageous to have negligible rotational diffusion (a small τ/θ ratio), so that homo-FRET can be identified.

For homo-FRET involving two or more fluorophores, and in the absence of any rotational diffusion, r(t) decays as a single exponential and is related to the homo-FRET transfer rate ω according to [10, 11, 140, 458]

$$r(t) = (r_0 - r_\infty)e^{-2\omega t} + r_\infty \qquad (3.22)$$

where r_0 is the initial anisotropy in the absence of rotation or energy transfer, as defined above, and r_∞ is the anisotropy at a long time after the excitation.[3] In the specific case of two fluorophores, $r_\infty = r_0/2$ and (3.22) reduces to

$$r(t) = \frac{r_0}{2}e^{-2\omega t} + \frac{r_0}{2} \qquad (3.23)$$

While hetero-FRET between different donors and acceptors to identify protein interaction can routinely be imaged with FLIM, mapping energy migration or homo-FRET to identify protein dimerisation requires polarization-resolved FLIM, i.e. TR-FAIM. The only way to detect homo-FRET where the fluorescence lifetime of the two fluorophores is the same, is by polarization measurements, because homo-FRET does not normally affect spectra or fluorescence lifetime [438]. If the fluorescence lifetimes of the two fluorophores are different, however, then FRET

---

[3] Equation (3.5) in [391] should be the same as (3.22) here.

can be identified by fluorescence lifetime measurements [327, 451]. This has, for example, been done in the case of tryptophan to tryptophan homo-FRET in barnase, where the tryptophans are located in different environments yielding different fluorescent lifetimes [448].

Homo-FRET studies are best done with fluorescent proteins, as their rotational diffusion during the excited state lifetime is negligible, especially when tethered to a protein of interest. The energy transfer can be extremely fast (a 2 ps transfer time has been quoted for YFP [203]), is an indicator of protein dimerisation or oligo-merisation, and polarization-resolved techniques can image it.

Anisotropy imaging can be performed as steady-state or time-resolved mea-surements in the time-domain or frequency-domain using scanning or wide-field methods [242, 387, 389], and has been combined with spectral imaging [115]. Photon counting approaches are particularly attractive because of their excellent signal-to-noise ratio and single photon sensitivity [116, 155, 315, 321, 405].

The combination of TIRF with time-resolved fluorescence anisotropy allows excitation with s- and p-polarized light and provides spatial information on the fluorescence depolarisation processes near an interface. This has, for example, facilitated the observation of the rotation of membrane dyes in- and out-of-plane [142, 371]. TIRF has indeed been combined with TR-FAIM, [57, 92] but time-resolved anisotropy imaging with s- and p-polarization of the excitation has not yet been demonstrated.

3.5.13 Phasor Analysis and Bayesian Analysis

Conventional FLIM data analysis in the time domain relies on Levenberg-Mar-quardt fitting algorithms to fit the experimental data to a mathematical model, i.e. compare data and theory [41]. This is a standard procedure that has been used in fluorescence spectroscopy for many decades [305]. However, improvements can be made, depending on the experimental conditions of the data acquisition, by max-imum-likelihood-estimator fitting. This approach has been reported to have a better performance for low photon counts than least square fitting (which in this case tends to underestimate the decay time) [267, 304, 404]. This is particularly relevant for time-resolved single molecule work [267]. Another way to obtain shot-noise limited lifetime accuracy down to extremely low photon numbers is to use the first moment of the photon distribution [16].

The recent development of phasor analysis [77, 170, 330] for FLIM [94] allows the visualization of the decay data without a specific mathematical model (but it does require a calibration measurement with a known reference sample). Although originally developed for data analysis in the frequency domain, it is equally well applicable for data recorded in the time domain, in particular for TCSPC FLIM. Essentially, the fluorescence decay is Fourier transformed at the frequency of the excitation modulation (laser repetition rate in time domain or modulation frequency in frequency domain), and the real part is plotted versus the imaginary part for each pixel. The resulting data cloud (or clouds) is on the universal semicircle for single exponential decays, and various quenching processes result in trajectories within (or even outside) this universal semicircle.

The recent development of Bayesian analysis is particularly relevant for fluorescence decays with a low number of photons [12]. The Bayesian approach allows a decay time estimation with a much narrower confidence limit than Levenberg-Marquardt or maximum likelihood estimator fitting if low photon numbers are involved [340]—as is the case more often than not in many FLIM experiments. If the photon numbers are high enough, then Bayesian fitting does not offer any advantages over conventional Levenberg-Marquardt fitting.

3.6 Summary and Outlook

The power of fluorescence-based optical imaging to drive major discoveries in cell biology is universally recognized. It offers two principal advantages: light microscopy allows the observation of structures inside a living sample in real time, and cellular components or compartments may be observed through specific fluorescence labelling. The key point of FLIM lies in the ability to monitor the environment of a fluorophore independent of its concentration—so in addition to the position of the fluorophore, its biophysical environment can be sensed via the lifetime.

There are various implementations of FLIM, and, depending on the application, each has its advantages and drawbacks. The ideal fluorescence microscope would acquire the entire multidimensional fluorescence emission contour of intensity, position, lifetime, wavelength and polarization in a single measurement, with single photon sensitivity, maximum spatial resolution and minimum acquisition time (Fig. 3.1). Needless to say, there is presently no technology with this unique combination of features, and to build one remains a challenge for instrumentation developers.

Acknowledgments We would like the UK's MRC, BBSRC and EPSRC for funding. Thanks also to Dylan Owen of the Physics Department and Randall Division of Cell and Molecular Biophysics at King's College London for valuable comments on the manuscript.

References

1. Z. Ahmed, C.C. Lin, K.M. Suen, F.A. Melo, J.A. Levitt, K. Suhling, J.E. Ladbury, Grb2 controls phosphorylation of FGFR2 by inhibiting receptor kinase and Shp2 phosphatase activity. J. Cell Biol. **200**, 493–504 (2013)
2. W.B. Amos, J.G. White, How the confocal laser scanning microscope entered biological research. Biol. Cell **95**, 335–342 (2003)

3. F. Arginelli, M. Manfredini, S. Bassoli, C. Dunsby, P. French, K. König, C. Magnoni, G. Ponti, C. Talbot, S. Seidenari, High resolution diagnosis of common nevi by multiphoton laser tomography and fluorescence lifetime imaging. Skin Res. Technol. **19**, 194–204 (2013)
4. J. Arlt, D. Tyndall, B.R. Rae, D.D.U. Li, J.A. Richardson, R.K. Henderson, A study of pile-up in integrated time-correlated single photon counting systems. Rev. Sci. Instrum. **84**, 103105 (2013)
5. E. Auksorius, B.R. Boruah, C. Dunsby, P.M.P. Lanigan, G. Kennedy, M.A.A. Neil, P.M.W. French, Stimulated emission depletion microscopy with a supercontinuum source and fluorescence lifetime imaging. Opt. Lett. **33**, 113–115 (2008)
6. D. Axelrod, Carbocyanine dye orientation in red cell membrane studied by microscopic fluorescence polarization. Biophys. J. **26**, 557–573 (1979)
7. D. Axelrod, Fluorescence polarization microscopy. Methods Cell Biol. **30**, 333–352 (1989)
8. B.J. Bacskai, J. Skoch, G.A. Hickey, R. Allen, B.T. Hyman, Fluorescence resonance energy transfer determinations using multiphoton fluorescence lifetime imaging microscopy to characterize amyloid-beta plaques. J. Biomed. Opt. **8**, 368–375 (2003)
9. A.N. Bader, S. Hoetzl, E.G. Hofman, J. Voortman, P.M.P.V.E. Henegouwen, G. van Meer, H.C. Gerritsen, Homo-FRET imaging as a tool to quantify protein and lipid clustering. ChemPhysChem **12**, 475–483 (2011)
10. A.N. Bader, E.G. Hofman, P.M.P.V.E. Henegouwen, H.C. Gerritsen, Imaging of protein cluster sizes by means of confocal time-gated fluorescence anisotropy microscopy. Opt. Express **15**, 6934–6945 (2007)
11. A.N. Bader, E.G. Hofman, J. Voortman, P.M.P.V.E. Henegouwen, H.C. Gerritsen, Homo-FRET imaging enables quantification of protein cluster sizes with subcellular resolution. Biophys. J. **97**, 2613–2622 (2009)
12. P.R. Barber, S.M. Ameer-Beg, S. Pathmananthan, M. Rowley, A.C.C. Coolen, A Bayesian method for single molecule, fluorescence burst analysis. Biomed. Opt. Express **1**, 1148–1158 (2010)
13. W.L. Barnes, Fluorescence near interfaces: the role of photonic mode density. J. Mod. Opt. **45**, 661–699 (1998)
14. A. Battisti, S. Panettieri, G. Abbandonato, E. Jacchetti, F. Cardarelli, G. Signore, F. Beltram, R. Bizzarri, Imaging intracellular viscosity by a new molecular rotor suitable for phasor analysis of fluorescence lifetime. Anal. Bioanal. Chem. **405**, 6223–6233 (2013)
15. W. Becker, *Advanced Time-Correlated Single Photon Counting Techniques, Springer Series in Chemical Physics*, vol. 81 (Springer, Berlin, Heidelberg, New York, 2005)
16. W. Becker, The bh TCSPC Handbook, 5th edition (2012)
17. W. Becker, Fluorescence lifetime imaging—techniques and applications. J Micros **247**, 119–136 (2012)
18. W. Becker, A. Bergmann, Lifetime-resolved imaging in nonlinear microscopy, in *Handbook of Biomedical Nonlinear Optical Microscopy*, ed by B.R. Masters, P.T.C. So (Oxford University Press, Oxford, 2008)
19. W. Becker, A. Bergmann, C. Biskup, Multispectral fluorescence lifetime imaging by TCSPC. Microsc. Res. Tech. **70**, 403–409 (2007)
20. W. Becker, A. Bergmann, C. Biskup, T. Zimmer, N. Klöcker, K. Benndorf, Multiwavelength TCSPC lifetime imaging *SPIE Proc 4620* (2002), pp. 79–84
21. W. Becker, A. Bergmann, E. Haustein, Z. Petrášek, P. Schwille, C. Biskup, L. Kelbauskas, K. Benndorf, N. Klöcker, T. Anhut, I. Riemann, K. König, Fluorescence lifetime images and correlation spectra obtained by multidimensional time-correlated single photon counting. Microsc. Res. Tech. **69**, 186–195 (2006)
22. W. Becker, V. Shcheslavskiy, S. Frere, I. Slutsky, Spatially resolved recording of transient fluorescence-lifetime effects by line-scanning TCSPC. Microsc. Res. Tech. **77**, 216–224 (2014)
23. W. Becker, V. Shcheslavskiy, Fluorescence lifetime imaging with near-infrared dyes, SPIE Proc 8588, (2013) 85880R

24. W. Becker, B. Su, A. Bergmann, K. Weisshart, O. Holub, Simultaneous fluorescence and phosphorescence lifetime imaging, SPIE Proc 7903, (2011), 790320
25. W. Becker, B. Su, O. Holub, K. Weisshart, FLIM and FCS detection in laser-scanning microscopes: increased efficiency by GaAsP hybrid detectors. Microsc. Res. Tech. **74**, 804–811 (2011)
26. M.J. Behne, J.W. Meyer, K.M. Hanson, N.P. Barry, S. Murata, D. Crumrine, R.W. Clegg, E. Gratton, W.M. Holleran, P.M. Elias, NHE1 regulates the stratum corneum permeability barrier homeostasis. microenvironment acidification assessed with fluorescence lifetime imaging. J. Biol. Chem. **277**, 47399–47406 (2002)
27. B. Belardi, A. delaZerda, D.R. Spiciarich, S.L. Maund, D.M. Peehl, C.R. Bertozzi, Imaging the glycosylation state of cell surface glycoproteins by two-photon fluorescence lifetime imaging microscopy. Angewandte Chemie Int. Ed. **52**, 14045–14049 (2013)
28. M.A. Bennet, P.R. Richardson, J. Arlt, A. McCarthy, G.S. Buller, A.C. Jones, Optically trapped microsensors for microfluidic temperature measurement by fluorescence lifetime imaging microscopy. Lab Chip **11**, 3821–3828 (2011)
29. R.K.P. Benninger, O. Hofmann, J. McGinty, J. Requejo-Isidro, I. Munro, M.A.A. Neil, A. J. deMello, P.M.W. French, Time-resolved fluorescence imaging of solvent interactions in microfluidic devices. Opt. Express **13**, 6275–6285 (2005)
30. R.K.P. Benninger, O. Hofmann, B. Önfelt, I. Munro, C. Dunsby, D.M. Davis, M.A.A. Neil, P.M.W. French, A.J. deMello, Fluorescence lifetime imaging of DNA-due interactions within continuous flow microfluidic systems. Angewandte Chemie Int. Ed. **46**, 2228–2231 (2007)
31. R.K.P. Benninger, Y. Koc, O. Hofmann, J. Requejo-Isidro, M.A.A. Neil, P.M.W. French, A. J. deMello, Quantitative 3D mapping of fluidic temperatures within microchannel networks using fluorescence lifetime imaging. Anal. Chem. **78**, 2272–2278 (2006)
32. R.K.P. Benninger, B. Önfelt, M.A.A. Neil, D.M. Davis, P.M.W. French, Fluorescence imaging of two-photon linear dichroism: cholesterol depletion disrupts molecular orientation in cell membranes. Biophys. J. **88**, 609–622 (2005)
33. M.Y. Berezin, S. Achilefu, Fluorescence lifetime measurements and biological imaging. Chem. Rev. **110**, 2641–2684 (2010)
34. O. Berezovska, P. Ramdya, J. Skoch, M.S. Wolfe, B.J. Bacskai, B.T. Hyman, Amyloid precursor protein associates with a nicastrin-dependent docking site on the presenilin 1-gamma-secretase complex in cells demonstrated by fluorescence lifetime imaging. J. Neurosci. **23**, 4560–4566 (2003)
35. M. Berndt, M. Lorenz, J. Enderlein, S. Diez, Axial nanometer distances measured by fluorescence lifetime imaging microscopy. Nano Lett. **10**, 1497–1500 (2010)
36. A. Biess, E. Korkotian, D. Holcman, Barriers to diffusion in dendrites and estimation of calcium spread following synaptic inputs. PLoS Comput. Biol. **7**, e1002182 (2011)
37. C.E. Bigelow, D.L. Conover, T.H. Foster, Confocal fluorescence spectroscopy and anisotropy imaging system. Opt. Lett. **28**, 695–697 (2003)
38. C.E. Bigelow, H.D. Vishwasrao, J.G. Frelinger, T.H. Foster, Imaging enzyme activity with polarization-sensitive confocal fluorescence microscopy. J. Microsc. **215**, 24–33 (2004)
39. D.J.S. Birch, Fluorescence detections and directions. Meas. Sci. Technol. **22**, 052002 (2011)
40. D.J.S. Birch, Multiphoton excited fluorescence spectroscopy of biomolecular systems. Spectrochim. Acta A **57**, 2313–2336 (2001)
41. D.J.S. Birch, R.E. Imhof, Time-domain fluorescence spectroscopy using time-correlated single photon counting, in *Topics in Fluorescence Spectroscopy: Techniques*, ed by J.R. Lakowicz (Plenum Press, New York, 1991)
42. D.K. Bird, K.M. Agg, N.W. Barnett, T.A. Smith, Time-resolved fluorescence microscopy of gunshot residue: an application to forensic science. J. Microsc.-Oxf. **226**, 18–25 (2007)
43. D.K. Bird, K.W. Eliceiri, C.H. Fan, J.G. White, Simultaneous two-photon spectral and lifetime fluorescence microscopy. Appl. Opt. **43**, 5173–5182 (2004)

44. D.K. Bird, L. Yan, K.M. Vrotsos, K.W. Eliceiri, E.M. Vaughan, P.J. Keely, J.G. White, N. Ramanujam, Metabolic mapping of MCF10A human breast cells via multiphoton fluorescence lifetime imaging of the coenzyme NADH. Cancer Res. **65**, 8766–8773 (2005)
45. R.H. Bisby, S.W. Botchway, G.M. Greetham, J.A. Hadfield, A.T. McGown, A.W. Parker, K. M. Scherer, M. Towrie, Time-resolved nanosecond fluorescence lifetime imaging and picosecond infrared spectroscopy of combretastatin A-4 in solution and in cellular systems. Meas. Sci. Technol. **23** (2012) 084001.
46. R.H. Bisby, S.W. Botchway, J.A. Hadfield, A.T. McGown, A.W. Parker, K.M. Scherer, Fluorescence lifetime imaging of E-combretastatin uptake and distribution in live mammalian cells. Eur. J. Cancer **48**, 1896–1903 (2012)
47. C. Biskup, T. Gensch, Fluorescence lifetime imaging of ions in biological tissues, in *Fluorescence Lifetime Spectroscopy and Imaging. Principles and Applications in Biomedical Diagnostics*, ed by D. Elson, P.W.M. French, L. Marcu (Taylor & Francis, Boca Raton, 2014)
48. C. Biskup, L. Kelbauskas, T. Zimmer, K. Benndorf, A. Bergmann, W. Becker, J. P. Ruppersberg, C. Stockklausner, N. Klöcker, Interaction of PSD-95 with potassium channels visualized by fluorescence lifetime-based resonance energy transfer imaging. J. Biomed. Opt. **9**, 753–759 (2004)
49. C. Biskup, T. Zimmer, K. Benndorf, FRET between cardiac Na$^+$ channel subunits measured with a confocal microscope and a streak camera. Nat. Biotechnol. **22**, 220–224 (2004)
50. M.J. Booth, T. Wilson, Low-cost, frequency-domain, fluorescence lifetime confocal microscopy. J. Micros. **214**, 36–42 (2004)
51. J.W. Borst, M.A. Hink, A. Hoek, A.J. Visser, Effects of refractive index and viscosity on fluorescence and anisotropy decays of enhanced cyan and yellow fluorescent proteins. J. Fluores. **15**, 153–160 (2005)
52. J.W. Borst, A.J.W.G. Visser, Fluorescence lifetime imaging microscopy in life sciences. Meas. Sci. Technol. **21**, 102002 (2010)
53. S.W. Botchway, A.M. Lewis, C.D. Stubbs, Development of fluorophore dynamics imaging as a probe for lipid domains in model vesicles and cell membranes. Eur. Biophys. J. Biophy. **40**, 131–141 (2011)
54. A.S. Brack, B.D. Brandmeier, R.E. Ferguson, S. Criddle, R.E. Dale, M. Irving, bifunctional rhodamine probes of myosin regulatory light chain orientation in relaxed skeletal muscle fibers. Biophys. J. **86**, 2329–2341 (2004)
55. H.G. Breunig, M. Weinigel, R. Bückle, M. Kellner-Höfer, J. Lademann, M.E. Darvin, W. Sterry, K. König, Clinical coherent anti-Stokes Raman scattering and multiphoton tomography of human skin with a femtosecond laser and photonic crystal fiber. Laser Phys. Lett. **10**, 025604 (2013)
56. S.Y. Breusegem, M. Levi, N.P. Barry, Fluorescence correlation spectroscopy and fluorescence lifetime imaging microscopy. Nephron J. **103**, e41–e49 (2006)
57. T. Bruns, W.S.L. Strauss, H. Schneckenburger, Total internal reflection fluorescence lifetime and anisotropy screening of cell membrane dynamics. J. Biomed. Opt. **13**, 041317 (2008)
58. I. Bugiel, K. König, H. Wabnitz, Investigation of cells by fluorescence laser scanning microscopy with subnanosecond time resolution. Lasers Life Sci. **3**, 47–53 (1989)
59. G.S. Buller, R.J. Collins, Single-photon generation and detection. Meas. Sci. Technol. **21**, 012002 (2010)
60. E.P. Buurman, R. Sanders, A. Draaijer, H.C. Gerritsen, J.J.F. van Ween, P.M. Houpt, Y.K. Levine, Fluorescence lifetime imaging using a confocal laser scanning microscope. Scanning **14**, 155–159 (1992)
61. N.I. Cade, G. Fruhwirth, S.J. Archibald, T. Ng, D. Richards, A cellular screening assay using analysis of metal-modified fluorescence lifetime. Biophys. J. **98**, 2752–2757 (2010)
62. N.I. Cade, G.O. Fruhwirth, T. Ng, D. Richards, Plasmon-assisted super-resolution axial distance sensitivity in fluorescence cell imaging. J. Phys. Chem. Lett. **4**, 3402–3406 (2013)

63. Z. Cao, C.C. Huang, W. Tan, Nuclease resistance of telomere-like oligonucleotides monitored in live cells by fluorescence anisotropy imaging. Anal. Chem. **78**, 1478–1484 (2006)
64. K. Carlsson, A. Liljeborg, Simultaneous confocal lifetime imaging of multiple fluorophores using the intensity-modulated multiple-wavelength scanning (IMS) technique. J. Micros. **191**, 119–127 (1998)
65. K. Carlsson, A. Liljeborg, R.M. Andersson, H. Brismar, Confocal pH imaging of microscopic specimens using fluorescence lifetimes and phase fluorometry: influence of parameter choice on system performance. J Micros **199**, 106–114 (2000)
66. A. Celli, S. Sanchez, M. Behne, T. Hazlett, E. Gratton, T. Mauro, The epidermal Ca^{2+} gradient: measurement using the phasor representation of fluorescent lifetime imaging. Biophys. J. **98**, 911–921 (2010)
67. F.T.S. Chan, C.F. Kaminski, G.S. Kaminski Schierle, HomoFRET fluorescence anisotropy imaging as a tool to study molecular self-assembly in live cells. ChemPhysChem **12**, 500–509 (2011)
68. A. Chatterjee, B. Maity, D. Seth, Torsional dynamics of thioflavin T in room-temperature ionic liquids: an effect of heterogeneity of the medium. ChemPhysChem **14**, 3400–3409 (2013)
69. Y. Chen, J.D. Mills, A. Periasamy, Protein localization in living cells and tissues using FRET and FLIM. Differentiation **71**, 528–541 (2003)
70. S.N. Cheng, R.M. Cuenca, B.A. Liu, B.H. Malik, J.M. Jabbour, K.C. Maitland, J. Wright, Y.S.L. Cheng, J.A. Jo, Handheld multispectral fluorescence lifetime imaging system for in vivo applications. Biomed. Opt. Express **5**, 921–931 (2014)
71. A.I. Chizhik, J. Rother, I. Gregor, A. Janshoff, J. Enderlein, Metal-induced energy transfer for live cell nanoscopy. Nat. Photon **8**, 124–127 (2014)
72. H. Choi, D.S. Tzeranis, J.W. Cha, P. Clemenceau, S.J.G. de Jong, L.K. van Geest, J.H. Moon, I.V. Yannas, P.T.C. So, 3D-resolved fluorescence and phosphorescence lifetime imaging using temporal focusing wide-field two-photon excitation. Opt. Express **20**, 26219–26235 (2012)
73. D. Chorvat, A. Chorvatova, Multi-wavelength fluorescence lifetime spectroscopy: a new approach to the study of endogenous fluorescence in living cells and tissues. Laser Phys. Lett. **6**, 175–193 (2009)
74. D. Chorvat, A. Chorvatova, Spectrally resolved time-correlated single photon counting: a novel approach for characterization of endogenous fluorescence in isolated cardiac myocytes. Eur. Biophys. J. Biophy. **36**, 73–83 (2006)
75. D.M. Chudakov, M.V. Matz, S. Lukyanov, K.A. Lukyanov, Fluorescent proteins and their applications in imaging living cells and tissues. Physiol. Rev. **90**, 1103–1163 (2010)
76. R. Cicchi, F.S. Pavone, Non-linear fluorescence lifetime imaging of biological tissues. Anal. Bioanal. Chem. **400**, 2687–2697 (2011)
77. A.H. Clayton, Q.S. Hanley, P.J. Verveer, Graphical representation and multicomponent analysis of single-frequency fluorescence lifetime imaging microscopy data. J. Micros. **213**, 1–5 (2004)
78. A.H.A. Clayton, Q.S. Hanley, D.J. Arndt-Jovin, V. Subramaniam, T.M. Jovin, Dynamic fluorescence anisotropy imaging microscopy in the frequency domain (rFLIM). Biophys. J. **83**, 1631–1649 (2002)
79. P.B. Coates, Origins of afterpulses in photomultipliers. J. Phys. D-Appl. Phys. **6**, 1159–1166 (1973)
80. M.J. Cole, J. Siegel, S.E.D. Webb, R. Jones, K. Dowling, P.M.W. French, M.J. Lever, L.O.D. Sucharov, M.A.A. Neil, R. Juškaitis, T. Wilson, Whole-field optically sectioned fluorescence lifetime imaging. Opt. Lett. **25**, 1361–1363 (2000)
81. D. Comelli, C. D'Andrea, G. Valentini, R. Cubeddu, C. Colombo, L. Toniolo, Fluorescence lifetime imaging and spectroscopy as tools for nondestructive analysis of works of art. Appl. Opt. **43**, 2175–2183 (2004)

82. D. Comelli, G. Valentini, R. Cubeddu, L. Toniolo, Fluorescence lifetime imaging and fourier transform infrared spectroscopy of Michelangelo's David. Appl. Spectrosc. **59**, 1174–1181 (2005)

83. M.W. Conklin, P.P. Provenzano, K.W. Eliceiri, R. Sullivan, P.J. Keely, Fluorescence lifetime imaging of endogenous fluorophores in histopathology sections reveals differences between normal and tumor epithelium in carcinoma in situ of the breast. Cell Biochem. Biophys. **53**, 145–157 (2009)

84. M. Cotlet, J. Hofkens, M. Maus, T. Gensch, M. van der Auweraer, J. Michiels, G. Dirix, M. van Guyse, J. Vanderleyden, A.J.W.G. Visser, F.C. de Schryver, Excited state dynamics in the enhanced green fluorescent protein mutant probed by picosecond time-resolved single photon counting spectroscopy. J. Phys. Chem. B **105**, 4999–5006 (2001)

85. S. Cox, G.E. Jones, Imaging cells at the nanoscale. Int. J. Biochem. Cell B **45**, 1669–1678 (2013)

86. R. Cubeddu, A. Pifferi, P. Taroni, A. Torricelli, G. Valentini, F. Rinaldi, E. Sorbellini, Fluorescence lifetime imaging: an application to the detection of skin tumors. IEEE J. Sel. Top. Quantum Electron. **5**, 923–929 (1999)

87. T.T. Cui-Wang, C. Hanus, T. Cui, T. Helton, J. Bourne, D. Watson, K.M. Harris, M.D. Ehlers, Local zones of endoplasmic reticulum complexity confine cargo in neuronal dendrites. Cell **148**, 309–321 (2012)

88. R.E. Dale, J. Eisinger, W.E. Blumberg, The orientational freedom of molecular probes. Biophys. J. **26**, 161–194 (1979)

89. Y. Dancik, A. Favre, C.J. Loy, A.V. Zvyagin, M.S. Roberts, Use of multiphoton tomography and fluorescence lifetime imaging to investigate skin pigmentation in vivo. J. Biomed. Opt. **18**, 26022 (2013)

90. P. De Beule, D.M. Owen, H.B. Manning, C.B. Talbot, J. Requejo-Isidro, C. Dunsby, J. McGinty, R.K.P. Benninger, D.S. Elson, I. Munro, M.J. Lever, P. Anand, M.A.A. Neil, P.M.W. French, Rapid hyperspectral fluorescence lifetime imaging. Microsc. Res. Tech. **70**, 481–484 (2007)

91. H. Deschout, K. Raemdonck, J. Demeester, S. De Smedt, K. Braeckmans, FRAP in pharmaceutical research: practical guidelines and applications in drug delivery. Pharm. Res. **31**, 255–270 (2014)

92. V. Devauges, C. Marquer, S. Lecart, J.C. Cossec, M.C. Potier, E. Fort, K. Suhling, S. Lévêque-Fort, Homodimerization of amyloid precursor protein at the plasma membrane: a homoFRET study by time-resolved fluorescence anisotropy imaging. PLoS ONE **7**, e44434 (2012)

93. D.L. Dexter, A theory of sensitized luminescence in solids. J. Chem. Phys. **21**, 836–850 (1953)

94. M.A. Digman, V.R. Caiolfa, M. Zamai, E. Gratton, The phasor approach to fluorescence lifetime imaging analysis. Biophys. J. **94**, L14–L16 (2008)

95. J.A. Dix, A.S. Verkman, Mapping of fluorescence anisotropy in living cells by ratio imaging. Applications to cytoplasmic viscosity. Biophys. J. **57**, 231–240 (1990)

96. J.M. Dixon, M. Taniguchi, J.S. Lindsey, PhotochemCAD 2: a refined program with accompanying spectral databases for photochemical calculations. Photochem. Photobiol. **81**, 212–213 (2005)

97. C.G. Dos Remedios, P.D.J. Moens, Fluorescence resonance energy transfer spectroscopy is a reliable "ruler" for measuring structural changes in proteins. Dispelling the problem of the unknown orientation factor. J. Struct. Biol. **115**, 175–185 (1995)

98. K. Dowling, S.C.W. Hyde, J.C. Dainty, P.M.W. French, J.D. Hares, 2-D fluorescence lifetime imaging using a time-gated image intensifier. Opt. Commun. **135**, 27–31 (1997)

99. H. Du, R.C.A. Fuh, J. Li, L.A. Corkan, J.S. Lindsey, PhotochemCAD: a computer-aided design and research tool in photochemistry. Photochem. Photobiol. **68**, 141–142 (1998)

100. R.R. Duncan, Fluorescence lifetime imaging microscopy (FLIM) to quantify protein-protein interactions inside cells. Biochem. Soc. Trans. **34**, 679–682 (2006)

101. C. Dunsby, P.M.P. Lanigan, J. McGinty, D.S. Elson, J. Requejo-Isidro, I. Munro, N. Galletly, F. McCann, B. Treanor, B. Önfelt, D.M. Davis, M.A.A. Neil, P.M.W. French, An electronically tunable ultrafast laser source applied to fluorescence imaging and fluorescence lifetime imaging microscopy. J. Phys. D-Appl. Phys. **37**, 3296–3303 (2004)

102. A.M. Edmonds, M.A. Sobhan, V.K.A. Sreenivasan, E.A. Grebenik, J.R. Rabeau, E.M. Goldys, A.V. Zvyagin, Nano-Ruby: a promising fluorescent probe for background-free cellular imaging. Part Part Syst. Char. **30**, 506–513 (2013)

103. A. Ehn, O. Johansson, J. Bood, A. Arvidsson, B. Li, M. Alden, Fluorescence lifetime imaging in a flame. P. Combust. Inst. **33**, 807–813 (2011)

104. A. Einstein, Strahlungsemission und -absorption nach der Quantentheorie. Berichte der Deutschen Physikalischen Gesellschaft **13–14**, 3128–3323 (1916)

105. A. Einstein, Zur Quantentheorie der Strahlung. Physikalische Zeitschrift **18**, 121–128 (1917)

106. M.D. Eisaman, J. Fan, A. Migdall, S.V. Polyakov, Invited review article: single-photon sources and detectors. Rev. Sci. Instrum. **82**, 071101 (2011)

107. J. Eisinger, R.E. Dale, Interpretation of intramolecular energy transfer experiments. J. Mol. Biol. **84**, 643–647 (1974)

108. A.D. Elder, C.F. Kaminski, J.H. Frank, φ^2 FLIM: a technique for alias-free frequency domain fluorescence lifetime imaging. Opt. Express **17**, 23181–23203 (2009)

109. A.D. Elder, S.M. Matthews, J. Swartling, K. Yunus, J.H. Frank, C.M. Brennan, A.C. Fisher, C.F. Kaminski, The application of frequency-domain fluorescence lifetime imaging microscopy as a quantitative analytical tool for microfluidic devices. Opt. Express **14**, 5456–5467 (2006)

110. D. Elson, J. Requejo-Isidro, I. Munro, F. Reavell, J. Siegel, K. Suhling, P. Tadrous, R. Benninger, P. Lanigan, J. McGinty, C. Talbot, B. Treanor, S. Webb, A. Sandison, A. Wallace, D. Davis, J. Lever, M. Neil, D. Phillips, G. Stamp, P. French, Time-domain fluorescence lifetime imaging applied to biological tissue. Photoch. Photobio. Sci. **3**, 795–801 (2004)

111. D.S. Elson, I. Munro, J. Requejo-Isidro, J. McGinty, C. Dunsby, N. Galletly, G.W. Stamp, M.A.A. Neil, M.J. Lever, P.A. Kellett, A. Dymoke-Bradshaw, J. Hares, P.M.W. French, Real-time time-domain fluorescence lifetime imaging including single-shot acquisition with a segmented optical image intensifier. New J. Phys. **6**, 180 (2004)

112. E.L. Elson, Fluorescence correlation spectroscopy: past, present, future. Biophys. J. **101**, 2855–2870 (2011)

113. V. Emiliani, D. Sanvitto, M. Tramier, T. Piolot, Z. Petrášek, K. Kemnitz, C. Duneux, M. Coppey-Moisan, Low-intensity two-dimensional imaging of fluorescence lifetimes in living cells. Appl. Phys. Lett. **83**, 2471–2473 (2003)

114. A. Esposito, Beyond range: innovating fluorescence microscopy. Remote Sens-Basel **4**, 111–119 (2012)

115. A. Esposito, A.N. Bader, S.C. Schlachter, D.J. van den Heuvel, G.S. Kaminski Schierle, A.R. Venkitaraman, C.F. Kaminski, H.C. Gerritsen, Design and application of a confocal microscope for spectrally resolved anisotropy imaging. Opt. Express **19**, 2546–2555 (2011)

116. A. Esposito, H.C. Gerritsen, F.S. Wouters, Optimizing frequency-domain fluorescence lifetime sensing for high-throughput applications: photon economy and acquisition speed. J. Opt. Soc. Am. A **24**, 3261 (2007)

117. A.D. Estrada, A. Ponticorvo, T.N. Ford, A.K. Dunn, Microvascular oxygen quantification using two-photon microscopy. Opt. Lett. **33**, 1038–1040 (2008)

118. O. Faklaris, V. Joshi, T. Irinopoulou, P. Tauc, M. Sennour, H. Girard, C. Gesset, J.C. Arnault, A. Thorel, J.P. Boudou, P.A. Curmi, F. Treussart, Photoluminescent diamond nanoparticles for cell labeling: study of the uptake mechanism in mammalian cells. ACS Nano **3**, 3955–3962 (2009)

119. S.M. Fernandez, R.D. Berlin, Cell-surface distribution of lectin receptors determined by resonance energy-transfer. Nature **264**, 411–415 (1976)

120. F. Festy, S.M. Ameer-Beg, T. Ng, K. Suhling, Imaging proteins in vivo using fluorescence lifetime microscopy. Mol. BioSyst. **3**, 381–391 (2007)

121. O.S. Finikova, A.Y. Lebedev, A. Aprelev, T. Troxler, F. Gao, C. Garnacho, S. Muro, R.M. Hochstrasser, S.A. Vinogradov, Oxygen microscopy by two-photon-excited phosphorescence. ChemPhysChem **9**, 1673–1679 (2008)
122. E. Fiserova, M. Kubala, Mean fluorescence lifetime and its error. J. Lumines. **132**, 2059–2064 (2012)
123. J.J. Fisz, Another look at magic-angle-detected fluorescence and emission anisotropy decays in fluorescence microscopy. J. Phys. Chem. A **111**, 12867–12870 (2007)
124. J.J. Fisz, Another treatment of fluorescence polarization microspectroscopy and imaging. J. Phys. Chem. A **113**, 3505–3516 (2009)
125. J.J. Fisz, Fluorescence polarization spectroscopy at combined high-aperture excitation and detection: application to one-photon-excitation fluorescence microscopy. J. Phys. Chem. A **111**, 8606–8621 (2007)
126. D. Fixler, Y. Namer, Y. Yishay, M. Deutsch, Influence of fluorescence anisotropy on fluorescence intensity and lifetime measurement: theory, simulations and experiments. IEEE Trans. Biomed. Eng. **53**, 1141–1152 (2006)
127. T. Förster, Energiewanderung und Fluoreszenz, Naturwissenschaften **33**, 166–175, translated into English by K. Suhling, J. Biomed. Opt. **17**, 011002 (2012)
128. T. Förster, G. Hoffmann, Die Viskositätsabhängigkeit der Fluoreszenzquantenausbeuten einiger Farbstoffsysteme. Zeitschrift für Physikalische Chemie Neue Folge **75**, 63–76 (1971)
129. E. Fort, S. Gresillon, Surface enhanced fluorescence. J. Phys. D-Appl. Phys. **41**, 013001 (2008)
130. T.H. Foster, B.D. Pearson, S. Mitra, C.E. Bigelow, Fluorescence anisotropy imaging reveals localization of meso-tetrahydroxyphenyl chlorin in the nuclear envelope. Photochem. Photobiol. **81**, 1544–1547 (2005)
131. T. French, P.T.C. So, D.J. Weaver, T. Coelho-Sampaio, E. Gratton, E.W. Voss, J. Carrero, Two-photon fluorescence lifetime imaging microscopy of macrophage-mediated antigen processing. J. Micros. **185**, 339–353 (1997)
132. G.O. Fruhwirth, S. Ameer-Beg, R. Cook, T. Watson, T. Ng, F. Festy, Fluorescence lifetime endoscopy using TCSPC for the measurement of FRET in live cells. Opt. Express **18**, 11148–11158 (2010)
133. K. Funk, A. Woitecki, C. Franjic-Wurtz, T. Gensch, F. Mohrlen, S. Frings, Modulation of chloride homeostasis by inflammatory mediators in dorsal root ganglion neurons. Mol. Pain **4**, 32 (2008)
134. B.M. Gadella, T.W.J. Gadella, B. Colenbrander, L.M.G. Van Golde, Visualization and quantification of glycolipid polarity dynamics in the plasma membrane of the mammalian spermatozoon. J. Cell Sci. **107**, 2151–2163 (1994)
135. B.M. Gadella, M. Lopez-Cardozo, L.M.G. Van Golde, B. Colenbrander, Glycolipid migration from the apical to the equatorial subdomains of the sperm head plasma membrane precedes the acrosome reaction. Evidence for a primary capacitation event in boar spermatozoa. J. Cell Sci. **108**, 935–945 (1995)
136. T.W.J. Gadella, T.M. Jovin, R.M. Clegg, Fluorescence lifetime imaging microscopy (FLIM) —spatial resolution of structures on the nanosecond timescale. Biophys. Chem. **48**, 221–239 (1993)
137. C.G. Galbraith, J.A. Galbraith, Super-resolution microscopy at a glance. J. Cell Sci. **124**, 1607–1611 (2011)
138. N.P. Galletly, J. McGinty, C. Dunsby, F. Teixeira, J. Requejo-Isidro, I. Munro, D.S. Elson, M.A.A. Neil, A.C. Chu, P.M.W. French, G.W. Stamp, Fluorescence lifetime imaging distinguishes basal cell carcinoma from surrounding uninvolved skin. Brit. J. Dermatol. **159**, 152–161 (2008)
139. S. Ganguly, A.H.A. Clayton, A. Chattopadhyay, Fixation alters fluorescence lifetime and anisotropy of cells expressing EYFP-tagged serotonin(1A) receptor. Biochem. Biophys. Res. Commun. **405**, 234–237 (2011)

140. I. Gautier, M. Tramier, C. Durieux, J. Coppey, R.B. Pansu, J.C. Nicolas, K. Kemnitz, M. Coppey-Moisan, Homo-FRET microscopy in living cells to measure monomer-dimer transition of GFP-tagged proteins. Biophys. J. **80**, 3000–3008 (2001)

141. E. Gaviola, Die Abklingungszeiten der Fluoreszenz von Farbstofflösungen. Ann. Phys.-Berlin **386**, 681–710 (1926)

142. M.L. Gee, L. Lensun, T.A. Smith, C.A. Scholes, Time-resolved evanescent wave-induced fluorescence anisotropy for the determination of molecular conformational changes of proteins at an interface. Eur. Biophys. J. Biophy. **33**, 130–139 (2004)

143. H.C. Gerritsen, N.A.H. Asselbergs, A.V. Agronskaia, W.G.J.H.M. Van Sark, Fluorescence lifetime imaging in scanning microscopes: acquisition speed, photon economy and lifetime resolution. J. Micros. **206**, 218–224 (2002)

144. H.C. Gerritsen, R. Sanders, A. Draaijer, C. Ince, Y.K. Levine, Fluorescence lifetime imaging of oxygen in living cells. J. Fluores. **7**, 11–16 (1997)

145. K.P. Ghiggino, J.A. Hutchison, S.J. Langford, M.J. Latter, M.A.P. Lee, P.R. Lowenstern, C. Scholes, M. Takezaki, B.E. Wilman, Porphyrin-based molecular rotors as fluorescent probes of nanoscale environments. Adv. Funct. Mater. **17**, 805–813 (2007)

146. V.V. Ghukasyan, F.J. Kao, Monitoring cellular metabolism with fluorescence lifetime of reduced nicotinamide adenine dinucleotide. J. Phys. Chem. C **113**, 11532–11540 (2009)

147. D. Gilbert, C. Franjic-Wurtz, K. Funk, T. Gensch, S. Frings, F. Mohrlen, Differential maturation of chloride homeostasis in primary afferent neurons of the somatosensory system. Int. J. Dev. Neurosci. **25**, 479–489 (2007)

148. A.V. Gohar, R.F. Cao, P. Jenkins, W.Y. Li, J.P. Houston, K.D. Houston, Subcellular localization-dependent changes in EGFP fluorescence lifetime measured by time-resolved flow cytometry. Biomed. Opt. Express **4**, 1390–1400 (2013)

149. E.M. Goldys, *Fluorescence Applications in Biotechnology and Life Sciences* (Wiley-Backwell, Hoboken, 2009)

150. O. Golfetto, E. Hinde, E. Gratton, Laurdan fluorescence lifetime discriminates cholesterol content from changes in fluidity in living cell membranes. Biophys. J. **104**, 1238–1247 (2013)

151. A.H. Gough, D.L. Taylor, Fluorescence anisotropy imaging microscopy maps calmodulin-binding during cellular contraction and locomotion. J. Cell Biol. **121**, 1095–1107 (1993)

152. C.C. Gradinaru, D.O. Marushchak, M. Samim, U.J. Krull, Fluorescence anisotropy: from single molecules to live cells. Analyst **135**, 452–459 (2010)

153. E.M. Graham, K. Iwai, S. Uchiyama, A.P. de Silva, S.W. Magennis, A.C. Jones, Quantitative mapping of aqueous microfluidic temperature with sub-degree resolution using fluorescence lifetime imaging microscopy. Lab Chip **10**, 1267–1273 (2010)

154. D.M. Grant, W. Zhang, E.J. McGhee, T.D. Bunney, C.B. Talbot, S. Kumar, I. Munro, C. Dunsby, M.A.A. Neil, M. Katan, P.M.W. French, Multiplexed FRET to image multiple signaling events in live cells. Biophys. J. **95**, L69–L71 (2008)

155. E. Gratton, S. Breusegem, J. Sutin, Q. Ruan, N. Barry, Fluorescence lifetime imaging for the two-photon microscope: time-domain and frequency-domain methods. J. Biomed. Opt. **8**, 381–390 (2003)

156. H.E. Grecco, K.A. Lidke, R. Heintzmann, D.S. Lidke, C. Spagnuolo, O.E. Martinez, E.A. Jares-Erijman, T.M. Jovin, Ensemble and single particle photophysical proper-ties (Two-Photon excitation, anisotropy, FRET, lifetime, spectral conversion) of commercial quantum dots in solution and in live cells. Microsc. Res. Tech. **65**, 169–179 (2004)

157. M. Green, Semiconductor quantum dots as biological imaging agents. Angew. Chem. **43**, 4129–4131 (2004)

158. M. Green, P. Howes, C. Berry, O. Argyros, M. Thanou, Simple conjugated polymer nanoparticles as biological labels. Proc. R. Soc. A Math. Phys. Eng. Sci. **465**, 2751–2759 (2009)

159. K. Greger, M.J. Neetz, E.G. Reynaud, E.H.K. Stelzer, Three-dimensional fluorescence lifetime imaging with a single plane illumination microscope provides an improved signal to noise ratio. Opt. Express **19**, 20743–20750 (2011)

160. J. Gu, C.Y. Fu, B.K. Ng, S. Gulam Razul, S.K. Lim, Quantitative diagnosis of cervical neoplasia using fluorescence lifetime imaging on haematoxylin and eosin stained tissue sections, J. Biophotonics **7**, 483–491 (2013)
161. K.I. Gutkowski, M.L. Japas, P.F. Aramendia, Fluorescence of dicyanovinyl julolidine in a room-temperature ionic liquid. Chem. Phys. Lett. **426**, 329–333 (2006)
162. T. Ha, T.A. Laurence, D.S. Chemla, S. Weiss, Polarization spectroscopy of single fluorescent molecules. J. Phys. Chem. B **103**, 6839–6850 (1999)
163. R.H. Hadfield, Single-photon detectors for optical quantum information applications. Nat. Photonics **3**, 696–705 (2009)
164. M. Haidekker, T.P. Brady, D. Lichlyter, E.A. Theodorakis, A ratiometric fluorescent viscosity sensor. J. Am. Chem. Soc. **128**, 398–399 (2006)
165. M.A. Haidekker, T. Ling, M. Anglo, H.Y. Stevens, J.A. Frangos, E.A. Theodorakis, New fluorescent probes for the measurement of cell membrane viscosity. Chem. Biol. **8**, 123–131 (2001)
166. M.A. Haidekker, M. Nipper, A. Mustafic, D. Lichlyter, M. Dakanali, E.A. Theodorakis, Dyes with segmental mobility: molecular rotors, in *Advanced Fluorescence Reporters in Chemistry and Biology I. Fundamentals and Molecular Design*, ed by A.P. Demchenko (Springer, Berlin, 2010), pp. 267–308
167. M.A. Haidekker, E.A. Theodorakis, Molecular rotors-fluorescent biosensors for viscosity and flow. Org. Biomol. Chem. **5**, 1669–1678 (2007)
168. M.A. Haidekker, A.G. Tsai, T. Brady, H.Y. Stevens, J.A. Frangos, E. Theodorakis, M. Intaglietta, A novel approach to blood plasma viscosity measurement using fluorescent molecular rotors, Am. J. Physiol.-Heart Circul. Physiol. **282**, H1609–H1614 (2002)
169. Q.S. Hanley, D.J. Arndt-Jovin, T.M. Jovin, Spectrally resolved fluorescence lifetime imaging microscopy. Appl. Spectrosc. **56**, 155–166 (2002)
170. Q.S. Hanley, A.H. Clayton, AB-plot assisted determination of fluorophore mixtures in a fluorescence lifetime microscope using spectra or quenchers. J. Micros. **218**, 62–67 (2005)
171. Q.S. Hanley, V. Subramaniam, D.J. Arndt-Jovin, T.M. Jovin, Fluorescence lifetime imaging: multi-point calibration, minimum resolvable differences, and artifact suppression. Cytometry **43**, 248–260 (2001)
172. K.M. Hanson, M.J. Behne, N.P. Barry, T.M. Mauro, E. Gratton, R.M. Clegg, Two-photon fluorescence lifetime imaging of the skin stratum corneum pH gradient. Biophys. J. **83**, 1682–1690 (2002)
173. A.G. Harpur, F.S. Wouters, P.I. Bastiaens, Imaging FRET between spectrally similar GFP molecules in single cells. Nat. Biotechnol. **19**, 167–169 (2001)
174. H. Harris, *The Birth of the Cell* (Yale University Press, New Haven, 1999)
175. E.N. Harvey, *A History of Luminescence from the Earliest Times Until 1900* (American Philosophical Society, Philadelphia, 1957)
176. J. Hedstrom, S. Sedarus, F.G. Prendergast, Measurements of fluorescence lifetimes by use of a hybrid time-correlated and multifrequency phase fluorometer. Biochemistry **27**, 6203–6208 (1988)
177. A.A. Heikal, S.T. Hess, W.W. Webb, Multiphoton molecular spectroscopy and excited-state dynamics of enhanced green fluorescent protein (EGFP): acid-base specificity. Chem. Phys. **274**, 37–55 (2001)
178. R. Heintzmann, G. Ficz, Breaking the resolution limit in light microscopy. Briefings Funct. Genomics Proteomics **5**, 289–301 (2006)
179. B. Herman, P. Wodnicki, S. Kwon, A. Periasamy, G.W. Gordon, N. Mahajan, W. Xue Feng, Recent developments in monitoring calcium and protein interactions in cells using fluorescence lifetime microscopy. J. Fluores. **7**, 85–92 (1997)
180. A.M. Heskes, C.N. Lincoln, J.Q.D. Goodger, I.E. Woodrow, T.A. Smith, Multiphoton fluorescence lifetime imaging shows spatial segregation of secondary metabolites in Eucalyptus secretory cavities. J. Micros. **247**, 33–42 (2012)

181. C. Hille, M. Berg, L. Bressel, D. Munzke, P. Primus, H.G. Löhmannsröben, C. Dosche, Time-domain fluorescence lifetime imaging for intracellular pH sensing in living tissues. Anal. Bioanal. Chem. **391**, 1871–1879 (2008)

182. S. Hirayama, D. Phillips, Correction for refractive index in the comparison of radiative lifetimes in vapour and solution phases. J. Photochem. **12**, 139–145 (1980)

183. L.M. Hirvonen, F. Festy, K. Suhling, Wide-field time-correlated single-photon counting (TCSPC) lifetime microscopy with microsecond time resolution. Opt. Lett. **39**, 5602–5605 (2014)

184. L.M. Hirvonen, T.A. Smith, Imaging on the nanoscale: super-resolution fluorescence microscopy. Aust. J. Chem. **64**, 41–45 (2011)

185. N.A. Hosny, D.A. Lee, M.M. Knight, Single photon counting fluorescence lifetime detection of pericellular oxygen concentrations. J. Biomed. Opt. **17**, 016007 (2012)

186. N.A. Hosny, G. Mohamedi, P. Rademeyer, J. Owen, Y. Wu, M.X. Tang, R.J. Eckersley, E. Stride, M.K. Kuimova, Mapping microbubble viscosity using fluorescence lifetime imaging of molecular rotors. Proc. Natl. Acad. Sci. USA **110**, 9225–9230 (2013)

187. B. Hötzer, R. Ivanov, S. Altmeier, R. Kappl, G. Jung, Determination of copper(II) ion concentration by lifetime measurements of green fluorescent protein. J. Fluoresc. **21**, 2143–2153 (2011)

188. B. Hötzer, R. Ivanov, T. Brumbarova, P. Bauer, G. Jung, Visualization of Cu^{2+} uptake and release in plant cells by fluorescence lifetime imaging microscopy. FEBS J. **279**, 410–419 (2012)

189. S.S. Howard, A. Straub, N.G. Horton, D. Kobat, C. Xu, Frequency-multiplexed in vivo multiphoton phosphorescence lifetime microscopy. Nat. Photonics **7**, 33–37 (2013)

190. P. Howes, M. Green, J. Levitt, K. Suhling, M. Hughes, Phospholipid encapsulated semiconducting polymer nanoparticles: their use in cell imaging and protein attachment. J. Am. Chem. Soc. **132**, 3989–3996 (2010)

191. P.D. Howes, R. Chandrawati, M.M. Stevens, Colloidal nanoparticles as advanced biological sensors. Science **346**, 1247390 (2014)

192. S. Hrabetova, Extracellular diffusion is fast and isotropic in the stratum radiatum of hippocampal CA1 region in rat brain slices. Hippocampus **15**, 441–450 (2005)

193. G. Hungerford, A. Allison, D. McLoskey, M.K. Kuimova, G. Yahioglu, K. Suhling, Monitoring Sol-to-Gel transitions via fluorescence lifetime determination using viscosity sensitive fluorescent probes. J. Phys. Chem. B **113**, 12067–12074 (2009)

194. G. Hungerford, D.J.S. Birch, Single-photon timing detectors for fluorescence lifetime spectroscopy. Meas. Sci. Technol. **7**, 121–135 (1996)

195. J. Hunt, A.H. Keeble, R.E. Dale, M.K. Corbett, R.L. Beavil, J. Levitt, M.J. Swann, K. Suhling, S. Ameer-Beg, B.J. Sutton, A.J. Beavil, A fluorescent biosensor reveals conformational changes in human immunoglobulin E Fc. implications for mechanisms of receptor binding, inhibition and allergen recognition. J. Biol. Chem. **287**, 17459–17470 (2012)

196. A.A. Istratov, O.F. Vyvenko, Exponential analysis in physical phenomena. Rev. Sci. Instrum. **70**, 1233–1257 (1999)

197. A. Jablonski, Über den Mechanismus der Photolumineszenz von Farbstoffphosphoren. Z. Phys. **94**, 38–46 (1935)

198. O. Jagutzki, A. Cerezo, A. Czasch, R. Dörner, M. Hattass, M. Huang, V. Mergel, U. Spillmann, K. Ullmann-Pfleger, T. Weber, H. Schmidt-Böcking, G.D.W. Smith, Multiple hit readout of a microchannel plate detector with a three-layer delay-line anode. IEEE Trans. Nucl. Sci. **49**, 2477–2483 (2002)

199. D.M. Jameson, J.A. Ross, Fluorescence polarization/anisotropy in diagnostics and imaging. Chem. Rev. **110**, 2685–2708 (2010)

200. E.A. Jares-Erijman, T.M. Jovin, FRET imaging. Nat. Biotechnol. **21**, 1387–1396 (2003)

201. L. Joosen, M.A. Hink, T.W.J. Gadella, J. Goedhart, Effect of fixation procedures on the fluorescence lifetimes of Aequorea victoria derived fluorescent proteins. J. Micros. **256**, 166–176 (2014)

202. T.M. Jovin, D.J. Arndt-Jovin, FRET microscopy: digital imaging of fluorescence resonance energy transfer, in *Cell structure and function by microspectrofluorometry*, ed by E. Kohen, J.G. Hirschberg, J.S. Ploem (Academic Press, London, 1989), pp. 99–117

203. G. Jung, Y.Z. Ma, B.S. Prall, G.R. Fleming, Ultrafast fluorescence depolarisation in the yellow fluorescent protein due to its dimerisation. ChemPhysChem **6**, 1628–1632 (2005)

204. C. Jüngst, M. Klein, A. Zumbusch, Long-term live cell microscopy studies of lipid droplet fusion dynamics in adipocytes. J. Lipid Res. **54**, 3419–3429 (2013)

205. G.S. KaminskiSchierle, C.W. Bertoncini, F.T.S. Chan, A.T. van der Goot, S. Schwedler, J. Skepper, S. Schlachter, T. van Ham, A. Esposito, J.R. Kumita, E.A.A. Nollen, C.M. Dobson, C.F. Kaminski, A FRET sensor for non-invasive imaging of amyloid formation in vivo. Chemphyschem **12**, 673–680 (2011)

206. H. Kaneko, I. Putzier, S. Frings, U.B. Kaupp, T. Gensch, Chloride accumulation in mammalian olfactory sensory neurons. J. Neurosci. **24**, 7931–7938 (2004)

207. M. Kasha, Characterization of electronic transitions in complex molecules. Discuss. Faraday Soc. **9**, 14–19 (1950)

208. V. Katsoulidou, A. Bergmann, W. Becker, How fast can TCSPC FLIM be made? in *SPIE Proc 6771*, (2007), B7710

209. S. Kawata, Y. Inouye, T. Ichimura, Near-field optics and spectroscopy for molecular nano-imaging. Sci. Prog. **87**, 25–49 (2004)

210. S.M. Keating, T.G. Wensel, Nanosecond fluorescence microscopy. Emission kinetics of fura-2 in single cells. Biophys. J. **59**, 186–202 (1991)

211. M. Köllner, J. Wolfrum, How many photons are necessary for fluorescence-lifetime measurements? Chem. Phys. Lett. **200**, 199–204 (1992)

212. K. König, H. Schneckenburger, R. Hibst, Time-gated in vivo autofluorescence imaging of dental caries. Cell. Mol. Biol. **45**, 233–239 (1999)

213. M. Koshioka, K. Sasaki, H. Masuhara, Time-dependent fluorescence depolarization analysis in 3-dimensional microspectroscopy. Appl. Spectrosc. **49**, 224–228 (1995)

214. S.V. Koushik, S.S. Vogel, Energy migration alters the fluorescence lifetime of Cerulean: implications for fluorescence lifetime imaging Förster resonance energy transfer measurements. J. Biomed. Opt. **13**, 031204 (2008)

215. M. Kress, T. Meier, R. Steiner, F. Dolp, R. Erdmann, U. Ortmann, A. Rück, Time-resolved microspectrofluorometry and fluorescence lifetime imaging of photosensitizers using picosecond pulsed diode lasers in laser scanning microscopes. J. Biomed. Opt. **8**, 26–32 (2003)

216. R.V. Krishnan, E. Biener, J.H. Zhang, R. Heckel, B. Herman, Probing subtle fluorescence dynamics in cellular proteins by streak camera based fluorescence lifetime imaging microscopy. Appl. Phys. Lett. **83**, 4658–4660 (2003)

217. R.V. Krishnan, A. Masuda, V.E. Centonze, B. Herman, Quantitative imaging of protein-protein interactions by multiphoton fluorescence lifetime imaging microscopy using a streak camera. J. Biomed. Opt. **8**, 362–367 (2003)

218. R.V. Krishnan, H. Saitoh, H. Terada, V.E. Centonze, B. Herman, Development of a multiphoton fluorescence lifetime imaging microscopy system using a streak camera. Rev. Sci. Instrum. **74**, 2714–2721 (2003)

219. K.V. Kuchibhotla, C.R. Lattarulo, B.T. Hyman, B.J. Bacskai, Synchronous hyperactivity and intercellular calcium waves in astrocytes in Alzheimer mice. Science **323**, 1211–1215 (2009)

220. M.K. Kuimova, Mapping viscosity in cells using molecular rotors. Phys. Chem. Chem. Phys. **14**, 12671–12686 (2012)

221. M.K. Kuimova, Molecular rotors image intracellular viscosity. Chimia **66**, 159–165 (2012)

222. M.K. Kuimova, S.W. Botchway, A.W. Parker, M. Balaz, H.A. Collins, H.L. Anderson, K. Suhling, P.R. Ogilby, Imaging intracellular viscosity of a single cell during photoinduced cell death. Nat. Chem. **1**, 69–73 (2009)

223. M.K. Kuimova, G. Yahioglu, J.A. Levitt, K. Suhling, Molecular rotor measures viscosity of live cells via fluorescence lifetime imaging. J. Am. Chem. Soc. **130**, 6672–6673 (2008)

224. F. Kukita, Solvent effects on squid sodium channels are attributable to movements of a flexible protein structure in gating currents and to hydration in a pore. J. Physiol.-London **522**, 357–373 (2000)

225. C.E. Kung, J.K. Reed, Fluorescent molecular rotors—a new class of probes for tubulin structure and assembly. Biochemistry **28**, 6678–6686 (1989)

226. C.E. Kung, J.K. Reed, Microsviscosity measurements of phospholipid bilayers using fluorescent dyes that undergo torsional relaxation. Biochemistry **25**, 6114–6121 (1986)

227. Y. Kuo, T.-Y. Hsu, Y.-C. Wu, J.-H. Hsu, H.-C. Chang, Fluorescence lifetime imaging microscopy of nanodiamonds in vivo, in *SPIE Proc 8635* (2013), 863503

228. E.S. Kwak, T.J. Kang, D.A.V. Bout, Fluorescence lifetime imaging with near-field scanning optical microscopy. Anal. Chem. **73**, 3257–3262 (2001)

229. M. Lahn, C. Dosche, C. Hille, Two-photon microscopy and fluorescence lifetime imaging reveal stimulus-induced intracellular Na^+ and Cl^- changes in cockroach salivary acinar cells. Am. J. Physiol.-Cell Physiol. **300**, C1323–C1336 (2011)

230. J.R. Lakowicz, *Principles of Fluorescence Spectroscopy*, 3rd edn. (Springer, New York, 2006)

231. J.R. Lakowicz, Radiative decay engineering 5: metal-enhanced fluorescence and plasmon emission. Anal. Biochem. **337**, 171–194 (2005)

232. J.R. Lakowicz, H. Szmacinski, K. Nowaczyk, K.W. Berndt, M. Johnson, Fluorescence lifetime imaging. Anal. Biochem. **202**, 316–330 (1992)

233. J.R. Lakowicz, H. Szmacinski, K. Nowaczyk, M.L. Johnson, Fluorescence lifetime imaging of calcium using Quin-2. Cell Calcium **13**, 131–147 (1992)

234. J.R. Lakowicz, H. Szmacinski, K. Nowaczyk, M.L. Johnson, Fluorescence lifetime imaging of free and protein-bound NADH. Proc. Natl. Acad. Sci. USA **89**, 1271–1275 (1992)

235. J.R. Lakowicz, H. Szmacinski, K. Nowaczyk, W.J. Lederer, Fluorescence lifetime imaging of intracellular calcium in COS cells using Quin-2. Cell Calcium **15**, 7–27 (1994)

236. K.Y. Law, Fluorescence probe for micro-environments—a new probe for micelle solvent parameters and premicellar aggregates. Photochem. Photobiol. **33**, 799–806 (1981)

237. M.D. Lesoine, S. Bose, J.W. Petrich, E.A. Smith, Supercontinuum stimulated emission depletion fluorescence lifetime imaging. J. Phys. Chem. B **116**, 7821–7826 (2012)

238. S. Lévêque-Fort, D.N. Papadopoulos, S. Forget, F. Balembois, P. Georges, Fluorescence lifetime imaging with a low-repetition-rate passively mode-locked diodepumped Nd:YVO_4 oscillator. Opt. Lett. **30**, 168–170 (2005)

239. J.A. Levitt, P.H. Chung, D.R. Alibhai, K. Suhling, Simultaneous measurements of fluorescence lifetimes, anisotropy and FRAP recovery curves, in *SPIE Proc 7902* (2011), 79020Y

240. J.A. Levitt, P.H. Chung, M.K. Kuimova, G. Yahioglu, Y. Wang, J.L. Qu, K. Suhling, Fluorescence anisotropy of molecular rotors. Chemphyschem **12**, 662–672 (2011)

241. J.A. Levitt, M.K. Kuimova, G. Yahioglu, P.H. Chung, K. Suhling, D. Phillips, Membrane-bound molecular rotors measure viscosity in live cells via fluorescence lifetime imaging. J. Phys. Chem. C **113**, 11634–11642 (2009)

242. J.A. Levitt, D.R. Matthews, S.M. Ameer-Beg, K. Suhling, Fluorescence lifetime and polarization-resolved imaging in cell biology. Curr. Opin. Biotechnol. **20**, 28–36 (2009)

243. Q. Li, T. Ruckstuhl, S. Seeger, Deep-UV laser-based fluorescence lifetime imaging microscopy of single molecules. J. Phys. Chem. B **108**, 8324–8329 (2004)

244. W. Li, Y. Wang, H.R. Shao, Y.H. He, H. Ma, Probing rotation dynamics of biomolecules using polarization based fluorescence microscopy. Microsc. Res. Tech. **70**, 390–395 (2007)

245. G. Liaugaudas, A.T. Collins, K. Suhling, G. Davies, R. Heintzmann, Luminescence-lifetime mapping in diamond. J. Phys.: Condens. Matter **21**, 364210 (2009)

246. G. Liaugaudas, G. Davies, K. Suhling, R.U.A. Khan, D.J.F. Evans, Luminescence lifetimes of neutral nitrogen-vacancy centres in synthetic diamond containing nitrogen. J. Phys.-Condes. Matter **24**, 435503 (2012)

247. D.S. Lidke, P. Nagy, B.G. Barisas, R. Heintzmann, J.N. Post, K.A. Lidke, A.H.A. Clayton, D.J. Arndt-Jovin, T.M. Jovin, Imaging molecular interactions in cells by dynamic and static fluorescence anisotropy (rFLIM and emFRET). Biochem. Soc. Trans. **31**, 1020–1027 (2003)

248. H.J. Lin, P. Herman, J.R. Lakowicz, Fluorescence lifetime-resolved pH imaging of living cells. Cytometry **52A**, 77–89 (2003)

249. L.L. Lin, J.E. Grice, M.K. Butler, A.V. Zvyagin, W. Becker, T.A. Robertson, H.P. Soyer, M.S. Roberts, T.W. Prow, Time-correlated single photon counting for simultaneous monitoring of zinc oxide nanoparticles and NAD(P)H in intact and barrier-disrupted volunteer skin. Pharm. Res. **28**, 2920–2930 (2011)

250. P.Y. Lin, S.S. Lee, C.S. Chang, F.J. Kao, Long working distance fluorescence lifetime imaging with stimulated emission and electronic time delay. Opt. Express **20**, 11445–11450 (2012)

251. P.Y. Lin, Y.C. Lin, C.S. Chang, F.J. Kao, Fluorescence lifetime imaging microscopy with subdiffraction-limited resolution. Jpn. J. Appl. Phys. **52**, 028004(2013)

252. L. Liu, J. Qu, Z. Lin, L. Wang, Z. Fu, B. Guo, H. Niu, Simultaneous time- and spectrum-resolved multifocal multiphoton microscopy. Appl. Phys. B-Lasers Opt. **84**, 379–383 (2006)

253. P. Loison, N.A. Hosny, P. Gervais, D. Champion, M.K. Kuimova, J.M. Perrier-Cornet, Direct investigation of viscosity of an atypical inner membrane of Bacillus spores: a molecular rotor/FLIM study. Biochim. Biophys. Acta **1828**, 2436–2443 (2013)

254. R.O. Loutfy, Fluorescence probes for polymer free-volume. Pure Appl. Chem. **58**, 1239–1248 (1986)

255. R.O. Loutfy, B.A. Arnold, Effect of viscosity and temperature on torsional relaxation of molecular rotors. J. Phys. Chem. **86**, 4205–4211 (1982)

256. J. Lu, C.L. Liotta, C.A. Eckert, Spectroscopically probing microscopic solvent properties of room-temperature ionic liquids with the addition of carbon dioxide. J. Phys. Chem. A **107**, 3995–4000 (2003)

257. K. Luby-Phelps, S. Mujumdar, R. Mujumdar, L.A. Ernst, W. Galbraith, A.S. Waggoner, A novel fluorescence ratiometric method confirms the low solvent viscosity of the cytoplasma. Biophys. J. **65**, 236–242 (1993)

258. Y.J. Ma, P. Rajendran, C. Blum, Y. Cesa, N. Gartmann, D. Bruhwiler, V. Subramaniam, Microspectroscopic analysis of green fluorescent proteins infiltrated into mesoporous silica nanochannels. J. Colloid Interface Sci. **356**, 123–130 (2011)

259. S.W. Magennis, E.M. Graham, A.C. Jones, Quantitative spatial mapping of mixing in microfluidic systems. Angew. Chem. Int. Ed. **44**, 6512–6516 (2005)

260. M. Magzoub, H. Zhang, J.A. Dix, A.S. Verkman, Extracellular space volume measured by two-color pulsed dye infusion with microfiberoptic fluorescence photodetection. Biophys. J. **96**, 2382–2390 (2009)

261. M. Malley, A heated controversy on cold light. Arch. Hist. Exact. Sci. **42**, 173–186 (1991)

262. S. Mao, R.K.P. Benninger, Y.L. Yan, C. Petchprayoon, D. Jackson, C.J. Easley, D.W. Piston, G. Marriott, Optical lock-in detection of FRET using synthetic and genetically encoded optical switches. Biophys. J. **94**, 4515–4524 (2008)

263. L. Marcu, P.M.W. French, D. Elson, *Fluorescence Lifetime Spectroscopy and Imaging: Principles and Applications in Biomedical Diagnostics* (CRC Press, Boca Raton, 2014)

264. B.R. Masters, P.T.C. So, E. Gratton, Multiphoton excitation fluorescence microscopy and spectroscopy of in vivo human skin. Biophys. J. **72**, 2405–2412 (1997)

265. D.R. Matthews, L.M. Carlin, E. Ofo, P.R. Barber, B. Vojnovic, M. Irving, T. Ng, S.M. Ameer-Beg, Time-lapse FRET microscopy using fluorescence anisotropy. J. Microsc. **237**, 51–62 (2010)

266. A.L. Mattheyses, A.D. Hoppe, D. Axelrod, Polarized fluorescence resonance energy transfer microscopy. Biophys. J. **87**, 2787–2797 (2004)

267. M. Maus, M. Cotlet, J. Hofkens, T. Gensch, F.C. De Schryver, J. Schaffer, C.A.M. Seidel, An experimental comparison of the maximum likelihood estimation and nonlinear least-squares fluorescence lifetime analysis of single molecules. Anal. Chem. **73**, 2078–2086 (2001)

268. A. Mayevsky, B. Chance, Oxidation-reduction states of NADH in vivo: from animals to clinical use. Mitochondrion **7**, 330–339 (2007)

269. G. McConnell, J.M. Girkin, S.M. Ameer-Beg, P.R. Barber, B. Vojnovic, T. Ng, A. Banerjee, T.F. Watson, R.J. Cook, Time-correlated single-photon counting fluorescence lifetime confocal imaging of decayed and sound dental structures with a white-light supercontinuum source. J. Microsc. **225**, 126–136 (2007)

270. B.J. McCranor, H. Szmacinski, H.H. Zeng, A.K. Stoddard, T. Hurst, C.A. Fierke, J.R. Lakowicz, R.B. Thompson, Fluorescence lifetime imaging of physiological free Cu(II) levels in live cells with a Cu(II)-selective carbonic anhydrase-based biosensor. Metallomics **6**, 1034–1042 (2014)

271. J. McGinty, D.W. Stuckey, V.Y. Soloviev, R. Laine, M. Wylezinska-Arridge, D.J. Wells, S. R. Arridge, P.M.W. French, J.V. Hajnal, A. Sardini, In vivo fluorescence lifetime tomography of a FRET probe expressed in mouse. Biomed. Opt. Express **2**, 1907–1917 (2011)

272. D.A. Mendels, E.M. Graham, S.W. Magennis, A.C. Jones, F. Mendels, Quantitative comparison of thermal and solutal transport in a T-mixer by FLIM and CFD. Microfluid. Nanofluid. **5**, 603–617 (2008)

273. X. Michalet, A. Cheng, J. Antelman, M. Suyama, K. Arisaka, S. Weiss, Hybrid photodetector for single-molecule spectroscopy and microscopy, in *SPIE Proc 6862* (2008), 68620F

274. X. Michalet, R.A. Colyer, J. Antelman, O.H.W. Siegmund, A. Tremsin, J.V. Vallerga, S. Weiss, Single-quantum dot imaging with a photon counting camera. Curr. Pharm. Biotechnol. **10**, 543–558 (2009)

275. X. Michalet, R.A. Colyer, G. Scalia, A. Ingargiola, R. Lin, J.E. Millaud, S. Weiss, O.H.W. Siegmund, A.S. Tremsin, J.V. Vallerga, A. Cheng, M. Levi, D. Aharoni, K. Arisaka, F. Villa, F. Guerrieri, F. Panzeri, I. Rech, A. Gulinatti, F. Zappa, M. Ghioni, S. Cova, Development of new photon-counting detectors for single-molecule fluorescence microscopy. Philos. Trans. R. Soc. B **368**, 20120035 (2013)

276. X. Michalet, F.F. Pinaud, L.A. Bentolila, J.M. Tsay, S. Doose, J.J. Li, G. Sundaresan, A.M. Wu, S.S. Gambhir, S. Weiss, Quantum dots for live cells, in vivo imaging, and diagnostics. Science **307**, 538–544 (2005)

277. X. Michalet, O.H.W. Siegmund, J. Vallerga, P. Jelinsky, J.E. Millaud, S. Weiss, Detectors for single-molecule fluorescence imaging and spectroscopy. J. Mod. Opt. **54**, 239–281 (2007)

278. X. Michalet, O.H.W. Siegmund, J.V. Vallerga, P. Jelinsky, J.E. Millaud, S. Weiss, Photon-counting H33D detector for biological fluorescence imaging. Nucl. Instrum. Methods Phys. Res. Sect. A **567**, 133–136 (2006)

279. M. Micic, D.H. Hu, Y.D. Suh, G. Newton, M. Romine, H.P. Lu, Correlated atomic force microscopy and fluorescence lifetime imaging of live bacterial cells. Colloid Surf. B **34**, 205–212 (2004)

280. M.Y. Min, D.A. Rusakov, D.M. Kullmann, Activation of AMPA, kainate, and metabotropic receptors at hippocampal mossy fiber synapses: role of glutamate diffusion. Neuron **21**, 561–570 (1998)

281. T. Minami, S. Hirayama, High quality fluorescence decay curves and lifetime imaging using an elliptical scan streak camera. J. Photochem. Photobiol. A **53**, 11–21 (1990)

282. A.C. Mitchell, J.E. Wall, J.G. Murray, C.G. Morgan, Direct modulation of the effective sensitivity of a CCD detector: a new approach to time-resolved fluorescence imaging. J. Micros. **206**, 225–232 (2002)

283. A.C. Mitchell, J.E. Wall, J.G. Murray, C.G. Morgan, Measurement of nanosecond time-resolved fluorescence with a directly gated interline CCD camera. J. Micros. **206**, 233–238 (2002)

284. A. Miyawaki, J. Llopis, R. Helm, J.M. McCaffery, J.A. Adams, M. Ikura, R.Y. Tsien, Fluorescent indicators for Ca^{2+} based on green fluorescent proteins and calmodulin. Nature **388**, 882–887 (1997)

285. N. Mohan, C.S. Chen, H.H. Hsieh, Y.C. Wu, H.C. Chang, In vivo imaging and toxicity assessments of fluorescent nanodiamonds in caenorhabditis elegans. Nano Lett. **10**, 3692–3699 (2010)
286. H. Mojzisova, J. Olesiak, M. Zielinski, K. Matczyszyn, D. Chauvat, J. Zyss, Polarization-sensitive two-photon microscopy study of the organization of liquid-crystalline DNA. Biophys. J. **97**, 2348–2357 (2009)
287. C.G. Morgan, A.C. Mitchell, J.G. Murray, Nanosecond time-resolved fluorescence microscopy: principles and practice. Proc. Roy. Microscop. Soc. **1**, 463–466 (1990)
288. G.A. Morton, H.M. Smith, R. Wasserman, Afterpulses in photomultipliers. IEEE Trans. Nucl. Sci. **14**, 443–448 (1967)
289. H. Murakoshi, H. Wang, R. Yasuda, Local, persistent activation of Rho GTPases during plasticity of single dendritic spines. Nature **472**, 100–104 (2011)
290. S. Murata, P. Herman, J.R. Lakowicz, Texture analysis of fluorescence lifetime images of AT- and GC- rich regions in nuclei. J. Histochem. Cytochem. **49**, 1443–1451 (2001)
291. S. Murata, P. Herman, J.R. Lakowicz, Texture analysis of fluorescence lifetime images of nuclear DNA with effect of fluorescence resonance energy transfer. Cytometry **43**, 94–100 (2001)
292. S. Murata, P. Herman, H.J. Lin, J.R. Lakowicz, Fluorescence lifetime imaging of nuclear DNA: effect of fluorescence resonance energy transfer. Cytometry **41**, 178–185 (2000)
293. T. Nakabayashi, I. Nagao, M. Kinjo, Y. Aoki, M. Tanaka, N. Ohta, Stress-induced environmental changes in a single cell as revealed by fluorescence lifetime imaging. Photoch. Photobio. Sci. **7**, 671–674 (2008)
294. T. Nakabayashi, H.P. Wang, M. Kinjo, N. Ohta, Application of fluorescence lifetime imaging of enhanced green fluorescent protein to intracellular pH measurements. Photoch. Photobio. Sci. **7**, 668–670 (2008)
295. F. Neugart, A. Zappe, F. Jelezko, C. Tietz, J.P. Boudou, A. Krueger, J. Wrachtrup, Dynamics of diamond nanoparticles in solution and cells. Nano Lett. **7**, 3588–3591 (2007)
296. T. Ng, A. Squire, G. Hansra, F. Bornancin, C. Prevostel, A. Hanby, W. Harris, D. Barnes, S. Schmidt, H. Mellor, P.I. Bastiaens, P.J. Parker, Imaging protein kinase C alpha activation in cells. Science **283**, 2085–2089 (1999)
297. T.A. Nguyen, P. Sarkar, J.V. Veetil, S.V. Koushik, S.S. Vogel, Fluorescence polarization and fluctuation analysis monitors subunit proximity, stoichiometry, and protein complex hydrodynamics. PLoS ONE **7**, e38209 (2012)
298. T. Ni, L.A. Melton, Two-dimensional gas-phase temperature measurements using fluorescence lifetime imaging. Appl. Spectrosc. **50**, 1112–1116 (1996)
299. C. Nicholson, J.M. Phillips, A.R. Gardner-Medwin, Diffusion from an iontophoretic point source in the brain—role of tortuosity and volume fraction. Brain Res. **169**, 580–584 (1979)
300. C. Nicholson, L. Tao, Hindered diffusion of high-molecular-weight compounds in brain extracellular microenvironment measured with integrative optical imaging. Biophys. J. **65**, 2277–2290 (1993)
301. T.A. Nielsen, D.A. DiGregorio, R.A. Silver, Modulation of glutamate mobility reveals the mechanism underlying slow-rising AMPAR EPSCs and the diffusion coefficient in the synaptic cleft. Neuron **42**, 757–771 (2004)
302. M.E. Nipper, M. Dakanali, E.A. Theodorakis, M.A. Haidekker, Detection of liposome membrane viscosity perturbations with ratiometric molecular rotors. Biochimie **93**, 988–994 (2010)
303. M.E. Nipper, S. Majd, M. Mayer, J.C. Lee, E.A. Theodorakis, M.A. Haidekker, Characterization of changes in the viscosity of lipid membranes with the molecular rotor FCVJ. Biochim. Biophys. Acta **1778**, 1148–1153 (2008)
304. G. Nishimura, M. Tamura, Artefacts in the analysis of temporal response functions measured by photon counting. Phys. Med. Biol. **50**, 1327–1342 (2005)
305. D.V. O'Connor, D. Phillips, *Time-Correlated Single-Photon Counting* (Academic Press, London, 1984)

306. S. Ogikubo, T. Nakabayashi, T. Adachi, M.S. Islam, T. Yoshizawa, M. Kinjo, N. Ohta, Intracellular pH sensing using autofluorescence lifetime microscopy. J. Phys. Chem. B **115**, 10385–10390 (2011)

307. P.R. Ogilby, Singlet oxygen: there is indeed something new under the sun. Chem. Soc. Rev. **39**, 3181–3209 (2010)

308. K. Okabe, N. Inada, C. Gota, Y. Harada, T. Funatsu, S. Uchiyama, Intracellular temperature mapping with a fluorescent polymeric thermometer and fluorescence lifetime imaging microscopy. Nat. Commun. **3**, 705 (2012)

309. A. Orte, J.M. Alvarez-Pez, M.J. Ruedas-Rama, Fluorescence lifetime imaging microscopy for the detection of intracellular pH with quantum dot nanosensors. ACS Nano **7**, 6387–6395 (2013)

310. D.M. Owen, P.M.P. Lanigan, C. Dunsby, I. Munro, D. Grant, M.A.A. Neil, P.M.W. French, A.I. Magee, Fluorescence lifetime imaging provides enhanced contrast when imaging the phase-sensitive Dye di-4-ANEPPDHQ in model membranes and live cells. Biophys. J. **90**, L80–L82 (2006)

311. D.M. Owen, C. Rentero, A. Magenau, A. Abu-Siniyeh, K. Gaus, Quantitative imaging of membrane lipid order in cells and organisms. Nat. Protoc. **7**, 24–35 (2012)

312. J.M. Paredes, M.D. Giron, M.J. Ruedas-Rama, A. Orte, L. Crovetto, E.M. Talavera, R. Salto, J.M. Alvarez-Pez, Real-Time phosphate sensing in living cells using fluorescence lifetime imaging microscopy (FLIM). J. Phys. Chem. B **117**, 8143–8149 (2013)

313. R. Patalay, C. Talbot, Y. Alexandrov, M.O. Lenz, S. Kumar, S. Warren, I. Munro, M.A.A. Neil, K. König, P.M.W. French, A. Chu, G.W.H. Stamp, C. Dunsby, Multiphoton multispectral fluorescence lifetime tomography for the evaluation of basal cell carcinomas. PLoS ONE **7**, e43460 (2012)

314. A. Paul, A. Samanta, Free Volume dependence of the internal rotation of a molecular rotor probe in room temperature ionic liquids. J. Phys. Chem. B **112**, 16626–16632 (2008)

315. S. Pelet, M.J.R. Previte, P.T.C. So, Comparing the quantification of Förster resonance energy transfer measurement accuracies based on intensity, spectral, and lifetime imaging. J. Biomed. Opt. **11**, 034017 (2006)

316. X. Peng, Z. Yang, J. Wang, J. Fan, Y. He, F. Song, B. Wang, S. Sun, J. Qu, J. Qi, M. Yan, Fluorescence ratiometry and fluorescence lifetime imaging: using a single molecular sensor for dual mode imaging of cellular viscosity. J. Am. Chem. Soc. **133**, 6626–6635 (2011)

317. R. Pepperkok, A. Squire, S. Geley, P.I.H. Bastiaens, Simultaneous detection of multiple green fluorescent proteins in live cells by fluorescence lifetime imaging microscopy. Curr. Biol. **9**, 269–272 (1999)

318. D. Perrais, N. Ropert, Altering the concentration of GABA in the synaptic cleft potentiates miniature IPSCs in rat occipital cortex. Euro. J. Neurosci. **12**, 400–404 (2000)

319. M. Peter, S.M. Ameer-Beg, Imaging molecular interactions by multiphoton FLIM. Biol. Cell **96**, 231–236 (2004)

320. Z. Petrášek, H.J. Eckert, K. Kemnitz, Wide-field photon counting fluorescence lifetime imaging microscopy: application to photosynthesizing systems. Photosynth. Res. **102**, 157–168 (2009)

321. J. Philip, K. Carlsson, Theoretical investigation of the signal-to-noise ratio in fluorescence lifetime imaging. J. Opt. Soc. Am. A **20**, 368–379 (2003)

322. J.C. Pickup, Z.L. Zhi, F. Khan, T. Saxl, D.J.S. Birch, Nanomedicine and its potential in diabetes research and practice. Diabetes-Metab. Res. **24**, 604–610 (2008)

323. R. Piet, L. Vargova, E. Sykova, D.A. Poulain, S.H.R. Oliet, Physiological contribution of the astrocytic environment of neurons to intersynaptic crosstalk. Proc. Natl. Acad. Sci. U. S. A. **101**, 2151–2155 (2004)

324. A. Pietraszewska-Bogiel, T.W.J. Gadella, FRET microscopy: from principle to routine technology in cell biology. J. Micros. **241**, 111–118 (2011)

325. D.W. Piston, Fluorescence anisotropy of protein complexes in living cells. Biophys. J. **99**, 1685–1686 (2010)

326. A. Pliss, L.L. Zhao, T.Y. Ohulchanskyy, J.L. Qu, P.N. Prasad, Fluorescence lifetime of fluorescent proteins as an intracellular environment probe sensing the cell cycle progression. ACS. Chem. Biol. **7**, 1385–1392 (2012)

327. G.B. Porter, Reversible energy-transfer. Theor. Chim. Acta **24**, 265–270 (1972)

328. P.P. Provenzano, K.W. Eliceiri, P.J. Keely, Multiphoton microscopy and fluorescence lifetime imaging microscopy (FLIM) to monitor metastasis and the tumor microenvironment. Clin. Exp. Metastas. **26**, 357–370 (2009)

329. J.L. Qu, L.X. Liu, D.N. Chen, Z.Y. Lin, G.X. Xu, B.P. Guo, H.B. Niu, Temporally and spectrally resolved sampling imaging with a specially designed streak camera. Opt. Lett. **31**, 368–370 (2006)

330. G.I. Redford, R.M. Clegg, Polar plot representation for frequency-domain analysis of fluorescence lifetimes. J. Fluores. **15**, 805–815 (2005)

331. G.I. Redford, Z.K. Majumdar, J.D.B. Sutin, R.M. Clegg, Properties of microfluidic turbulent mixing revealed by fluorescence lifetime imaging. J. Chem. Phys. **123**, 224504 (2005)

332. J.E. Reeve, A.D. Corbett, I. Boczarow, T. Wilson, H. Bayley, H.L. Anderson, Probing the orientational distribution of dyes in membranes through multiphoton microscopy. Biophys. J. **103**, 907–917 (2012)

333. A. Rei, G. Hungerford, M.I.C. Ferreira, Probing local effects in silica sol-gel media by fluorescence spectroscopy of p-DASPMI. J. Phys. Chem. B **112**, 8832–8839 (2008)

334. J. Requejo-Isidro, J. McGinty, I. Munro, D.S. Elson, N.P. Galletly, M.J. Lever, M.A.A. Neil, G.W.H. Stamp, P.M.W. French, P.A. Kellett, J.D. Hares, A.K.L. Dymoke-Bradshaw, High-speed wide-field time-gated endoscopic fluorescence-lifetime imaging. Opt. Lett. **29**, 2249–2251 (2004)

335. U. Resch-Genger, M. Grabolle, S. Cavaliere-Jaricot, R. Nitschke, T. Nann, Quantum dots versus organic dyes as fluorescent labels. Nat. Methods **5**, 763–775 (2008)

336. M.A. Rizzo, D.W. Piston, High-contrast imaging of fluorescent protein FRET by fluorescence polarization microscopy. Biophys. J. **88**, L14–L16 (2005)

337. M.J. Roberti, T.M. Jovin, E. Jares-Erijman, Confocal fluorescence anisotropy and FRAP imaging of alpha-synuclein amyloid aggregates in living cells. PLoS ONE **6**, e23338 (2011)

338. T. Robinson, P. Valluri, H.B. Manning, D.M. Owen, I. Munro, C.B. Talbot, C. Dunsby, J.F. Eccleston, G.S. Baldwin, M.A.A. Neil, A.J. de Mello, P.M.W. French, Three-dimensional molecular mapping in a microfluidic mixing device using fluorescence lifetime imaging. Opt. Lett. **33**, 1887–1889 (2008)

339. C. Rodenbücher, T. Gensch, W. Speier, U. Breuer, M. Pilch, H. Hardtdegen, M. Mikulics, E. Zych, R. Waser, K. Szot, Inhomogeneity of donor doping in SrTiO3 substrates studied by fluorescence-lifetime imaging microscopy. Appl. Phys. Lett. **103**, 162904 (2013)

340. M.I. Rowley, P.R. Barber, A.C.C. Coolen, B. Vojnovic, Bayesian analysis of fluorescence lifetime imaging data, SPIE Proc 7903, (2011), 790325

341. C. Rumble, K. Rich, G. He, M. Maroncelli, CCVJ is not a simple rotor probe. J. Phys. Chem. A **116**, 10786–10792 (2012)

342. D.A. Rusakov, L.P. Savtchenko, K.Y. Zheng, J.M. Henley, Shaping the synaptic signal: molecular mobility inside and outside the cleft. Trends Neurosci. **34**, 359–369 (2011)

343. A.G. Ryder, T.J. Glynn, M. Przyjalgowski, B. Szczupak, A compact violet diode laser-based fluorescence lifetime microscope. J. Fluores. **12**, 177–180 (2002)

344. Y. Sakai, S. Hirayama, A Fast deconvolution method to analyze fluorescence decays when the excitation pulse repetition period is less than the decay times. J. Lumines. **39**, 145–151 (1988)

345. W.Y. Sanchez, C. Obispo, E. Ryan, J.E. Grice, M.S. Roberts, Changes in the redox state and endogenous fluorescence of in vivo human skin due to intrinsic and photo-aging, measured by multiphoton tomography with fluorescence lifetime imaging. J. Biomed. Opt. **18**, 034016 (2012)

346. W.Y. Sanchez, T.W. Prow, W.H. Sanchez, J.E. Grice, M.S. Roberts, Analysis of the metabolic deterioration of ex vivo skin from ischemic necrosis through the imaging of intracellular NAD(P)H by multiphoton tomography and fluorescence lifetime imaging microscopy. J. Biomed. Opt. **15**, 046008 (2010)

347. R. Sanders, A. Draaijer, H.C. Gerritsen, P.M. Houpt, Y.K. Levine, Quantitative pH imaging in cells using confocal fluorescence lifetime imaging microscopy. Anal. Biochem. **227**, 302–308 (1995)

348. R. Sanders, H.C. Gerritsen, A. Draaijer, P.M. Houpt, Y.K. Levine, Fluorescence lifetime imaging of free calcium in single cells. Bioimaging **2**, 131–138 (1994)

349. F. Santamaria, S. Wils, E. De Schutter, G.J. Augustine, Anomalous diffusion in Purkinje cell dendrites caused by spines. Neuron **52**, 635–648 (2006)

350. M. Sauer, J. Hofkens, J. Enderlein, *Handbook of Fluorescence Spectroscopy and Imaging* (Wiley-VCH, Weinheim, 2011)

351. L.P. Savtchenko, D.A. Rusakov, Extracellular diffusivity determines contribution of high-versus low-affinity receptors to neural signaling. Neuroimage **25**, 101–111 (2005)

352. L.P. Savtchenko, S. Sylantyev, D.A. Rusakov, Central synapses release a resource-efficient amount of glutamate. Nat. Neurosci. **16**, 10–16 (2013)

353. T. Saxl, F. Khan, M. Ferla, D. Birch, J. Pickup, A fluorescence lifetime-based fibre-optic glucose sensor using glucose/galactose-binding protein. Analyst **136**, 968–972 (2011)

354. T. Saxl, F. Khan, D.R. Matthews, Z.L. Zhi, O. Rolinski, S. Ameer-Beg, J. Pickup, Fluorescence lifetime spectroscopy and imaging of nano-engineered glucose sensor microcapsules based on glucose/galactose-binding protein. Biosens. Bioelectron. **24**, 3229–3234 (2009)

355. K.C. Schuermann, H.E. Grecco, flatFLIM: enhancing the dynamic range of frequency domain FLIM. Opt. Express **20**, 20730–20741 (2012)

356. D. Schweitzer, M. Hammer, F. Schweitzer, R. Anders, T. Doebbecke, S. Schenke, E.R. Gaillard, E.R. Gaillard, In vivo measurement of time-resolved autofluorescence at the human fundus. J. Biomed. Opt. **9**, 1214–1222 (2004)

357. D. Schweitzer, S. Schenke, M. Hammer, F. Schweitzer, S. Jentsch, E. Birckner, W. Becker, A. Bergmann, Towards metabolic mapping of the human retina. Microsc. Res. Tech. **70**, 410–419 (2007)

358. A.D. Scully, A.J. Mac Robert, S. Botchway, P. O'Neill, A.W. Parker, R.B. Ostler, D. Phillips, Development of a laser-based fluorescence microscope with subnanosecond time resolution. J. Fluores. **6**, 119–125 (1996)

359. S. Seidenari, F. Arginelli, C. Dunsby, P. French, K. König, C. Magnoni, M. Manfredini, C. Talbot, G. Ponti, Multiphoton laser tomography and fluorescence lifetime imaging of basal cell carcinoma: morphologic features for non-invasive diagnostics. Exp. Dermatol. **21**, 831–836 (2012)

360. N. Sergent, J.A. Levitt, M. Green, K. Suhling, Rapid wide-field photon counting imaging with microsecond time resolution. Opt. Express **18**, 25292–25298 (2010)

361. N.C. Shaner, G.H. Patterson, M.W. Davidson, Advances in fluorescent protein technology. J. Cell Sci. **120**, 4247–4260 (2007)

362. N.C. Shaner, P.A. Steinbach, R.Y. Tsien, A guide to choosing fluorescent proteins. Nat. Methods **2**, 905–909 (2005)

363. P. Sharma, R. Varma, R.C. Sarasij, Ira, K. Gousset, G. Krishnamoorthy, M. Rao, S. Mayor, Nanoscale organization of multiple GPI-anchored proteins in living cell membranes. Cell **116**, 577–589 (2004)

364. T. Shibuya, The refractive index correction to the radiative rate constant. Chem. Phys. Lett. **103**(1), 46–48 (1983)

365. J. Siegel, D.S. Elson, S.E.D. Webb, K.C.B. Lee, A. Vlandas, G.L. Gambaruto, S. Lévêque-Fort, M.J. Lever, P.J. Tadrous, G.W.H. Stamp, Studying biological tissue with fluorescence lifetime imaging: microscopy, endoscopy, and complex decay profiles. Appl. Opt. **42**, 2995–3004 (2003)

366. A. Sillen, Y. Engelborghs, The correct use of "average" fluorescence parameters. Photochem. Photobiol. **67**, 475–486 (1998)
367. M.C. Skala, K.M. Riching, D.K. Bird, A. Gendron-Fitzpatrick, J. Eickhoff, K.W. Eliceiri, P. J. Keely, N. Ramanujam, In vivo multiphoton fluorescence lifetime imaging of protein-bound and free nicotinamide adenine dinucleotide in normal and precancerous epithelia. J. Biomed. Opt. **12**, 024014 (2007)
368. M.C. Skala, K.M. Riching, A. Gendron-Fitzpatrick, J. Eickhoff, K.W. Eliceiri, J.G. White, N. Ramanujam, In vivo multiphoton microscopy of NADH and FAD redox states, fluorescence lifetimes, and cellular morphology in precancerous epithelia. Proc. Natl. Acad. Sci. USA. **104**, 19494–19499 (2007)
369. A.D. Slepkov, A. Ridsdale, H.N. Wan, M.H. Wang, A.F. Pegoraro, D.J. Moffatt, J. P. Pezacki, F.J. Kao, A. Stolow, Forward-collected simultaneous fluorescence lifetime imaging and coherent anti-Stokes Raman scattering microscopy. J. Biomed. Opt. **16**, 021103 (2011)
370. G.E. Smith, The invention and early history of the CCD. Nucl. Instrum. Meth. A **607**, 1–6 (2009)
371. T.A. Smith, M.L. Gee, C.A. Scholes, Time-resolved evanescent wave-induced fluorescence anisotropy measurements, in *Reviews in Fluorescence*, ed by C.D.Geddes, J.R. Lakowicz (Springer, New York, 2005), pp. 245–271
372. J.A. Spitz, V. Polard, A. Maksimenko, F. Subra, C. Baratti-Elbaz, R. Meallet-Renault, R.B. Pansu, P. Tauc, C. Auclair, Assessment of cellular actin dynamics by measurement of fluorescence anisotropy. Anal. Biochem. **367**, 95–103 (2007)
373. J.A. Spitz, R. Yasukuni, N. Sandeau, M. Takano, J.J. Vachon, R. Meallet-Renault, R.B. Pansu, Scanning-less wide-field single-photon counting device for fluorescence intensity, lifetime and time-resolved anisotropy imaging microscopy. J. Micros. **229**, 104–114 (2008)
374. A. Squire, P.J. Verveer, O. Rocks, P.I.H. Bastiaens, Red-edge anisotropy microscopy enables dynamic imaging of homo-FRET between green fluorescent proteins in cells. J. Struct. Biol. **147**, 62–69 (2004)
375. V.K.A. Sreenivasan, A.V.Zvyagin, E.M. Goldys, Luminescent nanoparticles and their applications in the life sciences. J. Phys.-Condes. Matter **25**, 194101 (2013)
376. R. Steinmeyer, G.S. Harms, Fluorescence resonance energy transfer and anisotropy reveals both hetero- and homo-energy transfer in the pleckstrin homology-domain and the parathyroid hormone-receptor. Microsc. Res. Tech. **72**, 12–21 (2009)
377. O. Stern, M. Volmer, Über die Abklingungszeit der Fluoreszenz. Physikalische Zeitschrift **20**, 183–188 (1919)
378. G.G. Stokes, On the change of refrangibility of light. Philos. Trans. R. Soc. London **142**, 463–562 (1852)
379. G.G. Stokes, On the change of refrangibility of light II. Phil. Trans. R. Soc. London **143**, 385–396 (1853)
380. M. Straub, S.W. Hell, Fluorescence lifetime three-dimensional microscopy with picosecond precision using a multifocal multiphoton microscope. Appl. Phys. Lett. **73**, 1769–1771 (1998)
381. S.J. Strickler, R.A. Berg, Relationship between absorption intensity and fluorescence lifetime of molecules. J. Chem. Phys. **37**, 814–820 (1962)
382. C. Stringari, A. Cinquin, O. Cinquin, M.A. Digman, P.J. Donovan, E. Gratton, Phasor approach to fluorescence lifetime microscopy distinguishes different metabolic states of germ cells in a live tissue. Proc. Natl. Acad. Sci. USA **108**, 13582–13587 (2011)
383. C. Stringari, R.A. Edwards, K.T. Pate, M.L. Waterman, P.J. Donovan, E. Gratton, Metabolic trajectory of cellular differentiation in small intestine by phasor fluorescence lifetime microscopy of NADH. Sci. Rep. **2**, 568 (2012)
384. L. Stryer, Fluorescence energy transfer as a spectroscopic ruler. Annu. Rev. Biochem. **47**, 819–846 (1978)
385. L. Stryer, R.P. Haugland, Energy Transfer: A spectroscopic ruler. Proc. Natl. Acad. Sci. USA **58**, 719–726 (1967)

386. H. Studier, K. Weisshart, O. Holub, and W. Becker, Megapixel FLIM, (SPIE Proc 8948, (2014), 89481 K
387. K. Suhling, Fluorescence lifetime imaging, in *Cell Imaging*, ed by D. Stephens (Scion, Bloxham, 2006), pp. 219–245
388. K. Suhling, R.W. Airey, B.L. Morgan, Optimisation of centroiding algorithms for photon event counting imaging. Nucl. Instrum. Methods A **437**, 393–418 (1999)
389. K. Suhling, P.M.W. French, D. Phillips, Time-resolved fluorescence microscopy. Photochem. Photobiol. Sci. **4**, 13–22 (2005)
390. K. Suhling, G. Hungerford, R.W. Airey, B.L. Morgan, A position-sensitive photon event counting detector applied to fluorescence imaging of dyes in sol-gel matrices. Meas. Sci. Technol. **12**, 131–141 (2001)
391. K. Suhling, J. Levitt, P.H. Chung, Time-resolved fluorescence anisotropy imaging, in *Methods in Molecular Biology,* ed by Y. Engelborghs, A.J.W.G. Visser (Springer Science + Business Media, New York, 2014), pp. 503–519
392. K. Suhling, D. McLoskey, D.J.S. Birch, Multiplexed single-photon counting. II. The statistical theory of time-correlated measurements. Rev. Sci. Instrum. **67**, 2238–2246 (1996)
393. K. Suhling, J. Siegel, P.M.P. Lanigan, S. Lévêque-Fort, S.E.D. Webb, D. Phillips, D.M. Davis, P.M.W. French, Time-resolved fluorescence anisotropy imaging applied to live cells. Opt. Lett. **29**, 584–586 (2004)
394. K. Suhling, J. Siegel, D. Phillips, P.M.W. French, S. Lévêque-Fort, S.E.D. Webb, D.M. Davis, Imaging the environment of green fluorescent protein. Biophys. J. **83**, 3589–3595 (2002)
395. Y.H. Sun, N. Hatami, M. Yee, J. Phipps, D.S. Elson, F. Gorin, R.J. Schrot, L. Marcu, Fluorescence lifetime imaging microscopy for brain tumor image-guided surgery. J. Biomed. Opt. **15**, 056022 (2010)
396. K. Svoboda, D.W. Tank, W. Denk, Direct measurement of coupling between dendritic spines and shafts. Science **272**, 716–719 (1996)
397. R. Swaminathan, C.P. Hoang, A.S. Verkman, Photobleaching recovery and anisotropy decay of green fluorescent protein GFP-S65T in solution and cells: Cytoplasmic viscosity probed by green fluorescent protein translational and rotational diffusion. Biophys. J. **72**, 1900–1907 (1997)
398. E. Sykova, C. Nicholson, Diffusion in brain extracellular space. Physiol. Rev. **88**, 1277–1340 (2008)
399. H. Szmacinski, J.R. Lakowicz, Fluorescence lifetime characterization of magnesium probes: improvement of Mg^{2+} dynamic range and sensitivity using phase-modulation fluorometry. J. Fluores. **6**, 83–85 (1996)
400. H. Szmacinski, J.R. Lakowicz, Potassium and sodium measurements at clinical concentrations using phase-modulation fluorometry. Sens. Actuator B-Chem. **60**, 8–18 (1999)
401. H. Szmacinski, J.R. Lakowicz, Sodium Green as a potential probe for intracellular sodium imaging based on fluorescence lifetime. Anal. Biochem. **250**, 131–138 (1997)
402. P.J. Tadrous, J. Siegel, P.M.W. French, S. Shousha, E.N. Lalani, G.W. Stamp, Fluorescence lifetime imaging of unstained tissues: early results in human breast cancer. J. Pathol. **199**, 309–317 (2003)
403. R. Teixeira, P.M.R. Paulo, A.S. Viana, S.M.B. Costa, Plasmon-enhanced emission of a phthalocyanine in polyelectrolyte films induced by gold nanoparticles. J. Phys. Chem. C **115**, 24674–24680 (2011)
404. J. Tellinghuisen, C.W. Wilkerson, Bias and precision in the estimation of exponential decay parameters from sparse data. Anal. Chem. **65**, 1240–1246 (1993)
405. C. Thaler, S.V. Koushik, P.S. Blank, S.S. Vogel, Quantitative multiphoton spectral imaging and its use for measuring resonance energy transfer. Biophys. J. **89**, 2736–2749 (2005)
406. C. Thaler, S.V. Koushik, H.L. Puhl, P.S. Blank, S.S. Vogel, Structural rearrangement of CaMKII alpha catalytic domains encodes activation. Proc. Natl. Acad. Sci. USA **106**, 6369–6374 (2009)

407. R.G. Thorne, C. Nicholson, In vivo diffusion analysis with quantum dots and dextrans predicts the width of brain extracellular space. Proc. Natl. Acad. Sci. USA **103**, 5567–5572 (2006)
408. P. Tinnefeld, D.P. Herten, M. Sauer, Photophysical dynamics of single molecules studied by spectrally-resolved fluorescence lifetime imaging microscopy (SFLIM). J. Phys. Chem. A **105**, 7989–8003 (2001)
409. J. Tisler, G. Balasubramanian, B. Naydenov, R. Kolesov, B. Grotz, R. Reuter, J.P. Boudou, P.A. Curmi, M. Sennour, A. Thorel, M. Börsch, K. Aulenbacher, R. Erdmann, P.R. Hemmer, F. Jelezko, J. Wrachtrup, Fluorescence and spin properties of defects in single digit nanodiamonds. Acs Nano **3**, 1959–1965 (2009)
410. D.M. Togashi, R.I.S. Romao, A.M.G. da Silva, A.J.F.N. Sobral, S.M.B. Costa, Self-organization of a sulfonamido-porphyrin in Langmuir monolayers and Langmuir-Blodgett films. Phys. Chem. Chem. Phys. **7**, 3875–3884 (2005)
411. L. Tolosa, H. Szmacinski, G. Rao, J.R. Lakowicz, Lifetime-based sensing of glucose using energy transfer with a long lifetime donor. Anal. Biochem. **250**, 102–108 (1997)
412. D. Toptygin, Effects of the solvent refractive index and its dispersion on the radiative decay rate and extinction coefficient of a fluorescent solute. J. Fluores. **13**, 201–219 (2003)
413. D. Toptygin, R.S. Savtchenko, N.D. Meadow, S. Roseman, L. Brand, Effect of the solvent refractive index on the excited-state lifetime of a single tryptophan residue in a protein. J. Phys. Chem. B **106**, 3724–3734 (2002)
414. K. Torno, B.K. Wright, M.R. Jones, M.A. Digman, E. Gratton, M. Phillips, Real-time analysis of metabolic activity within lactobacillus acidophilus by phasor fluorescence lifetime imaging microscopy of NADH. Curr. Microbiol. **66**, 365–367 (2013)
415. M. Tramier M. Coppey-Moisan, Fluorescence anisotropy imaging microscopy for homo-FRET in living cells, in *Methods in Cell Biology*, ed by F.S. Kevin (Academic Press, San Diego, 2008), pp. 395–414
416. M. Tramier, K. Kemnitz, C. Durieux, J. Coppey, P. Denjean, R.B. Pansu, M. Coppey-Moisan, Restrained torsional dynamics of nuclear DNA in living proliferative mammalian cells. Biophys. J. **78**, 2614–2627 (2000)
417. B. Treanor, P.M. Lanigan, K. Suhling, T. Schreiber, I. Munro, M.A. Neil, D. Phillips, D.M. Davis, P.M.W. French, Imaging fluorescence lifetime heterogeneity applied to GFP-tagged MHC protein at an immunological synapse. J. Micros. **217**, 36–43 (2005)
418. C. Tregidgo, J.A. Levitt, K. Suhling, Effect of refractive index on the fluorescence lifetime of green fluorescent protein. J. Biomed. Opt. **13**, 031218 (2008)
419. P. Urayama, M.-A. Mycek, Fluorescence lifetime imaging microscopy of enodogenous biological fluorescence, in *Handbook of Biomedical Fluorescence*, ed by M.-A. Mycek, B. W. Pogue (Marcel Dekker, New York, 2003)
420. M.A. Uskova, J. Borst, M.A. Hink, A. van Hoek, A. Schots, A.L. Klyachko, A.J.W.G. Visser, Fluorescence dynamics of green fluorescent protein in AOT reversed micelles. Biophys. Chem. **87**, 73–84 (2000)
421. B.M. Uzhinov, V.L. Ivanov, M.Y. Melnikov, Molecular rotors as luminescence sensors of local viscosity and viscous flow in solutions and organized systems. Russ. Chem. Rev. **80**, 1179–1190 (2011)
422. B. Valeur, Pulse and phase fluorometries: an objective comparison, in *Fluorescence Spectroscopy in Biology*, ed by M. Hof, R. Hutterer, V. Fidler (Springer, Berlin, 2005), pp. 30–48
423. B. Valeur, M. Berberan-Santos, A brief history of fluorescence and phosphorescence before the emergence of quantum theory. J. Chem. Edu. **88**, 731–738 (2011)
424. B. Valeur, M.N. Berberan-Santos, *Molecular Fluorescence. Principles and Applications*, 2nd ed. (Wiley, Weinheim, 2012)
425. B.W. van der Meer, Kappa-squared: from nuisance to new sense. J. Biotechnol. **82**, 181–196 (2002)

426. T.J. van Ham, A. Esposito, J.R. Kumita, S.-T.D. Hsu, G.S. Kaminski Schierle, C.F. Kaminski, C.M. Dobson, E.A.A. Nollen, and C. W. Bertoncini, Towards multiparametric fluorescent imaging of amyloid formation: studies of a YFP model of α-Synuclein aggregation. J. Mol. Biol. **395**, 627–642 (2010)

427. H.J. van Manen, P. Verkuijlen, P. Wittendorp, V. Subramaniam, T.K. van den Berg, D. Roos, C. Otto, Refractive index sensing of green fluorescent proteins in living cells using fluorescence lifetime imaging microscopy. Biophys. J. **94**, L67–L69 (2008)

428. E.B. van Munster, T.W.J. Gadella, φ-FLIM: a new method to avoid aliasing in frequency-domain fluorescence lifetime imaging microscopy. J. Micros. **213**, 29–38 (2004)

429. E.B. van Munster, T.W.J. Gadella, Suppression of photobleaching-induced artifacts in frequency-domain FLIM by permutation of the recording order. Cytometry Part A **58**, 185–194 (2004)

430. M. van Zandvoort, C.J. de Grauw, H.C. Gerritsen, J.L.V. Broers, M. Egbrink, F.C.S. Ramaekers, D.W. Slaaf, Discrimination of DNA and RNA in cells by a vital fluorescent probe: lifetime imaging of SYTO13 in healthy and apoptotic cells. Cytometry **47**, 226–235 (2002)

431. C.J.R. Vanderoord, H.C. Gerritsen, F.F.G. Rommerts, D.A. Shaw, I.H. Munro, Y.K. Levine, Microvolume time-resolved fluorescence spectroscopy using a confocal synchrotron-radiation microscope. Appl. Spectrosc. **49**, 1469–1473 (1995)

432. R. Varma, S. Mayor, GPI-anchored proteins are organized in submicron domains at the cell surface. Nature **394**, 798–801 (1998)

433. B.D. Venetta, Microscope phase fluorometer for determining the fluorescence lifetimes of fluorochromes. Rev. Sci. Instrum. **30**, 450–457 (1959)

434. H.D. Vishwasrao, P. Trifilieff, E.R. Kandel, In vivo imaging of the actin polymerization state with two-photon fluorescence anisotropy. Biophys. J. **102**, 1204–1214 (2012)

435. M. Vitali, F. Picazo, Y. Prokazov, A. Duci, E. Turbin, C. Gotze, J. Llopis, R. Hartig, A.J.W. G. Visser, W. Zuschratter, Wide-field multi-parameter FLIM: long-term minimal invasive observation of proteins in living cells. PLOS ONE **6**, e15820 (2011)

436. M. Vitali, M. Reis, T. Friedrich, H.-J. Eckert, A wide-field multi-parameter FLIM and FRAP setup to investigate the fluorescence emission of individual living cyanobacteria, in *SPIE Proc 7376*, (2010), 737610

437. S.S. Vogel, T.A. Nguyen, B.W. van der Meer, P.S. Blank, The impact of heterogeneity and dark acceptor states on FRET: implications for using fluorescent protein donors and acceptors. PLoS ONE **7**, e49593 (2012)

438. S.S. Vogel, C. Thaler, P.S. Blank, S.V. Koushik, Time-Resolved Fluorescence Anisotropy, in *FLIM Microscopy in Biology and Medicine*, ed by A. Periasamy, R.M. Clegg (Chapman & Hall, Taylor & Francis Group, Boca Raton, 2010), pp. 245–288

439. M. Wachsmuth, Molecular diffusion and binding analyzed with FRAP. Protoplasma **251**, 373–382 (2014)

440. H. Wallrabe, A. Periasamy, Imaging protein molecules using FRET and FLIM microscopy. Curr. Opin. Biotechnol. **16**, 19–27 (2005)

441. B. Wandelt, P. Cywinski, G.D. Darling, B.R. Stranix, Single cell measurement of micro-viscosity by ratio imaging of fluorescence of styrylpyridinium probe. Biosens. Bioelectron. **20**, 1728–1736 (2005)

442. B. Wandelt, A. Mielniczak, P. Turkewitsch, G.D. Darling, B.R. Stranix, Substituted 4-[4-(dimethylamino)styryl] pyridinium salt as a fluorescent probe for cell microviscosity. Biosens. Bioelectron. **18**, 465–471 (2003)

443. H.W. Wang, V. Ghukassyan, C.T. Chen, Y.H. Wei, H.W. Guo, J.S. Yu, F.J. Kao, Differentiation of apoptosis from necrosis by dynamic changes of reduced nicotinamide adenine dinucleotide fluorescence lifetime in live cells. J. Biomed. Opt. **13**, 054011 (2008)

444. X.-D. Wang, O.S. Wolfbeis, Optical methods for sensing and imaging oxygen: materials, spectroscopies and applications. Chem. Soc. Rev. **43**, 3666–3761 (2014)

445. X.F. Wang, A. Periasamy, B. Herman, D.M. Coleman, Fluorescence lifetime imaging microscopy (FLIM): instrumentation and applications. Crit. Rev. Anal. Chem. **23**, 369–395 (1992)
446. X.F. Wang, T. Uchida, D.M. Coleman, S. Minami, A 2-dimensional fluorescence lifetime imaging system using a gated image intensifier. Appl. Spectrosc. **45**, 360–366 (1991)
447. X.F. Wang, T. Uchida, S. Minami, A fluorescence lifetime distribution measurement system based on phase-resolved detection using an image dissector tube. Appl. Spectrosc. **43**, 840–845 (1989)
448. K. Willaert, R. Loewenthal, J. Sancho, M. Froeyen, A. Fersht, Y. Engelborghs, Determination of the excited-state lifetimes of the tryptophan residues in barnase, via multifrequency phase fluorometry of tryptophan mutants. Biochemistry **31**, 711–716 (1992)
449. C.D. Wilms, J. Eilers, Photo-physical properties of Ca^{2+}-indicator dyes suitable for two-photon fluorescence-lifetime recordings. J. Micros. **225**, 209–213 (2007)
450. C.D. Wilms, H. Schmidt, J. Eilers, Quantitative two-photon Ca^{2+} imaging via fluorescence lifetime analysis. Cell Calcium **40**, 73–79 (2006)
451. P. Woolley, K.G. Steinhäuser, B. Epe, Förster-type energy-transfer—simultaneous forward and reverse transfer between unlike fluorophores. Biophys. Chem. **26**, 367–374 (1987)
452. F.S. Wouters, The physics and biology of fluorescence microscopy in the life sciences. Contemp. Phys. **47**, 239–255 (2006)
453. S.F. Wuister, C. de Mello Donega, A. Meijerink, Local-field effects on the spontaneous emission rate of CdTe and CdSe quantum dots in dielectric media. J. Chem. Phys. **121**, 4310–4315 (2004)
454. F.R. Xiao, C. Nicholson, J. Hrabe, S. Hrabetova, Diffusion of flexible random-coil dextran polymers measured in anisotropic brain extracellular space by integrative optical Imaging. Biophys. J. **95**, 1382–1392 (2008)
455. E. Yaghini, F. Giuntini, I.M. Eggleston, K. Suhling, A.M. Seifalian, A.J. MacRobert, Fluorescence lifetime imaging and FRET-induced intracellular redistribution of tat-conjugated quantum dot nanoparticles through interaction with a phthalocyanine photosensitiser. Small **10**, 782–792 (2014)
456. Y.L. Yan, G. Marriott, Fluorescence resonance energy transfer imaging microscopy and fluorescence polarization imaging microscopy, in *Methods in Enzymology, Biophotonics, Pt A* (Academic Press, London, 2003), pp. 561–580
457. C.M. Yengo, C.L. Berger, Fluorescence anisotropy and resonance energy transfer: powerful tools for measuring real time protein dynamics in a physiological environment. Curr. Opin. Pharmacol. **10**, 731–737 (2010)
458. E.K.L. Yeow, A.H.A. Clayton, Enumeration of oligomerization states of membrane proteins in living cells by homo-FRET spectroscopy and microscopy: theory and application. Biophys. J. **92**, 3098–3104 (2007)
459. Q.R. Yu, A.A. Heikal, Two-photon autofluorescence dynamics imaging reveals sensitivity of intracellular NADH concentration and conformation to cell physiology at the single-cell level. J. Photochem. Photobiol. B-Biol. **95**, 46–57 (2009)
460. Y. Zeng, Y. Wu, D. Li, W. Zheng, W.X. Wang, J.N.Y. Qu, Two-photon excitation chlorophyll fluorescence lifetime imaging: a rapid and noninvasive method for in vivo assessment of cadmium toxicity in a marine diatom Thalassiosira weissflogii. Planta **236**, 1653–1663 (2012)
461. H. Zhang, A.S. Verkman, Microfiberoptic measurement of extracellular space volume in brain and tumor slices based on fluorescent dye partitioning. Biophys. J. **99**, 1284–1291 (2010)
462. Q. Zhang, D.W. Piston, R.H. Goodman, Regulation of corepressor function by nuclear NADH. Science **295**, 1895–1897 (2002)
463. Q.L. Zhao, I.T. Young, J.G.S. de Jong, Photon budget analysis for fluorescence lifetime imaging microscopy. J. Biomed. Opt. **16**, 086007 (2011)
464. N.G. Zhegalova, G. Gonzales, M.Y. Berezin, Synthesis of nitric oxide probes with fluorescence lifetime sensitivity. Org. Biomol. Chem. **11**, 8228–8234 (2013)

465. K. Zheng, J. A. Levitt, K. Suhling, D.A. Rusakov, Monitoring nanoscale mobility of small molecules in organized brain tissue with time-resolved fluorescence anisotropy imaging, in *Neuromethods 84: Nanoscale Imaging of Synapses, New Concepts and Opportunities*, ed by U.V. Nägerl, A. Triller (Springer Science + Business Media, New York, 2014) pp. 125–143
466. K.Y. Zheng, A. Scimemi, D.A. Rusakov, Receptor actions of synaptically released glutamate: the role of transporters on the scale from nanometers to Microns. Biophys. J. **95**, 4584–4596 (2008)
467. L.-L. Zhu, D.-H. Qu, D. Zhang, Z.-F. Chen, Q.-C. Wang, H. Tian, Dual-mode tunable viscosity sensitivity of a rotor-based fluorescent dye. Tetrahedron **66**, 1254–1260 (2010)
468. M. Zimmer, Green fluorescent protein (GFP): applications, structure, and related photophysical behavior. Chem. Rev. **102**, 759–781 (2002)
469. H. Zollinger, *Color Chemistry: Syntheses, Properties, and Applications of Organic Dyes And Pigments* (Helvetica Chimica Acta, Zurich, 2003)

Chapter 4
Determination of Intracellular Chloride Concentrations by Fluorescence Lifetime Imaging

Thomas Gensch, Verena Untiet, Arne Franzen, Peter Kovermann and Christoph Fahlke

Abstract Fluorescence microscopy with membrane-permeable ion-sensitive fluorophores allows the non-invasive determination of intracellular ion concentrations. Chloride is the major anion of intra- and extracellular fluids influencing a great number of physiological processes. The dysfunction of chloride transporters and channels leads to disturbance in chloride homeostasis that can result in diseases of different parts of the body. The different existing chloride sensitive fluorophores and their usefulness in fluorescence lifetime imaging are put forward in this chapter. Fluorescence lifetime imaging of a chloride sensitive quinolinium dye (MQAE) has been established as an elegant tool to determine intracellular chloride concentrations of different cells in living biological tissue. Details of the experimental procedure are described and two case studies—chloride transport by a chloride transporter (KCC2) across the cell membrane and determination of intracellular chloride concentration in glia cells (EAAT1-positive astrocytes)—are presented.

T. Gensch (✉) · V. Untiet · A. Franzen · P. Kovermann · C. Fahlke
Institute of Complex Systems 4 (ICS-4, Cellular Biophysics),
Forschungszentrum Jülich, 52428 Jülich, Germany
e-mail: t.gensch@fz-juelich.de

V. Untiet
e-mail: v.untiet@fz-juelich.de

A. Franzen
e-mail: a.franzen@fz-juelich.de

P. Kovermann
e-mail: p.kovermann@fz-juelich.de

C. Fahlke
e-mail: c.fahlke@fz-juelich.de

© Springer International Publishing Switzerland 2015
W. Becker (ed.), *Advanced Time-Correlated Single Photon Counting Applications*,
Springer Series in Chemical Physics 111, DOI 10.1007/978-3-319-14929-5_4

4.1 Introduction

Chloride is the main physiological anion with great impact on the properties of biological fluids. This is true for both intracellular (cytosol and subcellular compartments) as well as extracellular (gastric and pancreatic juices, sweat, urine) fluids. Chloride transport across the plasma membrane of cells is an essential part of many physiological processes like synaptic transmission, pH and cell volume regulation, or transport processes in epithelial tissues. A considerable number of chloride channels, pumps and (co-)transporter proteins do exist in nature permitting regulation of the intracellular chloride concentration ($[Cl^-]_{int}$). Patch clamp and microelectrode based techniques have contributed a great deal to investigate chloride transport and chloride accumulation or deprivation in cells. However, fluorescence microscopy using chloride sensitive fluorophores has proven to be even more versatile to study such processes. This technique is non-invasive and allows fast excess to tissues as well as cells making it suitable for high-throughput screening. In particular, fluorescence microscopy in combination with multi-photon excitation—to a lesser extent also fluorescence confocal microscopy with one-photon excitation—permits addressing physiological questions not just in cultured cells but also in living tissue under native conditions. In the past 10 years fluorescence lifetime imaging microscopy (FLIM) has been established to study absolute chloride concentrations as well as relative changes in neuronal and other biological tissues. This chapter gives an overview on the methodology of FLIM and fluorophores, which are suitable for physiological investigations. Furthermore it contains examples from literature and presents some investigations using FLIM for the determination of chloride concentrations in glial cells.

4.2 Importance of Chloride Ions in Disease-Relevant Physiological Processes

Chloride is the major anion in intra- and extra-cellular solutions of multicellular organisms. Effective salt and solute transport requires the simultaneous movement of cations and anions. Therefore, chloride ion movement is crucial for the regulation of cellular salt content and volume. Chloride channels play an important role in synaptic transmission. Activation of ligand-gated anion channels such as GABA or glycine receptors inhibit neurons by increasing the membrane conductance and causing membrane hyper-polarization. Moreover chloride is co-transported in multiple secondary-active transporters, and there are classes of neurotransmitter transporters that exhibit associated chloride conductance [95]. Nature seems to utilize this potential of chloride ions. There are marked differences in the cytoplasmic chloride concentrations of diverse cell-types, fine-tuning the effects of such chloride channels and chloride co-transporters. Furthermore, chloride is not uniformly distributed over different cell organelles. In particular organelle acidification is associated with increased $[Cl^-]_{int}$ [26], and proper endo- and exo-cytosis seems to require defined chloride transport processes.

Changed chloride concentrations will result in altered transport rates or transport directions of chloride channels or Cl^--coupled co-transporters. Even slight changes in $[Cl^-]_{int}$ might impair inhibitory synaptic currents or change neurotransmitter release significantly. Volume regulation processes and the steady-state concentrations of other cellular constituents will be affected significantly as well. The physiological importance of regulated intracellular chloride concentration is highlighted by human diseases that are caused by changes in chloride transport proteins and altered chloride concentrations in the cytoplasm or in certain cellular organelles. Examples are disease states like hypertension, hepatic encephalopathy, neuropathic pain, and epilepsy [43, 65, 78, 100].

The high physiological importance of chloride concentrations is in marked difference to our understanding of the regulation of this important cellular parameter. There is still a lively debate about the processes involved in setting cellular chloride concentrations. For many intracellular organelles we have only rough estimates of their $[Cl^-]_{int}$. A prerequisite for studying this biological process is the availability of accurate techniques to determine steady-state as well as dynamic chloride concentration. Such techniques must leave physiological concentrations unaffected and work at high spatial resolution. Thus, optical methods seem to be uniquely suited for measuring $[Cl^-]_{int}$ [37, 55, 68, 92]. However, whereas multiple optical techniques have been developed for measuring the concentrations of all types of substrates this task appears to be difficult to solve for anions for reasons described below.

4.3 Fluorescence Imaging of Chloride—Historical

Fluorescence microscopy in cells relies on the excitation of a fluorophore and the detection of the emitted light. A good signal-to-noise ratio is only achieved if the excitation and emission probability of the fluorophore organic is high assuring that the detected signal is well above the thermal background signal of the detector. The emission signal (autofluorescence) generated by the excitation of endogenous fluorophores of the cell needs to be small compared to the signal of the fluorescent chloride sensor. Such imaging conditions can be achieved by choosing a specific excitation wavelength (laser wavelength, LED peak wavelength or lamp in combination with a bandpass filter) with high excitation probability and suitable optical filters for the emission optical pathway. The fluorescent sensor needs to be enriched within the cell to obtain sufficiently high fluorescence signals. Care has to be taken that scattered excitation light is well suppressed by optical filters.

A huge variety of fluorescence microscopy modalities have been developed since the invention of the first fluorescence microscope more than a 100 years ago [41, 66]. A number of readout modes do exist, which use different parameters of the emitted light to obtain contrast as for example: Fluorescence intensity, fluorescence intensity decay time, fluorescence anisotropy, fluorescence anisotropy decay or fluorescence correlation time [13, 27, 84, 85]. Some popular readout modes are secondary physical quantities, e.g. amplitudes of multi-exponential fluorescence intensity decay analysis (or their ratios) as well as ratios of fluorescence intensities using either two excitation and one emission wavelengths or one excitation and two emission wavelengths. In case of fluorescent chloride imaging three methods are of relevance: intensity, decay time and ratio.

Two different classes of fluorescent chloride sensors have been developed—organic fluorescent dyes based on collisional quenching and genetically encoded sensors based on fluorescent proteins (FP) with chloride-sensitive fluorescence or absorption properties. While both sensor types have been used in intensity-based fluorescent chloride imaging, the ratio approach is nearly exclusively applied for the latter type of sensors. In contrast, the fluorescence intensity decay analysis is almost solely used with organic fluorescent dyes. These different approaches have their origin in the sensing process and the resulting signal contrast one can achieve with the best and most commonly used fluorescent chloride sensors from each class (see below).

4.3.1 Genetically Encoded Fluorescent Sensors for Chloride Based on FPs

FP-based Cl^- sensors have a long history. A recent review gives a nice overview about the different sensors and their applications [16]. Early on in the generation of spectral variants of wt-GFP, a mutant with red-shifted absorption and emission spectra was generated (YFP T203Y/S65G/V68L/S72A; [94]), which is nowadays widely used as acceptor in FRET applications with CFP as donor. It was found that YFP was strongly pH and halide sensitive both affecting its performance as a FRET partner. Soon, pH- and halide-insensitive variants of YFP were produced. In parallel, the development of genetically encoded Cl^- sensors had been started [33, 48, 93]. In all those variants, binding of the Cl^- anion leads to a shift of the protonation equilibrium of the YFP chromophore toward the non-emissive, neutral form. The concentration of the anionic form with absorption around 515 nm is decreased upon Cl^- binding and the sensor gets dimmer. These sensors found great applications, for instance, in screening mutants of Cl^- transporters and channels. The Cl^- binding, however, can be described like a static quenching, and therefore, does not influence the fluorescence lifetime, which makes these sensors unsuitable for FLIM.

A FRET-based genetically encoded ratiometric fluorescent sensor for Cl^- has been described by Kuner and Augustine in 2000 [61], which they refer to as a "Clomeleon." To make their sensor ratiometric, they fused YFP with another, Cl^--insensitive GFP analogue CFP in order to create a FRET pair, whereby a ratio between the FRET-dependent emissions can be analyzed in presence as well as in absence of Cl^- ions. Some impressive examples of imaging physiological processes in vital brain or retina slices of transgenic "Clomeleon" mice have been reported [11, 24, 25, 37]. The group of Kuner produced other variants with different linkers or other YFP mutants, but without significant improvement.

While fluorescence lifetime contrast was observed in cultured neurons, no such application in neuronal or other tissues has been reported. Besides the relatively small dynamic range of "Clomeleon" near physiological pH, the very complicated fluorescence decay and the pH dependence of its donor (CFP) [35, 51, 76, 91] may be responsible for the lack of such studies. In 2008, an improved FRET-based Cl^--sensor named "Cl-sensor" has been introduced with a higher Cl^- affinity compared to "Clomeleon" [71]. It is much more chloride sensitive at physiological pH and shows a good performance in neuronal tissue [75, 96]. While the dynamic range appears to be good enough, it still uses CFP as a donor and no FLIM application has been performed until now.

Recently, Augustine and coworkers published a new "Clomeleon" variant ("SuperClomeleon" [40]) enhancing the Cl^--affinity almost by a factor of six (from 119 to 22.6 mM) by introducing four point mutations and a shorter linker. The replacement of the original donor (ECFP) by a less pH-dependent cyan GFP variant (Cerulean) lowers the pH sensitivity of this sensor. "SuperClomeleon" outperforms the original "Clomeleon" by far. Using "SuperClomeleon" inhibitory synaptic transmission of single neurons could be observed with good signal to noise ratio, which was non-detectable by "Clomeleon".

The major deficit all GFP-based Cl^- sensors have in common is that their fluorescence is chloride as well as pH dependent. In fact, the fluorescence properties of the FP are not influenced by Cl^- itself but rather by its protonation state, which is also affected by the environmental pH. As a consequence, all sensors have high pH-dependence of their Cl^- affinity. For quantitative measurements, the intracellular pH has to be determined independently in every single cell to calculate the Cl^- concentration. Beltram and coworkers generated an elegant sensor, which might circumvent this problem. They fused a dual-colour fluorescent GFP (E^2GFP) that shows a pH-dependent fluorescence [2] in a range, where neutral and anionic forms are both fluorescent, to a red emitting FP whose fluorescence is neither pH nor Cl^- dependent [3]. In addition to its pH dependence, the fluorescence of E^2GFP (neutral and anionic) is also statically quenched by Cl^- without changing the protonation state of the E^2GFP chromophore [2]. In this case, intracellular pH and intracellular Cl^- concentration can be determined simultaneously. Unfortunately, the static quenching mechanism does not lead to fluorescence lifetime contrasts.

It has to be concluded that the currently available Cl^- sensors are not suitable for relative or absolute Cl^- concentration measurements based on FLIM, neither in cultured cells nor in tissue. The FRET-based sensors need to be improved to get a larger dynamic range and a less complicated fluorescence decay of the FRET donor to allow FLIM. The pH dependence, however, will remain a problem.

4.3.2 Fluorescent Organic Dyes Sensing Chloride Based on Quinolinium Dyes

For most of the quinolinium dyes like SPQ, MEQ or MQAE, the fluorescence lifetime has been determined and is very long (up to 25 ns, see, e.g. [90] and Fig. 4.1a) compared to the fluorescence lifetime of classical fluorophore families (fluoresceins, rhodamines, oxazines, porphyrines, flavins, or cyanines). However, only one of these dyes (MQAE) is used in FLIM studies to determine $[Cl^-]_{int}$. Details of the properties and applications of the quinolinium dyes are well documented [14, 15, 34, 45, 46, 47, 49, 54, 81, 88, 89].

While the extinction coefficient is rather low, for both one- and two-photon excitation (2800 $cm^{-1} M^{-1}$ [350 nm] and 1.05 GM [750 nm]; hence 30 times smaller than those of fluorescein), the fluorescence quantum yield is high [42, 69, 90]. The fluorescence lifetime in the absence of Cl^- is extremely long (25 ns; [54, 90]). Noteworthy, the intracellular Stern–Volmer constant of MQAE determined in different cell types differs by almost one order of magnitude (see Table 4.1). Altogether, MQAE is a well-suited fluorescent dye for FLIM studies.

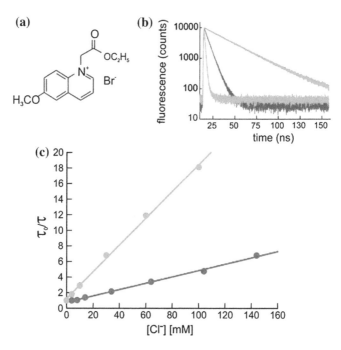

Fig. 4.1 **a** Chemical structure of MQAE; **b** fluorescence decay curves of MQAE (100 μM) dissolved in bi-distilled water in the presence of 0 mM (*green*), 20 mM (*pink*), and 100 mM (*cyan*) Cl^- (λ_{exc} = 380 nm; λ_{em} = 450 nm); **c** Stern–Volmer plot for MQAE in bi-distilled water for Cl^- (*green circles*; K_{SV} = 172 mM^{-1}) and a solution simulating intracellular ion concentrations (0–150 mM KCl, 150–0 mM Kalium gluconate, 2 mM $MgCl_2$, 20 mM HEPES buffer pH 7.2, 2 mM Na_2ATP, *red circles*; K_{SV} = 40.2 mM^{-1}). The *straight lines* are fits through the data points using (4.1), the slope equals K_{SV}

Table 4.1 Stern–Volmer constant K_{SV} and fluorescence lifetime τ_0 at 0 mM [Cl⁻] from intracellular calibrations of the fluorescent Cl⁻ sensor MQAE

Cell type	K_{SV} (mM⁻¹)	τ_0 (ns)	References
Phosphate buffer	185	25.1	[54]
Rat dorsal root ganglion (DRG) neurons	9.1	6.8	[54]
Perineuronal satellite cells from rat DRG	29.1	9.1	[54]
Rat olfactory sensory (OSN) neurons in olfactory epithelium	18	6.5	[55]
Mouse olfactory sensory (OSN) neurons in olfactory epithelium	13	4.9	[55]
Mice DRG neurons	3.05	3.9	[36]
Cockroach salivary duct cells	9.6	5.8	[42]
Cockroach salivary acinar peripheral cells	5.8	5.5	[63]
Cultured mice astrocytes (GLAST positive)	6.8	3.6	[87]

Only two research groups have been investigating Cl⁻ concentrations by FLIM in biological tissues so far—in neuronal tissue of rodents (Frings/Gensch) and in insect salivary glands (Dosche/Hille). In all studies, the Cl⁻ sensitive dye MQAE was used. The neuronal tissues contain neuron types (olfactory epithelium: olfactory sensory neurons (OSNs); dorsal root ganglion: nociceptors (DRG neurons)) for which elevated Cl⁻ concentrations have been proposed (based, e.g., on results of electrophysiological experiments). For the dendritic knobs in rat OSNs, values of 54 mM (50 mM extracellular Cl⁻) and 69 mM (150 mM extracellular Cl⁻) have been determined as well as a Cl⁻ gradient from the knob down the dendrite for high extracellular chloride concentration [55]. The knobs of OSNs in mice olfactory epithelium showed slightly lower resting Cl⁻ concentrations (37 at 50 mM extracellular Cl⁻).

The Cl⁻ concentration in DRG neurons [36] was very high in dorsal root ganglia from newborn mice (P1–P4, 77 mM) and lower in DRG neurons from mice in the third postnatal week (62 mM). These results represent a nice visualization of a physiological phenomenon in neuron maturation named the "chloride switch". Closer inspection revealed three groups of neurons with high, medium, and low intracellular Cl⁻ concentrations in the third postnatal week. The group of neurons with low Cl⁻ concentration cannot be found in dorsal root ganglia of newborn mice. It has to be noted that an intra-tissue calibration was not possible. Instead, the Stern–Volmer constant was determined in primary cultures of DRG neurons. In a follow-up study, the influence of inflammatory mediators on the Cl⁻ concentration of DRG neurons in intact, vital dorsal root ganglia was investigated [30]. An increase in a subset of DRG neurons was proven, but no absolute Cl⁻ concentrations are given due to the lack of intra-tissue calibration of MQAE.

The salivary gland system of insects serves as a model system to study the ion transport in epithelial tissue of higher organisms like mammals. Hille et al. [42] determined a high [Cl⁻]$_{int}$ of 59 mM in duct cells of the cockroach salivary glands. Later, the same group observed an increase in [Cl⁻]$_{int}$ after raising the extracellular chloride concentration [57] and determined [Cl⁻]$_{int}$ of a second cell type (acinar peripheral cells, 49 mM) of the salivary gland lobes [63]. The authors unwantedly highlighted the need for an intracellular calibration of the indicator dye sensitivity in the cell of interest. They first used the Stern–Volmer constant of MQAE and the fluorescence lifetime at 0 mM Cl$^-$ determined in duct cells (reasoning that those cells are from the same tissue and serving a similar function) and calculated a too high [Cl$^-$]$_{int}$ [62], which was corrected in a later publication [63]. Based on the intracellular Cl$^-$ concentration determined by FLIM, the Cl$^-$ transport in acinar peripheral cells is characterized with respect to blocking of anion transporters and dopamine stimulation.

4.4 Fluorescence Lifetime Imaging of Chloride—Practical

4.4.1 General Principle

As outlined above, fluorescence lifetime imaging of quinolinium dyes like MQAE (see Fig. 4.1a) has been established for the determination of absolute chloride concentrations in biological tissue. The fluorescence decay of MQAE in aqueous solutions of varying Cl$^-$ concentrations is large (in Fig. 4.1b the examples for 0, 20 and 100 mM chloride concentration in bi-distilled water are shown). The fluorescence lifetime can change by more than 40-fold for physiological Cl$^-$ concentrations.

The determination of Cl$^-$ concentrations is based on the mechanism of quenching of MQAE fluorescence by Cl$^-$ ions, namely the physical (also named dynamical) quenching. Quenching occurs upon collisions of Cl$^-$ ions and excited MQAE molecules that lead to an inverse proportionality dependence of the fluorescence lifetime to the Cl$^-$ concentration. The same dependence exists for the fluorescence intensity. Still, the fluorescence lifetime is the preferred readout parameter, since it is an intensive physical quantity that does not depend on e.g. the number of MQAE molecules or the power of the excitation light as it is true for the fluorescence intensity:

$$\tau/\tau_0 = 1 + K_{SV}[Cl^-] \tag{4.1}$$

K_{SV} (Stern–Volmer constant) and τ_0 (fluorescence lifetime in the absence of Cl$^-$) are parameters specific for MQAE depending on the surrounding environment (e.g. composition of solution or cell type). If possible, both parameters should be determined in that particular environment, although only K_{SV} is necessary. In Fig. 4.1c K_{SV} is determined for MQAE in bi-distilled water as well as in a solution mimicking intracellular conditions. The fluorescence lifetimes, τ, of MQAE obtained from analysis of fluorescence decays at different Cl$^-$ concentrations are set in relation to τ_0 and are plotted as a function of the chloride concentration. The slope of a linear fit through the data points yields K_{SV}.

The determination of the Stern–Volmer constant K_{SV} is a crucial step for intracellular chloride concentration measurements using FLIM of MQAE filled cells.

For the most commonly used method, the two-ionophore method, a Cl^-/OH^- antiporter, tributyltin, and a K^+/H^+ antiporter, nigericin, adjust the intra- and extracellular chloride concentration in presence of high potassium [59]. This method has been used for all experiments applying MQAE or SPQ. A number of other methods have been suggested and applied to calibrate genetically encoded sensors (Clomeleon and Cl-sensor), either by dialyzing different chloride concentrations through a patch pipette directly into the cell [24, 61] or by applying different detergents (Triton X-100 [32]; β-escin [96]). The only direct comparison of two calibration methods is found for the Cl^- sensor in CHO cells, where the K_{SV} values found by two-ionophore method or β-escin were almost identical [71, 96].

Table 4.1 summarizes all intracellular K_{SV} values of MQAE known from literature for different cell types. It is obvious that K_{SV} of MQAE is much smaller than the one in pure water, but six to sixties times lower depending on the different cell types. K_{SV} vary almost by a factor of 10—highlighting the necessity of a proper intracellular calibration of MQAE for every cell type studied. Once determined, the individual K_{SV} can be used to transform any measured MQAE fluorescence lifetime of the corresponding cell type into a Cl^- concentration using (4.1). The general procedure is shown with the example of GLAST-positive cultured astrocytes from mouse brain (Sect. 4.6 and Fig. 4.3).

Careful analysis of the fluorescence intensity decay of MQAE shows that even in pure bi-distilled water it cannot be described by a single mono-exponential function (data not shown). When using a bi-exponential function, both decay times are Cl^- dependent. An explanation on molecular scale for this behaviour might be two different interaction sites for collisions with Cl^-. However, if one limits the number of counts to typical values in FLIM experiments on cells (going from 500,000 to 5,000,000 counts in solution experiments to 50,000 or less in a fluorescence decay in a pixel of a FLIM image of cells, see Fig. 3 in [54]) a mono-exponential analysis is sufficient. Nevertheless, often there is the need to use a bi-exponential function for the analysis of MQAE FLIM images in cells. At least three possible explanations exist:

1. MQAE shows self-quenching in the low mM range [54].
2. Many other negatively charged small molecules or protein domains do quench the fluorescence of MQAE.
3. The hydrolyzed form of MQAE (much less membrane permeable) has a twofold lower K_{SV} compared to MQAE [54].
4. Inhomogeneous distribution of MQAE or the quenching entities as well as incomplete hydrolysis of MQAE [58] may therefore lead to the often observed multi-exponential nature of the fluorescence decay that usually can be described with a bi-exponential model function. In such cases, the amplitude weighted average fluorescence lifetime [8] is used for analysis (see e.g. [63]).

4.4.2 FLIM Measurement Setup

In fluorescence lifetime images the contrast is generated by plotting the fluorescence lifetime of the excited fluorophore (here MQAE) in spatial coordinates (x- and y-, more rarely also in z-direction). Since the fluorescence lifetime of suitable fluorophores is in the range of a few picoseconds to 10 ns one needs a fluorescence detection scheme with a picosecond time-resolution. To date, different measurement principles have been used to monitor the fluorescence decay data in combination with imaging: Gated image intensifiers [1, 23], streak cameras [12, 60], modulated image intensifiers [31, 64], modulation techniques with point scanning [29], scanning with analog-signal recording by fast digitisers [20], gated photon counting in several parallel time gates [18], conventional TCSPC with slow scanning [17], and multi-dimensional time correlated single photon counting (TCSPC) with fast point scanning [4, 5, 8]. The techniques differ in their time resolution, capabilities of resolving multi-exponential decay profiles, possibility of optical sectioning, multi-wavelength capabilities, photon efficiencies, acquisition times, and intensities they can be used at, see [10, 39, 84] and Chap. 3 for a comparison of different techniques. It is well accepted that TCSPC delivers the highest photon efficiency and thus the most accurate results at lower signal intensities. This fact is very important in biological experiments, since it allows the use of lower excitation intensities leading to less phototoxicity and longer observation times.

TCSPC-based FLIM requires an ultra-short pulsed excitation laser (50 fs–50 ps pulse width) with high repetition rate (10–80 MHz) that is scanned over the sample in x- and y-direction. Excitation can be obtained by the classic one-photon excitation process or by multi-photon excitation.

One-photon excitation excites fluorescence in a double-cone extending through the full depth of the sample. To obtain clear images from the focal plane (and thus avoid contamination of the fluorescence decay data with out-of-focus components) confocal detection is used. The fluorescence light passes back through the scanner and is focused into a pinhole on the upper focal plane of the microscope lens. Only light from the focal plane can pass the pinhole efficiently, out-of-focus light is suppressed. The principle is shown in Chap. 2, Fig. 2.6. The disadvantage of one-photon excitation is that emission light scattered on the way out of the sample does not pass the pinhole. One-photon excitation with confocal detection is therefore not suitable for thick-tissue imaging. Nevertheless, the suppression of scattered light and out-of-focus light is a huge advantage over wide-field imaging, see Chap. 2, Fig. 2.3.

Multiphoton excitation uses the simultaneous absorption of several (usually two) photons to excite the sample [38]. The use of multiphoton excitation in laser scanning microscope has been suggested by Wilson and Sheppard [98] and practically introduced by Denk et al. [21].

To obtain reasonable multiphoton absorption high spatial and temporal photon densities are required. Multiphoton microscopes therefore use femtosecond lasers as excitation sources. The most commonly employed lasers are Titanium-Sapphire lasers which have the additional advantage of being tuneable. The peak power of the pulses is high enough to obtain multiphoton excitation without the need of excessively high average power. Due to the nonlinearity of multiphoton absorption noticeable excitation is obtained only in the focus of the laser. Multiphoton excitation has several advantages over one-photon excitation: The excitation wavelength is in the near infrared where the (one-photon) absorption and scattering coefficients are lower than in the visible and ultraviolet range. Multiphoton excitation therefore penetrates deeper into biological tissue, see Chap. 2, Fig. 2.6. It also causes less phototoxicity: There is virtually no direct absorption of the laser light by tissue constituents, and no absorption outside the focal plane. This is an advantage especially for Cl^- imaging by MQAE. One-photon excitation would require a wavelength 350–400 nm which is strongly absorbed by NAD(P)H and FAD. The equivalent two-photon excitation wavelength (700–800 nm) is far less harmful to cells. Live samples can therefore be imaged for many minutes without impairing viability [54].

Another advantage of multiphoton excitation is that it does not require a pinhole to suppress out-of-focus light. The fluorescence can therefore be diverted from the excitation beam directly behind the microscope lens and transferred to a large-area detector. This 'non-descanned detection' (NDD) principle enables to record also photons which are scattered on the way out of the sample and assign them to the correct pixel of the image. The result is high detection efficiency even in deep layers of a tissue sample. Multiphoton excitation with non-descanned detection is described in Chap. 2, Fig. 2.7. An overview on FLIM applications is given in Chap. 3.

The combination of TCSPC with laser scanning is described in Chap. 1, Sect. 1.4. 5, and Chap. 2. TCSPC-based FLIM uses the multi-dimensional features of advanced TCSPC: The photon distribution is recorded not only over the times of the photons in the fluorescence decay but also over the coordinates of the scan [5, 8, 10]. The result can be interpreted as an array of pixels, each containing a full fluorescence decay curve in a large number of time channels. The recording process is independent of the scan rate. In particular, it works at the fast scan rates of modern laser scanning microscopes. The recording is simply continued over as many frames of the scan as necessary to obtain a sufficient signal-to-noise ratio of the result. At high scan rates the pixel rate can even be higher than the photon rate. The recording is than rather a probing of the sample by randomly detected photons than a point-by-point acquisition of decay data. The high scan rate has also other advantages: A fast preview function is available to adjust sample position and focus, and heat generation and excessive triplet accumulation in the illuminated spot are avoided.

TCSPC FLIM can be used to record more complex photon distributions, e.g. multi-wavelength FLIM data [5, 6], Z stack data, spatial mosaic data, time-series data [83], and fast temporal changes in the fluorescence decay functions [9], see also Chaps. 2 and 5.

TCSPC FLIM can be used with various photon counting detectors. Early systems used fast PMTs or MCP PMTs [4, 5]. Single-photon avalanche photodiodes (SPADs) have been introduced later but are limited to confocal detection systems because of their small area. Recent TCSPC FLIM systems almost exclusively use hybrid detectors. The advantage of these detectors is that they have high detection efficiency, clean instrument response functions, no afterpulsing, FCS capability, and an area large enough for non-descanned detection in multiphoton microscopes [7]. For multi-wavelength recording multi-anode PMTs with routing electronics are used [5, 6], see Chap. 1, Sect. 1.4.5.2, Chap. 2, Sect. 2.4.2, and Chap. 3, Sect. 3.5.11.

A frequent concern against TCSPC FLIM is that pile-up can occur at high count rates. TCSPC is based on the assumption that, on average, less than one photon is detected per excitation pulse. The detection of a possible second photon in the same pulse period must be small enough that it can be neglected. Pile up was a severe problem in early TCSPC systems working at pulse repetition rates in the kHz range. With excitation rates in the 50–80 MHz range as it is commonly used for FLIM pile-up is no longer a real problem. The pile-up error as a function of the count rate is shown in Fig. 1.5, Chap. 1. The lifetime error induced by pile up remains smaller than 2.5 % up to a detection rate of 10 % of the laser repetition rate [5, 8]. For a titanium-sapphire laser running at 80 MHz this corresponds to 8 MHz detection rate. This is far more than the count rates normally obtained from live samples [8].

TCSPC FLIM data are usually analysed by fitting the decay data with a single, double, or triple-exponential decay model. The fit procedure delivers the lifetimes and the amplitudes of the decay components, or amplitude or intensity-weighted lifetimes. Also the first moment of the decay data in the pixels can be used to determine the lifetime of a single-exponential approximation. A second way of data analysis is the 'Phasor' method. The phasor method transforms the data into the frequency domain and determines the phase and the modulation degree. These numbers are used as a signature of the fluorescence decay. Time-domain data analysis is described in detail in [8], phasor analysis in [22]. A brief description of TCSPC FLIM data analysis can be found in Chap. 15, Sect. 15.4.

4.5 Reversible Change of $[Cl^-]_{int}$ in a KCC2 Stable Cell Line

Chloride channels and transporters are broadly expressed in brain, kidney or lung and appear to play an important role in regulating $[Cl^-]_{int}$. KCC2 [77, 80] is a member of the SLC12 gene family. These cation-chloride co-transporters play an important role for the functioning of the central nervous system (CNS) [73]. A rare variant of KCC2 was recently shown to increase the risk of epilepsy [44, 53, 79]. In neurons, its expression is age dependent. KCC2 facilitates an electro-neutral ion transport with K^+ and Cl^- in a 1:1 stoichiometry. The sum of the ion gradients determines the ion transport direction. KCC2 (see Fig. 4.2a) exists as monomer as well as oligomer, but its functional form has not been identified unequivocally.

We decided to generate a cell line that expresses a chloride transporting protein constitutively by stably transfecting HEK293 cells (see e.g. [86]; source: European Collection of Cell Cultures) with the cDNA encoding the potassium-chloride

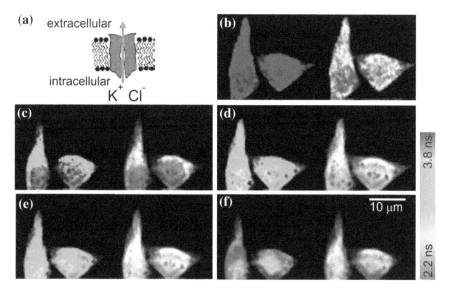

Fig. 4.2 a Schematic illustration of the potassium-chloride symporter KCC2; **b–f** FLIM image (*left*) and fluorescence intensity image (*right*) of HEK293 cells (two-photon excitation; λ_{exc} = 750 nm; λ_{obs} < 500 nm) expressing KCC2 (KCC2-HEK293) in aqueous solutions of 140 mM Na^+, 1 mM Mg^{2+}, 2 mM Ca^{2+}, 5 mM HEPES (pH 7.2) and varying K^+, Cl^-, Na^+ and gluconate concentrations: **b** 140 mM Na^+, 4 mM K^+, 150 mM Cl^- and 0 mM gluconate; **c** 0 mM Na^+, 144 mM K^+, 150 mM Cl^- and 0 mM gluconate (2 min incubation); **d** 0 mM Na^+, 144 mM K^+, 150 mM Cl^- and 0 mM gluconate (10 min incubation); **e** 0 mM Na^+, 144 mM K^+, 10 mM Cl^- and 140 mM gluconate (20 min incubation); **f** 0 mM Na^+, 144 mM K^+, 10 mM Cl^- and 140 mM gluconate (60 min incubation)

co-transporter KCC2 by a modified $CaPO_4$ co-precipitation method according to Chen and Okayama [19]. The cell line was dubbed KCC2-HEK293. Notably, HEK293 cells do not express KCC2 endogenously [32].

The fluorescence intensity and lifetime micrographs depicted in Fig. 4.2b–f have been acquired with two-photon excitation of MQAE-filled KCC2-HEK293. The series demonstrates a nearly reversible Cl^- import and subsequent export governed by the K^+ and Cl^- ion gradients between intracellular space and extracellular medium. Since the calibration of MQAE in KCC2-HEK293 has not been performed the following data are discussed quantitatively by using the fluorescence lifetime as a relative measure for $[Cl^-]_{int}$. Fluorescence decays had to be fitted with a bi-exponential function and the average fluorescence lifetime was used as readout parameter for internal chloride concentrations (see section General principle and [63]). Notably, the MQAE intensity signal was inhomogenously distributed in KCC2-HEK293 cells—especially at the beginning of the experiment—as it is the case for many other cell types studied so far. These organelle structures as well as the apparent differences in lifetime in the nucleus compared to the cytosol are not further discussed.

Under physiological conditions (4 mM K^+/140 mM Na^+/150 mM Cl^-) the fluorescence lifetime is long (>3.8 ns) corresponding to a low chloride concentration (Fig. 4.2b). Due to the high intracellular K^+ concentration (140 mM) KCC2 acts as Cl^- exporter and keeps $[Cl^-]_{int}$ around 4 mM [13, 67]. This is manifested by the long fluorescent lifetime measured under physiological conditions in Fig. 4.2b.

In the next step the composition of the extracellular solution was changed to high K^+ and Cl^- (144 mM K^+/150 mM Cl^-; Fig. 4.2c, d; cation concentration was kept constant by removing Na^+). Within 2 min the fluorescence lifetime decreased by around 1 ns (2.2–3 ns) with an inhomogeneous distribution (Fig. 4.2c). Obviously, the inversed Cl^- gradient drives the transport of Cl^- through KCC2 into the cells, due to the disappearance of the K^+ gradient. Within 8 min the fluorescence lifetime distribution became uniform with an average fluorescence lifetime of 2.7–3.1 ns (Fig. 4.2d).

At that point, the extracellular solution was replaced again, keeping the extracellular K^+ concentration high and lowering extracellular Cl^- concentration to 10 mM (144 mM K^+/10 mM Cl^-; Fig. 4.2e, f; osmolarity was kept constant by adding 140 mM gluconate). Again, as shown in Fig. 4.2e, f, the direction of ion transport through KCC2 was inverted, i.e., Cl^- and K^+ were transported out of the cell. This happened much slower compared to the first solution exchange. After 20 min the fluorescence lifetime is between 2.8 and 3.4 ns (Fig. 4.2e). In the next 40 min ion transport out of the cell has continued and $[Cl^-]_{int}$ reached values in the same range as at the beginning of this measurement with values from 3.4 to 4 ns (Fig. 4.2f). The K^+ concentration was kept high in the third solution so that no K^+ gradient but only the inversed Cl^- gradient would act on the KCC2-HEK293 cells. The observed reversed change of the fluorescence lifetime is a strong hint that indeed Cl^- concentration changes are observed in the experiment depicted in Fig. 4.2. The somewhat shorter fluorescence lifetimes of MQAE at the end of the experiment (Fig. 4.2f) compared to the beginning (Fig. 4.2b) is reasonable, since the minimal intracellular Cl^- concentration should not under-run that of the extracellular solution (10 mM Cl^-) applied in Fig. 4.2e, f. The K^+ ions that are transported over the plasma membrane by KCC2 at the various stages of the experiment will be presumably transported fast in the opposite direction by various endogenous (e.g. voltage-gated) K^+ channels of the KCC2-HEK293 cells [50, 101], so that most likely the intracellular K^+ concentration stays throughout the experiment near the value of 144 mM known from unperturbed mammalian cells [67].

The reversible change of $[Cl^-]_{int}$ presented in Fig. 4.2 is a good training example of how time-resolved fluorescence microscopy with two-photon excitation of MQAE can be used to monitor different $[Cl^-]_{int}$ that are varied by manipulating extracellular ion concentrations (here due to the action of the K^+/Cl^- co-transporter KCC2). It proves the usefulness of the method for observing such intracellular Cl^- changes and absolute concentrations occurring in more complicated physiological processes in cells where many Cl^- channels and transporters are active. On the other hand, the KCC2-HEK293 cell line is a good tool to study the intracellular behaviour of different Cl^- sensitive fluorescent dyes, since it offers the opportunity to control $[Cl^-]_{int}$.

4.6 Intracellular Cl⁻ Concentration in GLAST-Positive Astrocytes

The family of excitatory amino acid transporter (EAAT) is broadly expressed in the central nervous system. Its different subtypes can be found in brain neurons (EAAT3 and 4) and glial cells (EAAT1 and 2) and the retina as well as in peripheral tissue (EAAT5). Their major role is the clearance of glutamate from the synaptic cleft; thereby they allow for rounds of synaptic transmission and at once keep the extracellular glutamate concentration below a neurotoxic level. EAATs are secondary active transporters but they function as anion channels as well [74]. While the glutamate transport is well described the physiological impact of the associated anion channel function is still unknown.

For in vitro investigations of EAAT1 anion channel function primary cell culture from whole brain was used. Astrocytes were isolated from this cell culture by magnetic-beads using an anti-GLAST (the mouse homologue of the human EAAT1) antibody as described by Jungblut et al. [52]. These cells—from now on named pc-antiGLAST-AC—were cultured in serum-free media with heparin-binding epidermal growth factor [28].

Primary cell cultures are often used as model systems to study physiological properties. These cells mimic the in vivo situation better than stable cell lines. EAAT1 (GLAST) and many other Cl^- transporting membrane proteins are expressed in astrocytes and the $[Cl^-]_{int}$ may vary considerably compared to other cells. In the following the determination of $[Cl^-]_{int}$ of pc-antiGLAST-AC cells loaded with MQAE is described using fluorescence microscopy with two-photon excitation and FLIM readout.

The calibration of pc-antiGLAST-ACs is shown in Fig. 4.3. It uses the two-ionophore method described before. As in the case of the KCC2-HEK293 cells, the fluorescence intensity decays as a function of time were fitted satisfactorily with a bi-exponential function. A mono-exponential function was not sufficient to describe the data (see section *General principle* and [63]). The average fluorescence lifetime changed (see Fig. 4.3, top) from around 3.5 ns at the lowest Cl^- concentration (4 mM) to about 2 ns at the highest Cl^- concentration (100 mM). Around 10 pc-antiGLAST-ACs were measured for each Cl^- concentration, and the mean value was calculated. Figure 4.3, bottom left, shows the Stern–Volmer plot generated from these $\tau_{average}$ values (see 4.1) that displays an evidently linear relationship. From the slope the K_{SV} of MQAE in pc-antiGLAST-ACs hs been calculated (6.8 mM^{-1}).

Using this value, the FLIM images of a large number of pc-antiGLAST-ACs were transformed into chloride images as the one depicted in Fig. 4.4. The distribution of $[Cl^-]_{int}$ (shown in Fig. 4.3, bottom right) determined in pc-antiGLAST-ACs is relatively broad with arithmetical mean and median values equal to 22.8 and 17.4 mM, respectively. The asymmetry of the distribution may arise from measuring different types of GLAST-expressing astrocytes.

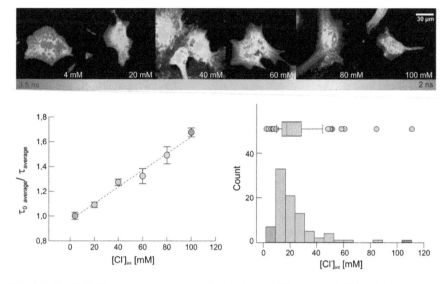

Fig. 4.3 *Top* FLIM images (two-photon excitation; λ_{exc} = 750 nm; λ_{obs} < 500 nm) of astrocytes loaded with MQAE in two-ionophore calibration solution (140 mM K^+, 10 mM Na^+, 10 mM HEPES, 4–140 mM Cl^-, 10–146 mM gluconate, 10 μM nigericin, 10 μM tributyltin, pH 7.4 adjusted with KOH) and varying Cl^- concentrations. Astrocytes are a primary culture from whole brain, isolated by antiGLAST (mouse EAAT1) magnetic-beads and cultured in serum-free media with heparin-binding epidermal growth factor. *Bottom left* Stern–Volmer analysis of average fluorescence lifetimes determined from intracellular calibration data of MQAE in cultured astrocytes (K_{SV} = 6.8 mM^{-1}). *Bottom right* Histogram of $[Cl^-]_{int}$ determined for 89 pc-antiGLAST-AC (DIV 5–DIV 26; isolated from two wild-type C57 B6 mice (10 to 16 days after birth): arithmetical mean = 22.8 mM; median = 17.4 mM. The box-whisker plot at the *top* visualizes that the central 50 % of $[Cl^-]_{int}$ values are within the *box* (*range* 14.8–28.48 mM) and the central 80 % of $[Cl^-]_{int}$ values are within the *whiskers* (*range* 9.7–44.4 mM)

4.7 Outlook

While fluorescence lifetime imaging of MQAE loaded cells and cell tissue has great opportunities and works in principle well (as shown in the examples and in the literature studies discussed in this chapter) there are a few limitations that prevented until now a more general usage of the method in physiological measurements both in the chloride-sensitive fluorescent sensors as well as in technical aspects of the method.

A complication is the long fluorescence lifetime of MQAE. Under physiological conditions, lifetimes on the order of 5 ns are obtained. Compared to the laser pulse period (12.5 ns) this is relatively long. The result is that the fluorescence does not entirely decay within the pulse period of the laser. One way to avoid this would be a reduction of the laser repetition rate by a pulse picker. However, lower repetition rate increases possible pile-up errors an thus reduces the maximum count rate the system can be used at. It is therefore better to run the laser at its full repetition rate

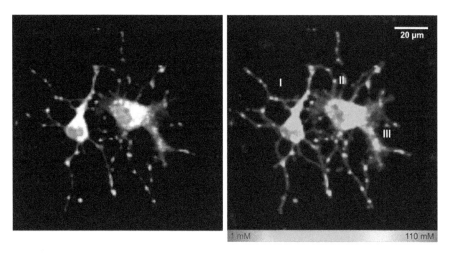

Fig. 4.4 Fluorescence intensity image (*left*) Cl⁻ map (*right*) of three pc-antiGLAST-AC with intracellular Cl⁻ concentrations of 17.5 mM (cell I), 17.7 mM (cell II) and 12.7 mM (cell III) derived from a FLIM image using the intracellular K_{SV} (6.8 mM^{-1}) determined in Fig. 4.3

and account for the incomplete decay by using an appropriate fit model in the data analysis [8].

One of the most severe constrictions of our system is the relatively long acquisition time of FLIM (typically 1–3 min). The acquisition speed can probably be reduced by using hybrid detectors [7]. Hybrid detectors typically have a 5–8 times higher photon detection efficiency, and deliver a better timing accuracy than conventional PMTs. As a result, the acquisition time can be reduced, usually by a factor of 10, even if the count rate is kept below the critical level of 5–8 MHz to avoid pile-up errors (see Chap. 1, Sect. 1.2). For a general discussion of the acquisition time of TCSPC FLIM see Chap. 2, Sect. 2.2.4.

Other ways to speed up the acquisition are scanning with lower pixel numbers [56], avoiding read-out times during subsequent measurements [56, 83], scanning only regions of interest [97] (which avoids wasting of time by scanning dark pixels) or intelligent binning in the data analysis, i.e. by using the binning function of SPCImage [8] or the inherent binning features of phasor analysis [22, 32, 82]. Other lifetime estimation techniques than nonlinear least-square fitting, like maximum likelihood estimation [72, 102] or the average photon arrival time histogram [70] have been applied successfully for the analysis of fluorescence decay with 50–300 photons. Such methods may allow faster acquisitions since they need less photons compared to least square fitting. However, for fluorescence data with more than a few 100 photons per pixel least-square fitting comes close to the theoretical accuracy of $1/N^{1/2}$ [8], so that possible improvements by these techniques are limited.

The problem of recording transient changes in the Cl⁻ concentration can be solved by advanced FLIM techniques which include the experiment time (the time after a stimulation) in the buildup of the photon distribution. Temporal mosaic imaging records transient changes with a resolution down to the frame time of the scanner. This is in the range of 40 ms to a few 100 ms. FLITS records transient changes along a line scan with a resolution down to the line time of the scanner, which is on the order of 1 ms. For both techniques, the total acquisition time is decoupled from the time resolution: The experiment is simply continued until enough photons have been recorded. Please see Chaps. 1, 2 and 5 of this book.

Another problem is the limited depth in a MQAE-stained biological tissue, where one can acquire FLIM images due to too low MQAE concentrations. Two-photon excitation allows to record FLIM images from regions 250 µm and deeper in the tissue, but MQAE staining is limited to the upper 50 µm to—at best—100 µm. Genetically encoded chloride sensors on the other hand do not have this limitation, but are not yet suited for FLIM imaging (see section Genetically encoded fluorescent sensors of chloride). Future chloride sensor development should take into account this issue and develop improved fluorophores solving this issue.

The determination and comparison of intracellular chloride concentrations will be relevant for solving pathomechanisms of diseases related to impaired or enhanced chloride transport (in e.g. brain, kidney, lung, muscle). Using cells and tissues of disease models and pharmacological blocking FLIM-based chloride concentration measurements helps not only understanding malfunctions occurring in diseases like ataxia or epilepsy (see e.g. [99]), but also could lead to the development of new therapeutic approaches.

References

1. A.V. Agronskaia, L. Tertoolen H.C. Gerritsen, High frame rate fluorescence lifetime imaging. J. Phys. D: Appl. Phys. **36**, 1655–1662 (2003)
2. D. Arosio, G. Garau, F. Ricci, L. Marchetti, R. Bizzarri, R. Nifosi, F. Beltram, Spectroscopic and structural study of proton and halide ion cooperative binding to GFP. Biophys. J. **93**, 232–244 (2007)
3. D. Arosio, F. Ricci, L. Marchetti, R. Gualdani, L. Albertazzi, F. Beltram, Simultaneous intracellular chloride and pH measurements using a GFP-based sensor. Nat. Meth. **7**, 516–518 (2010)
4. W. Becker, A. Bergmann, M.A. Hink, K. König, K. Benndorf, C. Biskup, Fluorescence lifetime imaging by time-correlated single photon counting. Microsc. Res. Technol. **63**, 58–66 (2004)
5. W. Becker, *Advanced Time-Correlated Single Photon Counting Techniques* (Springer, Berlin, 2005)
6. W. Becker, A. Bergmann, C. Biskup, Multi-spectral fluorescence lifetime imaging by TCSPC. Microsc. Res. Technol. **70**, 403–409 (2007)
7. W. Becker, B. Su, O. Holub, K. Weisshart, FLIM and FCS detection in laser-scanning microscopes: increased efficiency by GaAsP hybrid detectors. Microsc. Res. Technol. **74**, 804–811 (2011)
8. W. Becker, *The bh TCSPC Handbook*, 5th edn (Becker & Hickl GmbH, Berlin, 2012)

9. W. Becker, V. Shcheslavkiy, S. Frere, I. Slutsky, Spatially resolved recording of transient fluorescence-lifetime effects by line-scanning TCSPC. Microsc. Res. Technol. **77**, 216–224 (2014)
10. W. Becker, Fluorescence lifetime imaging—techniques and applications. J. Microsc. **247**, 119–136 (2012)
11. K. Berglund, W. Schleich, H. Wang, G. Feng, W.C. Hall, T. Kuner, G.J. Augustine, Imaging synaptic inhibition throughout the brain via genetically targeted Clomeleon. Brain Cell Biol. **36**, 101–118 (2008)
12. C. Biskup, T. Zimmer, K. Benndorf, FRET between cardiac Na + channel subunits measured with a confocal microscope and a streak camera. Nat. Biotechnol. **22**, 220–224 (2004)
13. C. Biskup, T. Gensch, in *Fluorescence Lifetime Imaging of Ions in Biological Tissues*, ed. by D. Elson, P. W. M. French and L. Marcu. Fluorescence Lifetime Spectroscopy and Imaging. Principles and Applications in Biomedical Diagnostics (Taylor & Francis, Boca Raton 2014)
14. J. Biwersi, A.S. Verkman, Cell-permeable fluorescent indicator for cytosolic chloride. Biochemistry **30**, 7879–7883 (1991)
15. J. Biwersi, B. Tulk, A.S. Verkman, Long-wavelength chloride-sensitive fluorescent indicators. Anal. Biochem. **219**, 139–143 (1994)
16. P. Bregestovski, D. Arosio, in *Green Fluorescent Protein-Based Chloride Ion Sensors for In Vivo Imaging*, ed. by G. Jung. Fluorescent Proteins II, Springer Ser. Fluoresc., vol. 12 (Springer, Berlin Heidelberg, 2012), pp. 99–124
17. I. Bugiel, K. Koenig, H. Wabnitz, Investigation of cells by fluorescence laser scanning microscopy with subnanosecond time resolution. Las. Life Sci. **3**, 47–53 (1989)
18. E.P. Buurman, R. Sanders, A. Draaijer, H.C. Gerritsen, J.J.F. van Veen, P.M. Houpt, Y.K. Levine, Fluorescence lifetime imaging using a confocal laser scanning microscope. Scanning **14**, 155–159 (1992)
19. C. Chen, H. Okayama, High-efficiency transformation of mammalian cells by plasmid DNA. Mol. Cell. Biol. **7**, 2745–2752 (1987)
20. S. Cheng, R.M. Cuenca, B. Liu, B.H. Malik, J.M. Jabbour, K.C. Maitland, J. Wright, Y.S. Cheng, J.A. Jo, Handheld multispectral fluorescence lifetime imaging system for in vivo applications. Biomed. Opt. Express **5**, 921–931 (2014)
21. W. Denk, J.H. Strickler, W.W. Webb, Two-photon laser scanning fluorescence microscopy. Science **248**, 73–76 (1990)
22. M.A. Digman, V.R. Caiolfa, M. Zamai, E. Gratton, The phasor approach to fluorescence lifetime imaging analysis. Biophys. J. **94**, L14–L16 (2008)
23. K. Dowling, M.J. Dayel, M.J. Lever, P.M.W. French, J.D. Hares, A.K.L. Dymoke-Bradshaw, Fluorescence lifetime imaging with picosecond resolution for biomedical applications. Opt. Lett. **23**, 810–812 (1998)
24. J. Duebel, S. Haverkamp, W. Schleich, G. Feng, G.J. Augustine, T. Kuner, T. Euler, Two-photon imaging reveals somatodendritic chloride gradient in retinal ON-type bipolar cells expressing the biosensor Clomeleon. Neuron **49**, 81–94 (2006)
25. V.I. Dzhala, K.V. Kuchibhotla, J.C. Glykys, K.T. Kahle, W.B. Swiercz, G. Feng, T. Kuner, G.J. Augustine, B.J. Bacskai, K.J. Staley, Progressive NKCC1-dependent neuronal chloride accumulation during neonatal seizures. J. Neurosci. **30**, 11745–11761 (2010)
26. V. Faundez, H.C. Hartzell, E.M. Adler, Teaching resources. Chloride concentration and pH along the endosomal pathway. Science's STKE : signal transduction knowledge environment, tr2 (2004)
27. F. Festy, S.M. Ameer-Beg, T. Ng, K. Suhling, Imaging proteins in vivo using fluorescence lifetime microscopy. Mol. BioSyst. **3**, 381–391 (2007)
28. L.C. Foo, N.J. Allen, E.A. Bushong, P.B. Ventura, W.S. Chung, L. Zhou, J.D. Cahoy, R. Daneman, H. Zong, M.H. Ellisman, B.A. Barres, Development of a method for the p purification and culture of rodent astrocytes. Neuron **71**, 799–811 (2011)
29. T. French, P.T.C. So, D.J. Weaver, T. Coelho-Sampaio, E. Gratton, E.W. Voss, J. Carrero, Two-photon fluorescence lifetime imaging microscopy of macrophage-mediated antigen processing. J. Microsc. **185**, 339–353 (1997)

30. K. Funk, A. Woitecki, C. Franjic-Würtz, T. Gensch, F. Möhrlen, S. Frings, Modulation of chloride homeostasis by inflammatory mediators in dorsal root ganglion neurons. Mol. Pain **4**, 32 (2008)

31. T.W.J. Gadella, T.M. Jovin, R.M. Clegg, Fluorescence lifetime imaging microscopy (FLIM)—spatial resolution of structures on the nanosecond timescale. Biophys. Chem. **48**, 221–239 (1993)

32. M. Gagnon, M.J. Bergeron, G. Lavertu, A. Castonguay, S. Tripathy, R.P. Bonin, J. Perez-Sanchez, D. Boudreau, B. Wang, L. Dumas, I. Valade, K. Bachand, M. Jacob-Wagner, C. Tardif, I. Kianicka, P. Isenring, G. Attardo, J.A.M. Coull, Y. De Koninck, Chloride extrusion enhancers as novel therapeutics for neurological diseases. Nat. Med. **19**, 1524–1528 (2013)

33. L.J. Galietta, P.M. Haggie, A.S. Verkman, Green fluorescent protein-based halide indicators with improved chloride and iodide affinities. FEBS Lett. **499**, 220–224 (2001)

34. C.D. Geddes, Optical halide sensing using fluorescence quenching: Theory, simulations and applications—a review. Meas. Sci. Technol. **12**, R53–R88 (2001)

35. A. Geiger, L. Russo, T. Gensch, T. Thestrup, S. Becker, K.-P. Hopfner, C. Griesinger, G. Witte, O. Griesbeck, Correlating calcium binding, FRET and conformational change in the biosensor TN-XXL. Biophys. J. **102**, 2401–2410 (2012)

36. D. Gilbert, C. Franjic-Wuertz, K. Funk, T. Gensch, S. Frings, F. Moehrlen, Differential maturation of chloride homeostasis in primary afferent neurons of the somatosensory system. Int. J. Dev. Neurosci. **25**, 479–489 (2008)

37. J. Glykys, V. Dzhala, K. Egawa, T. Balena, Y. Saponjian, K.V. Kuchibhotla, B.J. Bacskai, K.T. Kahle, T. Zeuthen, K.J. Staley, Local impermeant anions establish the neuronal chloride concentration. Science **343**, 670–675 (2014)

38. M. Göppert-Mayer, Über Elementarakte mit zwei Quantensprüngen. Ann. Phys. **9**, 273–294 (1931)

39. E. Gratton, S. Breusegem, J. Sutin, Q. Ruan, N. Barry, Fluorescence lifetime imaging for the two-photon microscope: time-domain and frequency-domain methods. J. Biomed. Opt. **8**, 381–390 (2003)

40. J.S. Grimley, L. Li, W. Wang, L. Wen, L.S. Beese, H.W. Hellinga, G.J. Augustine, Visualization of synaptic inhibition with an optogenetic sensor developed by cell-free protein engineering automation. J. Neurosc. **33**, 16297–16309 (2013)

41. O. Heimstädt, Das Fluoreszenzmikroskop. Zeitschr. wiss. Mikrosk. **28**, 330–337 (1911)

42. C. Hille, M. Lahn, G.-H. Loehmannsroeben, C. Dosche, Two-photon fluorescence lifetime imaging of intracellular chloride in cockroach salivary glands. Photochem. Photobiol. Sci. **8**, 319–327 (2009)

43. G. Huberfeld, L. Winter, D. Clemenceau, M. Baulac, K. Kaile, R. Miles, C. Rivera, Perturbed chloride homeostasis and GABAergic signaling in human temporal lobe epilepsy. J. Neurosci. **27**, 9866–9873 (2007)

44. C.A. Hübner, The KCl-cotransporter KCC2 linked to epilepsy. EMBO Rep. **15**, 732–733 (2014)

45. N.P. Illsley, A.S. Verkman, Membrane chloride transport measured using a chloride-sensitive fluorescent probe. Biochemistry **26**, 1215–1219 (1987)

46. S. Jayaraman, L. Teitler, B. Skalski, A.S. Verkman, Long-wavelength iodide-sensitive fluorescent indicators for measurement of functional CFTR expression in cells. Am. J. Phys. Cell Phys. **277**, C1008–C1018 (1999a)

47. S. Jayaraman, A.S. Verkman, Quenching mechanism of quinolinium type chloride-sensitive fluorescent indicators. Biophys. Chem. **85**, 49–57 (2000)

48. S. Jayaraman, P. Haggie, R.M. Wachter, S.J. Remington, A.S. Verkman, Mechanism and cellular applications of a green fluorescent protein-based halide sensor. J. Biol. Chem. **275**, 6047–6050 (2000)

49. S. Jayaraman, J. Biwersi, A.S. Verkman, Synthesis and characterization of dual-wavelength Cl^--sensitive fluorescent indicators for ratio imaging. Am. J. Physiol. Cell Physiol. **276**, C747–C757 (1999b)

50. B. Jiang, X. Sun, K. Cao, R. Wang, Endogenous KV channels in human embryonic kidney (HEK-293) cells. Mol. Cell. Biochem. **238**, 69–79 (2002)
51. M. Jose, D.K. Nair, C. Reissner, R. Hartig, W. Zuschratter, Photophysics of clomeleon by FLIM: Discriminating excited state reactions along neuronal development. Biophys. J. **92**, 2237–2254 (2007)
52. M. Jungblut, M.C. Tiveron, S. Barral, B. Abrahamsen, S. Knöbel, S. Pennartz, J. Schmitz, M. Perraut, F.W. Pfrieger, W. Stoffel, H. Cremer, A. Bosio, Isolation and characterization of living primary astroglial cells using the new GLAST-specific monoclonal antibody ACSA-1. Glia **60**, 894–907 (2012)
53. K.T. Kahle, N.D. Merner, P. Friedel, L. Silayeva, B. Liang, A. Khanna, Y. Shang, P. Lachance-Touchette, C. Bourassa, A. Levert, P.A. Dion, B. Walcott, D. Spiegelman, A. Dionne-Laporte, A. Hodgkinson, P. Awadalla, H. Nikbakht, J. Majewski, P. Cossette, T.Z. Deeb, S.J. Moss, I. Medina, G.A. Rouleau, Genetically encoded impairment of neuronal KCC2 cotransporter function in human idiopathic generalized epilepsy. EMBO Rep. **15**, 766–774 (2014)
54. H. Kaneko, I. Putzier, S. Frings, T. Gensch, *Determination of Intracellular Chloride Concentration in Dorsal Root Ganglion Neurons by Fluorescence Lifetime Imaging*, ed. by C.M. Fuller. Calcium-Activated Chloride Channels, Book Series: Current Topics in Membranes, vol. 53 (Academic Press, San Diego, 2002), pp. 167–194
55. H. Kaneko, I. Putzier, S. Frings, U.B. Kaupp, T. Gensch, Chloride accumulation in mammalian olfactory sensory neurons. J. Neurosci. **24**, 7931–7938 (2004)
56. V. Katsoulidou, A. Bergmann, W. Becker, How fast can TCSPC FLIM be made? Proc. SPIE **6771**, 67710B-1–67710B-7 (2007)
57. F. Koberling, V. Buschmann, C. Hille, M. Patting, C. Dosche, A. Sandberg, A. Wheelock, R. Erdmann, in *Fast Raster Scanning Enables FLIM in Macroscopic Samples Up to Several Centimetres*, Multiphoton Microscopy in the Biomedical Sciences X, Proceedings of SPIE, vol. 7569 (2010), p. 756931
58. C. Koncz, J.T. Daugirdas, Use of MQAE for measurement of intracellular [Cl^-] in cultured aortic smooth muscle cells. Am. J. Physiol. **267**, H2114–H2123 (1994)
59. R. Krapf, C.A. Berry, A.S. Verkman, Estimation of intracellular chloride activity in isolated perfused rabbit proximal convoluted tubules using a fluorescent indicator. Biophys. J. **53**, 955–962 (1988)
60. R.V. Krishnan, H. Saitoh, H. Terada, V.E. Centonze, B. Herman, Development of a multiphoton fluorescence lifetime imaging microscopy (FLIM) system using a streak camera. Rev. Sci. Instrum. **74**, 2714–2721 (2003)
61. T. Kuner, G.J. Augustine, A genetically encoded ratiometric indicator for chloride: capturing chloride transients in cultured hippocampal neurons. Neuron **27**, 447–459 (2000)
62. M. Lahn, C. Hille, F. Koberling, P. Kapusta, C. Dosche, in *pH and Chloride Recordings in Living Cells Using Two-Photon Fluorescence Lifetime Imaging Microscopy*, A. Periasamy, P.T.C. So, K. Konig. Multiphoton Microscopy in the Biomedical Sciences X, Proceedings of SPIE vol. 7569, 75690U (2010)
63. M. Lahn, C. Dosche, C. Hille, Two-photon microscopy and fluorescence lifetime imaging reveal stimulus induced intracellular Na^+ and Cl^- changes in cockroach salivary acinar cells. Am. J. Physiol. Cell Physiol. **300**, C1323–C1336 (2011)
64. J.R. Lakowicz, H. Szmacinski, K. Nowaczyk, K.W. Berndt, M. Johnson, Fluorescence lifetime imaging. Anal. Biochem. **202**, 316–330 (1992)
65. J.J. Li, R. Ji, Y.Q. Shi, Y.Y. Wang, Y.L. Yang, K.F. Dou, Changes in expression of the chloride homeostasis-regulating genes, KCC2 and NKCC1, in the blood of cirrhotic patients with hepatic encephalopathy. Exp. Ther. Med. **4**, 1075–1080 (2012)
66. H. Lehmann, Das Lumineszenz-Mikroskop, seine Grundlagen und seine Anwendungen. Zeitschr. wiss. Mikrosk. **30**, 418–470 (1913)
67. H. Lodish, A. Berk, C.A. Kaiser, M. Krieger, A. Bretscher, H. Ploegh, A. Amon, M.P. Scott, *Molecular Cell Biology*, 7th edn (W.H. Freeman & Co, New York, 2013), p. 485

68. H.J. Luhmann, S. Kirischuk, W. Kilb, Comment on "local impermeant anions establish the neuronal chloride concentration". Science **345**, 1130 (2014)
69. M.K. Mansoura, J. Biwersi, M.A. Ashlock, A.S. Verkman, Fluorescent chloride indicators to assess the efficacy of CFTR cDNA delivery. Hum. Gene Ther. **10**, 861–875 (1999)
70. A. Margineau, J. Hoota, M. van der Aueraer, M. Ameloot, A. Stefan, D. Beljonne, Y. Engelborghs, A. Herrmann, K. Muellen, F.C. De Schryver, J. Hofkens, Visualization of membrane rafts using a perylene monoimide derivative and fluorescence lifetime imaging. Bioph J. **93**, 2877–2891 (2007)
71. O. Markova, M. Mukhtarov, E. Real, Y. Jacob, P. Bregestovski, Genetically encoded chloride indicator with improved sensitivity. J. Neurosci. Meth. **170**, 67–76 (2008)
72. M. Maus, M. Cotlet, J. Hofkens, T. Gensch, F.C. De Schryver, J. Schaffer, C.A.M. Seidel, An experimental comparison of the maximum likelihood estimation and nonlinear least-squares fluorescence lifetime analysis of single molecules. Anal. Chem. **73**, 2078–2086 (2001)
73. I. Medina, P. Friedel, C. Rivera, K.T. Kahle, N. Kourdougli, P. Uvarov, C. Pellegrino, Current view on the functional regulation of the neuronal K^+-Cl^- cotransporter KCC2. Front. Cell Neurosci. **8**, 1–18 (2014)
74. N. Melzer, A. Biela, C. Fahlke, Glutamate modifies ion conduction and voltage-dependent gating of excitatory amino acid transporter-associated anion channels. J. Biol. Chem. **278**, 50112–50119 (2003)
75. M. Mukhtarov, O. Markova, E. Real, Y. Jacob, S. Buldakova, P. Bregestovski, Monitoring of chloride and activity of glycine receptor channels using genetically encoded fluorescent sensors. Phil. Trans. A Math. Phys. Eng. Sci. **366**, 3445–3462 (2008)
76. D.K. Nair, M. Jose, T. Kuner, W. Zuschratter, R. Hartig, FRET-FLIM at nanometer spectral resolution from living cells. Opt. Expr. **14**, 12217–12229 (2006)
77. J.A. Payne, T.J. Stevenson, L.F. Donaldson, Molecular characterization of a putative K-Cl cotransporter in rat brain, A neuronal-specific isoform. J. Biol. Chem. **271**, 16245–16252 (1996)
78. T. Price, F. Cervero, M.S. Gold, D.L. Hammond, S.A. Prescott, Chloride regulation in the pain pathway. Brain Res. Rev. **60**, 149–170 (2009)
79. M. Puskarjov, P. Seja, S.E. Heron, T.C. Williams, F. Ahmad, X. Iona, K.L. Oliver, B.E. Grinton, L. Vutskits, I.E. Scheffer, S. Petrou, P. Blaesse, L.M. Dibbens, S.F. Berkovic, K. Kaila, A variant of KCC2 from patients with febrile seizures impairs neuronal Cl^- extrusion and dendritic spine formation. EMBO Rep. **15**, 723–729 (2014)
80. C. Rivera, J. Voipio, J.A. Payne, E. Ruusuvuori, H. Lahtinen, K. Lamsa, U. Pirvola, M. Saarma, K. Kaila, The K^+/Cl^- co-transporter KCC2 renders GABA hyperpolarizing during neuronal maturation. Nature **397**, 251–255 (1999)
81. O. Stern, M. Volmer, Über die Abklingungszeit der Fluoreszenz. Phys. Zeitschr. **20**, 183–188 (1919)
82. C. Stringari, A. Cinquin, O. Cinquinb, M.A. Digman, P.J. Donovan, E. Gratton, Phasor approach to fluorescence lifetime microscopy distinguishes different metabolic states of germ cells in a live tissue. PNAS **108**, 13582–13587 (2011)
83. H. Studier, W. Becker, Megapixel FLIM. Proc. SPIE 8948 (2014)
84. K. Suhling, P.M.W. French, D. Phillips, Time-resolved fluorescence microscopy. Photochem. Photobiol. Sci. **4**, 13–22 (2005)
85. K. Suhling, J. Levitt, P.H. Chung, in *Time-Resolved Fluorescence Anisotropy Imaging*, vol. 3, ed. by Y. Engelborghs, A.J.W.G. Visser. Fluorescence Spectroscopy and Microscopy: Methods and Protocols, Methods in Molecular Biology (Springer, Humana Press, New York City, 2014), pp. 503–519
86. P. Thomas, T.G. Smart, HEK293 cell line: A vehicle for the expression of recombinant proteins. J. Pharmac. Toxic. Meth. **51**, 187–200 (2005)
87. V. Untiet, C. Fahlke, P. Kovermann, T. Gensch. (2013) (unpublished)
88. A.S. Verkman, Development and biological applications of chloride-sensitive fluorescent indicators. Am. J. Cell Physiol. **259**, C375–C388 (1990)

89. A.S. Verkman, *Chemical and Gfp-Based Fluorescent Chloride Indicators*, Physiology and Pathology of Chloride Transporters and Channels in the Nervous System, Part 2: Current Methods for Studying Chloride Regulation, Chap. 6, pp. 111–123 (2009)

90. A.S. Verkman, M.C. Sellers, A.C. Chao, T. Leung, R. Ketcham, Synthesis and characterization of improved chloride-sensitive fluorescent indicators for biological applications. Anal. Biochem. **178**, 355–361 (1989)

91. A. Villoing, M. Ridhoir, B. Cinquin, M. Erard, L. Alvarez, G. Vallverdu, P. Pernot, R. Grailhe, F. Merola, H. Pasquier, Complex fluorescence of the cyan fluorescent protein: Comparisons with the H148D variant and consequences for quantitative cell imaging. Biochem. **47**, 12483–12492 (2008)

92. J. Voipio, W.F. Boron, S.W. Jones, U. Hopfer, J.A. Payne, K. Kaila, Comment on "local impermeant anions establish the neuronal chloride concentration". Science **345**, 1130 (2014)

93. R.M. Wachter, S.J. Remington, Sensitivity of the yellow variant of green fluorescent protein to halides and nitrate. Curr. Biol. **9**, R628–R629 (1999)

94. R.M. Wachter, M.A. Elsliger, K. Kallio, G.T. Hanson, S.J. Remington, Structural basis of spectral shifts in the yellow-emission variants of green fluorescent protein. Structure **6**, 1267–1277 (1998)

95. J.I. Wadiche, S.G. Amara, M.P. Kavanaugh, Ion fluxes associated with excitatory amino acid transport. Neuron **15**, 721–728 (1995)

96. T. Waseem, M. Mukhtarov, S. Buldakova, I. Medina, P. Bregestovski, Genetically encoded Cl-sensor as a tool for monitoring of Cl⁻ dependent processes in small neuronal compartments. J. Neurosci. Meth. **193**, 14–23 (2010)

97. C.D. Wilms, H. Schmidt, J. Eilers, Quantitative two-photon Ca^{2+} imaging via fluorescence lifetime analysis. Cell Calc. **40**, 73–79 (2006)

98. T. Wilson, C. Sheppard, *Theory and Practice of Scanning Optical Microscopy* (Academic Press, London, 1984)

99. N. Winter, P. Kovermann, C. Fahlke, A point mutation associated with episodic ataxia 6 increases glutamate transporter anion currents. Brain **135**, 3416–3425 (2012)

100. Z.Y. Ye, D.P. Li, H.S. Byun, L. Li, H.L. Pan, NKCC1 upregulation disrupts chloride homeostasis in the hypothalamus and increases neuronal activity-sympathetic drive in hypertension. J. Neurosci. **32**, 8560–8568 (2012)

101. S.-P. Yu, G.A. Kerchner, Endogenous voltage-gated potassium channels in human embryonic kidney (HEK293) cells. J. Neurosc. Res. **52**, 612–617 (1998)

102. Y. Zeng, L. Jiang, W. Zheng, D. Li, S. Yao, J.Y. Qu, Quantitative imaging of mixing dynamics in microfluidic droplets using two-photon fluorescence lifetime imaging. Opt. Lett. **36**, 2236–2238 (2011)

Chapter 5
Calcium Imaging Using Transient Fluorescence-Lifetime Imaging by Line-Scanning TCSPC

Samuel Frere and Inna Slutsky

Abstract We present a technique that records transient changes in the concentration of free Ca^{2+} in live neurons with spatial resolution along a one-dimensional scan. The technique is based on recording fluorescence lifetime changes of a Ca^{2+} probe by multi-dimensional TCSPC. The sample is scanned with a high-frequency pulsed laser beam, single photons of the fluorescence light are detected, and a photon distribution over the distance along the scan, the arrival times of the photons after the excitation pulses and the time after a periodical stimulation of the sample is built up. The maximum resolution at which lifetime changes can be recorded is given by the line scan period. Transient lifetime effects can thus be resolved at a resolution of about 1 ms.

5.1 Calcium Imaging

Calcium ions serve as an intracellular signalling pathway that regulates a large spectrum of cellular functions such as intracellular transport, membrane potential, muscle contraction, gene expression, cell shapes, cell differentiation and cell death. Under resting condition, cytoplasmic calcium concentration is kept low (<150 nM) and calcium signalling is initiated by a change in calcium permeability at the plasma membrane or at the membranes of organelles such as the endoplasmic reticulum or the mitochondria, which contain high levels of calcium (several hundreds of μM, compares to few mM in the extracellular medium). The elevations in cytoplasmic calcium levels is temporally and spatially controlled by calcium homeostatic systems composed of calcium buffers (ex: calbindin, parvalbumin,

S. Frere (✉) · I. Slutsky
Department of Physiology and Pharmacology, Sackler School of Medicine,
Tel Aviv University, Tel Aviv, Israel
e-mail: samuel.frere@gmail.com

I. Slutsky
e-mail: islutsky@post.tau.ac.il

© Springer International Publishing Switzerland 2015
W. Becker (ed.), *Advanced Time-Correlated Single Photon Counting Applications*,
Springer Series in Chemical Physics 111, DOI 10.1007/978-3-319-14929-5_5

213

calretinin) and calcium transporters (plasmalemmal $Na^+\backslash Ca^{2+}$ exchanger and calcium pumps, endoplasmic Ca^{2+} pumps) [1].

In neuronal tissue, calcium functions are particularly important at the synapses with two main mechanisms controlling calcium permeability.

Presynaptically, the opening of transmembrane voltage-gated calcium channels (particularly of N- and P/Q-types) is triggered by the arrival of axonal action potentials from the cell body to the presynaptic terminal. Calcium induces the fusion of synaptic vesicles with the plasma membrane and the release of neurotransmitter (such as glutamate, GABA or acetylcholine) in the synaptic cleft. Notably, calcium dynamic is one of the main parameter that controls the probability of synaptic vesicle release and short-term plasticity [2].

Postsynaptically, activation of calcium-permeable NMDA receptors following binding of glutamate, D-serine and membrane depolarization. Calcium influx via NMDA receptors has been shown to be crucial for the induction of some forms of long-term synaptic plasticity and, therefore, critically affects memory formation [3]. In somato-dendritic compartments, calcium permeability is also controlled by voltage-gated calcium channels and can be used to monitor cell activity qualitatively.

According to this primordial role of calcium in cell physiology, fluorescent probes have been developed to measure the concentration of free Ca^{2+} as early as in the 70s. In the 80s, Roger Y. Tsien's group developed several calcium indicators that are still in use nowadays [4–7]. In addition, the repertoire of calcium measurement tools has been enriched by the recent development of calcium-sensitive proteins (genetically-encoded calcium indicators or GECIs, also pioneered by Tsien's group among others) with variable affinities to calcium, emission spectra or with specific cell-type or organelle/subcellular targeting.

Chemical calcium probes can be classified as ratiometric and non-ratiometric probes. They are characterized by dissociation constant ranging from ~ 100 nM to 100 μM, by UV or visible excitation wavelengths and by emission spectra ranging from 400 to 600 nm. The choice of the calcium probes will depend on the cell type, the dynamics of the calcium signal studied and the equipment availability (for UV-excited probes). In neuronal cells, non-ratiometric dye such as Oregon-green BAPTA1 (OGB-1) is suitable for low activity or single-action potential studies due to its high affinity (~ 170 nM). For highest evoked or spontaneous activities, Fluo-3 or -4 would be preferred since they are characterized by a lower Kd (~ 350–390 nM). As a general rule, calcium indicators should be used in the extreme range from 0.1 to 10 times the Kd of the probe.

While absolute calcium levels are evaluated by the measure of the intensity (or intensity ratio) of the probes, it is also possible to measure the effect of calcium binding on the fluorescence lifetime of the fluorescent probes. Indeed, several calcium probes have been tested for FLIM measurements such as Quin-2 [4], indo-1 [8, 9], calcium green [10], fura-2 [11]. Lifetime measurements of OGB-1 with 2-photon microscopy and TCSPC systems were used to get absolute calcium concentration in active neurons in acute brain slices [12] and in the astrocytes of a mouse brain in vivo [13]. The fluorescence lifetime of a fluorophore depends on its molecular environment but, within reasonable limits, not on its concentration

[5, 14]. Several problems, such as dye loading efficiency, cell volume changes, dye compartmentalization and extrusion, which are inherent to some calcium probes are therefore deviated using fluorescence lifetime (FL) measurements.

In addition to the calcium affinity of the dye or the quantum yield, other factors are required to perform lifetime imaging and obviously, fluorescence lifetime change upon calcium binding is a prerequisite [4]. Wilms and Eilers tested several calcium probes for FLIM [15]. The authors found that calcium orange, calcium green-1, magnesium green, Oregon green-2 and -5 N showed large Ca^{2+}-dependent lifetime changes while fura-2, Fluo-3, BTC and calcein displayed intensity changes but no change in the lifetimes. The authors concluded that calcium binding to the first group of indicators results in a change in the extinction coefficient, but does not affect the quantum yield and thus the fluorescence lifetime of the fluorophore. The authors also describe the presence of impurities in the commercially available dyes, which can represent up to 20 % of the total fluorescence and complicate the evaluation of FL parameters [15].

A more likely reason why the second group of probes does not show lifetime changes has been stated by Lakowicz [4]: The dyes have a fluorescence Ca-bound form and a non-fluorescent unbound form. Only fluorescence from the bound form is observed, therefore the lifetime does not change.

For recording the fluorescence lifetimes of Calcium probes a number of different FLIM techniques are around. FLIM can be classified into time domain and frequency domain techniques, analog and photon counting techniques, wide-field and scanning techniques. Almost all combinations are in use. Different techniques differ in their photon efficiency, i.e. in the number of photons required to obtain a given lifetime accuracy, their ability to resolve multi-exponential decay profiles, and their compatibility with optical sectioning techniques, especially confocal and multi-photon laser scanning techniques [16].

The technique with the highest photon efficiency is the combination of laser scanning with multi-dimensional TCSPC [17], see Chaps. 1 and 2 of this book. FLIM by multidimensional TCSPC is based on raster-scanning a sample with a high-frequency pulsed laser beam, detecting single photons of the fluorescent light emitted, and building up a photon distribution over the coordinates of the scan area, x and y, and the arrival times, t, of the photons after the laser pulses. The results can be interpreted as an array of pixels, each containing photons in a large number of time channels for consecutive times after the excitation pulses [17, 18]. The recording process delivers near-ideal efficiency and extremely high time resolution, and resolves multi-exponential decay functions. Moreover, multi-dimensional TCSPC solves the problem that the pixel rates in scanning microscope are often higher than the photon detection rates. No matter of how fast the scanner is moving the acquisition process simply assigns the photons to the right pixels and time channels. The accumulation is continued over as many frames of the scan as needed to obtain the desired signal-to-noise ratio.

To record transient effects in the fluorescence lifetime a time series of FLIM recordings can be performed, and the data saved into consecutive data files [18, 19]. Time-series of FLIM data can be recorded at surprisingly high rate, especially if readout times are avoided by dual-memory recording [18, 19], see Chap. 1, Sect. 1.4.4, Fig. 1.21, or by temporal mosaic FLIM, see Chap. 2, Sect. 2.4.4. Nevertheless, each step of a time series requires at least one x–y scan of the sample to be completed. With the typical frame rates of galvanometer scanners lifetime changes can be recorded at a maximum resolution of 40–100 ms. Intensity and fluorescence-lifetime changes depend on the endogenous change of the calcium concentration (by diffusion, buffering and slow uptake/extrusion) and on the dissociation rate of the dye, with higher affinity resulting in lower dissociation rate. The rising phase is very sharp. In the case of single-pulse stimulation of neurons, calcium signals rise in a few ms and decay in several hundreds of ms. An acquisition rate of 40 ms per image is therefore sufficient to show the presence of the Ca^{2+} transient but not to accurately record its function of time. An example is shown in Chap. 2, Fig. 2.30.

The usual way in laser scanning microscopy to resolve effects faster than the frame time is line scanning. Typical galvanometer scanners easily reach line times around 1 ms, therefore transient effects can be recorded at a resolution on the order of milliseconds. Below we will show how multi-dimensional TCSPC can be combined with line scanning and how it can be implemented in existing TCSPC FLIM systems.

5.2 Principles of Fluorescence Lifetime-Transient Scanning (FLITS)

Figure 5.1, left, shows the photon distribution built up by TCSPC FLIM. It is a distribution of photon numbers over the scan coordinates, x, y, and the arrival time within the laser pulse period, t. Figure 5.1, right, shows the photon distribution built up by line-scanning TCSPC. The difference is that one spatial coordinate (y) has been replaced with a second time axis, the 'experiment time', T. T is the time after a stimulation of the sample, or after any other event that is temporally correlated to a lifetime change in the sample. X is the distance along a spatially one-dimensional scan. The result is a data array that contains fluorescence decay curves for the pixels along the line of the scan for different times after the stimulation (Fig. 5.1).

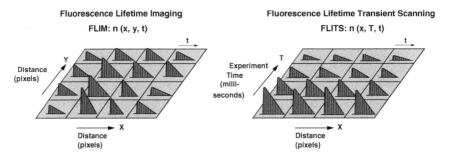

Fig. 5.1 *Left* Photon distribution built up by standard FLIM. *Right* Photon distribution built up by fluorescence lifetime-transient scanning

In principle, the result is the same as for a consecutive recording of the data of subsequent line scans. However, there is an important difference: Fluorescence Lifetime-Transient Scanning (FLITS) keeps the decay data for all times, T, and all distances, x, in the memory. Thus, the recording process can be made repetitive: The sample would be stimulated periodically, and the start of the recording along the T axis triggered by the stimulation. The recording then runs along the T axis periodically, and the photons are accumulated into one and the same photon distribution. With 'triggered accumulation', it is no longer necessary that each T step acquires enough photons to obtain a clear decay curve in each pixel. No matter when and from where a photon arrives, it is assigned to the right location in x, the right time, T, in the stimulation period, and the time, t, in the laser pulse period. The effect is the same as for FLIM at high scan rate: It is no longer necessary to wait in one x–y position (in one x–T position for FLITS) until enough photons have been recorded. A desired signal-to-noise-ratio is obtained by simply running the acquisition process for a sufficiently long time. Thus, the resolution of FLITS is only limited by the period of the line scan, which is about 1 ms for the commonly used scanners

FLITS can be performed in the commonly used FLIM modes of the Becker & Hickl (bh) TCSPC modules [18]. The synchronisation with the experiment is accomplished the usual way, via the pixel clock, line clock, and frame clock pulses. However, different than for FLIM, the frame clock for FLITS does not come from the scanner but from the stimulation of the sample. Switching between the two clock sources is controlled by the data acquisition software. The principle is illustrated in Fig. 5.2.

The SPC module thus records a photon distribution n (x, T, t) the coordinates of which are the distance, x, along the scanned line, the transient time, T, after the event that stimulates the sample, and the time, t, of the photon after the laser pulse. The transient-time, T, is given in multiples of the line time. The time scale of T can be varied either by varying the scan rate, or by defining 'line clock divider' values larger than one in the scan parameter section of the TCSPC system parameters [18].

The experiments described in this chapter were performed in a Zeiss LSM 7 MP multiphoton microscope with a Becker & Hickl Simple Tau 150 system and an HPM-100-40 hybrid detector. The general timing diagram of the FLITS experiment is shown in Fig. 5.3.

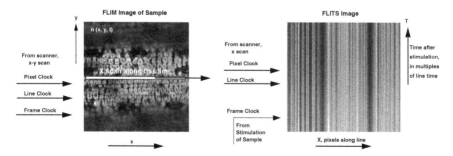

Fig. 5.2 Principle of recording transient lifetime effects by line scanning

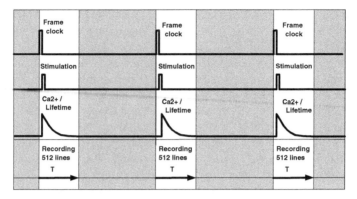

Fig. 5.3 Timing diagram for measurement of Ca^{2+} transients

5.3 Procedure

Hippocampal cultures were prepared from newborn rats, plated on coated cover slips and kept for 12–18 days in an incubator (37°, 5 % CO_2) in B-27 supplemented medium. For the loading of the cells with the calcium probes, two methods are routinely used. Permeant probes are commercially available (via acetoxymethyl ester or AM molecule). After 30–60 min incubations, this should result in a higher intracellular concentration of the probes compare to the concentration of the incubating solution. The AM-ester group is hydrophobic and allows the molecular complex to cross the lipidic plasma membrane. Once within the cell, the ester group is cleaved by endogenous esterases, which trap the probe inside the cells. In our experiments, primary neurons were loaded by an incubation of 30 min in Tyrode solution implemented with glutamatergic receptor blockers and containing 3 μM of cell-permeant OGB-1 AM [4] and Fluo-4 AM [6, 20]. Both dyes were prepared from 1 mM stocks diluted in DMSO. OGB-1 and Fluo-4 are two green fluorescent probes that are activated by a single wavelength in visible light (generally at 488 nm in one-photon microscopy). They present a broader absorption curve for two-photon microscopy and we used 920 nm as excitation wavelength. Alternatively, single neuron can be filled using an invasive method via passive diffusion of the dye from a filled patch-clamp pipette, which can take 20–30 min after that whole-cell configuration is obtained. Dye microinjection is not suitable for small cells such as cortical neurons.

After the incubation time, cover slips were first washed and kept for 10 min in the incubator to allow complete esterification of the dyes. Loaded cells were then transferred into a recording chamber (Warner Instruments, RC-49MF), placed on the stage of the microscope and imaged through a Zeiss 63×, NA = 1.0 water-immersion objective lens. The neurons were constantly perfused by Tyrode solution (for brain slices, oxygenated artificial cerebrospinal fluid is used). Optionally, the solution can be implemented with ionotropic glutamatergic blockers to prevent recurrent activation of the neuronal cultures and to record calcium dynamics in presynaptic compartments. Extracellular calcium and magnesium concentrations are generally in the range of 1.2–2.5 mM (1.2 mM here).

The experiment procedure was as described below:

1. To find an appropriate location of the line scan in the sample, an intensity image was taken first. Within this image, the location of the line scan was defined. Then the LSM 7 was put in the line scan mode. The line scan time was 1 ms, each two lines were combined by the line binning function of the FLIM system. The timing diagram shown in Fig. 5.3 presents the following steps of the lifetime measurements.
2. The scanning in the microscope and the measurement in the TCSPC system were started. This puts the TCSPC system in a state where it waits for a 'Frame Sync', which is, in this case, a pulse from the stimulation device.
3. The measurement by the TCSPC starts when it receives the CMOS signal provided by 1 of the 8 digital outputs of the Digidata 1440A (1440A, Molecular devices) commanded by the clampex software (line 1 of Fig. 5.3). The recording was performed over a total acquisition time of typically 2 min, i.e. photons from about 40 stimulation periods were accumulated.
4. Single stimulation or 50 Hz bursts of 5 stimulation pulses were applied to the cells periodically in intervals of 3 and 5 s, respectively. The pulses were 1 ms long and were applied 60 ms after the start of each FLITS cycle to get the baseline fluorescence (30 lines on the FLIM images). Electric stimulations were applied to the cells by an electric field created between two parallel platinum cylindrical electrodes connected to a stimulation unit (A385, World Precise Instruments). The stimulation unit was triggered by a CMOS signal provided by a different digital output of the Digidata (line 2 of Fig. 5.3).

5.4 Results

A result obtained with Oregon Green BAPTA 1-AM is shown in Fig. 5.4. The FLITS image is shown in Fig. 5.4, left. It can clearly be seen that the fluorescence lifetime changes when the stimulation induces a change in the Ca^{2+} concentration. The rise in the Ca^{2+} concentration occurs almost instantaneously. The concentration returns to resting concentration within the next 600–800 ms. A FLIM image taken after the FLITS experiment is shown in Fig. 5.4, right. The location of the line scan is indicated by a red line. The FLIM image shows that the fluorescence lifetime, and thus the Ca^{2+} concentration, after the FLITS experiment had returned to the resting level. No permanent lifetime changes or photobleaching were induced along the scanned line.

Time traces of the fluorescence lifetime and the fluorescence intensity extracted from the FLITS data are shown in Fig. 5.5.

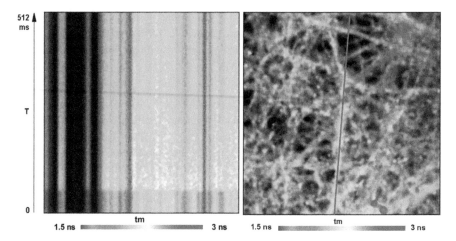

Fig. 5.4 Cultured neurons incubated with *Oregon Green* BAPTA 1-AM. *Left* FLITS image. *Right* FLIM image taken after the FLITS experiment. Lifetime analysis by bh SPCImage, amplitude-weighted lifetime of triple-exponential decay model

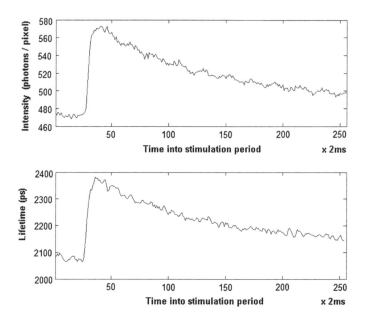

Fig. 5.5 Fluorescence intensity and fluorescence lifetime over time in the stimulation period

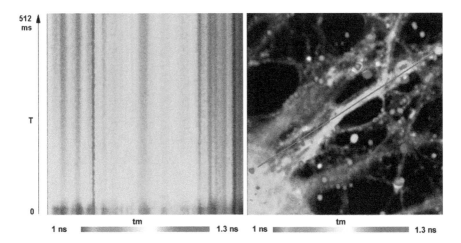

Fig. 5.6 Neurons incubated with Fluo-4 am. *Left* FLITS image. *Right* FLIM image taken after the FLITS experiment. Lifetime analysis by bh SPCImage, amplitude-weighted lifetime of triple-exponential decay model

Results obtained for neurons incubated with Fluo-4 AM are shown in Figs. 5.6 and 5.7. Fluo-4 AM [20] has a similar structure as Fluo-3AM [6] but has two chlorine atoms substituted with fluorine. This results in a higher absorption coefficient and higher fluorescence intensity. As can be seen from Figs. 5.6 and 5.7 Fluo-4 AM displays a large change in fluorescence intensity but virtually no change in the fluorescence lifetime. At first glance, this may be surprising: The mechanisms of the Ca^{2+} response are closely the same for Oregon Green and the Fluo family dye, both deriving from fluorescein as the fluorophore and BAPTA as the Ca^{2+} chelator [5, 7]. Both fluorophores have a Ca^{2+} bound form and an unbound form. The bound form has a higher fluorescence quantum yield, and thus a longer fluorescence lifetime than the unbound one. However, there is one difference: The ratio of the quantum efficiencies of the two forms is 10 for OGB-1 but 40 for Fluo-3 AM [5, 7]. It is possibly even larger for Fluo-4 AM [20]. That means the Ca^{2+}-free form of Fluo-4 AM is virtually non-fluorescent and, within the Ca^{2+} concentration range of interest, does not contribute to the emission. The detected fluorescence therefore comes essentially from the bound form. Consequently, also the fluorescence lifetime is that of the bound form, and does not change noticeably. The problem has already been stated by Lakowicz et al. [4]: A probe suitable for lifetime-based Ca^{2+} sensing must have detectable emission for both forms.

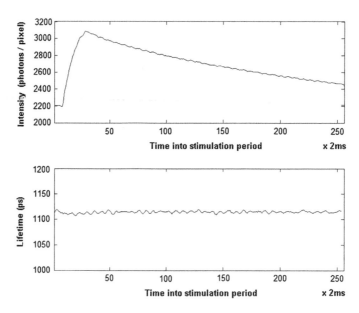

Fig. 5.7 Fluo-4 am, Fluorescence intensity and fluorescence lifetime over time in the stimulation period

5.5 Conclusions

FLITS is able to record transient changes in the fluorescence-lifetime of calcium probes at millisecond time resolution with spatially one-dimensional resolution. Technically, FLITS is obtained by line scanning and replacing the 'frame clock' of a TCSPC FLIM system with a trigger pulse that is synchronous with the event that stimulates the lifetime change in the sample. The technique can thus easily be implemented in confocal or multiphoton laser scanning microscopes, provided these are able to run a fast line scan. The technique works both with single-shot stimulation, or with periodic stimulation. For a given photon detection rate, the lifetime accuracy for single-shot stimulation decreases with decreasing time-channel width along the transient-time axis. This is not the case for periodic stimulation: Here the accuracy depends on the total acquisition time. Periodic stimulation is thus the key to high fluorescence-transient resolution. In the case of synaptic calcium imaging, caution should be taken that the frequency of stimulation does not result in some synaptic plasticity, which will affect calcium output along the experiments.

The advantage of lifetime detection over intensity measurement is that the result is independent of the unknown, and possibly variable, number of fluorescent molecules. Changes in the decay rates of the fluorophore can therefore be separated from changes in its concentration. Ca^{2+} imaging via the fluorescence lifetime has the additional advantage that absolute Ca^{2+} concentrations are obtained with non-ratiometric sensors after calibration [13]. In addition to OGB-1, Indo-1, which is excited by a single wavelength but present a shift in the wavelength of emission upon calcium binding (from 485 to 405 nm) would be suitable for multiphoton imaging and lifetime imaging [9]. One issue with Indo-1 is that over-excitation (at least in UV spectra) can lead to the formation of calcium-insensitive dyes but that are still able to emit photons, which will affect the correct amplitudes of bound- and unbound-dyes. This problem has not been described in the case of OGB-1.

It requires, however, that the sensor displays a clear lifetime change with the Ca^{2+} concentration. Lifetime changes occur only when both the Ca^{2+} bound and the Ca^{2+} unbound form of the sensor contribute to the sum fluorescence signal or when quantum yield is altered by binding calcium. This was found to be the case for OGB-1 AM but not for Fluo-4 AM and other Fluo family dyes.

References

1. M.J. Berridge, M.D. Bootman, H.L. Roderick, Calcium signalling: dynamics, homeostasis and remodelling. Nat. Rev. Mol. Cell Biol. **4**, 517–529 (2003)
2. T. Südhof, The synaptic vesicle cycle. Annu. Rev. Neurosci. **27**, 508–547 (2004)
3. K. Nakazawa, T.J. McHugh, M.A. Wilson, S. Tonegawa, NMDA receptors, place cells and hippocampal spatial memory. Nat. Rev. Neurosci. **5**, 361–372 (2004)
4. J.R. Lakowicz, H. Szmacinski, M.L. Johnson, Calcium imaging using fluorescence lifetimes an long-wavelength probes. J. Fluoresc. **2**, 47–62 (1992)
5. J.R. Lakowicz, *Principles of Fluorescence Spectroscopy*, 3rd edn. (Springer Science+Business Media LLC, Berlin, 2006)
6. A. Minta, J.P.Y. Kao, R.Y. Tsien, Fluorescent indicators for cytosolic calcium based on rhodamine and fluorescein chromophores. J. Biol. Chem. **264**, 8171–8178 (1989)
7. A. Takahashi, P. Camacho, J.D. Lechleiter, B. Herman, Measurement of intracellular calcium. Physiol. Rev. **79**, 1089–1124 (1999)
8. H. Szmacinski, I. Gryczynski, J.R. Lakowicz, Calcium-dependent lifetimes of Indo-1 for one- and two-photon excitation of fluorescence. Photochem. Photobiol. **58**, 341–345 (1993)
9. H. Szmacinski, I. Gryczynski, J.R. Lakowicz, Three-photon induced fluorescence of the calcium probe Indo-1. Biophys. J. **70**, 547–555 (1996)
10. R. Sanders, H.C. Gerritsen, A. Draaijer, P.M. Houpt, Y.K. Levine, Confocal fluorescence imaging of free calcium in single cells. J. Fluoresc. **4**, 291–294 (1994)
11. H. Szmacinski, J.R. Lakowicz, Possibility of simultaneous measuring low and high calcium concentrations using Fura-2 and lifetime-based sensing. Cell Calcium **18**, 64–75 (1995)
12. C.D. Wilms, H. Schmidt, J. Eilers, Quantitative two-photon calcium imaging via fluorescence lifetime analysis. Cell Calcium **49**, 73–79 (2006)
13. K.V. Kuchibhotla, C.R. Lattarulo, B.T. Hyman, B.J. Bacskai, Synchronous hyperactivity and intercellular calcium waves in astrocytes in Alzheimer mice. Science **323**, 1211–1215 (2009)
14. M.Y. Berezin, S. Achilefu, Fluorescence lifetime measurement and biological imaging. Chem. Rev. **110**, 2641–2684 (2010)
15. C.D. Wilms, J. Eilers, Photo-physical properties of calcium indicator dyes suitable for two-photon fluorescence-lifetime recordings. J. Microsc. **225**, 209–213 (2007)
16. W. Becker, Fluorescence Lifetime Imaging-Techniques and Applications. J. Microsc. **247**, 119–136 (2012)
17. W. Becker, Advanced time-correlated single-photon counting techniques (Springer, New York, 2005)

18. W. Becker, *The bh TCSPC Handbook*, 5th edn. (Becker & Hickl GmbH, Berlin, 2012)
19. V. Katsoulidou, A. Bergmann, W. Becker, How fast can TCSPC FLIM be made? Proc. SPIE **6771**, 67710B-1–67710B-7 (2007)
20. K.R. Gee, K.A. Brown, W.N. Chen, J. Bishop-Stewart, D. Gray, I. Johnson, Chemical and physiological characterization of fluo-4 Ca^{2+}-indicator dyes. Cell Calcium **27**, 97–106 (2000)

Chapter 6
Imaging Cell and Tissue O$_2$ by TCSPC-PLIM

James Jenkins, Ruslan I. Dmitriev and Dmitri B. Papkovsky

Abstract We describe a technique of imaging tissue oxygen using phosphorescence based probes and TCSPC-PLIM method. Included is a brief overview of the significance of biological oxygen imaging, the theory behind the phosphorescence quenching method, the main O$_2$ sensitive probes (mostly intracellular, cell-permeable) and imaging modalities currently available, highlighting their merits and limitations. In the practical part, the live cell microscopy imaging and TCSPC-PLIM hardware and software are described, along with the detailed experimental procedures of preparation of tissue samples, their staining with intracellular O$_2$ probes, acquisition of PLIM images and their processing to produce 2D and 3D maps of O$_2$ concentration. Several examples demonstrate practical use of O$_2$ imaging with different models of mammalian tissue, including cell mono-layers (2D model), multicellular spheroids, scaffold structures and tissue slices (3D models). Physiological experiments and multi-parametric analysis of these samples with some other fluorescent imaging probes are also presented.

James Jenkins and Ruslan I. Dmitriev these authors contributed equally to this work.

J. Jenkins
School of Biochemistry and Cell Biology, University College Cork,
Cork, Ireland
e-mail: 113224155@umail.ucc.ie

R.I. Dmitriev
School of Biochemistry and Cell Biology, University College Cork,
Western Road, Cork, Ireland
e-mail: r.dmitriev@ucc.ie

D.B. Papkovsky (✉)
Laboratory of Biophysics and Bioanalysis, School of Biochemistry and Cell Biology,
University College Cork, Cork, Ireland
e-mail: d.papkovsky@ucc.ie

© Springer International Publishing Switzerland 2015   225
W. Becker (ed.), *Advanced Time-Correlated Single Photon Counting Applications*,
Springer Series in Chemical Physics 111, DOI 10.1007/978-3-319-14929-5_6

6.1 Introduction

6.1.1 Tissue O_2: Simple Molecule, Complex Functions

Molecular oxygen (O_2) is a small, non-polar gaseous molecule essential for cell viability while simultaneously being capable of inducing detrimental effects upon living cells through energy stress, the creation of reactive oxygen species (ROS), hypoxia or hyperoxia [19]. O_2 is directly and indirectly involved in numerous cell functions inclusive of cell growth and differentiation [54], the production of ATP via oxidative phosphorylation [41], and it regulates the expression of proteins, necessary for the production of various metabolic and signalling molecules [49, 55]. In healthy mammalian tissues O_2 levels are maintained within specific physiological limits [63]. Deviations above or below the norm may result in drastic effects upon cell functions and behaviour such as perturbations of metabolism, changes in cell viability and growth rate, development of cancer, hypoxia induced resistance to cancer treatment and cell death through necrosis and apoptosis. Involvement of O_2 in a diverse range of cellular processes makes it a useful (though not very specific) indicator of many pathological conditions such as stroke, metabolic and neurological disorders and responses to drug stimulation [27, 34, 36, 52, 61].

From the analysis of cell and tissue O_2, the following parameters can be determined: local oxygenation state, the oxygen consumption rate, the oxygen gradients in respiring cells/tissue, changes of these parameters upon drug treatment or cell stimulation [12]. When combined with other relevant markers of cellular function, these measurements can provide a comprehensive picture of the physiological state of respiring samples.

6.1.2 2D and 3D Tissue Cultures

2D and 3D tissue cultures are used to model cell and tissue behaviour in their natural environment in vivo. They help reduce the use of animal or human material and problems of its availability, associated ethical issues and applicability of many experimental techniques [18]. 2D cultures (cell mono-layers) are widely used as simple models in basic mechanistic studies with cells, drug action and toxicity assays [25], but the limitations are unnatural flattened morphology of cells in such structures, lack of cell-to-cell interactions, paracrine action and signal transduction [18].

3D tissue models are more advanced as they mimic more closely tissue environment with prominent concentration gradients, diffusion limitations (O_2, nutrients, growth factors, drugs), heterogeneous populations of cells and their responses. Examples include scaffolds made of natural or synthetic polymers with seeded cells, multi-cellular spheroids and tissue slices [38]. These models are simpler than live animals and therefore more preferred for microscopy imaging [65].

Scaffolds aid the growth of cells in a 3D space, providing them support of varying stiffness and feeding components. Protein-derived scaffolds can function similar to the extracellular matrix (ECM), while rigid synthetic scaffolds promote cell invasion into the pores, interaction between neighbouring cells and formation of clusters, along with efficient gas exchange (O_2, CO_2) and access of nutrients to the cells [29]. Thus, microporous polystyrene-based scaffolds Alvetex™ (Amsbio) have pores of 35–40 μm and thickness of approximately 200 μm.

Spheroids comprise self-assembled clusters of cells with naturally formed ECM which is not present in 2D cultures. Free-floating spheroids can be formed in cultures of suspension cells, using serum-free media with growth factors which force the cells to aggregate. Although simple, this method is not very consistent and produces spheroids of variable size [26]. The hanging drop method uses specialized plates in which cells suspended in a 35–50 μL drop of media aggregate under gravity force at the bottom and form a spheroid. Although more expensive, this method produces individual spheroids of similar size and structure [60].

Excised slices are pieces of live tissue with native cyto-architecture, they can be cultured for some time (days to weeks) and used in model physiological experiments. Once detached from the vasculature, transfer of oxygen and nutrients to the whole tissue is greatly impeded (especially for deep regions), while the surface is exposed to high atmospheric O_2. Significant thickness of slices (hundreds of microns) complicates O_2 measurement, but also necessitates control of their oxygenation [65].

6.1.3 Phosphorescent Probes for O_2 Sensing and Imaging

Optical detection usually relies on dynamic quenching by O_2 molecules of long-lived excited states of certain photo-luminescent dyes or derived sensor materials. Luminescent signal relates to O_2 concentration as follows:

$$\frac{I_0}{I} = \frac{\tau_0}{\tau} = 1 + K_{sv}[O_2] \tag{6.1}$$

$$[O_2] = \frac{\tau_0 - \tau}{\tau * [K_{sv}]} = \frac{I_0 - I}{I * [K_{sv}]} \tag{6.2}$$

where I, I_0, τ and τ_0 are the luminescence intensity and lifetime values with and without the quencher respectively, K_{sv} is the Stern–Volmer quenching constant [51].

The phosphorescent Pt(II)- and Pd(II)-porphyrins, which have strong and characteristic absorption bands (370–420 and 500–550 nm regions), emission at around 630–700 nm, and lifetimes 20–100 μs for Pt(II)-porphyrins and 400–1000 μs for Pd(II)-porphyrins, are at the fore of oxygen sensing chemistry [12] (Table 6.1). Probes based on Pt-porphyrins are better suited for the whole physiological range (0–200 μM or 0–21 kPa O_2), while Pd-porphyrins—for the low O_2

Table 6.1 Phosphorescent characteristics of some common dyes used in O_2 sensing

Indicator dye[a]	Excitation optimum (nm)	Emission optimum (nm)	Lifetime τ_0, μs	Quantum yield/solvent	References
PtTFPP	392	650	70	$0.36/CHCl_3$	[11, 53]
PtTBP	416, 609	745	50	0.50/DMF	[22]
PdTPCPP	415, 524	690	~800	~0.1/water	[17]
PdTCPTBP	442, 632	790	240	0.12/water	[17]
$[Ru(bpy)_2(picH_2)]^{2+}$	460	607	0.7–0.9	0.07/water	[32, 43]

[a] *PtTFPP* Pt(II)-tetrakis(pentafluorophenyl)porphine; *PtTBP* Pt(II)-meso-tetrabenzoporphyrin, butyl octaester; *PdTPCPP* Pd(II)-meso-tetra-(4-carboxyphenyl)porphyrin; *PdTCPTBP* Pd(II)-meso-tetra-(4-carboxyphenyl)tetrabenzoporphyrindendrimer; $[Ru(bpy)_2(picH_2)]^{2+}$ [Ru(bpy)$_2$(2-(4-carboxyphenyl) imidazo-[4,5-*f*] [28, 41] phenanthroline)H$_2$)]$^{2+}$

range (<5 kPa). Phosphorescent tetrapyrrols with long wave spectral characteristics are represented by Pt-and Pd-benzoporphyrins, which are excitable at 590–650 nm and phosphoresce at 730–900 nm. Another important group is fluorescent complexes of Ru(II), which have shorter lifetimes (1–5 μs), reduced quenching and lower brightness. They allow for the more rapid signal acquisition rates but their low sensitivity to O_2 may be problematic [12]. Table 6.1 outlines some common O_2-sensitive dyes currently in use.

For biological applications and particularly imaging experiments, the sensor material is usually prepared in a soluble form as a probe. The probe can be a small molecule (free dye or its derivative), a supra molecular structure (e.g. dendrimer, protein or peptide conjugate) or nanoparticle formulation [31, 57]. The last two options allow the introduction of additional functionalities, such as two-photon antennae, vectors for targeted delivery, reference dye for ratiometric detection, as well as tuning of O_2 sensitivity and more specific delivery of the probe into the cell or tissue. The two main types of O_2 probes are cell-impermeable (or extra-cellular) and cell-permeable (intracellular or peri-cellular). Depending on the type and measurement task, the probe can be introduced in bulk to the respiring sample (e.g. cultured cells, tissue or spheroid), directly inside the cells or in live animal tissue.

Extra-cellular probes are useful for the measurement of oxygen consumption rate (OCR) in small biological samples, adding them to the media and analyzing the effects of drugs, culture conditions or cell metabolic state [12]. Such probes are also used in in vivo imaging via local or systemic intravenous administration [33], designed to remain in the vasculature, have low toxicity on cells (though organ toxicity and systemic effects can occur due to high doses used) [12]. However, extra-cellular probes are not very suitable for in vitro studies with respiring objects such as 3D tissue models as they do not stain the cells within tissue.

The measurement of O_2 in tissue samples with sub-cellular level of detail requires a probe that is capable of penetrating into the cells and across cell layers. Old methods for introducing probes into the cells (e.g. microinjection, electroporation, gene gun [21, 46]), are invasive, stressful, time consuming and poorly reproducible. Modern approaches use small molecule and nanoparticle probe structures which contain in their structure specific delivery vectors, such as cell penetrating/targeting peptides, carbohydrates, cationic polymers, chemical groups [16, 21, 35]. Such probes allow for efficient and passive self-loading, with minimal toxicity or effects on cellular function. Nanoparticle probes show promise due to their high brightness and photostability, stable calibration insensitive to the environment due to shielded reporter dye(s) [21]. Some cell-penetrating phosphorescent O$_2$ probes currently in use are presented in Table 6.2. Their luminescence spectra are shown in Fig. 6.1. The PtTFPP-based probes are normally excited with 405 nm laser (excitation at 532 or 546 nm can also be used), and emission is collected at 635–675 nm.

The choice of probe will depend upon the sample to be stained, detection mode and available measurement equipment. As can be seen, some probes can provide deep staining of tissue samples, while others are compatible with several different imaging modalities. The differences in brightness and photostability should also be considered when choosing appropriate probe for a particular experiment.

The Stern–Volmer equation predicts a linear relationship between the lifetime/intensity values of the probe and O$_2$ concentration in a homogeneous environment. However in many practical cases the indicator dye has heterogeneous micro-environment, which leads to non-linear Stern–Volmer plots. These effects can be accounted for with appropriate mathematical or physical models, such as the "two-site" model [20] and corresponding formula (6.3):

$$I_O/I = 1/(f_{(\alpha 1)}/(1 + K_{sv1}[Q] + f_{\alpha 2}/(1 + K_{sv2}[Q]))) \qquad (6.3)$$

where $f_{\alpha 1}$ and $f_{\alpha 2}$ are the two fractions, and K_{sv1} and K_{sv2}—their quenching constants [10].

6.1.4 O$_2$ Imaging Modalities

Phosphorescent O$_2$ probes allow for different detection modes: intensity, ratiometric and lifetime based detection, under one- or two-photon excitation. Simple intensity based imaging can be used to monitor *relative* changes in sample oxygenation, however, measured signal is a function of O$_2$, probe concentration and geometrical alignment. This brings disparities due to uneven probe loading, photo-bleaching, focusing, variations in the illumination strength across the sample and at different depths, and results in unstable calibration. Unless O$_2$ calibration is included in the experiment, this mode usually allows qualitative or semi-quantitative measurements. In ratiometric measurements the intensity signals of two spectrally distinct indicator dyes embedded in the probe (one being O$_2$-sensitive and another O$_2$-insensitive reference dye) are monitored simultaneously. This enables to account for many of the above instability factors and quantify the O$_2$ concentration from the ratio (I_{ref}/I_{O2}). Still this method cannot compensate for all the variables and interferences [51], and the system must be calibrated in a separate experiment and then re-checked frequently [20].

Table 6.2 O_2 sensing probes, tested in biological applications

Probe	Excitation optimum	Emission maxima/emission filter	Lifetimes (µs) (at 21 kPa and 0 kPa O_2)	Performed detection modes. Advantageous features	References
NanO2	390–405 nm	650 nm/635–675 nm	30–67	1-P excitation	[21]
	533, 546 nm			Quantitative 1-P PLIM	
				Intensity based O_2 sensing (relative changes)	
				Multiplexed detection	
MM2	390–405 nm	650 nm/635–675 nm	28–61	1-P and 2-P PLIM	[14, 30]
	760 nm (2-P)	Reference: 420–450 nm/ 438–458 nm		Quantitative ratiometric intensity based O_2 sensing (both 1-P and 2-P)	
NanO2-IR	614 nm	760 nm/750–810 nm	28–60	1-P PLIM	[59]
	440 nm			NIR excitation and emission	
				Intensity based O_2 sensing (relative changes)	
				Deep signal penetration in tissue. Multiplexing detection	
Pt-Glc	390–405 nm	650 nm/635–675 nm	20–57	1-P PLIM	[16]
	534 nm			Deep tissue staining of 3D cell models (spheroids, tissue slices)	
				Multiplexed detection	

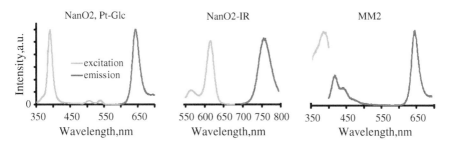

Fig. 6.1 Excitation (*green*) and emission (*red*) spectra of some commonly used O₂ probes

Phosphorescence lifetime imaging microscopy, PLIM (also called microsecond FLIM, phosphorescence quenching microscopy, PQM or two-photon phosphorescence lifetime microscopy, 2PLM) brings true quantitative context to O_2 imaging, since lifetime does not depend upon intensity values and geometrical alignment. Lifetime calibration is generally stable and minimally affected by measurement artefacts [56].

Luminescence lifetime measurements can be performed in the frequency domain or in the time domain, by wide-field or point-scanning techniques, and by analog or photon counting techniques. An overview can be found in [4] or [9].

The frequency domain method involves the excitation of the sample with intensity-modulated light (sine, square wave or pulsed) and the measurement of the delay of the probe emission signal (i.e. phase shift). The measured phase shift is then used to calculate the lifetime [51]:

$$\tau = \tan\frac{\varphi}{2\pi f}$$

where f is the modulation frequency of excitation and φ is the emission phase shift (degrees angle). Commercial frequency-domain FLIM systems use a heterodyne principle in combination with a single-pint detector and scanning (e.g. ISS) or with wide-field detection and a modulated image intensified camera (e.g. Lambert Instruments). The advantage of the camera is that it records the whole image simultaneously and thus has no problems with interference between the scan rate and the phosphorescence decay. However, wide-field imaging does not deliver any intrinsic depth resolution. To enable 3D imaging, the camera can be coupled with a confocal spinning disc accessory. However, this does not entirely solve the problem of light scattering inside the sample (see Chap. 2). Moreover, the photon efficiency for lifetime detection is less than ideal [50]. Accurate determination of lifetime values can be difficult, especially with low specific signals, multi-exponential decay, and significant optical and electronic background (mainly scattering, auto-fluorescence which bring additional phase shift components) [48].

Time-domain FLIM systems use short-pulse excitation and analysis of the decay of emission signal. Two different techniques are in use. Gated image intensifiers use wide-field excitation and detection. A gate is shifted over the decay curve, and a series of images is taken for consecutive gate delay. Time-gated detection can also be used suppress optical background (scattering, autofluorescence) and increase the S:N ratio and image contrast [51]. The disadvantage is that there is no intrinsic depth resolution, and a large influence of scattering inside thick samples, see Chap. 2, Fig. 2.1. Moreover, gating rejects most of the photons and thus results in poor photon efficiency. The problem has been addressed by employing a Rapid Lifetime Determination (RLD) method, whereby two intensity signals D1 and D2 are measured at two different delay times, t_1 and t_2, and the lifetime is calculated from the equation as follows [51]:

$$\tau = \frac{t_2 - t_1}{\ln \frac{D1}{D2}}$$

The main drawback is that only the lifetime of a single-exponential approximation of the decay function is obtained. Moreover, on imaging platforms with low signal-to-blank ratio (e.g. significant dark counts on the camera, long-lived optical background or poly-exponential emission decay), correct lifetime determination requires the collection and processing of multiple intensity frames with different delay times [21].

TCSPC-FLIM (Time Correlated Single Photon Counting) is arguably the most advanced and sensitive method for fluorescence and phosphorescence lifetime measurements in time domain. The system analyses individual photons emitted after a short surge of excitation, determines their arrival times, and re-constructs a complete decay curve from the times of the individual photons [39, 45], see Chap. 1 of this book. For FLIM and PLIM, TCSPC is combined with confocal or multi-photon laser scanning. A multi-dimensional TCSPC process is used which includes the position of the laser spot in the sample in the moment of the detection of a photon [2]. The result is an array of pixels, each containing a full fluorescence or phosphorescence decay curve in a large number of time bins. Operating with high-speed, large area photon counting detectors, the technique provides high detection sensitivity, high accuracy of lifetime determination, and optical sectioning capability. It does have some drawbacks such as a need for fast excitation sources, high-speed detectors and special TCSPC hardware, the increased time to scan the whole sample [51]. However, off-the-shelf TCSPC systems are quickly becoming commercially available. Importantly, TCSPC allows for simultaneous detection of FLIM and PLIM (see Chap. 1, Sects. 1.4.7 and 6.2 of this chapter) and multiplexing of O_2 imaging with measurement of other important fluorescent probes, cell biomarkers, auto-fluorescence (NADH and FAD), pH.

With the development of advanced O_2 probes and needs in tissue imaging, two-photon excitation systems are becoming increasingly popular, whereby a sample is excited by simultaneous absorption of two photons of lower energy. The excitation wavelength (approximately twice the wavelength of one-photon microscopy) is provided by a high-power femtosecond NIR laser (tunable range 650–900 nm). In result, two-photon microscopy enables deeper penetration of excitation into tissue, while excitation volume is restricted by the focal point, minimizing sample and probe photodamage, providing higher contrast and signal-to-noise ratio than one-photon microscopy [8, 42]. Standard two-photon imaging platforms usually rely on photon counting detectors, so they can be readily coupled/upgraded with TCSPC-FLIM hardware [8, 56], see Chap. 2, Sect. 2.2.2, Multi-Photon Microscopes.

6.2 Set-Up for O₂ Imaging by TCSPC-PLIM

Standard TCSPC-FLIM hardware (e.g. from Becker & Hickl GmbH) is compatible with conventional laser-scanning microscopes as well as with multi-photon excitation systems based on near-infrared femtosecond lasers [9]. For routine ex vivo work with cell and tissue models one-photon confocal PLIM systems are appropriate. Our confocal FLIM/PLIM system (Fig. 6.2) consists of an upright microscope (Axio Examiner Z1) with a motorized Z-stage and temperature-control (Carl Zeiss), to which a DSC-120 confocal scanner, TCSPC hardware and detectors (Becker & Hickl) are connected [7, 8]. The scanner has two optical channels for excitation, to which a 6 W picosecond super-continuum laser model SC400-4

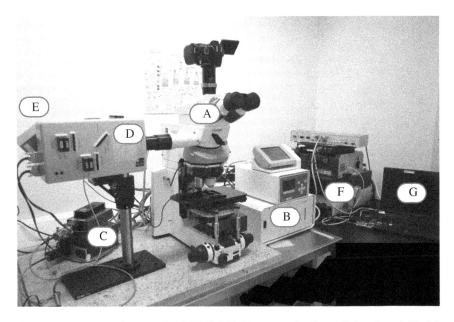

Fig. 6.2 Photograph of the confocal TCSPC-PLIM system for live cell imaging. *A* Upright microscope Axio Examiner Z1 with motorized heated stage (Zeiss) and Nikon D3100 digital camera for transmission light images (*top*). *B* Microscope and temperature controllers. *C* Picosecond 405 nm laser. *D* DCS-120 confocal scanner attached to the microscope with *E* Detector(s) connected from the back of scanner. *F* TCSPC and laser synchronization hardware. *G* Computer with instrument control software

(Fianium, UK) and a BDL-SMC 1 mW 405 nm picosecond diode laser (Becker & Hickl) are connected. The super-continuum laser provides tuneable excitation in a broad range of excitation wavelengths from 400 to 650 nm. However, it has low output power below 430 nm. The ps diode laser provides high excitation power at 405 nm. The Fianium laser is used for excitation of Pd- and Pt-benzoporphyrins (these have strong bands at around 440 and 614 nm), and a variety of standard fluorophores including GFP variants, rhodamines, Alexa Fluor, Cy dyes. The 405 nm ps diode laser is used to excite the Pt-porphyrins via their Soret bands, see Fig. 6.1. The excitation power is regulated manually by neutral density filter wheels in the DSC-120 scan head.

The DCS-120 scan head scans the excitation and detection beam by two fast galvanometer mirrors. Please see Chap. 2, Sect. 2.2.1 for details. The fluorescence and phosphorescence signals from the sample are collected back through the microscope lens, descanned by the galvanometer mirrors, split in two spectral or polarisation components, and focused into confocal pinholes. Each emission channel has individually selectable pinhole sizes and long pass and/or bandpass filters [7].

The two optical outputs of the DSC-120 scanner are connected to high-speed, high-sensitivity photon-counting hybrid detectors [5] (HPM-100-40 and -50, Becker & Hickl) These detectors are based on Hamamatsu R10467U-40 and -50 hybrid detector tubes which operate in the visible to red (400–700 nm, -40 version) and red to near-infrared (500–900 nm, -50 version) spectral regions. In addition to the high sensitivity and high speed, the advantage of the hybrid detectors is that they are virtually free of afterpulsing. Weak phosphorescence signals following a strong fluorescence signal are therefore not contaminated by afterpulsing.

The TCSPC FLIM technique used in the instrument is described in Chap. 1 of this book. It is based on detecting single photons of the light received from the sample, and determining the detection times of the photons with respect to the laser pulses. In addition to the detection times, the TCSPC module also determines the spatial position of the laser beam in the scan area in the moment of the photon detection. The recording process builds up a photon distribution over these parameters. The result is an image that contains a full fluorescence decay curve in each pixel [1, 2, 8, 9]. The process can be extended to record multi-wavelength FLIM data [3].

PLIM recording simultaneously with FLIM is achieved by on-off modulating the laser at the microsecond time scale synchronously with the scanning of the pixels, Chap. 1, Sect. 1.4.7, 'Phosphorescence Lifetime Imaging (PLIM)'. During the laser-on phases fluorescence is obtained, and phosphorescence built up. During the laser-off phases a pure phosphorescence decay is observed. FLIM and PLIM images are obtained by measuring the photon times both with respect to the laser pulses and with respect to the modulation period. The times in the laser pulse period deliver a FLIM image, the times in the modulation period a PLIM image [6, 8]. Except for the different time range, the data structure of PLIM is the same as for FLIM: Each pixel contains photon numbers in a large number of consecutive time channels, i.e. a full phosphorescence decay curve, see Fig. 6.3. With its two parallel TCSPC

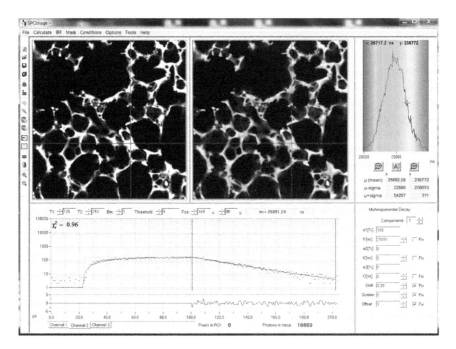

Fig. 6.3 Data processing window in the SPCImage data analysis software. PLIM-TCSPC measurement performed with a phosphorescent Alvetex O$_2$ scaffold at 21 % O$_2$: intensity image (*top left*) and PLIM image (*top right*). Phosphorescence decay (*bottom left*) and its fitting parameters (*bottom right*). *Blue* Photon numbers in subsequent time channels, *green* Instrument response function for PLIM, identical with laser-on phase. The fitting was restricted to the laser-off phase to exclude fluorescence components. The phosphorescence-lifetime colour range was set to 20–30 μs

channels, the system can record two FLIM and two PLIM images simultaneously for different wavelength. Please see Chap. 1, Sect. 1.4.7 for details.

The data acquisition software is able to repeat the measurement in different focal planes and thus record Z stacks of FLIM and PLIM images (see Chap. 2, Sect. 2.4, Mosaic FLIM). Time-lapse measurements can also be performed to study kinetic changes related to treatment with a drug or changing condition in the sample.

Importantly, throughout the whole imaging experiment respiring samples with cells and tissue must be maintained in a healthy and viable state, as close as possible to their natural functioning in vivo. Therefore, the imaging system must have stable and accurate temperature control at physiological levels (normally 37 °C for mammalian cells and tissues). Significant fluctuations of temperature, which affect probe calibration, respiration activity of cells and focusing of microscope optics, must be avoided. The media should also ensure normal physiological conditions, pH 7.2–7.4, necessary nutrients, serum and growth factors), through use of CO$_2$-bicarbonate or 10–20 mM HEPES buffer system [24]. Atmospheric control is highly desirable to establish *physiological* normoxia in the sample, or to modulate sample oxygenation and study cellular responses to hypoxia. Humidification of sample compartment minimizes evaporation of media during long-term imaging experiments. Such environmental control is usually achieved through the use of a mini-incubator on the microscope stage and controllers which maintain stable environment (temperature, O_2, CO_2, relative humidity).

Generated PLIM images can be used to assess oxygenation state of the sample and its different parts. higher lifetime values are indicative of reduced oxygen levels and vice versa. To convert measured lifetime values into O_2 concentration, calibration function (available from literature or experimentally [48]) should be applied. Such pixel-by-pixel conversion of the PLIM images allows generation of 2D and 3D maps of O_2 concentration for test samples. With ordinary biological specimens such as spheroids and brain tissue slices, this confocal PLIM-TCSPC system provides high sensitivity, light penetration depth of up to 100–300 μm and low photo-damage of the probe and cells during measurement [14, 16].

6.3 Experimental Procedure

A typical O_2 imaging experiment includes preparation of cells or tissue samples, staining them with a cell-permeable O_2 probe (or culturing them on O_2-sensitive scaffold—see below), assembling the sample on the microscope stage (equilibration, focusing the optics, pre-scans and selection of regions of interest, ROIs), image acquisition in TCSPC-PLIM mode and processing the data.

6.3.1 Sample Preparation

For high-resolution imaging live cells or tissue in the sample must be immobilized for duration of the measurements, to exclude any movement of objects being imaged. This can be achieved by cell attachment to coated surfaces (e.g. collagen, Matrigel^{TM}, poly-lysine), physical trapping in microfluidic biochips or growth on porous scaffolds such as Alvetex^{TM} or Millipore. Cultures of adherent cells do not require additional preparatory procedures, while multi-cellular spheroids and organoids grown in suspension for several days or even weeks, can be immobilized and prepared for imaging in about 2–6 h. The O_2 sensitive probe is selected based on its cell and tissue specificity, staining procedure (passive loading by simple addition to the media is preferred), acceptable doses and incubation times, achievable phosphorescent signals, sensitivity to O_2 in the studied range, overall performance of PLIM images and O_2 concentration maps produced. Importantly, the probe should have minimal cyto- and photo-toxicity and impact on cell and tissue function and metabolism. Preliminary optimization of staining with the O_2 probe (concentration added and incubation time) is usually required for each new type of cells or tissue.

O_2 imaging can also be combined with other fluorescence based probes and assays, e.g. molecular probes, fluorescent protein constructs, labelled antibodies.

Spectrally similar fluorescent probes which produce lifetime response to their analyte can be measured in the same spectral window simultaneously by nanosecond FLIM and PLIM [9]. Alternatively, spectrally different fluorescent dyes can be imaged in a different spectral channel in intensity mode (wavelength discrimination with appropriate filter and laser wavelengths). To facilitate visualization of cells, their morphology and sub-cellular compartments, test samples can be counterstained with appropriate fluorescent probes, such as nuclear (e.g. Hoechst 33342), cytosolic (Calcein Green) and mitochondrial (TMRM) stains, fluorescent antibodies or fluorescent protein constructs delivered by transfection. Such probes for multiparametric imaging should be tested in preliminary experiments, to ensure their compatibility and minimal cross-talk with the O$_2$ probe.

6.3.2 Measurement

O$_2$ imaging procedure by PLIM-TCSPC is described in more detail in [48, 56]. The sample is equilibrated at measurement conditions, scanned in preview mode and then imaged. The DCS-120 based confocal FLIM system requires manual setting of the neutral density filter and confocal aperture, and selection of either super-continuum or 405 nm diode laser. The other parameters, including scan rate, size of scanned region, pixel number (e.g. 128 × 128, 256 × 256, 512 × 512), time per pixel, PLIM time scale, and FLIM/PLIM data acquisition time, are controlled from the SPCM software [8] (Becker & Hickl). As the phosphorescence lifetime can vary over a very wide range these parameters may require optimization for the individual luminophores [14]. PLIM and FLIM intensity images (i.e. temporally integrated signals) can either be displayed directly or extracted from FLIM/PLIM data files. Transmission images are obtained with a standalone digital camera on the microscope port. For single-channel TCSPC systems multi-parametric imaging is performed by sequential scanning of sample (two or more scans). For a region 256 × 256 pixel scanning time in PLIM mode usually takes 1–2 min. The relatively long acquisition time is, in part, a result of the fact that the pixel time has to be longer than the phosphorescence decay time [8]. Nevertheless, the image acquisition time is kept as short as possible to prevent photodamage of the sample and probe, and to facilitate analysis of rapid changes in O$_2$ concentration over time. The laser power is selected as high as, but no higher than, to obtain a sufficient number of photons within the acquisition time dictated by the pixel number and phosphorescence decay time.

6.3.3 Data Analysis and Processing

Measured TCSPC-PLIM images are imported and processed in SPCImage software (Becker & Hickl), where decay fitting and precise calculations of fluorescence and phosphorescence lifetimes are carried out (Fig. 6.3). Importantly, photon distributions should be analyzed or at least inspected immediately after measurement, to verify that the image acquisition parameters, PLIM time scale, and laser-on time were optimal and ensure generation of sufficient quality decay curves which are easy to process and extract accurate lifetime values. After selecting an appropriate ROI and fitting parameters (shift, decay function order, offset and other parameters), lifetimes are calculated for the whole image (pixel-by-pixel transformation) and exported as a matrix. Binning can be used to improve photon distributions; however this reduces spatial resolution of the lifetime information [7, 8]. By applying O_2 calibration function (see below), lifetime images can then be converted into O_2 concentration images and used for graphical or numerical representation.

6.3.4 O_2 Calibration

When using a new PLIM set-up, probe, or imaging O_2 under non-standard conditions (temperature, sample type), it is necessary to calibrate the system to enable quantitative readout of O_2 concentration. This is achieved by loading cells or tissue sample with the probe, blocking respiration (e.g. with a mitochondrial inhibitor Antimycin A) to bring sample O_2 levels to atmospheric levels. Then the sample is exposed to different O_2 levels in the atmosphere, and after its equilibration PLIM images are acquired. Processing PLIM data and determining average lifetime values enables to work out the calibration function for given temperature. For the established O_2 probes calibrations are provided by vendors or can be found in literature [13, 14].

6.4 Practical Uses of 3D O_2 Imaging

6.4.1 Staining of Various Cell and Tissue Models with O_2 Probes

Cell-penetrating O_2 imaging probes provide efficient labelling of all cells in the sample and allow monitoring of local O_2 levels with sub-cellular spatial resolution. Different cell types and tissues differ in the structure. composition and fluidity of plasma membrane and transport mechanisms used for probe internalisation. For simple cultures of adherent cells (2D models) nanoparticle and small molecule probes (e.g. Pt-Glc, NanO2, MM2, see Table 6.2) usually provide good staining efficiency. Figure 6.4a shows intracellular staining with a small molecule probe, Pt-Glc (2.5 μM in the medium, overnight incubation), which localizes diffusely in the cytoplasm with partial enrichment in the lysosomes [16].

At the same time, 3D tissue models, even simple and relatively small multi-cellular aggregates and neurospheres, are more difficult to stain with fluorescent probes, expression constructs and antibodies, as such specimens possess strong diffusion limitations [37, 40]. For their efficient staining the probe should penetrate or bind to cell membrane, penetrate deep into tissue (which may contain cells of different type) and distribute across its volume. Only a few O$_2$ probes have been demonstrated to stain efficiently such 3D tissue models [16]. Thus nanoparticle based probes MitoImage^{TM} NanO2 and MM2 show efficient staining of HCT116 cell spheroids (Fig. 6.4), however they do not stain well pre-formed neurospheres [14]. Figure 6.4b–e, show different staining patterns for the different 3D tissue models and O$_2$ probes.

Organ and tissue explants (e.g. brain slices) are usually analyzed soon after isolation, but they can be maintained in culture for several days retaining their cyto-architecture and physiological behaviour. Staining of such samples is more

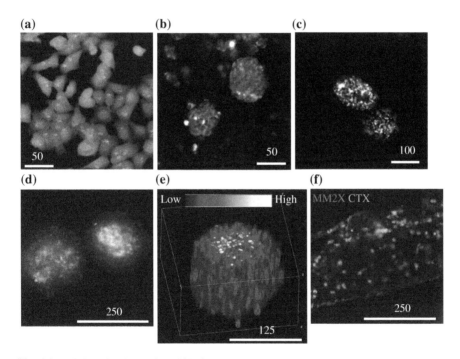

Fig. 6.4 Staining of various cell and 3D tissue models with the phosphorescent O$_2$ probes. **a** Live human colon cancer HCT116 cells stained with Pt-Glc probe (2.5 μM, 16 h incubation). Cell bodies are shown in *grey scale* and O$_2$ probe in *red*. **b** Live PC12 cell aggregates stained with Pt-Glc (2.5 μM, 16 h). **c** Live human colon cancer HCT116 tumor spheroids stained with NanO2 (5 μg/mL, 16 h). **d** Live neurospheres produced from rat primary cortical embryonic E18 cells stained with Pt-Glc (1 μM, 72 h). **e** 3D reconstruction (20 sections, 2 μm each) of fixed neurosphere stained with Pt-Glc (1 μM, 16 h). The intensity of probe staining is shown in *false-colour scale* provided on the *top*. **f** Live embryonic E16 rat brain slice (400 μm thick) stained with PA2 (25 μg/mL, *red*) and Cholera toxin-Alexa 488 conjugate (1 ng/mL, *green*) for 2 h. Sections **a–d**, **f** represent single optical sections (0.5–1 μm thick). Scale bar is in μm

challenging, as the continuous staining method used with spheroid cultures cannot be used. Many O_2 probes do not work with 3D tissue explants, but some are being developed specifically for such models [47, 62]. Figure 6.4f demonstrates efficient staining of embryonic brain slices with a new probe PA2 [16]. The small molecule Pt-Glc probe also provides very good staining of spheroids and tissue samples and allows mapping of O_2 concentration with sub-cellular spatial resolution.

Besides their staining efficiency, O_2 imaging probes must be evaluated for their effects on cell function and viability. Minimal probe doses should be used and possible toxicity effects evaluated prior to the start of complex physiological experiments. Non-specific and late markers of cell death such as propidium iodide (membrane integrity) are not sufficient, and preference should be given to more specific viability stains, such as the mitochondrial membrane potential, Ca^{2+}, OCR and ECA assays, tissue-specific and functional markers (see e.g. our recent study [14]). Overall, each tissue model requires careful selection of an O_2 imaging probe.

6.4.2 Analysis of Spheroid Oxygenation at Rest and upon Metabolic Stimulation

Until recently analysis of tissue O_2 was conducted with indirect or semi-quantitative markers of hypoxia, such as pimonidazole staining, HIF stabilization, protein markers (reviewed in [49]). The use of cell-penetrating phosphorescent O_2 probes and TCSPC-PLIM method allows the direct, quantitative and more detailed analysis of spatial and temporal heterogeneity of O_2 in tissue samples. It also provides valuable information on physiological status and metabolic activity of tissue and function of individual cells within.

Hypoxic regions of tumors have been shown to resist treatments by promoting angiogenesis and tumor progression and suppressing apoptosis [58, 64]. Many physiological conditions experienced by tumor cells, including the formation of cancer stem cells, can be modelled with tumor spheroids [40]. We analyzed spheroids produced from human colon cancer cells HCT116 with NanO2 probe efficient distribution of which across the spheroid allows the production of high quality PLIM images (see Fig. 6.5a, b). Line profiles across the spheroid (Fig. 6.5e, f) reveal clear trends for lifetime and O_2 concentration (though intensity signals are not changing much), and the presence of hypoxic core (approximately 60 μM O_2) inside the spheroid. These measurements were carried out in air atmosphere (21 % O_2).

In order to investigate cell death/viability within spheroids, we counter-stained them with CellTox Green (membrane integrity probe, binds to DNA in dead cells) and TMRM (stains polarized mitochondria of healthy cells) and analysed by multi-parametric imaging. Another spheroid was immunostained with BrdU to track proliferating cells [58]. The results in Fig. 6.5c, d, show that proliferating cells (BrdU-positive) and cells with active mitochondria (TMRM staining) resided at the

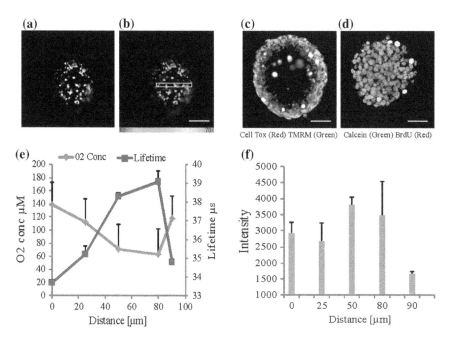

Fig. 6.5 Confocal TCSPC-FLIM based O₂ imaging of HCT116 tumor spheroids. **a** Intensity image of spheroid cross-section. **b** PLIM lifetime image which corresponds to **a** (the *colour palette* presents the lifetime range). **c** Fluorescent staining of HCT116 spheroids with TMRM (*green*) and Cell Tox green (*red*). **d** Fluorescent staining with calcein (*green*) and BrdU (*red*). **e** Lifetime values of the NanO2 probe (*red line*) and corresponding O₂ concentrations (*blue line*) for the cross-section shown in (**b**) as a *rectangle*. **f** Intensity values for the cross-sections shown in (**c**)

periphery, while dead cells (CellTox Green positive) were located mainly in spheroid hypoxic core regions where practically no cell division occurred.

In another experiment, we analysed neurospheres which comprise heterogeneous multi-cellular structures produced from embryonic neural tissue [14, 15]. Unlike tumor spheroids, hypoxic core of the neurosphere promotes survival and proliferation of neural stem cells, while its periphery is enriched with differentiating neuronal and glial cells. We stained neurospheres with Pt-Glc probe and analyzed them upon treatment with excitotoxic stimulant sodium glutamate which affects cell respiration and induces neuronal cell death [44]. Figure 6.6a, b, prove that glutamate reduces O₂ consumption in neurosphere core and increases its oxygenation. This effect can be attributed to direct toxic effect of glutamate on mitochondrial function of neural stem cells.

With another model, aggregates of PC12 cells [16], we performed stimulation with two model drugs [23]: mitochondrial uncoupler FCCP and inhibitor rotenone. Line profiles presented across the 100–200 μm aggregates (Fig. 6.6c) show average basal oxygenation level of 140 μM which quickly decreased to 20–40 μM by FCCP, but then reversed back by treatment with rotenone.

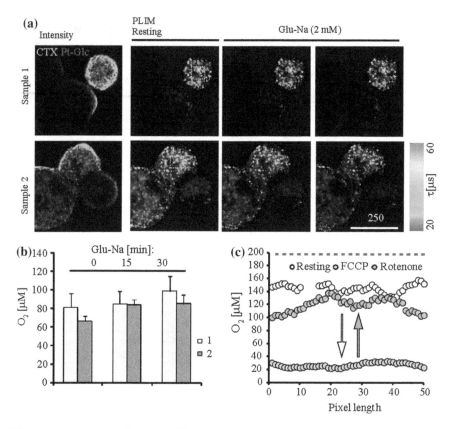

Fig. 6.6 a Responses to glutamate of live neurospheres (samples *1–2*) stained with Pt-Glc (1 μM, 72 h). *Left* Intensity images of Pt-Glc (*red*) and Cholera Toxin (*green*). *Right* Time-lapse PLIM images. **b** Calculated O_2 for selected ROI in different samples. **c** O_2 line profiles of PC12 multicellular aggregates, at rest and upon stimulation with FCCP (4 μM) and rotenone (2.5 μM). *Red dashed line* indicates air-saturated O_2 levels. Scale bar is in μm

Altogether, these examples show how TCSPC-PLIM can be used for quantitative monitoring of in situ oxygenation of spheroids and other 3D tissue models. This can be combined with drug treatments and multi-parametric analysis using other common probes and markers of cellular function, which require either live or fixed and immunostained cells.

6.4.3 Monitoring of Oxygenation of Cells Grown on Phosphorescent O_2-Sensitive Scaffolds

As an alternative to cell-penetrating probes, solid-state phosphorescent O_2-sensitive coatings and scaffold materials can also be used to study cell oxygenation [28, 49].

Thus, Alvetex scaffolds made of biocompatible porous polystyrene and having pores and voids of 36–40 µm, provide rigid support and efficient gas exchange for cultured cells and tissue slices. These scaffolds, which mimic natural cell environment in the tissue, can be converted into an O$_2$-sensitive material by impregnation with a phosphorescent dye such as PtTFPP. Figure 6.7a, shows confocal PLIM-TCSPC images of such scaffolds, their 3D reconstruction and response to changes in O$_2$ (Fig. 6.7b). The magnitude of lifetime changes for the scaffolds

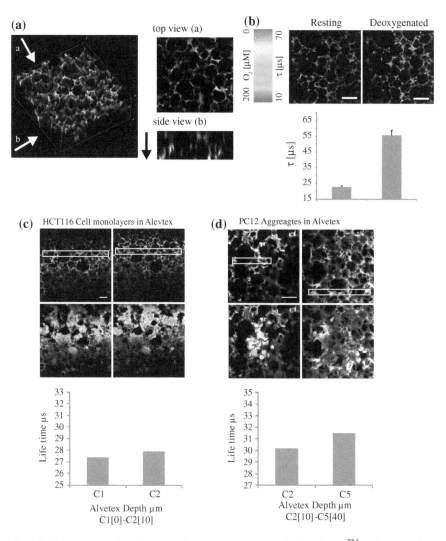

Fig. 6.7 Oxygenation of cells grown in phosphorescent scaffolds Alvetex^{TM} and analysed by TCSPC-PLIM. **a** *Top* (XY) and *side* (XZ) projections of reconstructed 3D image of the scaffold. **b** PLIM images of the scaffold at different O$_2$ levels. **c, d** Colon cancer cells (**c**, HCT116) and multi-cellular aggregates (**d**, PC12) in the scaffold imaged at different depths. Mean lifetime and O$_2$ values in selected ROI are shown underneath. Scale bar is 50 µm

(20–55 μs) is similar to NanO2, Pt-Glc probes based on the same dye, and response is fast due to their microporous structure and thin walls.

We applied suspensions of HCT116 cells on such phosphorescent Alvetex membranes and analyzed their oxygenation. At low cell density such cultures form monolayer structures which did not affect significantly the phosphorescent lifetimes of the scaffold, up to a depth of ~ 30 μm (Fig. 6.7c). However, the more dense aggregates of PC12 cells were seen to invade deep into the scaffold and deoxygenate surrounding areas, as reflected by increased lifetime values (Fig. 6.7d). This shows the usability of phosphorescent scaffolds for analysis of O_2 distribution and local gradients in cultured tissue-like structures. These materials are easy to make and use for the analysis of O_2 in any types of cells and tissues cultured on them. They possess high brightness and reproducibility resulting in high quality PLIM and O_2 images, do not require cell staining. O_2 imaging on scaffolds can also be coupled with other imaging probes and cell markers. Their drawback is variable and uncontrolled distance between the biological material and solid-state extracellular sensor.

6.5 Conclusions

Standard measurement set-up and experimental procedures for imaging O_2 concentration in 2D and 3D tissue models by TCSPC-PLIM were described. An overview of cell-permeable O_2-sensitive phosphorescent probes currently available and their main performance characteristics were provided. The use of these probes and imaging method were demonstrated in case studies with several different tissue models including spheroids, neurospheres and tissue slices. These examples provide a comprehensive guide for inexperienced users, to familiarize themselves with PLIM-TCSPC based O_2 imaging method and set it up on their own imaging platform. The on-going development of O_2 probes with improved bio-distribution in 3D tissue samples and compatibility with two-photon excitation, and technical advances in TCSPC and FLIM systems will further enhance the capabilities of this method, which has long-ranging applications in life and biomedical sciences.

Acknowledgments This work was supported by the Science Foundation Ireland, grant 12/RC/2276, the European Commission FP7 Program, grant FP7-HEALTH-2012-INNOVATION-304842-2.

References

1. W. Becker, A. Bergmann, M.A. Hink, K. König, K. Benndorf, C. Biskup, Fluorescence lifetime imaging by time-correlated single photon counting. Microsc. Res. Techn. **63**, 58–66 (2004)
2. W. Becker, *Advanced Time-Correlated Single-Photon Counting Techniques* (Springer, Berlin, 2005)

3. W. Becker, A. Bergmann, C. Biskup, Multispectral fluorescence lifetime imaging by TCSPC. Microsc. Res. Tech. **70**, 403–409 (2007)
4. W. Becker, A. Bergmann, Lifetime-Resolved Imaging in Nonlinear Microscopy, in *Handbook of Biomedical Nonlinear Optical Microscopy*, ed by B.R. Masters, P.T.C. So (Oxford University Press, Oxford, 2008)
5. W. Becker, B. Su, K. Weisshart, O. Holub, FLIM and FCS detection in laser-scanning microscopes: increased efficiency by GaAsP hybrid detectors. Microsc. Res. Tech. **74**, 804–811 (2011)
6. W. Becker, B. Su, A. Bergmann, K. Weisshart, O. Holub, Simultaneous fluorescence and phosphorescence lifetime imaging. Proc. SPIE **7903**, 790320 (2011)
7. DCS-120 Confocal Scanning FLIM Systems, user handbook, edition (Becker & Hickl GmbH, Berlin, 2012), available on www.becker-hickl.com
8. W. Becker, *The bh TCSPC Handbook*. 5th edn, (Becker & Hickl GmbH, Berlin, 2012), available on www.becker-hickl.com
9. W. Becker, Fluorescence lifetime imaging–techniques and applications. J. Microsc. **247**, 119–136 (2012)
10. E.R. Carraway, J.N. Demas, B.A. DeGraff, J.R. Bacon, Photophysics and photochemistry of oxygen sensors based on luminescent transition-metal complexes. Anal. Chem. **63**, 337–342 (1991)
11. M.R. Chatni, G. Li, D.M. Porterfield, Frequency-domain fluorescence lifetime optrode system design and instrumentation without a concurrent reference light-emitting diode. Appl. Opt. **48**, 5528–5536 (2009)
12. R.I. Dmitriev, D.B. Papkovsky, Optical probes and techniques for O2 measurement in live cells and tissue. Cell. Mol. Life Sci. **69**, 2025–2039 (2012)
13. R.I. Dmitriev, A.V. Zhdanov, G. Jasionek, D.B. Papkovsky, Assessment of cellular oxygen gradients with a panel of phosphorescent oxygen-sensitive probes. Anal. Chem. **84**, 2930–2938 (2012)
14. R.I. Dmitriev, A.V. Zhdanov, Y.M. Nolan, D.B. Papkovsky, Imaging of neurosphere oxygenation with phosphorescent probes. Biomaterials **34**, 9307–9317 (2013)
15. R.I. Dmitriev, A.V. Kondrashina, K. Koren, I. Klimant, A.V. Zhdanov, J.M.P. Pakan, K.W. McDermott, D.B. Papkovsky, Small molecule phosphorescent probes for O$_2$ imaging in 3D tissue models. Biomater. Sci. **2**, 853–866 (2014)
16. R. Dmitriev, S. Borisov, A. Kondrashina, J. P. Pakan, U. Anilkumar, J. M. Prehn, A. Zhdanov, K. McDermott, I. Klimant, D. Papkovsky, Imaging oxygen in neural cell and tissue models by means of anionic cell-permeable phosphorescent nanoparticles. Cell. Mol. Life Sci. (2014) doi: DOI:10.1007/s00018-014-1673-5.
17. I. Dunphy, S.A. Vinogradov, D.F. Wilson, Oxyphor R2 and G2: phosphors for measuring oxygen by oxygen-dependent quenching of phosphorescence. Anal. Biochem. **310**, 191–198 (2002)
18. N.T. Elliott, F. Yuan, A review of three-dimensional in vitro tissue models for drug discovery and transport studies. J. Pharm. Sci. **100**, 59–74 (2011)
19. T. Fenchel, B. Finlay, Oxygen and the spatial structure of microbial communities. Biol. Rev. Camb. Philos. Soc. **83**, 553–569 (2008)
20. Y. Feng, J. Cheng, L. Zhou, X. Zhou, H. Xiang, Ratiometric optical oxygen sensing: a review in respect of material design. Analyst **137**, 4885–4901 (2012)
21. A. Fercher, S.M. Borisov, A.V. Zhdanov, I. Klimant, D.B. Papkovsky, Intracellular O2 sensing probe based on cell-penetrating phosphorescent nanoparticles. ACS Nano **5**, 5499–5508 (2011)
22. O.S. Finikova, A.V. Cheprakov, S.A. Vinogradov, Synthesis and Luminescence of Soluble meso-Unsubstituted Tetrabenzo- and Tetranaphtho[2, 3]porphyrins. J. Org. Chem. **70**, 9562–9572 (2005)
23. K.A. Foster, F. Galeffi, F.J. Gerich, D.A. Turner, M. Müller, Optical and pharmacological tools to investigate the role of mitochondria during oxidative stress and neurodegeneration. Prog. Neurobiol. **79**, 136–171 (2006)

24. M.M. Frigault, J. Lacoste, J.L. Swift, C.M. Brown, Live-cell microscopy–tips and tools. J. Cell Sci. **122**, 753–767 (2009)
25. L.G. Griffith, M.A. Swartz, Capturing complex 3D tissue physiology in vitro. Nat. Rev. Mol. Cell Biol. **7**, 211–224 (2006)
26. X.-Y. Han, B. Wei, J.-F. Fang, S. Zhang, F.-C. Zhang, H.-B. Zhang, T.-Y. Lan, H.-Q. Lu, H.-B. Wei, Epithelial-mesenchymal transition associates with maintenance of stemness in spheroid-derived stem-like colon cancer cells, PLoS ONE **8**, e73341 (2013)
27. Y.-L. Hu, M. DeLay, A. Jahangiri, A.M. Molinaro, S.D. Rose, W.S. Carbonell, M.K. Aghi, Hypoxia-Induced autophagy promotes tumor cell survival and adaptation to antiangiogenic treatment in glioblastoma. Cancer Res. **72**, 1773–1783 (2012)
28. K. Kellner, G. Liebsch, I. Klimant, O.S. Wolfbeis, T. Blunk, M.B. Schulz, A. Göpferich, Determination of oxygen gradients in engineered tissue using a fluorescent sensor. Biotechnol. Bioeng. **80**, 73–83 (2002)
29. E. Knight, B. Murray, R. Carnachan, S. Przyborski, Alvetex(R): polystyrene scaffold technology for routine three dimensional cell culture. Methods Mol. Biol. **695**, 323–340 (2011)
30. A.V. Kondrashina, R.I. Dmitriev, S.M. Borisov, I. Klimant, I. O'Brien, Y.M. Nolan, A.V. Zhdanov, D.B. Papkovsky, A phosphorescent nanoparticle-based probe for sensing and imaging of (intra) cellular oxygen in multiple detection modalities. Adv. Funct. Mater. **22**, 4931–4939 (2012)
31. Y.E. Koo, Y. Cao, R. Kopelman, S.M. Koo, M. Brasuel, M.A. Philbert, Real-time measurements of dissolved oxygen inside live cells by organically modified silicate fluorescent nanosensors. Anal. Chem. **76**, 2498–2505 (2004)
32. J. Lakowicz, E. Terpetschnig, Z. Murtaza, H. Szmacinski, Development of long-lifetime metal-ligand probes for biophysics and cellular imaging. J. Fluoresc. **7**, 17–25 (1997)
33. A.Y. Lebedev, A.V. Cheprakov, S. Sakadzic, D.A. Boas, D.F. Wilson, S.A. Vinogradov, Dendritic phosphorescent probes for oxygen imaging in biological systems. ACS Appl. Mat. Interfaces **1**, 1292–1304 (2009)
34. K. Lee, R.A. Roth, J.J. LaPres, Hypoxia, drug therapy and toxicity. Pharm Ther **113**, 229–246 (2007)
35. Y.E. Lee, E.E. Ulbrich, G. Kim, H. Hah, C. Strollo, W. Fan, R. Gurjar, S. Koo, R. Kopelman, Near infrared luminescent oxygen nanosensors with nanoparticle matrix tailored sensitivity. Anal. Chem. **82**, 8446–8455 (2010)
36. L.U. Ling, K.B. Tan, H. Lin, G.N.C. Chiu, The role of reactive oxygen species and autophagy in safingol-induced cell death. Cell Death and Dis. **2**, e129 (2011)
37. J. Liu, J. Hilderink, T.A. Groothuis, C. Otto, C.A. Blitterswijk, J. Boer, Monitoring nutrient transport in tissue engineered grafts. J. Tissue Eng. Regen. Med. (2013). doi: 10.1002/term.1654
38. D.J. Maltman, S.A. Przyborski, Developments in three-dimensional cell culture technology aimed at improving the accuracy of in vitro analyses. Biochem. Soc. Trans. **38**, 1072–1075 (2010)
39. D. McLoskey, D. Campbell, A. Allison, G. Hungerford, Fast time-correlated single-photon counting fluorescence lifetime acquisition using a 100 MHz semiconductor excitation source. Meas. Sci. Technol. **22**, 067001 (2011)
40. G. Mehta, A.Y. Hsiao, M. Ingram, G.D. Luker, S. Takayama, Opportunities and challenges for use of tumor spheroids as models to test drug delivery and efficacy. J. Control Release **164**, 192–204 (2012)
41. R.L. Morris, T.M. Schmidt, Shallow breathing: bacterial life at low O(2). Nat. Rev. Microbiol. **11**, 205–212 (2013)
42. F.A. Navarro, P.T. So, R. Nirmalan, N. Kropf, F. Sakaguchi, C.S. Park, H.B. Lee, D.P. Orgill, Two-photon confocal microscopy: a nondestructive method for studying wound healing. Plast. Reconst. Surg. **114**, 121–128 (2004)
43. U. Neugebauer, Y. Pellegrin, M. Devocelle, R.J. Forster, W. Signac, N. Moran, T.E. Keyes, Ruthenium polypyridyl peptide conjugates: membrane permeable probes for cellular imaging. Chem. Commun. 5307–5309 (2008). doi: 10.1039/B810403D

44. D.G. Nicholls, L. Johnson-Cadwell, S. Vesce, M. Jekabsons, N. Yadava, Bioenergetics of mitochondria in cultured neurons and their role in glutamate excitotoxicity. J. Neurosci. Res. **85**, 3206–3212 (2007)
45. D.V. O'Connor, D. Phillips, *Time-Correlated Single Photon Counting* (Academic Press, London, 1984)
46. T.C. O'Riordan, A.V. Zhdanov, G.V. Ponomarev, D.B. Papkovsky, Analysis of intracellular oxygen and metabolic responses of mammalian cells by time-resolved fluorometry. Anal. Chem. **79**, 9414–9419 (2007)
47. F. Pampaloni, E.G. Reynaud, E.H. Stelzer, The third dimension bridges the gap between cell culture and live tissue. Nat. Rev. Mol. Cell Biol. **8**, 839–845 (2007)
48. D. Papkovsky, A.V. Zhdanov, A. Fercher, R.I. Dmitriev, J. Hynes, *Phosphorescent Oxygen-Sensitive Probes* (Springer, Berlin, 2012)
49. D.B. Papkovsky, R.I. Dmitriev, Biological detection by optical oxygen sensing. Chem. Soc. Rev. **42**, 8700–8732 (2013)
50. J.P. Philip, K. Carlsson, Theoretical investigation of the signal-to-noise ratio in fluorescence lifetime imaging. J. Opt. Soc. Am. **A20**, 368–379 (2003)
51. M. Quaranta, S.M. Borisov, I. Klimant, Indicators for optical oxygen sensors. Bioanal. Rev. **4**, 115–157 (2012)
52. M. Radisic, J. Malda, E. Epping, W. Geng, R. Langer, G. Vunjak-Novakovic, Oxygen gradients correlate with cell density and cell viability in engineered cardiac tissue. Biotechnol. Bioeng. **93**, 332–343 (2006)
53. S. Saharudin, K.M. Isha, Z. Mahmud, S.H. Herman, U.M. Noor, Performance evaluation of optical fiber sensor using different oxygen sensitive nano-materials, in Photonics (ICP). 2013 IEEE 4th International Conference on (IEEE2013), pp. 309–312
54. H. Sauer, M. Wartenberg, J. Hescheler, Reactive oxygen species as intracellular messengers during cell growth and differentiation. Cell Phys. and Biochem. **11**, 173–186 (2001)
55. G.L. Semenza, Hypoxia, clonal selection, and the role of HIF-1 in tumor progression. Crit. Rev. Biochem. Mol. Biol. **35**, 71–103 (2000)
56. J.A. Spencer, F. Ferraro, E. Roussakis, A. Klein, J. Wu, J.M. Runnels, W. Zaher, L.J. Mortensen, C. Alt, R. Turcotte, R. Yusuf, D. Cote, S.A. Vinogradov, D.T. Scadden, C.P. Lin, Direct measurement of local oxygen concentration in the bone marrow of live animals. Nature **508** (7495), 269–273 (2014)
57. E. Takahashi, T. Takano, Y. Nomura, S. Okano, O. Nakajima, M. Sato, In vivo oxygen imaging using green fluorescent protein. Am. J. Physiol. Cell Physiol. **291**, 31 (2006)
58. P. Taupin, BrdU immunohistochemistry for studying adult neurogenesis: paradigms, pitfalls, limitations, and validation. Brain Res. Rev. **53**, 198–214 (2007)
59. V. Tsytsarev, H. Arakawa, S. Borisov, E. Pumbo, R.S. Erzurumlu, D.B. Papkovsky, In vivo imaging of brain metabolism activity using a phosphorescent oxygen-sensitive probe. J. Neurosci. Methods **216**, 146–151 (2013)
60. Y.-C. Tung, A.Y. Hsiao, S.G. Allen, Y.-S. Torisawa, M. Ho, S. Takayama, High-throughput 3D spheroid culture and drug testing using a 384 hanging drop array. Analyst **136**, 473–478 (2011)
61. A.M. Weljie, F.R. Jirik, Hypoxia-induced metabolic shifts in cancer cells: Moving beyond the Warburg effect. Int. J. Biochem. Cell Biol. **43**, 981–989 (2011)
62. A. Williamson, S. Singh, U. Fernekorn, A. Schober, The future of the patient-specific Body-on-a-chip. Lab Chip **13**, 3471–3480 (2013)
63. D.F. Wilson, W.M.F. Lee, S. Makonnen, O. Finikova, S. Apreleva, S.A. Vinogradov, Oxygen pressures in the interstitial space and their relationship to those in the blood plasma in resting skeletal muscle. J. Appl. Physiol. **101**, 1648–1656 (2006)
64. W.R. Wilson, M.P. Hay, Targeting hypoxia in cancer therapy. Nat. Rev. Cancer **11**, 393–410 (2011)
65. K.M. Yamada, E. Cukierman, Modeling tissue morphogenesis and cancer in 3D. Cell **130**, 601–610 (2007)

Chapter 7
FRET Microscopy: Basics, Issues and Advantages of FLIM-FRET Imaging

Ammasi Periasamy, Nirmal Mazumder, Yuansheng Sun, Kathryn G. Christopher and Richard N. Day

Abstract Förster resonance energy transfer (FRET) is an effective and high resolution method to investigate protein–protein interaction in live or fixed specimens. The FRET technique is increasingly employed to evaluate the molecular mechanisms governing diverse cellular processes such as vesicular transport, signal transduction and the regulation of gene expression. For FRET to occur, protein moieties should be close together within 10 nm, the dipole moment of the fluorophore targeted to the proteins should have an appropriate orientation, and the spectral overlap of the donor emission with the acceptor absorption should be >30 %. FRET can be used to estimate the distance between interacting protein molecules in vivo or in vitro using light microscopy systems. Visible fluorescent proteins (VFPs) have been widely used as a FRET pair in addition to organic dyes. Light microscopy techniques including wide-field, confocal and multiphoton microscopy systems provide spatial information of the interacting proteins with nanometer resolution. For better interpretation and quantitation of the FRET signal the contaminations—also called spectral bleedthrough (SBT)—have to be removed.

A. Periasamy (✉)
W.M. Keck Center for Cellular Imaging, University of Virginia, Charlottesville, VA, USA
e-mail: ap3t@virginia.edu

N. Mazumder · K.G. Christopher
University of Virginia, Charlottesville, VA, USA
e-mail: mm3fd@virginia.edu

Y. Sun
University of Virginia, Biology, Charlottesville, VA 22904, USA
e-mail: yuansheng.sun@gmail.com

R.N. Day
Departments of Medicine and Cell Biology, University of Virginia, Charlottesville, VA, USA
e-mail: rnday@iupui.edu

© Springer International Publishing Switzerland 2015   249
W. Becker (ed.), *Advanced Time-Correlated Single Photon Counting Applications*,
Springer Series in Chemical Physics 111, DOI 10.1007/978-3-319-14929-5_7

Another imaging approach, fluorescence lifetime imaging microscopy (FLIM) also provides quantitative information with spatial and temporal details of protein-protein interactions. No algorithm is required here to remove any contamination, as in FLIM-FRET only the change in lifetime value of the donor without and with the acceptor molecules is monitored. The lifetime of the donor decreases at the occurrence of FRET. FLIM is sensitive to the local microenvironment of the molecule but insensitive to the change in fluorophore concentration or excitation intensity. The FLIM-FRET technique is ideal for dark acceptors and the investigation of NADH molecules such as NADH, FAD, Tryptophan, etc. FLIM-FRET techniques provide high temporal resolution of protein-protein interactions in live specimens.

7.1 Introduction

The historical precursors for the theory of Förster resonance energy transfer (FRET) date back to the 19th and beginning of 20th century with emerging understanding of electromagnetic and quantum mechanics. The first quantum mechanical theories of FRET were developed concurrently with the new theories of Heisenberg, Schrö-dinger and Dirac [1]. In the 1920s Jean-Baptiste Perrin and his son Francis Perrin explained the energy transfer process between two identical molecules in solution involving dipole-dipole intermolecular interactions. Perrin was the first to note that energy transfer is distance dependent and would occur over a specific range, which he calculated to be 150–250 Å—much larger than Förster's estimation [1, 2]. It was T. Förster in 1946 who established the correct distance (10–100 Å) over which the incoherent energy transfer (named FRET) would happen and provided the quantitative means to measure molecular distances with his now well-known equations. Förster's first paper on FRET was published in 1946 [3, 4]. More historical background about FRET is discussed in the literature [1].

The impact of great scientific insights can very often only be measured decades later. Even though the FRET technology is well established in cell biology and other research areas, Förster's FRET contribution is still evolving, still expanding and still offering new challenges in applying this great theory in the broader bio-medical sciences. Understanding the physics of FRET is important to implement it appropriately for various biological or clinical applications. FRET measurements typically require the donor and acceptor to be different fluorophores (called hetero-FRET), although the acceptor need not be fluorescent (e.g. dark quenchers) for measuring FRET based on the donor. FRET can also occur between identical fluorophores, called homo-FRET which can be measured by fluorescence anisot-ropy imaging [5]. Many FRET microscopy and spectroscopy techniques have been developed [6–12]—in this chapter we provide briefly the importance of microscopy techniques for FRET, intensity based FRET imaging and the issues associated with these techniques. We then discuss the advantages of the lifetime imaging FRET and it's calibration with FRET standards. We also talk about the time-correlated single photon counting (TCSPC) methodology to investigate protein-protein interactions in biological specimens.

7.2 Basics of FRET

FRET is a process by which radiationless transfer of energy occurs from a fluorophore molecule in the excited state to a molecule in close proximity in the ground state. The molecule donating the energy is called 'Donor' (D) and the molecule accepting the energy is called 'Acceptor' (A). When this occurs, the donor is said to be quenched and the acceptor is sensitized and the event becomes inter alia the basis for calculating proximities between molecules, providing a non-invasive approach to visualize the spatio-temporal dynamics of the interactions between protein partners in living specimens [1–3, 5–13]. There are three important conditions for FRET to occur, see Fig. 7.1. These are:

(i) The spectral overlap: FRET can only occur when the emission spectrum of a donor fluorophore significantly overlaps (>30 %) the absorption spectrum of an acceptor.

(ii) The distance between fluorophores attached to the molecules: The proximity between the donor and acceptor molecule should be within 10–100 Å (1–10 nm). The efficiency of energy transfer is inversely proportional to the sixth power of the radius, where the radius is the distance between the centres of the donor and acceptor dipoles.

(iii) Dipole moment orientation: The emission dipole of the donor and the acceptor absorption dipole must be oriented to each other. The magnitude of the relative orientation of the dipole-dipole coupling range is from 1 to 4. There is no FRET if it is oriented perpendicular to each other ($\kappa^2 = 0$). If the spectra are overlapped, the donor's oscillating emission dipole will look for a matching absorption dipole of an acceptor to oscillate in synchrony. The probability of these oscillations is higher if more acceptor molecules surround the donors. Each of the above three conditions have to be met; there is no FRET if any one of these conditions is not satisfied. Moreover, the maximum FRET signal occurs in the acceptor emission channel when the donor molecule is excited at peak absorption and the signal is collected at peak emission.

The energy transfer efficiency (E)—an expression of distance (in Angstrom) between interacting labelled molecules (r)—can be calculated using (7.1)

$$E = R_0^6/(Ro^6 + r^6) = 1 - (I_{DA}/I_D) \text{ or } r = R_0[(1/E) - 1]^{1/6} \qquad (7.1)$$

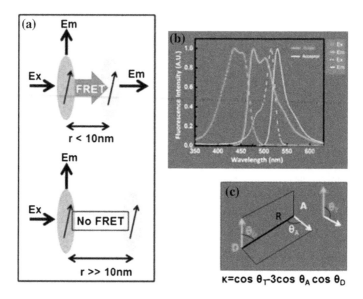

Fig. 7.1 Basic concepts of FRET. **a** FRET is the non-radiative energy transfer from an excited-state donor (D) to an acceptor (A) in close proximity (1–10 nm), via a long-range dipole-dipole coupling mechanism. The energy transfer efficiency (*E*) from D to A is dependent on the inverse of the sixth power of the distance between them (*r*), subject to the Förster distance (Ro) of the FRET pair, at which *E* is 50 %. Other than the D–A distance, FRET also requires two other conditions: **b** a significant spectral overlap between the donor emission and the acceptor absorption spectra; **c** a favourable dipole moment orientation κ^2 ranging from 1 to 4 (most favourable for FRET)

where R_0 is the Förster distance at which half of the excited-stated energy of the donor is transferred to the acceptor (E = 50 %); I_{DA} and I_D are the donor intensities in the presence and the absence of acceptor, respectively. The characteristic Förster distance (R_0) can be estimated based on (7.2).

$$Ro = 0.211 \cdot \{k^2 \cdot n^{-4} \cdot Q_D \cdot J\}^{\frac{1}{6}}, \quad \text{where } J = \in_A \frac{\int_0^\infty f_D(\lambda) f_A(\lambda) \lambda^4 d\lambda}{\int_0^\infty f_D(\lambda) d\lambda} \quad (7.2)$$

where κ^2 (dimensionless, ranging from 1 to 4) is the relative orientation between the dipoles of the donor emission and the acceptor absorption. For the calculation, the assumed dipole orientation for the random movement is $\kappa^2 = 2/3$ (see Fig. 7.1c); *n* is the medium refractive index; Q_D is the donor quantum yield; J (in units of $M^{-1} \times cm^{-1} \times nm^4$) expresses the degree of the overlap between the donor emission and the acceptor absorption spectra (see Fig. 7.1b). In detail, ε_A (in units of $M^{-1} \times cm^{-1}$) is the extinction coefficient of the acceptor at its peak absorption wavelength; λ is the wavelength in nanometer; both $f_D(\lambda)$ (donor emission spectrum) and $f_A(\lambda)$ (normalized acceptor absorption spectrum) are dimensionless.

7.3 FRET Pairs

Suitable fluorophore FRET partners are one of the keys for a successful FRET application [14–18]. FRET pairs can be selected from exogenous and endogenous fluorophores, the former being organic dyes, visible fluorescent proteins and quantum dots. An important criterion for the FRET pair selection is the Förster distance (Ro); a larger Ro will increase the likelihood of a FRET event. Additional positive factors are donors with a higher quantum yields, acceptors with a larger extinction coefficients, and FRET pairs with a larger spectral overlaps. On the other hand the acceptor molecule need not be fluorescent for FRET to occur and the donor to be quenched. Choosing an appropriate fluorophore depends on the target proteins or biological system under investigation. For example, visible fluorescent proteins (VFPs) may not be useful to investigate RNAs. A few selected FRET pairs are shown in Table 7.1.

7.3.1 Organic Fluorophores

The development of novel organic dyes that exhibit improved photo and pH stability, as well as excellent spectral characteristics, provides additional choices for FRET imaging. These organic dyes can be conjugated to ligands for live imaging to follow receptor-mediated cellular internalization [19]. Other applications use labelled antibodies to establish interactions between cellular components with FRET microscopy —albeit almost exclusively in fixed specimens [20]. Photophysical properties and Förster distances (Ro) of the FRET pairs can usually be obtained from manufacturers, e.g. Invitrogen (www.invitrogen.com) and Amersham Biosciences (www.gelifesciences.com). Nevertheless, the final selection of the right donor-acceptor

Table 7.1 Selected FRET pairs for FLIM

FRET pair	Donor lifetime (Unquenched) ns	Donor Ex/Em (nm)	Donor quantum yield
mCeruleun-EYFP	2.8	433/475	0.62
EGFP-mRFP1	2.4	488/525	0.77
EGFP-mCherry	2.4	488/525	0.77
mTurquoise-Venus	3.6	434/474	0.84
EGFP-EYFP	2.4	488/525	0.60
mCerulean-Venus	2.8	433/475	0.62
mCerulean-mCitrine	2.8	433/475	0.87
mTFP-Venus	2.65	462/492	0.85
mTFP-sREACH	2.65	462/492	0.85
mTFP-mKO2	2.65	462/492	0.85
Venus-tdTomato	2.8	514/530	0.65

pair should also include the actual biological question to be addressed, the type of biological specimen to be imaged and the instrument available to measure FRET.

7.3.2 Visible Fluorescent Proteins (VFPs)

Many visible fluorescent proteins (VFP) have been employed in combination with FRET microscopy to visualize dynamic protein interactions under physiological conditions [16]. After transfection, VFPs are directly expressed with the proteins of interest in live cells. The natural diversity of VFPs has provided scientists with a rich palette of variants with different biochemical and spectral characteristics, which represent a sizeable source of potentially powerful molecular tools for numerous applications in the study of complex biological systems. VFPs have provided dramatic new insights including contradicting previously used fluorescent techniques, however, as with all investigative approaches, any particular assay or experimental setup requires a specific choice or optimization of a VFP [14–18]. A few frequently used VFP FRET pairs are listed in Table 7.1.

After deciding to use FRET to study a protein of interest, fluorescent donor-acceptor pairs need to be generated. The most common approach is to obtain commercially available vectors with multiple cloning sites (e.g., BD Biosciences Clontech, Stratagene, Qbiogene, and other companies) to generate in-frame fusions with the VFP's genes. For example, GFP-type fusions can be generated at the N- or C-terminus of proteins of interest, or even internally, with the VFP being interspaced between domains of a protein coding sequence. It is generally a good idea to simultaneously prepare both N- and C-terminal fusions, hoping that at least one of these chimeras retain functional activity. However, to confirm the construction of functional GFP chimeras [17] certain functional tests are critical. After DNA sequencing to verify the constructs, DNA for mammalian cell transfections needs to be prepared with great care. In general, commercial kits (e.g., QiagenMaxiprep) work well. Many transfection reagents are routinely used to introduce plasmid DNA into living cells. Cerulean-Venus or mTFP-Venus is widely used FRET pairs for many biological applications and mTFP has an even better photo-stability and higher quantum yield than Cerulean [21].

Although the newer semiconductor nanocrystal quantum dots are still in the early application phases of biomedical FRET imaging, they have been successfully used as donors for in vitro FRET biological assays [22]. By utilizing the long-lifetime lanthanide chelates such as europium as the donor probe, time-resolved FRET approaches have been used for in vitro drug screening studies [23], where FRET significance is often quantified by the ratiometric FRET method. Since the europium probe has a much longer lifetime (microseconds to milliseconds) than organic compounds (nanoseconds), imaging in a time-resolved manner can easily eliminate background fluorescence from most compounds and dramatically increase the sensitivity of FRET signals [24].

The concentration (or the DNA amount) of the donor with respect to the acceptor has to be evaluated for a particular FRET pair. Most of the FRET pairs could work with equal amount of concentration but in many cases the acceptor concentration (or DNA amount) should ideally be higher than the donor; a higher level of acceptor may improve the probability of a favourable donor emission orientation dipole to the acceptor absorption dipole. It is also important to avoid over expression of the proteins of interest; over expression can cause homo FRET, i.e. donor-donor or acceptor-acceptor interactions.

7.4 Intensity-Based FRET Microscopy, Its Issues and Data Analysis

Monitoring protein-protein interactions using light microscopy techniques provides high spatial and temporal resolution. Even though the theoretical limitation of the microscopy is about 0.2 μm, one can image single molecule protein to ensemble of protein interactions using light microscopy by using appropriate optical configuration and detector. Any microscopy system including wide-field, confocal, two-photon and spectral imaging units can be used for FRET measurements by selecting appropriate filters and the excitation wavelengths for the selected FRET pairs. Here we compare and contrast some of the features available with the intensity based FRET microscopy systems.

Even though wide-field microscopy has successfully been used in FRET imaging over the last 30 years, the fluorescence signal in this system originates above and below the focal plane, which reduces the contrast due to the out-of-focus signal. This out-of-focus signal can be removed by using a digital deconvolution methods. This is a two-step process and it may not be useful in real time data collection from live specimens.

Laser scanning confocal microscopy is used to improve the signal to noise ratio and both, lateral and axial resolution. More importantly, the out-of-focus signals are rejected using a pinhole at the detector plane, which allows collecting time-lapse imaging at various optical sections of the specimen. Confocal microscopy allows imaging protein molecules at a distance of about 50–100 μm deep inside the specimen.

2-photon laser scanning microscopy is applied to imaging as deep as 1 mm or more inside the specimen. It uses infrared light for excitation, and the fluorophores are excited by a two-photon process [25]. Due to the nonlinearity of the excitation process efficient excitation is only obtained in the focal plane. Thus, no confocal pinhole is required to reject out-of-focus signals. The fluorescence photons can be sent directly to a non-descanned detector. Non-descanned detection results in high-efficiency detection from deep sample layers. In comparison to 1p confocal scanning, 2p microscopy produces high contrast images with less autofluorescence and photobleaching.

7.4.1 What Are the Issues Including Signal Bleed-Though in Intensity-Based FRET?

Intensity based microscopy systems are widely available and are used for FRET in many biomedical laboratories. The FRET signal obtained from these systems frequently contains several signal contaminations, listed below [26–28].

(i) Background signal—Background signals are generated by the biological specimen, detector noise, and light scattering. Usually, the level of these noise signals can be determined by using unlabeled biological specimens for subsequent subtraction.

(ii) Donor spectral bleedthrough or cross-talk—As shown in the Fig. 7.1b for the Cerulean- Venus FRET pair, the donor spectral bleed-through is due to the spectral overlap of donor emission and the acceptor absorption. To improve energy transfer efficiency, a > 30 % spectral overlap is desirable, which will increase the energy transfer efficiency at the price of increased spectral bleedthrough.

(iii) Acceptor spectral bleedthrough—The FRET signal is collected in the acceptor channel by exciting the donor molecule. While exciting the donor molecule, a certain percentage of the acceptor molecule can also be excited, due to the acceptor's absorption spectrum overlapping the donor excitation wavelength, undistinguishable from the FRET signal in the acceptor channel. Therefore, the contaminated FRET signal in the acceptor channel contains three components: the true FRET signal and the donor and acceptor bleed-through signals, which must be removed.

7.4.2 Processed FRET: PFRET Data Analysis

As mentioned above, FRET is distance-dependent event and a wealth of available information will be missed in a qualitative analysis. However, before proceeding to evaluate the data quantitatively, the above data correction steps have to be implemented: elimination of background noise and removal of donor and acceptor spectral bleedthrough (DSBT and ASBT). Another major problem with qualitative analyses is the fact that they do not differentiate random from controlled interactions. In contrast, quantitative analysis can discriminate between the two and confirms or otherwise controlled interactions of the molecules, dimerization, complex formation and extensive information of the molecular interaction with sensitivities as close as 5 % energy transfer efficiency. In the qualitative analysis approach, it is quite possible that one could miss a larger distance between the donor and acceptor molecules. How are the contaminations removed to extract the wealth of information provided by FRET microscopy techniques? The background signal due to optical light scattering, autofluorescence and detector noise signal is removed by background subtraction. The two types of spectral bleed-through (DSBT and ASBT) require some mathematical algorithm to remove the contaminations to isolate the FRET signal [10, 27–31].

There are various methods to assess the spectral bleed-through (SBT) contamination in FRET image acquisition [27–32]. Donor bleed-through can be calculated and corrected using the percentage of the spectral area of the donor emission spillover into the acceptor emission spectrum or FRET channel using a single-label donor. In the case of acceptor bleed-through it is difficult to determine the fraction of excitation of the acceptor by the donor wavelength and its emission in the acceptor or FRET channel [29]. Moreover, some of the available methods for FRET data analysis in the literature do not correct for any variation in the expression or concentration of the fluorophore labelled to the cells.

As described in the literature the PFRET [27] and spectral FRET (sFRET) [29, 32] algorithm approach works on the assumption that the double-labelled cells and single-labelled donor and acceptor specimens, imaged under the same conditions, exhibit the same SBT dynamics. This PFRET algorithm follows fluorescence levels pixel-by-pixel to establish the level of SBT in single-labelled cells, and then applies these values as a correction factor to the appropriate matching pixels of the double-labelled specimen. Two examples are shown in Fig. 7.2.

All these corrections are required for wide-field, confocal, and multiphoton microscopy systems. It is advantageous to use FRET standards (www.addgene.org) to verify the optical configuration of any system [33]. Figure 7.2 clearly demonstrates the FRET signals before (Panel FRET) and after correction (PFRET) in wide-field and confocal FRET. Using the PFRET algorithm the FRET signals are further interpreted by plotting the FRET data to show whether the protein interactions are a dimerized or random interaction [12, 19].

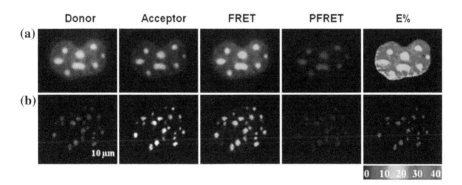

Fig. 7.2 Wide-field (**a**) and Confocal (**b**) FRET microscopy. PFRET image demonstrates that the PFRET algorithm removed the contamination in the FRET signal. The contamination includes donor and acceptor bleedthrough signals and the uneven expression in the cells. The Cerulean-Venus-C/EBPα protein is dimerized in the GHFT1 live cell nucleus. The colour bar indicates the energy transfer efficiency (E%)

In contrast, the above mentioned contamination correction is not required if fluorescence lifetime imaging microscopy is used to investigate protein-protein interactions by FRET [6, 8, 34]. Please see section below.

7.5 FLIM Microscopy

The fluorescence lifetime refers to the average time the molecule stays in its excited state before emitting a photon, which is an intrinsic property of a fluorophore. Fluorescence lifetime is sensitive to the local environment including pH, refractive index, temperature and insensitive to change in concentration and laser excitation intensity. There are different ways for measuring the fluorescence lifetime of a fluorophore: frequency domain (FD) and time domain (TD). Both methods can be combined with different optical imaging techniques and with different electronic signal recording principles. The techniques differ in time resolution, photon efficiency, [35–37], time resolution, capability to resolve multi-exponential decay profiles, the intensity range they can be applied to, acquisition speed, optical sectioning capabilities, and compatibility with different microscopy techniques. Please see [38, 39] for an overview.

Fluorescence lifetime measurements are performed both in single photon excitation and multiphoton excitation. Multiphoton microscopy provides several advantages in autofluorescence imaging than a single photon confocal microscopy, such as no photobleaching outside the focal plane [40, 41], less photodamage [42], and larger penetration depth for live cell and tissue imaging. Step-by-step instructions how to collect TD and FD FLIM images, advantages and disadvantages of TD and FD FLIM imaging are given in the literature [6–8, 39, 43–48]. Here we briefly describe some of the lifetime techniques used for FRET.

7.5.1 Frequency-Domain FLIM

Both TD and FD FLIM techniques are suitable for measuring the lifetime in wide-field or scanning mode, but the acquisition and analysis of the images differ. FD FLIM, the fluorophores are excited with periodically intensity-modulated light to determine the modulation in the emission signal [6–8, 48]. The fluorescence emits at the same modulation frequency but phase shifted due to the delay caused by the lifetime of the fluorophore relative to the excitation time. This delay is measured as a phase shift (ϕ_ω), where ω is the modulation frequency in radian/sec. The lifetime of the fluorophore also causes demodulation to the modulated excitation by a factor m_ω, as a function of the decay time and light modulation frequency. Suppose the modulation of the excitation is given by b/a, where 'a' is the average intensity and 'b' is the peak height of the incident light and the modulation of the emission is also defined as B/A. The modulation of the emission relative to the excitation is measured as, m = (B/A)/(b/a). The phase angle (ϕ_ω) and the modulation can be employed to calculate the lifetime using (7.3) and (7.4).

$$\tan \phi = \omega \tau_\phi \quad \tau_\phi = \omega^{-1} \tan \phi \tag{7.3}$$

$$m = \frac{1}{\sqrt{1 + \omega^2 \tau_m^2}} \quad \tau_m = \frac{1}{\omega} \left[\frac{1}{m^2} - 1 \right]^{\frac{1}{2}} \tag{7.4}$$

Note that, if the decay is a single exponential, both (7.3) and (7.4) yield the same and correct lifetime. If the decay is multi-exponential, then the apparent lifetimes in (7.3) and (7.4) represent a complex weighted average of the decay components [7]. Time-domain and frequency-domain are related to each other through Fourier transform. Time-domain data have intensity values in subsequent time channels, whereas, frequency-domain translate into amplitude and phase values at multiples of the signal repetition rate [6–8, 48].

7.5.2 Time-Domain FLIM

The time domain category includes FLIM based on gated cameras [8, 43, 46, 47], streak cameras [44, 45, 49], and FLIM by multi-dimensional time correlated single photon counting (TCSPC) [50–55], see Chap. 1 of this book.

7.5.2.1 Gated Camera

Ultrafast-gating image intensifiers coupled to a CCD camera is used to acquire time-resolved images of proteins in living cells. This camera allows operating the gate width from 300 ps to 1 ms and a repetition rate from single shots to 110 MHz (LaVision, Germany; Princeton instruments, USA). This gated image intensifier camera is synchronized with high-speed excitation laser pulses to trigger the camera gating pulse via a time-delay unit and synchronizing electronics. This gated camera can be coupled to any epi-fluorescence microscopy to implement FLIM-FRET imaging [43, 46]. A rapid lifetime determination (RLD) method is used for the gating camera-based FLIM imaging. RLD is a family of data analysis techniques for fitting experimental data that conform to single and double exponential decays with or without baseline contribution [47]. This allows us to calculate the decay parameters using the areas under different regions of the decay rather than recording a complete multipoint curve and analyzing the decay by the traditional least square methods. For quantitative analysis, it is important to estimate measurement precision in the presence of noise. The performance is evaluated over a wide range of experimental conditions in order to assess the optimum conditions and the theoretical limitations for contiguous and overlapped gating procedures for single- and double-exponential decay using Monte Carlo simulations [47].

7.5.2.2 Streak Camera

The principle the streak camera, its operation, and its application to FLIM-FRET measurements are explained in the literature [44, 45, 49, 56]. Briefly, the streak-scope consists of a photocathode surface, a pair of sweep electrodes, a micro-channel plate (MCP) to amplify photoelectrons coming off the photocathode and a phosphor screen to detect this amplified output of MCP. The streak camera operates by transforming the temporal profile of a light pulse into a spatial profile on a detector, by causing a time-varying deflection of the light across the width of the detector. The resulting image forms a "streak" of light, from which the duration of the light pulse can be inferred. An optical 2D image with spatial axes (x, y) is converted into a streak image with temporal information and with the axes (x, t). When a synchronous y-scanning is carried out on the region of interest, the above streak imaging process gives a complete stack of (x, y, t) streak images. This stack contains the complete information of optical intensity as well as the spatial and temporal information from the optical image. Numerical processing of all these streak images gives the final FLIM image. Every pixel in the FLIM image now contains the lifetime information.

7.5.2.3 TCSPC

TCSPC FLIM modules are widely used to acquire FLIM images for various biological applications [6–8, 51, 55, 57–62], please see [53] for more references. Technical details are described in Chaps. 1 and 2 of this book. The TCSPC FLIM board can be installed in a PC, or in an extension box of a laptop computer, as suggested by the manufacturer. The board receives the single-photon pulses from the detector, timing reference pulses from the laser, and scan synchronisation pulses from the scanner.

For every photon detected, the TCSPC module determines the time in the laser pulse period, t, and the location of the laser beam, x, y, in the scan area. The TCSPC module builds up a photon density histogram over these parameters, which represents an array of pixels, each containing fluorescence decay data in the form of photon numbers in consecutive time channels, see Chap. 1, Sect. 1.4.5. The procedure does not require that the scanner stays in one pixel until enough photons are recorded. The data can be acquired at any scan speed, by just running the acquisition over as many frames as needed to obtain the desired number of photons in the pixels.

The time resolution (the width of the instrument-response function, IRF) of the technique is given by the width of the laser pulse (which is negligible in a multi-photon microscope) and the transit time spread in the detector. The transit time spread is much smaller than the width of the single-photon response of the detector, see Chap. 1, Fig. 1.4. Therefore TCSPC FLIM reaches an excellent time resolution: Typical IRF widths are 25–30 ps for MCP PMTs [63], 120 ps for hybrid detectors [26], and 300 ps for conventional high-speed PMTs. Thus, the IRF width can be made much smaller than the decay times of the fluorophores, resulting in high lifetime accuracy and reproducibility.

TCSPC FLIM can be combined with multi-spectral detection [52]. The principle is described in Chap. 1, Sect. 1.4.5.2. Multi-spectral FLIM has been used to improve the reliability of FLIM-FRET measurements [64], to track the photoconversion of photosensitisers for PDT [65], and to obtain metabolic information from NADH/FAD fluorescence [66]. Please see also Chap. 13 of this book.

Data analysis software allows multi-exponential curve fitting of the acquired data on a pixel-by-pixel basis using a weighted least-squares numerical approach [53]. The sum of all time bins is equivalent to the intensity image and this is displayed to an image, pseudo-coloured according to the curve fit results. Therefore, each image can be easily displayed in a meaningful way to compare lifetimes within one or between different images.

7.5.3 Advantages of FLIM-FRET Imaging

As previously described, intensity based FRET methods are used for investigating the protein-protein interaction in biological systems. These techniques inherently have several disadvantages because of local probe concentration dependency and requirement of spectral bleedthrough corrections. Therefore, an alternative pixel by pixel analysis is FLIM which overcomes the problems of intensity-based FRET methods and make FLIM FRET an accurate method for FRET measurements. In FLIM-FRET measurement, only the fluorescence lifetime of donor molecule is determined. In case of FRET, the energy is transferred from donor to acceptor, and the donor fluorescence is quenched due to the interaction with the acceptor. Because the donor is losing its energy more rapidly its fluorescence lifetime decreases.

The energy transfer efficiency $(E = 1 - (\tau DA/\tau D))$ can be calculated by comparing the fluorescence lifetime of the donor in presence (τ_{DA}-the quenched lifetime) and in absence of acceptor (τ_D-unquenched lifetime) [6]. Note that τ_{DA} must be the amplitude-weighted lifetime [53] (7.7) to obtain accurate FRET efficiencies, see Sect. 7.6.2.

7.5.4 FLIM-FRET Standards

Appropriate selection of optical configuration (lens, excitation light, and detector) and the excitation or emission filter, based on FRET pair are essential to be successful in FRET imaging. It is difficult to calibrate or verify the optical configuration or filters in any FRET imaging system using a biological specimen. A corresponding FRET- standard pair can be created to calibrate both intensity and FLIM based FRET imaging system before using the same FRET pair for an experiment. For example, measurement of energy transfer efficiency for Cerulean linked by 5 amino acids with Venus (C5 V) should provide about 43 % energy

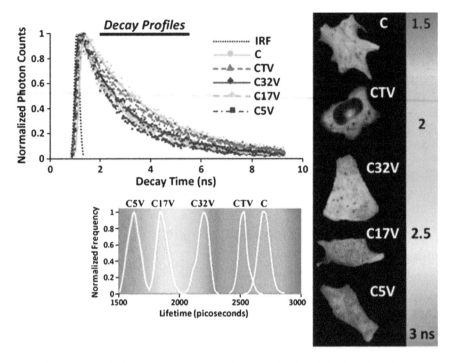

Fig. 7.3 FRET Standards. The data was collected using Becker &Hickl SPC 150 and HPM 100-40. FRET standards are available from www.addgene.org. E% for the following FRET Standards are C5 V-43 %; C17 V-38 %; C32 V-31 %; CTV-7 % [57, 61]

transfer efficiency. It is possible that the measurement will be within 3 % error. In Fig. 7.3 we have shown the FLIM-FRET measurements of various FRET standards prepared by the Vogel laboratory [33] see also www.addgene.org.

7.6 Instrumentation for FLIM-FRET Image Acquisition and Processing

7.6.1 FLIM-FRET Imaging System, Acquisition and Analysis

The imaging system is described in the literature [32, 48, 59]. Briefly, lifetime measurements in this study were made using a Zeiss(observer) epi-fluorescence microscope equipped with a 63x NA1.4 oil IR objective lens. This microscope was coupled to a Zeiss780 confocal/multi-photon system and a ChameleonVision II auto tuneable, (680–1080 nm) mode-locked ultrafast (80 MHz) pulsed (150 femtosecond) laser (Coherent, Inc.) with dispersion compensation. The Zeiss 780 system is controlled using the Zen software, and was configured to use the multi-photon laser to scan specimens. Emitted photons were collected using a bandpass emission filter by a fast hybrid PMT with a response time (IRF width) of approximately 120 picoseconds [26] (HPM-100-40, Becker &Hickl GmbH, Berlin, Germany). The usage of the photon-counting module board (SPC-150, Becker & Hickl GmbH) with a minimum temporal resolution of <4 ps rms records FLIM data with typically 256 × 256 or 512 × 512 pixels and 256 or 1024 time channels in each pixel [53]. Please see Chap. 1 of this book.

7.6.2 TCSPC FLIM Data Analysis

Lifetime calculation from the multi-exponential decay was implemented by fitting the experimental data to a mathematical convolution function of a decay model and the instrument response function (IRF) [53]. For example, the measured composite decays of two species can be modelled by convolution of an IRF (I_{instr}), with a double-exponential model function, defined in (7.5), with offset correction for the ambient light and/or dark noise (I_0), as shown by $I_c(t)$ in (7.6)

$$F(t) = a_1 e^{-t/\tau_1} + a_2 e^{-t/\tau_2},\tag{7.5}$$

$$I_c(t) = \int_{-\infty}^{\infty} I_{instru}(t)\{I_o + F(t)\}dt\tag{7.6}$$

where $a_1 e^{-t/\tau_1}$ and $a_2 e^{-t/\tau_2}$ denote the contributed fluorescence decays from short and long lifetime components, respectively; τ_1 and τ_2 represent their corresponding lifetime constants; and a_1 and a_2 refer to the corresponding initial amplitudes at zero time. The instrument response function, I_{instr}, can be estimated from the data itself [53] or measured experimentally depending upon the system (see Sect. 6.1.1). The average lifetime is calculated as an amplitude-weighted average of the two lifetime components:

$$\tau_m \equiv \frac{a_1\tau_1 + a_2\tau_2}{a_1 + a_2}.\tag{7.7}$$

The model parameters (i.e. a_i and τ_i) are typically derived by iteratively fitting the measured data $I_a(t)$ to $I_c(t)$, given by (7.6), while minimizing the goodness-of-fit function defined in (7.8), using the Levenberg-Marquardt algorithm,

$$\chi_R^2 = [\sum_{k=0}^{n}[I_a(t) - I_c(t)]^2/I_a(t)]/(n - p), \tag{7.8}$$

where n denotes the number of data (time) points, and p represents the number of model parameters. A good fit is characterized by an χ^2 close to 1 and residuals showing no noticeable systematic variations. Both lifetimes (τ_1 and τ_2) and amplitudes (a_1 and a_2—population sizes of molecules with the different decay rate) were obtained from fitting optimization software.

It is sometimes objected that the amplitude weighted lifetime, τ_m, is the 'wrong' lifetime to represent the 'apparent' lifetime of a multi-exponential decay, and the intensity-weighted lifetime,

$$\tau_i \equiv \frac{a_1\tau_1^2 + a_2\tau_2^2}{a_1\tau_1 + a_2\tau_2} \tag{7.9}$$

should be used. This is only correct if the closest approximation to the lifetime of a single-exponential fit is wanted. For obtaining FRET efficiencies from donor decays a lifetime definition has to be used that is proportional to the net quantum efficiency, or the net intensity of the donor. This is clearly (7.7), not (7.9). Interestingly, the fact that the amplitude-weighted lifetime has to be used for FRET calculation is rarely mentioned in the literature, although it has been correctly stated in [7].

The fact that (7.9) does not deliver the correct FRET intensity has consequences. Donor decay functions in presence of FRET are always multi-exponential [53]: Not all donor molecules have the correct orientation to interact with an acceptor, the acceptor labelling may be incomplete, or, in protein-interaction experiments, not all proteins may interact. If the FRET intensity under these circumstances is derived from a lifetime calculated by (7.9) or from an 'apparent' lifetime delivered by a FLIM technique unable to detect the double-exponential decay the FRET efficiency is determined too low. This may be one of the reasons of the discrepancies in the FRET efficiencies determined by different techniques and instruments.

7.6.3 Measuring the Instrument Response Function (IRF)

For time-domain FLIM measurements, fluorescence lifetimes are frequently comparable to both the excitation pulse width and the instrument response function (IRF) of a FLIM system. Therefore, to obtain fluorescence decays free from the instrumental distortions that result from the finite rise time, the width (changes due to optics) and the decay of the excitation pulse, and the detector and timing apparatus [32, 48, 51, 58], a deconvolution technique must be applied to extract the undistorted fluorescence decays that are convolved with the IRF. The IRF can either be estimated from fluorescence decay data or experimentally measured. The IRF should be representative of the experimental conditions, and thus is ideally

Table 7.2 Comparison of lifetime values for the measured and estimated IRF for different objective lens and the dispersion adjustment in the 2-photon excitation laser line

Objective	Process	With dispersion		Without dispersion	
20X	Estimated IRF	2.31 ± 0.07	$\chi^2 = 1.16$	2.35 ± 0.06	$\chi^2 = 1.15$
	Measured IRF	2.48 ± 0.10	$\chi^2 = 1.03$	2.49 ± 0.08	$\chi^2 = 1.01$
63X	Estimated IRF	2.38 ± 0.11	$\chi^2 = 1.19$	2.36 ± 0.12	$\chi^2 = 1.17$
	Measured IRF	2.47 ± 0.05	$\chi^2 = 1.01$	2.49 ± 0.05	$\chi^2 = 1.03$

The dispersion adjustment helps to reduce the pulse width expansion due to the microscope optics

measured under the same conditions used for the biological experiments. Conventionally, scattering nondairy coffee creamer can be used to record the IRF of a FLIM system for visible light excitation [46].

However, this does not work for the laser scanning microscope: Scattered excitation light is blocked by filters, and, in the case of a multiphoton microscope, the detector may not be sensitive to the laser wavelength.

In this study, we measured the IRF of the FLIM system through collecting the second-harmonic generation (SHG) signals emitted from urea crystals [53, 58]. SHG is an ultrafast nonlinear process that delivers a signal at one half of the excitation wavelength. To verify the utility of the measured IRFs, we used Coumarin6 dye dissolved in ethanol as a standard. This sample yields a single exponential decay and its fluorescence lifetime with measured IRF at room temperature is about 2.47 ns and a mean χ^2 of 1.01. In comparison, we also used the same routine to fit the data with the estimated IRF produced by the SPCImage software and obtained an average lifetime of about 2.38 ns and a mean χ^2 of 1.19. Although the excitation and emission wavelengths of the coumarin 6 sample were little different than those in the IRF measurements, it still produced better fitting results. Therefore, we used the measured results for all of our data processing. As shown in Table 7.2 the IRF was measured with 63x NA1.4 lens with dispersion adjustment. The IR beam pulse width expands about 10–20 % when it passes through microscope optics which influences the IRF. As shown in Table 7.2 we did not notice any difference in IRF with dispersion and without dispersion adjustment in our system.

7.7 Applications

The number of publications on FRET-FLIM has been increasing in the past few years [2, 6–8, 67]. The interaction and dynamics between SNAP 25 (donor) and rabphilin (acceptor) proteins were investigated during exocytosis in intact neuroendocrine cells [68, 69]. Protein interaction in sodium channels has been described in [70], interaction of SNARE proteins in [71].

There is a number of FRET applications which are directly related to clinical research [72]. The use of FRET to investigate the mechanism of infection of HeLa cells with enterovirus 71 was demonstrated by Ghukasyan et al. [73]. Bacskai et al. described the implementation of FLIM techniques for a spatially resolved FRET pair with senile plaques obtained from transgenic mouse brains of Alzheimer's disease [57]. Mechanisms of Huntington disease (a progressive neuro-degenerative disorder) were investigated in [74] and [75]. Please see [53] for more references.

In cancer research, gene expression profiling has provided information to determine the specific target in cancer therapies. Advanced in vivo two-photon excited fluorescence imaging with high spatial resolution and optical contrast investigates protein-protein interaction using FLIM-FRET technique [76]. Intravital FLIM was adopted for GFP and mRFP1 interactions by FRET in cancer cell in an animal model. FLIM-FRET analysis is well established not only for animal cells but also for plant cells and whole plants. In vivo protein interactions in plant cells were investigated by Boutant et al. [77] Christoph et al. showed that protoplast is expressed with mTurquoise-Venus FLIM/FRET pair and fitted with a bi-exponential model [54]. The fluorescence lifetime of donor (mTurquoise) is decreased from 3.6 to 2.6 ns after FRET events. In another study, Bucherl et al. reported the formation of BRI1 and SERK3 in the plasma membrane of epidermal cells tagged with GFP-mCherryas a FLIM/FRET pair [78]. It is even possible, and in some cases beneficial, to utilize non-fluorescent acceptors to measure FRET–FLIM [79–81]. The use of 'dark' acceptors in FLIM measurements allows a wider detection range for donor emission.

7.7.1 FRET with Dark Acceptor

For FLIM-FRET, we measure only the change in lifetime for the donor molecule. Selected FRET pairs are listed in Table 7.1. Recently, a non-fluorescent chromophore(non-radiative YFP variants) called resonance energy-accepting chromoprotein (REACh) has been used as an acceptor with matching donors, such as eGFP [79–81]. This non-radiative (quantum yield 0.04) REACh is a strong absorber and overcomes the problem of acceptor back-bleedthrough emission into the donor channel. On the other hand the absence of fluorescence from REACh means that the spectral window occupied by the acceptor is available for the detection of another fluorophore. As shown in Fig. 7.4 the sensitized emission is clearly shown in the acceptor channel in the case of Venus compared tosREACh. This indicates that there is no photon emission from sREACh. More importantly, REACh is a good acceptor for the FLIM-FRET microscopy since only the lifetime of the donor molecule is monitored and provides better signal-to-noise ratio of FLIM-FRET images.

Here we describe the usage of REACh as an acceptor with mTFP (Teal) as a donor. Teal is linked with 5 amino acid to sREACh (super-REACh). This FRET pair is expressed in GHFT1 live cells for FLIM-FRET imaging. Cells expressing Teal (mTFP)alone were used to determine the unquenched donor lifetime, while

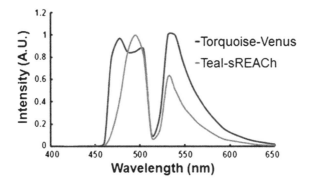

Fig. 7.4 The spectra of Torquoise-5aa-Venus and Teal (mTFP)-5aa-sREACh. The sREACh spectrum did not show any sensitized emission in the acceptor channel. The intensity peak in the acceptor channel is due to the donor spectral bleedthrough and no acceptor (sREACh) emission. Zeiss 780 NLO spectral imaging system; Ex 840 nm; Coherent Chameleon Vision II

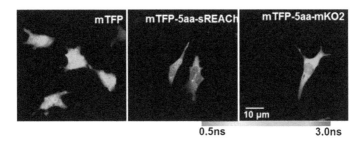

Fig. 7.5 Comparison of E% of sREACh(55 %) (the non-fluorescent acceptor) versus the fluorescent acceptor mKO2 (39 %). Donor for both acceptors is Teal (mTFP). As shown in Table 7.3, the E% value for sREACh is higher than the mKO2 acceptor. This may be due to the strong absorption property of sREACh

cells expressing the Teal-5aa-sReach construct were used to determine the quenched donor lifetime (Fig. 7.5).

The laser was tuned to 840 nm (the peak excitation of Teal) for multi-photon excitation and the emission signal from Teal was collected using a 475–503 nm emission filter. The laser power at the specimen plane was measured using a power meter (SSIM-VIS-IR, Coherent, Inc.) and was within the range of 0.7–1.5 mW. The data was typically acquired over 10–30 s, resulting in the accumulation of enough photon counts on the PMT for the analysis by either single or double exponential fitting. The lifetime results were then analyzed using the SPCImage software (Version 2.9.2.2989, Becker &Hickl GmbH), which allows multi-exponential curve fitting on a pixel-by-pixel basis using a weighted least-square numerical approach [53]. Please see Table 7.3.

Table 7.3 Comparison of E% for the selected FRET pair for measured and estimated IRF

	Donor mTFP	Donor-acceptor	
		mTFP-sREACh	mTFP-mKO2
Fluorescence lifetime (ns) (Measured IRF)	2.61 ± 0.03	1.15 ± 0.02	1.60 ± 0.03
E%		55 ± 3	39 ± 2
Fluorescence lifetime (ns) (Estimated IRF)	2.56 ± 0.08	1.29 ± 0.04	1.75 ± 0.09
E%		49 ± 2	33 ± 3

7.7.2 Live Versus Fixed Cells for FLIM-FRET Imaging

Fluorescence lifetime is sensitive to the probe environment and so it is advisable to use live cells for FLIM imaging. In the case of RNA-protein interaction, one has to use antibody-based probes to label RNA [82] with the requirement to fix cells. To explore the usage of fixed cell in FLIM-FRET imaging we used the FRET standards, cerulean-5aa-Venus (C5V) and compared the fitting analyses of the decay data sets obtained from the C5A (Cerulean-5aa-Amber) and C5V constructs expressed in live cells versus fixed (4 % para-formaldehyde) cells. We tested cells at different time duration post-fixing. The lifetime changes for fixed cells after 2–3 weeks post-fixation (Fig. 7.6).

Results showed that fixed cells kept in the refrigerator for at least a week show no change in lifetime. The cell line we used in our test is GHFT1 pituitary cells and we do not have further data on other cell lines, but believe that this may hold true for other cell lines, too. It is advisable for the user to test their own cell line and experiment protocol including the fixative.

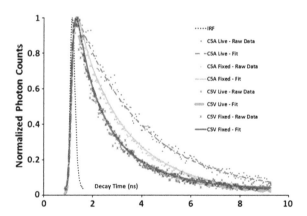

Fig. 7.6 Cerulean alone C5A construct (donor) had a shorter lifetime in the fixed (1.73 ns) cells than in the live (2.55 ns) cells; however, Cerulean lifetime in the double labelled (C-5aa-V) fixed cell was 1.2 ns compared to the live cells (1.33 ns). The difference in E% between the fixed (30.42 %) versus live (47.72 %) cells were about 17 % (Fixed cells were used after 2 weeks)

7.7.3 Homo-Dimerization of the CCAAT/Enhancer Binding Protein Alpha (C/EBPα) Transcription Factor

The biological model used here as an example is the basic region-leucine zipper (bZip) domain of the CCAAT/enhancer binding protein alpha (C/EBPα) transcription factor. The bZip family proteins form obligate dimers through their leucine-zipper domains, which positions the basic region residues for binding to specific DNA elements. Immuno cyto chemical staining of differentiated mouse adipocyte cells showed that endogenous C/EBPα was preferentially bound to satellite DNA-repeat sequences located in regions of centromeric heterochromatin [83–86]. When the C/EBPα bZip domain is expressed as a fusion to a fluorescent protein in cells of mouse origin, such as the pituitary GHFT1 cells used here (see Fig. 7.7), it is localized to the well-defined regions of centromeric heterochromatin in the cell nucleus.

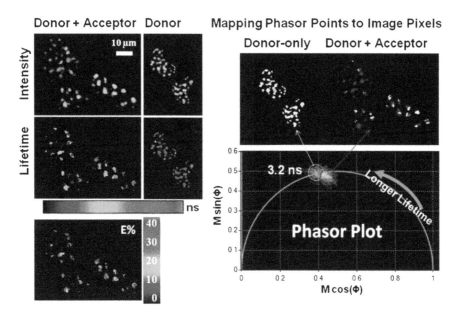

Fig. 7.7 Homo-dimerization of CCAAT/enhancer-binding protein alpha (C/EBPα) in live mouse pituitary cell nucleus using TCSPC FLIM-FRET microscopy. Cerulean-C/EBPα (Donor) and Venus-C/EBPα (Acceptor) were co-expressed in GHFT1 cell nucleus. In FLIM-FRET, the quenched (τ_{DA}) and unquenched (τ_D) donor lifetimes were measured from the doubly-expressing and the donor-alone singly-expressing cells, respectively, and the E% was determined by [1−(τ_{DA}/ τ_D)] Each image's pixel was plotted on the phasor plot [56, 87], demonstrating that the donor lifetimes decrease from the donor-alone to the double-label. The distribution of the donor alone is centred on the universal semicircle of the phasor plot, indicating that it has a single lifetime component. However, the phasor distribution of the double-label is located inside the universal semicircle of the phasor plot, indicating that it has more than one lifetime component

FRET microscopy has been demonstrated to be a perfect tool for detecting the homo-dimerization of C/EBPα in living cells [88]. A FRET system for this biological model can be built by fusing the C/EBPα bZip domain to two fluorescent proteins of a good FRET pair separately, e.g. Cerulean (FRET donor) and Venus (FRET acceptor) as used here. In FLIM, the donor fluorescence lifetimes are measured from cells co-expressing the donor (Cerulean-bZip) and the acceptor (Venus-bZip) as well as cells that only express the donor (Cerulean-bZip). Here, the Cerulean lifetimes in live GHFT1 cells expressing only Cerulean-bZip or both Cerulean-bZip and Venus-bZip (transfected by FuGENE 6) are measured by the TPE TCSPC FLIM approach, to demonstrate the quenching of Cerulean in doubly-expressed cells due to FRET between Cerulean-bZip and Venus-bZip (Fig. 7.7).

7.7.4 Endogenous Fluorophores for FRET (NADH-Tryptophan Interaction)

The most important autofluorescent constituents of cells and tissues are amino acids, the essential building blocks of proteins and enzymes, such as NADH and FAD, which regulate cell metabolism. The mitochondria are responsible for the energy supply of cells. Moreover, they are the main electron donor and acceptor in the biochemical process of oxidative phosphorylation, which is the main metabolic pathway used by biological systems to produce energy. This important feature enables the use of NADH fluorescence not only to investigate cellular morphology, but also to provide functional information about cellular metabolism, which can in turn be related to pathologies, e.g. cancerous or pre-cancerous conditions. It has been reported in the literature by biochemical methods that the tryptophan degradation was linked to breast cancer. A FRET assay using the autofluorescent properties of NADH and tryptophan can track the degradation of tryptophan as an alternative to biochemical methods.

The average prevalence of tryptophan (TRP) in 2000 proteins is about 1.4 %. Therefore, imaging tryptophan fluorescence in vivo can provide a new way to visualize cellular and subcellular protein contents at the molecular level. The distinction in lifetime between a free and protein bound NADH forms the basis of almost all fluorescence lifetime imaging techniques aimed at NADH. To avoid oxidation stress on cells caused by high-energy (365 nm) one-photon lifetime imaging, a three-photon excitation (740 nm) technique with a femtosecond Ti-sapphire laser has been used to image the interaction between TRP and bound NADH. The use of much milder irradiation minimizes the modification of cellular redox states and cell morphology. It also allows non-destructive fluorophore detection with high spatial and temporal resolution. To discriminate between free and protein-bound NAD(P)H, bi-exponential fluorescence lifetime imaging maps of cellular metabolism are typically generated (Fig. 7.8). However, many unresolved questions remain to be elucidated to fully understand this important metabolic pathway and to translate this knowledge into designing novel therapeutic strategies for diseases.

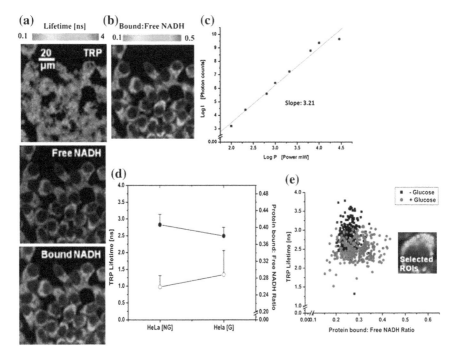

Fig. 7.8 Three-photon (3P) FLIM-FRET imaging of TRP-NADH interactions: **a** The representative lifetime images of TRP, free and protein bound NADH. **b** The corresponding protein bound: free NADH ratio image. **c** Effect of the average laser power [Log P] dependence on the emission intensity [Log I] of Tryptophan. **d** Averages and standard deviations of the TRP fluorescence lifetimes (*filled circles*) and the protein bound: free NADH ratios (*open circles*) as a function of perturbation with 5 mM glucose in live HeLa cells. **e** Scatter plot of 300 selected ROI's from both the absence [NG] and the presence [G] of glucose, clearly shows the decreased TRP lifetime with the increased protein bound: free NADH ratio in the presence of glucose in HeLa cells. Adapted from [89]

7.8 Summary

FRET microscopy techniques have been used for more than 30 years to provide spatial and temporal information of the protein molecules. FLIM is a unique technique with huge potential and application in biomedical field. Although, FLIM-FRET technique reveals the protein-protein interaction far beyond the limitation of optical resolution, imaging each protein molecule is still a challenge. On the other hand, the development of new fluorescent probes, advanced imaging techniques, detectors, optics and software, advanced FRET microscopy techniques to investigate protein-protein interactions in living specimens. For example, with high sensitivity detectors at $\sim 40\%$ quantum efficiency, lifetime time-lapse imaging can be implemented with few seconds of data acquisition time compared to a few minutes, see Chap. 2, Sect. 2.2.4. The acquisition time will also vary depending on cellular labelling. The phasor plot, a new algorithm for the FLIM data analysis, helps to detect different fluorophores at each pixel in the image frame. Application of the phasor plot approach helps to analyze the lifetime decay in great detail even if the number of fluorophores exceeds two, where the photon count level is not a serious issue. With the new invention of techniques and algorithms, FLIM can be a potential tool in clinical diagnosis, such as wound healing and cancer detections. Multimodality of FLIM-FRET with other imaging techniques, for example total internal reflection fluorescence microscopy (TIRF) and scanning near-field optical microscopy (SNOM) will provide new insights into protein-protein interactions at the cell surface. Additionally, the combination of high resolution microscopy systems such as stimulated emission depletion microscopy (STED), stochastic optical reconstruction microscopy (STORM), and photo-activated localization microscopy (PALM) will be added tools to investigate single molecular interactions.

Acknowledgments We acknowledge the funding from National Institutes of Health (HL101871 & OD016446) and the University of Virginia. We would like to thank Mr. Horst Wallrabe for his suggestions.

References

1. R.M. Clegg, The history of FRET: from conception through the labors of birth. in *Reviews in Fluorescence*, ed by C.D. Geddes, J.R. Lakowicz (Springer, New York, 2006), pp. 1–45
2. Y. Sun, H. Wallrabe, S.-A. Seo, A. Periasamy, FRET microscopy in 2010: the legacy of Theodor Förster on the 100th anniversary of his birth. Chem. Phys. Chem. **12**, 462–474 (2011)
3. Th. Förster, Energiewanderung und Fluoreszenz. Naturwissenschaften **6**, 166–175 (1946)
4. Th. Förster, Energy migration and fluorescence. Translated by Klaus Suhling. J. Biomed. Opt. **17**, 011002-1–011002-10
5. S.S. Vogel, C.T. Thaler, P.S. Blank, S.V. Koushik, Time-resolved fluorescence anisotropy. in *FLIM Microscopy in Biology and Medicine*, ed by A. Periasamy, R.M. Clegg (CRC Press (Taylor & Francis Group), New York, 2010), pp. 245–288
6. T.W.J. Gadella, *FRET and FLIM Techniques* (Elsevier, New York, 2009)
7. J.R. Lakowicz, *Principles of Fluorescence Spectroscopy*, 3rd edn. (Springer, Berlin, 2006)
8. A. Periasamy, R.M. Clegg, *FLIM Microscopy in Biology and Medicine* (CRC Press (Taylor & Francis Group), New York, 2010)
9. A. Periasamy, R.N. Day, *Molecular Imaging: FRET Microscopy and Spectroscopy* (Oxford University Press, New York, 2005)
10. Y. Sun, C. Rombola, V. Jyothikumar, A. Periasamy, Förster resonance energy transfer microscopy and spectroscopy to localize protein-protein interactions in live cells. Cytometry A **83A**(9), 780–793 (2013)
11. Y. Sun, H. Wallrabe, C. Booker, R.N. Day, A. Periasamy, Three-color spectral FRET microscopy localizes three interacting proteins in living cells. Biophys. J. **99**, 1274–1283 (2010)
12. H. Wallrabe, Y. Cai, Y. Sun, A. Periasamy, R. Luzes, X. Fang, H.-M. Kan, L.C. Cameron, D. A. Schafer, G.S. Bloom, IQGAP1 interactome analysis by In Vitro reconstitution and live cell 3-color FRET microscopy. Cytoskeleton **70**, 819–836 (2013)
13. S.S. Vogel, C. Thaler, S.V. Koushik, Fanciful FRET. Sci STKE **18**(331) (2006)

14. R.N. Day, M.W. Davidson, The fluorescent protein palette: tools for cellular imaging. ChemSoc Rev. **38**, 2887–2921 (2009)
15. R.H. Newman, M.D. Fosbrink, J. Zhang, Genetically encodable fluorescent biosensors for tracking signaling dynamics in living cells. Chem. Rev. **111**, 3614–3666 (2011)
16. G.H. Patterson, J. Lippincott-Schwartz, A photoactivatable GFP for selective photolabeling of proteins and cells. Science **297**, 1873–1877 (2002)
17. J. Sambrook, *Molecular Cloning: A Laboratory Manual* (Cold Spring Harbor Laboratory Press, New York, 1989)
18. J. Zhang, R.E. Campbell, A.Y. Ting, R.Y. Tsien, Creating new fluorescent probes for cell biology. Nat. Rev. Mol. Cell Biol. **3**, 906–918 (2002)
19. H. Wallrabe, M. Elangovan, A. Burchard, A. Periasamy, M. Barroso, Confocal FRET microscopy to measure clustering of ligand-receptor complexes in endocytic membranes. Biophys. J. **85**, 559–571 (2003)
20. Y. Chen, J.D. Mills, A. Periasamy, Protein localization in living cells and tissues using FRET and FLIM. Differentiation **71**, 528–541 (2003)
21. R.N. Day, C. Booker, A. Periasamy, The Characetrization of an improved donor fluorescent protein for Förster resonance energy transfer microscopy. J. Biomed. Opt. **13**, 031203 (2008)
22. U. Resch-Genger, M. Grabolle, S. Cavaliere-Jaricot, R. Nitschke, T. Nann, Quantum dots versus organic dyes as fluorescent labels. Nat. Methods **5**, 763–775 (2008)
23. Z. Xu, K. Nagashima, D. Sun, T. Rush, A. Northrup, J.N. Andersen, I. Kariv, E.V. Bobkova, Development of high-throughput TR-FRET and AlphaScreen assays for identification of potent inhibitors of PDK1. J. Biomol. Screen. **14**, 1257–1262 (2009)
24. K. Lundin, K. Blomberg, T. Nordstrom, C. indqvist, Development of a time-resolved fluorescence resonance energy transfer assay (cell TR-FRET) for protein detection on intact cells. Anal. Biochem. **299**, 92–97 (2001)
25. M. Göppert-Mayer, Über Elementarakte mit zwei Quantensprüngen. Ann. Phys. **9**, 273–294 (1931)
26. W. Becker, B. Su, K. Weisshart, O. Holub, FLIM and FCS detection in laser-scanning microscopes: increased efficiency by GaAsP hybrid detectors. Micr. Res. Tech. **74**, 804–811 (2011)
27. Y. Chen, M. Elangovan, A. Periasamy, FRET data analysis: the algorithm. in *Molecular Imaging: FRET Microscopy and Spectroscopy*, ed by A. Periasamy, R.N. Day (Elsevier, New York, 2005), pp. 126–145
28. Y. Sun, A. Periasamy, Additional correction for energy transfer efficiency calculation in filter-based Förster resonance energy transfer microscopy for more accurate result. J. Biomed. Opt. **15**(2), 020513 (2010)
29. Y. Chen, J.P. Mauldin, R.N. Day, A. Periasamy, Characterization of spectral FRET imaging microscopy for monitoring nuclear protein interactions. J. Microsc. **228**, 139–152 (2007)
30. G.W. Gordon, G. Berry, X.H. Liang, B. Levine, B. Herman, Quantitative fluorescence resonance energy transfer measurements using fluorescence microscopy. Biophys. J. **74**, 2702–2713 (1998)
31. A. Hoppe, K. Christensen, J.A. Swanson, Fluorescence resonance energy transfer-based stoichiometry in living cells. Biophys. J. **83**(6), 3652–3664 (2002)
32. Y. Sun, C.F. Booker, S. Kumari, R.N. Day, M. Davdison, A. Periasamy, Characterization of an orange acceptor fluorescent protein for sensitized spectral FRET microscopy using a white light laser. J. Biomed. Opt. **14**(5), 054009 (2009)
33. S.V. Koushik, H. Chen, C. Thaler, H.L. Puhl, S.S. Vogel, Cerulean, venus, and VenusY67C FRET reference standards. Biophys. J. **91**, L99–L101 (2006)
34. H. Wallrabe, A. Periasamy, Imaging protein molecules using FRET and FLIM microscopy. Curr. Opin. Biotechnol. **16**, 19–27 (2005)
35. H.C. Gerritsen, M.A.H. Asselbergs, A.V. Agronskaia, W.G.J.H.M. van Sark, Fluorescence lifetime imaging in scanning microscopes: acquisition speed, photon economy and lifetime resolution, J. Microsc. **206**, 218–224 (2002)

36. M. Köllner, J. Wolfrum, How many photons are necessary for fluorescence-lifetime measurements? Phys. Chem. Lett. **200**, 199–204 (1992)

37. J.P. Philip, K. Carlsson, Theoretical investigation of the signal-to-noise ratio in fluorescence lifetime imaging. J. Opt. Soc. Am. **A20**, 368–379 (2003)

38. W. Becker, A. Bergmann, in *Lifetime-Resolved Imaging in Nonlinear Microscopy*, ed by B.R. Masters, P.T.C. So, Handbook of Biomedical Nonlinear Optical Microscopy (Oxford University Press, Oxford, 2008)

39. W. Becker, Fluorescence lifetime imaging—techniques and applications. J. Microsc. **247**(2), 119–136 (2012)

40. G.H. Patterson, D.W. Piston, Photobleaching in two-photon excitation microscopy. Biophys. J. **78**(4), 2159–2162 (2000)

41. L.M. Tiede, M.G. Nichols, Photobleaching of reduced nicotinamide adenine dinucleotide and the development of highly fluorescent lesions in rat basophilic leukemia cells during multiphoton microscopy. Photochem. Photobiol. **82**, 656–664 (2006)

42. K. König, T.W. Becker, P. Fischer, I. Riemann, K.J. Halbhuber, Pulse-length dependence of cellular response to intense near-infrared laser pulses in multiphoton microscopes, Opt. Lett. 15; **24**(2), 113–115 (1999)

43. M. Elangovan, R.N. Day, A. Periasamy, Nanosecond fluorescence resonance energy transfer-fluorescence lifetime imaging microscopy to localize the protein interactions in a single living cell. J Microscopy. **205**, 3–14 (2002)

44. R.V. Krishnan, A. Masuda, V.E. Centonze, B. Herman, Quantitative imaging of protein-protein interactions by multiphoton fluorescence lifetime imaging microscopy using a streak camera. J. Biomed. Opt. **8**, 362–367 (2003)

45. R.V. Krishnan, H. Saitoh, H. Terada, V.E. Centonze, B. Herman, Development of a Multiphoton Fluorescence Lifetime Imaging Microscopy (FLIM) system using a streak camera. Rev. Sci. Instrum. **74**, 2714–2721 (2003)

46. A. Periasamy, P. Wodnicki, X.F. Wang, S. Kwon, G.W. Gordon, B. Herman, Time resolved fluorescence lifetime imaging microscopy using picosecond pulsed tunable dye laser system. Rev. Sci. Instrum. **67**, 3722–3731 (1996)

47. K.K. Sharman, A. Periasamy, H. Asworth, J.N. Demas, Error analysis of the rapid lifetime determination method for double-exponential decays and new windowing schemes. Anal. Chem. **71**, 947–952 (1999)

48. Y. Sun, R.N. Day, A. Periasamy, Investigating protein-protein interactions in living cells using fluorescence lifetime imaging microscopy. Nat. Protoc. **6**, 1324–1340 (2011)

49. C. Biskup, T. Zimmer, K. Benndorf, FRET between cardiac Na^+ channel subunits measured with a confocal microscope and a streak camera. Nat. Biotechnol. **22**(2), 220–224 (2004)

50. W. Becker, K. Benndorf, A. Bergmann, C. Biskup, K. König, U. Tirlapur, T. Zimmer, FRET measurements by TCSPC laser scanning microscopy. Proc. SPIE **4431**, 94–98 (2001)

51. W. Becker, *Advanced time-correlated single photon counting techniques* (Springer, Berlin, 2005)

52. W. Becker, A. Bergmann, C. Biskup, Multi-spectral fluorescence lifetime imaging by TCSPC. Micr. Res. Tech. **70**, 403–409 (2007)

53. W. Becker, The bh TCSPC handbook. 5th edition. Becker & Hickl GmbH (2012), www. becker-hickl.com

54. C.A. Bücherl, A. Bader, A.H. Westphal, S.P. Laptenok, J.W. Borst, FRET-FLIM applications in plant systems. Protoplasma **251**(2), 383–394 (2014)

55. R. Krahl, A. Bülter, F. Koberling, Performance of the Micro Photon Devices PDM 50CT SPAD detector with PicoQuant TCSPC systems, Technical Note (PicoQuant GmbH, 2005) (2005)

56. R.N. Day, Measuring protein interactions using Förster resonance energy transfer and fluorescence lifetime imaging microscopy. Methods **66**, 200–207 (2014)

57. B.J. Bacskai, J. Skoch, G.A. Hickey, R. Allen, B.T. Hyman, Fluorescence resonance energy transfer determinations using multiphoton fluorescence lifetime imaging microscopy to characterize amyloid-beta plaques. J. Biomed. Opt. **8**(3), 368–375 (2003)

58. W. Becker, Recording the Instrument Response Function of a Multiphoton FLIM System, Application Notes, www.becker-hickl.com (2007)
59. Y. Chen, A. Periasamy, Characterization of two-photon excitation fluorescence lifetime imaging microscopy for protein localization. Microscopy Research and Techniques. **63**, 72–80 (2004)
60. R.R. Duncan, A. Bergmann, M.A. Cousin, D.K. Apps, M.J. Shipston, Multi-dimensional time-correlated single-photon counting (TCSPC) fluorescence lifetime imaging microscopy (FLIM) to detect FRET in cells. J. Microsc. **215**, 1–12 (2004)
61. K.W. Eliceiri, C.-H. Fan, G.E. Lyons, J.G. White, Analysis of histology specimens using lifetime multiphoton microscopy. J. Biomed. Opt. **8**, 376–380 (2003)
62. K. König, I. Riemann, High-resolution multiphoton tomography of human skin with subcellular spatial resolution and picosecond time resolution. J Biomed Optics **8**, 432–439 (2003)
63. W. Becker, A. Bergmann, M.A. Hink, K. König, K. Benndorf, C. Biskup, Fluorescence lifetime imaging by time-correlated single photon counting. Micr. Res. Techn. **63**, 58–66 (2004)
64. C. Biskup, T. Zimmer, L. Kelbauskas, B. Hoffmann, N. Klöcker, W. Becker, A. Bergmann, K. Benndorf, Multi-dimensional fluorescence lifetime and fret measurements. Micr. Res. Tech. **70**, 403–409 (2007)
65. A. Rück, Ch. Hülshoff, I. Kinzler, W. Becker, R. Steiner, SLIM: A New Method for Molecular Imaging. Micr. Res. Tech. **70**, 403–409 (2007)
66. D. Chorvat, A. Chorvatova, Multi-wavelength fluorescence lifetime spectroscopy: a new approach to the study of endogenous fluorescence in living cells and tissues. Laser Phys. Lett. **6**, 175–193 (2009)
67. K. Abe, L. Zhao, A. Periasamy, X. Intes, M. Barroso, Non-invasive in vivo imaging of breast cancer cell internalization of transferrin by near infrared FRET. PLoS ONE **8**(11), e80269 (2013)
68. F.G. Cremazy, E.M. Manders, P.I. Bastiaens, G. Kramer, G.L. Hager, E.B. van Munster, P. J. Verschure, T.J. Gadella Jr, R. van Driel, Imaging in situ protein–DNA interactions in the cell nucleus using FRET–FLIM. Exp. Cell Res. **309**, 390–396 (2005)
69. J.D. Lee, Y.F. Chang, F.J. Kao, L.S. Kao, C.C. Lin, A.C. Lu, B.C. Shyu, S.H. Chiou, D.M. Yang, Detection of the interaction between SNAP25 and rabphilin in neuroendocrine PC12 cells using the FLIM/FRET technique. Microsc. Res. Tech. **71**(1), 26–34 (2008)
70. C. Biskup, L. Kelbauskas, T. Zimmer, K. Benndorf, A. Bergmann, W. Becker, J. P. Ruppersberg, C. Stocklausner, N. Klöcker, Interaction of PSD-95 with potassium channels visualized by fluorescence lifetime-based resonance energy transfer imaging. J. Biomed. Opt. **9**, 735–759 (2004)
71. C. Rickman, C.N. Medine, A.R. Dun, D.J. Moulton, O. Mandula, N.D. Halemani, S.O. Rizzoli, L.H. Chamberlain, R.R. Duncan, t-SNARE Protein Conformations Patterned by the Lipid Microenvironment. J. Biol. Chem. **285**, 13535–13541 (2010)
72. C.N. Medine, A. McDonald, A, Bergmann, R. Duncan, Time-correlated single photon counting FLIM: some considerations for Physiologists. Micr. Res. Tech. **70**, 421-425 (2007)
73. V. Ghukasyan, Y.-Y. Hsu, S.-H. Kung, F.-J. Kao, Application of fluorescence resonance energy transfer resolved by fluorescence lifetime imaging microscopy for the detection of enterovirus 71 infection in cells. J. Biomed. Opt. **12**, 024016-1–024016-8 (2007)
74. J.H. Fox, T. Connor, M. Stiles, J. Kama, Z. Lu, K. Dorsey, G. Lieberman, E. Sapp, R.A. Cherny, M. Banks, I. Volitakis, M. DiFiglia, O. Berezovska, A.I. Bush, S.M. Hersch, Cysteine oxidation within N-terminal mutant huntingtin promotes oligomerization and delays clearance of soluble protein. J. Biol. Chem. **286**, 18320–18330 (2011)
75. L. Munsie, N. Caron, R.S. Atwal, I. Marsden, E.J. Wild, J.R. Bamburg, S.J. Tabrizi, R. Truant, Mutant huntingtin causes defective actin remodeling during stress: defining a new role for transglutaminase 2 in neurodegenerative disease. Hum. Mol. Genet. **20**(10), 1937–1951 (2011)

76. M.T. Kelleher, G. Fruhwirth, G. Patel, E. Ofo, F. Festy, P.R. Barber, S.M. Ameer-Beg, B. Vojnovic, C. Gillett, A. Coolen, G. Kéri, P.A. Ellis, T. Ng, The potential of optical proteomic technologies to individualize prognosis and guide rational treatment for cancer patients, Targ. Oncol. **4**, 235–252 (2009)

77. E. Boutant, P. Didier, A. Niehl, Y. Mely, C. Ritzenthaler, M. Heinlein, Fluorescent protein recruitment assay for demonstration and analysis of in vivo protein interactions in plant cells and its application to Tobacco mosaic virus movement protein. Plant J. **62**, 171–177 (2010)

78. C.A. Bücherl, G.W. van Esse, A. Kruis, J. Luchtenberg, A.H. Westphal, J. Aker, A. van Hoek, C. Albrecht, J.W. Borst, C.S. de Vries, Visualization of BRI1 and BAK1(SERK3) membrane receptor heterooligomers during brassinosteroid signaling1[W][OPEN]. Plant Physiol. **162**(4), 1911–1925 (2013)

79. S. Ganesan, S.M. Ameer-Beg, T.T. Ng, B. Vojnovic, F.S. Wouters, A dark yellow fluorescent protein (YFP)-based Resonance Energy- Accepting Chromoprotein (REACh) for Förster resonance energy transfer with GFP. Proc. Natl. Acad. Sci. USA **103**, 4089–4094 (2006)

80. R.N. Day, M.W. Davidson, Fluorescent proteins for FRET microscopy: monitoring protein interactions in living cells. Bioassays **34**, 341–350 (2012)

81. H. Murakoshi, S.-J. Lee, R. Yasuda, Highly sensitive and quantitative FRET–FLIM imaging in single dendritic spines using improved non-radiative YFP. Brain Cell Biol. **36**, 31–42 (2008)

82. S .Rehman, J. T. Gladman, A. Periasamy, Y. Sun, M.S. van Mahade, Development of an AP-FRET based analysis for characterinzing RNA-protein interactions in myotonic dystrophy (DM1). PLoS ONE **9**(4), e95957

83. R.N. Day, T.C. Voss, J.F. Enwright, C.F. Booker, A. Periasamy, F. Schaufele, Imaging the localized protein interactions between Pit-1 and the CCAAT/enhancer binding protein alpha in the living pituitary cell nucleus. Mol. Endo. **17**(3), 333–345 (2003)

84. I.A. Demarco, A. Periasamy, C.F. Booker, R.N. Day, Monitoring dynamic protein interactions with photoquenching FRET. Nat. Methods **3**, 519–524 (2006)

85. Q.Q. Tang, M.D. Lane, Role of C/EBP homologous protein (CHOP-10) in the programmed activation of CCAAT/enhancer-binding protein-beta during adipogenesis. Proc. Natl. Acad. Sci. USA **97**, 12446–12450 (2000)

86. Q.Q. Tang, M.D. Lane, Activation and centromeric localization of CCAAT/enhancer-binding proteins during the mitotic clonal expansion of adipocyte differentiation. Genes Dev. **13**, 2231–2241 (1999)

87. E. Gratton, S. Breusegem, J. Sutin, Q. Ruan, N. Barry, Fluorescence lifetime imaging for the two-photon microscope: time-domain and frequency-domain methods. J. Biomed. Opt. **8**(3), 381–390 (2003)

88. R.N. Day, Measuring protein interactions using Förster resonance energy transfer and fluorescence lifetime imaging microscopy. Methods **66**, 200–207 (2014)

89. V. Jyothikumar, Y. Sun, A. Periasamy, Investigation of tyrptopahn-NADH in live human cells using 3-photon fluorescence lifetime imaging. J. Biomed. Opt. **18**(6), 060501 (2013)

Chapter 8
Monitoring HIV-1 Protein Oligomerization by FLIM FRET Microscopy

Ludovic Richert, Pascal Didier, Hugues de Rocquigny and Yves Mély

Abstract The majority of the human immunodeficiency virus type 1 (HIV-1) proteins are able to self assemble into oligomers. Since these oligomers generally exhibit functions that differ from those of their monomeric counterpart, the regulation of the monomer-oligomer equilibria plays a central role in the viral cycle. To characterize the oligomerization of these proteins in live cells, the combination of fluorescence lifetime imaging microscopy (FLIM) with Förster resonance energy transfer (FRET) has proven to be very powerful. In this review, we illustrate the application of FRET-FLIM on the characterization of the oligomerization of the Vpr, Vif and Pr55^{Gag} proteins of HIV-1 in fusion with eGFP and mCherry. For Vpr and Pr55^{Gag} proteins, very high levels of FRET leading to strong decreases in eGFP fluorescence lifetime are obtained, as a consequence of the rather small size of the viral proteins, the strong packing of the protomers and the presence of multiple acceptors for one donor. Analyzing the time-resolved decays by a two-component analysis further provides the possibility to discriminate monomers from oligomers and to monitor the spatiotemporal evolution of both populations in the cells. Though FRET-FLIM unambiguously reveals the oligomerization of a given protein, it hardly discloses the oligomer stoichiometry (number of protomers per oligomers). This parameter can be obtained by fluorescence correlation spectroscopy, which allows further interpreting the FRET-FLIM data. FRET-FLIM is also highly useful to identify the determinants of the oligomerization process and to investigate its regulation by other HIV-1 proteins and host proteins.

L. Richert · P. Didier · H. de Rocquigny · Y. Mély (✉)
UMR 7213 CNRS, Laboratoire de Biophotonique et Pharmacologie,
Faculté de Pharmacie, Illkirch, France
e-mail: yves.mely@unistra.fr

L. Richert
e-mail: ludovic.richert@unistra.fr

P. Didier
e-mail: pascal.didier@unistra.fr

© Springer International Publishing Switzerland 2015   277
W. Becker (ed.), *Advanced Time-Correlated Single Photon Counting Applications*,
Springer Series in Chemical Physics 111, DOI 10.1007/978-3-319-14929-5_8

8.1 Introduction

8.1.1 HIV-1 Proteins

The human immunodeficiency virus type 1 (HIV-1) is the causative agent of AIDS (acquired immunodeficiency syndrome). Its spherical viral particle has a diameter of 100–120 nm with an outer envelope originating from the lipid bilayer of the host cell membrane [91]. The two viral envelope glycoprotein subunits, the surface SUgp120 and the transmembrane TMgp41 proteins in the form of trimers, are anchored in the virion envelope [68]. The inner virion structure is filled by the matrix protein (MA) and contains the capsid protein (CA) organized as a conical core that contains the viral ribonucleoparticle (vRNP) [93]. This vRNP is composed of the genomic RNA of 9.6 kbases, in a dimeric form, coated by about 2000 copies of the nucleocapsid protein (NCp7) together with the viral protein R (Vpr) and viral enzymes, namely reverse transcriptase (RT), integrase (IN), and protease (PR). In addition, the virus contains also viral cofactors such as Tat and Rev, as well as the helper factors Vif, Vpr, Vpu, and Nef.

During viral infection by the human immunodeficiency virus type 1 (HIV-1), most of these viral proteins are in equilibrium between monomeric and various oligomeric forms. The latter exhibit frequently functions that differ from those of their monomeric counterpart, so that oligomerization appears as an optimized way for the virus to circumvent the limited size of its genome. For instance, viral enzymes such as proteases [33], integrases [3] and reverse transcriptase [56, 59] are generally active under their oligomeric form. Moreover, structural proteins, such as the capsid and the matrix proteins auto-assemble to form higher order multimers [7]. In addition, envelope proteins as well as accessory proteins are also concerned [66, 108]. The regulation of these monomer-oligomer equilibria plays a central role in the function of these proteins during the viral cycle. Therefore, drugs able to modulate these equilibria appear as potential therapeutic candidates in viral infections.

Significant progress was made on the understanding of the structures and properties of protein oligomers. NMR, Cryo-TEM and X-ray crystallography were extensively used to solve the structures of viral proteins in different oligomeric states (see reviews: [91, 97]). Moreover, the thermodynamics and kinetics of oligomerization were characterized by a number of biophysical techniques, including exclusion chromatography (SEC), analytical ultracentrifugation (AUC) and single molecule methods [27, 48, 61, 70]. However, these techniques are not well-suited to characterize protein oligomerization in live cells, and notably to monitor in real-time the dynamic properties of oligomer formation and disassembly.

8.1.2 FLIM-FRET Technique

In contrast, these questions can be addressed by FRET-based techniques that can be combined with both widefield and confocal microscopes. This allows the study of subcellular compartments, which can be particularly advantageous to separate the oligomerization that occurs during biosynthesis (in the endoplasmic reticulum (ER) and the Golgi) from that on the cell surface. However, several factors complicate the interpretation of FRET studies. First, the light used to activate the donor molecule may directly excite the acceptor as well. Second, the donor emission may leak into the channel used to detect the acceptor emission. Third, the FRET efficiency is affected by the expression levels of the donor and acceptor molecules, while autofluorescence and photobleaching can further complicate the measurements. To overcome these problems, appropriate correction factors must be used, which are obtained from separate measurements. However, this method is prone to errors, requiring multiple images of the same sample collected with various combinations of excitation and emission wavelengths, as well as measurements with model systems [49, 81, 88].

A much more convenient approach to monitor FRET is to use fluorescence lifetime imaging microscopy (FLIM) [10, 12, 14, 22, 78, 95]. This technique measures the fluorescence decay function of the FRET donor at each pixel of the FLIM image. From the decay functions, either an average fluorescence lifetime or the lifetimes and amplitudes of several decay components are derived. In contrast to fluorescence intensities, these parameters do not depend on the probe concentration, excitation intensity and, in first approximation, photobleaching. Therefore, they unambiguously and directly report on the amount of FRET between the donor and the acceptor, through a decrease of the fluorescence lifetime of the donor [11, 13, 16, 17, 23], see Chap. 7 of this book. There is also another advantage: there may be a non-interacting and an interacting fraction of the donor. The reason may be the orientation-dependence of FRET, or more interestingly, the presence of non-interacting and interacting protein fractions. There may be also donor populations with different FRET efficiency. Different donor fractions can be identified in the FLIM data by multi-exponential analysis of the decay functions. Multi-exponential decay analysis does, of course, require the decay functions to be recorded at high temporal resolution and in a large number of time channels. It also requires a large number of photons. At the same time, photoconversion or photobleaching of the donor or the acceptor must be avoided. Photoconversion changes the fluorescence decay times, photobleaching of the donor acts differently on the interacting and on the non-interacting donor fraction, and photobleaching of the acceptor decreases the apparent FRET efficiency [52]. That means high recording efficiency is required. These requirements are almost perfectly met by TCSPC FLIM [10, 12, 13, 17, 19]. The technique is based on scanning the sample with a high-frequency pulsed laser beam, detecting single photons of the fluorescence light, determining their times in the excitation pulse period and the location of the laser beam in the scan area in the moment of photo detection, and building up a three-dimensional histogram of the photon number over these parameters. Please see Chaps. 1 and 2 for details.

For the experiments described in this chapter we used a two-photon laser scanning microscope based on home-made confocal scanner optics, an Olympus IX70 microscope, a Spectra Physics Tsunami Ti: Sapphire laser, a Perkin-Elmer SPCM-AQR-14-FC SPAD detector, and a Becker & Hickl SPC-830 TCSPC FLIM module [13]. Mainly live-cell samples were used for the experiments. FLIM data were acquired for typically 30 s per sample. The data were analysed by Becker & Hickl SPCImage FLIM data analysis software [13].

To monitor the oligomerization of viral proteins by FRET-FLIM, the most common approach is to transfect cells with two plasmids in order to co-express the viral protein of interest tagged with a fluorescent protein (FP) acting as a FRET donor (plasmid 1) together with the viral protein tagged with a FP acting as an acceptor (plasmid 2). To increase the probability to observe the FRET process, viral proteins tagged with acceptor FP should be in excess over those tagged with the donor FP.

The observation of FRET through a decrease of the fluorescence lifetime of the donor critically depends on the distance between the donor and the acceptor [37, 38], see Chap. 7. Usually, FRET can be observed if the distances are below 10 nm. Since FPs exhibit a barrel-like shape with a diameter of 2 nm and a length of 4 nm, their size is comparable to the values of the Förster distance R_0, which prevents observing high FRET efficiencies and biases the determination of the distances between the oligomer sub-units. Moreover, due to steric hindrance between the labelled proteins, the possible orientations of the two FP dipoles involved in FRET are limited, which also impacts the determination of the inter-protein distances in the oligomers. Indeed, the orientations between the FPs affect κ^2, the orientation factor which is used to calculate the Förster distance. The values of κ^2 range between 0 and 4, and are assumed to be 2/3 in the case of dynamic isotropic averaging. In oligomers, due to steric constraints, only a limited range of κ^2 values is likely accessible, so that a significant deviation from the 2/3 value can be expected.

8.1.3 FRET to Multiple Acceptors

The stoichiometry of donors and acceptors in a molecular complex can also dramatically alter the FRET efficiency [86]. Each additional acceptor incrementally increases the FRET efficiency (Fig. 8.1a), so that the fluorescence lifetime of the donor will depend both on the number of acceptors and their relative arrangement in respect to the donor in the oligomer (Fig. 8.1b).

Other factors, such as the total concentration of fluorophores and the donor/acceptor ratio can influence the measured FRET efficiencies (Fig. 8.2). As modelized by Monte-Carlo simulations, high concentrations (mM range) in the cytoplasm, notably at high acceptor/donor ratios, will produce FRET signals as a result of random acceptor distributions [26]. Indeed, at these high acceptor concentrations, randomly placed donor will have a significant probability of being within 10 nm of at least one acceptor.

If donors and acceptors are confined in subcellular regions, such as in membranes, as few as 100 acceptors per μm^2 can result in "random FRET". Generally, any pair of integral membrane proteins labeled with a donor and an acceptor will generate approximately a 5 % random FRET efficiency when coexpressed in the

(a)   **(b)**

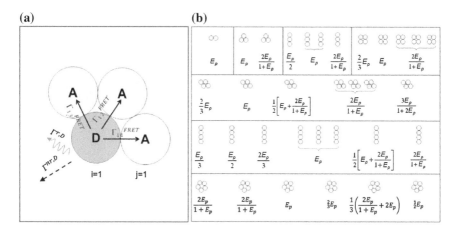

Fig. 8.1 Effect of multiple acceptors on FRET efficiency. **a** Pathways by which an excited donor can lose its electronic excitation when associated with multiple acceptors. i and j, counting indices for donors and acceptors, respectively; Γ denotes the rate constants of de-excitation; superscript r denotes a radiative process; superscript nr denotes a nonradiative process other than transfer to the acceptor. **b** Apparent FRET efficiencies for different sizes and configurations of oligomers. E_p corresponds to the energy transfer efficiency between a single donor and a single acceptor. Adapted from [86]

same membrane under normal imaging conditions. Therefore, determination of the fraction of FRET signal arising from specific protein-protein interactions as compared to that arising from random interaction due to overcrowding requires thus knowledge of the local abundance of acceptors in the sample.

To illustrate the use of FRET-FLIM in characterizing HIV-1 protein oligomerisation and obtaining key information on the viral life cycle, we will discuss its application for the investigation of the oligomerization of Vpr, Vif and Pr55^{Gag} proteins of HIV-1.

8.2 Vpr Oligomerization

8.2.1 Vpr Protein

Vpr is a 96 amino acid regulatory protein of HIV-1 that impacts the survival of the infected cells by causing a G2/M arrest and apoptosis [40, 80]. Its N-terminal domain plays a role in virion incorporation, nuclear localization and oligomerization, while its C-terminal domain is involved in the G2/M cell cycle arrest [46, 58] and apoptosis [57, 105]. The 3D structure of Vpr peptides and of full length Vpr in hydrophobic solvents or in the presence of micelles was solved by NMR [72, 90, 103]. However, the characterization of these 3D structures was tedious, as a consequence of the ability of Vpr to oligomerize via its leucine zipper like motifs

Fig. 8.2 Influence of donor/
acceptor ratio and total
fluorophore concentration on
the FRET efficiency for
fluorophores randomly
distributed in solution.
Adapted from [26]

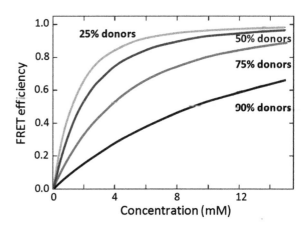

[20, 74]. This oligomerization is thought to constitute an intrinsic property of Vpr, since it was shown to be required for nuclear localization [17, 29, 40, 47, 48]. To further investigate and characterize Vpr oligomerization in cells, HeLa cells were transfected with plasmids coding for eGFP- and mCherry- tagged Vpr proteins, and monitored by two-photon FLIM measurements at 24 h post transfection. The eGFP tag was selected as a FRET donor, due to its high quantum yield (0.6) and its monoexponential time-resolved fluorescence decay. The mCherry tag was used as an acceptor due to the strong overlap between its absorption spectrum and the eGFP emission spectrum, resulting in a large Förster distance, R_0 (about 54 Å) [62, 69, 100]. The labeled Vpr proteins were found to accumulate mainly at the nuclear envelope, though some proteins were also expressed in the cytoplasm and the nucleus. By comparison with non labeled Vpr proteins, it was observed that fusion of eGFP or mCherry to the C-terminus of Vpr has a limited effect on Vpr localization in the cell. In contrast, fusion of these FPs to the N-terminus of Vpr induced a cellular redistribution, indicating that the FPs in this case may perturb the interaction of Vpr with the nuclear membrane [17, 29, 40, 67].

8.2.2 Fluorescence Lifetimes of Vpr-eGFP Fusion Proteins

We first analyzed the time-resolved decays obtained with our two-photon FLIM set-up [5, 25] using one-component decay analysis (Fig. 8.3 and Table 8.1).

The fluorescence lifetimes (τ) shown for the eGFP alone or linked to the N- or C-terminus of Vpr are the average values ($\pm$ standard deviation) for 10–35 cells. For each cell, measurements were performed at the nuclear envelope, in the nucleus and in the cytoplasm. The FRET efficiency (E) is given by

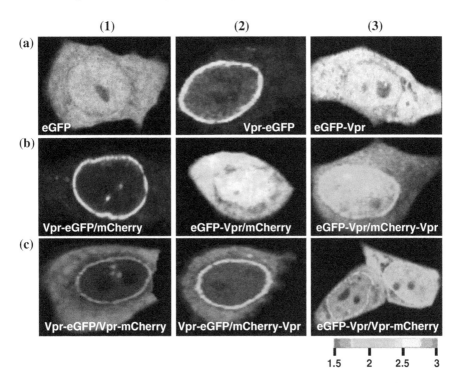

Fig. 8.3 FRET-FLIM monitoring of Vpr oligomerization. HeLa cells were transfected with plasmids coding for eGFP or eGFP-tagged Vpr alone or in combination with mCherry-tagged Vpr. The lifetimes (in ns) are represented using an arbitrary colour scale ranging from *blue* to *red*. *Panels A1–A3* show the FLIM images of cells expressing eGFP or eGFP-tagged Vpr alone. *Panels B1* and *B2* show cells co-expressing eGFP-tagged Vpr and mCherry; *Panels B3* and *C1–C3* correspond to cells co-expressing eGFP- and mCherry-tagged Vpr. Reprinted from [40]

$$E = 1 - \tau_{DA}/\tau_{D},$$

where τ_{DA} is the lifetime of the donor in the presence of the acceptor, and τ_{D} is the lifetime of the donor in the absence of the acceptor (see Chap. 7). The whole-cell FRET-efficiency (E) and the lifetime (τ) values represent the average values calculated over the entire cell.

Comparison with free eGFP (Fig. 8.3, panel A1) indicated that the eGFP fluorescence lifetime (2.4–2.5 ns) was not altered when fused to Vpr (Fig. 8.3, panels A2 and A3). Moreover, the fluorescence lifetime of eGFP-labeled Vpr was not affected by the co-expression of free mCherry (Fig. 8.3, panels B1 and B2), indicating that no short range interaction occurred with free mCherry [79, 95]. In sharp contrast, co-expression of eGFP-labeled Vpr and mCherry-labeled Vpr induced a sharp decrease in the eGFP lifetime, indicating that Vpr oligomerizes. The largest decrease in the eGFP lifetime (from 2.4 to 1.72 ns, Table 8.1) was observed at the nuclear rim when the C-terminally labeled Vpr proteins (Vpr-eGFP and

Vpr-mCherry) were co-expressed in the same cell (Fig. 8.3, panel B3), which corresponded to an apparent FRET efficiency of about 30 %. Vpr oligomerization was also observed in the cytoplasm and the nucleus, as shown by the 1.86 and 1.95 ns lifetimes (corresponding to 23 and 17 % apparent FRET efficiencies) measured in these compartments. Somewhat lower apparent FRET efficiencies (≤15 %) in all cell compartments were observed when a FP was at the N-terminus in one of the interacting partners (Fig. 8.3, panels C1–C3). Finally, co-expression of the two N-terminally labeled proteins (eGFP-Vpr and mCherry-Vpr) induced only a 8 % apparent FRET efficiency, suggesting that the N-terminal labelling of Vpr may alter its oligomerization.

8.2.3 Two-Component FRET Analysis

The lower lifetime values at the nuclear membrane compared to the cytoplasm and the nucleus suggests that either the eGFP and mCherry tags are closer to each other in the oligomers (giving a higher FRET efficiency) or that a higher fraction of Vpr proteins are in oligomeric form at the nuclear membrane. To address this question, we analyzed the time-resolved decays of Vpr-eGFP ± Vpr-mCherry with a two component analysis, using the model function [13]:

$$F(t) = \alpha_1 e^{-t/\tau_1} + \alpha_2 e^{-t/\tau_2}$$

As we transfected the plasmids coding for the eGFP- and mCherry-tagged Vpr at a 1:1 ratio and as a large fraction of Vpr proteins form dimers (see below), it is likely on a statistical basis that a significant number of eGFP-tagged proteins are not associated with mCherry-tagged proteins, and thus show a 2.4 ns lifetime, corresponding to the lifetime of free Vpr-eGFP. Therefore, we fixed the long-lived lifetime τ_2 at 2.4 ns in our two-component analysis. The result is shown in Fig. 8.4.

We found that the lifetime of the fast decay component, τ_1, was homogeneous all over the cell, with a mean value of 1.5 ns, that corresponds to a FRET efficiency of 37.5 % (Fig. 8.4a). In contrast, the amplitude, α_1, of the fast decay component was found to be much higher at the nuclear envelope (70 %) than in the cytoplasm or the nucleus (40 %) (Fig. 8.4a). Thus, the two component analysis clearly revealed the presence of two populations, one where eGFP-tagged Vpr proteins were associated with mCherry-tagged protein(s) and another one where eGFP-tagged Vpr proteins were either monomeric or associated with other eGFP-Vpr proteins. Moreover, as the FRET efficiency, and thus the value of τ_1, are expected to strongly depend on the size of the oligomers (Fig. 8.1), the homogeneity of the τ_1 values all over the cell suggests that the same oligomers are distributed all over the cell. Since the FRET efficiency is also dependent on the configuration of the oligomers and notably on the relative proportion of donor and acceptor in each oligomer (Fig. 8.1b), we repeated the FLIM-FRET experiments on cells transfected with an increased relative concentration of Vpr-mcherry (1:4 ratio) (Fig. 8.4). The decay

Table 8.1 Fluorescence lifetime and FRET efficiency of eGFP and eGFP-tagged Vpr in living cells

	Nuclear envelope		Cytoplasm		Nucleus		Whole cell	
	E (%)	τ (ns)	E (%)	τ (ns)	E (%)	τ (ns)	E (%)	τ (ns)
eGFP	–	–	–	2.50 (± 0.01)	–	2.50 (± 0.01)	–	2.50 (± 0.01)
Vpr-eGFP	–	2.36 (± 0.01)	–	2.40 (± 0.01)	–	2.41 (± 0.01)	–	2.39 (± 0.01)
eGFP-Vpr	–	2.47 (± 0.01)	–	2.46 (± 0.01)	–	2.47 (± 0.01)	–	2.47 (± 0.01)
Vpr-eGFP+mCherry	–	2.41 (± 0.02)	–	2.42 (± 0.01)	–	2.42 (± 0.01)	–	2.42 (± 0.01)
Vpr-eGFP+Vpr-mCherry	27	1.72 (± 0.02)	23	1.86 (± 0.03)	19	1.95 (± 0.03)	23	1.85 (± 0.03)
Vpr-eGFP+mCherry-Vpr	17	1.95 (± 0.02)	14	2.06 (± 0.02)	13	2.09 (± 0.02)	15	2.02 (± 0.03)
eGFP-Vpr+mCherry	–	2.43 (± 0.01)	–	2.43 (± 0.01)	–	2.43 (± 0.02)	–	2.43 (± 0.01)
eGFP-Vpr+Vpr-mCherry	13	2.14 (± 0.03)	9	2.25 (± 0.03)	6	2.32 (± 0.02)	9	2.25 (± 0.03)
eGFP-Vpr+mCherry-Vpr	13	2.14 (± 0.03)	7	2.28 (± 0.03)	6	2.31 (± 0.02)	8	2.28 (± 0.03)

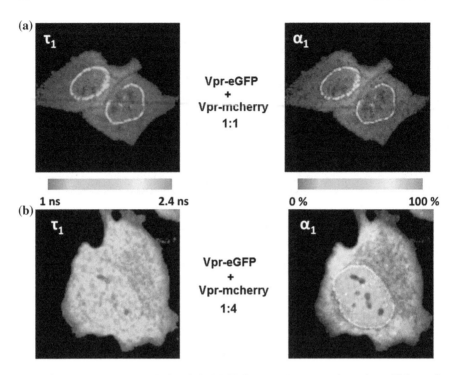

Fig. 8.4 Two-component analysis of FRET-FLIM experiments performed on HeLa cells transfected with DNA constructs encoding Vpr-eGFP and Vpr-mcherry in a 1:1 (**a**) and 1:4 ratio (**b**). The fluorescence decays were measured at each pixel and analysed by using a bi-exponential model with a long-lived component τ_2 fixed to the value of free eGFP (2.4 ns). In this case, the floating parameters were τ_1, α_1 and $\alpha_2 = 1-\alpha_1$. The τ_1 and α_1 values were converted into a colour code ranging from *blue* (1 ns, 0 %) to *red* (2.4 ns, 100 %). All images were acquired using a 50×50 μm scale and 128×128 pixels

curves were again analyzed using a two-components model with a τ_2 value fixed at 2.4 ns. As observed for the 1:1 ratio, the τ_1-component was homogenously distributed all over the cell, but with a decreased average value of 1.3 ns (FRET efficiency of 46 %). Moreover, the amplitudes α_1 of the fast decay component at the nuclear envelope and in the remaining part of the cell (Fig. 8.1b) were fully comparable to those observed at a 1:1 ratio (Fig. 8.1a).

8.2.4 Determination of Oligomer-Size by FCS

Though FLIM-FRET is highly useful to demonstrate Vpr oligomerization, it cannot be used to determine the number of protomers in the oligomers. To further characterize Vpr oligomerization in cells and interpret the data of the two components analysis by FRET-FLIM (Fig. 8.4), fluorescence correlation spectroscopy (FCS) was performed. This technique measures the fluctuations of the fluorescence intensity in the femtoliter volume defined by the focal volume of a confocal or a two-photon microscope [32, 87], see also Chap. 1, Sect. 1.5.2. These fluctuations mainly characterize the translational dynamics of the fluorescent molecules that diffuse through the focal volume. FCS can be performed in live cells, which allows the determination of several physical parameters, such as diffusion time, local concentration, and molecular brightness. The last parameter corresponds to the average number of photons emitted by a diffusing particle per second and increases proportionally with the number of fluorescent molecules inside a particle, so that the stoichiometry of the oligomers can be accessed.

Since no FCS measurement was possible at the nuclear envelope, as a result of strong eGFP photobleaching, FCS measurements were carried out in the cytoplasm and the nucleus. The diffusion time of eGFP (Fig. 8.5b) displays a narrow distribution centred around 0.4 ms [8]. Moreover, its anomalous diffusion coefficient α that accounts for the concentration, size, mobility and reactivity of the obstacles encountered by the diffusing species was around 1 (Fig. 8.5a), suggesting that eGFP freely diffuses as monomers in the cell [24, 96]. The distribution of the apparent diffusion time τ_A of Vpr-eGFP is shifted to 4 ms (Fig. 8.5e). As τ_A varies as the cubic root of the molecular mass of the diffusing species, this 1000-fold increase in the molecular mass of Vpr-eGFP oligomers as compared to eGFP, suggests that Vpr fusion proteins form large complexes in cells. Moreover, the α values of Vpr-eGFP are distributed around 0.75 showing that such complexes do not freely diffuse in the cell but interact with cellular components (Fig. 8.5d). While the brightness of eGFP monomers displays a narrow distribution centred at 1 kHz/particle (Fig. 8.5c), the brightness of Vpr-eGFP shows a major population around 2–3 kHz/particle and a minor population with a large distribution of brightness (Fig. 8.5f).

This suggests that Vpr-eGFP mainly forms dimers and trimers, as well as a smaller fraction of higher order oligomers. These small oligomers do not explain the aforementioned differences between the molar masses of eGFP and Vpr-eGFP complexes, indicating that Vpr oligomers probably interact with cellular proteins [80].

In light of the FCS results, the lifetime of the fast decay component measured in the two-component analysis (Fig. 8.4) may correspond to the mean value of the lifetimes exhibited by dimers containing Vpr-eGFP and Vpr-mCherry in a 1:1 ratio, and trimers containing both proteins at 2:1 and 1:2 ratio. In line with this hypothesis, the decrease in the lifetime of the fast decay component, τ_1, observed in Fig. 8.4b, as compared to Fig. 8.4a, could be explained by the increase in the proportion of trimers having two acceptors for one donor, as a result of the increased relative concentration of Vpr-mCherry. As shown in Fig. 8.1, these trimers are expected to be endowed with a higher FRET efficiency, and thus, with a lower lifetime value as compared to trimers having only one acceptor. Furthermore, as identical short-lived lifetime values were observed at the nuclear envelope and the remaining part of the cell for a given ratio of expressed Vpr-mCherry to Vpr-eGFP proteins (Fig. 8.4), we may reasonably conclude that the same oligomers are distributed all over the cell. Therefore, the increased α_1 value systematically

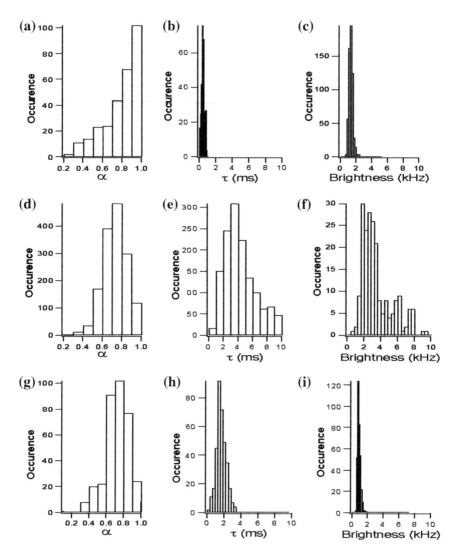

Fig. 8.5 Distribution histograms of anomalous diffusion coefficients, diffusion times and brightness of eGFP, Vpr-eGFP and ΔQ44 Vpr-eGFP. The anomalous diffusion coefficient (coefficient that accounts for the obstacles encountered by the diffusing species), diffusion times (average time needed to cross the focal volume) and brightness (count rates/species) determined by FCS are expressed as a function of the number of occurrences. **a–c** correspond to eGFP; **d–f** correspond to Vpr-eGFP; **g–i** correspond to ΔQ44 Vpr-eGFP. Reprinted from [40]

observed at the nuclear envelope as compared to the cytoplasm or the nucleus clearly indicates that dimers and trimers show a preferential binding to the nuclear envelope.

8.2.5 Mapping of Vpr Domains

In order to map the domains of Vpr involved in its oligomerization, point mutants of Vpr-eGFP and Vpr-mCherry were designed. Amino acids (L23, Q44, I60 and L67) located in the Vpr α-helices were changed to F (L23F) or A (I60A, L67A) or deleted (ΔQ44). Moreover, amino acids Q3, W54, R77 and R90 located outside of the α-helices were changed to R, G, Q and K, respectively. FLIM images obtained with these mutants are shown in Fig. 8.6.

The FLIM images of the Vpr-eGFP mutants expressed in the absence (Fig. 8.6, Column A) and in the presence of the corresponding Vpr-mCherry (Fig. 8.6, Column B) show that the four mutants in the α-helices (L23F, ΔQ44, I60A and L67A) have lost their ability to accumulate at the nuclear envelope. The very low transfer efficiency associated to these four mutants further indicated that they have lost their ability to oligomerize, indicating that the α-helices are involved in Vpr oligomerization and accumulation at the nuclear envelope. Noticeably, a small but significant FRET efficiency (6–7 %) was observed with the I60A mutant, indicating that this mutation is slightly less detrimental for Vpr oligomerization than the other three mutations. Interestingly, the brightness of ΔQ44 Vpr-eGFP measured by FCS is around 1.2 kHz

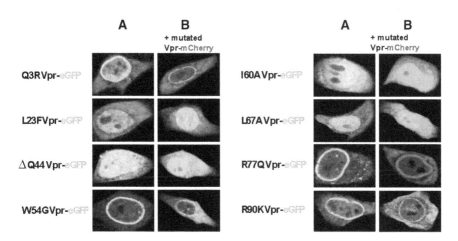

Fig. 8.6 Mapping of Vpr oligomerization by FRET-FLIM. HeLa cells were co transfected with point mutants of Vpr-eGFP and Vpr-mCherry. FLIM was carried out 24 h post transfection. *Column A* FLIM images of the Vpr-eGFP mutants alone. *Column B* FLIM images of cells co-expressing Vpr-eGFP and Vpr-mCherry mutants. A drastic reduction of Vpr oligomerisation is observed upon mutating residues L23, Q44, I60 and L67 (*column B*). Reprinted from [40]

(Fig. 8.5i), close to the value obtained for eGFP (Fig. 8.5c), confirming that the ΔQ44 Vpr-eGFP does not form oligomers. The diffusion coefficient τ_A for the Vpr ΔQ44 mutant is about 2 ms (Fig. 8.5h), a value in between that for eGFP (0.4 ms) and that for Vpr-eGFP (4 ms), suggesting that this Vpr mutant also interacts with host proteins. In sharp contrast with these mutants, the extrahelical Q3R, W54G, R77Q and R90K mutants showed an accumulation at the nuclear envelope and FRET efficiency levels comparable to those of the wild-type fusion protein. Thus, the Q3, W54, R77 and R90 residues located outside the α-helices appear to be not critical for Vpr intracellular localization and oligomerization.

Taken together, our data show that (i) Vpr oligomerizes when expressed in HeLa cells, (ii) Vpr oligomerization is required for its accumulation at the nuclear envelope, and (iii) Vpr oligomerization critically relies on the hydrophobic core formed by the three α helices, while the residues outside the helices are dispensable.

8.2.6 Interaction with Pr55^{Gag}

In a next step, our objective was to determine if Vpr oligomerization is required for its interaction with the Pr55^{Gag} polyprotein precursor, which plays a critical role in the recruitment of Vpr into the virus [6, 42, 65, 67]. This recruitment is thought to be mediated through interactions between Pr55^{Gag} and the first two α helices of Vpr [6, 60, 65]. To characterize the oligomerization state of Vpr on interaction with Pr55^{Gag}, eGFP- and mCherry-tagged Vpr proteins were coexpressed with non labeled Pr55^{Gag} (Fig. 8.7). Interestingly, the FRET signal provided by Vpr oligomers was observed mainly at the plasma membrane (Fig. 8.7A, image d), indicating that interaction with Pr55^{Gag} caused a relocation of Vpr oligomers at the plasma membrane. In addition, the relocation of Vpr oligomers was accompanied by a significant FRET increase (compare Fig. 8.7A, panels b and d; Fig. 8.7B), suggesting a Pr55^{Gag}-induced compaction of Vpr oligomers and/or a structural rearrangement of Vpr oligomers.

To further investigate the role of Vpr oligomerization for Pr55^{Gag} interaction, Vpr was mutated at the level of its non structured N- and C-termini (Q3R and R77Q) or in its hydrophobic α-helical core (L23F, ΔQ44, and L67A) (Fig. 8.8). For the Q3R and R77Q mutants that do not alter Vpr oligomerization, a high FRET efficiency was observed at the plasma membrane, indicating that these mutations marginally affect the interaction with Pr55^{Gag} (Fig. 8.8a, line 2 and Fig. 8.8b).

In contrast, mutations in the α helices (L23F, ΔQ44 and L67A), which abolish Vpr oligomerization and localization at the nuclear envelope (Fig. 8.8a, lines 3, 4 and 5 and Fig. 8.8b) were found to result in a loss of interaction with Pr55^{Gag} and redistribution at the PM. Thus, our data show that (i) Vpr proteins interacts with Pr55^{Gag} in their oligomeric form and (ii) Vpr oligomers are required for the redistribution of Vpr at the PM and probably, its incorporation into nascent viral particles.

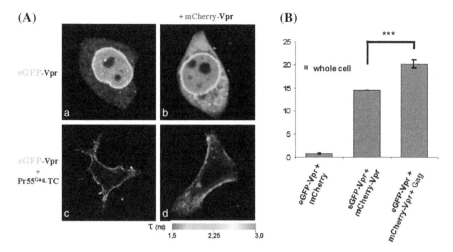

Fig. 8.7 FRET-FLIM investigation of the interaction of Vpr oligomers with Pr55^{Gag}. **A** FLIM of HeLa cells expressing eGFP-Vpr alone (*a*), eGFP-Vpr and mCherry-Vpr (*b*), eGFP-Vpr and Pr55^{Gag} (*c*), and both eGFP-Vpr/mCherry-Vpr and Pr55^{Gag} (*d*). **B** Histograms representing the FRET efficiencies between eGFP-Vpr and mCherry-Vpr in the absence or the presence of non-labelled Pr55^{Gag}. The FRET efficiency was calculated using the average lifetime values from at least 30 cells. Note that the *darker blue* in image *d* compared to the *blue* in image *b* corresponds to a significantly higher FRET efficiency between fluorescent proteins as determined by one-way ANOVA and Tukey's HSD test (*** $P < 10^{-3}$). Reprinted from [42]

8.3 Oligomerization of Vif Proteins

8.3.1 Self Association

The HIV-1 viral infectivity factor (Vif) is a small basic protein essential for viral fitness and pathogenicity [31]. Vif allows productive infection in non permissive cells, including most natural HIV-1 target cells, by counteracting the cellular cytosine deaminases APOBEC3G (apolipoprotein B mRNA-editing enzyme catalytic polypeptide-like 3G [A3G]) and A3F. In vitro, Vif was shown to self-associate in order to form dimers, trimers, and tetramers [4, 106]. Vif oligomerization is thought to be mediated by the region from positions 151 to 164, encompassing the conserved proline-rich region 161PPLP164 [106, 107] that is involved in the binding to A3G and is crucial for Vif function and viral infectivity [92, 106]. To check and characterize Vif oligomerization in living cells, eGFP- and mCherry-tagged Vif fusion proteins were expressed in HeLa cells and characterized by FRET-FLIM and FCS.

To investigate Vif multimerization, FLIM measurements were performed 24 h post transfection on cells co-expressing Vif proteins N-terminally labelled by eGFP and mCherry. Results are shown in Fig. 8.9.

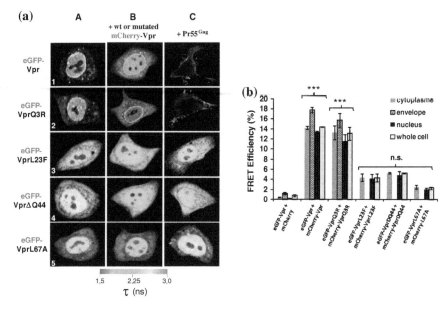

Fig. 8.8 Role of Vpr oligomerization in its interaction with Pr55^{Gag}. **a** FLIM experiments were carried out on cells expressing wild-type or mutant eGFP-Vpr alone (*column A*); wild-type (*wt*) or mutant eGFP-Vpr and their equivalent counterparts fused to mCherry (*column B*); wild-type or mutant eGFP-Vpr and Pr55^{Gag} (*column C*). **b** Histogram of the FRET efficiencies between eGFP-Vpr and mCherry-Vpr derivatives in the cytoplasm, at the nuclear envelope, at the plasma membrane, in the nucleus, and over the whole cell. The FRET efficiencies were calculated using the average lifetime values from at least 30 cells in three independent experiments. Multifactorial ANOVA and post hoc Dunnett tests were performed to compare the FRET efficiencies (* $P < 0.05$; *** $P < 10^{-3}$; *n.s.* nonsignificant). Reprinted from [42]

Control experiments (Fig. 8.9a, c) showed that the average lifetimes of eGFP-Vif expressed alone (2.53 ± 0.04 ns) or co-expressed with mCherry (2.48 ± 0.05 ns) were similar to the lifetime of eGFP alone [40, 79, 95], indicating that the eGFP fluorescence was not altered when fused to Vif and no interaction takes place between eGFP-Vif and free mCherry. In contrast, the eGFP-Vif fluorescence lifetime dropped to 2.21 ± 0.11 ns, when it was co-expressed with mCherry-Vif (Fig. 8.8b, c). This corresponded to a FRET efficiency of 12.2 % (Fig. 8.9d), that was substantially lower than that observed with Vpr (Fig. 8.3, Table 8.1). Nevertheless, as a threshold value of 5 % for FRET efficiency is generally considered to validate an interaction between two partners [102], the observed efficiency clearly supported the oligomerization of Vif proteins in living cells. To confirm this oligomerization, FLIM-FRET experiments were further performed on a mutant of Vif in which the 160PPLP164 motif was substituted for an oligomerization-defective AALA motif. The eGFP-Vif AALA mutant displayed an intermediate fluorescence lifetime of 2.35 ± 0.08 ns (Fig. 8.9c). The corresponding FRET efficiency of 6.6 % (Fig. 8.9d) indicated that the mutation reduced but not fully

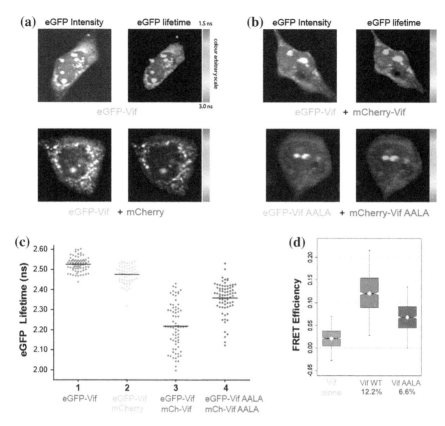

Fig. 8.9 FRET-FLIM investigation of the oligomerization of Vif and Vif AALA mutant. HeLa cells were transfected with 0.5 µg of plasmid coding for eGFP- and mCherry-labelled Vif proteins and were imaged by two photon FLIM 24 h post transfection. **a** and **b** Representative cells expressing eGFP-Vif or eGFP-VifAALA by fluorescence microscopy (*left*; greyscale) or by FLIM (*right*) using an arbitrary colour scale for the lifetimes. **c** Lifetime distribution of all analysed cells. Measurements were performed on more than 70 cells from at least 3 independent transfections. *Black bars* indicate the mean values. **d** *Box plots* depicting the FRET efficiency distributions. The mean FRET efficiency is represented by the *white diamonds*. Whiskers (*vertical dotted lines*) represent the interval containing 95 % of the FRET efficiencies. Reprinted from [9]

prevented Vif oligomerization. Interestingly enough, mutation of the PPLP motif of only one of the tagged Vif species was sufficient to reduce oligomerization, giving FRET efficiencies similar to the one observed when the mutation was present on both Vif partners.

Due to the rather low FRET values close to the 5 % FRET threshold, a careful statistical analysis was needed to strengthen our conclusions. Bayesian analyses of fluorescence lifetimes showed a 20–25 % reduction of Vif oligomerization when the PPLP motif was replaced by AALA. The distribution of the fluorescence lifetime of eGFP-Vif in the presence of mCherry-Vif was more dispersed than the one observed for eGFP-Vif alone or eGFP-Vif AALA (Fig. 8.9c), suggesting that wild-type Vif oligomers may be heterogeneous in the cell.

Thus, our data indicate that Vif oligomerizes in living cells and that the PPLP motif is involved in this oligomerization. Nevertheless, since oligomerization was not completely abrogated by mutation of the PPLP motif, other domains of Vif may also participate.

To determine the number of protomers in Vif oligomers, fluorescence correlation spectroscopy (FCS) was performed. The stoichiometry of the oligomers can be obtained through the determination of the molecular brightness that increases proportionally with the number of fluorescent molecules inside a particle. In the case of free eGFP in the cytoplasm of transfected HeLa cells, a Gaussian distribution centred at 0.16 kHz per particle was observed, reflecting a monomeric protein population (Fig. 8.10a, solid black circles). Using Vpr-eGFP as a positive control, we observed a major population presenting a twofold-higher brightness than free eGFP (0.35 kHz per particle) and a minor one showing a threefold higher brightness (about 0.55 kHz per particle), confirming that Vpr was able to form mainly dimers and trimers in cells (Fig. 8.10a, red triangles). The distribution curve

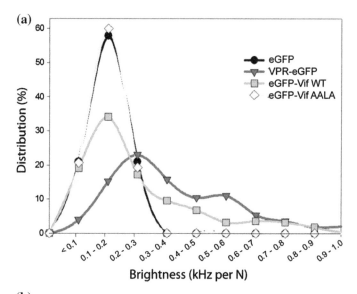

(b)

Group		eGFP	VPR-eGFP	eGFP-Vif WT	eGFP-Vif AALA
Brightness	Median	0.16	0.35	0.19	0.15
	25%	0.11	0.23	0.12	0.12
	75%	0.20	0.56	0.34	0.20

Fig. 8.10 FCS analysis of eGFP-Vif wild-type (*WT*) and AALA mutant. **a** Brightness distribution of free eGFP (*black circles*), Vpr-eGFP (*red triangles*), eGFP-Vif WT (*green squares*), and eGFP-Vif AALA (*yellow diamonds*). FCS measurements were performed 24 h post transfection in HeLa cells. Distributions were calculated from 100 autocorrelation curves. **b** Table with the brightness median for each condition, together with the values at the two quartiles. Reprinted from [9]

of eGFP-Vif was somewhat different, exhibiting a main broad population centred at 0.20 kHz per particle (as free eGFP) (Fig. 8.10a, green squares) and two minor populations centred at ∼0.4 kHz per particle and ∼0.8 kHz per particle. This distribution suggests that Vif exists as a mix of monomers and oligomers in cells, likely explaining the large distribution of eGFP-Vif lifetimes observed by FLIM (Fig. 8.9c). Finally, eGFP-Vif AALA exhibited a single population similar to the one obtained for free eGFP (Fig. 8.10a, yellow diamonds), suggesting that this mutant remains largely monomeric. Thus, the FCS results confirm that Vif oligomerizes in cells and confirm the involvement of the PPLP motif in the oligomerization process.

8.3.2 Interaction with Pr55^{Gag}

Next, we tested whether Vif oligomerization was required for its interaction with Pr55^{Gag}. FLIM data obtained on HeLa cells transfected with Vif fusion proteins and Pr55^{Gag} are shown in Fig. 8.11.

Figure 8.11a, left panels, shows that Pr55^{Gag} partially re-localizes Vif to the cell membrane, irrespective to the presence of the PPLP motif. Thus, it may be concluded that oligomerization of Vif is not critical for Pr55^{Gag} recognition. Moreover, the FRET efficiency (10.7 %) measured for eGFP-Vif in the presence of Pr55^{Gag} (Fig. 8.11b) was similar to the one observed in the absence of Pr55^{Gag} (12.2 %) (Fig. 8.9d), suggesting that Vif oligomerization is not affected by binding to Pr55^{Gag}. As in the absence of Pr55^{Gag}, the distribution of the fluorescence lifetime of eGFP-Vif was highly dispersed, confirming that Vif oligomers are heterogeneous in cells. By performing the same experiments using eGFP-Vif AALA and mCherry-Vif AALA mutants (Fig. 8.11a, b), we observed that in the presence of Pr55^{Gag}, the FRET efficiency of Vif AALA at the plasma membrane was decreased compared to that of wild-type Vif (Fig. 8.11d). Thus, our data show that Pr55^{Gag} proteins interact with Vif proteins and redistribute them at the plasma membrane. This interaction likely mediates Vif incorporation into nascent virions.

8.3.3 Binding to A3G

In a last step, our objective was to investigate whether the binding of A3G protein to Vif affects its oligomerization (Fig. 8.12a, b). Interestingly, the FRET efficiency between eGFP-Vif and mCherry-Vif dropped from 12.2 to 5.7 % (Fig. 8.12b) in the presence of A3G. In contrast, the FRET efficiency (10.4 %) in the presence of an A3G D128K mutant defective in Vif interaction [18, 89, 104] is similar to that observed in the absence of A3G. Furthermore, expression of Pr55^{Gag} was found to not affect the decrease in FRET efficiency induced by A3G. Thus, our data clearly show that A3G affects Vif oligomerization, independently of the presence of Pr55^{Gag}.

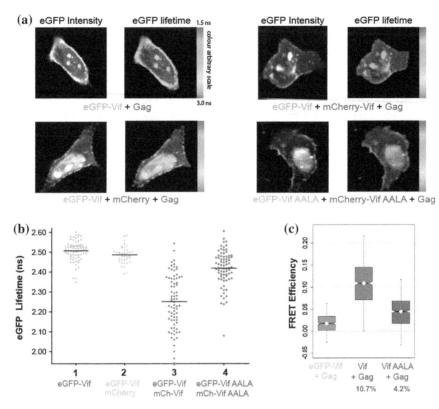

Fig. 8.11 FRET-FLIM investigation of Vif oligomerization in the presence of Pr55^{Gag} protein. **a** Effect of Pr55^{Gag} expression on eGFP fluorescence intensity (*greyscale*) and eGFP lifetime (*arbitrary colour scale*). **b** Lifetime distribution of all analyzed cells. The *horizontal black bars* indicate mean values. Controls in the absence of acceptor are in *green*, the Vif WT in *blue*, and the Vif AALA mutant in *red*. **c** *Box plots* of FRET efficiency distributions in the presence of Pr55^{Gag}. Mean FRET efficiency is represented by the *white diamonds*. Whiskers correspond to the interval containing 95 % of the FRET efficiencies. Reprinted from [9]

8.4 HIV-1 Assembly: Pr55^{Gag} Oligomerization

HIV-1 assembly is largely governed by Pr55^{Gag}. This polyprotein is formed of four structural domains that are from the N- to the C-terminus, the matrix (MA), capsid (CA), nucleocapsid (NC) and p6, which upon processing by the viral protease (PR) during virion maturation gives rise to MAp17, CAp24, NCp7 and p6 structural proteins present in infectious virions [71]. Pr55^{Gag} orchestrates HIV-1 virion formation [1, 7, 15] likely by initially undergoing oligomerization upon its binding to the genomic RNA acting as a scaffold [39, 61, 77]. The process of Pr55^{Gag} oligomerization is driven by series of homotypic interactions, involving the CA and SP1 regions and additionally, the MA region [28, 44, 45, 51, 73, 83, 84, 101]. Once Pr55^{Gag} oligomers are formed, the N-terminus of MA with its myristate becomes

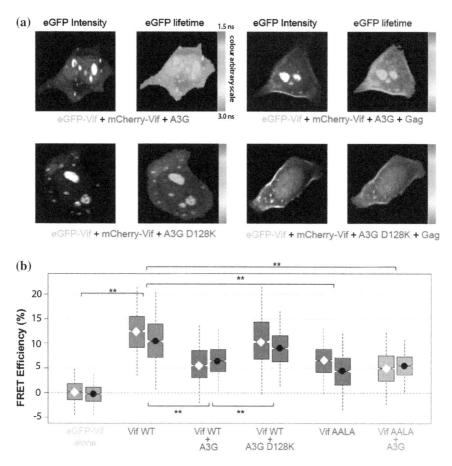

Fig. 8.12 FRET-FLIM analyses of the effect of A3G protein on Vif oligomerization. **a** Effect of A3G expression on eGFP fluorescence intensity (*greyscale*) and eGFP lifetime (*arbitrary colour scale*). **b** *Box plots* of FRET efficiency distributions in the absence (*white diamonds*) or presence (*black dots*) of Pr55^{Gag}. Whiskers represent the interval containing 95 % of the FRET efficiencies. *P* values for the different assays are indicated. ** <0.01. Reprinted from [9]

accessible, which allows the anchoring of the Pr55^{Gag} oligomers to the plasma membrane (see recent reviews: [1, 77]). To investigate Pr55^{Gag} oligomerization, pairs of Pr55^{Gag} proteins labelled at their C-terminus by eGFP and mCherry were first used [82]. Though these proteins were useful to monitor Pr55^{Gag} assembly by FRET [53, 63, 64], they lead to virus particles with aberrant morphology [50]. This drawback could be avoided by inserting the fluorescent proteins between the MA and CA domains [54, 55, 75] and expressing the chimeric proteins in the presence of an excess of non labelled Pr55^{Gag}. This protocol resulted in production of particles with wild-type morphology and infectivity [75].

In a first series of experiments, cells were transfected with Pr55^{Gag}/Pr55$^{Gag-eGFP}$ to determine the intrinsic fluorescence lifetime of eGFP fused to Pr55^{Gag} (Fig. 8.13a). A single fluorescence lifetime of 2.4 ns was obtained (Fig. 8.13b), indicating that the fusion to Pr55^{Gag} did not modify the photophysical properties of eGFP.

Next, cells expressing Pr55^{Gag}/Pr55$^{Gag-eGFP}$/Pr55$^{Gag-mCherry}$ in a 7/1/2 ratio were imaged by FLIM at 12 and 24 h post transfection. In a first attempt, a mono-exponential model was used to analyse the fluorescence decays in each pixel. At 12 h post transfection, Pr55^{Gag} polyproteins were found in the cytoplasm as well as at the plasma membrane (Fig. 8.13c). Moreover, the lifetime distribution showed clearly a bimodal distribution with peaks centred at 1.9 and 2.4 ns (Fig. 8.13d). While the 1.9 ns clearly evidenced the existence of Pr55^{Gag} oligomers, the 2.4 ns peak suggested that a fraction of Pr55^{Gag} proteins remained monomeric or was engaged in small oligomers where Pr55$^{Gag-mCherry}$ was absent. In contrast, at 24 h

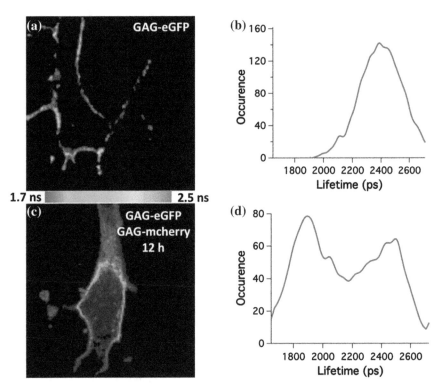

Fig. 8.13 One component analysis of FLIM images of cells expressing Pr55^{Gag}/Pr55$^{Gag-eGFP}$ (**a**, **b**) and Pr55^{Gag}/Pr55$^{Gag-eGFP}$/Pr55$^{Gag-mCherry}$ (**c**, **d**). The fluorescence decays were measured in each pixel and analysed with a mono-exponential function. The eGFP lifetime was then converted into a colour code ranging from *blue* (1.7 ns) to *red* (2.5 ns). The fluorescence lifetimes display an uni- (2.4 ns) and a bi-modal (1.9 and 2.4 ns) distribution for Pr55^{Gag}/Pr55$^{Gag-eGFP}$ and Pr55^{Gag}/Pr55$^{Gag-eGFP}$/Pr55$^{Gag-mCherry}$, respectively. All images were acquired using 50 × 50 μm scale and 128 × 128 pixels

post transfection, $Pr55^{Gag}$ molecules were found to accumulate at the plasma membrane with a lifetime distribution displaying a single peak centred at 1.9 ns (data not shown).

As the FLIM images at 12 h post transfection indicated a clear heterogeneity in the lifetime distribution, we then analysed the fluorescence decays with a two component model, where the long-lived lifetime τ_2 was fixed at the lifetime of free eGFP (2.4 ns). Through this analysis, it was possible to determine the distribution of the lifetime of the fast decay component, τ_1, and its respective population, α_1, giving thus access to the population of $Pr55^{Gag}$ oligomers undergoing FRET. As depicted in Fig. 8.14, at 12 h post transfection the distribution of τ_1 value was centred at 1.45 ns (corresponding to a FRET efficiency of 41 %) and homogeneously distributed within the cell as shown by the regular green staining [34]. These data indicated that

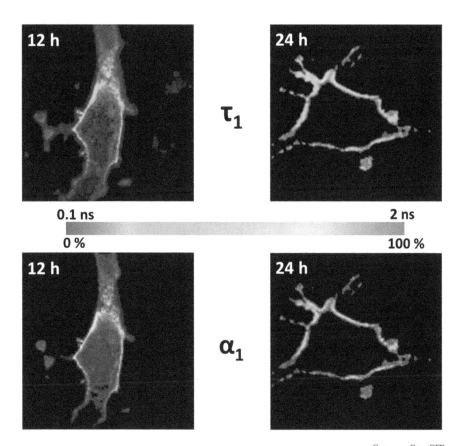

Fig. 8.14 Two component analysis of FLIM images of cells expressing $Pr55^{Gag}/Pr55^{Gag-eGFP}/Pr55^{Gag-mCherry}$ in a 7/1/2 ratio at 12 h or 24 h post transfection. The fluorescence decays in each pixel were analysed using a bi-exponential model, with a long-lived component τ_2 fixed to the lifetime of free eGFP (2.4 ns). The τ_1 and α_1 values were converted into a colour code ranging from *blue* (0.1 ns, 0 %) to *red* (2 ns, 100 %). All images were recorded using 50 × 50 μm scale and 128 × 128 pixels

Pr55^{Gag} oligomers already form in the cytoplasm. Moreover, as the FRET efficiency is very high, it is likely that the Pr55^{Gag} protomers are closely packed together in the oligomers, so that eGFP and mCherry can come in close contact. Interestingly, the amplitude of the short-lived component increased from 20 to 70 % from the cytoplasm to the plasma membrane, indicating that the concentration of Pr55^{Gag} oligomers progressively increased from the cytoplasm to the plasma membrane. At 24 h, the τ_1 component of Pr55$^{Gag\text{-}eGFP}$ was distributed around 1.2 ns and was thus, slightly below to that measured at 12 h post transfection, suggesting a further compaction of the Pr55^{Gag} protomers within the oligomers occurs with time. Moreover, the high amplitude (up to 75 %) of the short component showed that almost all Pr55$^{Gag\text{-}eGFP}$ proteins undergo FRET at the plasma membrane, in line with the well reported polymerization of Pr55^{Gag} polyproteins at the plasma membrane required to form new viral particles [30, 42, 53, 55].

To better evidence the gradient of Pr55^{Gag} oligomer concentration within the cytoplasm at 12 h post transfection, we increased the image sampling with a decreased field of view (256 × 256 pixels and 20 × 20 µm). In line with the data of Fig. 8.14, the lifetime of the fast decay component was homogeneously distributed within the cell, with an average value of 1.2 ns (Fig. 8.15). Due to the increased sampling, the increase in the concentration of Pr55^{Gag} oligomers from the cytoplasm (20 %) to the plasma membrane (75 %) was more clearly visible. In addition, bright internal dots characterized by a τ_1 lifetime of 1.1 ns with high amplitude could be clearly observed in the cytoplasm, suggesting that local assembly of a large number of Pr55^{Gag} proteins could occur in the cytoplasm.

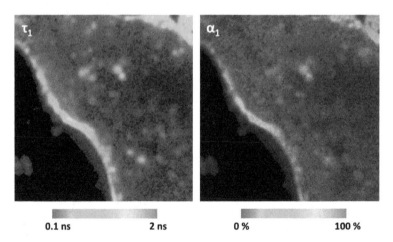

Fig. 8.15 Visualization of the gradient of Pr55^{Gag} oligomer concentration within the cytoplasm at 12 h post transfection. The FLIM image of a cell expressing Pr55^{Gag}/Pr55$^{Gag\text{-}eGFP}$/Pr55$^{Gag\text{-}mCherry}$ in a 7/1/2 ratio was recorded with increased sampling and decreased field of view (256 × 256 pixels and 20 × 20 µm). Analysis of data and colour code are as in Fig. 8.14. Taken together, our data indicate that formation of Pr55^{Gag} oligomers is initiated in the cytoplasm. The concentration of oligomers progressively increases from the cytoplasm to the plasma membrane, but without major changes in the organization and compaction of the Pr55^{Gag} protomers within the oligomers

8.5 Conclusion

Data on Vpr, Vif and Pr55^{Gag} proteins of HIV-1 in fusion with fluorescent proteins show that FRET-FLIM is a very powerful and sensitive tool to monitor their oligomerization in live cells. With both Vpr and Pr55^{Gag} proteins, very high levels of FRET were obtained, as a consequence of the rather small sizes of the viral proteins, the strong packing of the protomers and the presence of multiple acceptors for one donor. In the case of Vif, lower levels of FRET were observed, likely as a result of the less efficient oligomerization of this protein. For Vpr, the position of the fluorescent protein was found to be critical, with much higher levels of FRET being obtained when the fluorescent proteins are fused at the C-terminus of Vpr. In the case of Vpr and Pr55^{Gag} proteins, a two-component analysis could be used to analyze the time-resolved decays, giving the possibility to discriminate monomers from oligomers and monitor the spatiotemporal evolution of both populations in the cells. Though FLIM FRET can unambiguously reveal the oligomerization of a given protein, it can hardly be used to disclose the oligomer stoichiometry. The information on the number of protomers per oligomer can ideally be obtained by fluorescence correlation spectroscopy, by analyzing the brightness of the diffusing species. Moreover, FRET-FLIM was found to be also highly useful to identify the structural determinants of protein oligomerization, by using point mutants of the fluorescently labelled HIV-1 proteins. In addition, FRET-FLIM can be easily used to investigate the modulation of protein oligomerization by other HIV-1 proteins and host proteins. Of course, FRET-FLIM approaches are not limited to oligomerization of HIV-1 proteins and were already used to monitor the oligomerization of a number of other viral proteins [2, 19, 35, 76, 94] or mammalian proteins [14, 21, 36, 43, 85, 98, 99].

Acknowledgments We thank Salah Edin El Meshri for his technical help. This work was supported by the European Project THINPAD "Targeting the HIV-1 Nucleocapsid Protein to fight Antiretroviral Drug Resistance" (FP7—grant agreement 601969), Agence National Recherche sur le SIDA (ANRS) (2012-14. CSS2), SIDACTION (AI22-1-01963), Centre National Recherche Scientifique (CNRS), Université de Strasbourg and Institut National sur la Santé Et la Recherche Médicale (INSERM).

References

1. C.S. Adamson, E.O. Freed, Human immunodeficiency virus type 1 assembly, release, and maturation, *Advances in Pharmacology*, vol 55 (Elsevier, San Diego, Calif, 2007), pp. 347–387
2. K. Amari, E. Boutant, C. Hofmann, C. Schmitt-Keichinger, L. Fernandez-Calvino, P. Didier, A. Lerich, J. Mutterer, C.L. Thomas, M. Heinlein, Y. Mely, A.J. Maule, C. Ritzenthaler, A family of plasmodesmal proteins with receptor-like properties for plant viral movement proteins. PLoS Pathog. **6**, e1001119 (2010)
3. E. Asante-Appiah, A.M. Skalka, HIV-1 integrase: structural organization, conformational changes, and catalysis. Adv. Virus Res. **52**, 351–369 (1999)

4. J.R. Auclair, K.M. Green, S. Shandilya, J.E. Evans, M. Somasundaran, C.A. Schiffer, Mass spectrometry analysis of HIV-1 Vif reveals an increase in ordered structure upon oligomerization in regions necessary for viral infectivity. Proteins **69**, 270–284 (2007)
5. J. Azoulay, J.P. Clamme, J.L. Darlix, B.P. Roques, Y. Mely, Destabilization of the HIV-1 complementary sequence of TAR by the nucleocapsid protein through activation of conformational fluctuations. J. Mol. Biol. **326**, 691–700 (2003)
6. F. Bachand, X.J. Yao, M. Hrimech, N. Rougeau, E.A. Cohen, Incorporation of Vpr into human immunodeficiency virus type 1 requires a direct interaction with the p6 domain of the p55 gag precursor. J. Biol. Chem. **274**, 9083–9091 (1999)
7. M. Balasubramaniam, E.O. Freed, New insights into HIV assembly and trafficking. Physiology (Bethesda) **26**, 236–251 (2011)
8. D.S. Banks, C. Fradin, Anomalous diffusion of proteins due to molecular crowding. Biophys. J. **89**, 2960–2971 (2005)
9. J. Batisse, S.X. Guerrero, S. Bernacchi, L. Richert, J. Godet, V. Goldschmidt, Y. Mely, R. Marquet, H. de Rocquigny, J.C. Paillart, APOBEC3G impairs the multimerization of the HIV-1 Vif protein in living cells. J. Virol. **87**, 6492–6506 (2013)
10. W. Becker, A. Bergmann, M.A. Hink, K. König, K. Benndorf, C. Biskup, Fluorescence lifetime imaging by time-correlated single photon counting. Micr. Res. Techn. **63**, 58–66 (2004)
11. W. Becker, K. Benndorf, A. Bergmann, C. Biskup, K. König, U. Tirlapur, T. Zimmer, FRET measurements by TCSPC laser scanning microscopy. Proc. SPIE **4431**, 94–98 (2001)
12. W. Becker, *Advanced time-correlated single-photon counting techniques* (Springer, Heidelberg, New York, 2005)
13. W. Becker, The bh TCSPC Handbook, 5th edn. (2012). Printed copies available from Becker&Hickl GmbH, Electronic version available from www.becker-hickl.com
14. W. Becker, Fluorescence lifetime imaging—techniques and applications. J. Microsc. **247**, 119–136 (2012)
15. N.M. Bell, A.M. Lever, HIV Gag polyprotein: processing and early viral particle assembly. Trends Microbiol. **21**, 136–144 (2013)
16. C. Biskup, L. Kelbauskas, T. Zimmer, K. Benndorf, A. Bergmann, W. Becker, J. P. Ruppersberg, C. Stockklausner, N. Klöcker, Interaction of PSD-95 with potassium channels visualized by fluorescence lifetime-based resonance energy transfer imaging. J. Biomed. Opt. **9**, 735–759 (2004)
17. D.L. Bolton, M.J. Lenardo, Vpr cytopathicity independent of G2/M cell cycle arrest in human immunodeficiency virus type 1-infected CD4+T cells. J. Virol. **81**, 8878–8890 (2007)
18. H.P. Bogerd, B.P. Doehle, H.L. Wiegand, B.R. Cullen, A single amino acid difference in the host APOBEC3G protein controls the primate species specificity of HIV type 1 virion infectivity factor. Proc. Natl. Acad. Sci. U.S.A. **101**, 3770–3774 (2004)
19. K. Brandner, A. Sambade, E. Boutant, P. Didier, Y. Mely, C. Ritzenthaler, M. Heinlein, Tobacco mosaic virus movement protein interacts with green fluorescent protein-tagged microtubule end-binding protein 1. Plant Physiol. **147**, 611–623 (2008)
20. K. Bruns, T. Fossen, V. Wray, P. Henklein, U. Tessmer, U. Schubert, Structural characterization of the HIV-1 Vpr N terminus: evidence of cis/trans-proline isomerism. J. Biol. Chem. **278**, 43188–43201 (2003)
21. C.A. Bucherl, A. Bader, A.H. Westphal, S.P. Laptenok, J.W. Borst, FRET-FLIM applications in plant systems. Protoplasma **251**, 383–394 (2014)
22. N.S. Caron, L.N. Munsie, J.W. Keillor, R. Truant, Using FLIM-FRET to measure conformational changes of transglutaminase type 2 in live cells, PloS one **7**, e44159 (2012)
23. Chen, A. Periasamy, Characterization of two-photon excitation fluorescence lifetime imaging micros-copy for protein localization, Microsc. Res. Tech. **63**, 72–80 (2004)
24. Y. Chen, J.D. Muller, Q. Ruan, E. Gratton, Molecular brightness characterization of EGFP in vivo by fluorescence fluctuation spectroscopy, Biophys. J. **82**, 133–144 (2002)

25. J.P. Clamme, J. Azoulay, Y. Mely, Monitoring of the formation and dissociation of polyethylenimine/DNA complexes by two photon fluorescence correlation spectroscopy. Biophys. J. **84**, 1960–1968 (2003)

26. B. Corry, D. Jayatilaka, P. Rigby, A flexible approach to the calculation of resonance energy transfer efficiency between multiple donors and acceptors in complex geometries. Biophys. J. **89**, 3822–3836 (2005)

27. S. Dahmane, E. Rubinstein, P.E. Milhiet, Viruses and tetraspanins: lessons from single molecule approaches. Viruses **6**, 1992–2011 (2014)

28. S.A. Datta, L.G. Temeselew, R.M. Crist, F. Soheilian, A. Kamata, J. Mirro, D. Harvin, K. Nagashima, R.E. Cachau, A. Rein, On the role of the SP1 domain in HIV-1 particle assembly: a molecular switch? J. Virol. **85**, 4111–4121 (2011)

29. C. Depienne, P. Roques, C. Creminon, L. Fritsch, R. Casseron, D. Dormont, C. Dargemont, S. Benichou, Cellular distribution and karyophilic properties of matrix, integrase, and Vpr proteins from the human and simian immunodeficiency viruses. Exp. Cell Res. **260**, 387–395 (2000)

30. A. Derdowski, L. Ding, P. Spearman, A novel fluorescence resonance energy transfer assay demonstrates that the human immunodeficiency virus type 1 Pr55Gag I domain mediates Gag-Gag interactions. J. Virol. **78**, 1230–1242 (2004)

31. B.A. Desimmie, K.A. Delviks-Frankenberrry, R.C. Burdick, D. Qi, T. Izumi, V.K. Pathak, Multiple APOBEC3 restriction factors for HIV-1 and one Vif to rule them all. J. Mol. Biol. **426**, 1220–1245 (2014)

32. M.A. Digman, E. Gratton, Fluorescence correlation spectroscopy and fluorescence cross-correlation spectroscopy. Wiley Interdisc. Rev. Syst. Biol. Med. **1**, 273–282 (2009)

33. B.M. Dunn, M.M. Goodenow, A. Gustchina, A. Wlodawer, Retroviral proteases, Genome Biol. **3**, REVIEWS3006 (2002)

34. S.E. El Meshri, D. Dujardin, J. Godet, L. Richert, C. Boudier, J.L. Darlix, P. Didier, Y. Mely, H. de Rocquigny, Role of the nucleocapsid domain in HIV-1 Gag oligomerization and trafficking to the plasma membrane: A fluorescence lifetime imaging microscopy investigation. J. Mol. Biol. (2015, in press)

35. S. Engel, S. Scolari, B. Thaa, N. Krebs, T. Korte, A. Herrmann, M. Veit, FLIM-FRET and FRAP reveal association of influenza virus haemagglutinin with membrane rafts. Biochem. J. **425**, 567–573 (2010)

36. A.W. Fjorback, P. Pla, H.K. Muller, O. Wiborg, F. Saudou, J.R. Nyengaard, Serotonin transporter oligomerization documented in RN46A cells and neurons by sensitized acceptor emission FRET and fluorescence lifetime imaging microscopy. Biochem. Biophys. Res. Commun. **380**, 724–728 (2009)

37. Th. Förster, Zwischenmolekulare Energiewanderung und Fluoreszenz, Ann. Phys. (Serie 6) **2**, 55–75 (1948)

38. Th. Förster, Energy migration and fluorescence. Translated by Klaus Suhling. J. Biomed. Opt. **17**, 011002-1–10 (2012)

39. K.H. Fogarty, Y. Chen, I.F. Grigsby, P.J. Macdonald, E.M. Smith, J.L. Johnson, J.M. Rawson, L.M. Mansky, J.D. Mueller, Characterization of cytoplasmic Gag-gag interactions by dual-color z-scan fluorescence fluctuation spectroscopy. Biophys. J. **100**, 1587–1595 (2011)

40. J.V. Fritz, P. Didier, J.P. Clamme, E. Schaub, D. Muriaux, C. Cabanne, N. Morellet, S. Bouaziz, J.L. Darlix, Y. Mely, H. de Rocquigny, Direct Vpr-Vpr interaction in cells monitored by two photon fluorescence correlation spectroscopy and fluorescence lifetime imaging. Retrovirology **5**, 87 (2008)

41. J.V. Fritz, L. Briant, Y. Mely, S. Bouaziz, H. de Rocquigny, HIV-1 Viral Protein R: from structure to function. Future Virol. **5**, 607–625 (2010)

42. J.V. Fritz, D. Dujardin, J. Godet, P. Didier, J. De Mey, J.L. Darlix, Y. Mely, H. de Rocquigny, HIV-1 Vpr oligomerization but not that of Gag directs the interaction between Vpr and Gag. J. Virol. **84**, 1585–1596 (2010)

43. T.W. Gadella Jr, T.M. Jovin, Oligomerization of epidermal growth factor receptors on A431 cells studied by time-resolved fluorescence imaging microscopy. A stereochemical model for tyrosine kinase receptor activation. J. Cell Biol. **129**, 1543–1558 (1995)

44. T.R. Gamble, S. Yoo, F.F. Vajdos, U.K. von Schwedler, D.K. Worthylake, H. Wang, J. P. McCutcheon, W.I. Sundquist, C.P. Hill, Structure of the carboxyl-terminal dimerization domain of the HIV-1 capsid protein. Science **278**, 849–853 (1997)

45. B.K. Ganser-Pornillos, M. Yeager, W.I. Sundquist, The structural biology of HIV assembly. Curr. Opin. Struct. Biol. **18**, 203–217 (2008)

46. J. He, S. Choe, R. Walker, P. Di Marzio, D.O. Morgan, N.R. Landau, Human immunodeficiency virus type 1 viral protein R (Vpr) arrests cells in the G2 phase of the cell cycle by inhibiting p34cdc2 activity. J. Virol. **69**, 6705–6711 (1995)

47. N.K. Heinzinger, M.I. Bukinsky, S.A. Haggerty, A.M. Ragland, V. Kewalramani, M.A. Lee, H.E. Gendelman, L. Ratner, M. Stevenson, M. Emerman, The Vpr protein of human immunodeficiency virus type 1 influences nuclear localization of viral nucleic acids in nondividing host cells. Proc. Natl. Acad. Sci. U.S.A. **91**, 7311–7315 (1994)

48. P. Henklein, K. Bruns, M.P. Sherman, U. Tessmer, K. Licha, J. Kopp, C.M. de Noronha, W. C. Greene, V. Wray, U. Schubert, Functional and structural characterization of synthetic HIV-1 Vpr that transduces cells, localizes to the nucleus, and induces G2 cell cycle arrest. J. Biol. Chem. **275**, 32016–32026 (2000)

49. B. Herman, G. Gordon, N. Mahajan, V. Centonze, in *Methods of Cellular Imaging*, ed. by A. Periasamy. Measurement of fluorescence resonance energy transfer in the optical microscpe (Oxford University Press, New York, 2001)

50. L. Hermida-Matsumoto, M.D. Resh, Localization of human immunodeficiency virus type 1 Gag and Env at the plasma membrane by confocal imaging. J. Virol. **74**, 8670–8679 (2000)

51. C.P. Hill, D. Worthylake, D.P. Bancroft, A.M. Christensen, W.I. Sundquist, Crystal structures of the trimeric human immunodeficiency virus type 1 matrix protein: implications for membrane association and assembly. Proc. Natl. Acad. Sci. U.S.A. **93**, 3099–3104 (1996)

52. B. Hoffmann, T. Zimmer, N. Klöcker, L. Kelbauskas, K. König, K. Benndorf, C. Biskup, Prolonged irradiation of enhanced cyan fluorescent protein or Cerulean can invalidate Förster resonance energy transfer measurements. J. Biomed. Opt. 13(3), 031250-1 to -9 (2008)

53. I.B. Hogue, A. Hoppe, A. Ono, Quantitative fluorescence resonance energy transfer microscopy analysis of the human immunodeficiency virus type 1 Gag-Gag interaction: relative contributions of the CA and NC domains and membrane binding. J. Virol. **83**, 7322–7336 (2009)

54. W. Hubner, P. Chen, A. Del Portillo, Y. Liu, R.E. Gordon, B.K. Chen, Sequence of human immunodeficiency virus type 1 (HIV-1) Gag localization and oligomerization monitored with live confocal imaging of a replication-competent, fluorescently tagged HIV-1. J. Virol. **81**, 12596–12607 (2007)

55. W. Hubner, G.P. McNerney, P. Chen, B.M. Dale, R.E. Gordon, F.Y. Chuang, X.D. Li, D.M. Asmuth, T. Huser, B.K. Chen, Quantitative 3D video microscopy of HIV transfer across T cell virological synapses. Science **323**, 1743–1747 (2009)

56. A. Jacobo-Molina, J. Ding, R.G. Nanni, A.D. Clark, Jr., X. Lu, C. Tantillo, R.L. Williams, G. Kamer, A.L. Ferris, P. Clark et al., Crystal structure of human immunodeficiency virus type 1 reverse transcriptase complexed with double-stranded DNA at 3.0 A resolution shows bent DNA, Proc. Natl. Acad. Sci. USA **90**, 6320–6324 (1993)

57. E. Jacotot, K.F. Ferri, C. El Hamel, C. Brenner, S. Druillennec, J. Hoebeke, P. Rustin, D. Metivier, C. Lenoir, M. Geuskens, H.L. Vieira, M. Loeffler, A.S. Belzacq, J.P. Briand, N. Zamzami, L. Edelman, Z.H. Xie, J.C. Reed, B.P. Roques, G. Kroemer, Control of mitochondrial membrane permeabilization by adenine nucleotide translocator interacting with HIV-1 viral protein rR and Bcl-2. J. Exp. Med. **193**, 509–519 (2001)

58. J.B. Jowett, V. Planelles, B. Poon, N.P. Shah, M.L. Chen, I.S. Chen, The human immunodeficiency virus type 1 vpr gene arrests infected T cells in the G2+M phase of the cell cycle. J. Virol. **69**, 6304–6313 (1995)

59. L.A. Kohlstaedt, J. Wang, J.M. Friedman, P.A. Rice, T.A. Steitz, Crystal structure at 3.5 A resolution of HIV-1 reverse transcriptase complexed with an inhibitor. Science **256**, 1783–1790 (1992)
60. E. Kondo, F. Mammano, E.A. Cohen, H.G. Gottlinger, The p6gag domain of human immunodeficiency virus type 1 is sufficient for the incorporation of Vpr into heterologous viral particles. J. Virol. **69**, 2759–2764 (1995)
61. S.B. Kutluay, P.D. Bieniasz, Analysis of the initiating events in HIV-1 particle assembly and genome packaging. PLoS Pathog. **6**, e1001200 (2010)
62. A.J. Lam, F. St-Pierre, Y. Gong, J.D. Marshall, P.J. Cranfill, M.A. Baird, M.R. McKeown, J. Wiedenmann, M.W. Davidson, M.J. Schnitzer, R.Y. Tsien, M.Z. Lin, Improving FRET dynamic range with bright green and red fluorescent proteins. Nat. Methods **9**, 1005–1012 (2012)
63. D.R. Larson, M.C. Johnson, W.W. Webb, V.M. Vogt, Visualization of retrovirus budding with correlated light and electron microscopy. Proc. Natl. Acad. Sci. U.S.A. **102**, 15453–15458 (2005)
64. D.R. Larson, Y.M. Ma, V.M. Vogt, W.W. Webb, Direct measurement of Gag-Gag interaction during retrovirus assembly with FRET and fluorescence correlation spectroscopy. J. Cell Biol. **162**, 1233–1244 (2003)
65. C. Lavallee, X.J. Yao, A. Ladha, H. Gottlinger, W.A. Haseltine, E.A. Cohen, Requirement of the Pr55gag precursor for incorporation of the Vpr product into human immunodeficiency virus type 1 viral particles, J. Virol. **68**, 1926–1934 (1994)
66. J. Liu, A. Bartesaghi, M.J. Borgnia, G. Sapiro, S. Subramaniam, Molecular architecture of native HIV-1 gp120 trimers. Nature **455**, 109–113 (2008)
67. D. McDonald, M.A. Vodicka, G. Lucero, T.M. Svitkina, G.G. Borisy, M. Emerman, T. J. Hope, Visualization of the intracellular behavior of HIV in living cells. J. Cell Biol. **159**, 441–452 (2002)
68. A. Merk, S. Subramaniam, HIV-1 envelope glycoprotein structure. Curr. Opin. Struct. Biol. **23**, 268–276 (2013)
69. E.M. Merzlyak, J. Goedhart, D. Shcherbo, M.E. Bulina, A.S. Shcheglov, A.F. Fradkov, A. Gaintzeva, K.A. Lukyanov, S. Lukyanov, T.W. Gadella, D.M. Chudakov, Bright monomeric red fluorescent protein with an extended fluorescence lifetime. Nat. Methods **2007**(4), 555–557 (2007)
70. F. Michel, C. Crucifix, F. Granger, S. Eiler, J.F. Mouscadet, S. Korolev, J. Agapkina, R. Ziganshin, M. Gottikh, A. Nazabal, S. Emiliani, R. Benarous, D. Moras, P. Schultz, M. Ruff, Structural basis for HIV-1 DNA integration in the human genome, role of the LEDGF/P75 cofactor, EMBO J. **28**, 980–991 (2009)
71. G. Mirambeau, S. Lyonnais, R.J. Gorelick, Features, processing states, and heterologous protein interactions in the modulation of the retroviral nucleocapsid protein function. RNA Biol. **7**, 724–734 (2010)
72. N. Morellet, S. Bouaziz, P. Petitjean, B.P. Roques, NMR structure of the HIV-1 regulatory protein VPR. J. Mol. Biol. **327**, 215–227 (2003)
73. N. Morellet, S. Druillennec, C. Lenoir, S. Bouaziz, B.P. Roques, Helical structure determined by NMR of the HIV-1 (345-392) Gag sequence, surrounding p2: implications for particle assembly and RNA packaging. Protein Sci. **14**, 375–386 (2005)
74. N. Morellet, B.P. Roques, S. Bouaziz, Structure-function relationship of Vpr: biological implications. Curr. HIV Res. **7**, 184–210 (2009)
75. B. Muller, J. Daecke, O.T. Fackler, M.T. Dittmar, H. Zentgraf, H.G. Krausslich, Construction and characterization of a fluorescently labeled infectious human immunodeficiency virus type 1 derivative. J. Virol. **78**, 10803–10813 (2004)
76. A. Niehl, K. Amari, D. Gereige, K. Brandner, Y. Mely, M. Heinlein, Control of tobacco mosaic virus movement protein fate by CELL-DIVISION-CYCLE protein48. Plant Physiol. **160**, 2093–2108 (2012)
77. I.P. O'Carroll, F. Soheilian, A. Kamata, K. Nagashima, A. Rein, Elements in HIV-1 Gag contributing to virus particle assembly. Virus Res. **171**, 341–345 (2013)

78. S. Padilla-Parra, N. Auduge, H. Lalucque, J.C. Mevel, M. Coppey-Moisan, M. Tramier, Quantitative comparison of different fluorescent protein couples for fast FRET-FLIM acquisition. Biophys. J. **97**, 2368–2376 (2009)

79. R. Pepperkok, A. Squire, S. Geley, P.I. Bastiaens, Simultaneous detection of multiple green fluorescent proteins in live cells by fluorescence lifetime imaging microscopy. Curr. Biol. **9**, 269–272 (1999)

80. V. Planelles, S. Benichou, Vpr and its interactions with cellular proteins. Curr. Top. Microbiol. Immunol. **339**, 177–200 (2009)

81. A. Periasamy, *Methods in Cellular Imaging* (Oxford University Press, Oxford New York, 2001)

82. C. Perrin-Tricaud, J. Davoust, I.M. Jones, Tagging the human immunodeficiency virus gag protein with green fluorescent protein. Minimal evidence for colocalisation with actin. Virology **255**, 20–25 (1999)

83. O. Pornillos, B.K. Ganser-Pornillos, B.N. Kelly, Y. Hua, F.G. Whitby, C.D. Stout, W.I. Sundquist, C.P. Hill, M. Yeager, X-ray structures of the hexameric building block of the HIV capsid. Cell **137**, 1282–1292 (2009)

84. O. Pornillos, B.K. Ganser-Pornillos, M. Yeager, Atomic-level modelling of the HIV capsid. Nature **469**, 424–427 (2011)

85. A.S. Rinaldi, G. Freund, D. Desplancq, A.P. Sibler, M. Baltzinger, N. Rochel, Y. Mely, P. Didier, E. Weiss, The use of fluorescent intrabodies to detect endogenous gankyrin in living cancer cells. Exp. Cell Res. **319**, 838–849 (2013)

86. V. Raicu, D.R. Singh, FRET spectrometry: a new tool for the determination of protein quaternary structure in living cells. Biophys. J. **105**, 1937–1945 (2013)

87. J. Ries, P. Schwille, Fluorescence correlation spectroscopy, bioessays: news and reviews in molecular. Cell. Dev. Biol. **34**, 361–368 (2012)

88. H. de Rocquigny, S.E. El Meshri, L. Richert, P. Didier, J.L. Darlix, Y. Mely, Role of the nucleocapsid region in HIV-1 Gag assembly as investigated by quantitative fluorescence-based microscopy, Virus. Res. **193**, 78−88 (2014)

89. B. Schrofelbauer, D. Chen, N.R. Landau, A single amino acid of APOBEC3G controls its species-specific interaction with virion infectivity factor (Vif). Proc. Natl. Acad. Sci. U.S.A. **101**, 3927–3932 (2004)

90. W. Schuler, K. Wecker, H. de Rocquigny, Y. Baudat, J. Sire, B.P. Roques, NMR structure of the (52-96) C-terminal domain of the HIV-1 regulatory protein Vpr: molecular insights into its biological functions. J. Mol. Biol. **285**, 2105–2117 (1999)

91. W.I. Sundquist, H.G. Krausslich, HIV-1 assembly, budding, and maturation. Cold Spring Harb. Perspect. Med. **2**, a006924 (2012)

92. J.H. Simon, R.A. Fouchier, T.E. Southerling, C.B. Guerra, C.K. Grant, M.H. Malim, The Vif and Gag proteins of human immunodeficiency virus type 1 colocalize in infected human T cells. J. Virol. **71**, 5259–5267 (1997)

93. P.R. Tedbury, S.D. Ablan, E.O. Freed, Global rescue of defects in HIV-1 envelope glycoprotein incorporation: implications for matrix structure. PLoS Pathog. **9**, e1003739 (2013)

94. B. Thaa, A. Herrmann, M. Veit, Intrinsic cytoskeleton-dependent clustering of influenza virus M2 protein with hemagglutinin assessed by FLIM-FRET. J. Virol. **84**, 12445–12449 (2010)

95. M. Tramier, M. Zahid, J.C. Mevel, M.J. Masse, M. Coppey-Moisan, Sensitivity of CFP/YFP and GFP/mCherry pairs to donor photobleaching on FRET determination by fluorescence lifetime imaging microscopy in living cells. Microsc. Res. Tech. **69**, 933–939 (2006)

96. R.Y. Tsien, The green fluorescent protein. Annu. Rev. Biochem. **67**, 509–544 (1998)

97. B.G. Turner, M.F. Summers, Structural biology of HIV. J. Mol. Biol. **285**, 1–32 (1999)

98. Y. Ueda, S. Kwok, Y. Hayashi, Application of FRET probes in the analysis of neuronal plasticity. Frontiers in neural circuits **7**, 163 (2013)

99. E.B. van Munster, T.W. Gadella, Fluorescence lifetime imaging microscopy (FLIM). Adv. Biochem. Eng. Biotechnol. **95**, 143–175 (2005)

100. S.S. Vogel, B.W. van der Meer, P.S. Blank, Estimating the distance separating fluorescent protein FRET pairs. Methods **66**, 131–138 (2014)
101. U.K. von Schwedler, K.M. Stray, J.E. Garrus, W.I. Sundquist, Functional surfaces of the human immunodeficiency virus type 1 capsid protein. J. Virol. **77**, 5439–5450 (2003)
102. T.C. Voss, I.A. Demarco, R.N. Day, Quantitative imaging of protein interactions in the cell nucleus. Biotechniques **38**, 413–424 (2005)
103. K. Wecker, N. Morellet, S. Bouaziz, B.P. Roques, NMR structure of the HIV-1 regulatory protein Vpr in H2O/trifluoroethanol. Comparison with the Vpr N-terminal (1–51) and C-terminal (52–96) domains. Eur. J. Biochem. **269**, 3779–3788 (2002)
104. H. Xu, E.S. Svarovskaia, R. Barr, Y. Zhang, M.A. Khan, K. Strebel, V.K. Pathak, A single amino acid substitution in human APOBEC3G antiretroviral enzyme confers resistance to HIV-1 virion infectivity factor-induced depletion. Proc. Natl. Acad. Sci. U.S.A. **101**, 5652–5657 (2004)
105. X.J. Yao, A.J. Mouland, R.A. Subbramanian, J. Forget, N. Rougeau, D. Bergeron, E.A. Cohen, Vpr stimulates viral expression and induces cell killing in human immunodeficiency virus type 1-infected dividing Jurkat T cells. J. Virol. **72**, 4686–4693 (1998)
106. S. Yang, Y. Sun, H. Zhang, The multimerization of human immunodeficiency virus type I Vif protein: a requirement for Vif function in the viral life cycle. J. Biol. Chem. **276**, 4889–4893 (2001)
107. B. Yang, L. Gao, L. Li, Z. Lu, X. Fan, C.A. Patel, R.J. Pomerantz, G.C. DuBois, H. Zhang, Potent suppression of viral infectivity by the peptides that inhibit multimerization of human immunodeficiency virus type 1 (HIV-1) Vif proteins. J. Biol. Chem. **278**, 6596–6602 (2003)
108. G. Zanetti, J.A. Briggs, K. Grunewald, Q.J. Sattentau, S.D. Fuller, Cryo-electron tomographic structure of an immunodeficiency virus envelope complex in situ. PLoS Pathog. **2**, e83 (2006)

Chapter 9
Unraveling the Rotary Motors in F_oF_1-ATP Synthase by Time-Resolved Single-Molecule FRET

Michael Börsch

Abstract Detection of single fluorophore molecules was reported 25 years ago, at first in a crystalline matrix at cryogenic temperatures but quickly followed by single-molecule studies of biological machines at physiological conditions. Today, the Nobel Prize in Chemistry 2014 was awarded for 'breaking the resolution limit of optical microscopy', i.e. for super-resolution methods that are based on active control of the photophysical properties of single fluorophores. In this chapter, the development of conformational analysis of a single rotary double-motor F_oF_1-ATP synthase at work is reviewed. Analyzing single-molecule Förster resonance energy transfer (smFRET) in combination with confocal fluorescence lifetime (FLIM) recording of single enzymes has revealed stepsize and direction of the two distinct but coupled rotary motor parts, reversible elastic deformations for conformational energy storage within protein domains, and an internal mechanical regulatory mechanism to control the activity of the bacterial enzyme. TCSPC enabled the duty-cycle optimized multiplexed laser excitation schemes to control and correct the smFRET data analysis of this marvelous nanomachine which provides the cellular energy currency ATP.

9.1 Detecting Single Molecules for Two Decades

Counting single photons had been established more than 50 years ago. Precise correlation with high-frequency signals like the time of the exciting laser pulse enabled fluorescence lifetime spectroscopy with picosecond time resolution. Since then, enormous technical progress has been achieved. The excitation lasers became small, stable, 'turnkey'-like instruments offering a huge variety of wavelengths at affordable prices. Optical filters with very high transmission for fluorescence photons but nearly complete rejection of scattered laser light were developed.

M. Börsch (✉)
Single-Molecule Microscopy Group, Jena University Hospital,
Friedrich Schiller University Jena, Nonnenplan 2-4, 07743 Jena, Germany
e-mail: michael.boersch@med.uni-jena.de

© Springer International Publishing Switzerland 2015
W. Becker (ed.), *Advanced Time-Correlated Single Photon Counting Applications*,
Springer Series in Chemical Physics 111, DOI 10.1007/978-3-319-14929-5_9

Avalanche photodiodes and CCD cameras for single photon counting with high quantum efficiencies for photon detection were designed, and bulky recording electronics changed into computer cards which are controlled by software and store the photon time traces for subsequent analysis.

Optical detection of single molecules like fluorescent organic dyes in a crystalline matrix at low temperatures was reported 25 years ago [78, 79]. Nowadays, single-molecule detection is made easy by technological improvements. A variety of commercial microscopes is available that include different kinds of lasers, optics, detectors and data analysis software. Importantly, the research subjects have changed—from quantum phenomena of single impurities in solid state systems to the dynamics of individual biological machines at work in living cells and tissue.

Why are we interested in studying individual biomolecules? Single-molecule detection is the ultimate analysis approach at the lowest concentration limit. Specialized to identify very rare events, the optical signal of a single molecule has to overcome the background noise, for example autofluorescence and Raman scattering of solvents or the cytosol inside cells. Then, counting individual molecules is possible yielding quantitative numbers not only for a distribution maximum or mean values, but also for the complete description of a distribution, i.e. its width and shape. Diffusion properties of single molecules can unravel spatial heterogeneities in cells and also time-dependent fluctuations. Thereby, association phenomena like dimerization or oligomerization of biomolecules are revealed, even as short transient events. The control of photophysical properties of single fluorophores, for example using laser light for photoswitching molecules between fluorescent and non-fluorescent states [34], paved the road for 'breaking the resolution limit of optical diffraction' [1, 51] and resulted in the Nobel Prize in Chemistry 2014 awards to W.E. Moerner, E. Betzig and S.W. Hell (see www.nobelprize.org). With the introduction of single-molecule based superlocalization microscopies in 2006 [8, 53, 91], these imaging approaches allow to bridge the optical resolution gap between Förster resonance energy transfer (FRET) usable for distances less than 10 nm and the diffraction-limited resolution for two objects in an optical microscope in the range of 200 nm.

Biomedical applications of single-molecule microscopy are, for example, ultrasensitive pathogen detection or drug screening approaches. However, strong and rapidly increasing research areas are quantitative biophysical investigations on structure and function of molecular machines. This is the focus of the chapter which will review main aspects and possibilities of time correlated single photon counting (TCSPC) for single molecule analyses using examples from the author's research. For more than 15 years, we concentrate on experiments using confocal microscopes with point-like detectors which are used for fluorescence lifetime imaging (FLIM) with high time resolution. Widefield and total internal reflection (TIR) imaging of single biomolecules using EMCCD cameras are not discussed here. Because point-like detection with a confocal microscope has the disadvantage of being slow for imaging and cannot follow freely diffusing molecules for long time, novel experimental setups will be explained that aim at overcoming the limitations of a small, femtoliter-sized detection volume for a single biomolecule at work.

9.2 Binding and Diffusion of Single Molecules Through a Confocal Detection Volume

Aromatic amino acids like tryptophan or tyrosine are endogenous fluorophores in proteins. They cannot be used for single-molecule detection because low fluorescence quantum yields, limited photostabilities as well as absorbance in the UV or blue spectral range do not provide an appropriate signal-to-background ratio [72]. Instead endogenous fluorophores in proteins like chlorophylls in light harvesting complexes or allophycocyanines are bright enough to be detected in single proteins [46, 104]. Therefore, external labeling of the biological targets is required, either with organic dyes like rhodamines, the widely used fluorescent proteins directly fused to proteins of interest, or nanocrystals like Qdots or luminescent nanodiamonds [105]. Additional possibilities to attach fluorophores include specific protein tags, engineered cysteines and non-natural amino acids, and the chemistries for easy and quantitative reactions in aqueous solution to avoid non-specific labeling have been developed.

The principle of detecting a single molecule in a confocal microscope requires the knowledge of the size of the excitation and detection volume and the appropriate dilution of the molecule of interest. Given the wavelength-dependent resolution limit of about $\lambda/2$–200 nm [1], the likely threedimensional Gaussian intensity distribution of the exciting laser focus generated by the microscope objective has lateral dimensions of about 300 nm and an axial length of about 1 µm (Fig. 9.1a). In the emission pathway of the confocal microscope (Fig. 9.1b), a pinhole restricts the detection of fluorescence photons to a similar-sized detection volume. Accordingly, the confocal volume can be calculated. Depending on the wavelength, the laser beam diameter, the microscope objective parameters and the size of the pinhole, this volume can be adjusted between 0.2 and up to 10 fl. Relating this volume to the Avogadro constant $N_A = 6.022 \times 10^{23}$ mol^{-1} yields an upper concentration limit for the mean of a single-molecule in the detection volume at any time. As a rule of thumb, concentrations around 1 nM correspond to the confocal detection conditions in solution for a single molecule.

On the other hand, methods for getting a background-free buffer or "dark" cellular environment had to be found and are developed further to obtain the highest possible signal-to-noise ratio (SNR). In the confocal detection volume, the fluorescence signal of one single fluorophore has to overcome the background contributions of many water molecules and impurities. For example, a 1 nM rhodamine 110 solution in water corresponds to 1 fluorophore molecule in a 1 fl ($=10^{-15}$ L) volume surrounded by $\sim 5.5 \times 10^{10}$ water molecules. Due to the high fluorescence quantum yield of the dye, its fluorescence signal clearly exceeds the background as seen in Fig. 9.2a. In this time trajectory, recorded photons are binned to 1 ms intervals. The arbitrary Brownian motion of each dye molecule through the detection volume results in photon bursts from single molecules with different maximum intensities. The peaks in Fig. 9.2a reach about 40–55 counts/ms. In contrast, the background is found at 1–2 counts/ms, resulting in a SNR = 45/1.5 = 30.

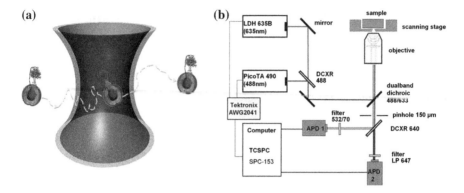

Fig. 9.1 a Principle of confocal single-molecule detection in solution. The fluorescence-labeled membrane protein F_oF_1-ATP synthase is reconstituted in a 120-nm liposome which diffuses freely through the laser focus. **b** Setup for confocal single-molecule FRET and FLIM. Two ps-pulsed lasers PicoTa490 and LDH 635B and the TCSPC electronics SPC153 are triggered by an arbitrary waveform generator Tektronix AWG 2041 with 1 ns time resolution. Two single-photon counting avalanche photodiodes APDs detect fluorescence of the FRET donor between 497 and 567 nm and of the FRET acceptor for wavelengths $\lambda > 647$ nm

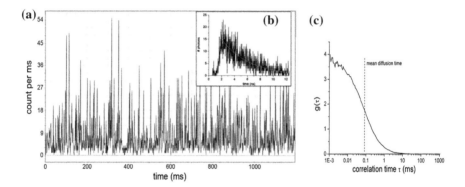

Fig. 9.2 a Time trace of single rhodamine 110 molecules freely diffusing in pure water. Excited with 488 nm, maximum count rates were 40–50 counts/ms with 1 ms binning time. **b** Fluorescence decay of a single rhodamine 110 molecule with simple monoexponential decay fit (*grey curve*) upon pulsed excitation with 80 MHz repetition rate at 488 nm. The single-molecule lifetime was $\tau_{rh} = 4.2$ ns. **c** Autocorrelation function (FCS) of the rhodamine 110 solution in water. The mean diffusion time was ~ 80 μs indicating a 2-fold extended detection volume compared to the standard diffraction-limited confocal volume

Using picosecond-pulsed laser excitation with high repetition rates (here 80 MHz, or 12.5 ns between the pulses, respectively) allows to extract the fluorescence lifetime of a single dye molecule during the transit time in the detection volume. Shown as an inset b in Fig. 9.2, the monoexponential decay fit to the photon histogram yields a lifetime $\tau \sim 4.2$ ns. This is in agreement with published data considering the simple fitting approach without appropriate deconvolution of the instrument response function (in the range of 600 ps).

The same TCSPC data with ns time resolution for each photon can be used to calculate the autocorrelation function for fluorescence correlation spectroscopy (FCS, see Chap. 1, Sect. 1.5.2 for calculation from TCSPC data). The autocorrelation function of the recorded time trace of rhodamine 110 in Fig. 9.2a is plotted in Fig. 9.2c. The prominent feature of the curve is a sigmoidal decay with a mean correlation time of ~ 80 μs. This is the average diffusion time of the small molecule rhodamine 110 through the confocal detection volume. Larger proteins or vesicles exhibit a right-shifted FCS curve towards longer correlation times. For example, the soluble F_1 part of $F_o F_1$-ATP synthase shows a ~ 10 times longer diffusion time compared to rhodamine 110 [14], and proteoliposomes with 120 nm diameter show ~ 100 time longer diffusion times than a single fluorophore molecule [16].

An extended autocorrelation function can be calculated—from ps to any times, as a so-called 'full correlation' function, see Chap. 1, Sect. 1.5.2. Then, all types of photophysical parameters are accessible within one single curve: fluorescence lifetime from ps to ns, rotational correlation times (or fluorescence anisotropies) from ten to hundreds of ns, photophysical processes including triplet state formation in μs, excited state protonation/deprotonation reactions, or cis/trans photoisomerization [48]. These fluorescence parameters can be exploited to measure biological and chemical reaction data, equilibrium constants and kinetics [52]. For example, binding of fluorescent small molecules like ligands or inhibitors to proteins will change their translational and rotational diffusion constants. Binding of ligands to enzymes might also change the effective hydrodynamic radius or shape of the protein by conformational changes which can be measured by FCS. Protein oligomerization and other aggregation phenomena are detectable by changes in diffusion properties and unravel binding or dissociation constants. Confocal measurements of photophysical properties in solution as well as in living cells by FCS might report local pH or pH changes, respectively, and triplet state dynamics can be used to monitor spatiotemporal O_2 concentration fluctuations. Reversible quenching of fluorescent reporters by photoinduced electron transfer (PET) can reveal conformational dynamics of proteins or nucleic acids, and averaging over many individual molecules in FCS might yield accurate time constants for these dynamics [95].

9.3 Analyzing Conformations of Single Molecules by FRET

A direct method of measuring bound *versus* unbound ligands to a single protein is based on Förster resonance energy transfer (FRET). FRET is known as a distance ruler for an approximate range between 2 and 10 nm [44]. One fluorescent reporter molecule is called the FRET donor which can transfer energy in its excited state to a nearby secondary reporter molecule called the FRET acceptor. Energy transfer results in an average loss of fluorescence intensity of the FRET donor molecule accompanied by a possible increase in fluorescence intensity of the FRET acceptor.

The dipole-dipole interaction depends strongly on the distance between the two dye molecules, orientational dynamics of transition dipoles, as well as their spectral properties. A more detailed description of the dependencies is reviewed elsewhere [27]. Energy transfer affects the lifetime of the excited FRET donor molecule, and shortening of FRET donor lifetime is distance-dependent. Three fundamental equations relating the FRET efficiency E_{FRET} to an intra- or intermolecular distance are given in the following:

$$E_{FRET} = I_A/(I_D + I_A) \tag{9.1}$$

$$E_{FRET} = 1 - \tau_{DA}/\tau_D \tag{9.2}$$

$$E_{FRET} = R_0^6/(R_0^6 + r_{DA}^6) \tag{9.3}$$

In (9.1) experimental fluorescence intensities of FRET donor (I_D) and acceptor (I_A) have to be corrected for background and spectral crosstalk from one to the other detection channel, as well as for relative quantum yields of the two dyes and detection efficiencies of donor and acceptor channel of the confocal microscope setup. To obtain the FRET efficiency E_{FRET} according to (9.2) the excited state lifetime of the donor fluorophore τ_D in the absence of an acceptor is measured independently from the actual single-molecule lifetime of a donor fluorophore in the presence of an acceptor τ_{DA}.

Once E_{FRET} is determined, the distance r_{DA} between donor and acceptor fluorophores is calculated based on the Förster radius R_0, i.e. the distance for 50 % energy transfer. The Förster radius comprises the spectral parameters of the fluorophores ('spectral overlap integral') and can be varied using different combinations of fluorophores. Typical values of R_0 are in the range of 5–6 nm for pairs of spectrally distinct rhodamine or cyanine dyes, or 3–4 nm for fluorescent proteins due to smaller absorption coefficients and lower quantum yields. Because the FRET distance dependence scales with the 6th power in (9.3), accurate distance measurements with sub-Å precision [7] are possible especially around distances similar to R_0.

Typical fluorophore spectra exhibit broad absorption and emission bands, with average spectral widths of tens of nm. For fluorophores with a small difference of a few nm between absorbance maximum and emission maximum, 'homo-FRET' between two identical fluorophores can occur. In this case, FRET donor and acceptor dyes cannot be discriminated spectrally. However, the mean fluorescence anisotropy of the two dyes can change due to homo-FRET, and, therefore, anisotropy measurements can be applied to identify co-localization of two labeled molecules within distances shorter than 10 nm [83]. Please see Chap. 12, Sect. 12.7.3.

Co-localization using FRET between two spectrally different fluorophores has become a standard approach, because simultaneous ratiometric measurement of the two intensities is easily possible, and necessary corrections for a quantitative distance calculation are feasible [56, 90]. Position-specific labeling with FRET donor and acceptor dyes within one protein is achievable in vitro using orthogonal labeling strategies. For example, fluorophore maleimides for attachment to introduced single cysteine residues can be combined with a genetic fusion construct of a fluorescent protein to another protein domain, or with other genetic fusion proteins like SNAP-, CLIP- or Halo-tags. Alternatively, reversible disassembly/assembly preparation procedures might exist for multisubunit proteins so that cysteines can be introduced to different residue positions and labeled specifically in the same protein. Strong binding of fluorescent ligands or inhibitors with nM dissociation constants are required for a FRET-based co-localization of these small molecules to the respective protein binding sites. However, specifically FRET-labeling single proteins in vivo is much more difficult to achieve.

Recently, the use of two fluorophores to monitor conformations by FRET was extended to multi-color FRET measurements [55, 71, 88, 101]. Thereby, ligand binding-induced conformational changes or changes in their respective dynamics in single proteins or nucleic acid complexes can be interpreted more accurately. Or the conformational changes of one protein/nucleic acid in a complex as induced by the interaction with another partner, can be simultaneously monitored its conformational changes. However, the accessible spectral range for multi fluorophore single-molecule FRET is limited by possible cross talk. In general, FRET between more than two dyes like in photonics wires, dendrimers or linear polymeric fluorophores [6] becomes more difficult to analyze quantitatively, and limits the use of these approaches.

Finally, the single-molecule FRET tool box can be combined with other single particle manipulation analysis techniques for further detailed insights or controlled changes in conformations, for example, by mechanical [49] or electrical stimulation [94], by electrochemically induced changes [13], and other options.

In the following our favorite membrane protein for single-molecule FRET studies, i.e. F_oF_1-ATP synthase from *Escherichia coli* and other bacteria, will be introduced and the single-molecule FRET experiments will be reviewed which revealed its conformational changes during catalysis, the suggested inhibitory control mechanisms, and transient elastic deformations of this highly efficient, rotary double-motor.

9.4 The Membrane-Bound Enzyme F_oF_1-ATP Synthase

Adenosine triphosphate (ATP) is the major "chemical energy currency" in biological systems. Upon hydrolysis of ATP, the standard free energy $\Delta G^{0'}$ (ATP) of -30 to -60 kJ mol^{-1} [5] can be used in the cell for a variety of transport and metabolic processes, and, as the result, adenosine diphosphate (ADP) and inorganic phosphate (P_i) are produced. The reversed process of ATP synthesis from ADP and P_i is accomplished by ubiquitous large membrane enzymes called ATP synthases, i.e. F_oF_1-ATP synthases in the bacterial plasma membrane, in the thylakoid membranes of chloroplasts or in the inner membrane of mitochondria, respectively. According to the 'chemiosmotic hypothesis' [77] (P.D. Mitchell, Nobel Prize in

Chemistry 1978) ATP synthesis is driven by the proton-motive force, PMF (or a sodium-motive force in some organisms [109]), comprising a concentration difference of protons, ΔpH, and an electric potential, $\Delta \Psi$, across the membrane.

The architecture, enzymatic rates and basic mechanistic properties of F_oF_1-ATP synthases have been reviewed [22, 24, 117, 121]. Crystallographic studies on F_oF_1-ATP synthase structure initially focused on the F_1 part from bovine heart mitochondria [2], but recently also the structure of the F_1 part from E. coli has been solved [26]. Bacterial F_oF_1-ATP synthases are composed of 9 different subunits. The F_1 part with subunit stoichiometry $\alpha_3\beta_3\gamma\delta\varepsilon$ is located outside the membrane and is housing three catalytic nucleotide binding sites. The membrane-embedded F_o part with subunit composition ab_2c_n (with n = 10 for the E. coli enzyme [65]) contains the proton translocating sites.

The mechanism by which proton translocation through F_o drives the catalytic reaction cycle within F_1 is yet to be defined at the atomic level. However, P. Boyer initially postulated a mechanism for this process in 1981 prior to the availability of structural information (reviewed in [22]). He suggested that F_oF_1-ATP synthase operated according to a 'binding change mechanism' where two of the three nucleotide binding sites within the α/β-subunit interface in F_1 either bind ADP and P_i, or generate ATP (see Fig. 9.3a). He also suggested that these reactions were both synchronized and dependent upon the position of a central asymmetric γ-subunit in F_1, which can change its orientation by rotation. Since chemical cross-linking data have shown a clear association between the γ-subunit (and the ε-subunit) of F_1 and the c-subunit ring in F_o [106], proton translocation through the F_o complex is the driving force behind the coupled γ-subunit rotation. Accordingly, a 'rotor' portion is formed in F_oF_1-ATP synthase by the rotating c-ring in F_o together with the γ- and ε-subunits in F_1. The ab_2-subunits of F_o together with the $\alpha_3\beta_3\delta$-subunits of F_1 then stabilize this stepwise-moving 'rotor' by acting as a non-moving 'stator', or external stalk, respectively.

9.4.1 Single-Molecule Methods to Monitor Rotary Catalysis of F_oF_1-ATP Synthase

Since the first partial structure of mitochondrial F_1 supporting this hypothesis of rotary catalysis was published in 1994 [2] resulting in the Nobel Prize in Chemistry 1997 to J.E. Walker and P.D. Boyer (and J.C. Skou for the distinct Na^+/K^+-ATPase), several research groups have attempted to reveal the molecular mechanisms of F_oF_1-ATP synthase function by applying a variety of biochemical [23, 36, 128] and spectroscopic techniques [92]. However, as the orientation of the rotary subunits cannot be synchronized with respect to the asymmetric peripheral stalk by subunits $ab_2\delta$ in F_oF_1-ATP synthase (Fig. 9.3a), only single-molecule techniques provide the ability to observe sequential conformational changes of the rotor as a function of time in individual enzymes. Fortunately, the bacterial F_oF_1-ATP

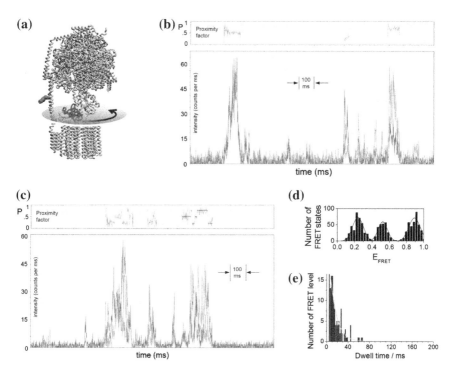

Fig. 9.3 **a** Model of F_oF_1-ATP synthase. Rotary subunits γ and the *c*-ring are shwon in *cyan* and ε in *blue*, stator subunits $α_3β_3δ$-ab_2 are colored *orange*. During ATP synthase, the rotary subunits move unidirectionally according to the *arrow*. For smFRET measurements, the donor fluorophore (*green*) is bound to and the acceptor (*red*) is bound to b_2. **b** Typical smFRET time trajectory showing photon bursts with three different but constant proximity factors P (*upper panel*) calculated from donor intensity (*blue trace*) and acceptor intensity (*green trace*). **c** Typical smFRET time trajectory showing photon bursts with three different but stepwise changing proximity factors P within each burst (*upper panel*) calculated from donor intensity (*blue trace*) and acceptor intensity (*green trace*). **d** FRET efficiency distribution from those FRET-labeled F_oF_1-ATP synthases showing FRET level fluctuations during ATP hydrolysis. **e** Dwell time distribution for the medium FRET level of F_oF_1-ATP synthases during ATP hydrolysis as shown in (**d**) (**d**, **e** [122], with permission)

synthase also catalyzes the opposite chemical reaction, i.e. ATP hydrolysis, which can be monitored more easily without the need for generating and maintaining a PMF required to synthesize ATP. Therefore, the experimental approaches to verify the rotary catalysis mechanism concentrated on the ATP hydrolysis reaction first.

In 1997, H. Noji, R. Yasuda, K. Kinosita Jr and M. Yoshida achieved this goal and reported the direct observation of subunit rotation in single F_1 parts driven by ATP hydrolysis [84]. They attached a fluorescent actin filament to the γ-subunit as the marker of rotation in the $α_3β_3γ$ sub-complex of the *Bacillus PS3* F_1 part. F_1 molecules were immobilized to a cover glass. In the presence of millimolar [ATP], these actin filaments rotated specifically in one direction, but stopped rotation after addition of azide N_3^-, an inhibitor of ATP hydrolysis. This spectacular demonstration of γ-subunit rotation in F_1 during ATP hydrolysis was followed up by similar experiments using *E. coli* F_1 [85], as well as for the other rotary subunit ε [69] and the *c*-ring [87, 93]. In addition, using magnetic beads instead of actin filaments allowed to manipulate—and even counterrotate—the attached subunit in individual F_1 parts by applying external magnetic fields [64, 89].

The tremendous success of these single-molecule observations is summarized in many reviews on the biophysical properties of γ-subunit rotation in the $α_3β_3$ fragment of the *Bacillus PS3* and of the *E. coli* F_1 parts [4, 9, 19, 61, 82, 115]. The findings include: unidirectional rotation of γ, 120° rotary steps at high [ATP] corresponding to stops at each of the three catalytic binding sites, 80°/40° substeps of γ at low [ATP] revealing that catalysis and substrate binding occur at different rotary angles [9, 120], product release of ADP and P_i at different angles in the full cycle [113], and the motor properties like [ATP]-dependent speed profiles [73, 81], torque [70, 85] and transient elastic properties [25, 67]. However, μm-long actin filaments or large beads represent a viscous load on the γ-subunit that dampen the details of rotational dynamics. Significantly smaller reporters have to be attached to minimize this load. Accordingly, the F_1 motor has been subsequently investigated using 40 nm gold nanobeads [9, 120] or 80 nm gold nanorods [73, 100] for light scattering, or the fluorescence anisotropy of single 1-nm-sized fluorophores [3, 47]. Most recently, subunit rotation in human mitochondrial F_1 revealed the existences of revealed two distinct substeps, one at 65° and the second one at 90° [102].

9.4.2 The TCSPC Approach to Monitor F_oF_1-ATP Synthase by Single-Molecule FRET

We have developed a single-molecule enzymology approach to investigate subunit rotation and individual catalytic rates by Förster resonance energy transfer within single F_oF_1-ATP synthases, starting in 1997 at the University of Freiburg in the group of P. Gräber. We prepared single F_oF_1-ATP synthases from *E. coli* reconstituted in freely diffusing lipid vesicles and monitored the orientation of the γ-subunit with respect to the peripheral stalk, i.e. the dimeric *b*-subunits [15–17]. Briefly the rotor subunit γ of F_oF_1 was labeled specifically with tetramethylrhodamine (TMR) to a genetically introduced cysteine residue as the FRET donor to yield F_1-γ106C-TMR. A bisfunctional cyanine 5 maleimide (Cy5bis) was synthesized as FRET acceptor in order to crosslink the dimeric stator subunits b_2, using two introduced cysteines and resulting in F_o-*b*64C-Cy5bis. To attach the fluorophores specifically to the enzyme, F_1 parts were purified and labeled as the first step [14]. F_o parts in liposomes were prepared separately, starting with from F_oF_1 enzymes that were labeled and reconstituted in liposomes. Selectivity of subunit labeling was examined by gel electrophoresis and showed that approximately 90 % of the two *b*-subunits in the F_o part were cross-linked by Cy5bis. Then, after biochemical removal of the F_1 part without any label, the labeled F_1-γ106C-TMR was reattached in the presence of Mg^{2+} on the remaining labeled F_o-b64C-Cy5bis parts. These FRET-labeled enzymes were embedded in lipid vesicles of about 120 nm in diameter. Biochemical measurements showed that the labeled enzymes were functional, with a mean turnover time $t_S \sim 50$ ms to synthesize one ATP molecule in one binding site of the enzyme, and $t_H \sim 20$ ms for the hydrolysis of one ATP molecule at room temperature.

Fluorescence of FRET donor and acceptor were measured with a confocal microscope setup for single-molecule detection. In the first experiments, a frequency-doubled continuous-wave Nd:YAG laser at 532 nm was attenuated to about 105 μW and focused into the buffer solution by a water immersion objective (UPLAPO, 40x, 1.15 N.A., Olympus). Fluorescence was measured in two spectral ranges for FRET donor and acceptor after passing the dichroic mirror (DCXR 540): TMR was measured in the range between 540 and 610 nm, and Cy5bis was measured at wavelengths above 665 nm. A 100-μm pinhole was used in the confocal detection pathway to reject out-of-focus fluorescence. Single photons were detected by two avalanche photodiodes (APD, SPCM AQR-15, Perkin Elmer) and counted by a multichannel scaler PC-card (PMS 300, Becker & Hickl, Germany) with a pre-defined time resolution of 1 ms/bin [15]. The signals were fed in parallel into a hardware correlator for fluorescence correlation spectroscopy (FCS). Visualization and analysis of the two-channel time trajectories were performed with the program 'BURST_ANALYZER' that was developed by N. Zarrabi at the University of Stuttgart since 2003 [122] (the latest version of the software is now available from Becker & Hickl, Berlin, Germany).

9.4.3 Different FRET Levels in Single F_oF_1-ATP Synthases

Typical time trajectories of FRET-labeled F_oF_1-ATP synthases traversing the confocal detection volume of a few femtoliters are shown in Fig. 9.3b, c. In the presence of 1 mM AMPPNP, a non-hydrolyzable ATP derivative, the rotor subunits γ or ε are expected to be locked into one of three orientations. Accordingly, three different FRET efficiencies can be expected, if the stopping positions of the label on the rotor are non-symmetrical with respect to the second fluorophore on the stator. In Fig. 9.3b lower panel, three photon bursts can be identified with fluorescence intensities for the FRET donor dye (blue trace) up to 65 counts/ms, for the FRET acceptor dye (green trace) up to 60 counts/ms, and a background noise of about 2–4 counts/ms. From these intensity data points per ms, a FRET efficiency E_{FRET} value is calculated for each time bin and plotted using (9.1). This trace is often called 'proximity factor' trace, because corrections for different detection efficiencies in the setup for the two fluorophores and their respective fluorescence quantum yields are not yet applied.

Luckily we had chosen the right positions for cysteine residues in either the rotary γ- or the ε-subunit in F_1 to enable an unambiguous discrimination of the three different FRET efficiencies or proximity factors. In the time trace in Fig. 9.3b, the first

FRET-labeled F_oF_1-ATP synthase exhibits a medium FRET efficiency of P ~ 0.5 corresponding to equal intensities of FRET donor and acceptor. The second photon burst shows a low FRET state around P ~ 0.3 characterized by higher count rates for the FRET donor (blue) than the FRET acceptor (green). In contrast, the last photon burst has higher count rates for the FRET acceptor, i.e. revealing a high FRET efficiency of P ~ 0.7. These different FRET efficiencies were clearly distinguishable.

During catalysis, the rotor will sequentially change its orientation with respect to the stator, and FRET efficiencies within a photon burst are expected to change sequentially as well. Shown in Fig. 9.3c are two typical photon bursts with stepwise changing FRET levels for active F_oF_1-ATP synthases. Whereas both fluorescence intensities fluctuate strongly due to Brownian motion of the proteoliposomes through the detection volume, the resulting proximity factor trace is simpler to analyze and displays FRET levels with stepwise changes. In the first photon burst, the active F_oF_1-ATP synthase starts with a very short high FRET level (H; proximity factor P ~ 0.8) followed by low FRET (L; P ~ 0.2) and medium FRET (M; P ~ 0.5), subsequently switching to L, then to H, and finally to the M level before leaving the detection volume. The second photon burst exhibits a more pronounced unidirectional sequence of FRET levels, i.e. in the order M $\rightarrow$ H $\rightarrow$ L $\rightarrow$ M $\rightarrow$ H $\rightarrow$ L.

At the beginning of our single-molecule FRET time trace analysis of F_oF_1-ATP synthases we assigned FRET levels within photon bursts manually. This approach was facilitated by the clear stepwise changes of FRET levels within less than 1 ms (switching time ~ 200 µs [130]) and the large differences between the discriminated proximity factors, and apparently constant proximity factors within each of the FRET levels. Assigned FRET levels were added to histograms. The FRET level distributions showed three maxima (or four in the presence of F_oF_1-ATP synthases without an FRET acceptor dye). Distributions were fitted by Gaussians. In Fig. 9.3d FRET efficiency histograms are shown for FRET-labeled F_oF_1-ATP synthase actively hydrolyzing ATP (i.e. 1 mM, in the presence of 2.5 mM $MgCl_2$). The enzyme was labeled with TMR on the rotary ε subunit and with Cy5bis crosslinking the static b_2 subunits. The fluorescence quantum yield $\Phi_D = 0.38$ for TMR bound at ε56C was measured relative to sulforhodamine 101; for the FRET acceptor Cy5bis the quantum yield $\Phi_A = 0.32$ was determined. The Förster radius was calculated $R_0 = 6.4$ nm assuming randomly distributed transition dipole moments, i.e. using an orientation factor $\kappa^2 = 2/3$. In the FRET distribution shown in Fig. 9.3d, a low-FRET level L corresponded to a FRET efficiency of $E_{FRET} = 0.23$, or a mean fluorophore distance of $r_{DA} = (7.9 \pm 0.2)$ nm (mean ± SD), respectively. The medium-FRET level was characterized by $E_{FRET} = 0.52$ or a fluorophore distance of $r_{DA} = (6.3 \pm 0.2)$ nm, and the high-FRET state exhibited $E_{FRET} = 0.86$ or a fluorophore distance of $r_{DA} = (4.3 \pm 0.3)$ nm, respectively [129]. Because we still do not know the exact position of the axis of rotation within the γ-ε-c_{10} rotor subunits and we do not know the position of the Cy5bis on the b_2 dimer (Fig. 9.3a is only a model based on partial information of structural details for the *E. coli* enzyme), these three FRET distances appeared plausible, assuming that the TMR fluorophore is located about 2–3 nm off-axis, and b64C positions for Cy5bis are located slightly outside and above the plane of TMR rotation. Then, a shortest TMR-Cy5bis distance of less than 2 nm and a maximum distance of up to 9 nm could be estimated for the rotating ε subunit.

9.4.4 Dwell Time Analysis of FRET Levels

Relevant information about catalysis is obtained from the analysis of the duration, or dwell time distribution, of the assigned FRET levels as shown in Fig. 9.3e. All FRET levels of F_oF_1 with proximity factors P between 0.15 and 0.39 were sorted to the L level, those with P between 0.4 and 0.69 comprised the M level, and those with P between 0.7 an 1.0 were assigned the H level. Note that F_oF_1 labeled with TMR but not Cy5bis, or with a photobleached Cy5bis, showed an apparent proximity factor P between 0.0 and 0.14. This level was assigned 'donor-only' and was omitted from analysis. Furthermore, the first and the last FRET level in a photon burst cannot be used because the real durations of these levels for F_oF_1 while entering or leaving the detection volume remain unknown. Therefore only the intermediary FRET levels were sorted by their FRET efficiency and binned in 2-ms intervals for the histogram of the M level in Fig. 9.3e. Dwell times were analyzed for L, M and H separately and slight differences were found. Fitting each distribution with a monoexponential decay resulted in dwell times for L ~ 17.6 ms, for M ~ 12.7 ms (Fig. 9.3e) and for H ~ 15.8 ms. In these dwell time histograms, a rising component at very short dwell times could not be analyzed due to a minimum duration of 4 ms for manually assigning a specific FRET level.

The dwell time differences indicated a small influence of the asymmetrically positioned peripheral stalk on the catalytic activity of the adjacent nucleotide binding sites in the F_1 part. Currently the structure of the bacterial F_oF_1-ATP synthase is not available at atomic resolution. However, the structure of the F_1 part of the mitochondrial enzyme has been solved at 2.8 Å resolution [2], and many details of the conformations of the three β-subunits with the catalytic nucleotide binding sites are known. Electron microscopic images of the bacterial [21] enzyme showed the shape of the ATP synthase and helped to establish the three-dimensional arrangement of the different subunits of F_oF_1. Biochemical cross-linking data improved the localization of the b_2-dimer at a non-catalytic interface between one α- and one β-subunit [75]. Thus, the conformational dynamics of two catalytic nucleotide binding sites could be affected by the attached b_2-dimer, and, because of the asymmetric nature of the subunits, could be affected differently. These dwell time differences were used later to identify the rotary orientation of the ε-subunit with respect to a new fluorophore position on the stator, i.e. to locate the C-terminus of the proton-translocating a-subunit of the F_o part by single-molecule FRET triangulation [37–39].

Dwell time analysis of FRET levels in photon bursts of single enzymes is a strong tool to discriminate conformational changes of protein domains from fluorophore movements [7] along the linker to the protein or the photophysics of the dyes. The ATP concentration dependence for dwell times of the rotary subunits of

F_oF_1-ATP synthase and F_1 parts have been analyzed in great details at high time resolution using the light scattering nanobeads or nanorods. For the FRET-labeled enzyme freely diffusing in solution, dwell time analysis is limited by both assignment of the shortest dwell time of a few ms due to limited photon counts, and by the maximum observation time of the enzyme in the detection volume. So far, single-molecule FRET was used to study rotation in F_oF_1 at maximum speed, i.e. at saturating high [ATP] for ATP hydrolysis or at very high PMF (with a starting ΔpH of ~ 4 and a large membrane potential) for ATP synthesis.

However, dwell time analysis can unravel the mechanism of competitive and non-competitive inhibitors or modulators, respectively. Single-molecule FRET measurements of F_oF_1 during ATP hydrolysis in the presence of aurovertin B, which likely binds to F_1 outside of the nucleotide binding sites [107], revealed that this non-competitive inhibitor did not affect the stopping angles of the rotor but prolonged the dwell times by an additional rising time component [66]. In the presence of ADP or P_i, products of ATP hydrolysis and competitive inhibitors, the rotor stopping angles differed and were identical to the stopping positions in the absence of any ATP [129], and fluctuations of FRET levels within single photon bursts were not found (and dwell times could not be measured). Using the single-molecule FRET approach to study other ATP-driven membrane transporters like P-glycoprotein (Pgp) we found that the competitive inhibitor VO_4^{3-}, which prevents ATP hydrolysis in biochemical assay and mimics P_i, did not suppress the conformational movements of the nucleotide binding domains completely, but altered and prolonged the dwell times [108, 127]. Similarly, the bacterial K^+ transporter KdpFABC, a P-type ATPase, showed ongoing conformational movements of the nucleotide binding domain in the presence of VO_4^{3-}, but much slower dynamics [50, 125].

The mechanism of another membrane transporter protein, the bacterial translocase YidC, which inserts a hydrophobic transmembrane α-helix into the lipid bilayer, was investigated by FRET between single YidC molecules in liposomes and its protein substrate and substrate mutants [40]. Dwell time analysis of FRET traces of the transiently associated substrates unraveled the fast kinetics in ms for this protein insertion process [119].

9.4.5 FRET Transition Density Plots for Sequential Conformational Changes

The strength of single-molecule biophysics is not only the possibility to resolve several distinct FRET levels of a single biomolecule, but also to disentangle the sequential order of its conformational transitions. For F_oF_1-ATP synthase, our first goal was to identify the direction of γ- and ε-subunit rotation during ATP synthesis with respect to ATP hydrolysis. As the chemical reaction was reversed, also the direction of rotation changed to the opposite [17, 35, 129]. These results were obtained by comparison of the sequences of manually assigned FRET levels, and the conclusions were based on predominances in the range of 80 % for the uni-directional sequence for each of the two contrary catalytic modes.

A different, more general approach for this FRET data analysis and a better way of finding rare conformational states is pattern recognition in the two dimensional histogram of FRET level transitions. This FRET transition density plot [74] relates FRET level i to level $i + 1$ and does not require additional a priori knowledge, for example the likely number of FRET levels (i.e. the three-stepped rotation in F_oF_1) or symmetries in sequential changes. In this histogram, major transitions appear more often, but also rare events and even missing conformational steps due to fast dynamics can be identified. We used the FRET transition density plot to reveal the step size of the proton-driven c-ring rotation in F_oF_1 during ATP synthesis [39]. For a three-stepped unidirectional rotation we had expected only three major popula-tions in the FRET transition density plot, but we found more and with smaller differences between the subsequent FRET efficiencies or FRET levels. This strongly supported a step size of 36° for a c-subunit, or, in other words, a proton-driven rotation of the c-ring in an one-after-another mode. Please see Chap. 1, Sect. 1.5.3, Fig. 1.59.

In addition, plotting FRET level i *versus* level $i + 2$ allows to identify how often conformations swivel back and forth. Especially for those single-molecule FRET data sets without well-defined FRET levels and with high uncertainties about the total number of conformations to be discriminated, the unbiased two dimensional plotting of assigned FRET levels is a recommended first step to begin the analysis. Currently we study the regulatory mechanism of bacterial F_oF_1-ATP synthase to has to prevent wasteful ATP hydrolysis in a living cell. Our goal is to monitor the mobile C-terminus of the rotating ε-subunit by single-molecule FRET [12, 20]. As the numbers of transitions, the FRET distances and the dwells are not known, unbiased methods to visualize FRET levels are required to observe the proposed mechanism of the C-terminus moving in an 'up'/'down' mode [60] and blocking ATP hydrolysis in an 'up' conformation as found in the X-ray structure of *E. coli* F_1 [26]. These methods include software-based searches in the time traces to find change-points [116] in intensities or in FRET efficiencies, or variable model-based approaches (for example, hidden Markov models, HMMs [74, 123, 127]).

9.4.6 Advanced TCSPC Methods for Time-Resolved and DCO-ALEX FRET

Additional controls are required to discriminate fluorophore photophysics and positional fluctuations from the conformational dynamics of the biomolecules. The alternative way to determine FRET efficiencies is based on measuring the FRET donor fluorophore lifetime in the presence of the acceptor r_{DA} according to (9.2). Picosecond-pulsed laserdiodes operating in the spectral range between 400 and 800 nm at fixed or adjustable repetition rates up to 80 MHz exist and can be easily integrated into confocal microscopes for single-molecule FRET measurements. For the single F_oF_1 data shown in Fig. 9.4 we used laser pulses of ~50 picoseconds at 530 nm provided by a frequency-doubled amplified laserdiode at 80 MHz [122]. Gaussian intensity distribution in the focus was achieved by passing the laser pulses through a polarization preserving single-mode fiber. Pulses were attenuated to an average power of 150 µW, and focused into the buffer solution by a water immersion objective (UPLANAPO, 60x, 1.2 N.A., Olympus). The fluorescence of FRET donor and acceptor was measured in two spectral regions by single photon counting APDs using the same filters, optics, and a 100-µm pinhole in the confocal detection pathway as described above. Photons were recorded in the parameter-tag mode of a TCSPC card with picosecond time resolution (SPC 630, Becker & Hickl, Germany) and channel-separated by a router (HRT-82). Data analysis including variable binning of the detected photons and the calculation of the autocorrelation functions was performed with the program 'BURST_ANALYZER'. The instrument response function of the setup was measured by the APD using back-scattered laser light, and resulted in a full-width-half-maximum instrument response function of 500 ps. This pulse shape was used for the deconvolution of the single-molecule lifetime data with a monoexponential decay function. As a reference, the fluorescence lifetime of single rhodamine 6G molecules (~100 pM) in water was measured yielding $\tau_{R6G} = 3.6 \pm 0.1$ ns (see Fig. 9.2b).

Shown in Fig. 9.4a, the F_oF_1 molecule exhibited a TMR lifetime of $\tau_{DA} = 2.1$ ns. A proximity factor P = 0.26 was calculated from the fluorescence intensities resulting in a L FRET level classification for this photon burst. Calculation was

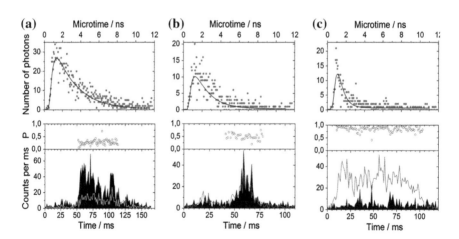

Fig. 9.4 **a** AMPPNP-trapped F_oF_1 in liposomes [122]. In the *upper diagrams*, the distribution of FRET donor TMR photons in the microtime interval between the laser pulses (80 MHz) is shown. The *lower diagram* shows the time trajectories of the fluorescence intensities of the FRET donor TMR, I_D (*black area*), and FRET acceptor Cy5, I_A (*gray line*), using 1 ms binning. The *middle trace* is the corresponding proximity factor $P = I_A/(I_D + I_A)$ calculated from the uncorrected dye intensities per 1 ms time bin. **a** Low-FRET state of F_oF_1, **b** medium-FRET state of F_oF_1, **c** high-FRET state of F_oF_1

started at 52 ms and stopped at 113 ms. In Fig. 9.4b, a TMR lifetime of τ_{DA} = 1.4 ns was measured for the F_oF_1 photon burst, and a proximity factor P = 0.48 was calculated starting at 41 ms and stopping at 78 based on intensities, corresponding to a M FRET level. The FRET-labeled F_oF_1 shown in Fig. 9.4c had a short TMR lifetime τ_{DA} = 0.5 ns. From the fluorescence intensities in the photon bursts a proximity factor P = 0.85 was calculated starting at 9 ms and stopping at 103 ms. This is the H FRET level orientation of the ε-subunit in F_oF_1.

In Fig. 9.5a, the F_oF_1 exhibited a longer TMR lifetime τ_D = 2.48 ns, and fluorescence intensities resulted in an apparent proximity factor P = 0.07, calculated with photons from 26 to 112 ms. Presumably, this F_oF_1 was labeled with TMR only. Therefore this fluorescence lifetime represented the FRET donor lifetime in the absence of an acceptor, τ_D (9.2). Accordingly, the TMR lifetime-based FRET efficiencies of the F_oF_1 in Fig. 9.4a–c were E_{FRET} = 0.15 for the L state, and E_{FRET} = 0.44 for the M state. For the H state a FRET efficiency E_{FRET} = 0.8 was found. The lifetime-based FRET efficiencies shown here were found in good agreement with the intensity-derived FRET efficiencies, supporting FRET causing the relative intensity changes and excluding photophysical artifacts.

Pulsed laserdiodes lasers can be easily combined by alternating excitation schemes to confirm the existence of the FRET acceptor dye on a single F_oF_1-ATP synthase, or to identify spectral shifts of the FRET donor in the absence of an acceptor, see Fig. 9.5b, c. These schemes are often called 'pulsed interleaved excitation' (PIE) [80] or 'alternating laser excitation' (ALEX) [68], with laser wavelength switching in ns oder μs. Since the first experiments 25 years ago,

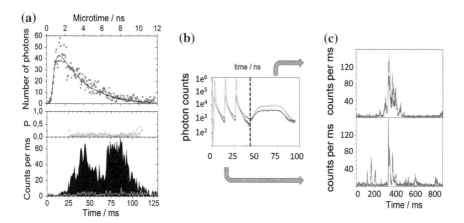

Fig. 9.5 a Donor-only state of F_oF_1 without FRET acceptor [122]. (**a–d** [122]). **b** Laser pulse scheme DCO-ALEX for two lasers to probe transient elastic deformation within the rotor subunits ε and c and simultaneous detection of rotational movement of the rotor with respect to a third label on the static a-subunit in a single F_oF_1-ATP synthase. Three pulses at 488 nm excite FRET donor EGFP an the stator (*blue decays*, and *blue time trace* in (**c**), *lower panel*) and FRET acceptors Alexa532 on ε and Cy5 on c (*green decays*, and *green* time trace in (**c**), *lower panel*), followed by a single AOM-switched 532 nm pulse to excite Alexa532 as new FRET donor on ε (*turquois decay* and time trace in (**c**)) and Cy5 on c as FRET acceptor in red (**b, c** [42], with permission)

single-molecule spectroscopy has revealed that fluorophores can switch their spectral properties [79], i.e. absorbance and emission spectra, quantum yield and fluorescence lifetime [112]. Using EGFP as the FRET donor in single F_oF_1-ATP synthase we showed that this fluorophore fused to the a-subunit exhibited different lifetimes of 2.8 and 2.1 ns, with the shorter lifetime corresponding to a red-shifted emission spectrum [37]. Therefore, alternating excitation to excite the FRET acceptor with a single ns pulse helped to exclude photon burst with apparent 'FRET' caused by transient spectral shifts of the donor fluorophore.

Simply alternating the two laser pulses in order to probe the existence of a FRET acceptor fluorophore by direct excitation with a second laser and to measure FRET by exciting the FRET donor with primary laser results in 1:1 division of the FRET-related photon recording, or a 50 % duty cycle. To improve the duty cycle of FRET recording, we apply different laser excitation schemes called duty-cycle optimized alternating laser excitation (DCO-ALEX) [124, 126]. An arbitrary waveform generator (AWG) with 1 ns time resolution and 8 programmable digital outputs is employed for sequential triggering the primary FRET laser with a sequence of up to six 12-ns-delayed pulses, followed by triggering a single pulse of the laser for direct acceptor test, and to synchronize the TCSPC recording electronics. Using the freely adjustable AWG, slower switching behavior of an acousto-optical modulator (AOM) with rise and decay times of about 5 ns for the direct FRET acceptor test can be compensated, and a FRET duty cycle of 75 % can still be achieved.

Single-molecule FRET [39] and gold nanorod imaging [63] revealed the 10-stepped rotation of the c-ring of F_o in F_oF_1-ATP synthases from $E.\ coli$ during ATP synthesis as well as ATP hydrolysis. The 10 steps per full rotation are coupled to 3-stepped rotation of γ- and ε-subunits in the F_1 part. Accordingly, twisting of elastic protein domains within the rotor subunits has to occur to compensate this symmetry mismatch and to maintain the coupling of the two rotary motors [25, 98, 99, 110]. To observe this elastic deformation within the rotary subunits and to ensure a catalytically active enzyme we labeled a single F_oF_1-ATP synthase with three fluorophores. The static a-subunit was labeled specifically by fusing the EGFP to its C-terminus, serving as the primary FRET donor fluorophore. When excited with a 488 nm pulse in a three-laser DCO-ALEX scheme, fluctuating FRET changes to either of two acceptor dyes, i.e. on the rotary ε- and on one rotary c-subunit, indicated an active enzyme [42, 126]. Simultaneously, i.e. by exciting the secondary FRET donor Alexa532 on the ε-subunit with an AOM-switched 532-nm laser pulse of about 10 ns, we measured FRET distance changes to the final FRET acceptor Cy5 on the c-ring. A final, single pulse of a 635-nm laserdiode in the AWG-controlled laser sequence enabled direct testing of the Cy5 dye being present in this F_oF_1-ATP synthase. However, in order to optimize the FRET duty cycle, the direct test of bound Cy5 could be omitted. A twisting within the rotary ε- and c-subunits of single active enzymes was shown, and the observed maximum twisting angles of more than 100° covered the necessary degree of elastic deformations required for the compensation of the two distinct rotary stepsizes [41].

9.4.7 Single-Molecule FRET Imaging of Immobilized F$_o$F$_1$-ATP Synthases

The dwell times of the rotary motors of F$_o$F$_1$-ATP synthases in liposomes determined in confocal smFRET experiments in solution were directly comparable to the cuvette experiments of the biochemical catalytic rates. However, experiments were limited by short observation times of a few hundred ms due to Brownian diffusion and the fluctuating sum of fluorescence intensities within a photon burst. Therefore, immobilizing the FRET-labeled enzyme was evaluated to prolong smFRET observation times. At first we incorporated a small amount of biotin-lipids to the proteoliposomes with F$_o$F$_1$ for subsequent surface attachment to streptavidin-modified cover glass [124]. After binding, the liposomes with a single FRET-labeled F$_o$F$_1$-ATP synthase remained as stable vesicles and did not spread out on the surface.

Scanning the surface using a x, y, z-piezo scanner (see Chap. 2, Sect. 2.8) resulted in single-molecule images with three different FRET efficiencies comparable to those obtained from freely diffusing enzymes in liposomes. However, after adding 1 mM ATP for ATP hydrolysis, only very few enzymes showed stepwise FRET efficiency fluctuations [18]. We concluded that the enzyme quickly diffuses all over the lipid bilayer of the proteoliposome within few ms, and eventually might contact the surface resulting in either denaturation or another inactivation process of the enzyme. Alternatively, restricted educt and product diffusion due to the local geometry of a thin buffer layer between glass surface and the curved liposome membrane might change the effective local concentrations of both ATP educt and ADP plus P$_i$ products, resulting in a very slow turnover. To avoid these geometrical constraints we also tried surface attachment of FRET-labeled F$_o$F$_1$-ATP synthase in detergent to the cover glass via Histags on the β-subunits in F$_1$. Using a maleimide-modified quantum dot bound covalently to the c-ring in F$_o$, we measured changing fluorescence anisotropies of a quantum dot as a marker to report c-ring rotation during ATP hydrolysis [45].

In collaboration with R. Iino and H. Noji (Tokyo University, Japan), we evaluated alternative labeling strategies for future in vivo smFRET measurements. Briefly, a CLIP-tag was fused to a shortened C-terminus of the rotary ε-subunit and labeled with FRET donor Alexa488 (or TMR for in vivo smFRET imaging), and a SNAP-tag was fused to the C-terminus of the a-subunit carrying the FRET acceptor Alexa647 [97]. Specificity of labeling the two fusion proteins was controlled by SDS-PAGE. Single-molecule FRET recording during ATP hydrolysis revealed three distinguishable FRET efficiencies, and mean dwell times of 12 ± 2 ms for all intermediary FRET levels indicated that these fusion constructs did not severely affect catalytic turnover of the liposome-reconstituted enzyme. However, many enzymes exhibited FRET changes oscillating between two FRET levels only.

Next, using this ε-CLIP/a-SNAP modified F$_o$F$_1$-ATP synthase we recorded smFRET/FLIM. Figure 9.6a shows the experimental setup of the surface-attached FRET-labeled F$_o$F$_1$ in detergent. Using piezo-driven sample scanning and pulsed

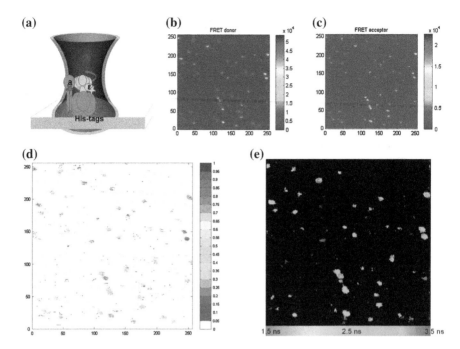

Fig. 9.6 **a** Setup for smFRET/FLIM to monitor subunit rotation in single surface-attached F_oF_1 in detergent. FRET donor Alexa488 was bound via a CLIP-tag fused to ε and FRET acceptor Alexa647 via a SNAP-tag fused to the *a*-subunit. Pulsed excitation at 488 nm with 60 ps and 80 MHz repetition rate for 4 ms/pixel. **b, c** False-colored fluorescence intensity images of FRET donor (**b**) and acceptor channel (**c**), with intensities in counts per s (*scale*: 256 pixel for x and y axis correspond to $20 \times 20 \ \mu m^2$). **d** FRET efficiency image as calculated from intensities in (**b**) and (**c**) per pixel. **e** False-colored FLIM image of the same area for individual F_oF_1-ATP synthases

excitation at 488 nm, intensity and lifetime-based FRET efficiencies of single enzymes on the surface were calculated. The false-colored, normalized intensity images of FRET donor Alexa488 in Fig. 9.4b and FRET acceptor Alexa647 in Fig. 9.4c revealed single F_oF_1 molecules on the surface, assured by spontaneous 'blinking' of the dyes as characterized by dark lines in the diffraction-limited image spot of the fluorophores. The width of individual fluorescence spots was found ~ 8 pixel in the 256×256 pixel images, corresponding to a FWHM of ~ 300 nm. The maximum photon count rates per pixel of FRET donor and FRET acceptor differed by a factor of 2.5, reflecting the different fluorescence quantum yields of Alexa488 (~ 0.85 according to the supplier) and Alexa647 (~ 0.3). FRET efficiencies in each pixel were calculated and shown in false-color coding in Fig. 9.4d. Dark blue single-molecule FRET spots correspond to a 'donor only' labeled F_oF_1, cyan pixels in the spots represent low FRET efficiency, green to yellow pixels indicate medium FRET, and orange to red pixels are related to high FRET efficiency values, or three rotary ε-subunit orientations with respect to the static *a*-subunit.

The corresponding fluorescence lifetime image of the FRET donor dye Alexa488 is shown in Fig. 9.4e. The lifetime information in the FLIM image is also false-colored. Pixels representing a 'donor only'-labeled F_oF_1 were characterized by lifetimes $\tau_D \sim 3.5$ ns in blue, cyan pixels indicate $\tau_{DA} \sim 3$ ns representing low FRET efficiencies, green pixels indicate $\tau_{DA} \sim 2.5$ ns or medium FRET, and orange to red pixels with $\tau_{DA} < 1.5$ ns are related to high FRET efficiency values. A reasonably good agreement between the intensity-based FRET efficiency assignment of single F_oF_1-ATP synthases and the lifetime-based FRET efficiencies was found. However, as the enzymes were bound directly to the glass surface, precise single-molecule lifetime determination was obstructed by additional background photons from the glass. As a consequence and stated for our smFRET/FLIM measurements in solution before [35, 37, 122], the lifetime information of the FRET donor fluorophore can be used as a support for the intensity-based FRET interpretation. Long-lasting FRET levels are more suitable for lifetime-based FRET efficiency calculation, as shown recently for the P-type ATPase SERCA using two-photon excitation and FRET lifetime trajectory analysis in vitro and smFRET/FLIM in vivo [86].

9.5 Perspectives

'Single-molecule active control microscopy' (SMACM [103]) has been phrased by W.E. Moerner (Stanford) as a basis of single-molecule super-localization microscopy to resolve cellular structures beyond the optical diffraction limit and related to other acronyms like STORM or PALM. Here, laser light is used to switch single fluorophores 'on' and 'off' reversibly, i.e. into a fluorescent or a non-fluorescent photophysical state.

Mechanical active control of individual F_1 motors of ATP synthase bound to a surface has been achieved since 10 years using a magnetic bead bound to the rotary γ-subunit [54, 62, 64, 89, 114, 115]. Changing the external magnetic fields allowed to counteract and stall the ATP-driven rotation at arbitrary angles, to push the rotary subunits backwards, and to mechanically drive the rotor in ATP synthesis direction for the first 'man-made' ATP molecules from ADP and P_i in the F_1 part of the enzyme. Thereby torque profiles of the motor and the angle-dependent catalytic steps of ATP binding, ATP hydrolysis and ADP and P_i release could be unraveled and quantitatively measured for a very detailed understanding of this part of the enzyme.

Single-molecule active control can be related also to keep an individual fluorophore at an arbitrarily chosen spatial position. A.E. Cohen and W.E. Moerner have developed such a control device called the 'Anti-Brownian electrokinetic trap' (ABEL trap [28–31, 33, 43]) to study single molecules in solution. This microfluidic device restricts diffusion to two dimensions in a less than 1 μm shallow transparent structure, and uses the position-dependent fluorescence of a single dye molecule to calculate its position and to push back the molecule to the center of the

ABEL trap. A combination of electrophoretic and electroosmotic forces is applied to compensate the Brownian motion with µs feedback, resulting in a apparently non-moving single molecule in solution.

Time-correlated single photon counting can be used to record several thousands of photons from a single molecule kept in solution by the ABEL trap and to observe conformational dynamics of intrinsically fluorescent proteins or specifically labeled proteins and nucleic acids, respectively, for long times of several tens of seconds [10, 11, 32, 46, 96, 111, 112]. So far, fluorescence lifetime analysis of freely diffusing single molecules in solution (see Fig. 9.4) were limited to a few tens or hundreds of photons, and observation times did not exceed a few hundred ms. However, with active position control we are enabled to measure single-molecule lifetimes much more precisely and to record long time trajectories of fluorescent lifetimes from a single biomolecule in solution without surface interference. In addition, keeping the single fluorophore at a given z-position and achieving homogeneous excitation in x and y allows to use the intensity information quantitatively. Transient changes of the photophysical properties of a single dye are identified as stepwise changes in the otherwise constant intensity, and single fluorophores can be cleary distinguished from multiple fluorophores.

In collaboration with A.E. Cohen and W.E. Moerner we have started to study the rotary motors of F_oF_1-ATP synthase in the ABEL trap. At first we studied the conformation of the regulatory ε-subunit in single FRET-labeled F_1 parts of the E. coli enzyme [12]. Labeling the moving C-terminus with Atto647N as the FRET acceptor and a rigid part of the γ-subunit with Atto488 as FRET donor, we could trap the 10 nm small soluble F_1 in buffer for several hundred ms instead of a mean transit time of 2–5 ms without the trap. Constant FRET levels but also switching between different FRET levels indicated conformational dynamics, and the FRET distance analysis confirmed the previous biochemical hypothesis, that in the presence of ATP most F_1 part have to be in an inactive conformation to prevent wasteful ATP hydrolysis. This finding is in agreement with the 10-fold stimulation of ATP hydrolysis rates in the presence of the detergent LDAO. LDAO activation is thought either to cause complete dissociation of ε from F_1 and to remove the mechanical blocking of γ-subunit rotation, or to somehow affect the C-terminus of ε so that it cannot insert back to the rotary axis. Single-molecule FRET of F_1 in the ABEL trap did not show any remaining double-labeled F1 parts in the presence of LDAO, which supports a complete dissociation of ε rather than a conformational change.

We evaluated an electrochemical active control for single F_oF_1-ATP synthases [13]. Using a small Pt-electrode and reducing H_2O for H_2 and OH^- generation at the tip of the electrode, we intended to establish a stable local pH gradient in the vicinity of the tip, so that freely diffusing F_oF_1 in liposomes should experience a pH change only while diffusing through the socalled 'chemical lens' created by the buffer around the electrode tip. After calibrating the tip-distance-dependent pH, placing the confocal excitation volume exactly in an area of pH > 8 in an otherwise acidic buffer at pH 5 for the proteoliposomes allowed to continuously record smFRET during ATP synthesis conditions. The data suggested that in the presence of ADP and P_i this approach could be used to detect single F_oF_1 rotating in ATP synthesis direction.

To overcome technical difficulties with this local electrochemistry approach like laser back scattering from the tip of the metallic electrode, light scattering due to H_2 gas bubble generation and luminescence from the glass capillary around the microelectrode, we now aim to apply confocal detection for smFRET/FLIM imaging of individual F_oF_1 enzymes in a planar, free-standing lipid bilayer ('Black Lipid Membrane', BLM). Novel microfluidic chips are commercially available which allow to position the BLM ~ 100 μm away from the cover glass. This stable z-position of the BLM enables precise single-photon counting applications like lipid and membrane protein diffusion measurements by either single-molecule tracking with EMCCD cameras [58, 59], or by confocal dual-focus FCS [118]. If this z-confinement is combined with a x, y-confinement within the BLM, confocal detection of FRET-labeled F_oF_1 using a fast laser focus pattern for homogenous illumination of a small 3×3 μm^2 area will enable constant intensities of single fluorophores like in the ABEL trap. In addition, electrophysiology controls and adjustable buffer compositions on both sides of the membrane will allow to control, maintain and change the driving force PMF for ATP synthesis, and to vary conditions over all parameters. The x, y confinement of F_oF_1 is achievable without hampering its rotational mobility within the membrane using the annexin V network which is built on top of the BLM [57]. Single-molecule localization, smFRET/FLIM and active control of individual enzyme activity will be merged.

Finally, one wants to obtain single-molecule active control for the analysis of FRET-labeled F_oF_1 in a living bacterium. At first the number of fluorescent ATP synthases has to be kept very low. Only one or two diffusing FRET-labeled F_oF_1 molecules can be followed at the diffraction limit and per time interval in a single small bacterium of 500 nm in diameter and 2 μm length. Future labeling strategies might involve photoactivatable fluorescent protein tags in combination with other bioorthogonal labeling strategies, or photobleaching down to single FRET-labeled molecules is required [97]. Microfluidics for stable attachment of bacteria to the cover glass and for simultaneous, continuous support with new growth media have to be used to actively control the physiological conditions of the cells. Thus we expect to continue our challenging endeavor to monitor single F_oF_1-ATP synthases at work by smFRET and FLIM, and to keep this extraordinary fundamental biological nanomachine [76] in focus.

Acknowledgments The author wants to thank the collaborators H. Noji (Tokyo University, Japan) and R. Iino (Okazaki Institute of Integrative Bioscience, Japan) for making the F_oF_1 constructs and single-molecule FRET measurements possible shown in Fig. 9.6. Briefly I want to thank all my co-workers over the years since 1997, first at the Institute for Physical Chemistry (Freiburg), namely M. Diez, B. Zimmermann, S. Steigmiller, R. Reuter, W. Wangler, R. Lehmann, M. Trost, J. Petersen, and F.-M. Boldt; then at the 3rd Institute of Physics (Stuttgart), namely M.G. Düser, N. Zarrabi, S. Ernst, T. Rendler, K. Seyfert, E. Hammann and A. Zappe; and now at the Single-Molecule Microscopy group (Jena) namely H. Sielaff, T. Heitkamp, S. Rabe, A. Renz, M. Renz, D. McMillan and B. Su for their careful single-molecule FRET experiments and analysis and their ongoing efforts for improvements. I am very grateful for the enduring collaborations with

S.D. Dunn (University of Western Ontario, London, Canada) and T.M. Duncan (SUNY Upstate Medical University, Syracuse, USA) for their strong biochemical F_oF_1 support and the appointment as 'adjunct assistant professor at Upstate', and with W.E. Moerner (Stanford) and A.E. Cohen (Harvard) to customize their ABEL trap concept for single-molecule FRET analysis of subunit rotation in a trapped F_oF_1-ATP synthase and for the invitation to Stanford as 'Visiting Scholar' from 2012 to 2014. I appreciate the ongoing generous support by J. Wrachtrup (Stuttgart) and by P. Gräber (Freiburg), and the long-term funding by the Baden-Württemberg Stiftung and the Deutsche Forschungsgemeinschaft (DFG grants BO 1891/8-1, BO 1891/10-1, BO 1891/10-2, BO 1891/15-1, BO 1891/16-1).

References

1. E. Abbe, Beiträge zur Theorie des Mikroskops und der mikroskopischen Wahrnehmung. Arch. f. Mikr. Anat. **9**, 413–420 (1873)
2. J.P. Abrahams, A.G. Leslie, R. Lutter, J.E. Walker, Structure at 2.8 Å resolution of F1-ATPase from bovine heart mitochondria. Nature **370**, 621–628 (1994)
3. K. Adachi, R. Yasuda, H. Noji, H. Itoh, Y. Harada, M. Yoshida, K. Kinosita Jr, Stepping rotation of F1-ATPase visualized through angle-resolved single-fluorophore imaging. Proc. Natl. Acad. Sci. USA **97**, 7243–7247 (2000)
4. K. Adachi, K. Oiwa, T. Nishizaka, S. Furuike, H. Noji, H. Itoh, M. Yoshida, K. Kinosita Jr, Coupling of rotation and catalysis in F(1)-ATPase revealed by single-molecule imaging and manipulation. Cell **130**, 309–321 (2007)
5. R.A. Alberty, Thermodynamics of the hydrolysis of adenosine triphosphate as a function of temperature, pH, pMg, and ionic strength. J. Phys. Chem. B **107**, 12324–12330 (2003)
6. F.E. Alemdaroglu, S.C. Alexander, D.M. Ji, D.K. Prusty, M. Börsch, A. Herrmann, Poly (BODIPY)s: a new class of tunable polymeric dyes. Macromolecules **42**, 6529–6536 (2009)
7. M. Antonik, S. Felekyan, A. Gaiduk, C.A. Seidel, Separating structural heterogeneities from stochastic variations in fluorescence resonance energy transfer distributions via photon distribution analysis. J. Phys. Chem. B **110**, 6970–6978 (2006)
8. E. Betzig, G.H. Patterson, R. Sougrat, O.W. Lindwasser, S. Olenych, J.S. Bonifacino, M.W. Davidson, J. Lippincott-Schwartz, H.F. Hess, Imaging intracellular fluorescent proteins at nanometer resolution. Science **313**, 1642–1645 (2006)
9. T. Bilyard, M. Nakanishi-Matsui, B.C. Steel, T. Pilizota, A.L. Nord, H. Hosokawa, M. Futai, R.M. Berry, High-resolution single-molecule characterization of the enzymatic states in Escherichia coli F1-ATPase. Philos. Trans. R. Soc. B: Biol. Sci. **368**, 20120023 (2013)
10. S. Bockenhauer, A. Furstenberg, X.J. Yao, B.K. Kobilka, W.E. Moerner, Conformational dynamics of single G protein-coupled receptors in solution. J. Phys. Chem. B **115**, 13328–13338 (2011)
11. S.D. Bockenhauer, W.E. Moerner, Photo-induced conformational flexibility in single solution-phase peridinin-chlorophyll-proteins. J. Phys. Chem. A **117**, 8399–8406 (2013)
12. S.D. Bockenhauer, T.M. Duncan, W.E. Moerner, M. Börsch, The regulatory switch of F1-ATPase studied by single-molecule FRET in the ABEL Trap. Proc. SPIE **8950**, 89500H (2014)
13. F.M. Boldt, J. Heinze, M. Diez, J. Petersen, M. Börsch, Real-time pH microscopy down to the molecular level by combined scanning electrochemical microscopy/single-molecule fluorescence spectroscopy. Anal. Chem. **76**, 3473–3481 (2004)
14. M. Börsch, P. Turina, C. Eggeling, J.R. Fries, C.A. Seidel, A. Labahn, P. Graber, Conformational changes of the H^+-ATPase from Escherichia coli upon nucleotide binding detected by single molecule fluorescence. FEBS Lett. **437**, 251–254 (1998)

15. M. Börsch, M. Diez, B. Zimmermann, R. Reuter, P. Graber, Monitoring γ-subunit movement in reconstituted single EFoF1 ATP synthase by fluorescence resonance energy transfer, in *Fluorescence spectroscopy, Imaging and Probes. New Tools in Chemical, Physical and Life Sciences*, ed. by R. Kraayenhof, A.J.W. Visser, H.C. Gerritsen (Springer, Berlin, 2002), pp. 197–207
16. M. Börsch, M. Diez, B. Zimmermann, R. Reuter, P. Graber, Stepwise rotation of the γ-subunit of EF(0)F(1)-ATP synthase observed by intramolecular single-molecule fluorescence resonance energy transfer. FEBS Lett. **527**, 147–152 (2002)
17. M. Börsch, M. Diez, B. Zimmermann, M. Trost, S. Steigmiller, P. Graber, Stepwise rotation of the γ-subunit of EFoF1-ATP synthase during ATP synthesis: a single-molecule FRET approach. Proc. SPIE **4962**, 11–21 (2003)
18. M. Börsch, J. Wrachtrup, The electromechanical coupling within a single molecular motor, in *Nanotechnology—Physics, Chemistry and Biology of Functional Nanostructures*, ed. by T. Schimmel, H. Lohneysen, C. Obermair, M. Barczewski (Landesstiftung Baden-Württemberg, Stuttgart, 2008), pp. 41–58
19. M. Börsch, Microscopy of single FoF1-ATP synthases—the unraveling of motors, gears, and controls. IUBMB Life **65**, 227–237 (2013)
20. M. Börsch, T.M. Duncan, Spotlighting motors and controls of single FoF1-ATP synthase. Biochem. Soc. Trans. **41**, 1219–1226 (2013)
21. B. Bottcher, I. Bertsche, R. Reuter, P. Graber, Direct visualisation of conformational changes in EF(0)F(1) by electron microscopy. J. Mol. Biol. **296**, 449–457 (2000)
22. P.D. Boyer, The binding change mechanism for ATP synthase—some probabilities and possibilities. Biochim. Biophys. Acta **1140**, 215–250 (1993)
23. V.V. Bulygin, T.M. Duncan, R.L. Cross, Rotation of the ε-subunit during catalysis by Escherichia coli FOF1-ATP synthase. J. Biol. Chem. **273**, 31765–31769 (1998)
24. R.A. Capaldi, R. Aggeler, Mechanism of the F(1)F(0)-type ATP synthase, a biological rotary motor. Trends Biochem. Sci. **27**, 154–160 (2002)
25. D.A. Cherepanov, A.Y. Mulkidjanian, W. Junge, Transient accumulation of elastic energy in proton translocating ATP synthase. FEBS Lett. **449**, 1–6 (1999)
26. G. Cingolani, T.M. Duncan, Structure of the ATP synthase catalytic complex (F(1)) from Escherichia coli in an autoinhibited conformation. Nat. Struct. Mol. Biol. **18**, 701–707 (2011)
27. R.M. Clegg, Fluorescence resonance energy-transfer and nucleic-acids. Methods Enzymol. **211**, 353–388 (1992)
28. A.E. Cohen, W.E. Moerner, An all-glass microfluidic cell for the ABEL trap: fabrication and modeling. Proc. SPIE **5930**, 59300S (2005)
29. A.E. Cohen, W.E. Moerner, The anti-Brownian electrophoretic trap (ABEL trap): fabrication and software. Proc. SPIE **5699**, 296–305 (2005)
30. A.E. Cohen, W.E. Moerner, Method for trapping and manipulating nanoscale objects in solution. Appl. Phys. Lett. **86**, 093109 (2005)
31. A.E. Cohen, W.E. Moerner, Suppressing Brownian motion of individual biomolecules in solution. Proc. Natl. Acad. Sci. USA **103**, 4362–4365 (2006)
32. A.E. Cohen, W.E. Moerner, Principal-components analysis of shape fluctuations of single DNA molecules. Proc. Natl. Acad. Sci. USA **104**, 12622–12627 (2007)
33. A.E. Cohen, W.E. Moerner, Controlling Brownian motion of single protein molecules and single fluorophores in aqueous buffer. Opt. Express **16**, 6941–6956 (2008)
34. R.M. Dickson, A.B. Cubitt, R.Y. Tsien, W.E. Moerner, On/off blinking and switching behaviour of single molecules of green fluorescent protein. Nature **388**, 355–358 (1997)
35. M. Diez, B. Zimmermann, M. Börsch, M. Konig, E. Schweinberger, S. Steigmiller, R. Reuter, S. Felekyan, V. Kudryavtsev, C.A. Seidel, P. Graber, Proton-powered subunit rotation in single membrane-bound FoF1-ATP synthase. Nat. Struct. Mol. Biol. **11**, 135–141 (2004)
36. T.M. Duncan, V.V. Bulygin, Y. Zhou, M.L. Hutcheon, R.L. Cross, Rotation of subunits during catalysis by Escherichia coli F1-ATPase. Proc. Natl. Acad. Sci. USA **92**, 10964–10968 (1995)

37. M.G. Düser, N. Zarrabi, Y. Bi, B. Zimmermann, S.D. Dunn, M. Börsch, 3D-localization of the a-subunit in FoF1-ATP synthase by time resolved single-molecule FRET. Proc. SPIE **6092**, 60920H (2006)

38. M.G. Düser, Y. Bi, N. Zarrabi, S.D. Dunn, M. Börsch, The proton-translocating *a* subunit of F0F1-ATP synthase is allocated asymmetrically to the peripheral stalk. J. Biol. Chem. **283**, 33602–33610 (2008)

39. M.G. Düser, N. Zarrabi, D.J. Cipriano, S. Ernst, G.D. Glick, S.D. Dunn, M. Börsch, 36 degrees step size of proton-driven c-ring rotation in FoF1-ATP synthase. EMBO J. **28**, 2689–2696 (2009)

40. S. Ernst, A.K. Schonbauer, G. Bar, M. Börsch, A. Kuhn, YidC-driven membrane insertion of single fluorescent Pf3 coat proteins. J. Mol. Biol. **412**, 165–175 (2011)

41. S. Ernst, M.G. Düser, N. Zarrabi, S.D. Dunn, M. Börsch, Elastic deformations of the rotary double motor of single FoF1-ATP synthases detected in real time by Förster resonance energy transfer, Biochimica et Biophysica Acta (BBA)—Bioenergetics, **1817**, 1722–1731 (2012)

42. S. Ernst, M.G. Düser, N. Zarrabi, M. Börsch, Three-color Förster resonance energy transfer within single FoF1-ATP synthases: monitoring elastic deformations of the rotary double motor in real time. J. Biomed. Opt. **17**, 011004 (2012)

43. A.P. Fields, A.E. Cohen, Electrokinetic trapping at the one nanometer limit. Proc. Natl. Acad. Sci. USA **108**, 8937–8942 (2011)

44. T. Förster, Energiewanderung Und Fluoreszenz. Naturwissenschaften **33**, 166–175 (1946)

45. E. Galvez, M. Düser, M. Börsch, J. Wrachtrup, P. Gräber, Quantum dots for single-pair fluorescence resonance energy transfer in membrane- integrated EFoF1. Biochem. Soc. Trans. **36**, 1017–1021 (2008)

46. R.H. Goldsmith, W.E. Moerner, Watching conformational- and photo-dynamics of single fluorescent proteins in solution. Nat. Chem. **2**, 179–186 (2010)

47. K. Hasler, S. Engelbrecht, W. Junge, Three-stepped rotation of subunits gamma and epsilon in single molecules of F-ATPase as revealed by polarized, confocal fluorometry. FEBS Lett. **426**, 301–304 (1998)

48. E. Haustein, P. Schwille, Fluorescence correlation spectroscopy: novel variations of an established technique. Annu. Rev. Biophys. Biomol. Struct. **36**, 151–169 (2007)

49. Y. He, M. Lu, J. Cao, H.P. Lu, Manipulating protein conformations by single-molecule AFM-FRET nanoscopy. ACS Nano **6**, 1221–1229 (2012)

50. T. Heitkamp, R. Kalinowski, B. Böttcher, M. Börsch, K. Altendorf, J.C. Greie, K(+)-Translocating KdpFABC P-type ATPase from Escherichia coli acts as a functional and structural dimer. Biochemistry **47**, 3564–3575 (2008)

51. S.W. Hell, J. Wichmann, Breaking the diffraction resolution limit by stimulated emission: stimulated-emission-depletion fluorescence microscopy. Opt. Lett. **19**, 780–782 (1994)

52. S.T. Hess, S. Huang, A.A. Heikal, W.W. Webb, Biological and chemical applications of fluorescence correlation spectroscopy: a review. Biochemistry **41**, 697–705 (2002)

53. S.T. Hess, T.P. Girirajan, M.D. Mason, Ultra-high resolution imaging by fluorescence photoactivation localization microscopy. Biophys. J. **91**, 4258–4272 (2006)

54. Y. Hirono-Hara, K. Ishizuka, K. Kinosita Jr, M. Yoshida, H. Noji, Activation of pausing F1 motor by external force. Proc. Natl. Acad. Sci. USA **102**, 4288–4293 (2005)

55. S. Hohng, C. Joo, T. Ha, Single-molecule three-color FRET. Biophys. J. **87**, 1328–1337 (2004)

56. S.J. Holden, S. Uphoff, J. Hohlbein, D. Yadin, L. Le Reste, O.J. Britton, A.N. Kapanidis, Defining the limits of single-molecule FRET resolution in TIRF microscopy. Biophys. J. **99**, 3102–3111 (2010)

57. T. Ichikawa, T. Aoki, Y. Takeuchi, T. Yanagida, T. Ide, Immobilizing single lipid and channel molecules in artificial lipid bilayers with annexin A5. Langmuir **22**, 6302–6307 (2006)

58. T. Ide, T. Yanagida, An artificial lipid bilayer formed on an agarose-coated glass for simultaneous electrical and optical measurement of single ion channels. Biochem. Biophys. Res. Commun. **265**, 595–599 (1999)
59. T. Ide, Y. Takeuchi, T. Aoki, T. Yanagida, Simultaneous optical and electrical recording of a single ion-channel. Jpn. J. Physiol. **52**, 429–434 (2002)
60. R. Iino, R. Hasegawa, K.V. Tabata, H. Noji, Mechanism of inhibition by C-terminal alpha-helices of the ε-subunit of Escherichia coli F_oF_1-ATP synthase. J. Biol. Chem. **284**, 17457–17464 (2009)
61. R. Iino, H. Noji, Operation mechanism of F(o)F(1)-adenosine triphosphate synthase revealed by its structure and dynamics. IUBMB Life **65**, 238–246 (2013)
62. Y. Iko, K.V. Tabata, S. Sakakihara, T. Nakashima, H. Noji, Acceleration of the ATP-binding rate of F1-ATPase by forcible forward rotation. FEBS Lett. **583**, 3187–3191 (2009)
63. R. Ishmukhametov, T. Hornung, D. Spetzler, W.D. Frasch, Direct observation of stepped proteolipid ring rotation in E. coli F_oF_1-ATP synthase. EMBO J. **29**, 3911–3923 (2010)
64. H. Itoh, A. Takahashi, K. Adachi, H. Noji, R. Yasuda, M. Yoshida, K. Kinosita, Mechanically driven ATP synthesis by F1-ATPase. Nature **427**, 465–468 (2004)
65. W. Jiang, J. Hermolin, R.H. Fillingame, The preferred stoichiometry of c subunits in the rotary motor sector of Escherichia coli ATP synthase is 10. Proc. Natl. Acad. Sci. USA **98**, 4966–4971 (2001)
66. K.M. Johnson, L. Swenson, A.W. Opipari Jr, R. Reuter, N. Zarrabi, C.A. Fierke, M. Börsch, G.D. Glick, Mechanistic basis for differential inhibition of the F(1)F(o)-ATPase by aurovertin. Biopolymers **91**, 830–840 (2009)
67. W. Junge, H. Sielaff, S. Engelbrecht, Torque generation and elastic power transmission in the rotary FOF1-ATPase. Nature **459**, 364–370 (2009)
68. A.N. Kapanidis, N.K. Lee, T.A. Laurence, S. Doose, E. Margeat, S. Weiss, Fluorescence-aided molecule sorting: Analysis of structure and interactions by alternating-laser excitation of single molecules. Proc. Natl. Acad. Sci. USA **101**, 8936–8941 (2004)
69. Y. Kato-Yamada, H. Noji, R. Yasuda, K. Kinosita Jr, M. Yoshida, Direct observation of the rotation of epsilon -subunit in F1-ATPase. J. Biol. Chem. **273**, 19375–19377 (1998)
70. K. Kinosita Jr, R. Yasuda, H. Noji, K. Adachi, A rotary molecular motor that can work at near 100 % efficiency. Philos. Trans. R. Soc. Lond. B Biol. Sci. **355**, 473–489 (2000)
71. J. Lee, S. Lee, K. Ragunathan, C. Joo, T. Ha, S. Hohng, Single-molecule four-color FRET. Angew. Chem. Int. Ed. Engl. **49**, 9922–9925 (2010)
72. M. Lippitz, W. Erker, H. Decker, K.E. van Holde, T. Basche, Two-photon excitation microscopy of tryptophan-containing proteins. Proc. Natl. Acad. Sci. USA **99**, 2772–2777 (2002)
73. J.L. Martin, R. Ishmukhametov, T. Hornung, Z. Ahmad, W.D. Frasch, Anatomy of F1-ATPase powered rotation. Proc. Natl. Acad. Sci. USA **111**, 3715–3720 (2014)
74. S.A. McKinney, C. Joo, T. Ha, Analysis of single-molecule FRET trajectories using hidden Markov modeling. Biophys. J. **91**, 1941–1951 (2006)
75. D.T. McLachlin, A.M. Coveny, S.M. Clark, S.D. Dunn, Site-directed cross-linking of b to the α, β, and a subunits of the Escherichia coli ATP synthase. J. Biol. Chem. **275**, 17571–17577 (2000)
76. T. Meier, J. Faraldo-Gomez, M. Börsch, ATP Synthase—A Paradigmatic Molecular Machine, in *Molecular Machines in Biology*, ed by J. Frank (Cambridge University Press: New York, 2012), pp. 208–238
77. P. Mitchell, Coupling of phosphorylation to electron and hydrogen transfer by a chemi-osmotic type of mechanism. Nature **191**, 144–148 (1961)
78. W.E. Moerner, L. Kador, Optical detection and spectroscopy of single molecules in a solid. Phys. Rev. Lett. **62**, 2535–2538 (1989)
79. W.E. Moerner, M. Orrit, Illuminating single molecules in condensed matter. Science **283**, 1670–1676 (1999)
80. B.K. Muller, E. Zaychikov, C. Brauchle, D.C. Lamb, Pulsed interleaved excitation. Biophys. J. **89**, 3508–3522 (2005)

81. M. Nakanishi-Matsui, S. Kashiwagi, H. Hosokawa, D.J. Cipriano, S.D. Dunn, Y. Wada, M. Futai, Stochastic high-speed rotation of Escherichia coli ATP synthase F1 sector: the ε-subunit-sensitive rotation. J. Biol. Chem. **281**, 4126–4131 (2006)

82. M. Nakanishi-Matsui, M. Sekiya, M. Futai, Rotating proton pumping Atpases: subunit/subunit interactions and thermodynamics. IUBMB Life **65**, 247–254 (2013)

83. T.A. Nguyen, P. Sarkar, J.V. Veetil, S.V. Koushik, S.S. Vogel, Fluorescence polarization and fluctuation analysis monitors subunit proximity, stoichiometry, and protein complex hydrodynamics. PLoS ONE **7**, e38209 (2012)

84. H. Noji, R. Yasuda, M. Yoshida, K. Kinosita Jr, Direct observation of the rotation of F1-ATPase. Nature **386**, 299–302 (1997)

85. H. Omote, N. Sambonmatsu, K. Saito, Y. Sambongi, A. Iwamoto-Kihara, T. Yanagida, Y. Wada, M. Futai, The γ-subunit rotation and torque generation in F1-ATPase from wild-type or uncoupled mutant Escherichia coli. Proc. Natl. Acad. Sci. USA **96**, 7780–7784 (1999)

86. S. Pallikkuth, D.J. Blackwell, Z. Hu, Z. Hou, D.T. Zieman, B. Svensson, D.D. Thomas, S.L. Robia, Phosphorylated phospholamban stabilizes a compact conformation of the cardiac calcium-ATPase. Biophys. J. **105**, 1812–1821 (2013)

87. O. Panke, K. Gumbiowski, W. Junge, S. Engelbrecht, F-ATPase: specific observation of the rotating c subunit oligomer of EF(o)EF(1). FEBS Lett. **472**, 34–38 (2000)

88. C. Ratzke, B. Hellenkamp, T. Hugel, Four-colour FRET reveals directionality in the Hsp90 multicomponent machinery. Nat. Commun. **5**, 4192 (2014)

89. Y. Rondelez, G. Tresset, T. Nakashima, Y. Kato-Yamada, H. Fujita, S. Takeuchi, H. Noji, Highly coupled ATP synthesis by F1-ATPase single molecules. Nature **433**, 773–777 (2005)

90. R. Roy, S. Hohng, T. Ha, A practical guide to single-molecule FRET. Nat. Methods **5**, 507–516 (2008)

91. M.J. Rust, M. Bates, X. Zhuang, Sub-diffraction-limit imaging by stochastic optical reconstruction microscopy (STORM). Nat. Methods **3**, 793–795 (2006)

92. D. Sabbert, S. Engelbrecht, W. Junge, Intersubunit rotation in active F-ATPase. Nature **381**, 623–625 (1996)

93. Y. Sambongi, Y. Iko, M. Tanabe, H. Omote, A. Iwamoto-Kihara, I. Ueda, T. Yanagida, Y. Wada, M. Futai, Mechanical rotation of the c subunit oligomer in ATP synthase (F0F1): direct observation. Science **286**, 1722–1724 (1999)

94. D.K. Sasmal, H.P. Lu, Single-molecule patch-clamp FRET microscopy studies of NMDA receptor ion channel dynamics in living cells: revealing the multiple conformational states associated with a channel at its electrical off state. J. Am. Chem. Soc. **136**, 12998–13005 (2014)

95. M. Sauer, H. Neuweiler, PET-FCS: probing rapid structural fluctuations of proteins and nucleic acids by single-molecule fluorescence quenching. Methods Mol. Biol. **1076**, 597–615 (2014)

96. G.S. Schlau-Cohen, Q. Wang, J. Southall, R.J. Cogdell, W.E. Moerner, Single-molecule spectroscopy reveals photosynthetic LH2 complexes switch between emissive states. Proc. Natl. Acad. Sci. USA **110**, 10899–10903 (2013)

97. K. Seyfert, T. Oosaka, H. Yaginuma, S. Ernst, H. Noji, R. Iino, M. Börsch, Subunit rotation in a single F[sub o]F[sub 1]-ATP synthase in a living bacterium monitored by FRET. Proc. SPIE **7905**, 79050K (2011)

98. H. Sielaff, H. Rennekamp, A. Wachter, H. Xie, F. Hilbers, K. Feldbauer, S.D. Dunn, S. Engelbrecht, W. Junge, Domain compliance and elastic power transmission in rotary F(O)F(1)-ATPase. Proc. Natl. Acad. Sci. USA **105**, 17760–17765 (2008)

99. H. Sielaff, M. Börsch, Twisting and subunit rotation in single FOF1-ATP synthase. Phil. Trans. R. Soc. B **368**, 20120024 (2013)

100. D. Spetzler, J. York, D. Daniel, R. Fromme, D. Lowry, W. Frasch, Microsecond time scale rotation measurements of single F1-ATPase molecules. Biochemistry **45**, 3117–3124 (2006)

101. I.H. Stein, C. Steinhauer, P. Tinnefeld, Single-molecule four-color FRET visualizes energy-transfer paths on DNA origami. J. Am. Chem. Soc. **133**, 4193–4195 (2011)

102. T. Suzuki, K. Tanaka, C. Wakabayashi, E. Saita, M. Yoshida, Chemomechanical coupling of human mitochondrial F1-ATPase motor. Nat. Chem. Biol. **10**, 930–936 (2014)
103. M.A. Thompson, M.D. Lew, W.E. Moerner, Extending microscopic resolution with single-molecule imaging and active control. Annu. Rev. Biophys. **41**, 321–342 (2012)
104. C. Tietz, F. Jelezko, U. Gerken, S. Schuler, A. Schubert, H. Rogl, J. Wrachtrup, Single molecule spectroscopy on the light-harvesting complex II of higher plants. Biophys. J. **81**, 556–562 (2001)
105. J. Tisler, G. Balasubramanian, B. Naydenov, R. Kolesov, B. Grotz, R. Reuter, J.P. Boudou, P.A. Curmi, M. Sennour, A. Thorel, M. Börsch, K. Aulenbacher, R. Erdmann, P.R. Hemmer, F. Jelezko, J. Wrachtrup, Fluorescence and spin properties of defects in single digit nanodiamonds. ACS Nano **3**, 1959–1965 (2009)
106. S.P. Tsunoda, R. Aggeler, M. Yoshida, R.A. Capaldi, Rotation of the c subunit oligomer in fully functional F1Fo ATP synthase. Proc. Natl. Acad. Sci. USA **98**, 898–902 (2001)
107. M.J. van Raaij, J.P. Abrahams, A.G. Leslie, J.E. Walker, The structure of bovine F1-ATPase complexed with the antibiotic inhibitor aurovertin B. Proc. Natl. Acad. Sci. USA **93**, 6913–6917 (1996)
108. B. Verhalen, S. Ernst, M. Börsch, S. Wilkens, Dynamic ligand induced conformational rearrangements in P-glycoprotein as probed by fluorescence resonance energy transfer spectroscopy. J. Biol. Chem. **287**, 1112–1127 (2012)
109. C. von Ballmoos, G.M. Cook, P. Dimroth, Unique rotary ATP synthase and its biological diversity. Annu. Rev. Biophys. **37**, 43–64 (2008)
110. A. Wachter, Y. Bi, S.D. Dunn, B.D. Cain, H. Sielaff, F. Wintermann, S. Engelbrecht, W. Junge, Two rotary motors in F-ATP synthase are elastically coupled by a flexible rotor and a stiff stator stalk. Proc. Natl. Acad. Sci. USA **108**, 3924–3929 (2011)
111. Q. Wang, R.H. Goldsmith, Y. Jiang, S.D. Bockenhauer, W.E. Moerner, Probing single biomolecules in solution using the anti-Brownian electrokinetic (ABEL) trap. Acc. Chem. Res. **45**, 1955–1964 (2012)
112. Q. Wang, W.E. Moerner, Lifetime and spectrally resolved characterization of the photodynamics of single fluorophores in solution using the anti-Brownian electrokinetic trap. J. Phys. Chem. B **117**, 4641–4648 (2013)
113. R. Watanabe, R. Iino, H. Noji, Phosphate release in F1-ATPase catalytic cycle follows ADP release. Nat. Chem. Biol. **6**, 814–820 (2010)
114. R. Watanabe, D. Okuno, S. Sakakihara, K. Shimabukuro, R. Iino, M. Yoshida, H. Noji, Mechanical modulation of catalytic power on F(1)-ATPase. Nat. Chem. Biol. **8**, 86–92 (2012)
115. R. Watanabe, H. Noji, Chemomechanical coupling mechanism of F(1)-ATPase: catalysis and torque generation. FEBS Lett. **587**, 1030–1035 (2013)
116. L.P. Watkins, H. Yang, Detection of intensity change points in time-resolved single-molecule measurements. J. Phys. Chem. B **109**, 617–628 (2004)
117. J. Weber, A.E. Senior, ATP synthase: what we know about ATP hydrolysis and what we do not know about ATP synthesis. Biochim. Biophys. Acta **1458**, 300–309 (2000)
118. K. Weiss, A. Neef, Q. Van, S. Kramer, I. Gregor, J. Enderlein, Quantifying the diffusion of membrane proteins and peptides in black lipid membranes with 2-focus fluorescence correlation spectroscopy. Biophys. J. **105**, 455–462 (2013)
119. S. Winterfeld, S. Ernst, M. Börsch, U. Gerken, A. Kuhn, Real time observation of single membrane protein insertion events by the Escherichia coli insertase YidC. PLoS ONE **8**, e59023 (2013)
120. R. Yasuda, H. Noji, M. Yoshida, K. Kinosita Jr, H. Itoh, Resolution of distinct rotational substeps by submillisecond kinetic analysis of F1-ATPase. Nature **410**, 898–904 (2001)
121. M. Yoshida, E. Muneyuki, T. Hisabori, ATP synthase—a marvellous rotary engine of the cell. Nat. Rev. Mol. Cell Biol. **2**, 669–677 (2001)
122. N. Zarrabi, B. Zimmermann, M. Diez, P. Graber, J. Wrachtrup, M. Börsch, Asymmetry of rotational catalysis of single membrane-bound F0F1-ATP synthase. Proc. SPIE **5699**, 175–188 (2005)

123. N. Zarrabi, M.G. Düser, R. Reuter, S.D. Dunn, J. Wrachtrup, M. Börsch, Detecting substeps in the rotary motors of FoF1-ATP synthase by Hidden Markov Models. Proc. SPIE **6444**, 64440E (2007)

124. N. Zarrabi, M.G. Düser, S. Ernst, R. Reuter, G.D. Glick, S.D. Dunn, J. Wrachtrup, M. Börsch, Monitoring the rotary motors of single FoF1-ATP synthase by synchronized multi channel TCSPC. Proc. SPIE **6771**, 67710F (2007)

125. N. Zarrabi, T. Heitkamp, J.-C. Greie, M. Börsch, Monitoring the conformational dynamics of a single potassium transporter by ALEX-FRET. Proc. SPIE **6862**, 68620M (2008)

126. N. Zarrabi, S. Ernst, M.G. Düser, A. Golovina-Leiker, W. Becker, R. Erdmann, S.D. Dunn, M. Börsch, Simultaneous monitoring of the two coupled motors of a single FoF1-ATP synthase by three-color FRET using duty cycle-optimized triple-ALEX. Proc. SPIE **7185**, 718505 (2009)

127. N. Zarrabi, S. Ernst, B. Verhalen, S. Wilkens, M. Börsch, Analyzing conformational dynamics of single P-glycoprotein transporters by Förster resonance energy transfer using hidden Markov models, Methods **65**, 168–179 (2014)

128. Y. Zhou, T.M. Duncan, V.V. Bulygin, M.L. Hutcheon, R.L. Cross, ATP hydrolysis by membrane-bound Escherichia coli F0F1 causes rotation of the gamma subunit relative to the beta subunits. Biochim. Biophys. Acta **1275**, 96–100 (1996)

129. B. Zimmermann, M. Diez, N. Zarrabi, P. Graber, M. Börsch, Movements of the ε-subunit during catalysis and activation in single membrane-bound H(+)-ATP synthase. EMBO J. **24**, 2053–2063 (2005)

130. B. Zimmermann, M. Diez, M. Börsch, P. Graber, Subunit movements in membrane-integrated EF0F1 during ATP synthesis detected by single-molecule spectroscopy. Biochim. Biophys. Acta **1757**, 311–319 (2006)

Chapter 10
Partitioning and Diffusion of Fluorescently Labelled FTY720 in Resting Epithelial Cells

Dhanushka Wickramasinghe, Randi Timerman, Jillian Bartusek and Ahmed A. Heikal

Abstract Sphingosine-1-phosphate (S1P) is a bioactive signalling molecule that mediates important cellular functions such as cell proliferation, cytoskeletal rearrangement, angiogenesis, mobilization of intracellular calcium, and immune cell trafficking. 2-Amino-2-[2-(4-octylphenyl) ethyl] propane-1,3-diol hydrochloride (also known as FTY720 or Fingolimod), a synthetic analog of sphingosine that has immunosuppressive properties, is the first oral drug approved by the U.S. FDA for treatment of multiple sclerosis. It is hypothesized that FTY720 is phosphorylated by sphingosine kinase 2 (SphK2) prior to binding to S1P receptor 1 (S1PR$_1$) followed by internalization and degradation of the receptor. Here we examined the cellular uptake, partitioning, and diffusion of two fluorescent analogs of this drug (namely, Bodipy-FTY720) in cultured C3H10T1/2 cells, derived from mouse embryos. Multichannel imaging and co-localization with the endoplasmic reticulum (ER) tracker (Glibenclamide-Bodipy-FL) indicates that Bodipy-FTY720 resides in the ER of C3H10T1/2 cells, where it is expected to be phosphorylated by SphK2, and is excluded from the nucleus. This conclusion is supported by the rotational and translational diffusion measurements of this FTY720 analog in the ER membrane of the cells. Our results also indicate a heterogeneous environment of the cellular Bodipy-FTY720 as revealed by two-photon fluorescence lifetime imaging. These studies represent a step forward towards elucidating the effects of Bodipy moiety on the action mechanism of FTY720 at the single-cell level.

D. Wickramasinghe
Department of Chemistry and Biochemistry, University of Minnesota Duluth,
Duluth, MN 55812 USA
e-mail: wickr012@d.umn.edu

R. Timerman · J. Bartusek · A.A. Heikal (✉)
Department of Chemistry and Biochemistry, College of Pharmacy,
University of Minnesota Duluth, Duluth, MN 55812 USA
e-mail: aaheikal@d.umn.edu

J. Bartusek
e-mail: bart0688@d.umn.edu

© Springer International Publishing Switzerland 2015
W. Becker (ed.), *Advanced Time-Correlated Single Photon Counting Applications*,
Springer Series in Chemical Physics 111, DOI 10.1007/978-3-319-14929-5_10

10.1 Introduction

Sphingosine-1-phosphate is a bioactive signalling molecule that mediates important cellular functions such as cell proliferation, cytoskeletal rearrangement, angiogenesis, mobilization of intracellular calcium, and immune cell trafficking [21, 30]. Sphingosine-1-phosphate is derived from the phosphorylation of Sphingosine, which is prevalent in mammalian cellular membranes. De novo biosynthesis of sphingolipids is initiated at the outer leaflet of the endoplasmic reticulum, ER, [17]. The first and rate-limiting step is catalyzed by serine palmitoyltransferase (SPT) to produce 3-ketodihydrosphinganine, which is then reduced rapidly to form dihydrosphingosine (DHS). Following N acylation, dihydroceramide is formed and then converted to ceramide, which can be further hydrolyzed by acid ceramidase to form sphingosine. Sphingosine can be phosphorylated and dephosphorylated by kinases and phosphatases, respectively. The phosphorylation of sphingosine, an ATP-dependent process, is carried out by sphingosine kinase 1 (SphK1) or sphingosine kinase 2 (SphK2) to produce sphingosine-1-phosphate (S1P). Dephosphorylation of S1P by Sphingosine phosphatase 1 or 2 (a magnesium dependent process) regulates the level of S1P in the cell. Another mechanism of S1P regulation is the irreversible degradation by sphingosine-1-phosphate lyase, which yields hexadecenal and phosphoethanolamine.

Sphingosine-1-phosphate acts as a ligand for five sphingosine-1-phosphate receptors (S1PR$_{1-5}$), which belong to the endothelial differentiation gene (EDG) family of G protein coupled receptors, GPCR, [32]. Sphingosine-1-phosphate acts as a bioactive metabolite in various cellular processes on the cell surface (e.g., binding to the five receptors S1PR$_{1-5}$) and uncharacterized intracellular targets. S1P is also transported to the extracellular milieu via ATP-binding cassette family transporters (ABC transporters), namely ABCC1 and ABCA1 [25, 29]. These S1PR$_{1-5}$ receptors have been discovered in almost every tissue tested with different expression levels, which suggest their importance in biological functions [17]. The phosphate group and the C3 hydroxyl group of S1P are believed to be significant for the binding affinity to Sphingosine-1-phosphate receptors [24]. Furthermore, the D-erythro configuration of S1P is essential for the high binding affinity to these receptors. However, in recent computational studies, it has been shown that the hydroxyl group of S1P has the least contribution in S1P binding recognition to these receptors except in S1PR$_3$ [28].

Sphingosine-1-phosphate has recently gained a lot of attention for its relation to the novel immunomodulator, FTY720, which is used to treat multiple sclerosis (MS). S1P plays a crucial role in MS-related processes that include inflammation and repair. During inflammation, S1P is released by platelets and can also be found in the serum at significant levels [9]. S1PR$_1$ is known to be widely expressed in lymphocytes and regulates the normal egress from the lymph nodes [17]. During such egress, T lymphocytes migrate across the S1P gradient between lymph nodes and the afferent lymph. Under a normal immune response, activation of T cells in the lymph nodes causes an internalization of the S1P$_1$ receptor. After activation of T cells, $S1PR_1$ is recycled and re-circulated by the T cells to peripheral sites. In patients with MS, circulating auto aggressive T cells cross the blood brain barrier into the central nervous system (CNS) causing inflammation. This leads to the destruction of the myelin sheath that surrounds the axons as well as the loss of oligodendrocytes in great numbers causing demyelination, which will ultimately cause the loss of axons and neurons. Although the CNS damage is usually self-repaired, the probability of recovery after repeated inflammation episodes is reduced and can lead to tissue damage and the development of MS.

FTY720 (Fingolimod) is an immunomodulator (Fig. 10.1) that is highly effective in animal model systems for transplantation and autoimmunity [10]. FTY720, a sphingosine analog [7, 8, 9], is a member of the sphingosine-1-phosphate receptor (S1PR) modulators [2] and the first oral drug to be approved by the US Food and Drug Administration (FDA) for MS treatment. Similar to sphingosine, FTY720 is also believed to be phosphorylated by SphK2 [27]. However, FTY720 is not phosphorylated by sphingosine kinase 1 and is believed to bind to $S1PR_1$ (highest affinity), $S1PR_3$, $S1PR_4$, and $S1PR_5$ of the GPCRs [11]. The biologically active form of FTY720 is found to be the (s)-enantiomer [1]. The phenyl ring between its polar head and the lipophilic tail has been shown to increase the agonism in $S1PR_5$, decrease of activity in $S1PR_2$ and loss of stereo-specificity in $S1PR_{1,3}$ [13]. The lipophilic tail of FTY720 (Fig. 10.2) is important for binding to the hydrophobic pocket of these receptors.

In a pharmacokinetics study, FTY720 is found in blood when administered intravenously; but not FTY720-P [31]. However, a higher level of FTY720-P was found in blood when administered orally. Since SphK2 is highly expressed in the liver, this suggests that the first-pass metabolism generates FTY720-P [31].

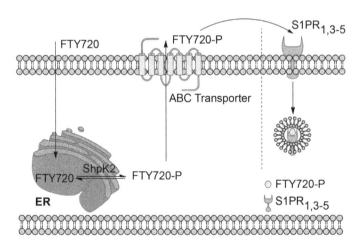

Fig. 10.1 Schematic sketch of the hypothesized mechanism of S1P-analog, FTY720. Following the cellular uptake, FTY720 is phosphorylated by SphK2 in the endoplasmic reticulum (ER). The FTY720-P is then transported to the extracellular milieu by an ABC transporter. The irreversible internalization of FTY720-P then takes place via S1P receptors ($S1PR_{1,3-5}$)

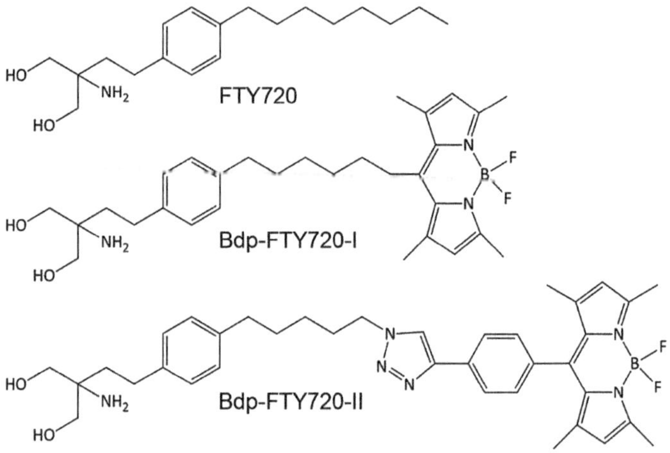

Fig. 10.2 The chemical structure of FTY720 and the two fluorescent analogs used in these studies, namely, Bodipy-FTY720-I and Bodipy-FTY720-II [7, 22]. In both Bdp-FTY720 derivatives, the Bodipy moiety is separated from the phosphorylation site of FTY720 [22], which is believed to be phosphorylated by SphK2 for the subsequent binding site to S1P receptors. The steady-state spectroscopy of Bdp-FTY720 (ethanol) reveals a maximum absorption and emission at 498 and 508 nm, respectively (data not shown). The corresponding maximum extinction coefficient at 500 nm is $\varepsilon = 72{,}000$ M^{-1} cm^{-1}. Similar spectra were observed for LZ570 under the same conditions with a negligible spectral shift

Eventually, FTY720 is metabolized in the liver by the cytochrome P450 CYP4F enzyme with a half-life of 5–6 days [20]. Although the phosphorylated FTY720 is an agonist to the S1P receptors, it induces internalization and degradation of the S1P receptor [32]. This interferes with the normal egress of the lymphocytes from the lymph nodes, thus inducing a prolonged state of immune system suppression. It takes about 2–8 days for full recovery to the normal expression levels of the $S1P_{1,\ 3-5}$ receptors. As a result, the activities of FTY720 due to short-term exposure lead to a prolonged immunosuppressant action.

Here we investigate the biophysical properties of two fluorescent analogs of the immunosuppressor FTY720, namely Bodipy-labelled FTY720 I and II (Fig. 10.2) at the single-cell level. Bodipy is a relatively hydrophobic fluorophore with a large extinction coefficient and fluorescence quantum yield [3, 14, 23, 33], which makes it ideal for fluorescence-based studies. The two Bodipy-FTY720 analogs are generous gifts from Dr. Robert Bittman (Queen's College, NY) and the synthesis was described elsewhere [23]. Our working hypothesis is that the fluorescence labelled Bdp-FTY720 (hereafter, Bdp-FTY720) is likely to follow the same mechanism as of the parent immune modulator (FTY720) as described in Fig. 10.1. To test this hypothesis, we employed multi-parametric fluorescence micro-spectroscopy methods of Bdp-FTY720 in adherent mouse embryo fibroblast C3H10T1/2 cells.

10.2 Materials and Methods

10.2.1 Sample Preparation

Stock solution of Bdp-FTY720 analogs were prepared in ethanol (1 mg/mL) and stored at 4 °C [23]. The steady-state spectroscopy of Bdp-FTY720 (ethanol), using Beckman spectrophotometer (DU800) and Fluorolog spectrofluorimeter (FL1000), reveals a maximum absorption ($\varepsilon = 72{,}000$ M^{-1} cm^{-1}) and emission at 500 and 508 nm, respectively.

ER Tracker Red^{TM} Dye (Molecular Probes) was used to label the endoplasmic reticulum [12] in adherent mouse embryo fibroblasts (C3H10T1/2 cells, CCL-226™) for co-localization studies. The cells were cultured in Eagle's Basal Medium (BME) Media (ATCC) containing 2 mM L-glutamine supplemented with 10 % heat inactivated Fetal Bovine Serum (FBS), 7.5 % w/v% Sodium Bicarbonate and Penicillin (100 U/mL medium)-Streptomycin (100 mg/mL medium). Cells were then plated in glass-bottom Petri dishes (MatTek) and incubated overnight before imaging. Bdp-FTY720 was added (~ 3 μM or 20 nM based on the type of experiments) to the cell culture media (10-min incubation), followed by three washes with Tyrode's buffer (135 mM NaCl, 5 mM KCl, 1 mM $MgCl_2$, 1.8 mM $CaCl_2$, 20 mM HEPES, and 5 mM glucose) prior to imaging.

10.2.2 FLIM System

The optical principle of the FLIM and micro-spectroscopy setup [18, 33] is shown in Fig. 10.3. The laser scanning microscope consists of an Olympus IX80 inverted microscope with an Olympus FV300 scan head. The excitation source is a Coherent Mira 900F femtosecond titanium-sapphire laser. For FLIM and anisotropy decay measurements, two photon (2P) excitation at the fundamental wavelength of the laser (950 nm, 76 MHz) was used. Alternatively, the second harmonic of the laser wavelength (475 nm, 4.2 MHz, Mira 9200, Coherent) can be used for complementary one-photon (1P) experiments.

The 2P-excited fluorescence signals are detected via a non-descanned (NDD) beam path. The fluorescence light is separated from the excitation light by a dichroic beamsplitter in the filter carousel of the microscope. The fluorescence light is projected out of a side port of the microscope. A field lens projects an image of the back aperture of the microscope lens on the detectors. Residual laser light is removed by a Chroma SP700 short pass filter. The light is then split into two polarisation or wavelength components by a polarising or dichroic beamsplitter. The fluorescence components are further cleaned up of by bandpass filters or polarisers in front of the detectors.

The detectors are Hamamatsu R3809U-50 multichannel PMTs (MCP-PMTs). In the TCSPC mode, the detectors deliver an IRF of about 30–50 ps FWHM [2, 4]. The fast IRF has advantages especially for anisotropy decay measurements: Differences

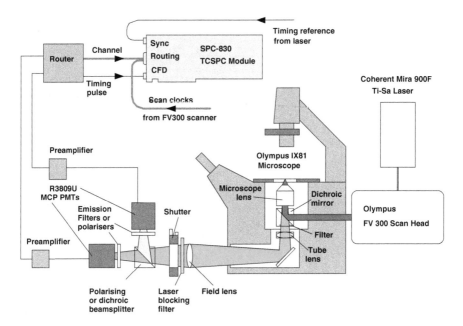

Fig. 10.3 Optical setup used for FLIM and micro-spectrometry. This figure is courtesy of Wolfgang Becker

in the IRF shape of the two detectors are negligible compared to the anisotropy decay time and need not be considered in the data analysis [4, 6]. However, operating MCP PMTs requires a few precautions. The detectors can easily be damaged by overload. Therefore the detectors are operated via a bh DCC-100 detector controller card. The DCC-100 controls the high voltage of the detector power supply. In case of overload, the preamplifiers detect the overload condition, and the DCC-100 shuts down the detector operating voltage and closes a shutter in the detector beam path [4, 6].

To obtain a reasonable count rate from the MCP PMTs the light has to be spread over the full cathode area. If the light is focused on a small spot, the microchannels in the illuminated area may saturate, and the IRF quickly degrades with increasing count rate [2, 4]. The focal length of the field lens and its position on the beam path was therefore selected to obtain an illuminated area of about 8 mm for the microscope lenses typically used. With these precautions, and by running the R3809Us at a reduced operating voltage of 3 kV (the maximum is 3.5 kV) the detectors can be operated at a count rate of several MHz [4, 6].

The single-photon pulses from the detectors are amplified by bh HFAC-26 preamplifiers. The amplified signals are connected to an HRT-41 router. The router combines the pulses of both detectors into a common timing pulse line and generates a routing signal that indicates which of the detectors delivered a particular photon pulse. These signals are processed in a single bh SPC-830 TCSPC FLIM module. The routing signal is used to obtain separate photon distributions for the photons detected in different detectors [2, 4, 6].

In contrast with the classic TCSPC [26], the FLIM mode of the SPC-830 uses a multi-dimensional recording process [2, 4], see Chap. 1. The result is a photon distribution over the arrival times of the photons after the laser pulses, the scan coordinates, and the detector channel number delivered by the router. To determine the spatial positions of the individual photons in the scan area the TCSPC module counts synchronisation pulses (frame, line and pixel clock) from the scanner. Unfortunately, the FV 300 does not deliver pixel clock pulses. The SPC-830 must therefore be operated with an internal pixel clock. That means the position within a line is not determined by counting pixel clocks from the scanner but by measuring the time after the last line clock [4].

The TCSPC/FLIM system is controlled by Becker & Hickl SPCM data acquisition software [4]. FLIM data were analysed by Becker & Hickl SPCImage data analysis software [4]. For a brief description please see Chap. 15, Sect. 15.4.

The same setup was used for time-resolved anisotropy, where the parallel $I_\parallel(t, x, y)$ and perpendicular $I_\perp(t, x, y)$ polarization fluorescence decays were measured simultaneously [18, 33]. The G-factor of our setup was measured using Rhodamine green (Invitrogen) in Phosphate Buffered Saline (PBS, pH 7.4, GIBCO) and the tail-matching approach [22]. Nonlinear least square fitting of time-resolved anisotropy was carried out using Origin 8.1 software.

10.2.3 FCS Recording

Fluorescence correlation spectroscopy (FCS) was used to quantify the translational diffusion of single Bdp-FTY720 molecules, excited at 488 nm, as they diffused through an open observation volume [19, 33]. In principle, the SPC-830 TCSPC module used in the FLIM and microspectrometer setup is also able to record FCS [4, 5]. However, the R3809U-50 detector in the FLIM path are not the most efficient FCS detectors. Moreover, 2P excitation and NDD often do not deliver the best possible signal-to-noise ratio for FCS. The reasons are probably that photobleaching in the focal plane is higher for 2P than for 1P excitation [15], and that daylight detection is hard to avoid in an NDD setup [4].

The fluorescence was therefore excited by 1P excitation and detected via a confocal beam path. The fluorescence light was focused on an optical fibre (50 μm in diameter), which acts as a confocal pinhole to reject out-of-focus fluorescence photons. The light was detected by a single-photon avalanche photodiode (SPAPD, SPCM CD-2969, Perkin-Elmer, Fremont, CA). The time-dependent fluorescence fluctuations, from different locals in living cells, were then autocorrelated using an external multiple-tau-digital correlator (ALV/6010-160). The recording process in the correlator is similar to the process described for TCSPC-based FCS recording, see Chap. 1, Sect. 1.5.2. The system is routinely calibrated using rhodamine green in PBS at room temperature (translational diffusion coefficient, $D_T = 2.8 \times 10^{-6}$ cm^2/s) and the autocorrelation curves were analyzed using Origin 8.1 (Origin®).

10.3 Results and Discussion

10.3.1 Internalization and Partitioning of Bodipy-FTY720 in C3H10T1/2 Cells

To examine the effect of the Bodipy moiety on the uptake of Bdp-FTY720 in mouse fibroblast C3H10T1/2 cells [27], we carried out time-lapse imaging using confocal and DIC microscopy. The Bdp-FTY720 (0.5 μM) was added at time zero to the on-stage cultured cells and representative time-lapse confocal images are shown in Fig. 10.4a. The integrated intracellular fluorescence of cellular Bdp-FTY720 (Fig. 10.4b) reached a plateau after ~20 min with a 50 % level after 6 ± 1 min. The minor modulation of the time-dependent integrated fluorescence was observed, regardless of the region of interest (squares) used for integration. At a later time following the internalization, Bdp-FTY720-labeled cytoplasmic vesicles appear throughout the cell. To examine the exocytosis of Bdp-FTY720-I, stained cells were also imaged following 24-h and 72-h incubations. The results show that the background fluorescence of both cellular Bdp-FTY720 derivatives diminishes with only remaining fluorescent vesicles throughout the cytoplasm of the cells. These results indicate an active exocytosis of both FTY720 analogs, presumably after phosphorylation in the ER by SphK2. The corresponding DIC images suggest viable cells with healthy morphology.

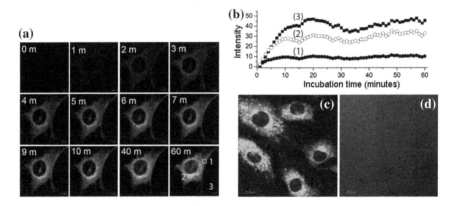

Fig. 10.4 Confocal and DIC images reveal the uptake and distribution of Bdp-FTY720 in living C3H10T1/2 cells. Representative time-lapse (0–60 min) imaging of Bdp-FTY720-II uptake by cultured C3H10T1/2 cells using on-stage incubation with 0.5 μM (**a**). Time-dependent fluorescence integration over different regions of interests (**b**, *squares*) shows that the uptake of the fluorescent immune modulator reaches a plateau after ~20 min with a 50 % point at ~6 min. Confocal (**c**) and DIC (**d**) images at a later time reveal the formation of Bdp-FTY720-II-labeled vesicles. As shown in those images, the FTY720 analogs are excluded from the nucleus. Confocal images of time-dependent incubation reveal a reduced background fluorescence signal and the density of the above-mentioned vesicles

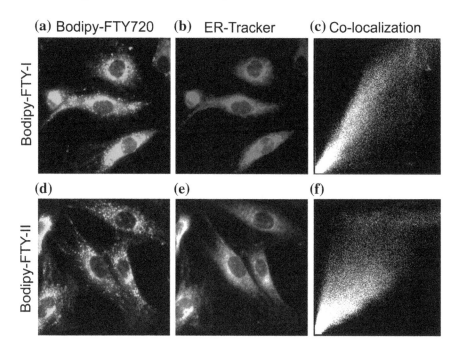

(a) Bodipy-FTY720   (b) ER-Tracker   (c) Co-localization

(d)   (e)   (f)

Fig. 10.5 Bdp-FTY720 analogs exhibit an affinity to the ER of cultured C3H10T1/2 cells. Doubly labelled cells with Bdp-FTY720 (**a, d**) and ER tracker-red, Glibenclamide-Bodipy-FL, (**b, e**) were imaged using two-channel (excited at 488 and 561 nm, respectively) confocal microscopy. The *green-red* correlation plots of these images (**c** and **f**) indicate co-localization of both Bdp-FTY720 analogs in the ER in C3H10T1/2 cells with an estimate correlation of 0.941

In order to test their partitioning, cultured C3H10T1/2 cells were also doubly stained with Bdp-FTY720-I and an ER marker (namely, Glibenclamide-Bodipy-FL), which were excited using 488-nm and 561-nm lasers, respectively. Two-channel confocal imaging of both fluorophores was recorded (Fig. 10.5a, b) and the corresponding correlation plot between the two dyes indicates the localization of Bdp-FTY720 in the ER of C3H10T1/2 cells (Fig. 10.5c). Similar results were obtained for the second Bdp-FTY720-II derivative (Fig. 10.5d–f). These results suggest that the Bodipy moiety does not interfere with the uptake of labelled FTY720 through the plasma membrane and its affinity to the ER membrane, similar to the parent immunosuppressor, where it is likely to be phosphorylated by SphK2.

10.3.2 2-Photon-FLIM Reveals Heterogeneity of Cellular Bodipy-FTY720

Representative 2P-FLIM Fluorescence (Fig. 10.6) reveals apparent heterogeneity of the fluorescence lifetime of Bdp-FTY720-II in adherent C3H10T1/2 cells with a

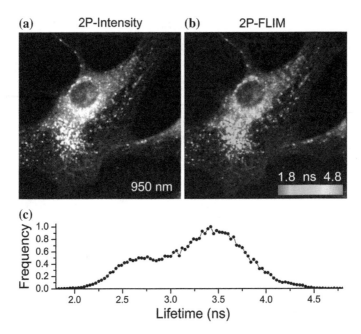

Fig. 10.6 Two-photon FLIM ofBdp-FTY720-II reveals heterogeneous conformations and local microenvironments in cultured C3H10T1/2 cells. Two-photon fluorescence intensity (**a**) and FLIM (**b**) images, excited at 950 nm, reveal different lifetimes of Bdp-FTY720-II in the ER membrane (longer lifetime) as compared with the labelled vesicles (*punctate-like features*). Similar measurements were carried out in Bdp-FTY720-I (data not shown)

lifetime-pixels histogram of a mean lifetime of 3.4 ± 0.5 ns. The histogram also reveals a short-lifetime shoulder, which suggests a double Gaussian profile throughout the image. In what seems to be the rough ER region, the fluorescence decays as a single exponential with a fluorescence lifetime of 4.4 ± 0.1 ns. Away from the nucleus and in the smooth ER regions, single exponential is adequate to describe the fluorescence decay with slightly shorter lifetime (4.1 ± 0.2 ns). In the apparent vesicle-like organelles observed in our confocal microscopy (Fig. 10.4), the fluorescence of BdpFTY720 decays as a bi-exponential (e.g., $\tau_1 = 1.29$ ns, $\alpha_1 = 0.52$, $\tau_2 = 4.52$ ns, $\alpha_2 = 0.47$) with a significantly shorter average lifetime (2.8 ± 0.1 ns).

In these FLIM images, the shorter fluorescence lifetime of Bdp-FTY720 observed in vesicles may be attributed to quenching (or fluorescence resonance energy transfer) between potential multiple copies of the fluorophore inside each vesicle. The fast component (τ_1) histogram of a given 2P-FLIM of Bdp-FTY720-II has a mean value of 2.3 ± 0.7 ns ($\alpha_1 = 0.62$) as compared with the slow-component histogram of 6.1 ± 0.7 ns ($\alpha_2 = 0.38$). As expected, the species with longer lifetime contribute most (64 %) to the detected fluorescence signal in these FLIM images (Fig. 10.6). The χ^2-histogram in these FLIM images has a mean value of 1.1 ± 0.1.

Complementary high temporal resolution, single-point 1P-fluorescence of Bdp-FTY720-II in the cytoplasm of C3H decays as a triple exponential ($\tau_1 = 1.61$ ns,

$\alpha_1 = 0.34$, $\tau_2 = 4.38$ ns, $\alpha_2 = 0.17$, $\tau_3 = 5.01$ ns, and $\alpha_3 = 0.49$) with an average lifetime of 3.76 ns. Under the same experimental conditions, the 1P-fluorescence of rhodamine green in a buffer decays as a single exponential with a lifetime of 3.99 ns. These FLIM results highlight the complex environmental and structural heterogeneity of these FTY720 derivatives in the cultured cells.

10.3.3 Rotational Diffusion, Fluidity, and Order of Bodipy-FTY720 in C3H10T1/2 Cells

Single-point, time-resolved anisotropy measurements (see Chap. 12) were carried out on randomly selected regions of interest in order to examine the degree of environmental restriction of Bdp-FTY720 in living C3H10T1/2 cells. In these measurements, the laser was focused randomly on regions of interest in the cytoplasm and away from the nucleus. The G-factor and calibration of our experimental setup was carried out using rhodamine green (PBS, pH 7.4), at room temperature, with an estimated rotational time of ~ 120 ps [16, 33]. Time-resolved anisotropy of Bdp-FTY720 (I and II, in ethanol) is best described with a bi-exponential decay with $\phi_1 \sim 80$ ps, $\beta_1 = 0.22$ and $\phi_2 \sim 550$ ps, $\beta_2 = 0.10$.

Representative anisotropy decay of Bdp-FTY720 in cultured C3H10T1/2 cells is shown in Fig. 10.7a. The anisotropy of cytosolic Bdp-FTY720 decays as a bi-exponential with a dominant, slow tumbling motion on the order of $\phi_1 = 20 \pm 5$ ns with amplitude $\beta_1 = 0.17 \pm 0.01$. The second fast component ($\phi_2 = 1.2 \pm 0.3$ ns) has amplitude (β_2) of 0.05 ± 0.01. The observed slow rotational time of Bdp-FTY720 is attributed to the association of this immune modulator with ER membranes in the intracellular milieu. With these fitting parameters of cellular Bdp-FTY720 anisotropy, we also estimate an order parameter (S) of 0.88 ± 0.05 [18], which indicates a degree of order among these fluorophores in the ER membrane. In addition, the lack of picoseconds rotational component indicates the lack of a freely tumbling FTY720 analog in the overall restrictive ER membrane.

To test whether Bdp-FTY720 will be secreted into the culture media, presumably after phosphorylation, we sampled the culture media of Bdp-FTY720-stained cells over 48 h of incubation. Representative anisotropy decays (Fig. 10.7a, curves 3 and 4) of both sampled Bdp-FTY720 species in the cultured media (24 h) seem to be distinct from the corresponding pure culture media incubating the unstained cells. The anisotropy decays of the likely secreted Bodipy-FTY720 species in the cultured medium is best described as bi-exponential with vastly different rotational times. As a control, similar culture media was sampled from cultured cells without the Bdp-FYTY720 staining, where the signal from pure culture medium was negligible (Fig. 10.7b, curve 2). The slight increase in the diffusion coefficient of Bdp-FTY720 may be attributed to its hydrophobic nature, which may lead to micelles formation.

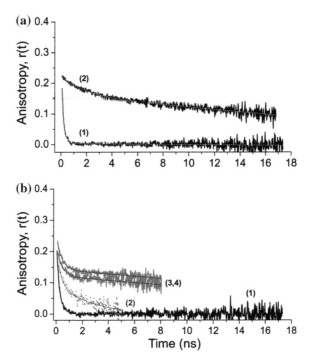

Fig. 10.7 The fluorescence anisotropy of Bdp-FTY720 is multi-exponential in both the cytosol and plasma membrane of cultured C3H10T1/2 cells. **a** The intracellular Bdp-FTY720-II anisotropy decays mostly as a bi-exponential with predominant slow component of $\phi_2 = 25 \pm 5$ ns ($\phi_2 = 0.19 \pm 0.02$) and a minor ($\phi_1 = 0.04 \pm 0.01$) fast component ($\phi_1 = 1.3 \pm 0.4$ ns) with an estimate initial anisotropy of $r_0 = 0.23 \pm 0.01$ (*curve 2*). As a control, the anisotropy of rhodamine green (PBS, pH 7.4) decays with a rotational time of 140 ± 10 ps (*curve 1*) and was used to determine the G-factor for our experimental setup. **b** Anisotropy decays of cultured medium, sampled as a function of time (here after 24 h incubation), from Bdp-FTY720-stained cells (*curves 3 and 4*). As a control, the culture media of non-stained cells yield very low fluorescence signal with distinct anisotropy decay (*curve 2*)

10.3.4 Translational Diffusion of Bodipy-FTY720 in C3H10T1/2 Cells

At the single-molecule level, FCS is the method of choice for quantifying the translational diffusion of Bdp-FTY720 in resting C3H10T1/2 cells. Changes in the hydrodynamic radius due to association with other biomolecules will affect the observed translational diffusion coefficient. The same approach provides insights into the restriction of Bdp-FTY720 in its local microenvironment in the living cell. The separation of both effects (i.e., association or environmental restriction) is rather a nontrivial task. The observation volume (radius ~ 260 nm) in our FCS setup was calibrated using rhodamine green in PBS buffer (pH 7.4, $\eta \sim 0.9$ cP) with known diffusion coefficient ($D = 2.8 \times 10^{-8}$ cm^2/s) at room temperature. As a control, the diffusion coefficient of free Bdp-FTY720 analogs in ethanol ($\eta \sim 1.1$ cP) is $\sim 2.1 \times 10^{-6}$ cm^2/s at room temperature.

In the plasma membrane (above the nucleus or at the periphery) of resting C3H10T1/2 cells, the fluorescence fluctuation autocorrelation of Bdp-FTY720-I decays as a two-dimensional anomalous diffusion with an anomalous coefficient (α) of 0.70 ± 0.06 (n = 10). The corresponding 2D-diffusion time is 25 ± 6 ms, which yields a diffusion coefficient of Bdp-FTY720 in the plasma membrane of $(11.7 \pm 0.5) \times 10^{-9}$ cm^2/s. These results indicate heterogeneity in the landscape of a restrictive plasma membrane in this cell line under resting conditions. When the autocorrelation curves were also satisfactorily fit using two-diffusing species in the 2D plasma membrane. Similar results were obtained for the other Bdp-FTY720-II analog.

Similar measurements were carried out on Bdp-FTY720 in the ER of C3H10T1/2 cells (Fig. 10.8a), where regions of interest were selected in the cytoplasm away from the nucleus. The fluorescence fluctuation autocorrelation of Bdp-FTY720 analogs in

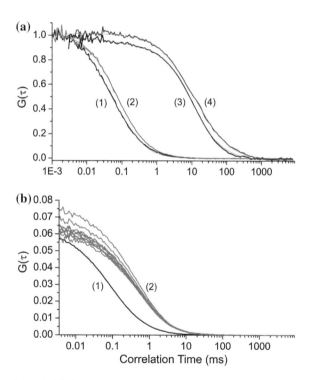

Fig. 10.8 Translational diffusion of Bdp-FTY720 in the ER and the plasma membrane of cultured C3H10T1/2 cells. **a** In the plasma membrane, Bdp-FTY720-I (*curve 3*) diffuses slightly faster than that in the ER (*curve 4*) of these cultured cells. For calibration, the autocorrelation functions of diffusing RhG (PBS, *curve 1*) and free Bdp-FTY720-II in ethanol (*curve 2*) at room temperature were used. **b** The fluorescence fluctuation autocorrelation of cultured media, sampled as a function of incubation time, is shown collectively as *curve 4*

the ER is best described by 3D, two-diffusing species model. Importantly, one of the diffusing species in the cytosol exhibits a significantly slower diffusion coefficient as compared with that in the plasma membrane. For example, a fraction (95 %) of the Bdp-FTY720 species diffuses at a rate of $(5.1 \pm 0.8) \times 10^{-8}$ cm^2/s as compared with 5 % of the population diffusing at a rate of $(2.69 \pm 0.04) \times 10^{-9}$ cm^2/s in the cytosol of resting C3H10T1/2 cells. Since the cytosolic regions of interests were chosen randomly, it is likely that some of the vesicles observed in our confocal images could be included in our observation volume, which might be assigned for $(2.69 \pm 0.04) \times 10^{-9}$ cm^2/s diffusing species.

It is worth noting that the diffusion coefficient of Bdp-FTY720 in the plasma membrane $(1.17 \pm 0.5 \times 10^{-8}$ cm^2/s) is about 4.3 times slower than that in the ER membrane $(5.1 \pm 0.8 \times 10^{-8}$ cm^2/s). Such distinct diffusion coefficients suggest a different structural mosaic of the ER membrane as compared with the plasma membrane of resting C3H10T1/2 cells. For example, the protein-to-lipid content ratio, lipid types, and cholesterol content would explain the observed difference in diffusion processes.

Both FCS and time-resolved anisotropy results were used to estimate the translational-to-rotational diffusion coefficient ratio ($D_T/D_R = 4a^2/3$) and therefore the hydrodynamic radius (a) of the intracellular Bdp-FTY720. Using Stokes-Einstein model, the D_T/D_R ratio is independent of the environmental viscosity surrounding the diffusing fluorophore. Our results indicate a hydrodynamic radius of ~ 14 nm, which is significantly larger than a free Bodip-FTY720 in a solution. In these calculations, we used the average rotational time along with the slow translational diffusion time of intracellular species.

For further control, we examined the culture media with and without Bdp-FTY720 (6 nM) under the same conditions. The signal from pure culture medium was negligible and without significant correlation amplitude. In addition, the autocorrelation function of free Bdp-FTY720 in a cultured medium (without cells) indicates a diffusion time of 0.21 ± 0.03 ms, which corresponds to a diffusion coefficient of 1.11×10^{-6} cm^2/s. The slight increase in the diffusion coefficient of Bdp-FTY720 may be attributed to its hydrophobic nature, which may lead to micelles formation. Representative fluorescence fluctuation autocorrelation curve of sampled Bdp-FTY720 species in the cultured media (24 h) is shown. The autocorrelation function of secreted Bodipy-FTY720 species in the cultured medium is best described with two diffusing species in 3D model. For example, the diffusion coefficient of the slowly diffusing Bdp-FTY720 species secreted in the culture medium is $(4.6 \pm 0.9) \times 10^{-7}$ cm^2/s at room temperature. Assuming the viscosity of water at room temperature, the hydrodynamic radius of such species is ~ 4.9 nm and therefore distinct from a free Bdp-FTY720 or its phosphorylated counterpart.

10.4   Conclusion

Using multi-parametric fluorescence micro-spectroscopy approaches, our results indicate that Bdp-FTY720 analogs are internalized through the plasma membrane of cultured, mouse embryonic C3H10T1/2 cells. The immune modulator analogs partition in the ER membrane, where the hypothesized phosphorylation takes place similar to the parent FTY720, which indicates a negligible effect of the Bodipy moiety on the cell uptake and partitioning in the ER. These conclusions are based on the double labelling experiments with ER Tracker (Glibenclamide-Bodipy-FL), the time-dependent formation of cellular Bdp-FTY720-labeled vesicles, and the slow 2D translational diffusion. Interestingly, the local environment (ER versus intracellular vesicles) of this fluorescent immune modulator differs according to FLIM measurements. The time-dependent sampling of the culture media of incubated cells reveals the presence of secreted Bdp-FTY720 that is different in size. Taken together, our results support, in part, the hypothesis that Bdp-FTY720 is likely to follow a similar mechanism as the parent immune modulator FTY720. These studies represent a step forward towards new insights into the action mechanism of FTY720 using its fluorescent derivative (Bdp-FTY720) at the single-cell level.

Acknowledgments   We are grateful to Dr. Robert Bittman (Queens College, NY) for his generous gift of both fluorescent analogs that were used in these studies. D.W. acknowledges the financial support (teaching assistantship) from the Department of Chemistry and Biochemistry, Swenson College of Science and Engineering, University of Minnesota Duluth. R.T. and J.B. were supported by Summer Undergraduate Research Program (SURP) and Undergraduate Research Opportunities Program (UROP) awards. Additional financial support was provided by Grant-In-Aid of Research, Artistry and Scholarship (University of Minnesota), NSF (MCB0718741), NSF/REU, and NIH (AG030949).

References

1. R. Albert, K. Hinterding, V. Brinkmann, D. Guerini, C. Muller-Hartwieg, H. Knecht, C. Simeon, M. Streiff, T. Wagner, K. Welzenbach, F. Zecri, M. Zollinger, N. Cooke, E. Francotte, Novel immunomodulator FTY720 is phosphorylated in rats and humans to form a single stereoisomer. Identification, chemical proof, and biological characterization of the biologically active species and its enantiomer. J. Med. Chem. **48**, 5373–5377 (2005)
2. F. Antochi, Update in neurology. Maedica (Buchar) **6**, 64 (2011)
3. F.S. Ariola, Z. Li, C. Cornejo, R. Bittman, A.A. Heikal, Membrane fluidity and lipid order in ternary giant unilamellar vesicles using a new bodipy-cholesterol Derivative. Biophys. J. **96**, 2696–2708 (2009)
4. W. Becker, *Advanced Time-Correlated Single-Photon Counting Techniques* (Springer, Berlin, Heidelberg, New York, 2005)
5. W. Becker, A. Bergmann, E. Haustein, Z. Petrasek, P. Schwille, C. Biskup, L. Kelbauskas, K. Benndorf, N. Klöcker, T. Anhut, I. Riemann, K. König, Fluorescence lifetime images and correlation spectra obtained by multi-dimensional TCSPC. Micr. Res. Tech. **69**, 186–195 (2006)

6. W. Becker, The bh TCSPC Handbook, 5th edn. (Becker & Hickl GmbH) (2012), electronic version available on www.becker-hickl.com, printed copies available from Becker & Hickl GmbH

7. T. Blom, N. Bäck, A.-L. Mutka, R. Bittman, Z. Li, A. de Lera, P.T. Kovanen, U. Diczfalusy, E. Ikonen, FTY720 Stimulates 27-hydroxycholesterol production and confers atheroprotective effects in human primary macrophages. Circ. Res. **106**, 720–729 (2010)

8. V. Brinkmann, A. Billich, T. Baumruker, P Heining, R. Schmouder, G. Francis, S. Aradhye, P. Burtin, Fingolimod (FTY720): discovery and development of an oral drug to treat multiple sclerosis. Nat. Rev. Drug. Discov. **9**, 883–897 (2010)

9. V. Brinkmann, K.R. Lynch, FTY720: targeting G-protein-coupled receptors for sphingosine 1-phosphate in transplantation and autoimmunity. Curr. Opin. Immunol. **14**, 569–575 (2002)

10. V. Brinkmann, D.D. Pinschewer, L. Feng, S. Chen, FTY720: altered lymphocyte traffic results in allograft protection. Transplantation **72**, 764–769 (2001)

11. V. Brinkmann, M.D. Davis, C.E. Heise, R. Albert, S. Cottens, R. Hof, C. Bruns, E. Prieschl, T. Baumruker, P. Hiestand, C.A. Foster, M. Zollinger, K.R. Lynch, The immune modulator fty720 targets sphingosine 1-phosphate receptors. J. Biol. Chem. **277**, 21453–21457 (2002)

12. S.S. Castillo, D. Teegarden, Ceramide conversion to sphingosine-1-phosphate is essential for survival in C3H10T1/2 cells. J. Nutr. **131**, 2826–2830 (2001)

13. J.J. Clemens, K.R. Lynch, T.L. Macdonald, Synthesis of para-alkyl aryl amide analogues of sphingosine-1-phosphate: discovery of potent S1P receptor agonists. Bioorg. Med. Chem. Lett. **13**, 3401–3404 (2003)

14. A.M. Davey, R.P. Walvick, Y. Liu, A.A. Heikal, E.D. Sheets, Membrane order and molecular dynamics associated with ige receptor cross-linking in mast cells. Biophys. J. **92**, 343–355 (2007)

15. P.S. Dittrich, P. Schwille, Photobleaching and stabilization of fluorophores used for single-molecule analysis with one- and two-photon excitation. Appl. Phys. B **73**, 829–837 (2001)

16. J.A. Dix, A.S. Verkman, Mapping of fluorescence anisotropy in living cells by ratio imaging. Application to cytoplasmic viscosity. Biophys J. **57**, 231–240 (1990)

17. H. Fyrst, J.D. Saba, An update on sphingosine-1-phosphate and other sphingolipid mediators. Nat. Chem. Biol. **6**, 489–497 (2010)

18. A.A. Heikal, Time-resolved fluorescence anisotropy and fluctuation correlation analysis of major histocompatibility complex class I proteins in fibroblast cells. Methods **66**, 283–291 (2014)

19. S.T. Hess, S. Huang, A.A. Heikal, W.W. Webb, Biological and chemical applications of fluorescence correlation spectroscopy: a review. Biochemistry **41**, 697–705 (2002)

20. Y. Jin, M. Zollinger, H. Borell, A. Zimmerlin, C.J. Patten, CYP4F enzymes are responsible for the elimination of fingolimod (FTY720), a novel treatment of relapsing multiple sclerosis. Drug Metab. Dispos. **39**, 191–198 (2011)

21. M. Jongsma, J. van Unen, P.B. van Loenen, M.C. Michel, S.L. Peters, A.E. Alewijnse, Different response patterns of several ligands at the sphingosine-1-phosphate receptor subtype 3 (S1P(3)). Br. J. Pharmacol. **156**, 1305–1311 (2009)

22. J.R. Lakowicz, *Principles of Fluorescence Spectroscopy*, (Springer 2006)

23. Z. Li, R. Bittman, Synthesis and spectral properties of cholesterol- and FTY720-containing boron dipyrromethene dyes. J. Org. Chem. **72**, 8376–8382 (2007)

24. H.S. Lim, Y.S. Oh, P.G. Suh, S.K. Chung, Syntheses of sphingosine-1-phosphate stereoisomers and analogues and their interaction with edg receptors. Bioorg. Med. Chem. Lett. **13**, 237–240 (2003)

25. P. Mitra, C.A. Oskeritzian, S.G. Payne, M.A. Beaven, S. Milstien, S. Spiegel, Role of ABCC1 in export of sphingosine-1-phosphate from mast cells. Proc. Natl. Acad. Sci. **103**, 16394–16399 (2006)

26. D.V. O'Connor, D. Phillips, *Time-Correlated Single Photon Counting* (Academic Press, London, 1984)

27. S.W. Paugh, S.G. Payne, S.E. Barbour, S. Milstien, S. Spiegel, The immunosuppressant FTY720 is phosphorylated by sphingosine kinase type 2. FEBS Lett. **554**, 189–193 (2003)

28. T.T.-C. Pham, J.I. Fells, Sr., D.A. Osborne, E.J. North, M.M. Naor, A.L. Parrill, Molecular recognition in the sphingosine 1-phosphate receptor family. J. Mol. Graph. Model. **26**, 1189–1201 (2008)
29. K. Sato, E. Malchinkhuu, Y. Horiuchi, C. Mogi, H. Tomura, M. Tosaka, Y. Yoshimoto, A. Kuwabara, F. Okajima, Critical role of ABCA1 transporter in sphingosine 1-phosphate release from astrocytes. J. Neurochem. **103**, 2610–2619 (2007)
30. S. Spiegel, S. Milstien, Functions of a new family of sphingosine-1-phosphate receptors. Biochim. Biophys. Acta **1484**, 107–116 (2000)
31. C.R. Strader, C.J. Pearce, N.H. Oberlies, Fingolimod (FTY720): a recently approved multiple sclerosis drug based on a fungal secondary metabolite. J. Nat. Prod. **74**, 900–907 (2011)
32. K. Takabe, S.W. Paugh, S. Milstein, S. Spiegel, "Inside-out" signaling of sphingosine-1-phosphate: therapeutic targets. Pharmacol. Rev. **60**, 181–195 (2008)
33. Q. Yu, M. Proia, A.A. Heikal, Integrated biophotonics approach for noninvasive and multiscale studies of biomolecular and cellular biophysics. J. Biomed. Opt. **13**, 041315 (2008)

Chapter 11
Probing Microsecond Reactions with Microfluidic Mixers and TCSPC

Sagar V. Kathuria and Osman Bilsel

Abstract Probing the in vitro kinetics and dynamics of macromolecules involved in biological processes is important for discerning their mechanism and function. These dynamics span the sub-microsecond to millisecond and longer timescales. In addition to resolving dynamics and kinetics, structural characterization of non-equilibrium intermediates over these time scales is often desired. In this chapter, we review recent advances in microfluidic mixing methods (both low-Reynolds laminar and chaotic/turbulent mixers) for initiating biochemical reactions and provide an overview of the interfacing of these techniques with time-correlated single photon counting (TCSPC) fluorescent detection methods. We focus on approaches in which both a kinetic reaction time axis and a TCSPC time axis are simultaneously monitored, often referred to as a "double kinetic" experiment. Methods for measurement and analysis of these experiments are presented in which the TCSPC time axis corresponds to fluorescence lifetimes, time-resolved FRET or time-resolved anisotropy. An overview of matrix methods, such as singular value decomposition, and maximum entropy methods for data analysis are also reviewed.

11.1 Introduction and Overview

A wide range of timescales are involved in biological processes, ranging from sub-microseconds for helix formation, to milliseconds for translation of an mRNA to a chain of amino acids and, in some cases, seconds for the folding of the chain to a functional specific three-dimensional structure. An overview is given in Fig. 11.1. A goal of the biophysicist is to understand the thermodynamics and mechanism of how these and related processes occur. A structural probe that a biophysicist can

S.V. Kathuria · O. Bilsel (✉)
Department of Biochemistry and Molecular Pharmacology, University of Massachusetts Medical School, Worcester, MA 01605, USA
e-mail: osman.bilsel@umassmed.edu

© Springer International Publishing Switzerland 2015
W. Becker (ed.), *Advanced Time-Correlated Single Photon Counting Applications*,
Springer Series in Chemical Physics 111, DOI 10.1007/978-3-319-14929-5_11

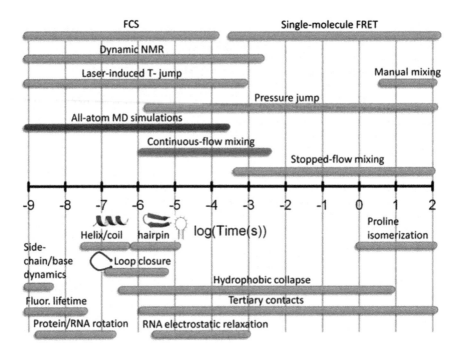

Fig. 11.1 Timescales for macromolecular dynamic processes (*lower half*) and techniques used to probe them (*upper half*). Copyright—2011 Wiley Periodicals, Inc. [31]

use with a time resolution to match the biological/biochemical process is a critical tool for revealing the underlying physics.

An expansive set of tools have been used to achieve a mechanistic and structural understanding of protein-protein, protein-RNA, and protein-DNA interaction, or protein folding and RNA folding. It is important to recognize the complementarity and synergy of many of these techniques. For example, in the case of protein folding, the design of a FRET pair for probing folding dynamics benefits significantly from the availability of a native crystal structure, radius of gyration measurements of the native and denatured state ensembles and residue level hydrogen-deuterium exchange protection information. The present overview of microsecond mixing interfaced with TCSPC is presented with this context in mind, that it can be a valuable tool in the toolkit of the biophysicist. It is also worth noting that numerous other complementary techniques are available with significantly greater time resolution than microfluidic mixing (Fig. 11.1), [31] such as temperature-jump, [14, 20, 24, 30, 44, 53, 67] NMR relaxation methods [28, 51] and FCS, [17, 18] but these have their own limitations with respect to initiation of the reaction and appropriateness for a given protein. The microfluidic methods discussed in this review are ideal for kinetic reactions initiated by a dilution reaction, as in the case of protein folding, [62] RNA folding, [50] or enzymatic reactions [34] in which a

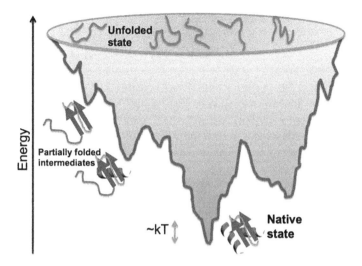

Fig. 11.2 Cartoon representation of the energy landscape for a protein. Some partially folded states are rarely accessed in equilibrium experiments. Non-equilibrium kinetic methods are therefore advantageous in studying them

reagent, such as ATP or an analogue, is added. For protein folding studies, the mixing studies provide a means for studying high-energy partially folded structures that are transiently populated as an approximately random-coil polypeptide folds to its native structure. The mixing approaches are able to achieve relatively high populations of these high-energy states, which may have too low of an equilibrium population for single-molecule studies for proteins with large stability, see Fig. 11.2.

11.1.1 Micromachining Advances Usher in a New Era

Although microsecond mixing devices have appeared in the literature for over three decades (without a dramatic improvement in mixing times) [33], they had not been widely adopted by the biophysics and enzymology researchers, who have favoured the more established stopped-flow method with its millisecond time resolution limitation. Over the past decade and a half, however, advances in micromachining methods (e.g., laser machining, photolithographic methods and DRIE etching) have enabled smaller micron-sized features and more intricate patterns to be machined. The devices have also been increasingly more robust, and ushered in new mixing techniques, such as laminar and chaotic mixing, in addition to the more traditional turbulent mixing methods [41, 54].

11.1.2 Concurrent Advances in TCSPC

Concurrent with developments of microfluidic based mixing devices in the late 1990s and early 2000s came the introduction of compact PCI-based high-speed TCSPC electronics. The higher counting rates possible with these PC-based TCSPC cards, their multi-dimensional recording capabilities [2, 3] (see Chap. 1) and high-repetition rate lasers made kinetic experiments, particularly of the 'double-kinetic' variety, [7, 9] (simultaneously performing reaction kinetics and TCSPC based time-resolved fluorescence kinetic) more accessible. These experiments, and their counterparts using camera-based detection [35], allowed researchers to probe millisecond timescale kinetic reactions using lifetime detected FRET. These developments allowed users to obtain not just an average end-to-end intramolecular distance for a given set of donor-acceptor labelled residues of a protein but to also obtain, via Laplace inversion, the time evolution of a distance distribution during a biochemical reaction [35].

11.1.3 Merging Microfluidics and TCSPC

The availability of continuous-flow microfluidic mixers opened the possibility of bringing the time-resolution of the 'double-kinetic' experiment into the microsecond time range. The availability of performing TCSPC-based lifetime-resolved FRET and anisotropy studies is significant in that many fundamental events, such as the formation of loops and hairpins of proteins [15, 26, 47, 67], and the folding of some globular proteins [37, 71, 72] occur on this timescale.

11.1.4 Matching Simulation and Experiment

The accessibility of the microsecond time regime in experiments has been paralleled over the past decade with complementary advances in computational molecular dynamics simulations. All-atom simulations can now be performed on proteins greater than 100 amino acids and extend into the millisecond timescale [40, 43]. The overlap between experiment and simulations allows validation of the simulations and provides atomistic insight into the mechanism governing the folding process and dynamics of a globular protein. The availability of distance distributions from lifetime-based FRET data [13] (see Chap. 7) obtained from TCSPC-based measurements in microfluidic devices, as discussed later in this chapter, can be a useful metric for comparing with simulations.

In this review, an overview of microfluidic devices interfaced with TCSPC will be presented, with an emphasis on experimental details and data analysis. This area is rapidly evolving, and the material presented here will perhaps foster new ideas and approaches for structural measurements of biological complexes in the microsecond time range.

11.2 Microfluidic Mixing Methods

Microfluidic mixers with microsecond time resolution have typically followed two general approaches (Fig. 11.3).

In one case, the mixer operates in the laminar regime and the mixing is accomplished by focusing of a central sheath of flow to a width of approximately 0.1 μm. Diffusion of the solvent molecules across this narrow sheath can occur on the microsecond timescale. These mixers are very efficient, allowing full experiments to be conducted with femtomoles of samples when used with a confocal microscope. The central 0.1 μm (or narrower) wide sheath is narrower than the sample volumes but this is not a serious limitation for fluorescence experiments, especially if a confocal instrument with a high numerical aperture objective is used. For coaxial laminar mixers, the small dimensions of the central sheath are an advantage as there is less out-of-focus light because the central sheath is less than the dimensions of the confocal volume [27]. By appropriate selection of the flow stream, a macroscopic observation volume is attainable for laminar flow mixers [22]. As shown in Fig. 11.4, the central sheath is selected and expanded in an effort to slow the flow rate sufficiently for single-molecule studies. Spectroscopic and kinetic studies have been performed with laminar mixers in the visible [22, 27, 52] and ultraviolet wavelength region [41, 49] (with extrinsic and intrinsic chromophores of proteins, respectively)

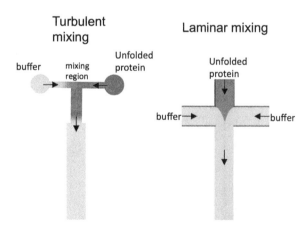

Fig. 11.3 Schematic overview of turbulent/chaotic and laminar mixing approaches to microsecond mixing. Channel widths are ~ 10 μm for laminar mixers and 30–100 μm for turbulent/chaotic mixers. The central sheath in the hydrodynamically focused laminar mixer is ~ 0.1 μm. Copyright © 2011 Wiley Periodicals, Inc. [32]

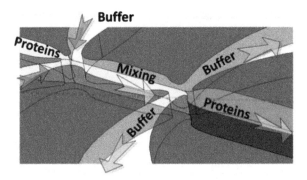

Fig. 11.4 A laminar mixer covering the 200 μs to seconds time range suitable for TCSPC and single-molecule studies. A 20 nm central sheath (*yellow*) is picked out of the flow and expanded (*orange*) to slow the flow rate. Reprinted by permission from [22]. Copyright (2011)

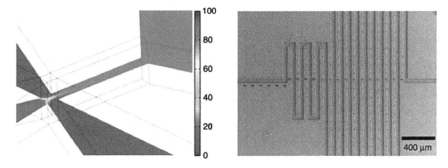

Fig. 11.5 An alternative simpler laminar mixer covering the ms to minutes time range suitable for TCSPC and single-molecule studies. The solutions mix in a 2.5 μm central channel (*yellow* to *red* transition) before expanding to ∼40 μm in the observation channel. The flow velocity is ∼1 μm/ms. Reprinted by permission from [70]. Copyright (2013)

with both ensemble and single-molecule experiments. However, these experiments have not been used with TCSPC detection.

Another popular approach to microsecond mixers relies on turbulent or chaotic mixing [33, 57]. Two solutions flowing at a fairly high flow rate (10–100 mL/min) in channels of 50–250 μm width are brought into contact in a way that generates Reynolds numbers on the order of several thousand (Fig. 11.5).

The solution breaks up such that the largest eddies are on the order of 0.1 μm, allowing diffusion, and mixing, to occur on the tens of microsecond timescale. These mixers tend to consume significantly more material than laminar mixers but allow a larger volume (and thus more signal) to be detected. The larger sample region makes these devices straightforward to interface with ensemble TCSPC setups. However, the high flow rates in these mixers give rise to very low residence times for sample molecules in the focal volume (∼ few microseconds) and make them less than ideal for single-molecule measurements.

The early capillary mixers used steady-state fluorescence detection [56, 60], but an early report by the Rousseau group showed that interfacing of a stainless steel mixer to a TCSPC detection setup could be readily accomplished [64]. An optical quality quartz cuvette mounted on the T-shaped mixer with 250×250 μm^2 channels achieved a mixing time of ~ 400 μs. A subsequent study using laser machined mixers constructed from 127 μm thick Kapton (polyimide) or PEEK (poly-ether-ether-ketone) used an epifluorescence detection arrangement with a TCSPC detection setup. Using this setup, approximately 100 TCSPC decay traces of horse heart cytocrome c were obtained over a 30–1200 μs range along the time range of the folding reaction [10].

A variety of mixers relying on split-and-recombine and/or flow obstructions that operate at Reynolds numbers in between the laminar and turbulent regimes have been reported [42, 59, 68]. While these "chaotic mixers" generally do not produce mixing on the same timescales as the turbulent mixing or hydrodynamic flow focusing techniques, they operate at relatively lower flow rates than the turbulent mixers and have larger detection volumes than the flow focused laminar mixers. These mixers can be easily interfaced with fluorescence detection using total intensity and TCSPC for both ensemble and single molecule techniques [21, 70]. An illustrative example is shown in Fig. 11.5, where mixing is achieved in a 2.5 μm wide channel before widening of the completely mixed solution. This approach slows the flow sufficiently for single-molecule experiments and also enables a large uniform volume to be used for various types of spectroscopy. Lastly, these mixers can achieve longer timescales ranging from a few hundred microseconds to minutes by introduction of a serpentine pattern [70]. While these timescales are easily accessible by most modern stopped-flow devices, the mixing efficiency of chaotic mixers is far superior to that of stopped-flow devices, consuming significantly less material. These devices also operate under lower turbulence and pressure regimes. Further, as the sample is in continuous flow mode, photobleaching is seldom a problem.

11.2.1 Materials

The last few years have seen significant advances in new mixer designs, facilitated in part due to user-friendly off-the-shelf simulation software and the availability of micromachining capabilities. Photolithography followed by etching is widely being used for micromachining of silica wafers [70] that are readily used in hydrodynamic flow focusing laminar mixer devices. A range of materials like quartz cover slips and silicon based polymer materials, PDMS, PMMS, POM, etc. can be used as window material on these chips for spectroscopic applications under low-pressures.

The higher pressures in turbulent flow mixing techniques prevents the use of delicate window materials and is generally circumvented by separating the mixing, which is performed in metal or PEEK mixers, from the observation channels. The optical quality of quartz along with its inert nature makes it ideal for most mixer types. The use of a femtosecond painting method for micro-etching quartz, developed by Bado and coworkers [8] has enabled the use of fused "single piece" quartz mixers [31, 33] with virtually no design limitations even under high pressures.

11.2.2 Optimization of Mixer Designs by Simulation

Robust commercially available computational fluid dynamics packages, e.g. COMSOL (Fig. 11.6) and ANSYS/Fluent, have led to systematic optimization of various mixer design parameters to achieve the highest mixing efficiencies at the lowest flow rates possible [16, 42, 46, 63]. Of the parameters optimized for various flow geometries, most groups have found that constricting the flow path at the point of mixing of two fluids in conjunction with an obstruction in the form of sharp turns

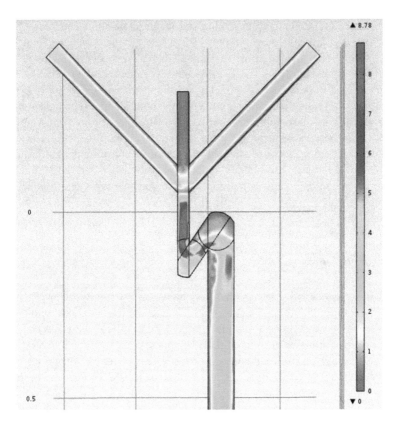

Fig. 11.6 Simulation using COMSOL. Fluid velocities at different positions along the channel are shown. The two side input channels (50 μm) are flowing at 2 m/s and the central channel (30 μm) flowing at 0.4 m/s. The flow velocity is uniform within 0.5 mm from the point of mixing

is the most efficient under non-laminar flow conditions. Under laminar conditions, the SAR, serpentine and herringbone patterned mixers appear to be more effective as they provide greater opportunity to increase the contact surface area between the two fluids and enhance passive diffusion.

One of the major limitations of continuous-flow mixers is that they have historically required either higher concentrations (laminar mixers) or amounts (turbulent mixers) of sample relative to stopped-flow experiments. Chaotic mixers can potentially overcome both of these limitations. Simulations play a significant role in developing geometries to optimize flow paths to achieve faster and more efficient mixing and to minimize dead spaces to reduce the priming sample volume required for uniform flow.

11.3 Interfacing Microfluidic Chips with TCSPC

11.3.1 Advantages of TCSPC

The motivation for interfacing microsecond mixing devices with TCSPC instrumentation [2, 3] stems from several factors. For FRET studies, the advantages offered by TCSPC can be substantial, see Chaps. 7 and 8 of this book. By collecting the time-resolved decay of the donor (and, where available, the acceptor), the user is able to record the average distance as well as the distribution of distances between a donor and acceptor fluorescent label on the molecule of interest. Therefore, by having the TCSPC axis, the user can distinguish whether the average FRET efficiency arises from a single population or multiple sub-populations. Additionally, the average FRET efficiency calculated from the donor lifetime is more accurate than that obtained through steady-state methods [13], see Chaps. 7 and 8. In steady-state methods, matching the concentrations of donor-only and donor-acceptor controls is a source of error in the average FRET efficiency determination. The fluorescence lifetime, however, is independent of concentration unless the concentration is sufficiently high that higher order oligomers, exciplexes or other quenching reactions are introduced. Furthermore, the theoretical standard deviation of a lifetime measurement is $\sqrt{N}/N$ for N photons [36], where N is on the order of 10^6. TCSPC comes very close to this ideal value. This latter point is helpful for the continuous-flow TCSPC experiment as the first moment of the lifetime decay can be used as a cross-check to confirm the intensity normalization, as will be discussed further below. Advantages of TCSPC detection also extend to anisotropy measurements, where the measurement of rotational correlation times can be very useful for assessing local structure and hydrodynamic size, see Chap. 12 of this book.

With the availability of compact high-repetition rate lasers over a broad spectral range and PC-based TCSPC cards, the adoption of TCSPC detection for microfluidic mixing devices is likely to increase. Below we present some of the instrumentation and design considerations in the practical application of continuous-flow microfluidic mixers with TCSPC detection.

11.3.2 Instrumentation

The basic setup for continuous-flow TCSPC experiments is shown in Fig. 11.7.

The setup is essentially that of an epi-fluorescence microscope equipped with an xy-scanning stage. The implementation currently in use at the University of Massachusetts Medical School utilizes the doubled or tripled output of a Ti:sapphire laser operating at 76 MHz repetition rate as the excitation source. The repetition rate is dropped to 3.8 MHz using an acousto-optical pulse picker for most experiments to allow more accurate determination of longer lifetimes (e.g., >4 ns), such as, experiments using Trp and cyano-Phe [1, 58, 65] fluorescence. The 3.8 MHz is due to the limitation of our pulse picker, in fact a repetition rate on the order of 20 MHz would be more appropriate. A laser power of several hundred µW is available but only a power of ∼ 100 µW is needed for excitation.

For experiments at 290–295 nm, a dichroic filter (FF310, Semrock Inc., Rochester, NY) is used to reflect the excitation to the sample. A single element plano-convex lens (35 mm focal length) is used to focus the excitation to an approximately 20 µm spot. Although a higher numerical aperture objective can be used, maintaining a constant intensity along the length of the 30 mm flow channel

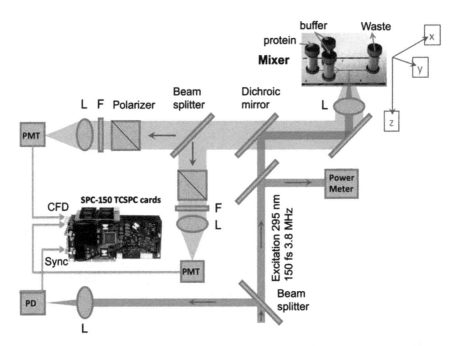

Fig. 11.7 Schematic of epifluorescence setup for continuous-flow TCSPC measurements. Abbreviations: *L* lens, *F* bandpass filter, *PMT* photomultiplier tube, *PD* fast photodiode. The flow channel is ∼ 50–70 µm wide, ∼ 50–100 µm deep and 30 mm long. The flow velocity is ∼ $(30 \text{ µs/mm})^{-1}$

becomes more challenging. A thin quartz plate is used to pick off a fraction of the excitation beam and directed into a power meter.

The microfluidic mixer is mounted horizontally onto a xy-stage (Ludl Biopoint 2). A positioning accuracy of a few microns is sufficient for channels that are >30 μm in width.

The fluorescence from the sample solution inside the channel is collimated, and transmitted by the dichroic mirror. The fluorescence light is further split in two components, which are projected onto UV sensitive PMT modules (PMH100-6) or microchannel plate (MCP) PMTs. The outputs of the PMTs are fed into two separate SPC150 TCSPC cards [3].

Wavelength selection is accomplished via bandpass filters (e.g., FF357/44 for tryptophan experiments), although the option of using a monochromator is available. The greater spectral width and excellent out-of-band rejection available with bandpass filters renders them very practical. A motorized Glan-Taylor polarizer at magic-angle is present in each detection path. The polarisers can be set to detect parallel and perpendicular polarisation components or for magic-angle detection. Magic-angle detection rejects contributions from chromophore reorientational dynamics from the decay data. Please note that the commonly used magic-angle of 54.7° strictly applies only for excitation and detection in parallel beams of light. The NA of our detection optics is 0.25, where the magic angle is still very close to the value for parallel beams. For excitation and detection at higher NA the magic angle decreases and approaches 45° for NA = 1 [19].

11.3.3 Alignment

The initial setup for the experiment consists of aligning the excitation beam, the detection path and in locating the start, stop and centre of the channel. Alignment of the excitation and detection paths is analogous to that of a confocal microscope without the pinhole. A concentrated dye solution is placed on the xy-stage instead of the mixer to view the fluorescent light path. The dichroic also transmits a small amount of the excitation light, which is useful for alignment purposes.

The precise location of the mixer flow channel is accomplished by placing an approximately 10-fold higher concentration of a fluorescent small molecule, appropriate for the experiment excitation and emission wavelengths, in the mixer channel (e.g., N-acetyl-tryptophanamide for tryptophan based experiments). The channel is point-scanned (at ∼5 μm resolution) perpendicular to the flow direction at approximately 1 mm intervals along the flow direction to locate the centre of the channel along the 30 mm flow path. A linear least squares fit to the measured centre positions yields the slope and intercept that allows calculation of a lookup table. With a lookup table in hand, an additional point-scan is recorded parallel to the flow channel along the centre. The mid-point of the rise of the signal provides the location of the start of the channel and provides an estimate of the zero-time of the experiment. Because the beam focus diameter is smaller than the channel width, the intensity variation along the flow channel is small, approximately 5 %, usually caused by the tolerance in thickness of the quartz plates used to construct the mixer. This intensity profile is recorded with each experiment and used to normalize the TCSPC traces at each point along the channel. The mixer is brought into focus in the z-direction by repeated scans perpendicular to the direction of flow while optimizing the signal intensity.

11.3.4 Data Acquisition Protocol

In a typical experiment the sample and control will be placed in separate sample loops. The blank solution (e.g., matching the sample in every respect except without protein) is placed in the pump pushing out the contents of the sample loop. The other pump contains the dilution buffer. The flow rate is on the order of 2–6 mL/min total, although efforts are underway to reduce this volume to below 1 mL/min. Control of the pumps and valves is under computer control.

A typical experiment uses 10 mL sample loops giving a data collection time of approximately 15–20 min for the sample. As the sample flows, a scan is made along the flow direction every minute, consisting of 60–100 points. Although the TCSPC cards can be operated in "continuous-flow" mode, recording photons without any gap, the double time mode tends to be more commonly employed. In the latter mode, at each step along the channel the xy-stage is positioned and TCSPC acquisition for approximately 0.5 or 1 s is individually triggered, keeping the stage stationary during acquisition. The duty cycle is therefore ~ 50–75 %.

Count rates for these experiments tend to be in the 5×10^4 to $<1 \times 10^5$ cps per detection channel. For an excitation pulse frequency of 3.8 MHz used in our experiments this is well within the range where pile-up is not an issue [2, 3]. Higher count rates may be available with more efficient detectors [4]. Even then, adverse pulse pile-up effects [11] can be avoided by using higher excitation pulse frequency, and by using multiple detectors and parallel TCSPC channels [2, 3].

Data acquisition consists of recording three separate measurements: a blank, sample and control. Each data set consists of approximately 15–20 scans with each scan comprising ~ 60-100 TCSPC traces with 4096 points in the excited state decay. The experiment is visualized by examining the integrals of the decays (Fig. 11.8).

Traces which overlay are summed. The control scans (e.g., NATA) are summed along the TCSPC axis since only the integrated intensity is used in the normalization. A corrected data set for the sample is therefore obtained according to the following relationship:

$$I_{corr}(t_{kin},\ t_{TCSPC}) = \frac{I(t_{kin},\ t_{TCSPC}) - Blank(t_{kin},\ t_{TCSPC})}{\sum_{t_{TCSPC}} (NATA(t_{kin},\ t_{TCSPC}) - Blank(t_{kin},\ t_{TCSPC}))}. \quad (11.1)$$

In the above expression the kinetic time axis (corresponding to distance along the flow direction in the channel) is denoted by t_{kin} and the TCSPC time axis is

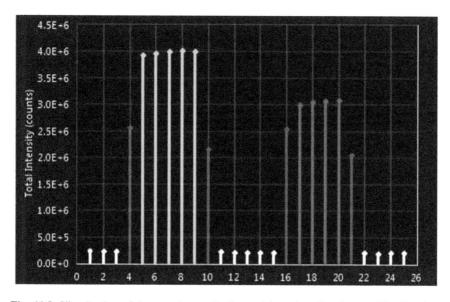

Fig. 11.8 Visualization of the experiment via the total intensity of each scan. The *blue* bars correspond to scans where the control is flowing through the mixer. The *magenta* bars correspond to the protein sample, the *white* bars are the blanks and the *grey* bars correspond to the traces where the sample flow was started or stopped and the entire channel has not reached flow-equilibrium

Fig. 11.9 Representative continuous-flow TCSPC data. Continuous-flow refolding of horse heart cytochrome c is shown over the 30–140 μs time range, where the collapse transition takes place. Reprinted from [32], Copyright (2014), with permission from Elsevier

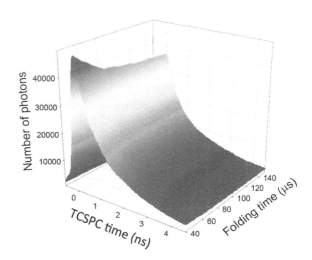

denoted by t_{TCSPC}. The former is in microseconds and the latter in ns. An analysis program with a graphical interface allows the user to analyze a data set in a few minutes, providing rapid feedback on the experiment. A representative data set, collected for horse heart cytochrome *c* is shown in Fig. 11.9.

With binning of several frames, the counts can be on par with what is necessary for distance distribution modelling and maximum entropy analysis, as is discussed further in this chapter.

11.3.5 Point Scanning TCSPC Versus CCD Based Steady-State Acquisition

The point scanning protocol used in these experiments has several pros and cons relative to the CCD based detection that is used by some groups [60]. Other than the advantage of TCSPC based acquisition, the point scanning allows full use of the incident beam intensity, which is very helpful in cases where the experiment is not limited by counting rate and is excitation flux limited. The CCD approach can be advantageous when excitation flux is not limiting. The CCD based approach provides the full kinetics simultaneously for all positions along the channel. The intensity variation along the channel tends to be larger for the CCD based approach but appears to not limit the data quality. The choice of using a CCD based approach or point scanning TCSPC depends on whether the fluorescence decay parameters are desired and the excitation flux available.

11.3.6 Alternative Approaches

Scanning along the flow channel can be avoided by using the multi-channel TCSPC architecture described in Chap. 1, Sect. 1.4.1. The entire flow channel would be illuminated simultaneously, an image of it projected on a multi-anode PMT, and the signals detected by the individual channels of the multi-anode PMT recorded simultaneously. The approach would increase the efficiency of the measurement enormously, and thus decrease the acquisition time and the amount of sample fluid needed for the experiment. A possible problem is that the number of data points along the channel would be limited by the number of PMT channels, typically 16. To spread the 16 data points reasonably over the protein folding time axis the flow speed must be adjusted to the expected folding time. An instrument using the multi-channel PMT technique has not been described yet.

Although the scope of this chapter is on the application of TCSPC, it is worth noting that other recording techniques can also be used for obtaining time-resolved fluorescence decays along a microfluidic microsecond mixer. One option is to use a fast digital oscilloscope (13 GHz bandwidth) in combination with an MCP-PMT, as implemented by Haas and coworkers [29]. An advantage of this approach is that multiple photons can be collected in a single detection channel per laser pulse. This is useful for experiments utilizing low-repetition rate lasers [55]. However, there is no need to use an extremely low repetition rate. For high repetition-rate lasers the

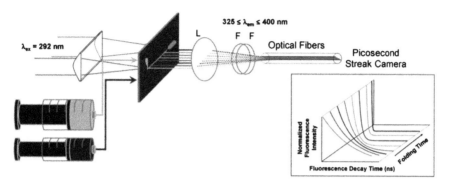

Fig. 11.10 Schematic of streak camera based fluorescence lifetime measurements. Copyright—2006 by The National Academy of Sciences of the USA [35]

problem of detecting multiple photons per laser pulse does not exist. The efficiency of analog recording is lower than for TCSPC, especially if the system is operated at low photon rates. The time resolution is limited by the single-electron response of the detector, not by the transit-time-spread as in the case of TCSPC, see Chap. 1, Fig. 1.4. It is about 300 ps for an MCP PMT, and 1–2 ns for a conventional PMT, compared to 30 and 200–300 ps in the case of TCSPC. Moreover, the current output of an MCP PMT tends to become a limiting factor: If several photons per excitation pulse are detected channel saturation almost certainly causes nonlinearity in the detected decay curves.

Lifetime-based FRET studies using a microsecond mixer were also performed using a streak camera by the Winkler group [35]. The picosecond streak camera allows the entire mixer to be imaged while providing excellent resolution along the excited state fluorescence decay (Fig. 11.10).

Their study provided high signal-to-noise picosecond resolved excited state decays at 25 points in the folding of a cytochrome c' four-helix bundle. The dynamic range of the camera allows the full excitation pulse (from an amplified laser) to be used. Other than the cost of the camera, the only limitation for this approach appears to be the coupling of the light into a fibre optic array bundle, which appears to have limited the earliest observation point. It seems likely that these limitations may be readily overcome with more recent microfluidic mixer designs.

11.4 Data Analysis

The continuous-flow TCSPC experiment presents the user with a rich data set, consisting of ~ 40 scans with ~ 100 kinetic time points and 4096 TCSPC time channels in each of the detection channels. One of the challenges is to rapidly evaluate the data set to gauge whether the experimental time window and signal-to-noise are sufficient and determine if a kinetic process is being observed. For time-resolved FRET and anisotropy experiments, this is done in several stages of analysis with each stage increasing in complexity and rigor. A brief overview of some of the analysis approaches is presented, with an emphasis on FRET and anisotropy data sets.

11.4.1 First Moment

The first moment of the histogram of the photon arrival times (often referred to as the "centre of mass") provides a quick overview of a data set. The first moment is known to a relative precision of $\sqrt{N}/N$ and is an easy to calculate metric that is also model independent and relatively insensitive to the number of photons. The first moment is calculated according to:

$$\langle \tau \rangle = \frac{\sum_i I_i(t_i - t_o)}{\sum_i I_i} \tag{11.2}$$

where I_i is the photon counts in the ith TCSPC channel, t_i is the TAC time of the ith channel and t_o is the excitation pulse arrival TAC time. This is mathematically equivalent to a quantum yield weighted average lifetime:

$$\langle \tau \rangle = \frac{\sum_i \alpha_i \tau_i^2}{\sum_i \alpha_i \tau_i} \tag{11.3}$$

where α_i is the amplitude of the ith exponential phase with relaxation time constant τ_i.

One potential drawback to the use of the first moment as a metric for kinetic analysis is that it is not rigorously proportional to mole fraction (concentration). Therefore, the time constant(s) of the kinetics can be biased to a lower or higher value depending on the relative quantum yields of the reactants. A workaround is to include the total intensity data $I(t_{kin})$, which has the quantum yield information, and fit it globally with the first moment. Assuming that the amplitudes in (11.3) are linearly proportional to concentration, $\alpha_i = q_i c_i$, for specie, i, the first moment at the reaction time, t_{kin}, can be written as:

$$\langle \tau \rangle (t_{kin}) = \frac{\sum_i q_i c_i(t_{kin}) \tau_i^2}{\sum_i q_i c_i(t_{kin}) \tau_i} \tag{11.4}$$

and the total intensity as:

$$I(t_{kin}) = \sum_i q_i c_i(t_{kin}) \tau_i. \tag{11.5}$$

For a two-state kinetic reaction ($i = 1, 2$),

$$c_1(t_{kin}) = c_1(0)e^{-k \cdot t}$$
$$c_2(t_{kin}) = c_2(0)(1 - e^{-k \cdot t}) \tag{11.6}$$

where k is the rate constant of the reaction, q_i are the relative quantum yields, and c_i, are the concentrations. The parameters q_i and τ_i are globally optimized simultaneously for the two kinetic traces. The approach can be generalized to an arbitrary kinetic mechanism.

This analysis of the first moment is analogous to that proposed previously for the analysis of anisotropy kinetic data [48]. It is worth noting that if the quantum yields are different for the two states the first moment and total intensity will not track each other. For the case where anisotropy and lifetime are collected, the global analysis would also include the steady-state anisotropy data:

$$r(t_{kin}) = \frac{\sum_i q_i c_i(t_{kin}) r_i}{\sum_i q_i c_i(t_{kin})} \tag{11.7}$$

where r_i are the steady state anisotropy values of the states in the kinetic mechanism obtained from the vertical and horizontally polarized data in the customary manner [48]. Similar, to the first moment, the parameters q_i and r_i are globally optimized. The first moment data calculated from the total intensity ($I = V + 2GH$) can also be included in the fitting to improve parameter estimation and kinetic model discrimination.

For the case where a FRET efficiency is sought, the average FRET efficiency, E, can be approximated (with the above caveats) from the average lifetimes of the donor-only, $\langle \tau_D \rangle$, donor-acceptor, $\langle \tau_{DA} \rangle$, and $\langle \tau_{DA} \rangle_{App}$, lifetimes:

$$\langle \tau_{DA} \rangle_{App} = x \cdot \langle \tau_{DA} \rangle + (1 - x) \cdot \langle \tau_D \rangle$$
$$E = 1 - \frac{\langle \tau_{DA} \rangle}{\langle \tau_D \rangle} = 1 - \frac{\langle \tau_{DA} \rangle_{App}}{x \cdot \langle \tau_D \rangle} + \frac{1 - x}{x} \tag{11.8}$$

where x represents the acceptor labelling efficiency. If x is less than unity, the measured lifetime of the donor in the presence of the acceptor, $\langle \tau_{DA} \rangle_{App}$, may be related to the true value, $\langle \tau_{DA} \rangle$, via the fractional labelling efficiency, x. An example of a first moment analysis for FRET data on a TIM barrel protein is shown in Fig. 11.11.

Use of the first moment data in practice requires subtraction of the pulse arrival time and consistency in the relative time range (relative pulse arrival time) used for the calculation. It should also be kept in mind that using the steady-state anisotropy and first moments provides an efficient method of determining the number of kinetic steps but comes at the expense of not fully utilizing the

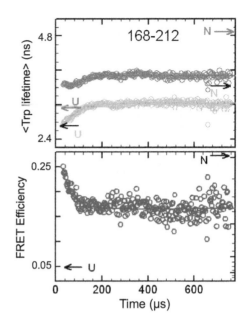

Fig. 11.11 First moment analysis of FRET data from a TIM barrel protein. The first moment of the donor-only (*red*) and donor-acceptor (*green*) labelled protein is shown in the *top panel*. The labelling efficiency corrected FRET efficiency (*blue*) is shown in the *bottom panel*. Copyright—2008 by The National Academy of Sciences of the USA [69]

information available in the time-resolved FRET and anisotropy data. This information can be obtained alongside the kinetic analysis via a full global analysis, as discussed below.

11.4.2 Global Analysis

Global analysis [6, 7, 9] is a more rigorous approach that involves direct fitting of the raw data. A global analysis allows more stringent testing of a model (e.g., kinetic model and/or distance distribution model) because the linkage of parameters over data sets reduces the number of parameters. The continuous-flow TCSPC data obtained with microfluidic mixers is ideal for this type of analysis although, depending on the extent of the global analysis, can be complicated to set up. In practice, the most robust parameter estimates are obtained if the full raw data set can be parameterized using a kinetic model for the continuous-flow time axis and using a distance distribution or anisotropy decay model for the TCSPC time axis. Below, an overview of a global analysis strategy of time-resolved FRET data is presented.

In one type of global analysis of tr-FRET data, decays from each kinetic time point (e.g., decays corresponding to donor-only and donor-acceptor samples) are globally analyzed using a distance distribution function. In practice, the global analysis is often performed stepwise: the donor-only decay is fit to a sum of exponentials, yielding the fractional amplitudes (α_i), total amplitude ($I_d(0)$) and decay constants (k_{d_i}).

$$I_d(t_{TCSPC}) = I_d(0) \cdot \sum_i^N \alpha_i e^{-k_{d_i} \cdot t_{TCSPC}} + const \qquad (11.9)$$

These parameters are then held fixed in the next step of the analysis, performed on the donor-acceptor decay, in which only the normalized distribution function parameters, $p(k_{ET})$, are allowed to vary:

$$I_{da}(t_{TCSPC}) = \int_0^\infty \sum_i^N \alpha_i e^{-k_{d_i} \cdot t_{TCSPC}} \cdot p(k_{ET}) \cdot e_{ET}^{-k_{ET} \cdot t_{TCSPC}} + const. \qquad (11.10)$$

In the above expressions, α_i is the fractional amplitude of the ith decay component with rate k_{di}, k_{ET} is the energy transfer rate and $p(k_{ET})$ is the energy transfer rate distribution. An assumption in this analysis is that, if the donor has multiple sub-populations with different donor decay rates (e.g. tryptophan), they are assumed to have the same energy-transfer rate distribution. (This assumption is not made in the 2D maximum entropy analysis discussed further below). The energy transfer rate distribution, $p(k_{ET})$, is related to the pair-distance distribution, $p(r)$, according to the Förster equation:

$$r^6 = R_0^6 \cdot \left(\frac{k_{Dave}}{k_{ET}} \right) \qquad (11.11)$$

where k_{Dave} is the inverse of the average lifetime of the donor in the absence of the acceptor. The distribution $p(k_{ET})$ is often described by a sum of Gaussian distribution functions:

$$p(r) = \sum_i \frac{a_i}{\sigma \sqrt{2\pi}} e^{-(r-\omega_i)^2 / 2\sigma_i^2} \qquad (11.12)$$

where a_i is the amplitude, ω_i the centre and σ_i the width of the ith Gaussian. The width of the Gaussian can be converted to the full-width at half-maximum by $FWHM = \sigma/2.354$. The adjustable parameters in the non-linear least squares optimization are the amplitude(s), a_i, width, σ_i, and centre, ω_i of the Gaussians and the offset (*const.*) [69]. Distribution functions other than Gaussians may be used. For example, if the distance distribution function of one of the components is expected to correspond to a random-coil, one of the Gaussian functions may be replaced by a worm-like chain model [66]:

$$p(r) = \frac{4\pi a N r^2}{l_c^2 \left(1 - \left(\frac{r}{l_c} \right)^2 \right)^{9/2}} \exp \left(\frac{-3 l_c}{4 l_p \left(1 - \left(\frac{r}{l_c} \right)^2 \right)} \right) \qquad (11.13)$$

where one fewer parameter is required (the amplitude, a, and the persistence length, l_p) to describe the functional form. The contour length, l_c, is obtained from the number of amino acids, N_{res} ($3.4 * N_{res}$) and is held fixed.

An extension of this analysis parameterizes both the time axis of the reaction kinetics and the TCSPC time axis in a combined analysis. This was performed, for example, for an associative anisotropy model analysis of the TCSPC axis together with a kinetic folding mechanism [9]. The analysis allows deconvolution of physical parameters (e.g., rotational correlation times, lifetimes, fundamental anisotropy etc.) for individual species in the reaction even if the species is not fully populated at 100 %. A similar approach can be applied to tr-FRET data as well. For example, for a kinetic reaction, where $A_1 \Leftrightarrow A_2 \Leftrightarrow \ldots A_n$, the distribution function at any given time is the population weighted linear combination of the distribution functions, $p_i(r)$, of the contributing species, Ai, $i = 1$, n:

$$p(r, t_{kin}) = \sum_i c_i(t_{kin}) \cdot p_i(r) \tag{11.14}$$

The populations (concentrations) can be obtained from the solution to the general rate equation:

$$\frac{d\vec{C}(t_{kin})}{dt} = \hat{S} \cdot \vec{C}(t_{kin}) = -\sum_{j=1,n;j\neq1} k_{ij} \cdot c_i(t_{kin}) + \sum_{j=1,n;j\neq1} k_{ji} \cdot c_j(t_{kin}) \tag{11.15}$$

where C is a vector of concentrations, S is the system matrix and t_{kin} denotes the time along the microsecond reaction kinetics axis (to distinguish from the TCSPC axis). The rate k_{ij} denotes the microscopic rate from species i to species j. In practice, signal-to-noise limits the number of distributions that can be reliably delineated.

11.4.3 Global Analysis with Diffusion of Donor and Acceptor

A further refinement of the global analysis described above takes into account the relative intramolecular diffusion of the donor and acceptor chromophores during the lifetime of the donor excited state [5]. For unfolded proteins and peptides, values of the diffusion constant have ranged from 5 to 15 $Å^2$/ns [25, 39]. Consideration of diffusion, therefore, becomes significant for highly flexible macromolecules probed with a donor having a long lifetime (>4 ns) and a short Förster distance. A straightforward implementation of diffusion begins with the distance distribution function, $p(r, t_{kin}, 0)$, and requires solving the diffusion equation for each of the times along the TCSPC time axis to obtain $p(r, t_{kin}, t_{TCSPC})$. If the diffusion coefficient is a constant, then, the distribution function at TCSPC time, t_{TCSPC}, is obtained according to a one-dimensional diffusion equation in the absence of a potential,

$$\frac{\partial p(r, t_{kin}, t_{TCSPC})}{\partial t} = D \cdot \frac{\partial^2}{\partial r^2} p(r, t_{kin}, t_{TCSPC}) \qquad (11.16)$$

that can be readily solved using finite difference numerical methods.

11.4.4 SVD: Singular Value Decomposition

SVD is a valuable data reduction strategy that provides a quick model-independent assessment of the number of distinguishable lifetime decays or "spectra" in the data. The SVD algorithm reduces the data into the minimum number of orthonormal basis vectors along both, the kinetic time axis, t_{kin}, and the TCSPC time axis, t_{TCSPC}. This is expressed as:

$$A = UWV^T \qquad (11.17)$$

where the columns of U represent the kinetic basis vectors, the columns of V are the TCSPC axis basis vectors and W is diagonal matrix containing the singular values. In practice, only the first few basis vectors of U and V are significant, with the remaining vectors being essentially random noise. SVD can, therefore, be used to filter data while knowing exactly what is being thrown out. The sum of the outer product of the first few significant vectors represents a least-squares approximation to the full data. The number of significant basis vectors also gives the user an indication of the minimum number of distinct species contributing to the data. In the example shown in Fig. 11.12, the SVD analysis suggested that only two dominant intermediates were present during the course of the collapse of cytochrome c, arguing for a concerted process [32].

SVD can also be used as an efficient route to a full global analysis. The parameters obtained from a global analysis of the SVD vectors (either the U or the V vectors) can be propagated using the coefficients of the other matrices to reconstruct a full parameter set. Application of SVD to trFRET and time-resolved anisotropy data sets is complicated by the requirements that two data sets need to be analyzed globally and examples of this have not appeared in the literature.

11.4.5 Maximum Entropy

Another model independent analysis approach suitable for fluorescence lifetime, anisotropy and FRET is the maximum entropy method (MEM). The MEM describes a data set (e.g., fluorescence excited state decay) with a distribution of rates, choosing the widest and smoothest distribution that is consistent with the data. MEM starts out with a flat a priori distribution and the amplitude of a given rate deviates from this a priori distribution only as warranted by the data.

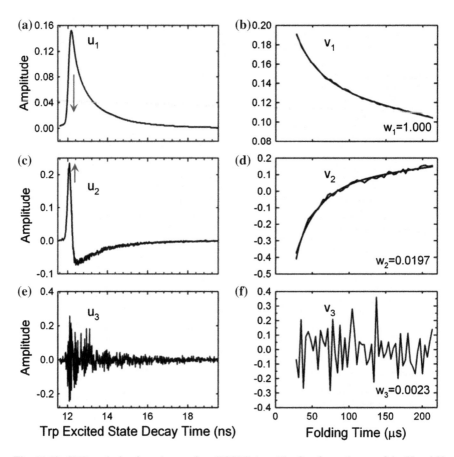

Fig. 11.12 SVD analysis of continuous-flow TCSPC data. The first few columns of the U and V matrices are shown. The remaining columns are essentially noise, similar to the third column. Data correspond to refolding of horse heart cytochrome c shown in Fig. 11.9. Reprinted from [32], Copyright (2014), with permission from Elsevier

The algorithm simultaneously minimizes χ^2 and maximizes the entropy of the distribution, the balance between the two determined by the Lagrange multiplier. An attractive feature of MEM is that it provides the most objective assessment of the true information content of the data [12, 45, 61]. MEM has been widely used in the TCSPC literature for both static and kinetic data [12, 35, 45]. For trFRET data one of the potential applications of MEM is to provide a model independent recovery of the donor-acceptor distance distribution function. Additionally, MEM offers the possibility of avoiding the assumption of the same distance distribution function for each sub-population of the donor excited state.

The ability of MEM to accurately recover the distribution of decay rates has been well established [38]. Extending MEM to time-resolved FRET involves analysis of both the donor and the donor-acceptor excited state decays. This approach parallels the MEM analysis of time-resolved anisotropy [23]. A concern about degeneracy in the 2D MEM distribution was raised; however, the MEM algorithm proposed by Kumar et al. [38] appears to not be susceptible to this, rendering the approach promising. The donor excited state decay is described according to (11.18):

$$I_d(t_{TCSPC}) = \int_0^\infty P(K_d) \cdot e^{-(k_d) \cdot t_{TCSPC}} dk_d$$

$$k_d \equiv 1/\tau_d$$

(11.18)

where k_d is defined, for convenience, as the inverse of the donor lifetime and $p(k_d)$ is the distribution of donor excited state decay rates. For the donor-acceptor labelled system the excited state decay is given as

$$I_{da}(t_{TCSPC}) = \int_0^\infty \int_0^\infty P(k_d, k_{ET}) \cdot e^{-(k_d + k_{ET}) \cdot t_{TCSPC}} dk_d dk_{ET}$$

(11.19)

where k_{ET} is the energy transfer rate given by the Förster equation. The two-dimensional distribution $p(k_d, k_{ET})$ describes the distribution of donor rates and energy-transfer rates. The distribution $p(k_d, k_{ET})$ is usually approximated in one-dimensional analyses as separate one-dimensional distributions giving rise to a "non-associative" model:

$$I_{da}(t_{TCSPC}) = \int_0^\infty P(k_d)e^{-k_d \cdot t_{TCSPC}} dk_d \cdot \int_0^\infty P(k_{ET})e^{-k_d \cdot t_{TCSPC}} dk_{ET}$$

(11.20)

This assumes that every subpopulation responsible for a different donor rate has the same energy-transfer rate distribution. The pair distance distribution is then calculated from the rate distribution according to the Förster equation [39]. Although this approximation results in significant computational advantages, the underlying assumption is not generally applicable. For example, a partially folded state and the unfolded state may be equally populated and the donor may exhibit different excited state lifetimes and a different donor-acceptor distance in each state. Scenarios such as this become especially likely at early times in folding reactions where marginally populated partially folded intermediates may interconvert rapidly. Discriminating these sub-populations is one of the goals of continuous-flow FRET studies. An analysis employing this 2D MEM approach was applied to the early steps in the folding of a TIM barrel protein (Fig. 11.13).

Although the FRET efficiencies were low and not ideal, the algorithm was able to discriminate two sub-populations. The main limitation of the 2D MEM approach is that the accuracy at low FRET efficiencies (e.g., 20 % and lower) is significantly reduced because of insufficient information in the data.

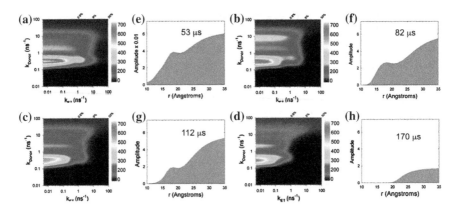

Fig. 11.13 2D MEM analysis at various time points during refolding of a TIM barrel protein. Used with permission [69]. Copyright—2008 by The National Academy of Sciences of the USA

One workaround to this limitation of the 2D MEM analysis [69] is to include an additional experimental axis or dimension such as the acceptor excited state decay. In preliminary tests this was found to yield significantly more reliable distribution functions in the low FRET efficiency tails of distributions. The computational cost of including the additional dimension scales non-linearly although the 3D MEM approach is within reach of current available computational power. The only potential drawback to MEM approaches is that these do not currently consider the diffusion of donor and acceptor chromophores during the excited state lifetime of the donor. However, for sufficiently large Förster distances and short lifetimes, this may still provide significant insights into the sub-populations present during a microsecond kinetic reaction.

11.5 Future Prospects

Significant advances in microfluidic technology have taken place over the past 5 years. Sample consumption, time resolution and temporal dynamic range have improved and the availability of simulation CFD software and microfabrication tools suggests that these advances will continue. These devices and TCSPC detection are used by a relatively small number of laboratories but the robustness of the devices and decreasing cost of TCSPC instrumentation point to increased adoption by the biochemistry community. These tools are likely to play an increasingly important role in FRET studies delineating enzyme mechanisms, protein-protein/RNA/DNA interactions and in enzymatic assays.

Acknowledgments The authors would like to thank McKenzie Davies for sharing representative data for the TCSPC analysis software figure, Brian Mackness and Jill Zitzewitz for helpful discussions and editing, and Bob Matthews for helpful discussions and support. This work was supported by NSF grants DBI1353942 to O. Bilsel and J.B. Perot and MCB1121942 to C.R. Matthews and O. Bilsel and NIH grant GM23303 to C.R. Matthews and O. Bilsel.

References

1. K.N. Aprilakis, H. Taskent, D.P. Raleigh, Use of the novel fluorescent amino acid p-cyanophenylalanine offers a direct probe of hydrophobic core formation during the folding of the N-terminal domain of the ribosomal protein L9 and provides evidence for two-state folding. Biochemistry **46**, 12308–12313 (2007)
2. W. Becker, *Advanced Time-Correlated Single-Photon Counting Techniques* (Springer, Berlin, 2005)
3. W. Becker, *The bh TCSPC Handbook*, 5th edn. Becker & Hickl GmbH (2012), www.becker-hickl.com, printed copies available from Becker & Hickl GmbH
4. W. Becker, B. Su, K. Weisshart, O. Holub, FLIM and FCS detection in laser-scanning microscopes: increased efficiency by GaAsP hybrid detectors. Microsc. Res. Tech. **74**, 804–811 (2011)
5. J.M. Beechem, E. Haas, Simultaneous determination of intramolecular distance distributions and conformational dynamics by global analysis of energy transfer measurements. Biophys. J. **55**, 1225–1236 (1989)
6. J.M. Beechem, Global analysis of biochemical and biophysical data. Methods Enzymol. **210**, 37–54 (1992)
7. J.M. Beechem, Picosecond fluorescence decay curves collected on millisecond time scale: direct measurement of hydrodynamic radii, local/global mobility, and intramolecular distances during protein-folding reactions. Methods Enzymol. **278**, 24–49 (1997)
8. Y. Bellouard, T. Colomb, C. Depeursinge, M. Dugan, A.A. Said, P. Bado, Nanoindentation and birefringence measurements on fused silica specimen exposed to low-energy femtosecond pulses. Opt. Express **14**, 8360–8366 (2006)
9. O. Bilsel, L. Yang, J.A. Zitzewitz, J.M. Beechem, C.R. Matthews, Time-resolved fluorescence anisotropy study of the refolding reaction of the alpha-subunit of tryptophan synthase reveals nonmonotonic behavior of the rotational correlation time. Biochemistry **38**, 4177–4187 (1999)
10. O. Bilsel, C. Kayatekin, L.A. Wallace, C.R. Matthews, A microchannel solution mixer for studying microsecond protein folding reactions. Rev. Sci. Instrum. **76**, 014302 (2005)
11. D.J. Birch, D. McLoskey, A. Sanderson, K. Suhling, A.S. Holmes, Multiplexed time-correlated single-photon counting. J Fluoresc **4**, 91–102 (1994)
12. J.C. Brochon, Maximum entropy method of data analysis in time-resolved spectroscopy. Methods Enzymol. **240**, 262–311 (1994)
13. Y. Chen, A. Periasamy, Characterization of two-photon excitation fluorescence lifetime imaging microscopy for protein localization. Microsc. Res. Tech. **63**, 72–80 (2004)
14. R.B. Dyer, E.B. Brauns, Laser-induced temperature jump infrared measurements of RNA folding. Methods Enzymol. **469**, 353–372 (2009)
15. W.A. Eaton, V. Munoz, S.J. Hagen, G.S. Jas, L.J. Lapidus, E.R. Henry, J. Hofrichter, Fast kinetics and mechanisms in protein folding. Annu. Rev. Biophys. Biomol. Struct. **29**, 327–359 (2000)
16. T. Egawa, J.L. Durand, E.Y. Hayden, D.L. Rousseau, S.R. Yeh, Design and evaluation of a passive alcove-based microfluidic mixer. Anal. Chem. **81**, 1622–1627 (2009)
17. E.L. Elson, Fluorescence correlation spectroscopy: past, present, future. Biophys. J. **101**, 2855–2870 (2011)
18. S. Felekyan, H. Sanabria, S. Kalinin, R. Kühnemuth, C.A. Seidel, Analyzing förster resonance energy transfer with fluctuation algorithms. Methods Enzymol. **519**, 39–85 (2013)

19. J.J. Fisz, Fluorescence polarization spectroscopy at combined high-aperture excitation and detection: application to one-photon-excitation fluorescence microscopy. J. Chem. Phys. A **111**, 8606–8621 (2007)

20. F. Gai, D. Du, Y. Xu, Infrared temperature-jump study of the folding dynamics of alpha-helices and beta-hairpins. Methods Mol. Biol. **350**, 1–20 (2007)

21. Y. Gambin, C. Simonnet, V. VanDelinder, A. Deniz, A. Groisman, Ultrafast microfluidic mixer with three-dimensional flow focusing for studies of biochemical kinetics. Lab Chip **10**, 598–609 (2010)

22. Y. Gambin, V. VanDelinder, A.C. Ferreon, E.A. Lemke, A. Groisman, A.A. Deniz, Visualizing a one-way protein encounter complex by ultrafast single-molecule mixing. Macmillan Publishers Ltd, Nat. Methods **8**, 239–241 (2011)

23. M. Gentin, M. Vincent, J.C. Brochon, A.K. Livesey, N. Cittanova, J. Gallay, Time-resolved fluorescence of the single tryptophan residue in rat alpha-fetoprotein and rat serum albumin: analysis by the maximum-entropy method. Biochemistry **29**, 10405–10412 (1990)

24. R. Gilmanshin, S. Williams, R.H. Callender, W.H. Woodruff, R.B. Dyer, Fast events in protein folding: relaxation dynamics of secondary and tertiary structure in native apomyoglobin. Proc. Natl. Acad. Sci. U.S.A. **94**, 3709–3713 (1997)

25. E. Haas, Ensemble FRET methods in studies of intrinsically disordered proteins. Methods Mol. Biol. **895**, 467–498 (2012)

26. S.J. Hagen, J. Hofrichter, A. Szabo, W.A. Eaton, Diffusion-limited contact formation in unfolded cytochrome c: estimating the maximum rate of protein folding. Proc. Natl. Acad. Sci. U.S.A. **93**, 11615–11617 (1996)

27. K.M. Hamadani, S. Weiss, Nonequilibrium single molecule protein folding in a coaxial mixer. Biophys. J. **95**, 352–365 (2008)

28. D.F. Hansen, H. Feng, Z. Zhou, Y. Bai, L.E. Kay, Selective characterization of microsecond motions in proteins by NMR relaxation. J. Am. Chem. Soc. **131**, 16257–16265 (2009)

29. E.B. Ishay, G. Hazan, G. Rahamim, D. Amir, E. Haas, An instrument for fast acquisition of fluorescence decay curves at picosecond resolution designed for "double kinetics" experiments: application to fluorescence resonance excitation energy transfer study of protein folding. Rev. Sci. Instrum. **83**, 084301 (2012)

30. K.C. Jones, C.S. Peng, A. Tokmakoff, Folding of a heterogeneous beta-hairpin peptide from temperature-jump 2D IR spectroscopy. Proc. Natl. Acad. Sci. U.S.A. **110**, 2828–2833 (2013)

31. S.V. Kathuria, L. Guo, R. Graceffa, R. Barrea, R.P. Nobrega, C.R. Matthews, T.C. Irving, O. Bilsel, Minireview: structural insights into early folding events using continuous-flow time-resolved small-angle X-ray scattering. Biopolymers **95**, 550–558 (2011)

32. S.V. Kathuria, C. Kayatekin, R. Barrea, E. Kondrashkina, R. Graceffa, L. Guo, R.P. Nobrega, S. Chakravarthy, C.R. Matthews, T.C. Irving, O. Bilsel, Microsecond barrier-limited chain collapse observed by time-resolved FRET and SAXS. J. Mol. Biol. **426**(9), 1980–1994 (2014)

33. S.V. Kathuria, A. Chan, R. Graceffa, R. Paul Nobrega, C. Robert Matthews, T.C. Irving, B. Perot, O. Bilsel, Advances in turbulent mixing techniques to study microsecond protein folding reactions. Biopolymers 99, 888–96 (2013)

34. B.A. Kelch, D.L. Makino, M. O'Donnell, J. Kuriyan, How a DNA polymerase clamp loader opens a sliding clamp. Science **334**, 1675–1680 (2011)

35. T. Kimura, J.C. Lee, H.B. Gray, J.R. Winkler, Site-specific collapse dynamics guide the formation of the cytochrome c' four-helix bundle. Proc. Natl. Acad. Sci. U.S.A. **104**, 117–122 (2007)

36. M. Köllner, J. Wolfrum, How many photons are necessary for fluorescence-lifetime measurements? Phys. Chem. Lett. **200**, 199–204 (1992)

37. J. Kubelka, W.A. Eaton, J. Hofrichter, Experimental tests of villin subdomain folding simulations. J. Mol. Biol. **329**, 625–630 (2003)

38. A.T.N. Kumar, L. Zhu, J.F. Christian, A.A. Demidov, P.M. Champion, On the rate distribution analysis of kinetic data using the maximum entropy method: applications to myoglobin relaxation on the nanosecond and femtosecond timescales. J. Phys. Chem. B **105**, 7847–7856 (2001)

39. J.R. Lakowicz, *Principles of Fluorescence Spectroscopy*, 3rd edn. (Springer, Berlin, 2006)
40. T.J. Lane, D. Shukla, K.A. Beauchamp, V.S. Pande, To milliseconds and beyond: challenges in the simulation of protein folding. Curr. Opin. Struct. Biol. **23**, 58–65 (2013)
41. L.J. Lapidus, S. Yao, K.S. McGarrity, D.E. Hertzog, E. Tubman, O. Bakajin, Protein hydrophobic collapse and early folding steps observed in a microfluidic mixer. Biophys. J. **93**, 218–224 (2007)
42. Y. Li, D. Zhang, X. Feng, Y. Xu, B.F. Liu, A microsecond microfluidic mixer for characterizing fast biochemical reactions. Talanta **88**, 175–180 (2012)
43. K. Lindorff-Larsen, S. Piana, R.O. Dror, D.E. Shaw, How fast-folding proteins fold. Science **334**, 517–520 (2011)
44. F. Liu, M. Nakaema, M. Gruebele, The transition state transit time of WW domain folding is controlled by energy landscape roughness. J. Chem. Phys. 131 (2009)
45. A.K. Livesey, J.C. Brochon, Analyzing the distribution of decay constants in pulse-fluorimetry using the maximum entropy method. Biophys. J. **52**, 693–706 (1987)
46. K. Malecha, L.J. Golonka, J. Bałdyga, M. Jasińska, P. Sobieszuk, Serpentine microfluidic mixer made in LTCC. Sens. Actuators, B **143**, 400–413 (2009)
47. V. Munoz, P.A. Thompson, J. Hofrichter, W.A. Eaton, Folding dynamics and mechanism of beta-hairpin formation. Nature **390**, 196–199 (1997)
48. M.R. Otto, M.P. Lillo, J.M. Beechem, Resolution of multiphasic reactions by the combination of fluorescence total-intensity and anisotropy stopped-flow kinetic experiments. Biophys. J. **67**, 2511–2521 (1994)
49. S.A. Pabit, S.J. Hagen, Laminar-flow fluid mixer for fast fluorescence kinetics studies. Biophys. J. **83**, 2872–2878 (2002)
50. S.A. Pabit, J.L. Sutton, H. Chen, L. Pollack, Role of ion valence in the submillisecond collapse and folding of a small RNA domain. Biochemistry **52**, 1539–1546 (2013)
51. A.G. Palmer, Nmr probes of molecular dynamics: overview and comparison with other techniques. Annu. Rev. Biophys. Biomol. Struct. **30**, 129–155 (2001)
52. H.Y. Park, X. Qiu, E. Rhoades, J. Korlach, L.W. Kwok, W.R. Zipfel, W.W. Webb, L. Pollack, Achieving uniform mixing in a microfluidic device: hydrodynamic focusing prior to mixing. Anal. Chem. **78**, 4465–4473 (2006)
53. C.M. Phillips, Y. Mizutani, R.M. Hochstrasser, Ultrafast thermally induced unfolding of RNase A. Proc. Natl. Acad. Sci. U.S.A. **92**, 7292–7296 (1995)
54. L. Pollack, M.W. Tate, N.C. Darnton, J.B. Knight, S.M. Gruner, W.A. Eaton, R.H. Austin, Compactness of the denatured state of a fast-folding protein measured by submillisecond small-angle x-ray scattering. Proc. Natl. Acad. Sci. U.S.A. **96**, 10115–10117 (1999)
55. V. Ratner, D. Amir, E. Kahana, E. Haas, Fast collapse but slow formation of secondary structure elements in the refolding transition of *E. coli* adenylate kinase. J. Mol. Biol. **352**, 683–699 (2005)
56. P. Regenfuss, R.M. Clegg, M.J. Fulwyler, F.J. Barrantes, T.M. Jovin, Mixing liquids in microseconds. Rev. Sci. Instrum. **56**, 283 (1985)
57. H. Roder, K. Maki, H. Cheng, Early events in protein folding explored by rapid mixing methods. Chem. Rev. **106**, 1836–1861 (2006)
58. J.M. Rogers, L.G. Lippert, F. Gai, Non-natural amino acid fluorophores for one- and two-step fluorescence resonance energy transfer applications. Anal. Biochem. **399**, 182–189 (2010)
59. F. Schonfeld, V. Hessel, C. Hofmann, An optimised split-and-recombine micro-mixer with uniform 'chaotic' mixing. Lab Chip **4**, 65–69 (2004)
60. M.C. Shastry, S.D. Luck, H. Roder, A continuous-flow capillary mixing method to monitor reactions on the microsecond time scale. Biophys. J. **74**, 2714–2721 (1998)
61. J. Skilling, R.K. Bryan, Maximum entropy image reconstruction: general algorithm. Mon. Not. R. Astron. Soc. **211**, 111–124 (1984)
62. T.R. Sosnick, D. Barrick, The folding of single domain proteins–have we reached a consensus? Curr. Opin. Struct. Biol. **21**, 12–24 (2011)
63. A.D. Stroock, S.K.W. Dertinger, A. Ajdari, I. Mezic, H.A. Stone, G.M. Whitesides, Chaotic mixer for microchannels. Science **295**, 647–651 (2002)

64. S. Takahashi, S.R. Yeh, T.K. Das, C.K. Chan, D.S. Gottfried, D.L. Rousseau, Folding of cytochrome c initiated by submillisecond mixing. Nat. Struct. Biol. **4**, 44–50 (1997)
65. H. Taskent-Sezgin, P. Marek, R. Thomas, D. Goldberg, J. Chung, I. Carrico, D.P. Raleigh, Modulation of p-cyanophenylalanine fluorescence by amino acid side chains and rational design of fluorescence probes of alpha-helix formation. Biochemistry **49**, 6290–6295 (2010)
66. D. Thirumalai, B.-Y. Ha, Statistical mechanics of semiflexible chains: a meanfield variational approach. arXiv.cond-mat/9705200 (1997)
67. P.A. Thompson, W.A. Eaton, J. Hofrichter, Laser temperature jump study of the helix <==> coil kinetics of an alanine peptide interpreted with a 'kinetic zipper' model. Biochemistry **36**, 9200–9210 (1997)
68. C.T. Wang, Y.C. Hu, T.Y. Hu, Biophysical micromixer. Sens. (Basel) **9**, 5379–5389 (2009)
69. Y. Wu, E. Kondrashkina, C. Kayatekin, C.R. Matthews, O. Bilsel, Microsecond acquisition of heterogeneous structure in the folding of a TIM barrel protein. Proc. Natl. Acad. Sci. U.S.A. **105**, 13367–13372 (2008)
70. B. Wunderlich, D. Nettels, S. Benke, J. Clark, S. Weidner, H. Hofmann, S.H. Pfeil, B. Schuler, Microfluidic mixer designed for performing single-molecule kinetics with confocal detection on timescales from milliseconds to minutes. Macmillan Publishers Ltd, Nat. Protoc. **8**, 1459–1474 (2013)
71. W.Y. Yang, M. Gruebele, Folding at the speed limit. Nature **423**, 193–197 (2003)
72. G. Zoldak, J. Stigler, B. Pelz, H. Li, M. Rief, Ultrafast folding kinetics and cooperativity of villin headpiece in single-molecule force spectroscopy. Proc. Natl. Acad. Sci. U.S.A. **110**, 18156–18161 (2013)

Chapter 12
An Introduction to Interpreting Time Resolved Fluorescence Anisotropy Curves

Steven S. Vogel, Tuan A. Nguyen, Paul S. Blank
and B. Wieb van der Meer

Abstract The decay in fluorescence anisotropy following an excitation pulse can be monitored using time-correlated single photon counting, and used to measure molecular rotation and homo-FRET in biomedical research. In this chapter we review the basis of polarized fluorescence emission, and how this emission can be detected and quantified to yield a time-resolved anisotropy curve. We then review chemical, instrumental and biological factors that can influence the shape and magnitude of anisotropy curves. Understanding the influence of these factors can then be used to aid in the interpretation of biomedical experiments using time-resolved anisotropy.

12.1 Introduction

Biophotonics is a branch of biophysics that analyzes changes in light to study biological processes [19]. Typically, the intensity or amplitude of light changes when interacting with a biological specimen as a result of absorption or scattering

S.S. Vogel (✉) · T.A. Nguyen
Division of Intramural Clinical & Biological Research, National Institutes of Health,
National Institute of Alcohol Abuse & Alcoholism, Bethesda, MD 20892-9411, USA
e-mail: stevevog@mail.nih.gov

T.A. Nguyen
e-mail: nguyenta2@mail.nih.gov

P.S. Blank
Eunice Kennedy Shriver National Institute of Child Health
and Human Development, Bethesda, MD, USA
e-mail: blankp@mail.nih.gov

B.W. van der Meer
Western Kentucky University, Department of Physics and Astronomy,
Kentucky, USA
e-mail: wieb.vandermeer@wku.edu

© Springer International Publishing Switzerland 2015
W. Becker (ed.), *Advanced Time-Correlated Single Photon Counting Applications*,
Springer Series in Chemical Physics 111, DOI 10.1007/978-3-319-14929-5_12

by molecules and molecular assemblies within the sample. For molecules that are intrinsically fluorescent, or fluorescent as a result of being labelled with a fluorophore, excitation light at one wavelength can be converted to another wavelength (longer wavelength for one photon excitation, and shorter wavelength for multiphoton excitation) [14]. This shift in the wavelength of light allows for very precise localization and quantification of fluorescent molecules, often with single molecule sensitivity. Biosensors have also been developed that show changes in their fluorescence intensity when bound to a specific ligand of biological interest. Typically, the intensity of either the bound or unbound state of the fluorophore is reduced or quenched. Ligand binding induces the molecule to switch to its alternate state thus transducing the binding of ligand into a change in fluorescence intensity (and often, fluorescence lifetime). The experimentalist can monitor this change to visualize the extent of ligand binding or the free concentration of ligand. In addition to the intensity (amplitude) and colour (wavelength) of light, another under-appreciated characteristic of light that can be monitored in biophotonics is the polarization or 'orientation' of light. Unlike changes in intensity and colour that are readily detected by most human eyes, the human eye is an extremely poor detector of the polarization of light [31]. Accordingly, even the notion that light has an orientation is often confusing, and the concept that the polarization of light can change with time is at best understood cerebrally rather than viscerally. It is our goal in this chapter to provide a more intuitive explanation of how and why the orientation of light changes when linear polarized light interacts with fluorescent biological samples, and in so doing provide a foundation for understanding how time-resolved fluorescence anisotropy measurements can be applied and interpreted to better understand biological processes.

12.2 The Molecular Basis for Orientation of Fluorescence Emission

A *fluorophore* is a molecule that can absorb a photon and after a short delay, typically within a few nanoseconds (10^{-9} s), emit a photon with a slightly longer wavelength (less energy). The absorption of a photon occurs within a specific range of wavelengths resulting in the excitation of one of the fluorophores electrons into an excited state. The wavelengths that can be absorbed by a specific fluorophore, its *excitation spectrum*, are dictated by the distribution of many discrete energy gaps defining the difference between ground and excited states present in the molecule. The energy differences between these states can be explained using quantum mechanics by accounting primarily for permitted electronic, vibrational and rotational states of the molecule. The absorption of a photon by a fluorophore is fast, occurring typically within a few femtoseconds (10^{-15} s). Over a period of a few hundred femtoseconds, the large selection of potential rotational and vibrational excited-state sublevels will decay into the lowest-energy singlet excited state.

This consolidation results from rotational- and vibrational-energy loss due to kinetic interactions of the excited fluorophore with surrounding molecules, and is in part responsible for the Stokes shift between the wavelengths of excitation and emission. An excited fluorophore may remain in the singlet state for a period lasting from picoseconds (10^{-12} s) to tens of nanoseconds (10^{-9} s). The amount of time any individual fluorophore and excitation event remains in the singlet state is stochastic. Nonetheless, fluorophores do have characteristic *fluorescence lifetimes* that describe the average time they will remain in the singlet-state. With time, a singlet state fluorophore will eventually decay to its ground state by either emitting a photon (fluorescence emission), by transferring energy to a nearby ground state fluorophore (Förster Resonance Energy Transfer, FRET), or the energy can be lost by the formation of a long-lived triplet state, collisional quenching, or some other non-radiative excited state reaction.

In polarization and anisotropy experiments, the orientation of photons emitted from a fluorophore is characterized relative to the orientation of the electric field of the excitation light. To understand the basis of fluorescence polarization we must appreciate that individual fluorophores have an 'orientation' in space that can change with time. Most importantly, their 'orientation', relative to the orientation of the excitation light electric field, can dramatically influence the probability that they absorb a photon to transition into an excited state. When a photon is absorbed, a ground-state electron in the fluorophore is boosted to a higher orbital. At that instant, the balance between lower-mass, fast-moving, negative charged electrons and higher-mass, slow-moving, positive charged protons in the molecule is perturbed establishing an oscillating dipole (a vector separation of positive and negative charges). This oscillating dipole is called a *Transition Dipole* or more specifically an *Absorption Dipole,* and has both magnitude and orientation. The highest probability for excitation is achieved when a fluorophore is illuminated with light at a wavelength appropriate for excitation, and if the electric field of the excitation light is parallel to the orientation of the fluorophores absorption dipole ($\theta = 0°$). In contrast, excitation does not occur if the electric field is perpendicular to the absorption dipole orientation ($\theta = 90°$). At intermediate orientations ($0° < \theta < 90°$) the probability for excitation can be calculated from the angle between the electric field orientation and that of the absorption dipole:

$$p_{excitation} \propto \cos^x \theta \tag{12.1}$$

where, for linearly polarized light, x is 2 for one-photon excitation and 4 for two-photon excitation. This preference for fluorophore excitation at low values of θ is called *photoselection* (see Fig. 12.1).

Once excited, the 'orientation' of a fluorophore can also influence the 'orientation' of a photon emitted as fluorescence. In biological experiments, there are three primary mechanisms by which the orientation of emitted fluorescence will be influenced. These are, as a result of: (1) conformational changes in the fluorophore while in the excited state, (2) molecular rotation of the fluorophore between the times it is excited and emits a photon, and (3) Förster Resonance Energy Transfer

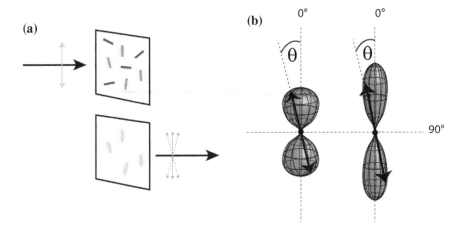

Fig. 12.1 Photoselection. Electric field orientation of linearly polarized light (*black arrow*) is depicted by a *double blue arrow* (**a**). Linear polarized light will selectively excite fluorophores whose absorption dipole orientation (depicted as *brown bars*) is similar (*light brown*) to that of the excitation lights electric field. Only these orientation 'selected' excited fluorophores will emit fluorescent light (**a**). The orientation of this emitted light (*green double arrows*), is correlated with the electric field orientation of the excitation light (*solid double arrow*), but is degenerate (*dashed green double arrows*) primarily because of the dependence of photoselection where θ is the angle between the electric field orientation of the excitation light and the absorption dipole orientation. Panel **b** depicts radial probability plots of the dependence of photoselection on θ ($p \propto \cos^x \theta \sin \theta$) where x = 2 for 1-photon excitation (*left*) or x = 4 for 2-photon excitation (*right*). In these plots the integrated area of each 3-dimentional plot (depicted in *pink*) is set to 1, and the length of a *double black arrow* (for a particular value of θ is proportional to its probability for excitation. Note that excitation is not possible when θ = 90°, and that 2-photon excitation will preferentially select fluorophores with θ values closer to 0°

(FRET) to a nearby fluorophore with a different transition dipole orientation. These will be discussed shortly, but first we must explain how polarization and anisotropy are measured, how these measurements differ, and why anisotropy measurements are usually preferred.

12.3 How Do We Measure the Anisotropy of Polarized Fluorescence Emissions?

How can a microscope be used to follow changes in the orientation of molecules in a biological context? Similarly, how can a microscope be used to monitor changes in the proximity of molecules within living cells? Conceptually, if not practically, changes in the orientation, and in some instances changes in the proximity of fluorescent molecules can be determined by measuring the intensity of emitted photons relative to the orientation of the electric vector of a linear polarized light source along three orthogonal axes.

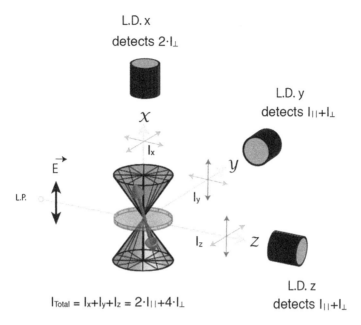

L.D. x
detects 2·I⊥

L.D. y
detects I∥+I⊥

$\mathcal{X}$

I$_x$

$\vec{E}$

$\mathcal{Y}$

L.P.

I$_y$

I$_z$

z

L.D. z
detects I∥+I⊥

I$_{Total}$ = I$_x$+I$_y$+I$_z$ = 2·I∥+4·I⊥

Fig. 12.2 Detecting polarization. The transition dipole (*blue double arrow*) of a fluorophore is excited by a linear polarized light source (L.P.). The electric field vector of the light source is shown as a *black double arrow*. The three dimensional orientation of the dipole can be characterized by 2 angles, θ and φ, where θ is the angle formed between the dipole and the X-axis (*pink double cone*), and φ is the angle formed between the projection of the dipole on the YZ plane (*green disk*) and the Z-axis. The light intensity emitted from this fluorophore will be proportional to the square of the dipole strength, and the dipole vector can be thought of as being composed of 3 directional components: x, y, and z. A signal proportional to the total intensity of light emitted by our fluorophore can be measured summing the signals detected by light detectors (L.D.) positioned on each of the three axes. The intensity information encoded in the x vector component by convention is called I∥ while the intensity information encoded in the y and z vector components is called I⊥. As a result of photoselection and the distribution symmetry of excited molecules formed around the X-axis for an isotropic solution, the y vector component is equal to the z vector component. For an isotropic solution of fluorophores excited with linearly polarized light whose electric vector is parallel to the X-axis, L.D.x will measure a signal whose intensity is proportional to 2·I⊥ (*crossed double-headed green arrows*), while L.D.y and L.D.z will each measure light signals whose intensity is proportional to I∥ + I⊥ (*crossed double-headed red* and *green arrows*). The total emission intensity will be proportional to 2·I∥ + 4·I⊥ or more simply I∥ + 2·I⊥. Note that the xy plane depicted here corresponds to the sample plane on a microscope, and L.D.z corresponds to a detector placed after the microscope condenser (or at an equivalent position on the epi-fluorescence path)

Figure 12.2 depicts a transition dipole (blue double arrow) of a fluorophore (from an isotropic solution) that is excited by linear polarized light (L.P.). The electric field orientation of this light source is shown as a black double arrow. The dipole orientation can be characterized by 2 angles, θ and φ. θ is the angle formed between the dipole and the X-axis in the XY-plane. φ is the angle formed between the projection of the dipole on the YZ-plane and the Z-axis.

The X-axis is an axis of symmetry as it is parallel to the electric vector of our light source. The Y and Z-axes are perpendicular to the light source electric vector and thus are not axes of symmetry. The fluorescent light intensity emitted from this fluorophore will be proportional to the square of its dipole length, and the dipole vector can be thought of as being composed of 3 directional components: x, y, and z. By placing light detectors (L.D.) on each of these three axes a signal proportional to the total light intensity emitted can be measured. The emitted light will be proportional to the sum of the three signals ($I_{total} = I_x + I_y + I_z$). L.D.x will detect light related to the yz vector components of the dipole. Similarly, L.D.y will detect light related to the xz vector components, and L.D.z will detect light related to the xy vector components. The intensity information encoded in the x vector component is called $I_{\parallel}$ while the intensity information encoded in the y and z vector components are called $I_{\perp}$. As a result of *photoselection* (the orientationally biased excitation of fluorophores by linearly polarized light) and the distribution symmetry of excited molecules formed around the X-axis for an *isotropic* (randomly orientated) solution, the y vector component will be equal to the z vector component. Accordingly, when an isotropic solution of fluorophores is excited with linearly polarized light whose electric vector is parallel to the X-axis, the L.D.x detector will measure an un-polarized light signal whose intensity is proportional to $2 \cdot I_{\perp}$. In contrast, L.D.y and L.D.z will each measure a polarized signal whose intensity is proportional to $I_{\parallel} + I_{\perp}$. If we sum the $I_{\parallel}$ and $I_{\perp}$ components along all three emission axes, we find that the total emission intensity will be proportional to $2 \cdot I_{\parallel} + 4 \cdot I_{\perp}$ or more simply $I_{\parallel} + 2 \cdot I_{\perp}$ [28].

As described above, a light detector placed along the Z-axis of a microscope will collect light proportional to $I_{\parallel} + I_{\perp}$. Furthermore, on most light microscopes it is relatively straightforward to position a light detector on this axis to detect changes in polarization. There are two general arrangements commonly used for measuring fluorescence polarization on a microscope, both require separating $I_{\parallel}$ and $I_{\perp}$ components of the fluorescence emission along the Z-axis, see Fig. 12.3.

The first design measures $I_{\parallel}$ and $I_{\perp}$ sequentially (panels a), while the second strategy measures $I_{\parallel}$ and $I_{\perp}$ at the same time (panel b). To measure $I_{\parallel}$ and $I_{\perp}$ sequentially, a linear polarizer (*Pol*) is positioned between the sample (*s*) and the light detector. When the linear polarizer is oriented with the electric field of the excitation source (0°), the detector will measure a signal proportional to $I_{\parallel}$ (Fig. 12.3a). Alternatively, when the polarizer is oriented orthogonal to the electric field of the excitation source (90°), the detector will measure a signal proportional to $I_{\perp}$. On a simple and inexpensive system the orientation of the linear polarizing filter can be changed manually. Alternatively to minimize the time interval between measuring $I_{\parallel}$ and $I_{\perp}$ signals, the rotation of the polarizing filter can be automated using either a mechanized rotating filter mount or by using a filter wheel to switch between two orthogonally oriented linear polarizing filters.

To accurately measure polarization in the schemes outlined above it is essential to accurately align the polarizing filters at 0° and 90°. In our laboratory, fluorescence emission filters are removed to allow linearly polarized excitation light to project directly onto the light detector. Next the orientation of the polarizing filter is rotated until the weakest transmitted signal is found (90°). The 0° orientation is then

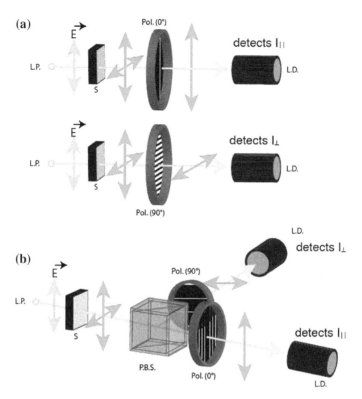

Fig. 12.3 Separating $I_\parallel$ and $I_\perp$. When linearly polarized light (L.P.) with electronic vector E is used to excite a sample (s) on a microscope the fluorescence emission along the z-axis (*yellow arrow*) will be comprised of $I_\parallel$ and $I_\perp$. The magnitude of these two intensity components can be measured either sequentially (panel **a**) or in parallel (panel **b**). The sequential configuration uses a single light detector (L.D.) and a linear polarizing filter (Pol.) that is first positioned parallel (panel **a** *top*; 0°) to the electronic vector of the light source to measure $I_\parallel$ and then positioned perpendicular (panel **a** *bottom*; 90°) to measure $I_\perp$. In the parallel detection configuration (panel **b**) a polarizing beam splitter (P.B.S.) is used in conjunction with two linear polarizing filters and two light detectors to measure $I_\parallel$ and $I_\perp$ simultaneously

a 90° rotational offset from the 90° position. This alignment procedure assumes that the detectors are insensitive to the polarization of the light. Side-on photo-multipliers are often very sensitive to polarization while end-on photomultipliers typically are not.

To measure $I_\parallel$ and $I_\perp$ signals in parallel, a polarizing beam splitter (*P.B.S.*) can be used to separate these signals simultaneously (Fig. 12.3b). For live cell mea-surements this scheme is preferable to the sequential arrangement because it eliminates artefacts caused by the motion of polarized cellular components during the time interval between measuring $I_\parallel$ and $I_\perp$ signals. By placing a polarizing beam splitter after the sample, $I_\parallel$ and $I_\perp$ fluorescent signals can be separated and measured by two separate dedicated light detectors. Typically, the $I_\perp$ signal is reflected by the beam splitter while the $I_\parallel$ signal is transmitted. Polarizing beam splitters can be wavelength dependent. Consequently, when adapting a microscope for polarization measurements a polarizing beam splitter that has a flat response over a wavelength range that is matched to the emission spectrum of the fluorophore of interest should be selected. It is also worth noting that the *contrast ratio* (the intensity ratio of the transmitted polarization state vs. the attenuated state) of polarizing beam splitters is reasonable in the transmitted pathway (typically >500:1), but typically poor in the reflected pathway (20:1). For this reason, we use two linear polarizing filters whose orientation is matched to the output of the beam splitter to augment the contrast ratio. Linear polarizing filters typically have contrast ratios that are at least 500:1. Finally, when measuring $I_{\parallel}$ and $I_{\perp}$ in parallel, each pathway downstream of the polarizing beam splitter will have its own dedicated photo-detector.

A photomultiplier tube would typically be used as the light detector for steady-state polarization measurements in conjunction with laser scanning microscopy such as confocal microscopy or two-photon microscopy. The use of photomultipliers, ideally with a short instrument response function (<200 ps) in conjunction with a pulsed laser light source allows time-resolved polarization measurements using time correlated single photon counting (TCSPC; [3]). Two fluorescence lifetime decay curves are generated, one representing the decay of $I_{\parallel}(t)$ and the other for $I_{\perp}(t)$. Ideally, the inverse of the laser repetition rate should be $\geq$ to 5 times the lifetime of the fluorophore being studied. For most fluorophores a repetition rate of 20–50 MHz is ideal. With two-photon excitation often a less than ideal laser repetition rate is used (80–90 MHz), and this compromise results in truncated time-resolved decay curves as well as the potential for temporal bleed-through artefacts. The microscope designs shown in panel A or B can also be adapted for steady state polarization imaging using a EMCCD camera as the light detector.

12.4 How Are Polarized Emissions Quantified?

To begin to understand the basis of time-resolved anisotropy measurements we will start with a 'simple' population of fluorophores in solution. We will assume that the fluorophores in this sample are randomly oriented (*isotropic*), that the absorption and emission dipoles of these fluorophores are collinear (have the same orientation), and that these fluorophores do not rotate while in the excited state. Once we understand this simple system we will then consider the impact of having non-collinear dipoles and molecular rotation. With the assumptions of this simple system the orientation of photons emitted from the excited fluorophores in this population will be highly correlated with the orientation of the linear polarized light used to excite them as dictated by the orientational dependence of photoselection (12.1). Essentially, in this population, fluorophores with low values of θ will be preferentially excited while those with θ values near $90°$ will not be excited. It must also be appreciated that even though the fluorophores in this population are randomly oriented, the number of molecules with θ values near $0°$ will be much less than those with θ values near $90°$ (for a more detailed explanation see [31]).

Ultimately, the probability of exciting any specific molecule in this isotropic population will be proportional to the product of the probability for excitation at a specific θ values (12.1) and the probability for finding a fluorophore with that θ value in the isotropic population ($\sin \theta$):

$$p \propto \cos^x \theta \cdot \sin \theta \tag{12.2}$$

where x is 2 for one-photon excitation and 4 for two-photon excitation with linearly polarized light. The key concept to understand is that photo selection with linearly polarized light transforms a randomly oriented population of ground-state fluorophores into an ordered population of excited state fluorophores whose dipole orientations are strongly correlated with the orientation of the excitation light electric field. This near instantaneous transformation into a population of ordered excited-state fluorophores as a result of a laser excitation pulse, and our ability to monitor the population orientation using polarization measurements as described above, forms the basis of time-resolved anisotropy measurements. At this point we should note that because this 'simple' population is static, it does not rotate during the excited state; polarization measurements should not change as a function of time after the laser pulse. In most biological samples, fluorophores can rotate to some extent during the excited state and so polarization measurements can and will change as a function of time.

The polarization state of a photon emitted by a fluorophore will be correlated with the orientation of the fluorophore (strictly speaking its emission dipole) at the instant of returning to the ground state [33]. As described above, for an isotropic population of static fluorophores with collinear absorption and emission dipoles excited by linearly polarized light, the polarization of emitted photons will also be strongly correlated with the electric field orientation of the polarized light source. Thus, under the conditions described, the polarization of photons emitted by an isotropic population of fluorophores can be used to determine how similar the orientation of the fluorophores are relative to the orientation of the electric field of the excitation source. We need to answer two more fundamental questions before we can more fully understand this correlation quantitatively and then apply it to biological questions; these are: (1) What is the relationship between the orientation of a fluorophore emission dipole and the probability of detecting its emitted photon through parallel or perpendicularly oriented linear polarizers? and (2) How can we parameterize the orientation of the emission from an isotropic population of fluorophores using the measured $I_{\parallel}$ and $I_{\perp}$ values?

Figure 12.4 depicts how the emission dipole orientation (double green arrow) of a fluorophore from an isotropic population of fluorophores excited with polarized light will affect the signal intensity measured through a linear polarizer oriented either parallel (Fig. 12.4a) or perpendicular (Fig. 12.4b) to the electric field of the light source.

A representation of the three-dimensional excited state probability distribution (see 12.2) is depicted in pink. The orientation of any single fluorophore from this excited state population can be described by two angles, θ (Fig. 12.4c) and ϕ (4D).

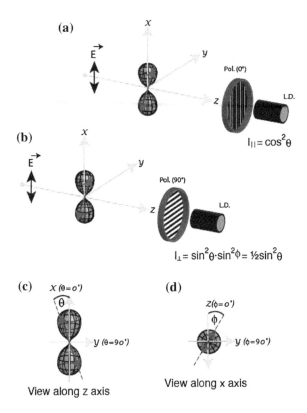

Fig. 12.4 The probability of detection through a polarizing filter. The probability that a photon emitted by a fluorophore will pass through a linear polarizing filter (Pol.) and detected by a light detector (L.D.) is a function of both the orientation of the fluorophores emission dipole (*green double arrow*), and the orientation of the filter. When the filter is situated along the Z-axis and is oriented at 0° relative to the electric field (*E*) of the light source the photomultiplier will detect $I_{\parallel}$ (**a**). When the filter is rotated to 90° relative to the electric field the photomultiplier will detect $I_{\perp}$(**b**). $I_{\parallel}$ will be proportional to $\cos^2 \theta$, where θ is the angle formed between electric field of the light source and the emission dipole of the fluorophore (panel **c**). $I_{\perp}$ will be proportional to $\sin^2 \theta \cdot \sin^2 \phi$, where ϕ is the angle formed between the emission dipole of the fluorophore and the Z-axis (panel **d**). For an isotropic distribution of fluorophores excited with linearly polarized light, the distribution of excited state dipole orientations (*pink* hour-glass shaped cloud) will have a symmetrical distribution of ϕ values around the X-axis (**d**). Due to this symmetry, the value of $\sin^2 \phi = \frac{1}{2}$. Accordingly, for an isotropic distribution of fluorophores $I_{\perp}$ will be proportional to $\frac{1}{2} \sin^2 \theta$. Notice that for an isotropic population of fluorophores values of $I_{\parallel}$ and $I_{\perp}$ are functions of θ alone

When the linear polarizer is oriented parallel to the electric field polarization (0°) the light intensity measured through the filter will be proportional to $\cos^2 \theta$ (where θ is the polar angle of the emitting molecule relative to the electric field polarization). Thus, for a population of fluorophores the measured $I_{\parallel}$ intensity will be proportional to an average of all the $\cos^2 \theta$ values weighted by their abundance. When the filter polarizer is oriented perpendicular to the orientation of the excitation electric field polarization (90°), the light intensity measured will be proportional to $\sin^2 \theta \cdot \sin^2 \phi$.

This equation can be simplified because the excited state distribution is symmetrical around the X-axis, and for the population, the average of $\sin^2 \phi = \frac{1}{2}$. Consequently, for the population of fluorophores, $I_\perp$ will be proportional to an abundance-weighted average of all $\frac{1}{2} \sin^2 \theta$ values. This equation illustrates that when a population of randomly oriented fluorophores is excited by linearly polarized light, both $I_\parallel$ and $I_\perp$ will be determined by the value of θ (the polar angle of the emitting molecules relative to the electric field polarization) alone.

Finally we must discus how $I_\parallel$ and $I_\perp$ values can be used to parameterize the orientation of fluorophore populations. There are two main conventions that have been used in the literature, the *polarization ratio* (p) and *emission anisotropy* (r). The polarization ratio is simply the intensity difference between $I_\parallel$ and $I_\perp$ divided by the intensity observed by a photo-detector placed along either the Y- or Z-axis ($I_\parallel + I_\perp$):

$$p = (I_\parallel - I_\perp)/(I_\parallel + I_\perp)$$

By setting $I_\perp$ or $I_\parallel$ to 0 the limiting values of polarization can be determined; the polarization ratio can range from -1 to 1. A polarization value of 1 indicates a perfect alignment of emission dipoles with the orientation of the light source electric field. A polarization value of -1 indicates a perfectly orthogonal orientation. Analogous to polarization, the numerator for calculating the emission anisotropy is the intensity difference between $I_\parallel$ and $I_\perp$. However, the denominator used for calculating the emission anisotropy is the emission intensity with parallel and perpendicular components proportional to the *total* intensity ($I_\parallel + 2 \cdot I_\perp$):

$$r = (I_\parallel - I_\perp)/(I_\parallel + 2 \cdot I_\perp) \tag{12.3}$$

Substituting 0 for either $I_\perp$ or $I_\parallel$ reveals that the range of possible values for emission anisotropy is from -0.5 to 1. For anisotropy measurements a value of 1 indicates a perfect alignment of emission dipoles with the orientation of the light source and a value of -0.5 indicates a perfectly orthogonal orientation. A population of randomly oriented excited-state fluorophores will have an anisotropy value of 0. Clearly, the polarization ratio and emission anisotropy are just different expressions used to parameterize the same phenomenon, the orientation of light emitted relative to the orientation of the linearly polarized light source electric field. The relationship between r and p is:

$$r = 2 \cdot p/(3 - p)$$

12.5 Calculating the Time-Resolved Anisotropy Curve

In this chapter we use emission anisotropy rather than polarization because in most biological applications anisotropy is more amenable to analysis. Specifically, we will describe time-resolved emission anisotropy, which monitors how anisotropy values change as a function of time after photo selection/excitation. In Fig. 12.5a, b we display $I_\parallel(t)$ and $I_\perp(t)$ decay curves for a sample containing the yellow fluorescent protein *Venus* in an aqueous buffer.

We will use this sample to introduce how time-resolved anisotropy can be used to analyze fluorescent biological samples. For these measurements the Venus fluorophore was excited using two-photon excitation at 950 nm. $I_\parallel(t)$ and $I_\perp(t)$ decay curves contain information about the fluorescence lifetime (τ) of Venus as well as the dipole orientation as Venus rotates while in its excited state. The Venus $I_\parallel(t)$ and $I_\perp(t)$ decay curves are well described by the following two equations:

$$I_\parallel(t) = G\tilde{F}_0 \cdot e^{(-t/\tau)} \left(2r_0 \cdot e^{(-t/\theta_{rot})} + 1 \right) \qquad (12.4)$$

$$I_\perp(t) = \tilde{F}_0 \cdot e^{(-t/\tau)} \left(-r_0 \cdot e^{(-t/\theta_{rot})} + 1 \right) \qquad (12.5)$$

where τ is the *fluorescence lifetime* of Venus the (average amount of time a Venus molecule remains in the excited state), θ_{rot} is the *rotational correlation time* of Venus in solution, r_0 is the *limiting anisotropy* (the anisotropy value at time equals

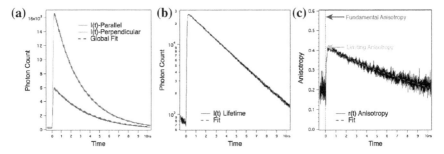

Fig. 12.5 The lifetime and time-resolved anisotropy of the *yellow* fluorescent protein Venus. A sample of the fluorescent protein venus was excited using 950 nm light pulses from a Ti: sapphire laser using two-photon photo selection and a fluorescence detection scheme similar to the layout depicted in Fig. 12.3b. The measured $I(t)_\parallel$ and $I(t)_\perp$ decay curves are plotted in panel **a**. These two traces were globally fit (*red dashed lines*) to (12.4) and (12.5) and yielded a Venus lifetime of 3.23 ns, a rotational correlation time of 14.31 ± 0.05 ns, and a limiting anisotropy of 0.41. The $I(t)_\parallel$ and $I(t)_\perp$ traces in panel **a** were next processed using (12.6) to yield the fluorescence lifetime decay, or (12.8) to yield the time-resolved anisotropy decay (panels **b** and **c** respectively). The *dashed red line* in panel **b** depicts a fit using a single exponential decay model and yielded a lifetime of 3.20 ns. The *dashed red line* in panel **c** also depicts a fit using a single exponential decay model and yielded a rotational correlation time for Venus of 14.9 ± 0.1 ns and a limiting anisotropy value of 0.41. The fundamental anisotropy value expected for an isotropic population of fluorophores with collinear absorption and emission dipoles excited by two-photon excitation is 0.57. The difference between this fundamental anisotropy value and the measured limiting anisotropy (0.41) can be accounted for (using 12.12) by multiplying the limiting anisotropy by the product of an instrumental depolarization factor for a NA 1.2 water objective (*dNA* = 0.81, see Table 12.1) and a depolarization factor for a β dipole angle of 15.4° (*dβ* = 0.89, see 12.11). We have previously measured an upper limit for the Venus dipole angle β of 16° using low NA objectives [31]

zero before any dipole rotation has occurred), G is an experimentally measured constant, particular for a specific microscope configuration, that accounts for differences in the sensitivity between $I_\perp$ or $I_\parallel$ measurements, and F_0 is an intensity-scaling factor, where $\tilde{F}_0 = F_0 \cdot (1 + 2G)/G$. If the sensitivity of both measurements is the same, then G will have a value of 1. In our experience using either the same, or matched photomultiplier tubes, G has a value close to 1 (in this example it had a value of 0.9). Notice, since (12.4) and (12.5) are functions of the same 4 variables, $I_\parallel(t)$ and $I_\perp(t)$ decay curves shown in Fig. 12.5a can be fit globally to (12.4) and (12.5) respectively to yield more accurate estimates of τ, τ_{rot} and r_0. Global analysis of the curves depicted in Fig. 12.5a yielded a Venus lifetime of 3.23 ns (± 0.00), a rotational correlation time of 14.3 $\pm$ 0.1 ns, and a limiting anisotropy of 0.41 (± 0.00). Importantly, (12.4) and (12.5) are only applicable if the fluorophore lifetime decays as a single exponential, and if the molecule is spherical, such that its rotational correlation times around its X-, Y-, and Z-axes are identical.

The Venus $I_\parallel(t)$ and $I_\perp(t)$ decay curves depicted in Fig. 12.5a can be used to calculate a fluorescence lifetime decay curve using (12.6) and is plotted in Fig. 12.5b:

$$I(t) = I_\parallel(t) + 2 \cdot G \cdot I_\perp(t) \tag{12.6}$$

By substituting (12.4) and (12.5) into (12.6) we observe that $I(t)$ decays as a single exponential with an intensity amplitude of $F_0(1 + 2G)$.

$$I(t) = F_0(1 + 2G) \cdot e^{-t/\tau} \tag{12.7}$$

The decay curve in Fig. 12.5b can be modelled as a single exponential with a tau of 3.2 ns (± 0.0) in excellent agreement with the lifetime predicted by global fitting of the decay curves in Fig. 12.5a as well as with values in the literature [20].

The Venus $I_\parallel(t)$ and $I_\perp(t)$ decay curves depicted in Fig. 12.5b can also be used to calculate an emission anisotropy decay curve using (12.8) and is plotted in Fig. 12.5c:

$$r(t) = \frac{I_\parallel(t) - G \cdot I_\perp(t)}{I_\parallel(t) + 2G \cdot I_\perp(t)} \tag{12.8}$$

By substituting (12.4) and (12.5) into (12.8) we can see that $r(t)$ for this example should decay as a single exponential:

$$r(t) = r_0 \cdot e^{-t/\theta_{rot}} \tag{12.9}$$

Analysis of the time resolved anisotropy curve in Fig. 12.5c results in a rotational correlation time of 14.9 $\pm$ 0.1 ns, in good agreement with the rotational correlation time previously reported [20] and a limiting anisotropy of 0.41 $\pm$ 0.00. These values were also in excellent agreement with the values obtained using global fitting of the decay curves in Fig. 12.5a. The major point is that (12.6) and (12.8) allow the extraction of either the fluorescence lifetime or the time-resolved anisotropy decay from $I_\parallel(t)$ and $I_\perp(t)$ decay curves. Equations (12.7) and (12.9) are applicable only to fluorophores whose lifetime decays as a single exponential and are spherical, having identical rotational correlation times around their X-, Y-, and Z-axes. In contrast, (12.6) and (12.8) can be applied to extract either the fluorescence lifetime or time-resolved anisotropy from $I_\parallel(t)$ and $I_\perp(t)$ decay curves obtained from more complex fluorescent samples having multiple lifetimes and/or multiple anisotropy decay components. For this reason $I_\parallel(t)$ and $I_\perp(t)$ decay curves are obtained from more complicated biological samples and are then processed using (12.8) to generate a time-resolved anisotropy decay curve. The remainder of this chapter will address how to interpret time-resolved anisotropy decay curves.

12.6 The Anatomy of Time-Resolved Anisotropy

A time-resolved anisotropy curve, such as the curve depicted in Fig. 12.5c, describes the ensemble anisotropy value of a population of fluorophores, and follows how this value changes as a function of elapsed time from photo selection with a short pulse of linearly polarized light at t = 0. The time-resolved anisotropy curve has two characteristics that are worth considering, the limiting anisotropy, r_0, which is the population anisotropy value at the instant of photo selection (t = 0), and a sequence of anisotropy values that describes how the population anisotropy changes as a function of time after photo selection. We will start with interpreting the value of the limiting anisotropy, r_0, and then discus how to analyze and interpret how anisotropy values can change with time.

The X-axis of time-resolved anisotropy curves represents elapsed time from the instant of photo selection (typically with a short pulse of linearly polarized laser light). The span of the X-axis is primarily dictated by the repetition rate of the pulsed laser used for photo-excitation, and typically ranges from t = 0 to 12.5 ns when an 80 MHz Ti:Sapphire laser is used for two-photon excitation, to 20 ns when a 50 MHz pulsed diode laser is used for one-photon excitation, and to 50 ns when a 20 MHz pulsed diode laser is used. The time resolution of time-correlated single photon counting (TCSPC) electronics used in these types of biophotonics measurements are typically better than a few picosecond, so any value measured for the limiting anisotropy should represent the value of the population anisotropy within ~ 5 ps of the instant that the population was photo-selected. Equation (12.2), in conjunction with a random number generator to produce values of θ from an isotropic distribution, can be used to run Monte Carlo simulations to predict what the anisotropy value should be for an isotropic population of fluorophores excited with 1-, 2-, or 3-photon excitation. These calculations predict anisotropy values of 0.40, 0.57, and 0.67 respectively [31]. These theoretical predictions represent the highest possible anisotropy values for a randomly oriented population of fluorophores excited by 1-, 2- or 3-photon linearly polarized light, and are called the *Fundamental Anisotropy*. In practice, the *actual* anisotropy value measured at t = 0, r_0, is typically less than the Fundamental Anisotropy and is called the *Limiting*

Anisotropy. Note that r_0 is the anisotropy at $t = 0$, not to be confused with the anisotropy at infinite time, r_∞. We can see this attenuation for the Venus anisotropy with two-photon excitation in Fig. 12.5c.

12.7 Depolarization Factors and Soleillet's Rule

An excellent starting point for learning how to interpret time-resolved anisotropy curves is to consider why the *limiting* anisotropy is almost always less than the *fundamental* anisotropy. *Depolarization factors* (*d*) are experimental influences that can account for a decrease in the measured anisotropy. One of the prime reasons for using anisotropy rather than polarization ratios is based on Soleillet's rule [22]. Soleillet's rule states that the anisotropy measured will be equal to the fundamental anisotropy multiplied by all applicable depolarization factors [13, 28]:

$$r = r_f \cdot \prod_i d_i \qquad (12.10)$$

For anisotropy experiments involving proteins or other cellular components labelled with a fluorophore, there are typically four depolarization factors that should be considered when attempting to account for any discrepancies between theory and a measured anisotropy value. These are: (1) depolarization due to instrumentation, (2) depolarization due to non-collinear absorption and emission dipoles, (3) depolarization due to molecular rotation, and (4) depolarization occurring as a result of Förster Resonance Energy Transfer (FRET). The first two of these factors occur very rapidly and, if applicable, would result in a near immediate drop from the fundamental anisotropy.

12.7.1 Instrumental Depolarization

By definition, the lenses used to build optical microscopes (primarily objectives and condensers) alter the orientation of light. Accordingly, one adverse effect of measuring anisotropy on a microscope (as opposed to a dedicated fluorimeter) is that the orientation of linearly polarized excitation light when it reaches the sample is no longer purely linearly polarized. Similarly, when the fluorescence emission is collected and relayed to the photo-detector, $I_{\parallel}$ signal can be contaminated with $I_{\perp}$ components and vice-a-versa. In general, the higher the numerical aperture (NA) of the optics, the greater the instrumental depolarization. The quantitative theory for predicting the impact of lens NA on depolarization (and the value of d_{NA}, see Table 12.1) for a particular microscope setup has a complex dependence on the numerical aperture of each lens used in the excitation and emission light paths, and the refractive index of the medium [1, 2]. It can be visualized empirically by

Table 12.1 Depolarization factor, d_{NA}, for different numerical apertures of the lens and different refractive indexes of the immersion medium

N.A.	Reference Index	d_{NA}
1.4	Oil	0.80
1.2	Oil	0.87
1.2	Water	0.81
1.0	Water	0.89
0.9	Water	0.92
0.75	Water	0.95
0.95	Air	0.77
0.9	Air	0.82
0.75	Air	0.89
0.65	Air	0.92
0.6	Air	0.94
0.5	Air	0.96
0.4	Air	0.97

time-resolved anisotropy measurements of the same sample while systematically altering the NA [31]. Simpler schemes for correcting for instrumental depolarization by altering the value of 2 in the denominator of (12.6) to a value ranging from 2 to 1 with increasing NA optics [3, 9, 26, 27] may have value in producing a lifetime that decays as a single exponential, but cannot be used in (12.8) to predict measured anisotropy values as it incorrectly predicts that anisotropy values will increase with higher NA objectives.

Depolarization due to non-collinear dipoles: while it is usually assumed that the absorption dipole is identical to the emission dipole of a fluorophore, this is not always true. If we designate the angle difference between the absorption and emission dipole of a fluorophore as β, a depolarization factor that accounts for non-collinear dipoles can be calculated [14]:

$$d_\beta = \frac{3}{2}\cos^2\beta - \frac{1}{2} \tag{12.11}$$

β is thought to be a static intrinsic property of a fluorophore that does not change during the course of a biological experiment. In contrast, some fluorophores do undergo a conformational change in response to absorbing a photon. These dynamic changes in fluorophore structure are typically extremely rapid. Thus, regardless of the mechanism of non-collinearity, non-zero values for β are thought to attenuate the fundamental anisotropy to yield a diminished limiting anisotropy. In most biological applications of time-resolved anisotropy, the objective is to observe how the anisotropy changes with time from the initial limiting anisotropy to explore the fluorophores ability to rotate, as well as if it is in close proximity to other fluorophores (Homo-FRET). Accordingly, when applying Soleillet's rule it is often assumed that the limiting anisotropy observed, r_0, is a product of both d_{NA} of the objective and d_β (see 12.11) without determining the absolute value of each individual depolarization factor:

$$r_0 = r_f \cdot d_{NA} \cdot d_\beta = r_f \cdot d_{NA} \cdot \left(\frac{3}{2} \cos^2 \beta - \frac{1}{2} \right) \qquad (12.12)$$

12.7.2 Depolarization Due to Molecular Rotation

Depolarization caused by the molecular rotation of fluorophores is typically much slower occurring on a time scale of hundreds of picoseconds to hundreds of nanoseconds. Accordingly molecular rotation typically manifests on a time-resolved anisotropy curve as a slow decay of anisotropy from the limiting anisotropy value. We have already shown in (12.9) that a spherical molecule that is free to rotate around its X-, Y-, and Z-axes will decay in time as a single exponential [13, 28]:

$$r(t) = r_0 \cdot e^{-t/\theta_{rot}}$$

where r_0 is the *limiting anisotropy*, and θ_{rot} is the *rotational correlation time* of the molecule. The rotational correlation time θ_{rot} is related to the *rotational diffusion coefficient*, D_r, and is a measure of how rapidly a molecule can rotate; the smaller the value of θ_{rot}, the faster the fluorophore can rotate. The relationship between θ_{rot} and D_r for a molecule in solution is [5, 13]:

$$\theta_{rot} = \frac{1}{6D_r} \qquad (12.13)$$

The Stokes-Einstein relationship [13] describes how D_r is a function of the absolute temperature (T), the viscosity (η), and the volume per molecule (V):

$$D_r = \frac{k_B T}{6V\eta} \qquad (12.14)$$

where k_B is Boltzmann's constant. Accordingly:

$$\theta_{rot} = \frac{V\eta}{k_B T} \qquad (12.15)$$

The time-resolved anisotropy decay of molecules that are asymmetrical can also be modelled but as a multi-exponential decay (requiring up to 5 decay components):

$$r(t) = r_0 \cdot \sum_i a_i \cdot e^{-t/\theta_i} \qquad (12.16)$$

where θ_i is the *i*th *rotational correlation time* of the molecule and a_i is the fractional amplitudes of the *i*th decay component. Often the signal to noise level of time resolved anisotropy measurements, particularly at times much longer than the fluorescence lifetime, preclude accurate fitting to multi-exponential decay models such as in (12.16). In practice, even non-spherical molecules can be fit well to (12.9).

12.7.3 Depolarization Due to Homo-FRET
A.K.A. Energy Migration

Another possible reason for observing a time-dependent depolarization when analyzing time-resolved anisotropy decay curves is if energy transfer is occurring between fluorophores in the sample [4, 6, 9, 12, 15, 16, 23–27, 31]. Förster Resonance Energy Transfer (FRET) is a physical phenomenon where excited state energy is transferred by a non-radiative mechanism from a photo-selected fluorophore (the donor) to a nearby ground state chromophore (the acceptor) [7, 8, 10, 11, 17, 18, 32]. There are two general types of FRET, Hetero-FRET (where the excitation and emission spectra of the donor and acceptor are different, and Homo-FRET where the donor and acceptor have identical spectra (they are distinct molecules of the same type of fluorophore). In this chapter we will only consider the impact of Homo-FRET on anisotropy measurements. When Homo-FRET occurs the donor returns to its ground state and the acceptor is concomitantly raised to its excited state. Clearly, if the dipole orientation of the acceptor is different from the dipole orientation of the donor, the emission anisotropy can be altered with Homo-FRET. For Homo-FRET to occur three conditions must me met: (1) The distance between a donor and acceptor fluorophores must be typically less than 10 nm. (2) The fluorophore must have a small Stokes shift such that its absorption spectrum has a large overlap with its emission spectrum. and (3) The acceptors absorption dipole must not be oriented perpendicular to the orientation of the electric field created by the donors oscillating dipole. In time-resolved anisotropy measurements biological samples are typically labelled with a single fluorophore, and the sample concentration is experimentally manipulated to keep it low enough so the mean distance between molecules is much greater than 10 nm. Despite these precautions to avoid Homo-FRET, many biological molecules can form multimers that may result in Homo-FRET. Equation (12.17) describes the anisotropy decay for a spherical complex having two fluorophores in close proximity (a dimer) separated by distance R:

$$r(t) = r_0 \cdot \left(a_{FRET} \cdot e^{-t/\phi} + a_{Rot} \cdot e^{-t/\theta} \right) \qquad (12.17)$$

where ϕ is the *Homo-FRET correlation time* of the complex, θ is the *rotational correlation time*, a_{FRET} is the fractional amplitude of the Homo-FRET decay component, a_{Rot} is the fractional amplitude of the rotational decay component. For energy migration between two fluorophores the Homo-FRET correlation time is a function of ω, the FRET transfer rate [21, 24, 29]:

$$\phi = \frac{1}{2\omega} \qquad (12.18)$$

If ω, the FRET transfer rate is much faster than the fluorescence emission rate $(1/\tau)$, then the amplitude of the Homo-FRET decay component for Homo-FRET between two fluorophores will be approximately 0.5, and:

$$a_{Rot} = 1 - a_{FRET} \qquad (12.19)$$

In general, a_{FRET} will be approximately $1 - (1/N)$ where N is the number of fluorophores in a complex that are transferring energy by Homo-FRET. Thus the a_{FRET} amplitude for a trimer transferring energy by Homo-FRET will be greater than for a dimer at the same separation [31]. The rate of energy transfer by FRET has an inverse sixth power dependence on the separation of the donor and acceptor:

$$\omega = \frac{1}{\tau} \cdot \left(\frac{R_0}{R}\right)^6 \qquad (12.20)$$

where R is the separation between the donor and acceptor, and R_0 is the Förster distance or the distance where the transfer rate (ω) is equal to the emission rate (Γ), the inverse of the lifetime (τ) in the absence of FRET. Equation (12.20) assumes that donors and acceptors in the sample population are randomly oriented and rotate rapidly during the excited state lifetime such that the dipole orientation factor κ^2 is equal to 2/3. If it is not valid to assume a κ^2 value of 2/3, for example, if fluorophores do not rotate rapidly during the excited state, then the transfer rate of an individual FRET pair can be calculated using (12.21) by specifying the specific value of κ^2 [30]:

$$\omega = \frac{3}{2} \cdot \frac{\kappa^2}{\tau} \left(\frac{R_0}{R}\right)^6 \qquad (12.21)$$

Under these circumstances it is possible to observe heterogeneity in the transfer rates of the population of FRET pairs despite having a homogenous separation between donors and acceptors [30]. Vis-à-vis Homo-FRET, having heterogeneous κ^2 values can result in an attenuated Homo-FRET decay component amplitude (a_{FRET}) as well as observing multiple Homo-FRET correlation times (as manifest by a multi-exponential Homo-FRET decay component).

12.7.4 Complex Time-Resolved Anisotropy Curves

Time-resolved anisotropy curves typically decay from a limiting anisotropy (r_0) to an anisotropy value of zero if fluorophores are free to rotate in all directions (see 12.16). Similarly, if Homo-FRET is occurring, a faster decay component associated with FRET will be observed in addition to the rotational anisotropy decay (for example see 12.17). For small fluorophores that can rotate rapidly, it may be difficult to differentiate between depolarization caused by fast rotation and depolarization caused by FRET. In some samples the ability of fluorophores to rotate will be hindered. This is often observed for fluorophores that partition into lipidic membranes that align either perpendicular or parallel to the plane of the membrane. The time-resolved anisotropy curve for samples with 'hindered' rotation will decay from the limiting anisotropy value to a non-zero asymptote, r_∞, the anisotropy value at infinite time:

$$r(t) = (r_0 - r_\infty) \cdot \left(\sum_i a_i \cdot e^{-t/\theta_i} \right) + r_\infty \tag{12.22}$$

Even more complicated time-resolved anisotropy curves can be observed if more than one type of fluorophore is present and if the different fluorophores have different lifetimes and time-resolved anisotropies. With multiple fluorophores the average time-resolved anisotropy of a population of fluorophores is [14]:

$$\bar{r}(t) = \sum_i f(t)_i \cdot r(t)_i \tag{12.23}$$

where $f(t)_i$ is the time-dependent fractional intensity of the ith fluorescent component and $r(t)_i$ is the time-resolved anisotropy of the ith fluorescent component. The time-dependent fractional intensity, $f(t)_i$, is the abundance-weighted lifetime of a single fluorophore in the population divided by the sum of the abundance weighted lifetimes of all fluorescent species in the population:

$$f(t)_x = \frac{a_x \cdot I(t)_x}{\sum_i a_i \cdot I(t)_i} \tag{12.24}$$

where a_x is the abundance of fluorophore x in the population and $I(t)_x$ is the lifetime of fluorophore x. Equation (12.24) indicates that the *anisotropy* of a population of fluorophores is simply the intensity weighted sum of the anisotropy values of the individual fluorophores in the population. Interpretation of ensemble anisotropy values from populations with more than one type of fluorophore (each having unique anisotropy and lifetime values) can be challenging (for example see Fig. 12.6).

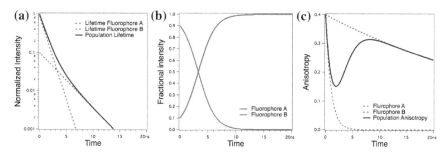

Fig. 12.6 A complex biphasic time-resolved anisotropy curve. In samples having two or more types of fluorophores having different lifetimes and rotational correlation times, complex multiphasic time-resolved anisotropy curves are possible. To illustrate, we use (12.24) to simulate the ensemble time-resolved anisotropy of a population composed of two fluorophores, A and B. Fluorophore A has a lifetime of 3 ns (panel **a**) and initially (at t = 0) accounts for 10 % of the fluorescence detected from the sample (panel **b**). In contrast, fluorophore B has a shorter lifetime of 1 ns (panel **a**) but accounts for 90 % of the fluorescence at t = 0 (Panel **b**). Because fluorophore B has a shorter lifetime than fluorophore A, with time its fractional intensity will decline while the fractional intensity of fluorophore B will increase reciprocally (panel **b**). The rotational correlation time of fluorophore A is 30 ns, while the rotational correlation time of fluorophore B is 1 ns (panel **c**, the limiting anisotropy was set at 0.4). The *black* trace in panel **c** depicts a biphasic anisotropy decay curve. Such a complex time-resolved anisotropy curve could be observed in cells expressing a protein of interest genetically-tagged with a fluorescent protein if the cell has a high level of auto-fluorescence background

Acknowledgments The intramural programs of the National Institutes of Health, National Institute on Alcohol Abuse and Alcoholism, Eunice Kennedy Shriver National Institute of Child Health and Human Development, Bethesda, MD 20892 supported this work.

References

1. D. Axelrod, Carbocyanine dye orientation in red cell membrane studied by microscopic fluorescence polarization. Biophys. J. **26**, 557–573 (1979)
2. D. Axelrod, Fluorescence polarization microscopy. Methods Cell Biol. **30**, 333–352 (1989)
3. W. Becker, *Advanced Time-Correlated Single Photon Counting Techniques* (Springer, Berlin, 2005)
4. S.M. Blackman, D.W. Piston, A.H. Beth, Oligomeric state of human erythrocyte band 3 measured by fluorescence resonance energy homotransfer. Biophys. J. **75**, 1117–1130 (1998)
5. C.R. Cantor, P.R. Schimmel, *Biophysical Chemistry Part II: Techniques for the Study of Biological Structure and Function* (W.H. Freeman & Co., New York, 1980)
6. A.H. Clayton, Q.S. Hanley, D.J. Arndt-Jovin, V. Subramaniam, T.M. Jovin, Dynamic fluorescence anisotropy imaging microscopy in the frequency domain (rFLIM). Biophys. J. **83**, 1631–1649 (2002)
7. R.M. Clegg, Förster resonance energy transfer—FRET what it is, why do it, and how it's done, in *FRET and FLIM Techniques*, ed. T.W.J. Gadella (Elsevier, New York, 2009), pp. 1–57
8. T.W.J. Gadela, *FRET and FLIM Techniques* (Elsevier, New York, 2009)

9. I. Gautier, M. Tramier, C. Durieux, J. Coppey, R.B. Pansu, J.C. Nicolas, K. Kemnitz, M. Coppey-Moisan, Homo-FRET microscopy in living cells to measure monomer-dimer transition of GFP-tagged proteins. Biophys. J. **80**, 3000–3008 (2001)

10. E.A. Jares-Erijman, T.M. Jovin, FRET imaging. Nat. Biotechnol. **21**, 1387–1395 (2003)

11. E.A. Jares-Erijman, T.M. Jovin, Imaging molecular interactions in living cells by FRET microscopy. Curr. Opin. Chem. Biol. **10**, 409–416 (2006)

12. T.M. Jovin, Fluorescence polarization and energy transfer: theory and application, in *Flow Cytometry and Sorting*, ed. by M.R. Melamed, P.F. Mullaney, M.L. Mendelsohn (Wiley, New York, 1979), pp. 137–165

13. J.R. Lakowicz, *Principles of Fluorescence Spectroscopy* (Plenum, New York, 1999)

14. J.R. Lakowicz, *Principles of Fluorescence Spectroscopy* (Springer, New York, 2006)

15. D.S. Lidke, P. Nagy, B.G. Barisas, R. Heintzmann, J.N. Post, K.A. Lidke, A.H. Clayton, D. J. Arndt-Jovin, T.M. Jovin, Imaging molecular interactions in cells by dynamic and static fluorescence anisotropy (rFLIM and emFRET). Biochem. Soc. Trans. **31**, 1020–1027 (2003)

16. T.A. Nguyen, P. Sarkar, J.V. Veetil, S.V. Koushik, S.S. Vogel, Fluorescence polarization and fluctuation analysis monitors subunit proximity, stoichiometry, and protein complex hydrodynamics. PLoS ONE **7**, e38209 (2012)

17. A. Periasamy, R.N. Day (eds.), *Molecular Imaging: FRET Microscopy and Spectroscopy* (Oxford University Press, Oxford, 2005)

18. D.W. Piston, G.J. Kremers, Fluorescent protein FRET: the good, the bad and the ugly. Trends Biochem. Sci. **32**, 407–414 (2007)

19. P.N. Prasad, *Introduction to Biophotonics* (Wiley, Hoboken, 2003)

20. P. Sarkar, S.V. Koushik, S.S. Vogel, I. Gryczynski, Z. Gryczynski, Photophysical properties of Cerulean and Venus fluorescent proteins. J. Biomed. Opt. **14**, 034047 (2009)

21. P. Sharma, R. Varma, R.C. Sarasij, Ira, K. Gousset, G. Krishnamoorthy, M. Rao, S. Mayor, Nanoscale organization of multiple GPI-anchored proteins in living cell membranes. Cell **116**, 577–589 (2004)

22. P. Soleillet, Sur les parametres caracterisant la polarisation partielle de la lumierer dans les phenomenes de fluorescence. Ann. Phys. (Paris) **12**, 23–97 (1929)

23. K. Suhling, J. Siegel, P.M. Lanigan, S. Leveque-Fort, S.E. Webb, D. Phillips, D.M. Davis, P. M. French, Time-resolved fluorescence anisotropy imaging applied to live cells. Opt. Lett. **29**, 584–586 (2004)

24. F. Tanaka, N. Mataga, Dynamic depolarization of interacting fluorophores. Effect of internal rotation and energy transfer. Biophys. J. **39**, 129–140 (1982)

25. C. Thaler, S.V. Koushik, H.L. Puhl 3rd, P.S. Blank, S.S. Vogel, Structural rearrangement of CaMKIIalpha catalytic domains encodes activation. Proc Natl Acad Sci USA **106**, 6369–6374 (2009)

26. M. Tramier, M. Coppey-Moisan, Fluorescence anisotropy imaging microscopy for homo-FRET in living cells. Methods Cell Biol. **85**, 395–414 (2008)

27. M. Tramier, T. Piolot, I. Gautier, V. Mignotte, J. Coppey, K. Kemnitz, C. Durieux, M. Coppey-Moisan, Homo-FRET versus hetero-FRET to probe homodimers in living cells. Methods Enzymol. **360**, 580–597 (2003)

28. B. Valeur, *Molecular Fluorescence* (Wiley, Weinheim, 2002)

29. R. Varma, S. Mayor, GPI-anchored proteins are organized in submicron domains at the cell surface. Nature **394**, 798–801 (1998)

30. S.S. Vogel, T.A. Nguyen, B.W. van der Meer, P.S. Blank, The impact of heterogeneity and dark acceptor states on FRET: implications for using fluorescent protein donors and acceptors. PLoS ONE **7**, e49593 (2012)

31. S.S. Vogel, C. Thaler, P.S. Blank, S.V. Koushik, Time Resolved Fluorescence Anisotropy, in *FLIM Microscopy in Biology and Medicine*, ed. by A. Periasamy, R.M. Clegg (Taylor & Francis, Boca Raton, 2010), pp. 245–290

32. S.S. Vogel, C. Thaler, S.V. Koushik, Fanciful FRET, Sci STKE 2006, re2 (2006)

33. G. Weber, Polarization of the fluorescence of macromolecules. I. Theory and experimental method. Biochem. J. **51**, 145–155 (1952)

Chapter 13
Time-Resolved Spectroscopy of NAD(P)H in Live Cardiac Myocytes

Alzbeta Marcek Chorvatova

Abstract Monitoring cell and tissue physiological parameters, such as metabolic state in real time and in their true environment is a continuous challenge. In the last decades, advanced photonics techniques were developed, combining fluorescence spectroscopy with time-resolved and imaging techniques, thus opening completely new opportunities for investigation of fluorescence parameters in living cells. This is particularly true in the case of evaluation of endogenous fluorescence or auto-fluorescence (AF) of living cells, derived from nicotinamide dinucleotide (phosphate) (NAD(P)H). We have pioneered the application of time-resolved fluorescence spectroscopy to evaluate changes in metabolic oxidative state directly in living cardiac cells by means of endogenous NAD(P)H fluorescence. NAD(P)H fingerprinting was investigated in living cardiac myocytes isolated from left ventricle (LV) of rats by spectrally-resolved lifetime detection using spectrally-resolved time-correlated single photon counting (TCSPC). Metabolic modulation was employed to evaluate individual NAD(P)H fluorescence components. Advanced data analysis leading to development of techniques and analytical approaches aimed at precise separation of individual fluorescence components from the recorded AF signals was also performed. Spectral decomposition of time-resolved NAD(P)H fluorescence signals by linear unmixing approach was successfully achieved. Gathered results demonstrate that combined approaches between time-resolved, spectroscopic and imaging systems open new possibilities for understanding the precise role of mitochondria in complex pathophysiological conditions and for finding new non-invasive clinically-relevant diagnostic applications.

A. Marcek Chorvatova (✉)
Department of Biophotonics, International Laser Centre, Bratislava, Slovakia
e-mail: alzbeta.chorvatova@ilc.sk

© Springer International Publishing Switzerland 2015
W. Becker (ed.), *Advanced Time-Correlated Single Photon Counting Applications*,
Springer Series in Chemical Physics 111, DOI 10.1007/978-3-319-14929-5_13

407

13.1 Non-invasive Investigation of Metabolic Oxidative State in Living Cardiac Myocytes

Metabolism is crucial for cardiac cell contraction. The heart is a pump converting chemical energy into mechanical work and the power for this work is gathered almost entirely from oxidation of carbon fuels and, to a great extent, these fuels are provided by coronary (myocardial) blood flow. In the heart, oxidative metabolism is primarily the function of mitochondria in the process of oxidative phosphorylation by the respiratory chain [17]. This process is coupled with the oxidation of NADH, the principal electron donor for the electrochemical gradient that is indispensable for oxidative energy metabolism. The first step in this process, which accounts for 95 % of ATP generation needed for cardiomyocyte contraction, is the dehydrogenation of NADH by flavoproteins of Complex I of the mitochondrial respiratory chain. Because of the high oxidative metabolism, heart cells have enhanced oxidative capacity, demonstrated by their ultrastructure [4]: 25–35 % of total cardiomyocyte volume is occupied by mitochondria, distributed in stripes, where oxidative phosphorylation takes place. Energy generated by the network of cardiac cells serves for the main function of the heart—its contraction—allowing the pumping of the blood. Mitochondrial role in cellular bioenergetics predestine them to play a pivotal role in various human diseases—the implication of the mitochondrial dysfunction in CV disease was largely reviewed [2]. Possibility to study, non-invasively, rapid changes in metabolic oxidative state is therefore crucial for understanding of processes in living cardiac cells and their alterations in pathophysiological conditions.

In the CVS, endogenous fluorescence of NAD(P)H and flavins has long been used for non-invasive monitoring of processes, such as modifications in metabolic oxidative state, or change in myocardial tissue status (reviewed in [28]). Dehydrogenation of NADH, the principal electron donor for the electrochemical gradient, by flavoproteins of the mitochondrial respiratory chain is indispensable for ATP generation and thus oxidative energy metabolism. In addition, during oxidative stress, cellular NADPH content is modulated through the release of peroxides and various by-products that have been shown to decrease the activity of several enzymes, such as NADP-isocitrate dehydrogenase. In the last decades, advanced photonics techniques have been developed to monitor endogenous fluorescence in living cells and tissues [25, 30].

13.1.1 Experimental Set-up for Time-Resolved NAD(P)H Recordings on Living Cardiac Cells

The experimental setup is shown in Fig. 13.1 It consists of a Zeiss Axiovert 200 microscope, a Becker & Hickl 375 nm picosecond diode laser, a Proscan Solar 100 spectrograph, and a PML-16 multichannel detector and an SPC-830 TCSPC module from Becker & Hickl.

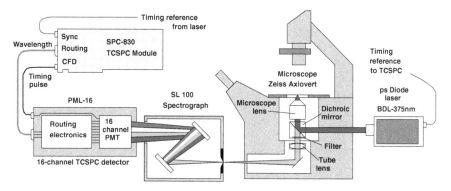

Fig. 13.1 Experimental setup for time-resolved fluorescence spectroscopy. A 375-nm picosecond laser diode (Becker & Hickl) was used as an excitation source with ~1 mW output power. The emitted cell autofluorescence, separated from laser excitation by 395 nm dichroic and 397 nm long-pass filter in the Axiovert 200 M (Zeiss, Canada), was spectrally dispersed by an imaging spectrograph (Solar SL 100 M, Proscan, Germany), detected by a PML 16 16-channel TCSPC detector and recorded by an SPC-830 TCSPC module, both Becker & Hickl, Germany. Described in detail in [20]

The collimated beam of the picosecond diode laser is sent through the back port of the microscope. There it is reflected by a dichroic mirror in a standard beamsplitter cube in the filter carousel of the microscope, and focused into the sample by the microscope lens. It creates a slightly defocused elliptical spot with typical dimensions of 10 × 20 µm, chosen in regard to average the fluorescence signal over the width of a single myocyte [4]), see inset in Figs. 13.4, 13.7.

The fluorescence is detected back through the microscope lens, and separated from the excitation light by the dichroic mirror in the filter cube. A filter at the output of the filter cube suppresses scattered excitation light. The tube lens of the microscope focuses the light from the illuminated area into the output image plane at the side port of the microscope.

The spectrograph is put into this plane with its input slit. The spectrograph projects a spectrum of the fluorescence signal on the cathode of an R5099-L16 16-channel PMT inside the PML-16 detector module. The detector module contains routing electronics which provides, for every photon detected, a timing pulse and a routing signal. The routing signal indicates the PMT channel in which the photon was detected. The timing pulse and the routing signal are fed into the SPC-830 TCSPC module which builds up a photon distribution over the times of the photons after the laser pulses and the detector channel, see Chap. 1, Sect. 1.4.1 for details. The result is a 2-dimensional data array that contains individual decay curves for the 16 wavelengths. Because the decay curves are built up simultaneously, motion in the cardiomyocytes or photobleaching do not cause distortion in the recorded spectra. The setup and its application to the study of cardiac myocytes are described in [20, 29, 31].

For the results presented here, fluorescence decays were measured for 30 s in the 16 spectral channels simultaneously over a wavelength range of 385–675 nm.

The half-width of the instrument response function (IRF) was estimated to be 0.2 ns [20] with the temporal resolution of the system set to 24.4 ps/channel. Cells were mounted on an inverted microscope and studied at room temperature in 4-well chambers with UV-proof, coverslip-based slides (LabTech, Canada). Data were always evaluated from the first measurement of each cell (measured for 30 s) to avoid artefacts induced by photobleaching.

13.1.2 Time-Resolved Fluorescence Spectroscopy of NAD(P)H in Vitro

Measurement of the fluorescence of endogenous fluorophores in vitro is a prerequisite for demonstrating their presence in living cells, as well as for having reference spectra for separation of individual components. NADH and NADPH are endogenously fluorescent molecules [30]; their fluorescence spectra and fluorescence lifetimes (Fig. 13.2A) were studied in vitro in the intracellular media-mimicking solution (pH 7.25) [1, 29]. Spectral intensity was linearly dependent on the NADPH concentration (Fig. 13.2Aa), as illustrated at the maximum emission wavelength of 450 nm (Fig. 13.2Ac) with no modification in the fluorescence decays

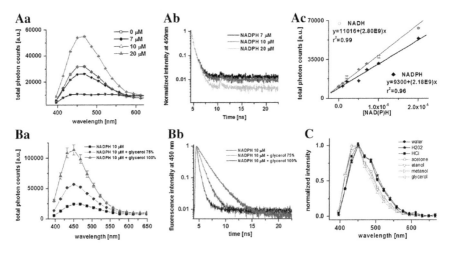

Fig. 13.2 NAD(P)H fluorescence in vitro. **A** Concentration-dependence of NADPH fluorescence in the intracellular media-mimicking solution (pH 7.25), excitation 375 nm. Background-corrected emission spectra (**a**), normalized fluorescence decays at 450 nm (**b**) and the total photon counts at 450 nm (**c**) as a function of the NADH and NADPH concentration. **B** Viscosity-dependence of the NADPH (10 μM) fluorescence in intracellular solution. Background-corrected emission spectra (**a**) and normalized fluorescence decays at 450 nm (**b**). Data are shown as mean ± SEM (n = 5 for all but 0 μM for which n = 3). **C** Normalized spectra of NADH (20 μM) in inorganic (*grey*) and organic (*black*) solvents (all 100 %). Published in [1, 29]

(Fig. 13.2Ab). Normalized spectra for NADPH concentrations from 1 to 20 μM overlaid perfectly (data not shown), confirming the same molecular origin. Comparable results were obtained for NADH, in accordance with the fact that NADH and NADPH are inseparable by fluorescence spectra, or fluorescence lifetimes, only with slightly higher quantum yield for NADH than for NADPH (Fig. 13.2Ac) [1, 89].

To test the sensitivity to molecular environment, we have recorded NADH and NADPH in different organic and inorganic solvents (Fig. 13.2B and C). As illustrated by the addition of different concentrations of glycerol versus H_2O, the total photon counts, measured at the emission maximum (λ_{em} = 450 nm) revealed exponential rise of the peak of NADPH fluorescence intensity (Fig. 13.2Ba) [29]. This result was due to the prolongation of the fluorescence lifetimes in a highly viscous environment (Fig. 13.2Bb) and is in agreement with previous findings [46] showing that the sensitivity of NADPH fluorescence intensity to changes in its microenvironment can be related to changes in the molecular mobility and/or modification in the conformation of NADPH molecules in more viscous milieu. Importantly, we also noted a slight (10 nm) blue spectral shift when NADH or NADPH fluorescence was recorded in organic solvents, as opposed to the inorganic ones (Fig. 13.2C). These experiments also demonstrated sensitivity of the constructed experimental set-up for recording physiological concentrations of these endogenous fluorophores.

13.1.3 Time-Resolved NAD(P)H Fluorescence Spectroscopy in Cardiac Mitochondria

Cardiac mitochondria are crucial for the heart oxidative metabolic status, as they contain respiratory chain responsible for the ATP production in cardiac cells. To demonstrate that, in cardiomyocytes, the presence of endogenous fluorescence derived from NAD(P)H primarily related to these organelles, confocal images of unstained myocytes were gathered by the 2 photon excitation induced by 777 nm light to verify NAD(P)H fluorescence distribution (Fig. 13.3a). Recorded images clearly demonstrated that endogenous NAD(P)H fluorescence in cells was distributed in stripes.Verification of mitochondrial distribution using mitochondrial fluorescence probe Tetramethylrhodamine (TMRM) confirmed stripe-like distribution of mitochondria [4]. These results are in agreement with 2 photon microscopy and spectroscopy recording used previously to evaluate spatial distribution and intensity changes of NAD(P)H fluorescence in living cardiac myocytes [59]. AF excitation-emission spectra in heart mitochondria isolated from rat hearts by differential centrifugation [23] demonstrated presence of both NAD(P)H and flavin fluorescence (Fig. 13.3b). This result revealed comparable peaks to those recorded in cardiac myocytes [23] and was also in agreement with previous observations in pigeon heart mitochondria that identified excitation/emission spectra of oxidized

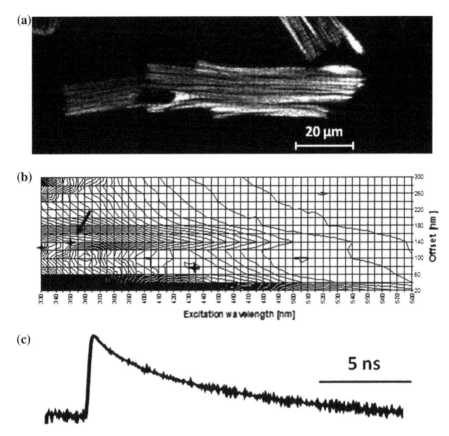

Fig. 13.3 Mitochondrial NAD(P)H fluorescence of rat cardiac cells. **a** Confocal image of NAD(P) H distribution in a cardiac cell (777 nm two-photon excitation, HFT KP 700/488 dichroic filter and 450–470 nm spectral range for emission detection). **b** Excitation-emission matrix recorded in isolated cardiac mitochondria (*arrow* points to the NAD(P)H peak) by spectro-fluorimetry. **c** Example of the NAD(P)H fluorescence decay recorded in isolated rat cardiac mitochondria after excitation by 375 nm laser by time-resolved fluorescence. For more details, see [23, 58]

flavoproteins and reduced pyridine nucleotide fluorescence [16]. Time-resolved measurements from isolated mitochondria after excitation at 375 nm showed at least a double exponential decay of 0.47 and 1.32 ns (Fig. 13.3c) [58]. These results correspond to data in intact porcine heart mitochondria, where 3 NAD(P)H fluorescence lifetime pools have been identified [10]: a free pool with a lifetime of 0.2–0.4 ns, an intermediate pool (1–2 ns), and a long lifetime pool (3–8 ns). The intermediate and long lifetime pools were proposed to be resulting from protein binding of NAD(P)H.

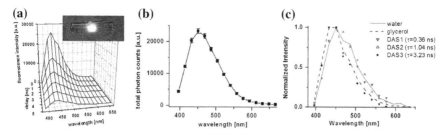

Fig. 13.4 Spectrally- and time-resolved AF of cardiac myocytes. **a** Original recording of the time-resolved spectroscopy measurements of NAD(P)H fluorescence in living cardiac cells, excitation at 375 nm (mean of 10 measurements; in the inset, representative transmission image of the single cardiomyocyte illumination). **b** Total photon counts of NAD(P)H fluorescence in rat cardiac myocytes in control conditions (Data are shown as mean ± SEM, n = 70 cells). **c** Decay associate spectra for estimated fluorescence lifetimes. Published in [20, 33, 39]

13.1.4 Time-Resolved NAD(P)H Fluorescence Spectroscopy in Isolated Cardiac Myocytes

Recording of time- and spectrally-resolved fluorescence from isolated living LV cardiomyocytes was highly reproducible for both NAD(P)H (Fig. 13.4a) [21, 29, 31] and flavins [23, 24]. Cardiac myocytes were isolated from LV of female rats after retrograde perfusion of the heart with proteolytic enzymes, as previously reported [5, 32]. Steady-state emission spectra of the cardiomyocyte NAD(P)H fluorescence, calculated from the total photon counts on each spectral channel, had a spectral maximum at 450 nm (Fig. 13.4b). Analysis of the AF decay recorded in cardiomyocytes showed acceptable chi-square values ($\chi^2 < 1.2$) and flat plots of weighted residuals when using at least a 3-exponential model with lifetimes (Table 13.1). Gathered data are in good agreement with lifetime values estimated in cardiomyocyte mitochondria [10], as well as in human cardiac myocytes [18, 19].

13.2 Analysis of Individual NAD(P)H Fluorescence Components

Finding an appropriate analytical approach for analysis of time-resolved fluorescence data, particularly when performed in living cells and tissues, is a prerequisite to insure that physiological significance of the recorded data is correctly recovered [64]. At the same time, it is also one of the main limitation of fluorescence lifetime imaging (FLIM) recordings, in regard to large amount of complex data, as well as in regard to availability of the sufficient number of photons for each measurement point. The physically relevant models need to include the general knowledge of the investigated system, together with the established statistical properties of the residuals [82].

Table 13.1 Fluorescence parameters of cardiomyocyte AF ($\lambda_{excitation/emission}$ = 375 nm/450 nm)

	Photon counts (a.u.)	a_1	τ_1 (ps)	a_2	τ_2 (ps)	a_3	τ_3 (ps)	χ^2
Control (70/13)	23231 ± 786	69.29 ± 1.04	685.68 ± 10.80	27.63 ± 0.92	2028.23 ± 52.09	3.14 ± 0.23	12678.87 ± 776.34	1.02 ± 0.01
Rotenone (28/6)	58940 ± 1833[*]	71.04 ± 1.63	563.92 ± 10.28[*+]	26.25 ± 1.48	1696.32 ± 56.78[*]	2.95 ± 0.22[+]	8757.18 ± 649.73	1.05 ± 0.01
Cyanide (10/2)	55384 ± 7002[*]	68.80 ± 1.33	485.71 ± 12.62[*+]	28.80 ± 1.21	1582.34 ± 55.32[*+]	2.47 ± 0.18[+]	9431.74 ± 1274.96	1.01 ± 0.01
Rotenone/cyanide (10/2)	55603 ± 3556[*]	72.3 ± 1.80	555.54 ± 16.42	25.00 ± 1.56	1688.47 ± 91.04	2.71 ± 0.34	8481.59 ± 701.50	1.05 ± 0.02
DNP (15/3)	10824 ± 990[*#]	67.29 ± 1.65	901.42 ± 54.61[*#]	27.07 ± 1.72	2162.46 ± 118.57[#]	5.59 ± 0.80[*#]	15828.98 ± 1828.51[#]	1.01 ± 0.01
BHB/AcAc (20:1) (27/4)	29308 ± 1553[*&]	66.3 ± 1.64	598.31 ± 22.38	31.05 ± 1.50	1801.64 ± 64.66	2.78 ± 0.26	12597.52 ± 1427.89	1.01 ± 0.01
BHB/AcAc (2:1) (26/4)	21544 ± 1099	64.96 ± 1.99	657.52 ± 21.66	31.38 ± 1.92	1855.63 ± 65.37	3.72 ± 0.25	10848.68 ± 683.22	1.00 ± 0.00
Lactate/pyruvate (15/3)	24066 ± 1861	65.73 ± 2.66	655.47 ± 23.61	30.87 ± 2.32	1893.02 ± 117.35	3.63 ± 0.47	10380.86 ± 2009.34	1.01 ± 0.00

	P_{max} (a.u.)	$a_1\tau_1$ (a.u.) ($*10^3$)	r_1	$a_2\tau_2$ (a.u.) ($*10^3$)	r_2	$a_3\tau_3$ (a.u.) ($*10^3$)	r_3	$\langle\tau\rangle$ (ns)
Control (70/13)	1546 ± 68	48.00 ± 1.26	0.36 ± 0.01	53.57 ± 1.00	0.40 ± 0.01	32.69 ± 1.71	0.24 ± 0.01	1.34 ± 0.03
Rotenone (28/6)	3565 ± 160[*]	40.46 ± 1.53	0.36 ± 0.02	42.54 ± 0.96[*+]	0.40 ± 0.01	23.13 ± 1.05	0.22 ± 0.01	1.06 ± 0.01[*]
Cyanide (10/2)	3638 ± 468[*]	33.54 ± 1.42[*]	0.33 ± 0.01	45.16 ± 1.31	0.45 ± 0.01	21.65 ± 1.00[+]	0.22 ± 0.01	1.00 ± 0.02[*]
Rotenone/cyanide (10/2)	3762 ± 230[*]	40.42 ± 2.09	0.39 ± 0.02	40.99 ± 0.72[*]	0.40 ± 0.01	21.19 ± 1.38	0.21 ± 0.02	1.03 ± 0.01[*]
DNP (15/3)	771 ± 95[*#]	60.51 ± 3.69[*#]	0.32 ± 0.02	59.12 ± 5.21[#]	0.31 ± 0.02[*#]	74.62 ± 7.20[*#]	0.38 ± 0.02[*#]	1.94 ± 0.12[*#]
BHB/AcAc (20:1) (27/4)	1982 ± 95[§]	40.06 ± 2.20	0.32 ± 0.02	54.38 ± 1.31	0.44 ± 0.01	30.06 ± 1.22	0.24 ± 0.01	1.24 ± 0.02
BHB/AcAc (2:1) (26/4)	1386 ± 74	43.62 ± 2.56	0.32 ± 0.02	55.71 ± 2.10	0.41 ± 0.02	37.46 ± 1.76	0.27 ± 0.01	1.37 ± 0.02
Lactate/pyruvate (15/3)	1480 ± 93	43.75 ± 2.95	0.34 ± 0.02	55.19 ± 1.54	0.43 ± 0.01	30.54 ± 2.68	0.23 ± 0.02	1.29 ± 0.02

Total photon counts, fluorescence lifetimes (τ_1–τ_3) and their relative amplitudes (a_1–a_3) of single cardiomyocytes in control conditions and in the presence of 1 μmol/L rotenone, 4 mmol/L cyanide, 50 μmol/L DNP, 3 mmol/L BHB with 150 or 1.5 mmol/L AcAc (ratio 20:1 and 2:1, respectively), lactate (1 mmol/L) in the presence of pyruvate (100 μmol/L), or octanoate (1 mmol/L). In grey, maximum AF emission (P_{max}: time-resolved at Δt = 0 ns), calculated relative intensities and relative fractions for each component as well as average lifetime. Data are shown as mean ± SEM (number of cells/number of animals); *p < 0.05 versus control, #p < 0.05 versus rotenone, &p < 0.05 versus BHB/AcAc (ratio 2:1).

The separation and quantification of individual components underlying multi-exponential decays of autofluorescing species in biological cells and tissues still represent tough scientific challenges that have not yet been fully resolved. Searching for the best approach for precise separation of fluorescence from individual fluorophores directly in living cells and tissues is therefore one of the main tasks to solve.

13.2.1 Global Analysis and Decay Associated Spectra (DAS)

Typical approach to analyze fluorescence decay (illustrated in Fig. 13.3c) is to fit the decay curve, while establishing discrete components—fluorescence lifetimes—for each fluorescence decay. This approach allows to calculate relative intensities and relative fractions for each component, as well as an average fluorescence lifetime for each decay (Table 13.1). However, although this approach can serve to distinguish between distinct cellular states, it has several limitations. On one hand, such components do not necessarily match the fluorescence lifetime distribution for such system. On the other hand, values thus gathered cannot easily be related to one precise fluorescence component. One solution is to estimate spectral shapes of underlying components by recording the fluorescence decays on multiple wavelengths (Fig. 13.4a), which is useful to link lifetimes with variable amplitudes. In such case, spectral coordinates can be processed to create a set of decay-associated spectra (DAS), corresponding to individually-resolved lifetimes, estimated by global analysis [61, 88]. The DAS for cardiomyocyte NAD(P)H fluorescence (Fig. 13.4c) were constructed for each lifetime pool as a product of wavelength-dependent fractions of the NAD(P)H fluorescence emission of each lifetime pool with respect to total fluorescence, multiplied by total photon counts [10]. The resulting DAS showed that, for cardiomyocytes in control conditions, the 1st and 2nd lifetime pools with sub-nanosecond decay kinetics had similar spectra with maximum around 450–470 nm and red-shifted shoulder at 490 nm, while the 3rd lifetime pool with the longest lifetime has a blue-shifted peak with the maximum around 430–450 nm. Comparison of the first two spectra observable in the cells to the spectral profiles of NADH showed almost perfect fit of the component measured in water (and in inorganic solvents in vitro, Fig. 13.2c, in general), with the first two DAS components, as well as a very good fit of the glycerol- (and organic solvents in general)-like component to the third DAS component. These results uncover the presence of two spectrally-distinct components in UV-excited AF of cardiac cells with 3 fluorescent lifetimes. However, although this result gives an indication of underlying fluorescence species and their spectral shapes, it is of a limited use in complex cellular environment.

13.2.2 Time-Resolved (TRES) and Area-Normalized (TRANES) Emission Spectra

To estimate the number of individual spectral components present in the fluorescence decays recorded in complex environment of living cardiac cells, an approach proposed for semi-quantitative analysis of the multi-fluorophore mixtures [62] was applied. Series of consecutive spectral profiles decaying in time, time-resolved emission spectra (TRES) sequences were computed from original multi-wavelength TCSPC recordings of NAD(P)H fluorescence by summing the data in several consecutive temporal windows (Fig. 13.5a), as previously described for flavin fluorescence [20]. TRES analysis offers an alternative approach to DAS by estimating spectral components in complex sample. In addition, TRANES—area normalized TRES—method [62] where the spectra are further background subtracted and normalized, indicates the number of the most pronounced fluorescence species present in the sample. This approach allowed to precisely identify the ns changes of the spectra during the fluorescence decay of NAD(P)H in cardiomyocytes. Without any a priori knowledge of excited-state kinetics, the presence of each isoemissive point indicates the existence of two emitting species in the sample [62]. TRANES analysis of the cell AF in control conditions (Fig. 13.5b) revealed one isoemissive point at ∼460 nm (see grey arrow at the Fig. 13.5b), pointing to two dominant spectral components with different decay kinetics, peaking around 430 nm (component 1) and 470 nm (component 2). In both cases, the component at longer wavelength rapidly faded in the first couple of ns, suggesting the presence of the process with the corresponding lifetime at the order of 1 ns or less; while the peak with shorter emission maximum was readily observable even after 5 ns of the decay. Most important components in the data were then identified by principal component analysis (PCA), while the component's most probable positions (spectral maxima) and thus their spectral profiles were determined by a target transformation technique (see [20] for details) following analysis of cell responses to metabolic modulators. With this approach, in addition to 2 components already present in control conditions (Fig. 13.4c): the 1st component peaking at 430–

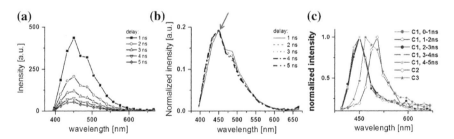

Fig. 13.5 Time-resolved, area-normalized emission spectroscopy (TRANES) of NAD(P)H fluorescence in a cardiac myocyte. (**a**, **b**) TRES and TRANES of cardiac cell AF in control conditions; *arrow* points to the isoemissive point. **c** Principal components resolved by PCA in living cardiac cells. Published in [33]

450 nm and the 2nd one with 450–470 nm spectral maximum, we also resolved 3rd significant spectral component with 510–530 nm spectral maximum (Fig. 13.5c), noted in some metabolic conditions only.

13.2.3 Separation of Individual NAD(P)H Fluorescence Components by Spectral Unmixing

To separate individual components in time-resolved NAD(P)H fluorescence, an original approach of time-resolved spectral decomposition was applied. Based on spectral decomposition of flavin fluorescence from spectrally-resolved TCSPC data [27], performed using spectral unmixing algorithms for multispectral imaging [37, 97], individual spectral components of NAD(P)H fluorescence recorded by multi-wavelength fluorescence lifetime spectroscopy were resolved in isolated cardiomyocytes [21, 31]. As featured at Fig. 13.6a, presence of three different components was confirmed with largest contribution of C1 and C2 to the overall fluorescence amplitude, as illustrated by spectrally unmixed fluorescence intensity for each component (Fig. 13.6b). A 0.5 ns steps served for the reconstitution of the fluorescence decay of each component using the sequential PCA over different time delays. Mono- or multi-exponential fit (Fig. 13.6c) was then applied to gather information on the fluorescence lifetime and amplitude of each resolved component: 1st (Fig. 13.6-C1), 2nd (Fig. 13.6-C2), and 3rd (Fig. 13.3-C3), respectively [31, 33]. Both components derived from NAD(P)H in organic and inorganic solvents presented mono-exponential decay. The 3rd component, corresponding to flavin fluorescence, was best fitted with double exponential decay; with its maximum at 520 nm, this component had little contribution to the overall photon counts at 450 nm. In addition to the Phasor (lifetime, spectral, or combined) approach [38, 42, 43, 83], this method thus represents a new tool to separate individual components in time-resolved fluorescence spectroscopy data, based on distinct spectral and lifetime characteristics, which are corresponding to precise molecules species.

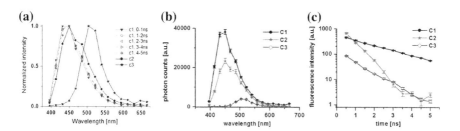

Fig. 13.6 Spectral decomposition of NAD(P)H fluorescence components in living cardiac cells. **a** Reference spectra selected as a base for spectral decomposition. **b** Photon counts calculated for each resolved component by area integration. **c** Fluorescence kinetics of the three spectral components (*C1*-1st component, *C2*-2nd component, *C3*-3rd component), estimated by spectral unmixing. The number of analyzed cells: n = 120; data are presented as mean ± SEM. Published in [21, 33]

13.2.4 Responses to Metabolic Modulation

In order to identify the nature of the resolved spectra, metabolic approach—similar to the chemiometric one—was employed. The sensitivity of the resolved components to changes in respiratory chain, in NADH production, as well as in lipid composition was tested in their cellular environment by evaluating the cell responsiveness to known metabolic regulators (Fig. 13.7 and Table 13.1) [33]. First, the sensitivity of the components to modulation of the mitochondrial respiratory chain was verified. On one hand, enhancement in the mitochondrial NADH content following restriction of the respiratory chain by Rotenone (1 μM), the Complex I inhibitor, significantly increased the fluorescence amplitude of both components (Fig. 13.7Ba, Bb), without notable effect on the fluorescence lifetimes (Fig. 13.7Aa, Ab). The addition of Na-cyanide (4 mM), the inhibitor of Complex IV, alone, or in the presence of Rotenone maintained higher fluorescence amplitudes of the two components. On the other hand, uncoupling ATP synthesis by DNP (50 μmol/L) or FCCP (200 nmol/L) lowered NAD(P)H content in cells, in accordance with a higher NADH dehydrogenation rate. This action was accompanied by change in the

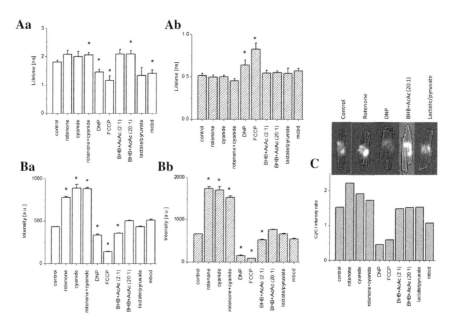

Fig. 13.7 Metabolic modulation of unmixed components. Estimated fluorescence lifetimes (**A**), together with calculated photon counts (**B**) of the resolved fluorescence component 1 (**a**) and component 2 (**b**) at 450 nm during metabolic modulation. (**C**) Ratio of the component 2 and component 1 amplitude with representative transmission image of the cell illumination in each condition. Data are shown as mean ± SEM, Control conditions (n = 120), Rotenone (1 μM; n = 30), Na-cyanide (4 mM; n = 10), NaCN+Rotenone (n = 15), DNP (50 μM; n = 15), fccp (200 nM; n = 20), after application of 20:1 (n = 35) or 2:1 BHB/AcAc (n = 18), lactate (2 mM) and pyruvate (100 μM; n = 10), MCβD (1 %; n = 10), *p < 0.05 versus control. Published in [33]

value of the fluorescence lifetime: decrease in the 1st, but increase in the 2nd resolved NAD(P)H component. These effects of modulators had repercussions also on the "free/bound" component amplitude ratio (Fig. 13.7c), which was proposed to correspond to NADH/NAD$^+$ reduction/oxidation pair [9, 75] and is thus an important indicator of mitochondrial metabolic status modifications.

Second, to promote mitochondrial NADH production as opposed to the one in the cytosol, BHB (3 mmol/L) was administered to cardiomyocytes in basic extra-cellular solution in the presence of different AcAc concentrations (Fig. 13.7): 1.5 mmol/L (ratio 2:1), closer to physiological conditions [76], and 150 μmol/L (ratio 20:1), to favour NADH production. As expected, increasing the BHB/AcAc ratio from 2:1 to 20:1, a condition favourable to NADH production, evoked rise in the amplitude of both 1st and 2nd component, in accordance with a higher NADH concentration in cardiomyocyte mitochondria, without affecting fluorescence life-times, or the component amplitudes ratio and thus mitochondrial redox potential. Third, to promote rise in the cytosolic NADH production, lactate (2 mmol/L) was administered in the presence of pyruvate (100 μmol/L) in concentrations found in the blood of studied animals [4]. This action had no significant effect on the fluorescence, indicating that the analyzed fluorescence signal is principally derived from the AF of mitochondria. Finally, the sensitivity of the resolved components to the presence of cholesterol in the cell membranes was also verified (Fig. 13.7). We noted that depletion of the cholesterol content with MCßD (1 %) [68] increased the fluorescence lifetime of the 1st component. These experiments demonstrated that the resolved components are selectively sensitive to changes in metabolic state, while pointing to higher specificity of the 1st lifetime component to lipid environment.

13.2.5 Perspectives in the Component Analysis: Data Classification

Presented analytical approach, based on spectral linear unmixing of individual component demonstrates how to resolve and determine precise molecular species present in a complex sample. However, the use of such procedure may not be always advantageous when data classification rather than evaluation of underlying molecular origins is required, as is often the case when rapid decision-making is needed, for example in clinical diagnostics. In this case, other methods for data classification can be considered. Estimation of spectral and/or time-resolved patterns associated with healthy versus diseases states is another mean to achieve classification of specific states and allow diagnostics of metabolic states in living cells and tissues. To classify metabolic state of living cells, classical approach based on spectral analysis can be employed, where the individual spectral components contributing to cell AF are estimated by blind source separation using non-negative matrix factorization [73]. This approach was demonstrated, for example, for flavin fluorescence images of isolated cardiac cells in respect to responses to metabolic modulators [26]. However, in order to use an analytical approach better suited for fast data evaluation needed in situations such as clinical diagnostics, the classic classification method can be replaced by an advanced data analysis: the machine learning approach [70]. This approach is based on support vector machine with the set of the automatically calculated features from recorded spectral profile of spectral AF images. Our first results showed that machine learning can effectively classify the spectrally resolved flavin fluorescence images without the need of detailed knowledge about the sources of AF and their spectral properties [71]. This approach can also prove to be suitable for analysis of clinical data recorded by time-resolved fluorescence spectroscopy in the future.

13.3 Time-Resolved NAD(P)H Spectrometry in CVS Physiology

To introduce AF-assisted examination in cardiac myocytes for the CVS physiology studies, verification of NAD(P)H recordings in close-to-physiological conditions needs to be performed. In the case of living cardiac myocytes, this includes the modulation of AF during cells contraction and verification of the extent of photobleaching. In addition, to evaluate the methodology for investigation of metabolic changes during physiological LV remodelling, two conditions were chosen: NAD(P)H changes were investigated in the natural condition of pregnancy, while the effect of pharmaceutical drug was tested on a model drug—the Na pump inhibitor—ouabain.

13.3.1 Recordings in Contracting Myocytes

To evaluate changes of the metabolic oxidative state under physiological conditions, recording of NAD(P)H fluorescence was tested during cardiac cell contraction. Such recording is a complex technology problem. To solve this issue, we have designed a set-up [22] that allowed us to record rapid changes in AF during cell contraction stimulated by external platinum electrodes, incorporated in a home-made bath and triggered by a pulse generator at a frequency of 0.5 Hz (to stabilize sarcoplasmic reticulum loading), or 5 Hz (the rat heart rate) [4]. The basic timing sequence of the experiment is shown in Fig. 13.8, left.

The myocyte is stimulated periodically by an electrical pulse. After every stimulation pulse it contacts for a short period of time and then relaxes. Synchronously with the stimulation the ps diode laser is turned on for 100 ms via its 'laser on' signal. During the turn-un time the laser excites fluorescence and sends synchronisation pulses to the SPC-830 TCSPC module. The SPC-830 records only

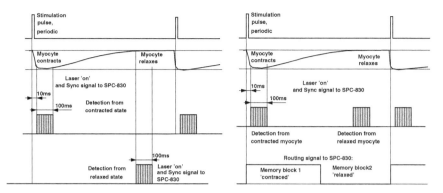

Fig. 13.8 Autofluorescence recording during cardiomyocyte contraction. *Left* Experiment timing, selection of myocyte state via temporal position of the Laser 'on' pulse in the stimulation period. *Right* Recording of contracted and relaxed state in the same stimulation period. Data are recorded into different memory blocks of TCSPC module by routing signal

when the synchronisation pulses are present. Fluorescence from different contraction states of the myocyte can thus be recorded by shifting the 'laser-on' pulse within the stimulation period.

Alternatively, the principle can be modified to record fluorescence from the contracted and the relaxed state in the same stimulation period, see Fig. 13.8, right. In that case, the laser is turned on both at the beginning (in the contracted state) and at the end (in the relaxed state) of the stimulation period. The data from the two states are routed into different memory blocks of the SPC-830 module by a (TTL) routing signal. Please see Chap. 1, Sects. 1.4.1 and 1.4.2 for routing principle.

We selected the relatively tiny illumination (laser-on) time due to the short length of the contraction cycle of a field-stimulated cardiac cell [32]. However, recording of fluorescence during such rapid process is problematic, as the fluorescence signal in 1 contraction cycle is insufficient to gather analysable data. We therefore accumulated the data over a large number of stimulation periods. Typically, we used a total collection time of 100 s (corresponding to the effective photon counting time of 100 s $\times$ 0.5 s^{-1} $\times$ 100 ms = 5 s). This approach allowed us to record a sufficient amount of emitted fluorescence in the 100-ms range, which is required to be able to study physiological behaviour of cardiomyocytes during their contraction.

Employing this setup, experiments can be performed at variable temperature (from room 24 °C to physiological 35 °C). The applicability of the newly developed setup was demonstrated by recording calcium transients at maximum contraction in cells loaded with the Fluo-3 fluorescent probe (excited by 475 nm pulsed picosecond diode laser) [22]. When time-resolved spectroscopy signals of flavin [22] or NAD(P)H [34] were evaluated during cell contraction at frequency 0.5 Hz, no significant change in the fluorescence at the peak of contraction and at rest was found. These results are in accordance with previously published observation that, during cell contraction, NADH levels have been shown not to change in single rat

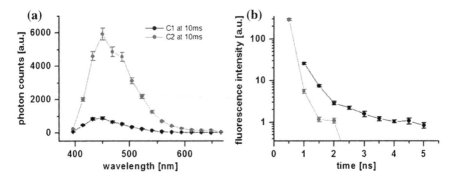

Fig. 13.9 Autofluorescence recording during cardiomyocyte contraction. *Left* NAD(P)H fluorescence photon counts calculated for the resolved component 1 and component 2 by area integration. *Right* Fluorescence kinetics of the two spectral components estimated by spectral unmixing in contracting cardiac myocytes (375 nm excitation, 0.5 Hz stimulation rate, room temperature, recorded at the peak of contraction, 10–110 ms, n = 15 cells). Data are shown as mean ± SEM. For more details, see [22, 34]

myocytes [49, 50, 90], while in trabeculae, a decrease in the fluorescence was seen [13]. Interestingly, as opposed to steady-state measurements, we noted more important contribution of the component 2 ("free NAD(P)H") when compared to the component 1 during contraction at 10 ms, see Fig. 13.9.

These experiments demonstrate the applicability of the time-resolved spectroscopy AF measurements under physiological conditions. They also confirmed that the use of repetitive scanning, synchronized to cell contraction, is a safe approach to gather dynamic information from these cells.

13.3.2 Photobleaching

Photobleaching, recorded following repetitive scanning, is a phenomenon that can significantly hamper fluorescence recordings and has to be monitored. It is considered to be closely related to two phenomena: photodamage and phototoxicity [63]. The level of photobleaching of NAD(P)H fluorescence was verified under our standard recording conditions (as described in Sect. 13.1.1). In these experiments, as illustrated in Fig. 13.10, NAD(P)H fluorescence was measured every minute for 30 s for the duration of 7 min, while the laser power was not modified. In order to precisely identify the effect of photobleaching, individual components of the NAD (P)H fluorescence were resolved by linear unmixing of the time-resolved spectroscopy signals, as described in Sect. 13.2 [33]. We identified comparable level of photobleaching of the two components [34] with the maintenance of the component amplitude ratio, which is a prerequisite for the correct estimation of metabolic state modifications. Low photobleaching rate is also a requirement for investigation of NAD(P)H changes during repetitive fluorescence recording performed in

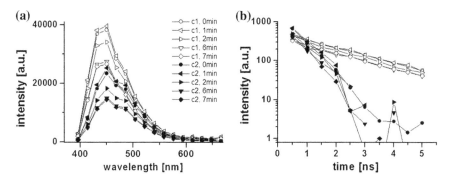

Fig. 13.10 Photobleaching of the two individual NAD(P)H fluorescence components. **a** Photon counts calculated by area integration and **b** fluorescence decays for each resolved component. Photobleaching was induced by excitation of a defocused elliptical spot with a 375 nm picosecond laser (~1 mW output power) for 30 s repeated every 60 s for 7 min. Detailed in [34]

experiments where cells were electrically stimulated to induce contractions (discussed in the Sect. 13.3.1). These results demonstrate that the application of time-resolved spectroscopy recording has much lesser effect on photobleaching than the confocal microscopy, for example, in agreement with previous observations that FLIM by TCSPC is less prone to bleaching than wide-field time-gating of frequency-domain FLIM [63], for example. These results confirmed that time-resolved spectroscopy recording by TCSPC is a safe method for monitoring metabolic oxidative state by means of recordings of endogenous fluorescence from live cells.

13.3.3 Cardiac Remodeling in Pregnancy

To investigate metabolic modifications under physiological conditions, a highly interesting physiological condition of pregnancy has been chosen as a model system. Normal pregnancy (P) induces significant CVS changes, associated with haemodynamic and endocrine modifications that contribute to maternal volume expansion and are necessary for fetal homeostasis and well being [60]. The heart remodeling observed during P resembles that seen in women during long-distance exercise training [48, 51], but there are important differences. Unlike endurance training [12], plasma hormones during P are modified and could affect cardiac function. Importantly, marked metabolic changes occur in this condition. Indeed, since the fetus has an absolute need for glucose, late P is associated with significant changes in carbohydrate metabolism [45], providing continuous nutrient availability to the developing fetus. This results in low blood glucose concentration, as well as in augmented levels of triglycerides and lactate in maternal blood [55, 56, 66]. In addition, in late P, basal oxidative metabolism is increased, leading to enhanced ATP demand with an elevated rate of mitochondrial NADH and oxidation [91]. P is characterized by metabolic remodeling of the mother's heart and by adaptations of haemodynamic and

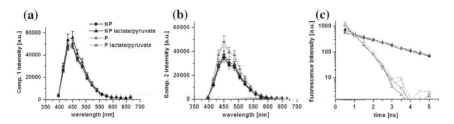

Fig. 13.11 NAD(P)H fluorescence in pregnancy. Photon counts calculated by area integration for the **a** component 1 and **b** component 2 resolved by linear unmixing of time-resolved spectroscopy data. Data are shown as mean ± SEM. **c** fluorescence decays for each resolved component. Presence of glucose (10 mmol/L) only (*circles*, n = 78 for NP, n = 26 for P), or in combination with lactate (2.0 mmol/L) and pyruvate (100 μmol/L) (*triangles*, n = 15 for NP, n = 15 for P) in NP (*black symbols*) versus P (*grey symbols*). For more details, see [3]

cardiac parameters. We demonstrated an adaptive hypertrophic cardiac remodeling during pregnancy, which includes cardiomyocyte dimensions and functions [3, 4]. Despite significant modifications in metabolism, we observed no modifications in the NAD(P)H fluorescence in control conditions in P (Fig. 13.11). Interestingly, spectrally-resolved lifetime detection of NAD(P)H fluorescence revealed modifications in metabolic oxidative state in the presence of substrates lactate/pyruvate that are naturally increased in P [4], mainly affecting the 2nd "free" NAD(P)H component (Fig. 13.11b). This result is in agreement with metabolic switch from glucose to lactate/pyruvte taking place in P.

13.3.4 Monitoring the Effect of the Pharmaceutical Drug Ouabain

AF has a great potential to monitor cellular responses to treatment by pharmaceutical drugs: non-invasive AF evaluation was employed by Hanley et al. [52] to measure reaction of NADH to volatile anesthetics halothane, isoflurane and sevoflurane. Rodrigo et al. [78] studied protective effects of 9,10-dinitrophenol on the cellular damage induced by metabolic inhibition and reperfusion in freshly isolated rat ventricular myocytes. We have tested the applicability of the method to evaluate the effect of pharmaceutical drug ouabain—the Na pump inhibitor—on the mitochondrial metabolic state [31, 39]. Ouabain was reported to decrease the NAD(P)H fluorescence in rat cardiac cells [67], but little information was available on mechanisms underlying this action. We demonstrated (Fig. 13.12) that the decrease in the time- and spectrally-resolved NAD(P)H fluorescence by ouabain was due to decrease in the "free" NAD(P)H fluorescence component leading to lowering of the free/bound NAD(P)H ratio, accompanied by reduced % of oxidized nucleotides, but increased NADH production in living cardiac myocytes [31]. Gathered findings suggest that ouabain induces a decrease in the cell oxidation. This information is

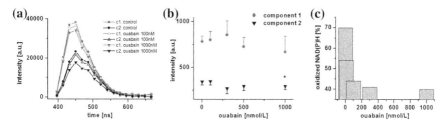

Fig. 13.12 Effect of pharmaceutical drug ouabain on metabolic state. **a** Photon counts calculated for each resolved NAD(P)H fluorescence component by area integration with rising ouabain concentrations. **b** Comparison of the effect of rising concentrations of ouabain on the integral NAD (P)H fluorescence intensity calculated for the component 1 (*black triangle*) and 2 (*grey circles*) at 450 nm spectral channel as the component amplitude multiplied by the unitary reference spectrum (data are shown as mean ± SEM, control (0 nM), n = 120; ouabain 100 nM, n = 36; ouabain 300–1000 nM, n = 15 *p < 0.05 vs. 0nM). **c** Concentration-dependent effect of ouabain on percentage of oxidized nucleotides is evaluated as [F(fully reduced) − Fcontrol]/[F(fully reduced) − F(fully oxidized)], where the fully reduced state is induced in the presence of Rotenone and the fully oxidized state is induced in the presence of DNP. Published in [31]

crucial for understanding of the effect of this pharmaceutical drug, particularly in regard to relationship between metabolic state and cardiac cell contractility [3].

The effect of ouabain was also studied in P, as circulating levels of cardiotonic steroids—endogenous Na pump inhibitors (which include ouabain)—are known to correlate with cardiac hypertrophy and/or changes in blood pressure [80], and are significantly elevated in pregnancy-induced hypertension (PIH) in women [69] as well as in experimental PIH (ePIH) in rats [41]. In normal P, inhibition of Na$^+$/K$^+$-ATPase has been suggested, namely via ouabain, to help plasma volume expansion [69]. In addition, Na$^+$/K$^+$-ATPase is known to be regulated by angiotensin and aldosterone, increased in normal P, but blunted in PIH and ePIH [72]. In cardiac myocytes, Na pump inhibition regulates not only excitation-contraction coupling via [Na$^+$]$_i$-stimulated reverse sodium/calcium exchange [93], but also cardiomyocyte metabolism, including the Ras-Rac-NAD(P)H oxidase cascade [94]. We demonstrated that P affected the action of ouabain on NAD(P)H fluorescence [39]. Altogether, these results demonstrated that the time-resolved spectroscopy of NAD(P)H can be highly valuable also for testing the effect of pharmaceutical drugs on cellular metabolism in physiological conditions.

13.4 Time-Resolved NAD(P)H Spectrometry in CVS Pathology

Numerous diseases of the heart, including hypertension and diabetes, are often linked to alterations in mitochondrial energy metabolism associated with the mitochondrial dysfunction [36]. Chronic alterations of fuel metabolism and oxidative stress status are factors that could impair the capacity of mitochondria to fulfil their crucial role in energy production [65] and thereby contribute to the activation of pathways governing cell death and/or disease [11, 35]. This creates room for numerous applications of non-invasive AF measurements in identification and study of CVS pathology. Effects of oxidative stress were tested together with the study of cardiomyocyte AF under conditions of pathological pregnancy. Clinical applications of time-resolved spectroscopy of NAD(P)H in cardiac myocyte was demonstrated in the case of diagnostics of early stages of cardiac allograft rejection.

13.4.1 Evaluation of the Effect of Oxidative Stress

Oxidative stress is a condition resulting from imbalance between oxidized and reduced species [79]. Lipid peroxidation is a major biochemical consequence of the oxidative deterioration of polyunsaturated lipids in cell membranes and causes damage to membrane integrity and loss of protein function. One of the most reactive products of n-6 polyunsaturated fatty acid peroxidation of membrane phospholipids is 4-hydroxy-2-nonenal (HNE), the primary α,β-unsaturated hydroxyalkenal formed in cells by LPO process [40]. HNE is generated in the peroxidation of lipids containing polyunsaturated omega-6 acyl groups, such as arachidonic or linoleic acids, and of the corresponding fatty acids. In cardiac myocytes, HNE has been shown capable of affecting NADPH production by inactivating mitochondrial $NADP^+$-isocitrate dehydrogenase activity, an important enzyme that controls redox and energy status [7, 8, 96]. The sensitivity of NAD(P)H fluorescence to oxidative stress was tested and proved using stressors such as hydrogen peroxide (H_2O_2) or HNE in living LV myocytes (see Supplement in [3] for details). The effect of HNE was evaluated on the NAD(P)H content in order to better understand mechanisms underlying the lipid peroxidation action. In these experiments, cells were pre-heated to 35 °C to mimic effects at physiological temperatures. Gathered results demonstrated that HNE reduces cell NAD(P)H content (Fig. 13.13) via decrease in the production of NADH and stimulation of the NAD(P)H use by flavoprotein complexes [29]. Consequently, HNE provoked an important cell oxidation by dehydrogenation of both free and bound NAD(P)H molecules in living cardiac cells. This action is likely to affect the overall energy production and use in the heart. These findings not only shed a new light on the effect of HNE on the regulation of NAD(P)H in living cardiac cells, but also demonstrated that oxidative stress via NADPH pathway can have significant effect on the AF measurement and, consequently, has to be taken into account in the analysis of AF signals.

13.4.2 Pathological Pregnancy

Heart disease remains an important cause of women's mortality [84]. Pregnancy induced hypertension was proved to be a predictor of CV disease for women later in

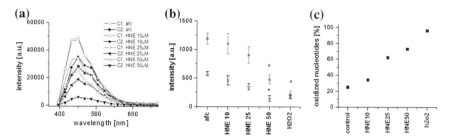

Fig. 13.13 Effect of HNE on NAD(P)H fluorescence. **a** Photon counts calculated for each resolved component by area integration in the presence of rising HNE concentrations. **b** Concentration-dependent effect of HNE on the integral NAD(P)H fluorescence intensity of the component 1 (*black asterisks*) and 2 (*grey squares*), calculated at the 450 nm spectral channel as the component amplitude multiplied by the unitary reference spectrum; data are shown as mean ± SEM, control, n = 78 cells; HNE, 25 μM, n = 71; HNE, 50 μM, n = 19; *p < 0.05 versus control. **c** Percentage of oxidized nucleotides, calculated as fluorescence at (fully reduced − control)/(fully reduced − fully oxidized) state in different HNE concentrations and/or in the presence of H_2O_2. Published in [29]

their lives [81, 92]. In so far as cardiac changes are concerned, maternal heart rate is one of the earliest to increase in P, followed by augmented blood volume. Hypervolemia is the result of active sodium and water retention because of alterations in osmoregulation and the renin-angiotensin-aldosterone system [45], accompanied by decreased sodium serum levels of about 3–4 mM in humans [53] as well as in rats [6]. Subsequently, cardiac output rises together with stroke volume in response to elevated heart rate and reaches maximum values in the final stages of delivery, placing increased volume load on the heart. Blood volume expansion leads to myocardial adaptations mainly affecting the LV, which is more susceptible to heightened load. In contrast to pathological conditions, these alterations are associated with physiological reduction of blood pressure and are reversible. Our work demonstrated that while adaptive compensated cardiomyocyte remodeling is present in normal P in rats, maladaptive components were identified in the model of experimental pregnancy-induced-hypertension [4].

In this setting, mineralocorticoid hormones, including aldosterone of the renin-angiotensin-aldosterone chain, play an important but poorly understood role. Aldosterone is markedly elevated during P [14, 15]; this steroid hormone controls blood volume, and its rise contributes to maternal volume expansion [60]. Mineralocorticoid receptors are expressed in the heart of humans and rodents [77], indicating possible involvement on cardiac metabolic adaptations. Glucocorticoid-mineralocorticoid complexes have been proposed to be activated by lowered NADH level, a determinant of the redox state, and/or by reactive oxygen species [44]. We have tested the effect of mineralocorticoid inhibitor canrenoate in P (P_{can}). These experiments demonstrated that responsiveness of cardiomyocytes to lactate/pyruvate (Fig. 13.2), was modified by canrenoate (Fig. 13.14), (see also Supplementary material in [3]) due to regulation of the "free" NAD(P)H, pointing to the possible role of this hormone in metabolic remodeling that takes place in

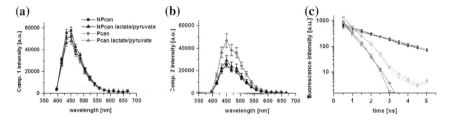

Fig. 13.14 Effect of lactate/pyruvate in MR-Inhibitor canrenoate treated rats. **a** NAD(P)H fluorescence component 1 and **b** component 2 resolved by linear unmixing of time-resolved spectroscopy data. Data are shown as mean ± SEM. **c** Fluorescence decays for each resolved component. Presence of glucose (10 mmol/L) only (*squares*, n = 20 for NP$_{can}$, n = 20 for P$_{can}$), or in combination with lactate (2.0 mmol/L) and pyruvate (100 μmol/L) (*asterisks*, n = 10 for NP$_{can}$, n = 25 for P$_{can}$) in NP$_{can}$ (*black symbols*) versus NP$_{can}$ (*grey symbols*). For more details, see [3]

P [3]. Gathered finding, together with the ones on the Na pump inhibitor ouabain opened a completely new chapter of investigation of P as a highly interesting metabolic condition, also recognized by the Judith Heiny's editorial in Experimental Physiology [54].

13.4.3 Monitoring Rejection of Transplanted Hearts

Transfer of gathered knowledge to clinics is a crucial part of the technology application. We have tested possibility of such knowledge transfer to evaluate early stages of cardiac rejection in paediatric patients with transplanted hearts. Cardiac allograft rejection involves several processes, including activation and proliferation of T-lymphocyte subsets (leading to lymphocyte infiltration and immune destruction of graft tissue), as well as graft vessel injury and thrombosis [57]. Although molecular mechanisms of the rejection remain unclear, alloantigen-dependent and -independent factors are known contributors [87]. Ischemia-reperfusion injury is the most influential alloantigen-independent factor [47], resulting in cardiac cell hypoxia. Such modifications, which also include alterations in the cell oxidative metabolism, often develop rapidly. Some observations suggest that cardiac cells undergo modifications in their oxidative state with the progression of cardiac rejection [85, 86], namely as a result of cell hypoxia, following ischemic changes of cardiac cells. Evaluation of the oxidative metabolism can thus serve as an early indication of the rejection of transplanted hearts.

Time-resolved NAD(P)H fluorescence spectroscopy was therefore attempted in examination of transplanted tissues. A strong correlation between changes in AF spectra and the rejection grade were found in rat heart allograft model [74], but more difficulties were encountered using human tissues [95], possibly due to use of frozen fractions. When time-resolved NAD(P)H spectroscopy method was applied to human cardiac cells freshly isolated from endomyocardial biopsy tissue

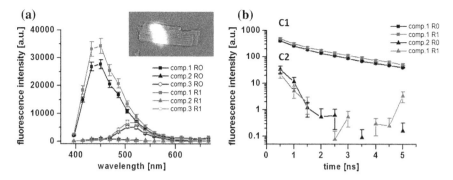

Fig. 13.15 Effect of rejection on NAD(P)H fluorescence in transplanted hearts. **a** NAD(P)H emission spectra determined as total photon counts in human cardiomyocytes isolated from cardiac biopsy of heart transplanted patients presenting no rejection (*R0*) or mild rejection (*R1*) (recording time 30 s, excitation at 375 nm; in the inset, illustrative picture of the recorded human cardiac myocyte: the *bright* spot corresponds to a fluorescence excited by a defocused laser beam used for recording). **b** NAD(P)H fluorescence decays of the component 1 (*C1*) and component 2 (*C2*), resolved after linear unmixing in human cardiomyocytes and compared in *R0* and *R1*. Data are shown as mean ± SEM; R0, n = 48 cells; R1, n = 35 cells. Published in [21]

(Fig. 13.15a), an increase in NAD(P)H fluorescence in the hearts of transplanted paediatric patients was detected (Fig. 13.15b) [19, 21]. Spectral unmixing revealed that from the two NAD(P)H components present in human cardiomyocytes [21], mild rejection of transplanted hearts in paediatric patients affected primarily the lifetimes of the first C1 component (Fig. 13.15-C1, C2). This result not only explained higher NAD(P)H fluorescence intensities, but also proved the applicability of the time-resolved AF spectroscopy as a non-invasive diagnostic tool directly in clinics. At the same time, it also showed difficulties to resolve individual components in the clinical data and prompted search for other advanced data classification approaches, as discussed in Sect. 13.2.

Acknowledgements Supported by Integrated Initiative of European Laser Infrastructures LaserLab Europe III (EC's FP7 under grant agreement no 284464) and the Research grant agency of the Slovak Research and Development Agency APVV-0242-11. Part of this work was presented as doctoral dissertation thesis of AC. I would like to thank Y. Cheng for the preparation of Table 13.1.

References

1. S. Aneba, Y. Cheng, A. Mateasik, B. Comte, D. Chorvat, A. Chorvatova, Probing of cardiomyocyte metabolism by spectrally resolved lifetime detection of NAD(P)H fluorescence. Comput. in Cardiol. **34**, 349–352 (2007)
2. S.W. Ballinger, Mitochondrial dysfunction in cardiovascular disease. Free Radic. Biol. Med. **38**, 1278–1295 (2005)

3. V. Bassien-Capsa, F.M. Elzwiei, S. Aneba, J.C. Fouron, B. Comte, A. Chorvatova, Metabolic remodelling of cardiac myocytes during pregnancy: the role of mineralocorticoids. Can. J. Cardiol. **27**, 834–842 (2011)

4. V. Bassien-Capsa, J.C. Fouron, B. Comte, A. Chorvatova, Structural, functional and metabolic remodeling of rat left ventricular myocytes in normal and in sodium-supplemented pregnancy. Cardiovasc. Res. **69**, 423–431 (2006)

5. M.C. Battista, E. Calvo, A. Chorvatova, B. Comte, J. Corbeil, M. Brochu, Intra-uterine growth restriction and the programming of left ventricular remodelling in female rats. J. Phys. Lond. **565**, 197–205 (2005)

6. A. Beausejour, K. Auger, J. St. Louis, M. Brochu, High-sodium intake prevents pregnancy-induced decrease of blood pressure in the rat. Am. J. Physiol. Heart Circ. Physiol. **285**, H375–H383 (2003)

7. M. Benderdour, G. Charron, B. Comte, R. Ayoub, D. Beaudry, S. Foisy, D. Deblois, C. Des Rosiers, Decreased cardiac mitochondrial NADP+-isocitrate dehydrogenase activity and expression: a marker of oxidative stress in hypertrophy development. Am. J. Physiol. Heart Circ. Physiol. **287**, H2122–H2131 (2004)

8. M. Benderdour, G. Charron, D. Deblois, B. Comte, C. Des Rosiers, Cardiac mitochondrial NADP+-isocitrate dehydrogenase is inactivated through 4-hydroxynonenal adduct formation: an event that precedes hypertrophy development. J. Biol. Chem. **278**, 45154–45159 (2003)

9. D.K. Bird, L. Yan, K.M. Vrotsos, K.W. Eliceiri, E.M. Vaughan, P.J. Keely, J.G. White, N. Ramanujam, Metabolic mapping of MCF10A human breast cells via multiphoton fluorescence lifetime imaging of the coenzyme NADH. Cancer Res. **65**, 8766–8773 (2005)

10. K. Blinova, S. Carroll, S. Bose, A.V. Smirnov, J.J. Harvey, J.R. Knutson, R.S. Balaban, Distribution of mitochondrial NADH fluorescence lifetimes: steady-state kinetics of matrix NADH interactions. Biochemistry **44**, 2585–2594 (2005)

11. V. Borutaite, G.C. Brown, Mitochondria in apoptosis of ischemic heart. FEBS Lett. **541**, 1–5 (2003)

12. R.W. Braith, M.A. Welsch, M.S. Feigenbaum, H.A. Kluess, C.J. Pepine, Neuroendocrine activation in heart failure is modified by endurance exercise training. J. Am. Coll. Cardiol. **34**, 1170–1175 (1999)

13. R. Brandes, D.M. Bers, Increased work in cardiac trabeculae causes decreased mitochondrial NADH fluorescence followed by slow recovery. Biophys. J. **71**, 1024–1035 (1996)

14. M. Brochu, J.P. Gauvin, J. St. Louis, Increase of aldosterone secretion in adrenal cortex suspensions derived from pregnant rats. Proc. Soc. Exp. Biol. Med. **212**, 147–152 (1996)

15. M. Brochu, J.G. Lehoux, S. Picard, Effects of gestation on enzymes controlling aldosterone synthesis in the rat adrenal. Endocrinology **138**, 2354–2358 (1997)

16. B. Chance, B. Schoener, R. Oshino, F. Itshak, Y. Nakase, Oxidation-reduction ratio studies of mitochondria in freeze-trapped samples. NADH and flavoprotein fluorescence signals. J. Biol. Chem. **254**, 4764–4771 (1979)

17. B. Chance, B. Thorell, Fluorescence measurements of mitochondrial pyridine nucleotide in aerobiosis and anaerobiosis. Nature **184**, 931–934 (1959)

18. Y. Cheng, N. Dahdah, N. Porier, J. Miro, D. Chorvat Jr., A. Chorvatova, Spectrally and time-resolved study of NADH autofluorescence in cardiac myocytes from human biopsies. Proc. SPIE: Int. Soc. Opt. Eng. **6771**, 677104-1–677104-13 (2007)

19. Y. Cheng, A. Mateasik, N. Dahdah, N. Poirier, J. Miro, D. Jr. Chorvat, A. Chorvatova, Analysis of NAD(P)H fluorescence components in cardiac myocytes from human biopsies: a new tool to improve diagnostics of rejection of transplanted patients. Proc. SPIE: Int. Soc. Opt. Eng. **7183**, 718319-1–718319-8 (2009)

20. D. Chorvat Jr., A. Chorvatova, Spectrally resolved time-correlated single photon counting: a novel approach for characterization of endogenous fluorescence in isolated cardiac myocytes. Eur. Biophys. J. Biophys. Lett. **36**, 73–83 (2006)

21. D. Chorvat Jr., A. Mateasik, Y. Cheng, N. Poirier, J. Miro, N.S. Dahdah, A. Chorvatova, Rejection of transplanted hearts in patients evaluated by the component analysis of multi-wavelength NAD(P)H fluorescence lifetime spectroscopy. J. Biophotonics **3**, 646–652 (2010)

22. D. Chorvat Jr., S. Abdulla, F. Elzwiei, A. Mateasik, A. Chorvatova, Screening of cardiomyocyte fluorescence during cell contraction by multi-dimensional TCSPC. Proc. SPIE: Int. Soc. Opt. Eng. **6860**, 686029-1–686029-12 (2008)
23. D. Chorvat Jr., V. Bassien-Capsa, M. Cagalinec, J. Kirchnerova, A. Mateasik, B. Comte, A. Chorvatova, Mitochondrial autofluorescence induced by visible light in single rat cardiac myocytes studied by spectrally resolved confocal microscopy. Laser Phys. **14**, 220–230 (2004)
24. D. Chorvat Jr., A. Chorvatova, Study of flavins in living cardiomyocytes using spectrally resolved time correlated single photon counting. J. Mol. Cell. Cardiol. **39**, 187 (2005)
25. D. Chorvat Jr., A. Chorvatova, Multi-wavelength fluorescence lifetime spectroscopy: a new approach to the study of endogenous fluorescence in living cells and tissues. Laser Phys. Lett. **6**, 175–193 (2009)
26. D. Chorvat Jr., J. Kirchnerova, M. Cagalinec, J. Smolka, A. Mateasik, A. Chorvatova, Spectral unmixing of flavin autofluorescence components in cardiac myocytes. Biophys. J. **89**, L55–L57 (2005)
27. D. Chorvat Jr., A. Mateasik, J. Kirchnerova, A. Chorvatova, Application of spectral unmixing in multi-wavelength time-resolved spectroscopy. Proc. SPIE: Int. Soc. Opt. Eng. **6771**, 677105-1–677105-12 (2007)
28. A.M. Chorvatova, in *Natural Biomarkers for Cellular Metabolism: Biology, Techniques and Applications,* ed. by A.A. Heikal, V. Ghukasyan. Autofluorescence-assisted examination of cardiovascular system physiology and pathology (Taylor and Francis, Boca Raton, 2014), pp. 245–271
29. A. Chorvatova, S. Aneba, A. Mateasik, D. Chorvat, B. Comte, Time-resolved fluorescence spectroscopy investigation of the effect of 4-hydroxynonenal on endogenous NAD(P)H in living cardiac myocytes. J. Biomed. Opt. **18**, 67009 (2013)
30. A. Chorvatova, D. Chorvat, Jr., Tissue fluorophores and their spectroscopic characteristics, in *Fluorescence Lifetime Spectroscopy and Imaging for Tissue Biomedical Diagnostics*, ed. by L. Marcu, P. French, D. Elson, (CRC Press, Boca Raton, 2014), pp. 47–84
31. A. Chorvatova, F. Elzwiei, A. Mateasik, D. Chorvat Jr., Effect of ouabain on metabolic oxidative state in living cardiomyocytes evaluated by time-resolved spectroscopy of endogenous NAD(P)H fluorescence. J. Biomed. Opt. **17**, 101505 (2012)
32. A. Chorvatova, G. Hart, M. Hussain, Na(+)/Ca(2+) exchange current (I(Na/Ca)) and sarcoplasmic reticulum Ca(2+) release in catecholamine-induced cardiac hypertrophy. Cardiovasc. Res. **61**, 278–287 (2004)
33. A. Chorvatova, A. Mateasik, D. Chorvat Jr., Spectral decomposition of NAD(P)H fluorescence components recorded by multi-wavelength fluorescence lifetime spectroscopy in living cardiac cells. Laser Phys. Lett. **10**, 125703 (2013)
34. A. Chorvatova, A. Mateasik, D. Chorvat Jr., Laser-induced photobleaching of NAD(P)H fluorescence components in cardiac cells resolved by linear unmixing of TCSPC signals, pp. 790326-1–790326-9 (2011)
35. S. Cortassa, M.A. Aon, E. Marban, R.L. Winslow, B. O'Rourke, An integrated model of cardiac mitochondrial energy metabolism and calcium dynamics. Biophys. J. **84**, 2734–2755 (2003)
36. N.S. Dhalla, N. Afzal, R.E. Beamish, B. Naimark, N. Takeda, M. Nagano, Pathophysiology of cardiac dysfunction in congestive heart failure. Can. J. Cardiol. **9**, 873–887 (1993)
37. M.E. Dickinson, G. Bearman, S. Tilie, R. Lansford, S.E. Fraser, Multi-spectral imaging and linear unmixing add a whole new dimension to laser scanning fluorescence microscopy. Biotechniques **31**, 1272–1276 (2001)
38. M.A. Digman, V.R. Caiolfa, M. Zamai, E. Gratton, The phasor approach to fluorescence lifetime imaging analysis. Biophys. J. **94**, L14–L16 (2008)
39. F. Elzwiei, V. Bassien-Capsa, J. St-Louis, A. Chorvatova, Regulation of the sodium pump during cardiomyocyte adaptation to pregnancy. Exp. Physiol. **98**, 183–192 (2013)
40. H. Esterbauer, R.J. Schaur, H. Zollner, Chemistry and biochemistry of 4-hydroxynonenal, malonaldehyde and related aldehydes. Free Radic. Biol. Med. **11**, 81–128 (1991)

41. O.V. Fedorova, N.I. Kolodkin, N.I. Agalakova, A.R. Namikas, A. Bzhelyansky, J. St. Louis, E.G. Lakatta, A.Y. Bagrov, Antibody to marinobufagenin lowers blood pressure in pregnant rats on a high NaCl intake, J. Hypertens. **23**, 835–842 (2005)

42. F. Fereidouni, A.N. Bader, H.C. Gerritsen, Spectral phasor analysis allows rapid and reliable unmixing of fluorescence microscopy spectral images. Opt. Express **20**, 12729–12741 (2012)

43. F. Fereidouni, K. Reitsma, H.C. Gerritsen, High speed multispectral fluorescence lifetime imaging. Opt. Express **21**, 11769–11782 (2013)

44. J.W. Funder, RALES, EPHESUS and redox. J. Steroid Biochem. Mol. Biol. **93**, 121–125 (2005)

45. S.G. Gabbe, J.R. Niebyl, J.L. Simpson, in *Obstetrics. Normal and Problem Pregnancies*, ed. by S. G. Gabbe, J.R. Niebyl, J.L. Simpson,(Churchill Livingstone, A Harcourt Health Science Company, U.S.A., Philadelphia, Pennsylvania, 2002)

46. A. Gafni, L. Brand, Fluorescence decay studies of reduced nicotinamide adenine dinucleotide in solution and bound to liver alcohol dehydrogenase. Biochemistry **15**, 3165–3171 (1976)

47. P.B. Gaudin, B.K. Rayburn, G.M. Hutchins, E.K. Kasper, K.L. Baughman, S.N. Goodman, L.E. Lecks, W.A. Baumgartner, R.H. Hruban, Peritransplant injury to the myocardium associated with the development of accelerated arteriosclerosis in heart transplant recipients. Am. J. Surg. Pathol. **18**, 338–346 (1994)

48. L.E. Ginzton, R. Conant, M. Brizendine, M.M. Laks, Effect of long-term high intensity aerobic training on left ventricular volume during maximal upright exercise. J. Am. Coll. Cardiol. **14**, 364–371 (1989)

49. E.J. Griffiths, H. Lin, M.S. Suleiman, NADH fluorescence in isolated guinea-pig and rat cardiomyocytes exposed to low or high stimulation rates and effect of metabolic inhibition with cyanide. Biochem. Pharmacol. **56**, 173–179 (1998)

50. E.J. Griffiths, S.K. Wei, M.C. Haigney, C.J. Ocampo, M.D. Stern, H.S. Silverman, Inhibition of mitochondrial calcium efflux by clonazepam in intact single rat cardiomyocytes and effects on NADH production. Cell Calcium **21**, 321–329 (1997)

51. M. Guazzi, F.C. Musante, H.L. Glassberg, J.R. Libonati, Detection of changes in diastolic function by pulmonary venous flow analysis in women athletes. Am. Heart J. **141**, 139–147 (2001)

52. P.J. Hanley, J. Ray, U. Brandt, J. Daut, Halothane, isoflurane and sevoflurane inhibit NADH: ubiquinone oxidoreductase (complex I) of cardiac mitochondria. J. Physiol. **544**, 687–693 (2002)

53. A.P. Heenan, L.A. Wolfe, G.A. Davies, M.J. McGrath, Effects of human pregnancy on fluid regulation responses to short-term exercise. J. Appl. Physiol. **95**, 2321–2327 (2003)

54. J. Heiny, Pumped up and pregnant. Exp. Physiol. **98**, 48 (2013)

55. E. Herrera, M.A. Lasuncion, M. Palacin, A. Zorzano, B. Bonet, Intermediary metabolism in pregnancy. First theme of the Freinkel era. Diabetes **40**(2), 83–88 (1991)

56. E. Herrera, H. Ortega, G. Alvino, N. Giovannini, E. Amusquivar, I. Cetin, Relationship between plasma fatty acid profile and antioxidant vitamins during normal pregnancy. Eur. J. Clin. Nutr. **58**, 1231–1238 (2004)

57. F.M. Hoffman, Outcomes and complications after heart transplantation: a review. J. Cardiovasc. Nurs. **20**, S31–S42 (2005)

58. J. Horilova, Z. Tomaskova, M. Grman, A. Ilesova, M. Bucko, A. Vikartovska, P. Gemeiner, V. Stefuca, I. Lajdova, D. Chorvat, A. Chorvatova, Monitoring metabolic oxidative state in living cells and systems using time-resolved fluorescence and spectroscopy techniques. Proc. ADEPT 131–134 (2013)

59. S. Huang, A.A. Heikal, W.W. Webb, Two-photon fluorescence spectroscopy and microscopy of NAD(P)H and flavoprotein. Biophys. J. **82**, 2811–2825 (2002)

60. E. Jensen, C. Wood, M. Keller-Wood, The normal increase in adrenal secretion during pregnancy contributes to maternal volume expansion and fetal homeostasis. J. Soc. Gynecol. Investig. **9**, 362–371 (2002)

61. J.R. Knutson, J.M. Beechem, L. Brand, Simultaneous analysis of multiple fluorescence decay curves: a global approach. Chem. Phys. Lett. **102**, 501–507 (1983)

62. A.S.R. Koti, M.M. Krishna, N. Periasamy, Time-resolved area-normalized emission spectroscopy (TRANES): a novel method for confirming emission from two excited states. J. Phys. Chem. A **105**, 1767–1771 (2001)
63. S. Kumar, C. Dunsby, P.A. De Beule, D.M. Owen, U. Anand, P.M. Lanigan, R.K. Benninger, D.M. Davis, M.A. Neil, P. Anand, C. Benham, A. Naylor, P.M. French, Multifocal multiphoton excitation and time correlated single photon counting detection for 3-D fluorescence lifetime imaging. Opt. Express **15**, 12548–12561 (2007)
64. J.R. Lakowicz, On spectral relaxation in proteins. Photochem. Photobiol. **72**, 421–437 (2000)
65. E.J. Lesnefsky, S. Moghaddas, B. Tandler, J. Kerner, C.L. Hoppel, Mitochondrial dysfunction in cardiac disease: ischemia–reperfusion, aging, and heart failure. J. Mol. Cell. Cardiol. **33**, 1065–1089 (2001)
66. A. Leturque, S. Hauguel, J.P. Revelli, A.F. Burnol, J. Kande, J. Girard, Fetal glucose utilization in response to maternal starvation and acute hyperketonemia. Am. J. Physiol. **256**, E699–E703 (1989)
67. T. Liu, D.A. Brown, B. O'Rourke, Role of mitochondrial dysfunction in cardiac glycoside toxicity. J. Mol. Cell. Cardiol. **49**, 728–736 (2010)
68. T. Loftsson, P. Jarho, M. Masson, T. Jarvinen, Cyclodextrins in drug delivery. Expert Opin. Drug Deliv. **2**, 335–351 (2005)
69. D.A. Lopatin, E.K. Ailamazian, R.I. Dmitrieva, V.M. Shpen, O.V. Fedorova, P.A. Doris, A.Y. Bagrov, Circulating bufodienolide and cardenolide sodium pump inhibitors in preeclampsia. J. Hypertens. **17**, 1179–1187 (1999)
70. J. Luts, F. Ojeda, R. Van de Plas, M.B. De, H.S. Van, J.A. Suykens, A tutorial on support vector machine-based methods for classification problems in chemometrics. Anal. Chim. Acta **665**, 129–145 (2010)
71. A. Mateasik, D. Chorvat, A. Chorvatova, Analysis of spectrally resolved autofluorescence images by support vector machines. Proc. SPIE **8588**, 85882J-1–85882J-10 (2013)
72. A.S. Mihailidou, H. Bundgaard, M. Mardini, P.S. Hansen, K. Kjeldsen, H.H. Rasmussen, Hyperaldosteronemia in rabbits inhibits the cardiac sarcolemmal Na(+)- K(+) pump. Circ. Res. **86**, 37–42 (2000)
73. A.S. Montcuquet, L. Herve, F. Navarro, J.M. Dinten, J.I. Mars, Nonnegative matrix factorization: a blind spectra separation method for in vivo fluorescent optical imaging. J. Biomed. Opt. **15**, 056009 (2010)
74. D.C. Morgan, J.E. Wilson, C.E. MacAulay, N.B. MacKinnon, J.A. Kenyon, P.S. Gerla, C. Dong, H. Zeng, P.D. Whitehead, C.R. Thompson, B.M. McManus, New method for detection of heart allograft rejection: validation of sensitivity and reliability in a rat heterotopic allograft model. Circulation **100**, 1236–1241 (1999)
75. R. Niesner, B. Peker, P. Schlusche, K.H. Gericke, Noniterative biexponential fluorescence lifetime imaging in the investigation of cellular metabolism by means of NAD(P)H autofluorescence. ChemPhysChem **5**, 1141–1149 (2004)
76. L.H. Opie, P. Owen, Effects of increased mechanical work by isolated perfused rat heart during production or uptake of ketone bodies. Assessment of mitochondrial oxidized to reduced free nicotinamide-adenine dinucleotide ratios and oxaloacetate concentrations. Biochem. J. **148**, 403–415 (1975)
77. P. Pearce, J.W. Funder, High affinity aldosterone binding sites (type I receptors) in rat heart. Clin. Exp. Pharmacol. Physiol **14**, 859–866 (1987)
78. G.C. Rodrigo, C.L. Lawrence, N.B. Standen, Dinitrophenol pretreatment of rat ventricular myocytes protects against damage by metabolic inhibition and reperfusion. J. Mol. Cell. Cardiol. **34**, 555–569 (2002)
79. C.X. Santos, N. Anilkumar, M. Zhang, A.C. Brewer, A.M. Shah, Redox signaling in cardiac myocytes. Free Radic. Biol. Med. **50**, 777–793 (2011)
80. W. Schoner, Endogenous digitalis-like factors. Clin. Exp. Hypertens. A **14**, 767–814 (1992)
81. E.W. Seely, C.G. Solomon, Insulin resistance and its potential role in pregnancy-induced hypertension. J. Clin. Endocrinol. Metab. **88**, 2393–2398 (2003)

82. M. Straume, S.G. Frasier-Cadoret, M.L. Johnson, Least-squares analysis of fluorescence data, in *Topics in Fluorescence Spectroscopy*, ed. by J.R. Lakowicz, (Plenum Press, New York, 1991), pp. 177–240

83. C. Stringari, J.L. Nourse, L.A. Flanagan, E. Gratton, Phasor fluorescence lifetime microscopy of free and protein-bound NADH reveals neural stem cell differentiation potential. PLoS ONE **7**, e48014 (2012)

84. J. Tan, Cardiovascular disease in pregnancy. Curr. Obstet. Gynaecol. **11**, 137–145 (2001)

85. M. Tanaka, G.K. Mokhtari, R.D. Terry, L.B. Balsam, K.H. Lee, T. Kofidis, P.S. Tsao, R.C. Robbins, Overexpression of human copper/zinc superoxide dismutase (SOD1) suppresses ischemia-reperfusion injury and subsequent development of graft coronary artery disease in murine cardiac grafts. Circulation **110**, II200–II206 (2004)

86. M. Tanaka, S. Nakae, R.D. Terry, G.K. Mokhtari, F. Gunawan, L.B. Balsam, H. Kaneda, T. Kofidis, P.S. Tsao, R.C. Robbins, Cardiomyocyte-specific Bcl-2 overexpression attenuates ischemia-reperfusion injury, immune response during acute rejection, and graft coronary artery disease. Blood **104**, 3789–3796 (2004)

87. G. Vassalli, A. Gallino, M. Weis, W. von Scheidt, L. Kappenberger, L.K. von Segesser, J.J. Goy, Alloimmunity and nonimmunologic risk factors in cardiac allograft vasculopathy. Eur. Heart J. **24**, 1180–1188 (2003)

88. P.J. Verveer, A. Squire, P.I. Bastiaens, Global analysis of fluorescence lifetime imaging microscopy data. Biophys. J. **78**, 2127–2137 (2000)

89. M. Wakita, G. Nishimura, M. Tamura, Some characteristics of the fluorescence lifetime of reduced pyridine nucleotides in isolated mitochondria, isolated hepatocytes, and perfused rat liver in situ. J. Biochem. (Tokyo) **118**, 1151–1160 (1995)

90. R.L. White, B.A. Wittenberg, NADH fluorescence of isolated ventricular myocytes: effects of pacing, myoglobin, and oxygen supply. Biophys. J. **65**, 196–204 (1993)

91. R.L. White, B.A. Wittenberg, Mitochondrial NAD(P)H, ADP, oxidative phosphorylation, and contraction in isolated heart cells. Am. J. Physiol. Heart Circ. Physiol. **279**, H1849–H1857 (2000)

92. M. Wolf, C.A. Hubel, C. Lam, M. Sampson, J.L. Ecker, R.B. Ness, A. Rajakumar, A. Daftary, A.S. Shakir, E.W. Seely, J.M. Roberts, V.P. Sukhatme, S.A. Karumanchi, R. Thadhani, Preeclampsia and future cardiovascular disease: potential role of altered angiogenesis and insulin resistance. J. Clin. Endocrinol. Metab. **89**, 6239–6243 (2004)

93. Z. Xie, T. Cai, Na+-K+-ATPase-mediated signal transduction: from protein interaction to cellular function. Mol. Interv. **3**, 157–168 (2003)

94. Z. Xie, P. Kometiani, J. Liu, J. Li, J.I. Shapiro, A. Askari, Intracellular reactive oxygen species mediate the linkage of Na+/K+-ATPase to hypertrophy and its marker genes in cardiac myocytes. J. Biol. Chem. **274**, 19323–19328 (1999)

95. M.H. Yamani, S.W. van de Poll, N.B. Ratliff, B.E. Kuban, R.C. Starling, P.M. McCarthy, J.B. Young, Fluorescence spectroscopy of endomyocardial tissue post-human heart transplantation: does it correlate with histopathology? J. Heart Lung Transplant. **19**, 1077–1080 (2000)

96. J.H. Yang, E.S. Yang, J.W. Park, Inactivation of NADP + -dependent isocitrate dehydrogenase by lipid peroxidation products. Free Radic. Res. **38**, 241–249 (2004)

97. T. Zimmermann, J. Rietdorf, R. Pepperkok, Spectral imaging and its applications in live cell microscopy. FEBS Lett. **546**, 87–92 (2003)

Chapter 14
Fluorescence Lifetime Measurements of NAD(P)H in Live Cells and Tissue

Alex J. Walsh, Amy T. Shah, Joe T. Sharick
and Melissa C. Skala

Abstract Autofluorescence intensity and lifetime imaging of NAD(P)H yields quantitative, non-invasive measurements of cellular metabolism. NAD(P)H is a coenzyme involved in cellular metabolism processes including glycolysis and oxidative phosphorylation. The NAD(P)H fluorescence lifetime includes a short and long lifetime component due to the two possible physiological conditions of NAD(P)H, free or protein-bound (to an enzyme and/or substrate). Fluorescence lifetimes of NAD(P)H have been imaged in cells, ex vivo tissues, and in vivo tissues to investigate cellular metabolism at basal conditions and with perturbations. In particular, NAD(P)H fluorescence lifetimes are altered in pre-malignant and malignant cells and tissues compared with non-malignant cells and tissues across several cancers including head and neck cancers, breast cancer, and skin cancer. Additionally, NAD(P)H fluorescence lifetimes decrease in cancer cells and tumors following drug treatment and therefore, these metabolic endpoints show potential for drug monitoring and screening.

14.1 Introduction

The reduced form of nicotinamide adenine dinucleotide (NAD(P)H) is a co-enzyme in cellular metabolism. It functions as the main electron donor in oxidative phosphorylation, and also in the reactions that feed into oxidative phosphorylation (e.g.

A.J. Walsh · A.T. Shah · J.T. Sharick · M.C. Skala (✉)
Department of Biomedical Engineering, Vanderbilt University,
Nashville, TN 37235, USA
e-mail: m.skala@vanderbilt.edu

A.J. Walsh
e-mail: alexandra.j.walsh@vanderbilt.edu

A.T. Shah
e-mail: a.shah@vanderbilt.edu

J.T. Sharick
e-mail: joseph.t.sharick@vanderbilt.edu

© Springer International Publishing Switzerland 2015
W. Becker (ed.), *Advanced Time-Correlated Single Photon Counting Applications*,
Springer Series in Chemical Physics 111, DOI 10.1007/978-3-319-14929-5_14

glycolysis, citric acid cycle, etc.). NAD(P)H is autofluorescent, with an excitation maximum at 350 nm and an emission maximum at 460 nm [46]. Intracellular NAD (P)H is localized in the cytoplasm and mitochondria. Early optical imaging studies focused on mapping NAD(P)H intensities in cells and tissues, in order to gain non-invasive insight into relative cellular metabolic rates [11]. The fluorescence lifetime of NAD(P)H provides further information on the relative amounts of free and protein-bound NAD(P)H [40]. The fluorescence lifetime of free NAD(P)H ($\sim$300 ps) is distinctly different from that of protein-bound NAD(P)H ($\sim$2.5 ns) due to quenching from the adenine moiety of the molecule when it is in its free state.

The fluorescence decay function of NAD(P)H is an advantageous measure compared with fluorescence intensity because the lifetime provides additional information on protein-binding activity whereas the NAD(P)H fluorescence intensity spectrum is similar for both the free and protein-bound conformations. Additionally, the decay parameters are independent of concentration and are therefore a self-referenced measure which require less system calibration compared with intensity measurements. The fluorescence lifetimes and the relative amounts of bound and unbound NAD(P)H have been characterized in cell culture, applied to animal models to understand disease progression, and piloted in human tissues both ex vivo and in vivo. These studies have provided unique insight into the role of cellular metabolism in pathology and in response to therapy. Future technology development will exploit the fluorescence lifetime of NAD(P)H to streamline disease diagnosis, improve treatment monitoring, and accelerate meaningful drug development.

14.2 NAD(P)H Fluorescence Lifetime in Cells

The fluorescence decay of NAD(P)H is typically fit by a double-exponential decay model, where the short component represents free NAD(P)H and the long component represents protein-bound NAD(P)H. The amplitude-weighted (mean) lifetime is calculated by $\tau_m = \alpha_1\tau_1 + \alpha_2\tau_2$, where α_1 and α_2 represent the relative contributions from free and protein-bound NAD(P)H, respectively ($\alpha_1 + \alpha_2 = 1$), and τ_1 and τ_2 represent the lifetimes of the free and protein-bound components, respectively. In order to record the fast lifetime component correctly the time-channel width should be no larger than about 1/5 of its decay time. The number of time channels should be large enough to cover a time interval of about 5 times the slow decay component [34]. That means the time channel width should be about 50 ps, and the number of time channels about 250. These numbers are in the range where TCSPC delivers near-ideal photon efficiency [4]. Moreover, correct double-exponential decay components are only obtained if the data are free of out-of-focus fluorescence. Therefore, an imaging technique with optical sectioning capability must be used. These requirements are almost perfectly met by TCSPC FLIM in combination with confocal or multiphoton laser scanning, see Chap. 2. Most of the data shown in this section were recorded by TCSPC FLIM in combination with two-photon laser scanning microscopes. Our own system uses a Becker & Hickl

SPC-150 TCSPC FLIM module [6] with a Hamamatsu H7422P-40 PMT in combination with a Prairie Technology two-photon laser scanning microscope.

The fluorescence lifetime of cellular NAD(P)H has been characterized for several metabolic perturbations, including treatment with metabolic inhibitors. The NAD(P)H lifetime has also been characterized for cancer cells expressing particular mutations and in response to anti-cancer treatment. Additionally, the NAD(P)H lifetime has been investigated to differentiate mechanisms of cell death and to measure stem cell differentiation.

Metabolic perturbations have been shown to affect the fluorescence lifetime of NAD(P)H, see also Chap. 13. In particular, these perturbations include adding inhibitors of oxidative phosphorylation or glycolysis. The electron transport chain is a series of oxidation and reduction reactions that create a proton gradient across the mitochondrial membrane, and this gradient drives ATP synthesis. During the electron transport chain, NAD(P)H is oxidized to NAD+, but this oxidation can be disrupted by compounds like cyanide. Cyanide treatment causes a decrease in the mean NAD(P)H lifetime (τ_m) in cell lines from the breast (MCF10A) and oral cavity (OKF6) [8, 55, 60]. This decrease in τ_m reflects an increase in the contribution from free NAD(P)H (α_1).

Oxidative phosphorylation inhibition achieved with cobalt chloride ($CoCl_2$) treatment in MCF10A cells (Fig. 14.1) and with rotenone treatment in BKEz-7 endothelial cells from calf aorta cause similar shifts in NAD(P)H τ_m and α_1 [51]. Inhibition of glycolysis in MCF10A cells caused an increase in the NAD(P)H protein-bound lifetime (τ_2) and a decrease in the relative amount of protein-bound NAD(P)H (α_2) (Fig. 14.1). Glycolysis inhibition with deoxyglucose treatment in the BKEz-7 cells caused a slight increase in NAD(P)H τ_m, reflecting an increase in NAD(P)H α_2, but statistical significance was not tested [51].

In addition to metabolic inhibitors, serum starvation and cellular confluency affect cellular metabolism and induce changes in the NAD(P)H lifetime. Serum levels in media can affect cellular metabolism because serum contains substrates that are used in glycolysis and oxidative phosphorylation. Serum-starvation in MCF10A cells causes increased NAD(P)H α_1 and decreased NAD(P)H τ_m, which are similar effects as those observed due to treatment with cyanide [8]. Serum-starvation and oxidative phosphorylation inhibition affect the NAD(P)H lifetime

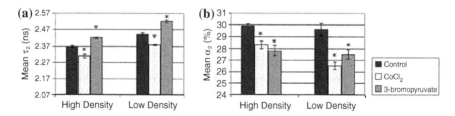

Fig. 14.1 The long-lifetime component of NAD(P)H decreases with $CoCl_2$ treatment and increases with 3-boropyruvate treatment in high-density and low-density MCF10A cells (**a**). The contribution from the long-lifetime component decreases with $CoCl_2$ and 3-bromopyruvate treatment in MCF10A cells (**b**). Reproduced with permission from [56]

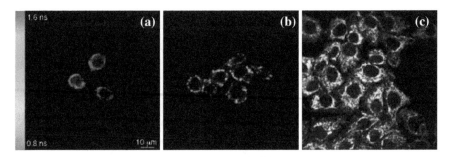

Fig. 14.2 Representative images show MCF10A cells at a higher confluency have a lower NAD(P)H fluorescence lifetime. Reproduced with permission from [8]

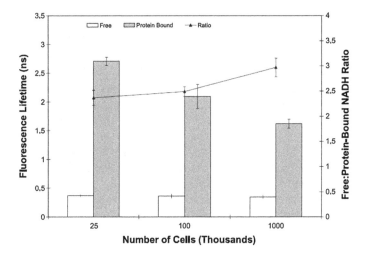

Fig. 14.3 MCF10A cells that have been serum-starved show increased free-to-protein-bound NAD(P)H ratio plated at 10,000 cells per dish compared with at 25 and 100 cells per dish. Reproduced with permission from [8]

similarly since both perturbations slow cellular metabolism. Confluency also affects cellular metabolism. A higher confluency increases NAD(P)H α_1 and decreases NAD(P)H τ_2, causing a decreased NAD(P)H τ_m, see Figs. 14.2 and 14.3.

The fluorescence lifetime of NAD(P)H has been shown to distinguish malignant versus non-malignant cells and as well as different sub-types of malignant cells based on receptor status. Common phenotypes in breast cancers include overexpression of the estrogen receptor (ER) and human epidermal growth factor receptor 2 (HER2). Walsh et al. measured the NAD(P)H fluorescence lifetime of malignant breast cancer cells including MCF7, which express ER, as well as SKBr3, MDA-MB-361, and BT474, which express HER2 (Fig. 14.4). The NAD(P)H lifetime of malignant cells was greater than that of the non-malignant cell line MCF10A [60]. This increase was attributed to a decreased NAD(P)H α_1. Furthermore, the NAD(P)H mean lifetime of

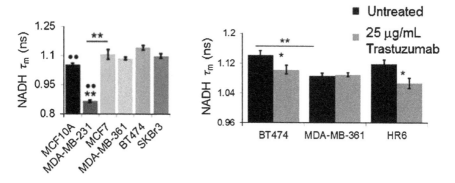

Fig. 14.4 *Left* NAD(P)H lifetime is decreased in triple negative breast cancer (MDA-MB-231) compared to non-malignant breast cells (MCF10A) and increased in estrogen receptor positive (MCF7) and HER2 positive cells (MDA-MB-361, BT474, SKBr3). *Right* NAD(P)H lifetime decreases in BT474 cells with trastuzumab treatment, is unchanged in MDA-MB-361 cells, and is decreased in HR6 cells with trastuzumab treatment. Reproduced with permission from [60]

the MDA-MB-231 breast cancer cells, which do not overexpress ER or HER2, was lower than that of the MCF10A cells. Additionally, squamous cell carcinoma cells from the oral cavity, including SCC25 and SCC61, show an increased NAD(P)H α_1 compared with the nonmalignant cell line OKF6 [55].

Walsh et al. showed that the NAD(P)H lifetime can resolve response to anti-cancer treatment. These studies were tested in breast cancer cells with treatment using trastuzumab, which is a clinically-used antibody that targets HER2 and prevents downstream signalling. The cell lines studied included a trastuzumab-responsive cell line, BT474, and a partly responsive cell line, MDA-MB-361. Additionally, a cell line derived from a BT474 xenograft with an acquired resistance to trastuzumab, HR6, was also measured. Trastuzumab treatment for 24 h caused a decrease in the mean NAD(P)H lifetime in BT474, but not MDA-MB-361 cells (Fig. 14.4). This change in NAD(P)H lifetime in BT474 cells was attributed to an increased contribution from free NAD(P)H. Interestingly, the HR6 cells also showed a decreased mean NAD(P)H lifetime, which could reflect altered internal signalling from trastuzumab treatment.

Shah et al. tested therapeutic response in oral cancer cells treated with cetuximab, an antibody that targets the epidermal growth factor receptor (EGFR), BGT226, an investigational small molecular inhibitor that targets phosphoinositide 3-kinase (PI3K) and mammalian target of rapamycin (mTOR), and cisplatin, standard chemotherapy. Treatment with these drugs for 24 h was done in SCC25 and SCC61, two EGFR-overexpressing cell lines (Figs. 14.5 and 14.6). SCC61 cells also have upregulated PI3K activation. For both cell lines, the contribution of free NAD(P)H was shown to decrease with BGT226 and cisplatin treatment (Fig. 14.6). However, SCC61 cells also show a decreased contribution from free NAD(P)H with cetuximab treatment, which could reflect altered metabolic pathways in response to treatment. These results indicate that the NAD(P)H decay parameters are a sensitive measure that can resolve early response to anti-cancer treatment.

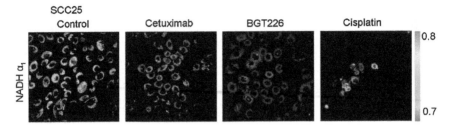

Fig. 14.5 Representative images show NAD(P)H α_1 of SCC25 cells after treatment with cetuximab, BGT226, or cisplatin for 24 h. Reproduced with permission from [55]

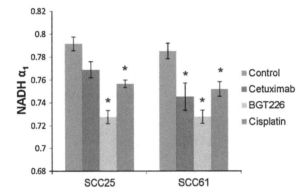

Fig. 14.6 NAD(P)H α_1 decreases with BGT226 and cisplatin treatment in SCC25 and SCC61 cells after 24 h of treatment. Additionally, NAD(P)H α_1 decreases with cetuximab treatment in SCC61 cells. Reproduced with permission from [55]

The NAD(P)H fluorescence lifetime has been used to characterize cell death by apoptosis or necrosis [61]. Wang et al. treated HeLa cells and 143 osteosarcoma cells with staurosporine (STS) to induce apoptosis or hydrogen peroxide (H_2O_2) to induce necrosis. STS treatment caused an initial increase in NAD(P)H lifetime 15 min after treatment and then a gradual decrease for up to two hours (Fig. 14.7). These effects are attributed to an increased contribution of protein-bound NAD(P)H or an increased lifetime of protein-bound NAD(P)H, reflecting NAD(P)H binding to different enzymes in apoptosis. Necrosis caused no change in NAD(P)H fluorescence lifetime over one hour (Fig. 14.8). To validate the process of apoptosis, STS treatment showed increased caspase 3 activity compared with H_2O_2 treatment after two hours. These results indicate that the NAD(P)H lifetime has potential to be a label-free method to distinguish methods of cell death, which could provide beneficial information for optimizing treatments that cause apoptosis instead of necrosis.

Stem cell differentiation has been characterized using NAD(P)H fluorescence lifetime [28, 39]. Guo et al. incubated human mesenchymal stem cells (hMSCs) in osteogenic induction media, causing osteogenic differentiation. Over 21 days,

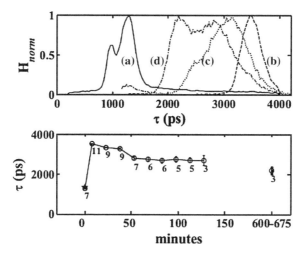

Fig. 14.7 The *top graph* shows distributions of lifetimes from (**a**) before treatment with STS and after treatment for 0–15 min (**b**), 30–45 min (**c**), and 60–75 min (**d**). The *bottom graph* shows the average lifetime, τ, over time after STS treatment. Reproduced with permission from [61]

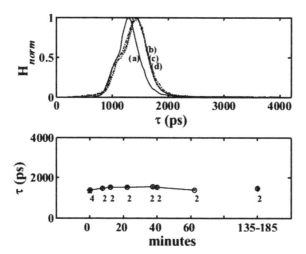

Fig. 14.8 The *top graph* shows distributions of lifetimes from (**a**) before treatment with H_2O_2 and after treatment for 0–15 min (**b**), 30–50 min (**c**), and 50–70 min (**d**). The *bottom graph* shows the average lifetime, τ, over time after H_2O_2 treatment. Reproduced with permission from [61]

osteogenic differentiation caused an increase in the mean lifetime of NAD(P)H and a decrease in the ratio between free and protein-bound NAD(P)H.

Overall, the fluorescence lifetime of NAD(P)H is a sensitive measure of metabolic processes, particularly for applications for characterizing cancer cells and response to anti-cancer treatment, measure the effect of inhibitors targeting metabolic pathways, characterizing cell death, and monitoring stem cell differentiation.

14.3 NAD(P)H Fluorescence Lifetime in Preclinical Models

Preclinical models of human cancers remain a vital step in the drug discovery and development process. Understanding the pathophysiology of cancer, studying the mechanisms of intrinsic and acquired resistance to therapies, and identifying novel therapeutic targets and agents rely on the use of a wide spectrum of animal models [49]. Both transgenic and carcinogen-promoter-induced tumor models will be discussed in this section. The former are similar to human cancers in their phenotype, histology, and genetic makeup, and allow for the development of tumors in their appropriate microenvironment, but may develop asynchronously and are time and labour intensive. The latter are similar to human cancers in their phenotype, histology, and biochemistry, and are highly reproducible, but require repeated applications of carcinogens and require a long time frame for tumor development [49]. While animal models are imperfect representations of genetically heterogeneous human cancers, they are indispensable for the testing of new diagnostic imaging modalities and devices prior to translation to a clinical setting.

The ability of optical metabolic imaging to distinguish precancerous cells from normal cells has been investigated in vivo using the hamster cheek pouch model of oral cancer, which mimics the development of squamous epithelial cancer in the human oral cavity [1]. Skala et al. demonstrated the use of multiphoton microscopy to simultaneously image NAD(P)H lifetime and subcellular morphology in this model, which was generated by treatment with 0.5 % DMBA, a powerful carcinogen [57]. A significant decrease in the contribution and lifetime of protein-bound NAD(P)H was reported in epithelial precancers versus normal tissue (Fig. 14.9), along with a significant increase in intracellular variability of NAD(P) Hfluorescence lifetimes.

Neoplastic cells, like the ones present in epithelial precancers, favour glycolysis over oxidative phosphorylation as a means of generating ATP [27], which is consistent with the observed decrease in contribution of protein-bound NAD(P)H

Fig. 14.9 Volume-averaged NAD(P)H lifetime variables in normal, low-grade precancerous, and high-grade precancerous animals obtained from in vivo multiphoton images. Reproduced with permission from [57]

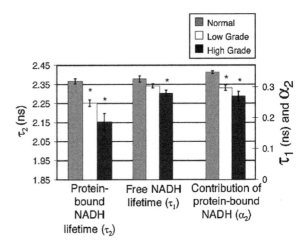

lifetime in these cells. The decrease in protein-bound NAD(P)H lifetime is attributed to a possible increase in dynamic quenching [41] or a change in distribution of NAD(P)H enzyme binding sites in neoplastic tissues [2]. This study demonstrates an in vivo application of multiphoton lifetime imaging for high-resolution metabolic mapping, guiding development of similar time-resolved and steady state fluorescence schemes for clinical detection of human precancers and cancers.

Jabbour et al. has developed a multimodal optical system for imaging of oral precancer that takes advantage of this ability of multiphoton FLIM to probe the metabolic activity in possibly precancerous tissue [30]. The integrated system uses FLIM to generate real-time wide-field images of the biochemical makeup of the specimen, which then guide reflectance confocal microscopy (RCM) imaging to suspicious sites within the field of view (FOV). RCM allows clinicians to probe subcellular morphology and is able to distinguish between normal, precancerous, and cancerous oral tissue [18], but has a limited FOV and thus requires guidance, usually by visual inspection. The multimodal optical system was built using an automated translation stage which moved a sample between the FOVs of two separate FLIM and RCM subsystems. The FOV of the FLIM subsystem was 16×16 mm^2 with a lateral resolution of 62.5 μm, while the RCM subsystem FOV was measured to have a 400 μm diameter with a lateral resolution of 0.97 μm. In the FLIM subsystem, three emission collection spectral bands were generated, including one corresponding to NAD(P)H (452 ± 22.5 nm). The illumination source for the RCM subsystem was a near infrared continuous wave diode-pumped solid state laser with $\lambda = 1064$ nm (power at sample <45 mW).

The ability of the multimodal optical system to differentiate between normal and precancerous tissue in vivo was tested using the same hamster cheek pouch model of oral cancer described above. For imaging, the hamster was anesthetized, and the check pouch pulled and clamped to a mount. FLIM imaging was performed first, and spatial features of the FLIM intensity and lifetime images were used to choose locations for RCM (15 per treated pouch) (Fig. 14.10).

As expected, NAD(P)H fluorescence intensity and lifetime is relatively consistent within healthy tissue. The RCM images taken from the centre of the FOV in Fig. 14.10b, c, g shows keratin scattering at the surface, as well as epithelial nuclei beneath the outer layer. FLIM and RCA images from a DMBA-treated hamster cheek pouch show significant variability in fluorescence intensities, lifetimes, and cell morphologies (Fig. 14.11). The NAD(P)H fluorescence lifetime was found to be much shorter (2.62 ± 0.79 ns) compared to normal tissue (4.60 ± 0.25 ns) and low-grade dysplasia (4.29 ± 0.29 ns) (Region 2). This is expected due to an increased fluorescence contribution from bound NAD(P)H. Corresponding RCM images show larger cell nuclei in Region 2 versus Region 1, consistent with the respective diagnoses of low-grade dysplasia and cytologic atypia by histopathology. The classification ability of this dual-modality system could have clinical relevance in assessing whether a sample is normal, benign, premalignant, or malignant.

For situations in which in vivo optical measurements are not feasible, it has been shown that FLIM measurements in excised tissue maintained in chilled tissue media

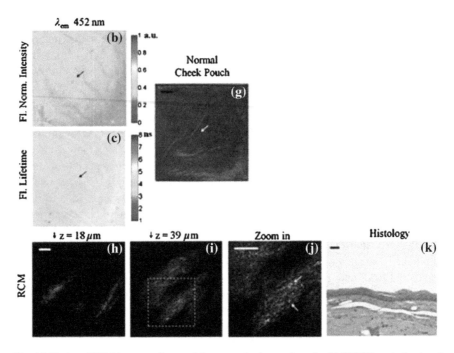

Fig. 14.10 b, c FLIM images of normal hamster cheek pouch at the NAD(P)H emission band (normalized intensity and lifetime). **g** Photograph of FLIM imaging area. *Arrows* indicate RCM imaging region. **h–i** RCM images taken at indicated depths. **j** Zoomed in image taken from dotted square in (**i**). *White arrows* indicate nuclei. **k** H&E histology image from RMC region. *Scale bars* **a** and **g** 2 mm, and **h**, **j**, and **k** 50 μm. Reproduced with permission from [30]

for up to 8 h can represent in vivo metabolic states [50, 61]. Walsh et al. compared the NAD(P)H lifetime in hamster cheek epithelia in vivo, in live cultured biopsies up to 48 h, and in flash-frozen and thawed samples [59]. The mean lifetime of NAD (P)H in cultured tissue appears to have a lifetime similar to that of in vivo tissue through 4 h, while the frozen-thawed sample is 13 % greater, a significant difference ($p < 0.001$). By 12 h, the mean lifetime of NAD(P)H decreases significantly by 10 % and remains significantly lower than the in vivo value (Fig. 14.12).

These measurements suggest that optical metabolism measurements from tissue that has been cultured for up to 8 h represent the metabolic state of the in vivo tissue. The metabolic state of frozen-thawed tissue does not represent that of in vivo tissue. Cells undergoing stress, such as apoptosis, have also shown increased mean NAD(P)H fluorescence lifetimes, suggesting that freezing and thawing tissue exerts more stress on cells than live culture [61]. While optical metabolic measurements varied from in vivo values after 8 h of live culture, cell morphology and histological analysis did not show any changes over the entire time course of the experiment. This suggests that NAD(P)H lifetime measurements are more sensitive to molecular changes than these standard analysis methods.

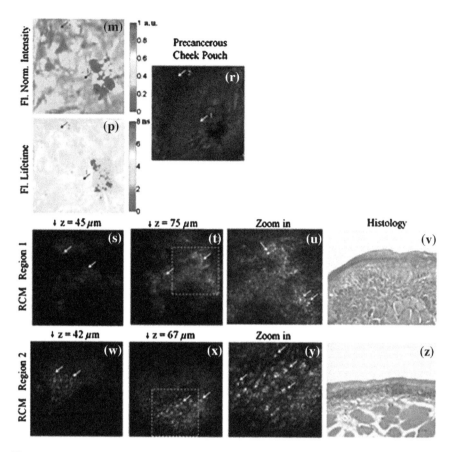

Fig. 14.11 m, p FLIM images of DMBA-treated hamster cheek pouch at the NAD(P)H emission band (normalized intensity and lifetime). **r** Photograph of FLIM imaging area. *Arrows* indicate RCM imaging regions (**s–t, w–x**) RCM images taken at indicated depths and regions. **u, y** Zoomed in images taken from *dotted square* in (**t, x**). *White arrows* indicate nuclei. **v, z** H&E histology image from RMC region. Scale same as Fig. 14.10. Reproduced with permission from [30]

Conklin et al. demonstrated that fluorescence measurements of endogenous NAD(P)H can be made accurately in fixed tissue samples such as classic histopathology slides [6]. Histology allows a pathologist to identify morphological changes that provide a high degree of accuracy in tumor staging and prediction of patient outcome. The authors demonstrate that mouse mammary tumors can be distinguished from normal epithelium by multiphoton FLIM measurements in unstained histopathology slides, allowing detection of changes in the metabolic state of tumor cells. Staining with hematoxylin and eosin (H&E) was shown to interfere with endogenous fluorescence measurements. The mice in this study were transgenic for the polyomavirus middle-T (PyVMT) oncogene under the control of the mammary specific MMTV promoter to generate primary tumors. This breast tumor model reliably demonstrates the progression from hyperplasia to adenoma to

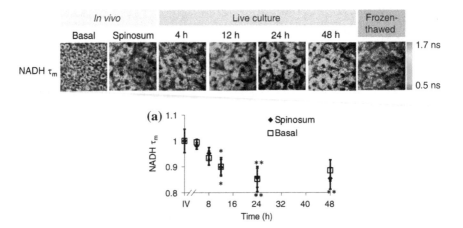

Fig. 14.12 *Top* Representative NAD(P)H τ_m images from each location and time point. *Bottom* NAD(P)H τ_m relative to in vivo value (*p < 0.05; **p < 0.01). Reproduced with permission from [59]

Table 14.1 Components of NAD(P)H fluorescence lifetime in tumor epithelium versus normal epithelium

	τ_1 (ps)		τ_2 (ps)		a_1	
	Normal	Tumor	Normal	Tumor	Normal	Tumor
780 nm excitation	456	546	2360	2538	47	49
Average shift	(90)		(178)		(2)	
[n = 10]	*					

Reproduced with permission from [19]
* denotes that the tumor value is significant compared to the normal value, $p<0.05$ for a Student's t-test

carcinoma in a manner closely resembling human ductal carcinoma in situ, and is dependably invasive and metastatic [43].

The study focused on the carcinoma in situ (CIS) stage of disease progress, which is similar to human ductal carcinoma in situ (DCIS), and is of interest for seeking optical biomarkers of cancer because it contains normal epithelium contiguous with tumor neoplasia. A mouse mammary tumor was sectioned and imaged unstained with both normal and tumor epithelium visible. The short component of NAD(P)H lifetime was found to undergo a statistically significant increase in tumor epithelium versus normal tissue (Table 14.1). Finally, the section was stained with H&E to show the ability to subsequently investigate cell morphology.

This data suggests that combining fluorescence imaging of unstained slides with subsequent classic bright-field imaging of H&E stains can give a more advanced diagnosis of disease state than either alone. Surprisingly, endogenous fluorescence of cells is preserved throughout histological processes including fixing, sectioning, and de-paraffinizing.

Preclinical models of human cancers and other disease states are invaluable in the development and testing of new devices that measure the fluorescence lifetime of NAD(P)H in tissues both in vivo and ex vivo. Despite imperfections in the ability of these models to accurately represent genetically and phenotypically heterogeneous human diseases, their use will continue to be an important step in the translation of NAD(P)H lifetime technologies to the clinical setting.

14.4 Clinical Applications

Due to the promising results obtained in cells and tissues, time-resolved fluorescence imaging and spectroscopy has the potential for a multitude of clinical applications. These applications necessarily include also other endogenous fluorophores, such as FAD, lipofuscin, melanin, and porhyrines [17, 22, 35, 52, 53]. Clinical instruments are on the market for ophthalmology [23, 52–54] and dermatology [36, 37], see Chaps. 15 and 16 of this book.

Current efforts of NAD(P)H FLIM translation are predominately in accessible tissues—the skin, and the oral cavity [9, 10, 45]; however, studies of biopsied tissues include also interrogation of brain, colon, and breast tissues [13, 20, 25, 43, 62]. Clinically, time resolved NAD(P)H fluorescence endpoints are used to differentiate normal from malignant and pre-malignant tissues, and identify tumor margins.

Early experiments of time-resolved NAD(P)H fluorescence were performed with single-point spectroscopy measurements on excised tissues. The NAD(P)H fluorescence lifetime of excised tissues recapitulates the in vivo state as long as the tissue is hydrated and imaged within several hours [50, 59]. De Beule et al. used a hyperspectral fluorescence lifetime probe for skin cancer diagnosis [20]. This system used two picosecond lasers emitting light at 355 and 440 nm and collected spectrally resolved fluorescence emission between 390 and 600 nm. This system utilized multi-wavelength TCSPC [4–6] at 16 emission channels in combination with laser multiplexing [4–6] (see Chap. 1, Sects. 1.4.1 and 1.4.2) to simultaneously obtain spectrally resolved lifetime data at two excitation wavelengths. De Beule et al. found no significant difference between the fluorescence lifetimes of basal cell carcinoma and normal, surrounding tissue when excited with 355 nm light [20]. However, it is important to note that the point measurement provides an average value over the probed area and the fluorescence signal from the tissue excited at 355 nm may include fluorescence emission from not only NAD(P)H, but also collagen, keratin an melanin. Significant improvement was obtained by single-point multi-wavelength TCSPC in a micro-spectrometric setup [14–16], see Chap. 13.

However, for clinical use fluorescence lifetime detection must be combined with imaging and, if possible, with spectrally resolved detection. Acquisition times should be in the range of a few seconds or below, the data should be recorded in a sufficiently large number of time channels to resolve double-exponential decay function into their components, and the laser power should be within the limits of laser safety standards.

Cheng et al. designed a FLIM system with a rigid probe of diameter 1.7 cm and a length of 14 cm [12]. Imaging was performed by scanning, time-resolved recording by using analog-signal detection and recording by a fast digitizer. A frequency-tripled Nd:YAG laser with excitation wavelength 355 nm and 1 ns pulse length was used for excitation. Images of 150 × 150 pixels were obtained within an acquisition time of 1 s. The endoscope was tested in vivo in healthy epithelium of a hamster cheek pouch model, and spatially uniform NAD(P)H autofluorescence lifetime data was measured as expected (Fig. 14.13, left). The time resolution of the detection system was 320 ps, deconvolution of the impulse response function was performed offline using an optimized Laguerre algorithm (CITE). It delivered a mean NAD(P)H lifetime of 4.44 ± 0.13 ns. This accuracy was considered sufficient for clinical imaging of oral cancer.

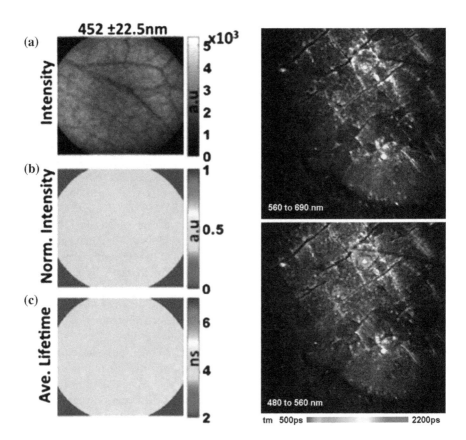

Fig. 14.13 Endoscope testing in vivo. *Left* Images obtained by Cheng et al. Maps of NAD(P)H (452 ± 22.5 nm emission). **a** Absolute intensity, **b** normalized intensity, and **c** average lifetime. 150 x 150 pixels, acquisition time 1 s. Reproduced with permission from [12]. *Right* Image obtained by TCSPC FLIM and scanning through an endoscope, see Chap. 2, Sect. 2.6. Keratome on human skin, 512 × 512 pixels, two wavelength channels, image area 5 × 5 mm, acquisition time 20 s

The advantage of this device is that it can achieve relatively high imaging speed without loss of temporal resolution. The sampling rate of the digitiser (one sample per 150 ps) allows for deconvolution of the non-ideal instrument response for accurate single-exponential lifetime analysis but not necessarily for double exponential analysis. The disadvantage is that the photon efficiency of analog recording is lower than for photon counting, and high excitation power is used. It should also be noted that the acquisition time of 1 s was obtained at a relative moderate number of pixels. For images of the same pixel number TCSPC FLIM is not significantly slower [6, 32].

The FLIM technique with the highest temporal resolution and with the highest photon efficiency is TCSPC FLIM. The principle is described in Chap. 1 of this book, the combination with optical imaging techniques in Chap. 2. For clinical imaging, TCSPC FLIM is often considered too slow in terms of acquisition time. This is not necessarily correct, as has been shown in [4, 5, 33]: Acquisition times on the order of one second can be achieved, see also Chap. 2, Sect. 2.2.4. TCSPC FLIM can be combined with multi-spectral detection [4–6], see Chap. 1, Sect. 1.4.5.2. Multi-spectral TCSPC FLIM in combination with a two-photon laser scanning microscope has been introduced [3] and the applicability to NAD(P)H detection been demonstrated by Becker et al. [4–6] and Rück et al. [48]. The technique has been used for NAD(P)H imaging [63], for identification of malignant melanoma in skin [22], and for diagnosis of squamous intraepithelial neoplasia [39]. A system with 8 fully parallel spectral channels working at an acquisition time of 5 s has been described in [6].

TCSPC FLIM systems are available not only for microscopy but also imaging of cm-sized objects. These are placed directly in the primary image plane of a confocal scanner [6], see Chap. 2, Sect. 2.5. Scanning has also been demonstrated through flexible [33] endoscopes. TCSPC FLIM by scanning through a rigid endoscope is described in Chap. 2, Sect. 2.6 of this book. The system delivered excellent spatial and temporal resolution and an acquisition time of 10–20 s for images of 512×512 pixels. The acquisition time can probably be reduced by a factor 16 for images of 128×128 pixels. Images obtained with this system is shown in Fig. 14.13, right.

14.4.1 Ex Vivo Tissue Studies

Time-resolved NAD(P)H fluorescence spectroscopy and imaging has been explored in a variety of tissues, including skin, brain, breast, head and neck tissues, and colon. However, it is often difficult to separate NAD(P)H autofluorescence from that of other endogenous fluorophores including collagen, keratin, and FAD in tissues. The NAD(P)H fluorescence signal can be separated from collagen and FAD by spectrally resolved emission data and the fluorescence lifetime [38, 40, 41, 63]. Collagen has a much longer lifetime ($\sim$ 5–6 ns) than NAD(P)H, and emission is typically at shorter wavelengths than NAD(P)H emission [41]. Many of the studies discussed herein excite tissue at a single wavelength and collect wave-length resolved emission spectrum to infer trends across samples for different fluorophores. NAD(P)H can be excited in the 330–400 nm range (660–800 nm two-photon) and emission is centred around 450 nm [29, 41].

While the probe-based single-point spectroscopy study of Kennedy et al. [33] did not find significant differences in the lifetime between basal cell carcinoma and normal tissue, an imaging study was able to delineate between the two tissue states. In a subsequent study, Galletly et al. used a wide-field macroscopic time-domain FLIM system to interrogate basal cell carcinoma and normal tissue [25]. This system excited tissue at 355 nm and collected emission at either 375 or 455 nm, with the assumption that the fluorescence signal collected at 375 nm is predominately from collagen while the emission from 455 nm is predominately from NAD(P)H. Fluorescence lifetime data was acquired using a CCD camera with a gated optical intensifier to obtain 25 time gates at 250 ps intervals. Galletly et al. report a significantly ($p < 0.0001$) reduced mean fluorescence lifetime (1.402 ± 0.127 ns basal cell carcinoma versus 1.554 ± 0.141 ns normal) of basal cell carcinomas in the 455 nm emission channel [25]. Due to this difference in fluorescence lifetimes, FLIM shows promise for in vivo detection and diagnosis of basal cell carcinoma. In addition to basal cell carcinoma, time-resolved fluorescence imaging has been investigated for melanoma detection and diagnosis. In a study by Dimitrow et al., fluorescence lifetime images of skin lesions were acquired both in vivo and from biopsies [22]. This study reports decreased short and long lifetime values with two-photon excitation at 760 nm in melanocytes compared to normal tissue but no lifetime changes in melanoma.

Time resolved fluorescence imaging of human tissue NAD(P)H is not limited to skin, but has also been performed on biopsies of colon, stomach, bladder, liver and pancreas. In this study by McGinty et al., a wide-field fluorescence lifetime imaging system detected fluorescence lifetime data by a gated optical intensifier coupled to a CCD camera which recorded images at 25 different time points with 250 ps intervals [44]. The tissue was excited at 355 nm and emission was collected with a 375 nm long-pass filter. McGinty et al. found that the fluorescence lifetime of colon tumors increased compared to normal colon tissue [44]. Additionally, they observed increased (but not statistically significant) fluorescence lifetimes of gastric cancer and bladder cancer [44]. Two precancerous colon lesions exhibited decreased mean fluorescence lifetimes [44]. Due to the 355 nm excitation and 375 nm long-pass filter, most of the fluorescence signal was most likely from NAD(P)H, however, there may be a contribution from collagen and/or keratin in this study. Additionally, McGinty et al. demonstrated an imaging system utilizing only two time gates to build a fluorescence decay curve which was capable of video rate (10 Hz) acquisition speeds, which provides real-time imaging for point-of-care or intraoperative use [44].

In addition to these abdominal cancers, NAD(P)H FLIM has been performed on slices of human brain biopsies. In the brain, multiphoton excited NAD(P)H fluorescence lifetimes are often fit to three lifetimes due to an additional long lifetime often attributed to an additional bound species [13]. In ex vivo measurements from brain glioma tissue samples, a spectroscopic time-resolved fluorescence probe excited tissue at 337 nm and a 460 nm bandpass filter isolated NAD(P)H emission [62]. The lifetime of 460 nm emission was shortest in cerebral cortex (1.1 ns), slightly greater in low grade glioma (1.15 ns), greater in high grade glioma (1.3 ns), and greatest in normal white matter (1.8 ns) [62]. While the lifetime at 460 nm alone was not sufficient to classify tissue as benign or malignant, additional information including the lifetime at 390 ns (collagen) and fluorescence intensities resulted in a classification algorithm with an accuracy above 90 % [62].

Kantelhardt et al. [31] and Leppert et al. [42] used two-photon excitation and TCSPC FLIM to record lifetime images of glioma and the surrounding brain tissue. They found a significantly increased mean (amplitude-weighted) lifetime for the glioma cells compared to the surrounding brain cells.

14.4.2 In Vivo Human Studies

In addition to investigating excised human tissues, time-resolved fluorescence imaging of NAD(P)H has been performed in vivo for diagnosis and margin assessment of brain and head and neck cancers. In the brain, time-resolved fluorescence spectroscopy measurements have identified longer NAD(P)H lifetimes in low grade glioma (τ = 1.38 ns) than normal white matter (τ = 1.19 ns), normal cortex (τ = 1.16 ns) and high grade glioma (τ = 1.13 ns) [9, 10]. Using the spectral and lifetime characteristics of NAD(P)H and collagen, this study classified low grade gliomas with 100 % sensitivity and 94 % specificity [10].

In head and neck cancers, time-resolved measurements of NAD(P)H fluorescence are explored as a means of diagnosing cancerous tissue. A non-invasive diagnostic tool is particularly important for head and neck cancers to preserve tissue and function of vital organs—such as the tongue, mouth, esophagus, and vocal box. When head and neck squamous cell carcinomas are excited at 337 nm and the emission at 460/25 nm collected, the lifetime of cancerous tissue is significantly shorter than normal tissue [45, 58]. A wide-field FLIM system with a 4 mm field of view quickly collects and graphically displays NAD(P)H lifetimes for intraoperative margin assessment [58]. Representative images are shown in Fig. 14.14, demonstrating the added contrast of the lifetime measurements over fluorescence intensity or bright field imaging alone [58].

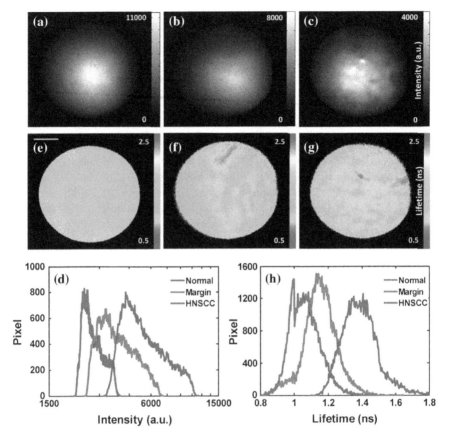

Fig. 14.14 Autofluorescence fluorescence lifetime imaging microscopy images of human buccal mucosa: (**a–c**) depict the intensity images, and (**e–g**) depict the average lifetime images from three areas: normal, tumor, and adjacent normal-tumor, their corresponding histograms are depicted in (**d**) for intensity and (**h**) for average lifetime. HNSCC, head and neck squamous cell carcinoma. Reproduced with permission from [58]

14.5 Conclusions

Fluorescence lifetime imaging of NAD(P)H provides specific information on the relative amounts of free and protein-bound NAD(P)H in a cell. This unique metabolic endpoint has been characterized with controlled experiments in cell culture, and has been applied to understand disease progression in animal models. Due to the non-invasive nature of these measurements, NAD(P)H fluorescence lifetime imaging has also been used to characterize human tissue in vivo, and several ex vivo human tissue studies also hold promise for future translation, see Chaps. 2, 13, 14–16. Recent technological developments have improved the acquisition time of fluorescence lifetime imaging techniques. TCSPC FLIM with laser scanning has been shown to record lifetime images at acquisition times of less than one second [32], and spectrally resolved lifetime images within 5 s [6]. Dynamic effects in the fluorescence lifetime of a sample can be recorded at a resolution of 1 ms by line scanning TCSPC [8], and at a resolution of 40 ms by temporal mosaic FLIM, see Chaps. 2 and 5 of this book. TCSPC FLIM can be performed in image areas in the cm^2 range, and through optical periscopes, see Chap. 2. Motion artefacts can be eliminated by parallel IR imaging and aligning subsequent frames online (see Chap. 16), probably also by recording individual frames by mosaic FLIM and aligning the frames on-line or off-line. Wide-field FLIM techniques by gated camera have been improved to record images in several time windows simultaneously, thus increasing the photon efficiency and reducing the acquisition time [21, 24, 26, 47, 58]. Thus, there is a wide range of technologies capable of capturing the relatively weak signals of NAD(P)H and using them for clinical applications.

References

1. S. Andrejevic, J.F. Savary, C. Fontolliet, P. Monnier, H. van Den Bergh, 7,12-dimethylbenz[a] anthracene-induced 'early' squamous cell carcinoma in the Golden Syrian hamster: evaluation of an animal model and comparison with 'early' forms of human squamous cell carcinoma in the upper aero-digestive tract. Int. J. Exp. Pathol. **77**, 7–14 (1996)
2. S. Banerjee, D.K. Bhatt, Histochemical studies on the distribution of certain dehydrogenases in squamous cell carcinoma of cheek. Indian J. Cancer **26**, 21–30 (1989)
3. W. Becker, A. Bergmann, C. Biskup, T. Zimmer, N. Klöcker, K. Benndorf, Multi-wavelength TCSPC lifetime imaging. Proc. SPIE **4620**, 79–84 (2002)
4. W. Becker, *Advanced Time-Correlated Single-Photon Counting Techniques* (Springer, Berlin, 2005)
5. W. Becker, A. Bergmann, C. Biskup, Multi-spectral fluorescence lifetime imaging by TCSPC. Microsc. Res. Technol. **70**, 403–409 (2007)
6. W. Becker, *The bh TCSPC Handbook*, 5th edn. (Becker & Hickl GmbH 2012). www.becker-hickl.com
7. W. Becker, V. Shcheslavkiy, S. Frere, I. Slutsky, Spatially resolved recording of transient fluorescence-lifetime effects by line-scanning TCSPC. Microsc. Res. Technol. **77**, 216–224 (2014)
8. D.K. Bird, L. Yan, K.M. Vrotsos, K.W. Eliceiri, E.M. Vaughan, P.J. Keely, J.G. White, N. Ramanujam, Metabolic mapping of MCF10A human breast cells via multiphoton fluorescence lifetime imaging of the coenzyme NADH. Cancer Res. **65**, 8766–8773 (2005)
9. P.V. Butte, Q. Fang, J.A. Jo, W.H. Yong, B.K. Pikul, K.L. Black, L. Marcu, Intraoperative delineation of primary brain tumors using time-resolved fluorescence spectroscopy. J. Biomed. Opt. **15**, 027008 (2010)
10. P.V. Butte, A.N. Mamelak, M. Nuno, S.I. Bannykh, K.L. Black, and L. Marcu, Fluorescence lifetime spectroscopy for guided therapy of brain tumors. NeuroImage *54* (Suppl 1), S125–135 (2011)
11. B. Chance, B. Schoener, R. Oshino, F. Itshak, Y. Nakase, Oxidation-reduction ratio studies of mitochondria in freeze-trapped samples. NADH and flavoprotein fluorescence signals. J. Biol. Chem. **254**, 4764–4771 (1979)
12. S. Cheng, R.M. Cuenca, B. Liu, B.H. Malik, J.M. Jabbour, K.C. Maitland, J. Wright, Y.S. Cheng, J.A. Jo, Handheld multispectral fluorescence lifetime imaging system for in vivo applications. Biomed. Opt. Express **5**, 921–931 (2014)

13. T.H. Chia, A. Williamson, D.D. Spencer, M.J. Levene, Multiphoton fluorescence lifetime imaging of intrinsic fluorescence in human and rat brain tissue reveals spatially distinct NADH binding. Opt. Express **16**, 4237–4249 (2008)

14. A. Chorvat, Chorvatova, Spectrally resolved time-correlated single photon counting: a novel approach for characterization of endogenous fluorescence in isolated cardiac myocytes. Eur. Biophys. J. **36**, 73–83 (2006)

15. D. Chorvat, A. Chorvatova, Multi-wavelength fluorescence lifetime spectroscopy: a new approach to the study of endogenous fluorescence in living cells and tissues. Laser Phys. Lett. **6**, 175–193 (2009)

16. D. Chorvat Jr., A. Mateasik, Y.G Cheng, N.Y Poirier, J. Miro, N.S. Dahdah, A. Chorvatova, Rejection of transplanted hearts in patients evaluated by the component analysis of multi-wavelength NAD(P)H fluorescence lifetime spectroscopy. J. Biophotonics. **3** 646–652 (2010)

17. A. Chorvatova, D. Chorvat, Tissue fluorophores and their spectroscopic characteristics, in *Fluorescence Lifetime Spectroscopy and Imaging*, ed. by L. Marcu, P.W.M. French, D.S. Elson (CRC Press, Boca Raton, 2015)

18. A.L. Clark, A. Gillenwater, R. Alizadeh-Naderi, A.K. El-Naggar, R. Richards-Kortum, Detection and diagnosis of oral neoplasia with an optical coherence microscope. J. Biomed. Opt. **9**, 1271–1280 (2004)

19. M.W. Conklin, P.P. Provenzano, K.W. Eliceiri, R. Sullivan, P.J. Keely, Fluorescence lifetime imaging of endogenous fluorophores in histopathology sections reveals differences between normal and tumor epithelium in carcinoma in situ of the breast. Cell Biochem. Biophys. **53**, 145–157 (2009)

20. P.A. De Beule, C. Dunsby, N.P. Galletly, G.W. Stamp, A.C. Chu, U. Anand, P. Anand, C.D. Benham, A. Naylor, P.M. French, A hyperspectral fluorescence lifetime probe for skin cancer diagnosis. Rev. Scientific Instrum. **78**, 123101 (2007)

21. P.A.B. De Beule, D.M. Owen, H.B. Manning, C.B. Talbot, J. Requejo-Isidro, C. Dunsby, J. McGinty, R.K.P. Benninger, D.S. Elson, I. Munro, M. John Lever, P. Anand, M.A.A. Neil, and P. M.W. French, Rapid hyperspectral fluorescence lifetime imaging. Microsc. Res. Technol. **70**, 481–484 (2007)

22. E. Dimitrow, I. Riemann, A. Ehlers, M.J. Koehler, J. Norgauer, P. Elsner, K. Konig, M. Kaatz, Spectral fluorescence lifetime detection and selective melanin imaging by multiphoton laser tomography for melanoma diagnosis. Exp. Dermatol. **18**, 509–515 (2009)

23. C. Dysli, G. Quellec, M Abegg, M.N. Menke, U. Wolf-Schnurrbusch, J. Kowal, J. Blatz, O. La Schiazza, A.B. Leichtle, S. Wolf, M.S. Zinkernagel, Quantitative analysis of fluorescence lifetime measurements of the macula using the fluorescence lifetime imaging ophthalmoscope in healthy subjects. IOVS. **55**, 2107–2113 (2014)

24. D.S. Elson, I. Munro, J. Requejo-Isidro, J. McGinty, C. Dunsby, N. Galletly, G.W. Stamp, M. A.A. Neil, M.J. Lever, P.A. Kellett, A. Dymoke-Bradshaw, J. Hares, P.M.W. French, Real-time time-domain fluorescence lifetime imaging including single-shot acquisition with a segmented optical image intensifier. New J. Phys. **6**, 180 (2004)

25. N.P. Galletly, J. McGinty, C. Dunsby, F. Teixeira, J. Requejo-Isidro, I. Munro, D.S. Elson, M. A. Neil, A.C. Chu, P.M. French, G.W. Stamp, Fluorescence lifetime imaging distinguishes basal cell carcinoma from surrounding uninvolved skin. Br. J Dermatol. **159**, 152–161 (2008)

26. D.M. Grant, J. McGinty, E.J. McGhee, T.D. Bunney, D.M. Owen, C.B. Talbot, W. Zhang, S. Kumar, I. Munro, P.M. Lanigan, G.T. Kennedy, C. Dunsby, A.I. Magee, P. Courtney, M. Katan, M.A.A. Neil, P.M.W. French, High speed optically sectioned fluorescence lifetime imaging permits study of live cell signaling events. Opt. Express **15**, 15656–15673 (2007)

27. C.J. Gulledge, M.W. Dewhirst, Tumor oxygenation: a matter of supply and demand. Anticancer Res. **16**, 741–749 (1996)

28. H.W. Guo, C.T. Chen, Y.H. Wei, O.K. Lee, V. Gukassyan, F.J. Kao, H.W. Wang, Reduced nicotinamide adenine dinucleotide fluorescence lifetime separates human mesenchymal stem cells from differentiated progenies. J. Biomed. Opt. **13**, 050505 (2008)

29. S. Huang, A.A. Heikal, W.W. Webb, Two-photon fluorescence spectroscopy and microscopy of NAD(P)H and flavoprotein. Biophys. J. **82**, 2811–2825 (2002)

30. J.M. Jabbour, S. Cheng, B.H. Malik, R. Cuenca, J.A. Jo, J. Wright, Y.S. Cheng, K.C. Maitland, Fluorescence lifetime imaging and reflectance confocal microscopy for multiscale imaging of oral precancer. J. Biomed. Opt. **18**, 046012 (2013)

31. S.R. Kantelhardt, J. Leppert, J. Krajewski, N. Petkus, E. Reusche, V.M. Tronnier, G. Hüttmann, A. Giese, Imaging of brain and brain tumor specimens by time-resolved multiphoton excitation microscopy ex vivo. Neuro-Onkology **95**, 103–112 (2007)

32. V. Katsoulidou, A. Bergmann, W. Becker, How fast can TCSPC FLIM be made? Proc. SPIE **6771**, 67710B-1–67710B-7 (2007)

33. G.T. Kennedy, H.B. Manning, D.S. Elson, M.A.A. Neil, G.W. Stamp, B. Viellerobe, F. Lacombe, C. Dunsby, P.M.W. French, A fluorescence lifetime imaging scanning confocal endomicroscope. J. Biophoton. **3**, 103–107 (2010)

34. M. Köllner, J. Wolfrum, How many photons are necessary for fluorescence-lifetime measurements? Phys. Chem. Lett. **200**, 199–204 (1992)

35. K. König, I. Riemann, High-resolution multiphoton tomography of human skin with subcellular spatial resolution and picosecond time resolution. J. Biomed. Opt. **8**, 432–439 (2003)

36. K. König, Clinical multiphoton tomography. J. Biophoton. **1**, 13–23 (2008)

37. K. König, A. Uchugonova, in *FLIM Microscopy in Biology and Medicine*, ed. by A. Periasamy, R.M. Clegg. Multiphoton Fluorescence Lifetime Imaging at the Dawn of Clinical Application, (CRC Press, Boca Raton, 2009)

38. K. König, A. Uchugonova, E. Gorjup, Multiphoton fluorescence lifetime imaging of 3d-stem cell spheroids during differentiation. Microsc. Res. Techn. **74**, 9–17 (2011)

39. S. Khoon Teh, W. Zheng, S. Li, D. Li, Y. Zeng, Y. Yang, J.Y. Qu, Multimodal nonlinear optical microscopy improves the accuracy of early diagnosis of squamous intraepithelial neoplasia. J. Biomed. Opt. **18**(3) 036001-1 to -11 (2013)

40. J.R. Lakowicz, H. Szmacinski, K. Nowaczyk, M.L. Johnson, Fluorescence lifetime imaging of free and protein-bound NADH. Proc. Natl. Acad. Sci. **89**, 1271–1275 (1992)

41. J. Lakowicz, *Principles of Fluorescence Spectroscopy*, 3rd edn. (Springer, Berlin, 2006)

42. J. Leppert, J. Krajewski, S.R. Kantelhardt, S. Schlaffer, N. Petkus, E. Reusche, G. Hüttmann, A. Giese, Multiphoton excitation of autofluorescence for microscopy of glioma tissue. Neurosurgery **58**, 759–767 (2006)

43. E.Y. Lin, J.G. Jones, P. Li, L. Zhu, K.D. Whitney, W.J. Muller, J.W. Pollard, Progression to malignancy in the polyoma middle T oncoprotein mouse breast cancer model provides a reliable model for human diseases. Am. J. Pathol. **163**, 2113–2126 (2003)

44. J. McGinty, N.P. Galletly, C. Dunsby, I. Munro, D.S. Elson, J. Requejo-Isidro, P. Cohen, R. Ahmad, A. Forsyth, A.V. Thillainayagam, M.A. Neil, P.M. French, G.W. Stamp, Wide-field fluorescence lifetime imaging of cancer. Biomed. Opt. Express **1**, 627–640 (2010)

45. J.D. Meier, H. Xie, Y. Sun, Y. Sun, N. Hatami, B. Poirier, L. Marcu, D.G. Farwell, Time-resolved laser-induced fluorescence spectroscopy as a diagnostic instrument in head and neck carcinoma, Otolaryngology–head and neck surgery. Am. J. Otolaryngol Head Neck Surg. **142**, 838–844 (2010)

46. N. Ramanujam, Fluorescence spectroscopy of neoplastic and non-neoplastic tissues. Neoplasia **2**, 89–117 (2000)

47. J. Requejo-Isidro, J. McGinty, I. Munro, D.S. Elson, N.P. Galletly, M.J. Lever, M.A.A. Neil, G.W.H. Stamp, P.M.W. French, P.A. Kellett, J.D. Hares, A.K.L. Dymoke-Bradshaw, High-speed wide-field time-gated endoscopic fluorescence-lifetime imaging. Opt. Lett. **29**, 2249–2251 (2004)

48. A. Rück, Ch. Hülshoff, I. Kinzler, W. Becker, R. Steiner, SLIM: a new method for molecular imaging. Microsc. Res. Technol. **70**, 403–409 (2007)

49. B.A. Ruggeri, F. Camp, S. Miknyoczki, Animal models of disease: pre-clinical animal models of cancer and their applications and utility in drug discovery. Biochem. Pharmacol. **87**, 150–161 (2014)

50. W.Y. Sanchez, T.W. Prow, W.H. Sanchez, J.E.Grice, M.S. Roberts, Analysis of the metabolic deterioration of ex-vivo skin, from ischemic necrosis, through the imaging of intracellular NAD(P)H by multiphoton tomography and fluorescence lifetime imaging microscopy (MPT-FLIM). J. Biomed. Opt. **15**(4), 046008 (2010)

51. H. Schneckenburger, M. Wagner, P. Weber, W.S. Strauss, R. Sailer, Autofluorescence lifetime imaging of cultivated cells using a UV picosecond laser diode. J. Fluoresc. **14**, 649–654 (2004)

52. D. Schweitzer, S. Schenke, M. Hammer, F. Schweitzer, S. Jentsch, E. Dirckner, W. Becker, Towards metabolic mapping of the human retina. Microsc. Res. Technol. **70**, 403–409 (2007)

53. D. Schweitzer, in *Fundus Autofluorescence*, ed. by N. Lois, J.V. Forrester. Quantifying Fundus Autofluorescence. (Wolters Kluwer, Lippincott Willams & Wilkins, 2009)

54. D. Schweitzer, in *Metabolic Mapping* ed. by F.G. Holz, R.F. Spaide. Medical Retina Essential in Opthalmology, (Springer, Berlin, 2010)

55. A.T. Shah, M. Demory Beckler, A.J. Walsh, W.P. Jones, P.R. Pohlmann, M.C. Skala, Optical metabolic imaging of treatment response in human head and neck squamous cell carcinoma. PLoS ONE. **9**, e90746 (2014)

56. M.C. Skala, K.M. Riching, D.K. Bird, A. Gendron-Fitzpatrick, J. Eickhoff, K.W. Eliceiri, P. J. Keely, N. Ramanujam, In vivo multiphoton fluorescence lifetime imaging of protein-bound and free nicotinamide adenine dinucleotide in normal and precancerous epithelia. J. Biomed. Opt. **12**, 024014 (2007)

57. M.C. Skala, K.M. Riching, A. Gendron-Fitzpatrick, J. Eickhoff, K.W. Eliceiri, J.G. White, N. Ramanujam, In vivo multiphoton microscopy of NADH and FAD redox states, fluorescence lifetimes, and cellular morphology in precancerous epithelia. Proc. Natl. Acad. Sci. U.S.A. **104**, 19494–19499 (2007)

58. Y. Sun, R. Liu, D.S. Elson, C.W. Hollars, J.A. Jo, J. Park, Y. Sun, L. Marcu, Simultaneous time- and wavelength-resolved fluorescence spectroscopy for near real-time tissue diagnosis. Opt. Lett. **33**, 630–632 (2008)

59. A.J. Walsh, K.M. Poole, C.L. Duvall, M.C. Skala, Ex vivo optical metabolic measurements from cultured tissue reflect in vivo tissue status. J. Biomed. Opt. **17**, 116015 (2012)

60. A.J. Walsh, R.S. Cook, H.C. Manning, D.J. Hicks, A. Lafontant, C.L. Arteaga, M.C. Skala, Optical metabolic imaging identifies glycolytic levels, subtypes, and early-treatment response in breast cancer. Cancer Res. **73**, 6164–6174 (2013)

61. H.W. Wang, V. Gukassyan, C.T. Chen, Y.H. Wei, H.W. Guo, J.S. Yu, F.J. Kao, Differentiation of apoptosis from necrosis by dynamic changes of reduced nicotinamide adenine dinucleotide fluorescence lifetime in live cells. J. Biomed. Opt. **13**, 054011 (2008)

62. W.H. Yong, P.V. Butte, B.K. Pikul, J.A. Jo, Q. Fang, T. Papaioannou, K. Black, L. Marcu, Distinction of brain tissue, low grade and high grade glioma with time-resolved fluorescence spectroscopy. Frontiers in bioscience : a journal and virtual library **11**, 1255–1263 (2006)

63. Y. Zeng, B. Yan, Q. Sun, S. Khoon Teh, W. Zhang, Z. Wen, Jianan Y. Qu, Label-free in vivo imaging of human leukocytes using two-photon excited endogenous fluorescence. J. Biomed. Opt. **18**(4), 040103-1–040103-3 (2013)

Chapter 15
Fluorescence Lifetime Imaging of the Skin

Washington Y. Sanchez, Michael Pastore, Isha N. Haridass,
Karsten König, Wolfgang Becker and Michael S. Roberts

Abstract Skin research and clinical dermatological diagnoses have traditionally relied on morphological assessment obtained by various imaging techniques. With the advent of various new technologies, including polarised, confocal, multiphoton, infrared, photoacoustic, sonographic and magnetic resonance techniques, there has been a major step forward in non-invasive intravital imaging—the measurement of alterations or aberrations in the skin below the surface, in vivo. An even greater understanding of the processes associated with morphological changes and/or as a result of exogenous/endogenous chemical changes in the skin are now also possible by the use of ancillary detection techniques such as reflectance, Raman, CARS, interferometric (optical coherence tomography), and fluorescence lifetime imaging (FLIM). In this chapter, we examine the developments in our understanding of

W.Y. Sanchez (✉) · I.N. Haridass · M.S. Roberts
Therapeutics Research Centre, Princess Alexandra Hospital, Brisbane,
Woolloongabba, QLD 4102, Australia
e-mail: w.sanchez1@uq.edu.au

I.N. Haridass
e-mail: isha.haridass@gmail.com

M.S. Roberts
e-mail: m.roberts@uq.edu.au

M. Pastore
Division of Health Sciences, University of South Australia, School of Pharmacy
and Medical Science, Adelaide, SA 5000, Australia
e-mail: michael.pastore@mymail.unisa.edu.au

K. König
Department of Biophotonics and Laser Technology, Saarland University,
66123 Saarbruecken, Germany
e-mail: k.koenig@blt.uni-saarland.de

W. Becker
Becker & Hickl GmbH, Berlin, Germany
e-mail: becker@becker-hickl.de

© Springer International Publishing Switzerland 2015
W. Becker (ed.), *Advanced Time-Correlated Single Photon Counting Applications*,
Springer Series in Chemical Physics 111, DOI 10.1007/978-3-319-14929-5_15

FLIM as applied to animal and human skin. We also consider the application of FLIM in using skin autofluorescence to define the metabolic and pathological state of the skin, quantifying transdermal drug delivery and the effect of cosmetics on the skin. Finally, we review the potential application of FLIM in dermatology, especially in the diagnosis of various skin conditions, including skin cancer.

15.1 Introduction

The skin is the largest organ in the body and one that protects us from our external environment. At the same time, skin facilitates the homeostasis of the body by controlling moisture content, temperature etc. as well as being a key organ in our surveillance of the environment and interaction with it through touch and feel. However, exposure to sunlight, chemicals, physical and heat stress as well as the process of aging can modify and impair this organ. Furthermore, this organ is often a window to the inner body in that many internal diseases are manifested by changes in the skin. Whilst a key goal in dermatology is to be able to diagnose and treat disorders, handheld dermoscopy is the primary means of non-invasively imaging patient skin to assist in diagnosis. Mechanistic studies have largely been defined by in vitro cell culture experiments from skin-derived cells, Franz cell studies to study cutaneous drug permeation, and histologically prepared biopsies for clinical diagnostics. Cosmetology also has a key focus on protecting and maintaining skin health by the use of appropriate cosmetics and other procedures. In pharmacology and pharmaceutical sciences, we need to know how compounds distribute in and respond in the different regions of the skin over time. In toxicology, we need to know if and when exposure may be harmful.

In order to address these questions, we need to know what occurs below the skin barrier. To visualise and quantify the internal components of skin, along with any exogenously applied compounds, the tissue is processed in a manner that typically distorts its structure. In particular, the main barrier to solute movement into and through the skin, the outermost stratum corneum, can flake-off during the fixation process in H&E stains (Fig. 15.1a) and in electron microscopy (Fig. 15.1b). Embedding skin in optical cutting temperature compound for cryosectioning preserve the tissue morphology if adequately hydrated, and staining with an exogenous fluorescent dye (e.g. acriflavine) identifies viable cells in the epidermis and dermis (Fig. 15.1c). However, this approach is not appropriate for measuring topical penetration of water soluble dyes, as it leads to artificial staining of the section by solubilising the dye present on the skin surface. While appropriate for nanoparticle penetration studies, cryosectioning requires destructive tissue processing and importantly eliminates viable epidermal tissue autofluorescence and thus a source of critical metabolic data that can be extracted via FLIM [108]. Non-invasive imaging preserves all layers of the skin intact and compact (including hexagonal structure corneocytes in the stratum corneum), has well defined furrows and dermatoglyphs,

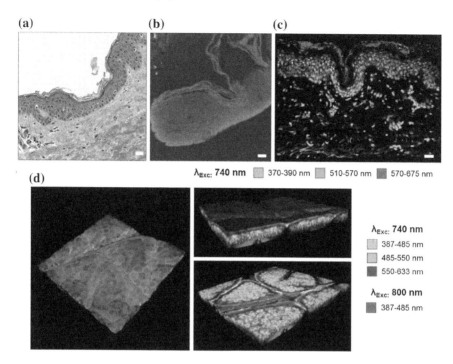

Fig. 15.1 Comparison of cross-sectional imaging of human skin. **a** Light microscopy of H&E stained of excised human skin. *White scale bar* represents a length of 20 μm. **b** Scanning electron microscopy of a section of excised human skin. *White scale bar* represents a length of 10 μm. **c** Multispectral multiphoton imaging of cryosectioned excised human skin (740 nm excitation) after the topical application of ZnO-NP (*blue-purple*), and post-stained with acriflavine (*green*). *White scale bar* represents a length of 20 μm. **d** Freshly excised human skin imaged non-invasively with multiphoton microscopy. The pseudo coloured optical sections are coloured according to the skin autofluorescence across multiple spectral channels at 740–800 nm excitation

and preserves the all of the natural autofluorescence of both excised and in vivo skin (Fig. 15.1d).

Non-invasive imagine along, however, is usually limited to the characterisation of skin morphology. This can also be achieved by a range of complementary technologies that reflect the responses occurring when light strikes an object it has been focused on. These include reflection of the light (confocal reflectance), fluorescence (wide angle, confocal and multiphoton fluorescence), sound (photoacoustic), scattering (Raman, optical coherence tomography). However, in the special case of fluorescence imaging, much more can be discerned about the biochemistry of the object and the presence of exogenous solutes by also analysing the spectroscopic emissions arising after light interacts with the object (Fig. 15.2).

In general, spectroscopic emissions associated with confocal and multiphoton microscopy can be examined by three methods [104, 105]. The first recognises that the emission spectra of various solutes differ in how they vary with wavelength. The

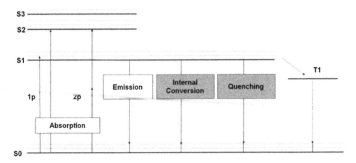

Fig. 15.2 Jablonski energy diagram demonstrating the various decay pathways of an excited molecule. From the ground state (*S0*), a molecule absorbs energy from an exogenous source to be excited to a higher energy level (*S1–S3*). Excited molecules within the higher energy level states (*S2–S3*) rapidly decay to the *S1* state from where it may further decay to the ground state via fluorescence emission of a photon, or via other pathways such as internal conversion or quenching. Adapted from [105]

quantification of and comparison of emissions for solutes for various wavelength bands is called *spectral imaging* and this technique is widely used to describe skin [94]. The second method recognises that each fluorescent compound has its own specific fluorescence lifetime profile that is also dependent on external factors such as the environment that the compound is in, and whether the compound is directly interacting with other molecules in the biological milieu. We can base our imaging on the lifetime distribution for a specific region in a tissue of interest, a process based on time-correlated single photon counting (TCSPC) and called fluorescence lifetime imaging (FLIM) [92]. Finally, we can assess solutes in terms of the way that they respond to polarised light, a process referred to as anisotropy. In principle, these modalities can be combined to yield even greater specificity and sensitivity of analysis for particular components and processes. Table 15.1 summarises the first two of these emission characteristics for the common endogenous components in the skin.

In this overview, we limit our discussion to the FLIM of the skin, without and with the other modalities. We examine the use of skin FLIM for a range of applications, including defining the redox state of the skin and the penetration of solutes and nanomaterials into the skin. As the fluorescence lifetime of fluorophores only alters with physio-chemical changes in the molecule itself, and not with intensity, FLIM can be used to assess both structural and functional changes non-invasively within skin. Consequently, it provides an additional level of data that can be extracted from non-invasive confocal and multiphoton microscopy for both skin research and clinical imaging. It can also complement in vitro cell culture experiments of skin-derived cells, Franz cell studies to study cutaneous drug permeation, and histologically prepared biopsies for clinical diagnostics. It has the potential to take the approach of non-invasive imaging of patients' skin using handheld dermoscopy for diagnostic analysis in clinical dermatology to another level. With the increasing availability of portable, affordable and cost-effective confocal and

Table 15.1 Fluorescence lifetime properties of endogenous fluorophores in skin

Autofluorescent fluorophore	Excitation (nm)	Emission (nm)	State	Lifetime (ns)	References
Stratum corneum					
Keratin	750–900	400–600		1.4	[11, 30, 100, 134]
Viable epidermis					
NAD(P)H	720–780 (2P)	400–550	Free	0.2–0.39, 1.14	[10, 11, 37, 73, 90, 124, 125, 134]
			Bound	2.2–2.5	
FAD	720–900 (2P)	480–650	Free	40 (monomer), 130 (dimer), 2.0–2.8, 5.2	[11, 22, 54, 62, 87, 115, 134]
			Bound	0.08, 0.7, 1.7	
Lipofuscin	366 (1P, max)	500–605 nm		0.3, 1.2, 4.8	[38, 55]
Melanin	720–880 (2P)	510–600		0.2, 0.5–1.9, 7.9	[11, 23, 62, 134]
Dermis					
Elastin	700–740	570–590		0.2–0.4, 2.3–2.5	[11, 62, 134, 136]
Collagen	800 (SHG; 2P)	400 (SHG)		0	[11, 62, 134]

multiphoton microscopes, a combination of these modalities with FLIM can enable improved specificity and sensitivity in non-invasive high resolution imaging that, at this time, is still in a phase of advanced development and has yet to become routine clinical practice.

15.1.1 The Skin, a Vital Organ

Figure 15.3 shows a diagrammatic illustration of the skin. It is the largest organ of the body at about 16 % of total body weight, has a large surface area of 1.5–2.0 m², is metabolically active, synthesizes vitamin D3 and assists in minimising thermal, physical, and mechanical injury, water loss and UV damage, as well as regulating body temperature, facilitating sensation and providing active immunological surveillance. It is a multilayer structure consisting of an insulation and energy storage layer, the subcutaneous fat; a support layer that provides cushioning, nutrition, innervation, surveillance, feel and excretion, the dermis (1.1–2.3 mm) and the barrier layer, the epidermis (0.05–1.50 mm thick). These layers are punctuated by appendages allowing the excretion of sweat and sebum, the latter, together with hair, providing an additional level of protection. The epidermis, in turn, consists of four sub-layers which are in a continual state of regeneration to produce a dead desquamating layer, the stratum corneum. This layer is also undulating, defined as

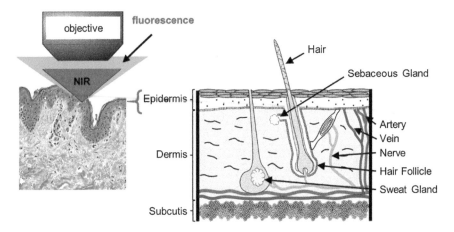

Fig. 15.3 Schematic diagram of skin structure

rete, to provide it with flexibility. The key structural cells of the epidermis in the remaining three layers are the stem, basal, spinous and stratum granulosum cells, which becoming increasingly flattened as the cells evolve into the stratum corneum. Associated with these cells are two protective skin cells, the melanocytes and the Langerhans cells. The melanocytes are responsible for the production of melanin and skin pigmentation, whereas the Langerhans cells are the skin's main dendritic cell with the primary goals of surveillance for foreign objects and communication with the immune system.

15.1.2 Challenges in Assessing Skin Health

One of the first laser scanning platforms developed was the confocal laser scanning microscope (CLSM). CLSM conveyed several advantages over traditional microscopy including significantly improved spatial resolution and the ability to obtain optical sections of cells and tissues [33, 102, 123]. CLSM could also be used in either fluorescence or reflectance modes. The development of multiphoton microscopy was a revolutionary improvement in laser scanning microscopes, conferring several advantages over CLSM including: narrow focal volume, eliminating the need for a pinhole to filter out non-specific fluorescence from another focal depth, improved resolution, deeper tissue imaging, and less photodamage due to the use of lower energy near-infrared photons to excite the tissue. Fundamentally, both CLSM and MPM only show the fluorescence intensity of a sample within a defined spectral range. While this data represents the relative quantity of a mix of fluorophores within the tissue, it cannot distinguish between fluorophores with an overlapping emission spectra or their physical and chemical states. Fortunately,

TCSPC FLIM has enabled an additional analytical layer on top of multi-spectral laser scanning microscopy to resolve the various physio-chemical states of fluorophores with overlapping emission spectra.

15.1.3 Non-invasive Imaging Below the Skin Surface— Limitations and Opportunities

Non-invasive microscopy of endogenous fluorophores in skin can be used for a variety of tasks including metabolic imaging of the viable epidermis, and visualisation of melanin distribution. However, there are some limitations to using this technique in assessing the penetration of actives into the skin and as a diagnostic tool. A non-negligible prerequisite is that the administered molecule is fluorescent. This greatly reduces the number of actives that can be studied with this technique. Furthermore, fluorescent actives must have an excitation wavelength greater than 350 nm to be able to be visualized by MPM and confocal microscopy.

Non-invasive multiphoton microscopy is able to provide sub-micrometer resolution of endogenous fluorophores to a depth of about 200 µm [26, 81]. It is important to note that as deeper strata are measured, the fluorescence intensity detected may decrease as light is loss due to scattering. When imaging live subjects, details that may seem minor, such as the quality of the surface being imaged and the subject's movements, may significantly reduce the quality images obtained.

The photons responsible for exciting the fluorophores within a sample can also cause their degradation. Photo-bleaching is one of the factors that may directly influence the data obtained through non-invasive imaging and FLIM, as it influences the intensity and lifetime of the photo-damaged fluorophore. In addition, photo-bleaching may result in the formation of free radicals that may affect the metabolic state of the cells. This, along with overlap of the spectral signals of different fluorophores, and fluorescence quenching, are some of the challenges that are presented with non-invasive and FLIM microscopy of the skin [105].

15.1.4 Current Range of Technologies to Define Skin Morphology

The several instruments used for imaging below the skin use radiation such as X rays, light and sound. In general, low wavelength modalities typically seen in confocal microscopy enable high resolution, but damage the skin and have poor penetration depth. There are several long wavelength modalities (i.e. near-infrared) that limit damage to the skin and have significantly greater tissue penetration depth. Depending on the modality, these techniques can vary in resolution and offer different ways to analyse skin structure, function and percutaneous penetration.

Wide-field fluorescence microscopy provides a two-dimensional imaging resolution of up to 50 nm, but has several drawbacks. It delivers poor image contrast and lacks any optical sectioning capability. Optical sectioning can be added at the expense of photon efficiency [21, 47]. This increases the rate of photo-bleaching, and as a result, only highly photo-stable fluorophores can be imaged with this technique.

In comparison, multiphoton microscopy allows imaging depths of up to 200 μm. The chances of photo-bleaching are greatly reduced with this technique, which confers the ability to image live cells and tissue with subcellular resolution. This also allows the precise visualization of the corneocytes of the stratum corneum on the surface as well as the collagen and elastin fibres of the dermis [82]. Structures below the skin surface can be observed at video rates using CLSM and the z-stacks obtained can be used to construct a three-dimensional representation of the skin. However, CLSM has a poor penetration depth compared to multiphoton microscopy (85 μm).

There are several other detection systems that can be used to analyse skin structure at different depths and resolutions. Optical coherence tomography (OCT) offers greater imaging depths (1.0–1.5 mm)compared to multiphoton microscopy and CLSM, although signal intensity decreases as imaging depth increases. As a result, only superficial layers of the skin, such as the stratum corneum can be visualized clearly, as subcellular resolution is lost in more profound layers [128]. One of the major limitations of OCT is the strong non-specific background signal that is detected, which deters the ability to distinguish between background noise and viable signal [128]. Photoacoustic microscopy allows a greater imaging depth in addition to the unique ability of imaging skin vasculature [79]. When using these techniques, one has to consider that the subject movement during imaging will result in decreased image quality.

Raman microscopic imaging detects vibrational changes in the protein and lipid components of the skin, which can be interpreted as structural changes occurring in different skin strata [135]. This imaging method has a resolution of approximately 80 μm. Signal intensity decreases with tissue depth as elastic light scattering increases [16]. Automated confocal Raman microspectrometry can be used to determine water concentrations in hydrated and non-hydrated *stratum corneum*, which shows the this method has great possibilities [15].

15.2 Fluorescence Lifetime Imaging Microscopy (FLIM)

15.2.1 Metabolic Imaging of Skin

Autofluorescence FLIM analysis is predominantly used for assessing the metabolic state of tissue in response to a change in the microenvironment, or the addition of an exogenous agent. As a result, NAD(P)H and FAD are the primary targets in autofluorescent FLIM analysis. Metabolic changes in tissue can be measured as changes in the fluorescence lifetime of NAD(P)H or FAD, and/or as changes in the redox ratio or ratio of free and protein-bound components of NAD(P)H [107–109, 115, 116, 124], see also Chaps. 13 and 14 of this book.

15.2.2 Fluorescence Lifetime Changes of NAD(P)H and FAD in the Viable Epidermis

Fluorescence lifetime changes do not offer any descriptive knowledge about the underlying causes of these changes. However, certain lifetime changes are reproducibly associated with cellular events such as apoptotic cell death and oxidative stress [108, 126, 134], and can be used assess skin viability and toxicological effects of topically applied agents in excised tissue and in vivo.

Fluorescence lifetime changes associated with apoptosis have been well described in vitro. The ATP-kinase inhibitor staurosporine was used to induced apoptosis within cultured cells in vitro and was associated with an increase in the fluorescence lifetime of NAD(P)H [45, 126, 134], which occurred prior to apoptotic morphological changes and caspase 3 activity. These changes were specific to the induction of apoptosis and not with necrotic cell death initiated by hydrogen peroxide [126]. Using these observations as a reference point, changes in skin viability can be examined.

Assessing and maintaining the viability of skin explants is essential for successful skin grafting. Traditionally, viability assays are used to measure the activity of NAD(P)H-dependent enzymes to reduce tetrazolium-based dyes into formazan, resulting in a colour change [43]. However, this suffers from several drawbacks in terms of assay time, cost, and tissue destruction as a result of obtaining a painful biopsy.

FLIM of excised tissue, stored under various temperature and nutrient conditions, demonstrated that spatial viability can be measured non-invasively using multiphoton microscopy [108]. As expected, the viability of excised skin stored at room temperature and 37 °C deteriorated more rapidly than skin kept at 4 °C. Metabolic deterioration of excised skin was associated with an increase in the fluorescence lifetime of NAD(P)H [108]. In contrast, the fluorescence lifetime of skin kept at 4 °C showed a fluorescence lifetime comparable to freshly excised and in vivo human skin. Using this approach, muliphoton FLIM could potentially be used to routinely monitor skin graft viability and for the early detection of skin graft failure.

Non-invasive multiphoton FLIM microscopy has also been used to measure lifetime changes associated with skin wound healing [25]. Similarly to the viability studies described above, the wound formation was associated with an increase in the fluorescence lifetime of NAD(P)H, which decreased linearly with wound healing over time [25]. These changes are hypothesised to be due to an increase in the proportion of intercellular protein-bound NAD(P)H in the wounded skin.

15.2.3 Oxidative Stress

In addition to apoptosis, oxidative stress is associated with defined fluorescence lifetime changes that can be examined in skin. Ultraviolet (UV) A exposure was used to induce oxidative stress in Chinese hamster ovary cells in vitro, which resulted in a decrease in the fluorescence lifetime of NAD(P)H [60]. Oxidative stress, as a result from ischemia-reperfusion injury, in the rat liver is associated with a decrease in the fluorescence lifetime of NAD(P)H [120].

In the context of skin research and clinical assessment, exposure to UV radiation increases cellular oxidative stress [103]. This effect was investigated using in vivo human skin, comparing NAD(P)H and FAD fluorescence lifetime differences between solar exposed and solar protected forearm skin of various age groups. Similar to the in vitro studies, a significant decrease in the NAD(P)H and FAD lifetime of the viable epidermis in the dorsal forearm (solar exposed), relative to the volar forearm (solar protected), was observed [107].

15.2.4 Redox Ratio Using FLIM Parameters

The redox ratio is an indicator of the redox state and viability of the cell [129]. Traditionally, the redox ratio was calculated by measuring either: (1) changes in NAD(P)H intensity [52], or (2) the fluorescence intensity ratio of FAD to NAD(P)H [18], as the reduced form of FAD and oxidised form NAD(P)H are both non-fluorescent [49]. However, this approach has several drawbacks, namely spectral overlap with other endogenous fluorophores [134].

FLIM analysis of the ratio of free to protein-bound NAD(P)H is an alternative approach to measuring the redox state of the cell [45, 73, 109]. Moreover, a more sophisticated analysis of the ratio of free and protein-bound components of NAD(P)H and FAD can be measured due to the precision of multi-exponential component analysis by FLIM [54, 57, 73]. Indeed, in vitro studies demonstrate that the increase in the relative abundance of free NAD(P)H (α_1) relative to protein-bound NAD(P)H (α_2) correlates with the increasing confluency of breast epithelial and HeLa cells [10, 45]. This observation is believed to be associated with a gradual decrease of available nutrients and oxygen in the culture media, and thus an increase in the NADH/NAD$^+$ ratio. This observation can be reproduced with cultured cells in vitro by serum starvation and the addition of various electron transport chain inhibitors to disrupt oxidative phosphorylation [10]. Overall, the increase in free-to-protein-bound NAD(P)H ratio, and corresponding decrease in the average NAD(P)H lifetime, is associated with the inhibition of oxidative phosphorylation in vitro [10, 45]. Thus it is considered that the α_1/α_2 ratio of NAD(P)H is inversely proportional to the metabolic rate of the cells.

In the aforementioned wound healing model, the free-to-bound ratio of NAD(P)H was lower near the wound relative to normal skin during the early stages of wound healing [25]. This data suggests that at the site of wound healing at this time, oxidative phosphorylation may be the primary bioenergetic pathway used. Over time throughout the wound healing process, non-invasive FLIM was performed daily to determine healing-associated changes in the fluorescence lifetime and free-to-bound ratio of NAD(P)H. This data can be used clinically to assess tissue recovery from injury.

15.3 FLIM Instrumentation for Clinical Use

FLIM is used routinely for both skin research and clinical dermatological applications. While the former is well-established in the literature, clinical FLIM intravital microscopy is a developing field with significant potential. Intravital FLIM applications in dermatology include: diagnostic analysis of suspicious skin lesions [28, 113], assessing skin viability [40, 108], analysis of skin aging [107], and measuring the liposome-mediated delivery of agents and their spatial distribution [41]. Despite the powerful analytical capacity of FLIM, relatively few systems exist for intravital FLIM microscopy for either research or clinical purposes.

There is a wide variety of different FLIM techniques. The techniques differ in the electronic and optical principles to resolve the fluorescence signals spatially and temporally [2, 5]. These approaches differ in their acquisition time and photon counting efficiency (i.e. the number of photons needed to obtain a given lifetime accurately), resolution of multi-exponential decay functions, optical sectioning, and the fluorescence intensities they can be used. In particular, there is constant controversy whether a system for clinical use should use wide-field imaging or scanning. It is certainly correct that a wide-field system is able to achieve shorter acquisition time than a scanning system. Wide-field FLIM is therefore used for a number of diagnostic applications, such as analysis of biopsies of human skin cancerous lesions [39, 84]. Autofluorescence lifetime differences were demonstrated between normal and lesional tissue, validating the utility of FLIM for diagnostic analysis [39]. However, a wide field systems also has a number of disadvantages.

The first one is that it requires a camera-like image sensor. There is currently no such sensor that acquires picosecond time-resolved data in a large number of time channels of all pixels simultaneously. Therefore, the time resolution must be provided by time-gating: a gated pulse is shifted over the decay curve, acquiring a number of images for different time after the excitation. Of course, gating reduces the photon efficiency, i.e. the number of photons needed to obtain a given lifetime accurately. To maintain a reasonable photon efficiency, the decay data are often acquired in only a few time gates. This, however, reduces the capability of resolving multi-exponential decay profiles into their decay components [2, 5].

The most significant problem of wide-field imaging in comparison to scanning is that the images are far more impaired by scattering: a wide-field system detects scattered light from all pixels of the image, a scanning system only from the currently excited pixel. Scattering is not a major problem when cell cultures or thin tissue sections are used. It may also have little effect on the image contrast when large-area images are taken directly from the surface of biological tissue. However, when images are recorded at μm resolution from planes inside the tissue, the differences in image contrast are dramatic, see Fig. 2.1 in Chap. 2. The problem of scattering is enhanced by the fact that a broad wide-field system does not provide any kind of out-of-focus suppression and optical sectioning capability. In principle, out-of-focus signals can be removed from the data by structured illumination techniques [21, 47], but this further reduces the photon efficiency.

The only technique that combines laser scanning with near perfect photon efficiency and time resolution is TCSPC FLIM. TCSPC with laser scanning was already used by Bugiel, König and Wabnitz in 1989 [13]. These early implementations used classic TCSPC and slow scanning. The slow scan rate was a serious limitation which possibly led to the misconception that TCSPC FLIM needs extremely long acquisition times. The limitation in scan rate was overcome with the introduction of multi-dimensional TCSPC by Becker & Hickl, see Chaps. 1 and 2 of this book. In this chapter, we exclusively used optical principles based on fast laser scanning in combination with FLIM by multi-dimensional TCSPC.

15.3.1 Confocal Scanning with One-Photon Excitation

Confocal scanning systems scan the sample with a focused laser beam, and detect the fluorescence back through the optical scanner and through a pinhole in an optical plane conjugate with the focal plane in the sample. Light from above or below the focal plane and light scattered in the sample is not focused into the pinhole and thus significantly suppressed, see Chap. 2, Fig. 2.3. Confocal scanning systems can easily be combined with TCSPC FLIM: All that is needed is a pulsed laser for excitation and a detector with single-photon sensitivity and picosecond timing resolution for the individual photons.

A commercially available instrument of this type with the capability of TCSPC FLIM for intravital imaging is the VivaScope® (Caliber Imaging & Diagnostics Inc., formerly Lucid Inc., Rochester, NY, USA). In its basic configuration the VivaScope® is a reflectance confocal microscope (RCM) system developed for intravital imaging of human patients for dermatological diagnostics [46, 98, 99, 111].

The VivaScope® enables the operator to obtain optical sections of excised or in vivo skin [53, 75, 110]. Both animal and human skin is imaged with the VivaScope® predominantly by RCM, which uses the inherent variations in refractive indices of skin structures to provide spatial imaging of the tissue [88]. Besides its application in cosmetic (e.g. photo-aging, pigmentation), the VivaScope® has been extensively used for the non-invasive diagnosis of both melanocytic and non-melanocytic skin lesions, such as basal cell carcinoma and actinic keratosis [50].

Together with the reflectance mode, the VivaScope® is also used as a (stead-state) fluorescence microscope either by exploiting the autofluorescence properties of the skin, or with the use of exogenous fluorophores [118]. Both hydrophilic molecules (e.g. fluorescein) and hydrophobic molecules (e.g. curcumin) can be used as contrasting dyes to distinguish between cellular and tissue structures. Moreover, these compounds are used to study skin penetration pathways and skin penetration kinetics (i.e. dermatopharmacokinetics).

The VivaScope® microscope platform can be equipped with pulsed confocal lasers and a fibre output for fluorescence excitation and FLIM detection. A sim-plified optical diagram of the Vivascope with RCM and FLIM is shown in Fig. 15.4. The light source of the RCM system is a 785 nm diode laser. The laser beam passes the beam splitters, BS1, BS2, BS3, and is deflected by a fast-rotating polygon mirror (x direction) and a galvanometer mirror (y direction). The polygon mirror provides ultra-fast scanning with line times on the order of 100 µs. The scan lens projects the pivot point of the beam on the microscope lens. The microscope lens focuses the laser beam into the sample. It also collects and collimates the light returned from the sample. The returned beam is de-scanned by the scan mirrors, and focused into a pinhole in front of the reflectance detector. Important to the function of the reflection system are the λ/4 plate behind the microscope lens and the polarising beamsplitter in front of the reflection detector. These elements have the effect that reflected light with the same polarisation as the laser is suppressed. Light directly reflected at the surface is polarised, light scattered at tissue structures inside the sample is not. The image contrast for deep-tissue structures is thus substantially improved.

The Vivascope is upgraded for FLIM by adding an input for a ps diode laser and a confocal output to a FLIM detector. A dichroic beamsplitter, BS1 reflects the FLIM laser wavelengths (405–488 nm) and transmits the reflectance laser (785 nm). BS3 is a wideband (50/50 %) beamsplitter, it transmits 50 % of the returned light to the reflectance PMT and reflects 50 % to the FLIM fibre port. The fibre forms the pinhole for confocal detection, and transmits the fluorescence light to the FLIM

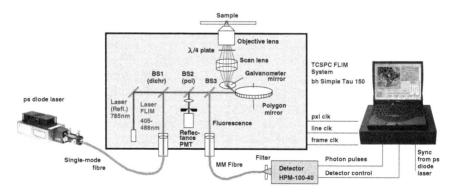

Fig. 15.4 Optical principle of the VivaScope® with TCSPC FLIM

detector. We use a Becker & Hickl HPM-100-40 hybrid GaAsP detector [4]. The FLIM system is a standard Becker & Hickl Simple Tau system [4]. The synchronisation with the scanning is obtained by scan clock pulses from the Vivascope scanner, pxl clk, line clk, and frame clk, the synchronisation with the laser pulses sequence by synchronisation pulses from the ps diode laser. The FLIM system acquires the FLIM data by building up a photon distributions over the coordinates of the scan and the times of the photons in the laser pulse period [4, 18]. The recording process is compatible with the ultra-fast scanning used in the Vivascope: the photons of a large number of frames are accumulated, and the acquisition is continued until the desired signal-to-noise ratio is achieved. Please see Chaps. 1 and 2 for details.

By combining RCM at NIR wavelengths with FLIM, the VivaScope® combines the recording high-contrast intensity imaging in deep-tissue layers with the recording of fluorescence-lifetime images. It is thus delivers both morphological and metabolic information. One drawback however, is that, FLIM cannot be obtained from the same depth as RCM images: The excitation wavelength of FLIM is shorter, so that the penetration depth is smaller. The combination of several imaging techniques in the same instrument also leads to a sub-optimal sensitivity of FLIM. As the instrument is designed now, a part of the excitation and fluorescence light is lost at the 50/50 beamsplitter, see Fig. 15.4. There is plenty of sensitivity reserved for FLIM with exogenous fluorophores but there is little sensitivity left when the instrument is used for autofluorescence FLIM.

Figure 15.5 contrasts the confocal reflectance and FLIM modes of a modified VivaScope® system. The data demonstrates that standard RCM yields little more than spatial information about the skin structure (Fig. 15.5b). When equipped with FLIM, the various physio-chemical components of autofluorescent and fluorescent compounds can be resolved with the FOV and imaging depth advantages that the VivaScope® provides (Fig. 15.5c, d).

15.3.2 Multiphoton Scanning with Non-descanned Detection

The quality of fluorescence images from deep tissue layers can substantially be improved by using two-photon excitation. The process of simultaneous absorption of two or more photons has been theoretically investigated 1931 by Göppert-Mayer [42], suggested for fluorescence excitation in laser scanning systems by Wilson and Sheppard [130] in 1984, and practically introduced by Denk and Strickler [26] in 1990. The use of multiphoton excitation in a laser scanning microscope has a number of advantages. First, the excitation wavelength for fluorophores with absorption bands in the visible range is in the NIR. Scattering and absorption coefficients in the NIR are lower than in the visible range. Therefore, a larger penetration depth is achieved. Second, the excitation happens predominantly in the focus of the laser. Multiphoton excitation therefore offers inherent depth resolution. No pinhole is required to suppress out-of-focus light. The fluorescence signal can

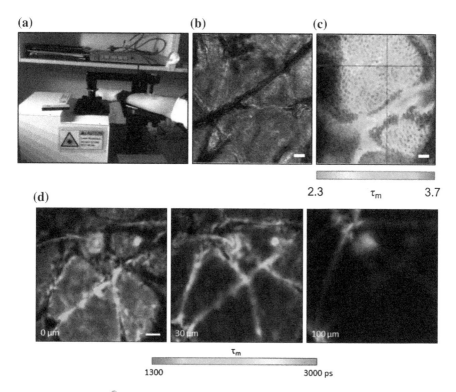

Fig. 15.5 VivaScope® reflectance confocal and fluorescence lifetime imaging microscopy of in vivo human skin. **a** The VivaScope® is equipped with a flexible arm with optics compatible for fluorescence lifetime imaging. **b** Reflectance confocal image of the stratum corneum surface of in vivo human skin. **c** Pseudo-coloured fluorescence lifetime image (according to the average weighted fluorescence lifetime, τm) of the stratum corneum of in vivo human skin after topical application of an acriflavine (1 mg/mL) solution, using an excitation wavelength of 488 nm. The *white scale bar* represents 40 μm. **d** FLIM images of acriflavine solution (1 mg/mL in ethanol) penetration in the hair follicle and skin furrows of in vivo human skin after topical application, using an excitation wavelength of 488 nm. The *yellow scale bar* represents 50 μm

therefore be diverted directly behind the microscope lens, and transferred to a large-area detector. The principle is called non-descanned detection (NDD). In comparison to confocal detection, the advantage of NDD is that photons scattered on the way out of the sample reach the detectors and contribute to the buildup of the image. The absence of excitation outside the focal plane has the additional advantage of reduced photobleaching and photodamage in the sample volume above and below the image plane.

The downside of multiphoton imaging is that it requires a femtosecond laser to obtain noticeable excitation efficiency. The need of a femtosecond laser makes a multiphoton system relatively expensive. One may also raise concerns about the use of femtosecond pulses with high peak power in clinical applications. It has, however, been shown that two-photon excitation is safe to human cells and tissue [34, 35, 60, 61, 65].

(a)   (b)

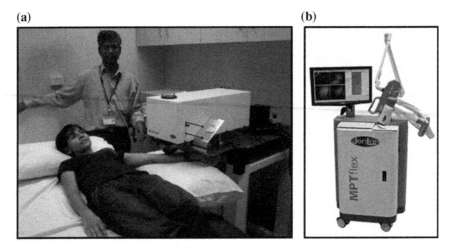

Fig. 15.6 DermaInspect® multiphoton microscopes first generation fixed (**a**), second and third generation mobile (**b**) systems. Fluorescence lifetime imaging modules are located internally and operate across multiple spectral channels. Adapted from [67]

Two-photon excited autofluorescence fluorescence imaging of human skin has been developed by Gratton, König and Masters [60, 62, 80–82], and commercially introduced by Jenlab GmbH, Germany [64]. The first system was the DermaInspect®. First generation DermaInspect® microscopes were fixed units, requiring human volunteers or patients to optimally position the scanning area on their body to the fixed objective head, see Fig. 15.6a. The next generation MPTflex® and MPTflex-CARS®systems [12] (Fig. 15.6b) have the microscope/scanner assembly mounted on a 360° flexible arm, allowing the volunteer/patient to remain in a comfortable position for intravital imaging [67]. With these improvements to mobility and usability, TCSPC FLIM remains an optional module to the MPTflex® platform.

From the beginning, DermaInspect® contained a FLIM option using Becker & Hickl TCSPC FLIM technique. The basic optical principle of the FLIM part of the DermaInspect® is shown in Fig. 15.7. The laser beam is scanned by galvanometer mirrors and projected into the plane of the microscope lens by the scan lens. The microscope lens focuses the laser beam into the sample. The fluorescence light is collected by the microscope lens, and separated from the excitation beam by a dichroic mirror. It is further split into several spectral components, and detected by several (up to four) detectors. The signals of the detectors were originally detected by a single Becker & Hickl SPC-830 TCSPC FLIM module via a router [4, 18], please see Chap. 1, Sect. 1.4.1. Later systems use parallel detection by several SPC-150 modules (Chap. 1, Sect. 1.4.5.1), as indicated in Fig. 15.7. A typical FLIM image recorded by this instrument is shown in Fig. 15.8.

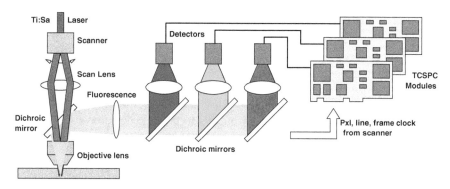

Fig. 15.7 Optical Principle of the FLIM part of the DermaInspect®

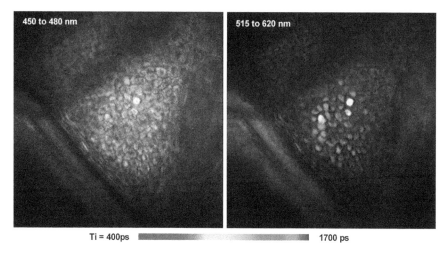

Fig. 15.8 Stratum granulosum of human forearm. Excitation 750 nm, detection from 450 to 480 nm (*left*) and 515 to 650 nm (*right*). Amplitude weighted lifetime of double-exponential decay

The DermaInspect® and MPTflex® systems were designed especially for dermatological research and clinical analysis and have been used for multiple intravital FLIM studies in human skin for drug penetration and toxicology [74, 76, 104], photo-ageing assessment [58, 107], characterisation and diagnostic differentiation between normal and cancerous lesions [28, 114]. The multiphoton tomograph MPTflex-CARS is the certified multiphoton tomograph of the third generation which provides optical biopsies with chemical information. The systems have delivered data for an impressive number of publications, see [12, 28, 41, 44, 56, 63, 66, 68–71, 74, 76, 78, 79, 89, 96, 101, 104, 105, 107, 108, 119, 122, 127].

15.4 FLIM Data Analysis

Below we give only a brief description, focusing especially on the features important to multiphoton FLIM data of human skin. TCSPC FLIM data are arrays of pixels each of which contains a (usually large) number of time channels. The time channels contain photon numbers for consecutive times after the excitation pulse. The data can be analysed either directly in the time domain, or transformed into the frequency domain and analysed by in the frequency space.

15.4.1 Time-Domain Analysis

To obtain fluorescence decay parameters from TCSPC an iterative convolution process is used. The function of a suitable decay model is convoluted with the instrument response function (IRF). The IRF is the response of the detection system to the excitation pulse itself. It is either measured or calculated from the fluorescence decay data. The fit procedure then optimises the model parameters until the best fit to the photon numbers in the time channels is achieved. The principle is in use for analysis of single fluorescence decay curves for many years [92]. For FLIM, the fit procedure has to be repeated for all pixels of the image [4].

The simplest decay model is a single exponential function. It is described by a single decay time. In most cases, however, the decay profiles have to be modelled by sums of two or three exponential functions, see (15.1) and (15.2). The model function, (15.1), is described by several decay times, τ_i, and amplitude coefficients, a_i:

$$F(t) = \sum_{i=1}^{n} a_i e^{-t/\tau_i}, n \geq 1 \tag{15.1}$$

where

$$\sum_{n=1}^{i} a_i = 1 \tag{15.2}$$

The final FLIM image is obtained by assigning the brightness to the total photon number in the pixel, and the colour to a selected decay parameter. This can be the lifetime of a single-exponential decay, an amplitude- or intensity-weighted average of the lifetimes in a multi-exponential decay, or a lifetime or amplitude ratio, see Fig. 15.9a.

Typical applications of using the decay parameters are images of the ratio of bound/unbound NADH (α_1/α_2) and FRET images (α_1/α_2 for amount of interacting donor, τ_1/τ_2 component for donor-acceptor distance).

(a)

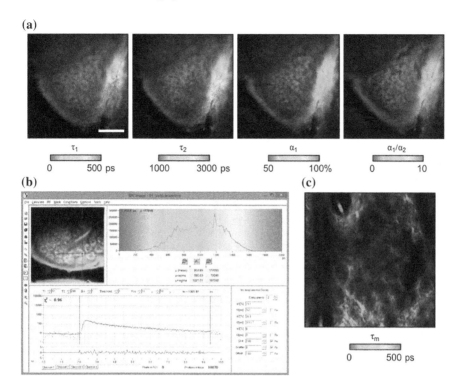

Fig. 15.9 TCSPC pseudo-coloured multiphoton FLIM image of in vivo (**a**) and (**b**) excised human skin, and the dermis of the excised mouse ear (**c**) at an excitation wavelength of 740 nm. **b** The fluorescence lifetime decay curve and lifetime parameters of excised human skin. **c** The FLIM pseudo-coloured image of the dermis in the excised mouse ear displays the ultra-fast lifetime component corresponding to collagen SHG (*blue*), which has a lifetime approaching zero, and the longer wavelength autofluorescence lifetime of keratin hair follicles (*red*)

For many applications it is sufficient to calculate a mean fluorescence lifetime, τ_m, from the parameters of the multi-exponential decay [116]. The mean lifetime is an average of the decay times of the decay components, weighted by their amplitudes:

$$\tau_m = \frac{\sum_{i=1}^{N} \alpha_1 \tau_1}{\sum_{i=1}^{N} \alpha_1} \tag{15.3}$$

A pseudo-coloured image of the mean (or amplitude-weighted) lifetime of a double-exponential decay model is shown in Fig. 15.9b. It should be noted that the mean lifetime is *not* identical with the intensity-weighted (average) lifetime, and with the lifetime of a single-exponential fit of the decay function [4].

It is also possible to extract SHG (second-harmonic generation) components from the FLIM data via their infinitely short lifetime or via time-gating the intensity data in the lifetime display, see Fig. 15.9c.

15.4.2 Frequency Domain (Phasor) Analysis

Frequency-Domain data contain two numbers in each pixel: a phase and a modulation degree. Such data can be analysed elegantly by the 'Phasor' approach [27]. Phasor analysis does not explicitly aim on determining fluorescence lifetimes or decay components for the pixels. Instead, it uses the phase and the modulation degree directly. For each pixel, a pointer (the phasor) is defined and displayed in a polar plot, see Fig. 15.7. The phase is used as the angle of the pointer, the modulation degree as the amplitude. This 'phasor plot' has several remarkable features:

- The phasors of pixels with single-exponential decay profiles end on a semicircle. The location on the semicircle depends on the fluorescence lifetime.
- Phasors of pixels with multi-exponential decay profiles, i.e. sums of several decay components, end inside the semicircle.
- Phasors of decay profiles which result from the convolution of several processes end outside the semicircle. An example is the emission of a FRET-excited acceptor, which is the convolution of the decay function of the interacting donor with the fluorescence decay of the acceptor.

Pixels with different signature in the phasor plot are then back-annotated in the image by giving them different colours. An example is shown in Fig. 15.10. It shows a phasor plot for several cells, B–F, shown at the top of the figure. As can be seen in the lower part of the figure, the fluorophores form different spots in the phasor plot, as does the autofluorescence of the untransfected cells. As can be seen from Fig. 15.10, phasor analysis is an excellent tool for distinguishing pixel areas with different decay signature. Please note also that all spots are inside the semicircle, i.e. none of the fluorophores present in the cells delivers a perfectly single-exponential decay in the biological environment.

15.5 FLIM of Skin Autofluorescence

Historically, functional and structural information about the skin was assessed either through in vitro cell culture, or histopathology of skin biopsies. Drug penetration studies were performed by placing excised skin within a Franz cell, measuring the flux of a compound from the donor solution through the skin (full thickness, split-thickness or separated epidermis) into the receptor solution [36]. For intravital imaging the dermatoscope, a handheld epiluminescence microscope, was the long-standing non-invasive diagnostic tool for assessing suspicious skin lesions [77]. Further analysis requires painful biopsies to be taken for histopathological assessment, a non-trivial time-consuming and costly procedure.

The emergence of confocal and especially multiphoton microscopy has stimulated a revolution in dermatological research and clinical diagnostics. Skin

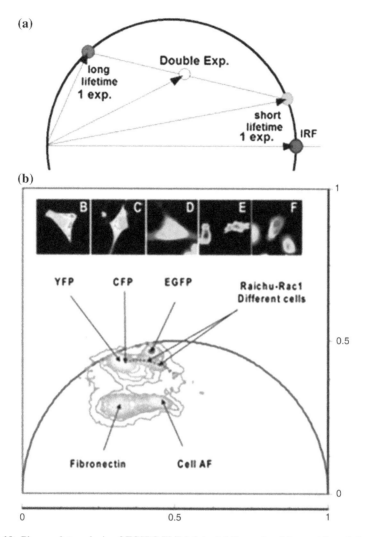

Fig. 15.10 Phasor plot analysis of TCSPC FLIM data. (**a**) Example of the position of single and multi-exponential lifetime components on a phasor plot. (**b**) Resolution of cellular autofluorescence and fluorescently-labelled proteins on a phasor plot extracted from lifetime data collected by TCSPC. Adapted from [27]

autofluorophores, which are endogenous molecules or proteins with fluorescence properties, are used for non-invasive functional, structural, penetration, and diagnostic studies within excised and in vivo skin. Autofluorescence offers a means for label-free analysis of structural and functional information within skin.

Autofluorescence intensity alone yields limited information and can be complicated by fluorophores with overlapping emission spectra, despite using multi-spectral confocal or multiphoton imaging. In addition, autofluorescence intensity only informs the investigator that the fluorophore is there and in its approximate concentration.

Fluorescence lifetime imaging adds another dimension to the data that can be obtained from autofluorescence imaging of skin. FLIM yields information of the physio-chemical state of the autofluorophore as a function of the direct binding interactions with nearby molecules and the local microenvironment [5]. As the fluorescence lifetime of a molecule is independent of its concentration, unlike intensity, FLIM is ideal for contrasting fluorophores with overlapping emission spectra. Moreover, lifetime changes in certain autofluorophores such as nicotin-amide adenine dinucleotide (phosphate) or NAD(P)H, which are involved in key intracellular metabolic pathways, are indicative of cell death, ischeamia reperfusion, and oxidative stress [108, 121, 126, 134].

15.5.1 Stratum Corneum Barrier Autofluorophores

15.5.1.1 Keratin

A major component of the stratum corneum is keratin, which is a family of fibrous structural proteins. Keratin monomers, produced in skin within the viable epidermis by keratinocytes, are arranged into bundles within the cytoplasm to form inter-mediary filaments. Intermediary filaments makeup the cytoskeletal backbone of the cell [91] and are involved in cell motility and division [95]. Within corneocytes in the stratum corneum, keratin intermediate filaments serve as a platform for fliggarin monomers to aggregate in order to flatten the cell and establish the intercellular barrier [91]. Using two-photon excitation, keratin autofluorescence is visible within a broad excitation range between 740 and 860 nm. Similarly, the fluorescence emission of keratin is broad ranging from 420 to 600 nm [134]. The fluorescence lifetime of keratin is approximately 1.4 ns [30], following a single component model Fig. 15.11.

15.5.2 Viable Epidermal Autofluorescence and Associated Fluorophores

15.5.2.1 NAD(P)H and FAD

There are two autofluorophores that play a central role in multiple cellular bioen-ergetic pathways: NAD(P)H and flavine adenine dinucleotide (FAD). NADH is an electron donor for the electron transport chain in oxidative phosphorylation, while NADPH is an intracellular reducing agent used as a coenzyme for anabolic

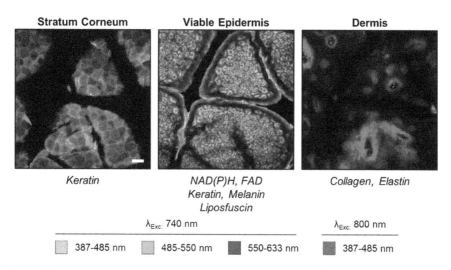

Fig. 15.11 In vivo cross-section of human skin imaged by multiphoton microscopy. Skin autofluorescence is pseudo-coloured *green*, while collagen second-harmonic generation is pseudo-coloured *red*

reactions and antioxidant [132]. NADH is used in various cell processes including biosynthesis, apoptosis and cell signalling [132]. As the spectral and lifetime properties of NADH and NADPH are essentially indistinguishable, they are aggregated and analysed together as NAD(P)H. NAD(P)H is excited between 720 and 780 nm (two-photon) and displays an emission range between 410 and 550 nm [134]. Using a bi-exponential decay model, the lifetimes of NAD(P)H are resolved into fast (~ 0.2–0.4 ns) and slow (~ 2.2–2.5 ns) components corresponding to free and protein-bound NAD(P)H (Table 15.1). FAD is an electron acceptor and redox cofactor, while in the reduced form it is a central electron transporter involved in oxidative phosphorylation [49]). FAD is excited by two-photon excitation between 720 and 880 nm, with an emission range of 510–600 nm [134]. Similarly to NAD (P)H, FAD has a free and protein-bound lifetime components corresponding to the slow (~ 2.0–2.8 ns) and fast (~ 0.7 ns) lifetimes.

15.5.2.2 Melanin

Melanin is a major component of the stratum basale and refers to a family of melanin compounds: pheomelanin, eumelanin, and neuromelanin. Phenotypically, melanin is the key factor for skin pigmentation while functionally it serves to protect cells from photodamage. Melanin protects cells from photodamage by rapidly converting absorbed ultraviolet light into heat as well as other chemical reactions [85]. Melanin is produced by melanocytes in skin, situated within the stratum basale, and packaged into lipid vesicles known as melanosomes that are distributed and endocytosed by other cells in the viable epidermis [131]. Melanin has a two-photon excitation range between ~720 and 880 nm and fluorescence emission greater than 510 nm [134]. The reported fluorescence lifetime of melanin varied between studies, but can be approximated to an ultra-fast component of ~0.15 ns and a second component ranging between 0.5 and 2 ns [23].

15.5.2.3 Lipofuscin

Liposfuscin is a yellow-brown pigment visible particularly in ageing skin and in the fundus of the eye, see Chap. 16. Lipofuscin is an aggregate term that refers to intercellular granules containing a heterogeneous mix of highly cross-linked components such as oxidized proteins (30–70 %), lipids (20–50 %), sugars and metals. Lipofuscin is believed to be formed in part from impaired proteasomal degradation of oxidized protein, as a function of age-related oxidative stress, which forms cross-links or hydrophobic bonds with other material [51]. Lipofuscin can be detected using a two-photon excitation wavelength of 800 nm and emission window between 500 and 550 nm. Multiple fluorescence lifetimes are reported for lipofusin (0.32, 1.2 and 4.8 ns) and should therefore by analysed by FLIM using a multiple exponential component model.

15.5.3 Dermal Autofluorescence and Its Fluorophores

15.5.3.1 Elastin

Elastin is a key structural protein within the dermis and conveys elastic recoil in skin [1]. The loss of elastin distribution in the dermis, and therefore elasticity of the skin, is considered as a major phenotypic characteristic associated with skin ageing [31]. This can be measured using multiphoton microscopy by measuring the ratio of elastin fluorescence to collagen SHG in the dermis [20, 107]. Elastin can be excited by two-photon microscopy within a broad range spanning 740–920 nm, with a similarly broad fluorescence emission range between 400 and 600 nm [134]. Elastin is reported to have a fluorescence lifetime of ~0.3–2 ns [64].

15.5.3.2 Collagen

Collagen is one of the major components of the dermis that, along with elastin, provides the skin's structural support. Collagen is synthesised and secreted into the dermis, forming bundles to provide the underlying structural strength of the skin [14]. Collagen emits both fluorescence (see Chap. 16) and second-harmonic-generation (SHG) signals (see Fig. 15.9c). Second-harmonic generation is a non-linear optical phenomenon whereby the excitation laser light is frequency-doubled after focusing on a sample (i.e. two photons of 800 nm is frequency-doubled to a single

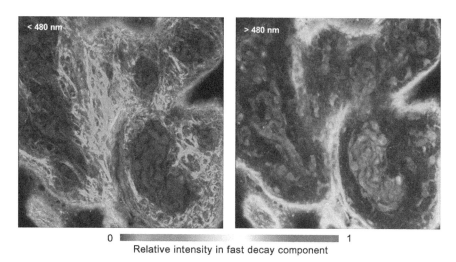

Fig. 15.12 Two-photon FLIM of pig skin, excitation 800 nm. *Colour* shows relative intensity in fast decay component. *Left* Wavelength channel <480 nm, the fast decay component represents the SHG signal from collagen. *Right* Wavelength channel >480 nm. The collagen structures are not seen in this channel because the SHG wavelength is outside the wavelength interval detected. LSM 710 NLO with bh Simple-Tau 152 dual-channel FLIM system, NDD detection. Data from [4], re-analysed

photon of 400 nm). The SHG light is emitted in forward direction, therefore scattering in the sample is required to detect it back through the microscope lens [83]. As the SHG process is near-instantaneous, the SHG signal has a near-zero fluorescence lifetime [83]. In two-photon FLIM data, the SHG signal contrasts collagen effectively from elastin and keratin [62, 105]. A few precautions must be observed to record SHG components effectively: a femtosecond laser excitation must be used, the SHG wavelength must be within detection wavelength interval, the sample must be thick enough to scatter the SHG light back into the microscope lens, and non-descanned detection must be used to detect the back-scattered signal. An example of SHG detection from collagen is shown in Fig. 15.12, left.

Caution is indicated when SHG sources other than collagen are present in the tissue. This may be the case in nanoparticle penetration studies [24, 74, 76] when SHG is used as an indicator of the presence of the particles. SHG signals from collagen may then interfere with the SHG signal of the nanoparticles.

15.6 Heterogeneity in Skin Fluorescence Lifetimes

While the fluorescence lifetime of autofluorophores is consistent in the absence of physio-chemical changes to the molecule itself, there is reported heterogeneity of the viable epidermal fluorescence lifetime between individuals and skin locations on the body. These differences are due to a number of factors including variation in the various proportions of autofluorophores within the viable epidermis. This is best exemplified by the presence of melanin [23].

15.6.1 Body Site

The average fluorescence lifetime of the viable epidermis was reported to increase with age in one study, which also demonstrated significant lifetime differences between volar and dorsal forearms, and thigh [9]. These differences (Fig. 15.13) were attributed to cell density and also the presence of melanin, with its ultra-short lifetime component, that was more abundant in solar-exposed dorsal forearm than other sites.

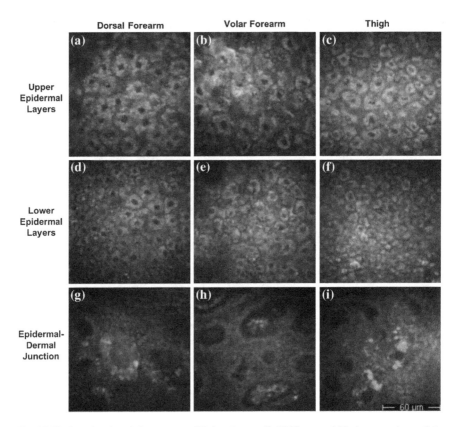

Fig. 15.13 Pseudocolored fluorescence lifetime image (0–2000 ps; *red-blue*) comparison of the dorsal forearm (**a, d, g**), volar forearm (**b, e, h**), and thigh (**c, f, i**) from the upper (**a–c**) and lower (**d–f**) epidermal layers, and epidermal-dermal junction (**g–i**). Jenlab DermaInspect with Becker & Hickl SPC-830 TCSPC FLIM module

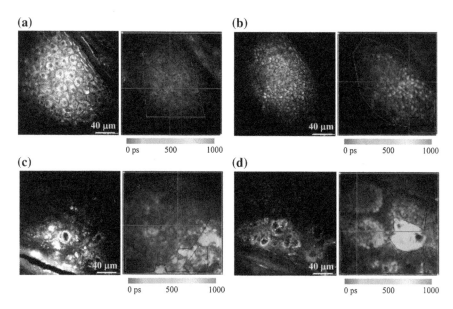

Fig. 15.14 Intravital multiphoton-FLIM microscopy (pseudo-coloured to τm) to observe the distribution of melanin within in vivo stratum granulosum (**a** and **c**) and stratum basale (**b** and **d**) of Asian (**a, b**) and African (**c, d**) volunteers. Jenlab DermaInspect with Becker & Hickl SPC-830 TCSPC FLIM module. Adapted from [23]

15.6.1.1 Melanin

The fluorescence lifetime of autofluorophores can be used to observe the relative quantity and distribution within skin. Melanin, with an ultra-fast lifetime component, is easily observed using multiphoton FLIM microscopy [23, 117], as demonstrated when comparing its relative abundance between Asian and African skin, as demonstrated in Fig. 15.14.

Using the lifetime properties of melanin, non-invasive clinical imaging of patients with melanocytic lesions [29, 113] demonstrate the potential for FLIM to enhance the diagnostic and prognostic assessment of these lesions before and after therapeutic treatment (Fig. 15.15).

15.6.2 Imaging, Harvesting and Storage Conditions

Due to the complex setup of FLIM-equipped microscopes in a laboratory setting, only certain types of samples can be imaged in skin research:

1. Epidermal or dermal cells, isolated from excised tissue, cultured in vitro
2. Freshly excised skin from animals or humans (donor or cadaveric tissue)
3. Histologically prepared fixed skin from animals or humans.

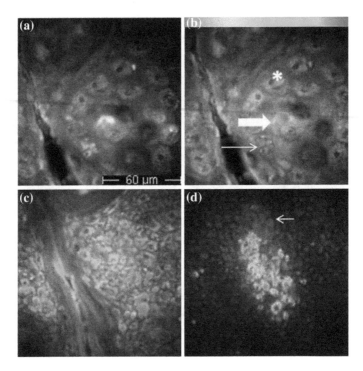

Fig. 15.15 Multiphoton intravital FLIM microscopy of melanocytic infiltration into the viable epidermis of in vivo human skin. Side-by-side multiphoton intensity (**a**) and pseudo-coloured FLIM (τ_m) images of the stratum granulosum of a melanoma lesion. *Arrows* in (**b**) demonstrate the short lifetime component (*red-yellow*) corresponding to melanin, whereas normal autofluorescence (*green-blue*) is indicated by a *star*. High concentrations of melanin, within normal melanocytes (*thin arrow*) and melanoma cells, are also visible FLIM images of the stratum spinosum (**c**) and stratum basale (**d**). Adapted from [112]

Confocal and multiphoton laser scanning microscopes equipped with FLIM on the laboratory bench are constrained to examining mouse and rat skin for intravital imaging. More specialised equipment is required for intravital human FLIM microscopy.

For analysis of excised tissue, tissue viability is an important factor to consider in lifetime changes of the viable epidermis. One study demonstrated that the progressive deterioration of excised skin, due to ischemic necrosis, is associated with increases in the average fluorescence lifetime of NAD(P)H [108]. This is an important factor to consider in ex vivo studies in order to attribute changes in the lifetime to their appropriate causes. In order to preserve the average fluorescence lifetime of the viable epidermis, excised skin should be stored at 4 °C for a period of no more than 1 week. Storage of the skin at –20 °C largely eliminates NAD(P)H and FAD autofluorescence and should be avoided if the skin imaging aims to measure the metabolic rate of tissue.

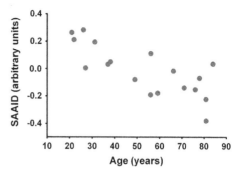

Fig. 15.16 Changes in the SAAID index according to age [45]

15.6.3 Normal and Solar-Induced Aging

Heterogeneity in the viable epidermal fluorescence lifetime can also be attributed to both natural and solar-induced photodamage. Non-invasive imaging to quantify the degree of skin aging measures the ratio of collagen SHG to skin autofluorescence (AF) in the dermis (i.e. elastin), an index known as SAAID:

$$SAAID = \frac{SHG - AF}{SHG + AF} \tag{15.4}$$

The SAAID index is reported to decrease with aging (see Fig. 15.16) and solar-induced photo-aging [59, 107], corresponding to the reduction of collagen and disruption of elastin fibre networks. These components, elastin and collagen, can be measured specifically in the dermis using FLIM to measure age and photodamage associated changes as reported [58].

Within the viable epidermis, the decrease in the average fluorescence lifetime tends to occur with solar-exposure, whereas age-related changes are associated with an increase in the fluorescence lifetime [107], as also found in an earlier study [9]. These studies demonstrate that considering the age and photodamage-related changes to the autofluorescence lifetime, in the viable epidermis and dermis, are critically important in large population FLIM studies.

15.7 Using FLIM to Assess the Topical Penetration of Actives into Skin

Franz cell diffusion studies are traditionally used for cutaneous delivery and trans-dermal penetration studies. While this technique is a proven and effective means of calculating the dermatopharmacokinetics of topically applied agents, the data generated fails to provide any spatial information. Spatial data is essential for determining the penetration route and assessing potential cellular uptake. Moreover, spatial data can be used to direct the optimisation of formulations and other penetration enhancers for targeted delivery and transport of compounds through the skin.

Histology can be used to determine the spatial distribution of a topically applied compound, but requires multiple skin samples for each time point, adding to the overall cost in time and resources. Although laser scanning microscopy permits non-invasive imaging of intact excised skin or in vivo skin for topical penetration studies, the effectiveness of this technique is limited to the ability to resolve background autofluorescence and the topically applied agent spectrally. If the agent and autofluorescence is spectrally distinguishable, little more than spatial data could be obtained by measuring the relative fluorescence intensity. With the introduction of TCSPC FLIM to laser scanning microscopes, an additional means of resolving various fluorophores can be achieved irrespective of their emission spectral overlap. Furthermore, the fluorescence lifetime of the topically applied agent (dye, drug, or nanoparticle) can be used to resolve various states of the compound (i.e. free/non-interacting versus binding/interacting) as it penetrates into the skin [105].

15.7.1 Fluorescent Dyes and Macromolecules

FLIM has been used to contrast, assess and quantify the penetration of topically applied fluorescent dyes in excised and in vivo human skin. One study used the fluorescence lifetime properties of acriflavine and fluorescein, both of which have an average lifetime greater than 2 ns, to isolate and quantify the presence of dye within viable epidermis of in vivo human skin as a measure of changes in barrier integrity associated with aging and photoaging [107]. Similarly, the VivaScope® FLIM system was used to image the penetration of topically applied fluorescein into in vivo human skin [106] and differentiate between tissue autofluorescence and topical fluorescein and 9-cyanoanthracene in porcine skin [89] (Fig. 15.17).

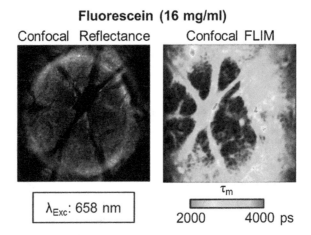

Fig. 15.17 Confocal reflectance (*left*) and pseudo-coloured fluorescence lifetime image (*right*) of fluorescein (16 mg/mlit's) distribution in the stratum corneum of in vivo human skin after topical application. Lucid Vivascope with Becker & Hickl Simple-Tau 150 TCSPC FLIM system

15.7.2 siRNA

Combined with FLIM, fluorescent tagging of siRNA can be used to assess the effectiveness of various nano-carriers systems, such as flexible liposomes, to deliver their payload into the viable epidermis. In a recent study we used FRET to show that fluorescein labelled siRNA delivered to viable epidermis with secosomes were actually delivered into viable skin after topical application, while there was no significant penetration of zinc oxide nanoparticles within the viable epidermis [41], see Fig. 15.18.

15.7.3 Nanoparticles

FLIM has best shown its utility in the study topical nanoparticle penetration into human skin. FLIM offers unprecedented advantages of regular confocal and multiphoton imaging as it enables an effective contrast between skin autofluorescence and nanoparticles, and metabolic imaging of the viable epidermis. In the latter case, nanotoxicological data can be extracted while simultaneously measuring the spatial map of nanoparticle distribution in the skin.

15.7.3.1 Zinc Oxide Nanoparticles

One of the most well studied nanoparticles is zinc oxide (ZnO-NP) as it is widely used is cosmetical products, sunscreens in particular. Due to the broad consumer adoption of nano-sunscreens, some public health concerns have been raised regarding the potential penetration of ZnO-NP into the viable epidermis to cause toxicity [32]. Using multiphoton FLIM, repeated in vivo human studies have demonstrated little to no penetration of ZnO-NP nanoparticles into the viable epidermis and the absence of any fluorescence lifetime changes in NAD(P)H or FAD associated with oxidative stress or cell death [74, 76, 104, 137]. Rather, ZnO-NP are predominantly localised with the stratum corneum, ridges and furrows of the skin (Fig. 15.19).

Zinc oxide nanoparticle signal can be separated from background autofluorescence lifetimes in two ways: (1) examining the amplitude coefficient of the short lifetime components, which predominantly occupy 90–100 % of the total lifetimes detected [74, 76], and (2) spectral isolation of the ZnO-NP SHG signal [24].

15.7.3.2 Gold and Silver Nanoparticles

The dual action of FLIM microscopy to measure nanoparticle penetration and the metabolic state of the tissue for nanotoxicology has also been reported for gold

(a)

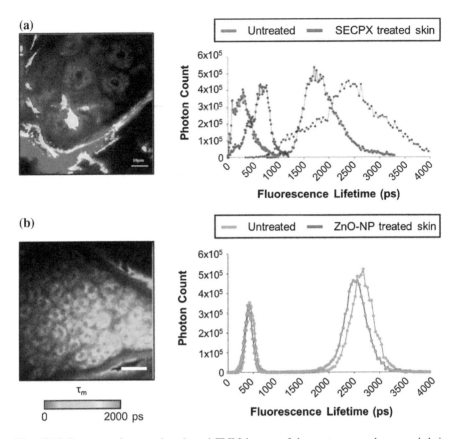

(b)

Fig. 15.18 Representative pseudo-coloured FLIM images of the stratum granulosum and their corresponding lifetime histograms, of the gated viable epidermis, after application of SECosomes (**a**) and zinc oxide nanoparticles (**b**) to human skin following MPM excitation at 740 nm. A shift in fluorescence lifetime observed for the NAD(P)H for the secosomes (**a**) and not zinc oxide nanoparticles (**b**) compared to normal untreated skin (*blue line*), suggesting that the fluorescein-labelled siRNA in secosomes have penetrated into the viable epidermis. In contrast, lifetime differences between normal and zinc oxide treated skin can be attributed to normal variability in imaging the viable epidermis and is absent of a strong ultra-fast short lifetime component from zinc oxide. The *white scale bar* represents a length of 10 μm (**a**) and 40 μm (**b**). FLIM image and histogram of (**a**) is adapted from [41]

nanoparticles. FLIM was used in parallel with dermoscopy, RCM, light microscopy histology, and transmission electron microscopy (TEM) to measure the level of penetration of topically applied gold nanoparticles (6 and 15 nm in size). In conjunction, FLIM was used to assess the viable state of the epidermal cells after topical treatment with gold nanoparticles in an aqueous and toluene penetration enhancer solution [72]. FLIM validated the other imaging techniques showing that non-polar 6 nm gold nanoparticles in a toluene vehicle penetrated into the viable epidermis of human skin. Moreover, the toluene vehicle caused a reduction in the

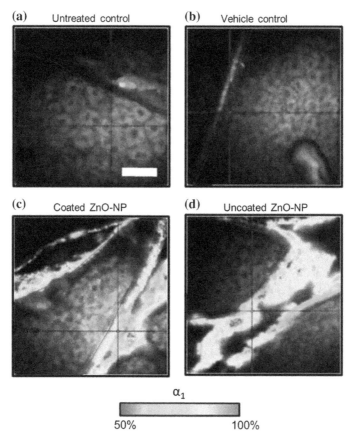

Fig. 15.19 Pseudocolored multiphoton FLIM image (amplitude coefficient of the short lifetime component, α1) of the distribution of zinc oxide nanoparticles (ZnO-NP) into in vivo human skin 6 h after topical application. The stratum granulosum of in vivo human skin was imaged by multiphoton microscopy with FLIM after topical application of coated (**c**) and uncoated (**d**) ZnO-NP, compared to the untreated (**a**) and vehicle (**b**; caprylic/capric triglyceride) controls. *White scale bar* is equivalent to 40 μm. Adapted from [74]

fluorescence intensity of NAD(P)H, suggesting necrosis of the epidermal cells. In contrast, 15 nm gold nanoparticles suspended in aqueous solution were found to accumulate within the superficial stratum corneum and furrows, without penetration into the viable epidermis. Interestingly, the fluorescence intensity of NAD(P)H trended lower than the aqueous vehicle control, but was unchanged relative to the vehicle controls in terms of the α_1/α_2 ratio [72].

Similarly, silver nanoparticles (<50 nm) were spatially resolved and contrasted according to their lifetimes with background autofluorescence in normal and barrier-disrupted excised human skin [101]. Barrier impairment was achieved by tape-stripping but was surprisingly found not to enhance topical penetration of silver nanoparticles into the skin. However, incomplete barrier-impairment via tape-stripping may account for differences between similar examinations with topical silver nanoparticle penetration into human skin [68]. Furthermore, while there was a non-statistically significant trend in α_1/α_2 ratio to increase with silver nanoparticle application, no changes in the metabolic state of the viable epidermis were detected [101].

As with zinc oxide [74, 76], both gold and silver nanoparticles can be distinguished from background autofluorescence by measuring the short lifetime amplitude coefficient between 90 and 100 % [72, 101].

15.7.3.3 Silica Nanoparticles

In addition to topical penetration, FLIM has also been used to measure the uptake of n-(6-aminohexyl)-aminopropyltrimethoxysilane (AHAPS) functionalized silicone dioxide nanoparticles by macrophages following a sub-cutaneous injection in mouse skin [93]. As with other nanoparticle studies, FLIM was used to effectively distinguish between the nanoparticle signal and background autofluorescence, irrespective of overlapping fluorescence emission spectra.

15.8 Evaluation of Effectiveness of Cosmetics with FLIM

Cosmeceutical products are designed to improve the barrier integrity, function, and health of human skin, as well as repair the effects of natural and sun-induced photoaging. In the pursuit of this goal, cosmeceutical development requires a more thorough understanding of the various properties of human skin in order to optimise formulations for topical application. Fluorescence lifetime imaging is positioned uniquely to inform an investigator of physio-chemical changes in autofluorophores and fluorescence lifetime reporters to determine skin properties such as pH, divalent metal cation distribution, and structural protein distribution.

15.8.1 pH

One of the earliest studies to employ fluorescence lifetime reporters used 2',7'-bis-(2-carboxyethyl)-5-(and-6)-carboxyfluorescein to spatially map the pH gradient within the stratum corneum of hairless mouse skin using two-photon non-invasive FLIM [48]. As lifetime changes in the reporter are not dependent on concentration, unlike fluorescence intensity, obstacles in calibrating pH measurements are eliminated using FLIM. The data demonstrated that the intracellular pH of the corneocytes was neutral in contrast to the narrow intercellular spaces between corneocytes [48], which can be used by cosmetic scientists to appropriately formulate and target products for the skin.

15.8.2 Calcium

FLIM has also been used to spatial map the distribution of calcium ions within skin using the fluorescence lifetime reporter dye calcium green 5N (CG5N) [8, 17]. Using FLIM, two major lifetime components can be resolved corresponding to the free and calcium-bound states of CG5N, which are distinct from the background autofluorescence. Interestingly, this research had identified that the majority of calcium is found intracellularly within the Golgi and endoplasmic reticulum, with very little overall calcium located in the extracellular spaces of the viable epidermis (2–7 %). Using the lifetime parameters of CG5N, the spatial concentration of calcium was also calculated to determine the concentration gradient within normal [8] and barrier impaired skin [17].

15.8.3 Assessing Cosmetic Effectiveness

Dermatological cosmetic products assert a variety of claims to improving the health and vitality of skin via the topical application of products containing anti-oxidant formulations (e.g. folic acid, CoQ10). As reviewed above, changes in the redox state of the viable epidermis can be assessed using non-invasive multiphoton FLIM. In order to assess some of these claims, cosmetic products were daily applied to in vivo human skin over a period of 6 days and imaged at the end of this period for changes in the redox state of the viable epidermis [101]. Figure 15.20 shows changes in the lifetime parameters of NAD(P)H in the stratum granulosum after daily application of folic acid, demonstrating a marked reduction in the α_1/α_2 ratio and elevated metabolic rate. Overall changes in NAD(P)H intensity, measured by FLIM photon counts, after the repeated application of these cosmetic products are shown in Fig. 15.21. Despite the various trends observed, significant changes in the relative quantity of NAD(P)H were observed from retinol application in the stratum spinosum after 2 days of treatment [101]. This study demonstrates the capacity to use FLIM to validate or challenge dermatological cosmeceutical claims safely using non-invasive multiphoton microscopy.

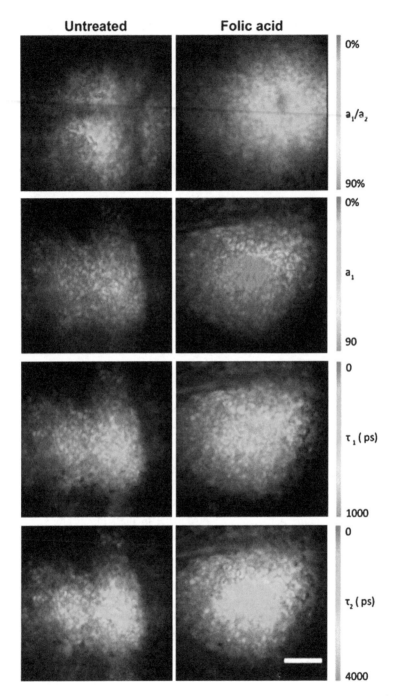

Fig. 15.20 Fluorescence lifetime parameters of in vivo human skin measured after repeated daily application of folic acid over 6 days. The *white scale bar* represents a length of 40 μm. Adapted from [101]

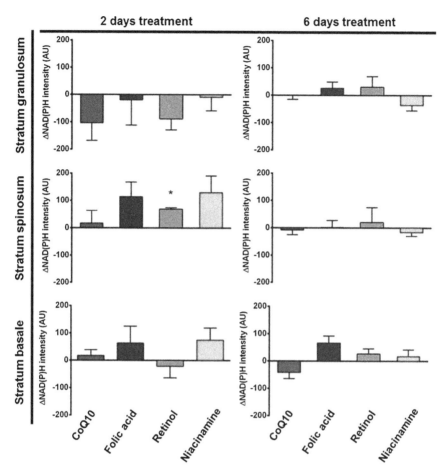

Fig. 15.21 Relative intensity of NAD(P)H (sum of α1 and α2 photon counts) of various skin strata, measured by multiphoton FLM, after repeated topical application of CoQ10, folic acid, retinol and niacinamine containing cosmetic products to in vivo human skin [101]. * p < 0.05 student's t-test

15.9 Skin FLIM in Dermatology

FLIM of lesional skin, due to psoriasis or atopic dermatitis, has largely investigated fluorescence lifetime differences in the viable epidermis, compared to normal skin, and the potential for topically applied agents (specifically nanoparticles) to penetrate into the viable layers.

FLIM of in vivo human skin has shown that the NAD(P)H lifetime of the viable epidermis increases with atopic dermatitis severity [67]. Successful topical treatments for atopic dermatitis led to the normalisation of the autofluorescence lifetime

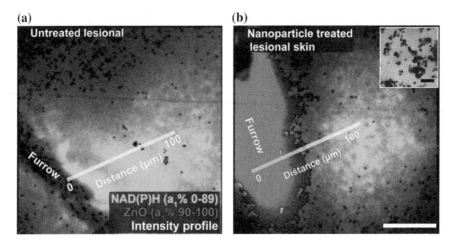

Fig. 15.22 Penetration of topically applied ZnO-NP from the furrows into lesional skin. Pseudocolored FLIM images of untreated (**a**) and ZnO-NP topically-treated lesional skin (**b**) according to the amplitude coefficient of the short lifetime. *Green* represents cellular autofluorescence (0–89 %), whereas ZnO-NP signal (90–100 %) is shown in *red*. The *white scale bar* represents a distance of 40 μm. Adapted from [101]

of the viable epidermis. Topical application of ZnO-NP to in vivo lesional skin (in psoriatic and atopic dermatitis patients) was measured by FLIM to quantify the level of penetration into the viable epidermis and assess the metabolic state of the cells [68, 76]. In these studies, topically applied ZnO-NP failed to penetrate into the viable epidermis, as shown in Fig. 15.22, despite possessing an impaired stratum corneum barrier [76]. Moreover, the metabolic rate was unaffected by the application of ZnO-NP to normal or lesional skin [68]. Non-invasive multiphoton FLIM can therefore be used to investigate the effectiveness of therapeutic treatments and assess potential nanotoxicological hazards in psoriasis and atopic dermatitis.

15.9.1 FLIM of Skin Cancer

One of the more profound differences between the normal and cancerous cells is the tendency for the glycolytic bioenergetic metabolic pathway to be dominant in tumours, despite a readily available supply of oxygen that would normally activate oxidative phosphorylation [86]. As differences between glycolysis and oxidative phosphorylation, and their resulting redox states within the cells, can be distinguished according to the fluorescence lifetimes of NAD(P)H and FAD, non-invasive cancer imaging in dermatology is a provocative emerging field of research with far reaching clinical implications.

The autofluorescence lifetime differences between normal and cancerous tissues were described in a landmark study using non-invasive multiphoton FLIM imaging of an oral epithelial cancer model within the hamster check pouch [115, 116]. These studies demonstrated that high grade pre-cancerous tissue displayed a lower average fluorescence lifetime and higher relative contribution of the short lifetime component than lower grade pre-cancerous tissue and normal epithelial cells. As these differences mirrored the inhibition of oxidative phosphorylation by the addition of CoCl$_2$, the lifetime changes associated with cancer progression were hypothesised to be the result of a suppression of oxidative phosphorylation and elevation of glycolysis as the predominant bioenergetic pathway [116]. This study has inspired similar investigations in dermatological assessment of skin lesions to determine if similar lifetime differences can be observed between normal and cancerous tissue.

15.9.2 Melanoma

Multiphoton FLIM studies of melanoma lesions in human skin demonstrated an overall decrease in the average fluorescence lifetime of the viable epidermis [19, 29, 43, 114]. This decrease is associated with the elevated distribution of melanin due to infiltration of melanoma cells into the upper layers of the viable epidermis (Fig. 15.23)

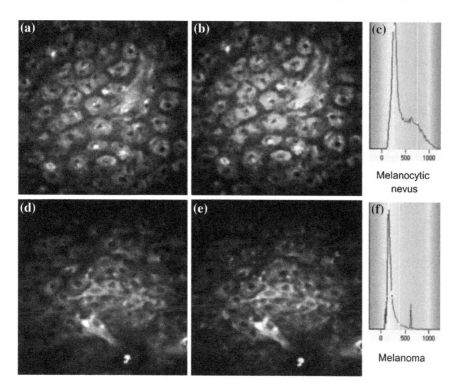

Fig. 15.23 Autofluorescence lifetime differences between of a melanocytic nevus (**a–c**) and melanoma (**b–f**) lesion. Multiphoton intensity (**a, d**) and pseudo-coloured lifetime images (**b, e**) show the spatial distribution of autofluorophore lifetimes, with the corresponding lifetime histograms of the image (**c, f**). Adapted from [29]

[29, 43, 114]. Interestingly, analysis of keratinocyte autofluorescence lifetimes show no significant differences [43], indicating the FLIM analysis of melanoma lesions highlights the presence of melanoma cells rather than underlying metabolic differences.

15.9.3 Basal Cell Carcinoma

One of the earliest studies to investigate potential autofluorescence lifetime differences between normal and basal cell carcinoma (BCC) lesions in human skin used wide-field FLIM [39]. Using a UV laser excitation source, they observed an overall decrease in the average fluorescence lifetime of the BCC lesion compared to normal skin. In contrast, multiphoton (740 nm; 2-photon) TCSPC FLIM analysis of excised BCC lesions from human patients observed a slightly elevated autofluorescence lifetime to normal skin [19], which was also confirmed by a separate study published shortly after [43].

In terms of distinguishing between types of skin lesions, one study showed significant fluorescence lifetime differences between dysplastic nevi and nodular BCC (nBCC) in freshly excised human skin samples [97]. The average fluorescence lifetime of dysplastic nevi was significantly lower than nBCC, which is most likely due to the presence of melanin's ultra-fast short lifetime component.

A large clinical study was performed using the DermaInspect® with FLIM to analyse morphological, structural and autofluorescence lifetime differences between normal and BCC skin lesions [113]. As with earlier studies, BCC cells demonstrate a significantly elevated autofluorescence lifetime compared to normal cells. Moreover, the lifetime properties of collagen and elastin were used to confirm the distribution of structural proteins surrounding and beneath BCC nodules. A similar study also demonstrated that the average fluorescence lifetime of BCC lesions was significantly elevated relative to normal skin across multiple spectral channels [96], as shown in Fig. 15.24.

15.9.4 Squamous Cell Carcinoma

Aside from morphological changes characteristic of squamous cell carcinoma (SCC), FLIM has thus far failed to demonstrate autofluorescence lifetime differences between normal and SCC lesions [105]. However, only one study has been performed to date, leaving this strand of non-invasive FLIM imaging open to new advances for SCC research diagnostics.

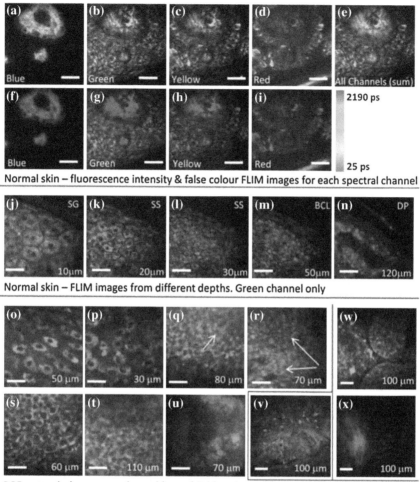

Normal skin – fluorescence intensity & false colour FLIM images for each spectral channel

Normal skin – FLIM images from different depths. Green channel only

BCC – morphology, green channel (o-u,w) & blue channel (v,x)

Fig. 15.24 Comparison of the autofluorescence lifetime of normal versus BCC lesional in vivo human skin. **a** Multiphoton intensity (**a–e**) and pseudo-coloured lifetime images (**f–i**) of spectral channels *blue* (**a, f**; 360–460 nm), *green* (**b, g**; 425–515 nm), *yellow* (**c, h**; 515–620 nm) and *red* (**d, i**; 620–640 nm). **j–n** 10 μm incremental Z-stack lifetime images of normal skin within the green spectral channel. **o–u** Lifetime images of BCC lesions at various depths measured with the green spectral channel. **v** Lifetime image of the dermis in a BCC lesion taken in the *blue* spectral channel. Lifetime images of the dermis in a BCC lesion taken in the *green* (**w**) and *blue* (**x**) spectral channels respectively. Adapted from [96]

15.10 Photoluminescence Lifetime Imaging Microscopy (PLIM)

One of the major problems associated with fluorescence imaging in biological specimens is the potential interference of autofluorescence with the fluorescence of biomarkers or other exogenous compounds to be located in the tissue. In 2008, we documented the well-known complex spectral and temporal behaviour of ZnO photoluminescence for uncoated and silane coated nanoparticles. We used a bh DCS-120 confocal scanning FLIM system [5, 104] in the simultaneous fluorescence/phosphorescence lifetime imaging (FLIM/PLIM) mode. Simultaneous FLIM/PLIM is based on TCSPC in combination with on/off modulation of a high-frequency pulsed laser. Photon times are determined both for the temporal location in the laser pulse period and in the modulation period [3, 4], see Fig. 15.25. The times in the pulse period are used to build up a FLIM image, the times in the modulation period to build up a PLIM image. Please see Chap. 1, Sect. 1.4.7 for technical details, and Chap. 6 for O_2-sensing applications.

The ZnO nanoparticles were excited by a 405 nm picosecond diode laser. The pulse period was 20 ns, the modulation period 100 μs. SPCImage [4, 5] was then used to fit the fluorescence and photoluminescence lifetimes as shown in Fig. 15.26. It is evident that both the coated and uncoated nano zinc oxide have a fast fluorescence decay profile relative to a slow luminescence decay profile seen for uncoated zinc oxide. Interestingly, the coated nano ZnO shows a lack of slow luminescence decay phenomena, consistent with photoluminescence being a surface phenomenon.

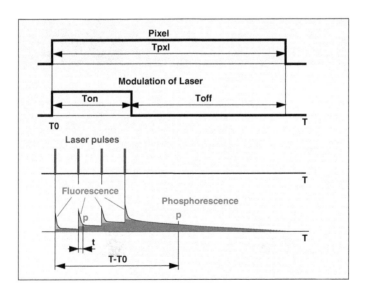

Fig. 15.25 Simultaneous FLIM and PLIM processes implemented in the TCSPC modules

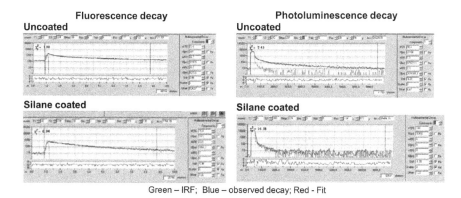

Fig. 15.26 Comparison of the fluorescence and photoluminescence lifetime decay profiles for uncoated and coated zinc oxide [104]

As an adjunct to our earlier FLIM analysis of the penetration of nano ZnO into human skin, we have now carried out pilot studies with the new FLIM/PLIM technique implemented by B&H in their FLIM systems for the Zeiss 710 NLO [6]. The system controls the Ti:Sa laser of the LSM 710 and records as shown in Fig. 15.25, yeilding lifetime images both at the picosecond and microsecond time scale.

A comparison of the FLIM and PLIM for normal human skin is shown in Fig. 15.27. It is evident that, after excitation at 740 nm, there is weak fluorescence for

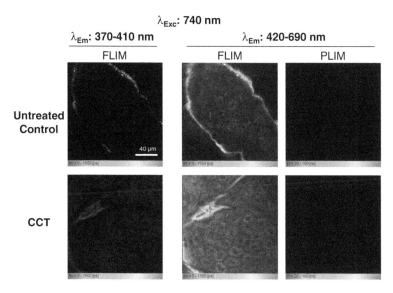

Fig. 15.27 FLIM and PLIM for normal human skin after excitation at 740 nm for emission wavebands of 370–410 and 420–690 nm

λ_{Exc}: 740 nm

λ_{Em}: 370-410 nm λ_{Em}: 420-690 nm

FLIM   FLIM   PLIM

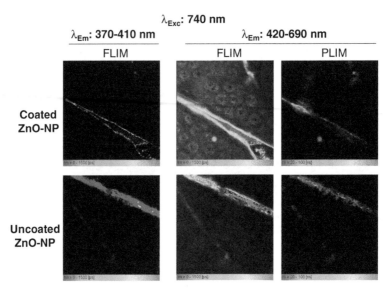

Fig. 15.28 FLIM and PLIM for normal human skin after excitation at 740 nm for emission wavebands of 370–410 and 420–690 nm

an emission waveband of 370–410 nm and strong fluorescence for the wavelength band 420–690 nm. In contrast, there is no PLIM for the former wavelength band.

The comparable images after application of nano ZnO are shown in Fig. 15.28. Here, it is evident that the characteristic ZnO images exist in the furrows for both the low fluorescence wavelength band and higher phosphorescence wavelength band. This preliminary data suggests that there is an opportunity to use PLIM to distinguish solutes and nanomaterials from autofluorescence when there is a very strong autofluorescence signal.

15.11 Summary

Fluorescence lifetime imaging leverages existing laser scanning microscope platforms to add a powerful layer of functional data. The advantage of FLIM is not just an enhanced ability to distinguish between fluorophores according to their lifetime, especially in combination with multi-spectral imaging, but also resolving physiochemical changes in these fluorophores. Combined with non-invasive imaging, the application of FLIM in skin research and clinical diagnostics has rapidly expanded. For skin penetration studies, FLIM allows a user to simultaneously monitor the spatial distribution of the topically applied agent in the skin and cellular metabolic response over time. In clinical diagnostics, FLIM uses the metabolic and phenotypic differences between normal and cancerous tissue to accurately resolve lesional tissue. In the future, FLIM is expected to be a mainstream technique for skin research and clinical practice.

Acknowledgments This work was supported by the National Health and Medical Research Council of Australia, the Australian Research Council and the Cancer Councils of South Australia and Queensland.

References

1. A.K. Baldwin, A. Simpson, R. Steer, S.A. Cain, C.M. Kielty, Elastic fibres in health and disease. Expert Rev. Mol. Med. **15**, e8 (2013)
2. W. Becker, A. Bergmann, Lifetime-resolved Imaging in Nonlinear Microscopy, in *Handbook of Biomedical Nonlinear Optical Microscopy* (Oxford University Press, New York, 2008)
3. W. Becker, B. Su, A. Bergmann, K. Weisshart, O. Holub, Simultaneous fluorescence and phosphorescence lifetime imaging. Proc. SPIE **7903**, 790320 (2011)
4. W. Becker, *The bh TCSPC handbook*, 5th edn. Becker & Hickl GmbH (2012), electronic version available on www.becker-hickl.com
5. W. Becker, Fluorescence lifetime imaging-techniques and applications. J. Microsc. **247**(2), 119–136 (2012)
6. Becker & Hickl GmbH, DCS-120 *Confocal Scanning FLIM Systems*, user handbook, edition 2012, available on www.becker-hickl.com
7. Becker & Hickl GmbH, *Modular FLIM systems for Zeiss LSM* 510 *and LSM* 710 *Family Laser Scanning Microscopes*, user handbook, 5th edn. Becker & Hickl GmbH (2012), available on www.becker-hickl.com
8. M.J. Behne, S. Sanchez, N.P. Barry, N. Kirschner, W. Meyer, T.M. Mauro, I. Moll, E. Gratton, Major translocation of calcium upon epidermal barrier insult: imaging and quantification via FLIM/Fourier vector analysis. Arch. Dermatol. Res. **303**(2), 103–115 (2011)
9. E. Benati, V. Bellini, S. Borsari, C. Dunsby, C. Ferrari, P. French, M. Guanti, D. Guardoli, K. Koenig, G. Pellacani, G. Ponti, S. Schianchi, C. Talbot, S. Seidenari, Quantitative evaluation of healthy epidermis by means of multiphoton microscopy and fluorescence lifetime imaging microscopy. Skin Res. Technol. **17**(3), 295–303 (2011)
10. D.K. Bird, L. Yan, K.M. Vrotsos, K.W. Eliceiri, E.M. Vaughan, P.J. Keely, J.G. White, N. Ramanujam, Metabolic mapping of MCF10A human breast cells via multiphoton fluorescence lifetime imaging of the coenzyme NADH. Cancer Res. **65**(19), 8766–8773 (2005)
11. H.G. Breunig, H. Studier, K. König, Multiphoton excitation characteristics of cellular fluorophores of human skin in vivo. Opt. Express **18**, 7857–7871 (2010)
12. H.G. Breunig, R. Bückle, M. Kellner-Höfer, M. Weinigel, J. Lademann, W. Sterry, K. König, Combined in vivo multiphoton and CARS imaging of healthy and disease-affected human skin. Microsc. Res. Tech. **75**, 492–498 (2012)
13. I. Bugiel, K. König, H. Wabnitz, Investigations of cells by fluorescence laser scanning microscopy with subnanosecond time resolution. Lasers Life Sci. **3**(1), 47–53 (1989)
14. T.M. Callaghan, K.-P. Wilhelm, A review of ageing and an examination of clinical methods in the assessment of ageing skin. Part I: cellular and molecular perspectives of skin ageing. Int. J. Cosmet. Sci. **30**(5), 313–322 (2008)
15. P.J. Caspers, G.W. Lucassen, H.A. Bruining, G.J. Puppels, Automated depth-scanning confocal Raman microspectrometer for rapid in vivo determination of water concentration profiles in human skin. J. Raman Spectrosc. **818**, 813–818 (2000)
16. P.J. Caspers, A.C. Williams, E.A. Carter, H.G.M. Edwards, B.W. Barry, H.A. Bruining, G.J. Puppels, Monitoring the penetration enhancer dimethyl sulfoxide in human stratum corneum in vivo by confocal Raman spectroscopy. Pharm. Res. **19**(10), 1577–1580 (2002)
17. A. Celli, S. Sanchez, M. Behne, T. Hazlett, E. Gratton, T. Mauro, The epidermal Ca(2+) gradient: measurement using the phasor representation of fluorescent lifetime imaging. Biophys. J. **98**(5), 911–921 (2010)

18. B. Chance, B. Schoener, R. Oshino, F. Itshak, Y. Nakase, Oxidation-reduction ratio studies of mitochondria in freeze-trapped samples. NADH and flavoprotein fluorescence signals. J. Biol. Chem. **254**(11), 4764–4771 (1979)
19. R. Cicchi, S. Sestini, V. De Giorgi, D. Massi, T. Lotti, F.S. Pavone, Nonlinear laser imaging of skin lesions. J. Biophotonics **1**(1), 62–73 (2008)
20. R. Cicchi, D. Kapsokalyvas, V. De Giorgi, V. Maio, A. Van Wiechen, D. Massi, T. Lotti, F. S. Pavone, Scoring of collagen organization in healthy and diseased human dermis by multiphoton microscopy. J. Biophotonics **3**(1–2), 34–43 (2010)
21. M.J. Cole, J. Siegel, S.E.D. Webb, R. Jones, K. Dowling, M.J. Dayel, D. Parsons-Karavassilis, P.M. French, M.J. Lever, L.O. Sucharov, M.A. Neil, R. Juskaitas, T. Wilson, Time-domain whole-field lifetime imaging with optical sectioning. J. Microsc. **203**, 246–257 (2001)
22. N.P. Damayanti, A.P. Craig, J. Irudayaraj, A hybrid FLIM-elastic net platform for label free profiling of breast cancer. Analyst **138**(23), 7127–7134 (2013)
23. Y. Dancik, A. Favre, C.J. Loy, A.V. Zvyagin, M.S. Roberts, Use of multiphoton tomography and fluorescence lifetime imaging to investigate skin pigmentation in vivo. J. Biomed. Opt. **18**(2), 26022 (2013)
24. M.E. Darvin, K. König, M. Kellner-Hoefer, H.G. Breunig, W. Werncke, M.C. Meinke, A. Patzelt, W. Sterry, J. Lademann, Safety assessment by multiphoton fluorescence/second harmonic generation/hyper-Rayleigh scattering tomography of ZnO nanoparticles used in cosmetic products. Skin Pharmacol. Physiol. **25**(4), 219–226 (2012)
25. G. Deka, W. Wu, F. Kao, In vivo wound healing diagnosis with second harmonic and fluorescence lifetime imaging. J. Biomed. Opt. **18**(6), 061222 (2013)
26. W. Denk, J.H. Strickler, W.W.W. Webb, Two-photon laser scanning fluorescence microscopy. Science **248**, 73–76 (1990)
27. M. Digman, V.R. Caiolfa, M. Zamai, E. Gratton, The phasor approach to fluorescence lifetime imaging analysis. Biophys. J. **94**(2), L14–L16 (2008)
28. E. Dimitrow, I. Riemann, A. Ehlers, M.J. Koehler, J. Norgauer, P. Elsner, K. König, M. Kaatz, Spectral fluorescence lifetime detection and selective melanin imaging by multiphoton laser tomography for melanoma diagnosis. Exp. Dermatol. **18**(6), 509–515 (2009)
29. E. Dimitrow, M. Ziemer, M. Koehler, J. Norgauer, K. König, P. Elsner, M. Kaatz, Sensitivity and specificity of multiphoton laser tomography for in vivo and ex vivo diagnosis of malignant melanoma. J. Invest. Dermatol. **129**, 1752–1758 (2009)
30. A. Ehlers, I. Riemann, M. Stark, K. König, Multiphoton fluorescence lifetime imaging of human hair. Microsc. Res. Tech. **70**(2), 154–161 (2007)
31. M. El-Domyati, S. Attia, F. Saleh, D. Brown, D.E. Birk, F. Gasparro, H. Ahmad, J. Uitto, Intrinsic aging vs. photoaging: a comparative histopathological, immunohistochemical, and ultrastructural study of skin. Exp. Dermatol. **11**(5), 398–405 (2002)
32. T. Faunce, Toxicological and public good considerations for the regulation of nanomaterial-containing medical products. Expert Opin. Drug Saf. **7**(2), 103–106 (2008)
33. R. Fink-Puches, R. Hofmann-Wellenhof, J. Smolle, H. Kerl, Confocal laser scanning microscopy: a new optical microscopic technique for applications in pathology and dermatology. J. Cutan. Pathol. **22**(3), 252–259 (1995)
34. F. Fischer, B. Volkmer, S. Puschmann, R. Greinert, W. Breitbart, J. Kiefer, R. Wepf, Risk estimation of skin damage due to ultrashort pulsed, focused near-infrared laser irradiation at 800 nm. J. Biomed. Opt. **13**(4), 041320 (2008)
35. F. Fischer, B. Volkmer, S. Puschmann, R. Greinert, W. Breitbart, J. Kiefer, R. Wepf, Assessing the risk of skin damage due to femtosecond laser irradiation. J. Biophoton. **1**, 470–477 (2008)
36. D.R. Friend, In vitro skin permeation techniques. J. Control. Release **18**(3), 235–248 (1992)

37. A. Gafni, L. Brand, Fluorescence decay studies of reduced nicotinamide adenine dinucleotide in solution and bound to liver alcohol dehydrogenase. Biochemistry **15**(15), 3165–3171 (1976)
38. E.R. Gaillard, S.J. Atherton, G. Eldred, J. Dillon, Photophysical studies on human retinal lipofuscin. Photochem. Photobiol. **61**(5), 448–453 (1995)
39. N.P. Galletly, J. McGinty, C. Dunsby, F. Teixeira, J. Requejo-Isidro, I. Munro, D.S. Elson, M.A.A. Neil, A.C. Chu, P.M.W. French, G.W. Stamp, Fluorescence lifetime imaging distinguishes basal cell carcinoma from surrounding uninvolved skin. Br. J. Dermatol. **159** (1), 152–161 (2008)
40. U. Gehlsen, A. Oetke, M. Szaszák, N. Koop, F. Paulsen, A. Gebert, G. Huettmann, P. Steven, Two-photon fluorescence lifetime imaging monitors metabolic changes during wound healing of corneal epithelial cells in vitro. Graefes Arch. Clin. Exp. Ophthalmol. **250**(9), 1293–1302 (2012)
41. B. Geusens, M. Van Gele, S. Braat, S.C. De Smedt, M.C.A. Stuart, T.W. Prow, W. Sanchez, M.S. Roberts, N.N. Sanders, J. Lambert, Flexible nanosomes (SECosomes) enable efficient siRNA delivery in cultured primary skin cells and in the viable epidermis of ex vivo human skin. Adv. Funct. Mater. **20**(23), 4077–4090 (2010)
42. M. Göppert-Mayer, Über Elementarakte mit zwei Quantensprüngen. Ann. Phys. **9**, 273–294 (1931)
43. M.M. Ghosh, S.G. Boyce, E. Freedlander, S. MacNeil, A simple human dermal model for assessment of in vitro attachment efficiency of stored cultured epithelial autografts. J. Burn Care Rehabil. **16**(4), 407–417 (1995)
44. V.D. Giorgi, D. Massi, S. Sestini, R. Cicchi, F.S. Pavone, T. Lotti, V. De Giorgi, Combined non-linear laser imaging (two-photon excitation fluorescence microscopy, fluorescence lifetime imaging microscopy, multispectral multiphoton microscopy) in cutaneous tumours: first experiences. J. Eur. Acad. Dermatol. Venereol. **23**(3), 314–316 (2009)
45. V.V Ghukasyan, F. Kao, Monitoring cellular metabolism with fluorescence lifetime of reduced nicotinamide adenine dinucleotide †. J. Phys. Chem. C **113**(27), 11532–11540 (2009)
46. S. González, Z. Tannous, Real-time, in vivo confocal reflectance microscopy of basal cell carcinoma. J. Am. Acad. Dermatol. **47**(6), 869–874 (2002)
47. M.G.L. Gustafsson, Nonlinear structured-illumination microscopy: wide-field fluorescence imaging with theoretically unlimited resolution. Proc. Natl. Acad. Sci. U. S. A. **102**, 13081–13086 (2005)
48. K. Hanson, M. Behne, N. Barry, T. Mauro, Two-photon fluorescence lifetime imaging of the skin stratum corneum pH gradient. Biophys. J. **83**(3), 1683–1690 (2002)
49. A.A. Heikal, Intracellular coenzymes as natural biomarkers for metabolic activities and mitochondrial anomalies. Biomark. Med. **4**(2), 241–63 (2010)
50. R. Hofmann-Wellenhof, G. Pellacani, J. Malvehy, H.P. Soyer (eds.), *Reflectance Confocal Microscopy for Skin Diseases* (Springer, Berlin, Heidelberg, 2012) pp. 0–500
51. A. Höhn, T. Grune, Lipofuscin: formation, effects and role of macroautophagy. Redox Biol. **1**(1), 140–144 (2013)
52. S. Huang, A.A. Heikal, W.W. Webb, Two-photon fluorescence spectroscopy and microscopy of NAD(P)H and flavoprotein. Biophys. J. **82**(5), 2811–2825 (2002)
53. M. Huzaira, F. Rius, M. Rajadhyaksha, R.R. Anderson, S. González, Topographic variations in normal skin, as viewed by in vivo reflectance confocal microscopy. J. Invest. Dermatol. **116**(6), 846–852 (2001)
54. M.S. Islam, M. Honma, T. Nakabayashi, M. Kinjo, N. Ohta, pH Dependence of the fluorescence lifetime of FAD in solution and in cells. Int. J. Mol. Sci. **14**(1) 1952–1963 (2013)
55. T. Jung, A. Höhn, T. Grune, Lipofuscin: detection and quantification by microscopic techniques. Methods Mol. Biol. **594**, 173–193 (2010)
56. M. Kaatz, A. Sturm, P. Elsner, K. König, R. Bückle, M.J. Koehler, Depth-resolved measurement of the dermal matrix composition by multiphoton laser tomography. Skin Res. Technol. **16**, 131–136 (2010)

57. Y.-T. Kao, C. Saxena, T.-F. He, L. Guo, L. Wang, A. Sancar, D. Zhong, Ultrafast dynamics of flavins in five redox states. J. Am. Chem. Soc. **130**(39), 13132–13139 (2008)

58. M.J. Koehler, A. Preller, P. Elsner, K. König, U.C. Hipler, M. Kaatz, Non-invasive evaluation of dermal elastosis by in vivo multiphoton tomography with autofluorescence lifetime measurements. Exp. Dermatol. **21**(1), 48–51 (2012)

59. M.J. Koehler, K. König, P. Elsner, R. Bückle, M. Kaatz, In vivo assessment of human skin aging by multiphoton laser scanning tomography. Opt. Lett. **31**(19), 2879–2881 (2006)

60. K. König, P.T. So, W.W. Mantulin, B.J. Tromberg, E. Gratton, Two photon excited lifetime imaging of autofluorescence in cells during UVA and NIR photostress. J. Microsc. **183**(Pt 3), 197–204 (1996)

61. K. König, in *Methods in Cellular Imaging*, ed by A. Periasamy. Cellular Response to Laser Radiation in Fluorescence Microscopes (Oxford University Press, New York, 2001) pp. 236–254

62. K. König, I. Riemann, High-resolution multiphoton tomography of human skin with subcellular spatial resolution and picosecond time resolution. J. Biomed. Opt. **8**, 432–439 (2003)

63. K. König, A. Ehlers, F. Stracke, I. Riemann, In vivo drug screening in human skin using femtosecond laser multiphoton tomography. Skin Pharmacol. Physiol. **19**(2), 78–88 (2006)

64. K. König, Clinical multiphoton tomography. J. Biophotonics **1**(1), 13–23 (2008)

65. K. König, in *Handbook of Biomedical Nonlinear Optical Microscopy*, ed. by B.R. Masters, P.T.C. So. Multiphoton-induced Cell Damage (Oxford University Press, New York, 2008)

66. K. König, M. Weinigel, D. Hoppert, R. Bückle, H. Schubert, M.J. Köhler, M. Kaatz, P. Elsner, Multiphoton tissue imaging using high-NA microendoscopes and flexible scan heads for clinical studies and small animal research. J. Biophoton. **1**, 506–513 (2008)

67. K. König, Hybrid multiphoton multimodal tomography of in vivo human skin. IntraVital **1**, 11–26 (2012)

68. K. König, A.P. Raphael, L. Lin, J.E. Grice, H.P. Soyer, H.G. Breunig, M.S. Roberts, T.W. Prow, Applications of multiphoton tomographs and femtosecond laser nanoprocessing microscopes in drug delivery research. Adv. Drug Deliv. Rev. **63**(4–5), 388–404 (2011)

69. K. König, M. Speicher, M.J. Köhler, R. Scharenberg, M. Kaatz, Clinical application of multiphoton tomography in combination with high-frequency ultrasound for evaluation of skin diseases. J. Biophotonics **3**, 759–773 (2010)

70. K. König, M. Speicher, R. Bückle, J. Reckfort, G. McKenzie, J. Welzel, M.J. Koehler, P. Elsner, M. Kaatz, Clinical optical coherence tomography combined with multiphoton tomography of patients with skin diseases. J. Biophotonics **2**, 389–397 (2009)

71. K. König, H.G. Breunig, R. Bückle, M. Kellner-Höfer, M. Weinigel, E. Büttner, W. Sterry, J. Lademann, Optical skin biopsies by clinical CARS and multiphoton fluorescence/SHG tomography, Laser Phys. Lett. **8**, 1–4 (2011). doi:10.1002/lapl.201110014

72. H.I. Labouta, D.C. Liu, L.L. Lin, M.K. Butler, J.E. Grice, A.P. Raphael, T. Kraus, L.K. El-Khordagui, H.P. Soyer, M.S. Roberts, M. Schneider, T.W. Prow, Gold nanoparticle penetration and reduced metabolism in human skin by toluene. Pharm. Res. **28**(11), 2931–2944 (2011)

73. J. Lakowicz, H. Szmacinski, K. Nowaczyk, M. Johnson, Fluorescence lifetime imaging of free and protein-bound NADH. Proc. Natl. Acad. Sci. **89**(4), 1271–1275 (1992)

74. V.R. Leite-Silva, M. Le Lamer, W.Y. Sanchez, D.C. Liu, W.H. Sanchez, I. Morrow, D. Martin, H.D.T. Silva, T.W. Prow, J.E. Grice, M.S. Roberts, The effect of formulation on the penetration of coated and uncoated zinc oxide nanoparticles into the viable epidermis of human skin in vivo. Eur. J. Pharm. Biopharm. **84**(2), 297–308 (2013)

75. Y. Li, S. Gonzalez, T.H. Terwey, J. Wolchok, Y. Li, I. Aranda, R. Toledo-Crow, A.C. Halpern, Dual mode reflectance and fluorescence confocal laser scanning microscopy for in vivo imaging melanoma progression in murine skin. J. Invest. Dermatol. **125**(4), 798–804 (2005)

76. L.L. Lin, J.E. Grice, M.K. Butler, A.V. Zvyagin, W. Becker, T.A. Robertson, H.P. Soyer, M.S. Roberts, T.W. Prow, Time-correlated single photon counting for simultaneous monitoring of zinc oxide nanoparticles and NAD(P)H in intact and barrier-disrupted volunteer skin. Pharm. Res. **28**(11), 2920–2930 (2011)

77. R.M. Mackie, C. Fleming, A.D. McMahon, P. Jarrett, The use of the dermatoscope to identify early melanoma using the three-colour test. Br. J. Dermatol. **146**(3), 481–484 (2002)
78. M. Manfredini, F. Arginelli, C. Dunsby, P. French, C. Talbot, K. König, G. Ponti, G. Pellacani, S. Seidenari, High-resolution imaging of basal cell carcinoma: a comparison between multiphoton microscopy with fluorescence lifetime imaging and reflectance confocal microscopy. Skin Res. Technol. **19**, e433–e443 (2013)
79. K. Maslov, G. Stoica, L.V. Wang, In vivo dark-field reflection-mode photoacoustic microscopy. Opt. Lett. **30**, 625–627 (2005)
80. B.R. Masters, P.T.C. So, E. Gratton, Multiphoton excitation fluorescence microscopy and spectroscopy of in vivo human skin. Biophys. J. **72**, 2405–2412 (1997)
81. B.R. Masters, P.T. So, E. Gratton, Multiphoton excitation microscopy of in vivo human skin. Functional and morphological optical biopsy based on three-dimensional imaging, lifetime measurements and fluorescence spectroscopy. Ann. N. Y. Acad. Sci. **838**, 58–67 (1998)
82. B.R. Masters, P. So, Confocal microscopy and multi-photon excitation microscopy of human skin in vivo. Opt. Express **8**, 2–10 (2001)
83. B.R. Masters, P.T.C. So, *Handbook of Biomedical Nonlinear Optical Microscopy* (Oxford University Press, New Yok, 2008)
84. J. McGinty, N.P. Galletly, C. Dunsby, I. Munro, D.S. Elson, J. Requejo-Isidro, P. Cohen, R. Ahmad, A. Forsyth, A.V. Thillainayagam, M.A.A. Neil, P.M.W. French, G.W. Stamp, Wide-field fluorescence lifetime imaging of cancer. Biomed. Opt. Express **1**(2), 627–640 (2010)
85. P. Meredith, J. Riesz, Radiative relaxation quantum yields for synthetic eumelanin. Photochem. Photobiol. **79**(2), 211–216 (2007)
86. E.C. Nakajima, B. Van Houten, Metabolic symbiosis in cancer: refocusing the Warburg lens. Mol. Carcinog. **52**(5), 329–337 (2013)
87. N. Nakashima, K. Yoshihara, F. Tanaka, K. Yagi, Picosecond fluorescence lifetime of the coenzyme of D-amino acid oxidase. J. Biol. Chem. **255**(11), 5261–5263 (1980)
88. K.S. Nehal, D. Gareau, M. Rajadhyaksha, Skin imaging with reflectance confocal microscopy. Semin. Cutan. Med. Surg. **27**, 37–43 (2008)
89. Z. Nie, R. An, J.E. Hayward, T.J. Farrell, Q. Fang, Hyperspectral fluorescence lifetime imaging for optical biopsy. J. Biomed. Opt. **18**(9), 096001 (2013)
90. R. Niesner, B. Peker, P. Schlüsche, K.-H. Gericke, Noniterative biexponential fluorescence lifetime imaging in the investigation of cellular metabolism by means of NAD(P)H autofluorescence. ChemPhysChem **5**(8), 1141–1149 (2004)
91. K. Nishifuji, J.S. Yoon, The stratum corneum: the rampart of the mammalian body. Vet. Dermatol. **24**(1), 60–72.e15–6 (2013)
92. D.V. O'Connor, D. Phillips, *Time-correlated single photon counting* (Academic Press, London, 1984)
93. A. Ostrowski, D. Nordmeyer, A. Boreham, R. Brodwolf, L. Mundhenk, J.W. Fluhr, J. Lademann, C. Graf, E. Rühl, U. Alexiev, A.D. Gruber, Skin barrier disruptions in tape stripped and allergic dermatitis models have no effect on dermal penetration and systemic distribution of AHAPS-functionalized silica nanoparticles. Nanomedicine **10**(7):1571–1581 (2014)
94. J. Palero, H. Bruijn, A. Ploeg van den Heuvel, H. Sterenborg, H. Gerritsen, Spectrally resolved multiphoton imaging of in vivo and excised mouse skin tissues. Biophys. J. **93**(3), 992–1007 (2007)
95. X. Pan, R.P. Hobbs, P.A. Coulombe, The expanding significance of keratin intermediate filaments in normal and diseased epithelia. Curr. Opin. Cell Biol. **25**(1), 47–56 (2013)
96. R. Patalay, C. Talbot, Y. Alexandrov, M.O. Lenz, S. Kumar, S. Warren, I. Munro, M.A.A. Neil, K. König, P.M.W. French, A. Chu, G.W.H. Stamp, C. Dunsby, Multiphoton multispectral fluorescence lifetime tomography for the evaluation of basal cell carcinomas. PLoS ONE **7**(9), e43460 (2012)
97. R. Patalay, C. Talbot, Y. Alexandrov, I. Munro, M.A.A. Neil, K. König, P.M.W. French, A. Chu, G.W. Stamp, C. Dunsby, Quantification of cellular autofluorescence of human skin

using multiphoton tomography and fluorescence lifetime imaging in two spectral detection channels. Biomed. Opt. Express **2**(12), 3295–3308 (2011)

98. G. Pellacani, A.M. Cesinaro, S. Seidenari, Reflectance-mode confocal microscopy of pigmented skin lesions—improvement in melanoma diagnostic specificity. J. Am. Acad. Dermatol. **53**(6), 979–985 (2005)

99. G. Pellacani, P. Guitera, C. Longo, M. Avramidis, S. Seidenari, S. Menzies, The impact of in vivo reflectance confocal microscopy for the diagnostic accuracy of melanoma and equivocal melanocytic lesions. J. Invest. Dermatol. **127**(12), 2759–2765 (2007)

100. A.-M. Pena, M. Strupler, T. Boulesteix, M.-C. Schanne-Klein, Spectroscopic analysis of keratin endogenous signal for skin multiphoton microscopy. Opt. Express **13**(16), 6268 (2005)

101. T.W. Prow, J.E. Grice, L.L. Lin, R. Faye, M. Butler, W. Becker, E.M.T. Wurm, C. Yoong, T. A. Robertson, H.P. Soyer, M.S. Roberts, Nanoparticles and microparticles for skin drug delivery. Adv. Drug Deliv. Rev. **63**(6), 470–491 (2011)

102. M. Rajadhyaksha, M. Grossman, D. Esterowitz, R.H. Webb, R.R. Anderson, In vivo confocal scanning laser microscopy of human skin: melanin provides strong contrast. J. Invest. Dermatol. **104**(6), 946–952 (1995)

103. L. Rittié, G.J. Fisher, UV-light-induced signal cascades and skin aging. Ageing Res. Rev. **1**(4), 705–720 (2002)

104. M.S. Roberts, M.J. Roberts, T.A. Robertson, W. Sanchez, C. Thörling, Y. Zou, X. Zhao, W. Becker, A.V. Zvyagin, In vitro and in vivo imaging of xenobiotic transport in human skin and in the rat liver. J. Biophotonics **1**(6), 478–493 (2008)

105. M.S. Roberts, Y. Dancik, T.W. Prow, C.A. Thorling, L.L. Lin, J.E. Grice, T.A. Robertson, K. König, W. Becker, Non-invasive imaging of skin physiology and percutaneous penetration using fluorescence spectral and lifetime imaging with multiphoton and confocal microscopy. Eur. J. Pharm. Biopharm. **77**(3), 469–488 (2011)

106. T.A. Robertson, F. Bunel, M.S. Roberts, Fluorescein derivatives in intravital fluorescence imaging. Cells **2**(3), 591–606 (2013)

107. W.Y. Sanchez, C. Obispo, E. Ryan, J.E. Grice, M.S. Roberts, Changes in the redox state and endogenous fluorescence of in vivo human skin due to intrinsic and photo-aging, measured by multiphoton tomography with fluorescence lifetime imaging. J. Biomed. Opt. **18**(6), 61217 (2013)

108. W.Y. Sanchez, T.W. Prow, W.H. Sanchez, J.E. Grice, M.S. Roberts, Analysis of the metabolic deterioration of ex vivo skin from ischemic necrosis through the imaging of intracellular NAD(P)H by multiphoton tomography and fluorescence lifetime imaging microscopy. J. Biomed. Opt. **15**(4), 046008 (2010)

109. H. Schneckenburger, Fluorescence decay kinetics and imaging of NAD(P)H and flavins as metabolic indicators. Opt. Eng. **31**(7), 1447 (1992)

110. A. Scope, U. Mahmood, D.S. Gareau, M. Kenkre, J.A. Lieb, K.S. Nehal, M. Rajadhyaksha, In vivo reflectance confocal microscopy of shave biopsy wounds: feasibility of intraoperative mapping of cancer margins. Br. J. Dermatol. **163**(6), 1218–1228 (2010)

111. S. Segura, S. Puig, C. Carrera, J. Palou, J. Malvehy, Development of a two-step method for the diagnosis of melanoma by reflectance confocal microscopy. J. Am. Acad. Dermatol. **61** (2), 216–229 (2009)

112. S. Seidenari, F. Arginelli, S. Bassoli, J. Cautela, P.M.W. French, M. Guanti, D. Guardoli, K. König, C. Talbot, C. Dunsby, Multiphoton laser microscopy and fluorescence lifetime imaging for the evaluation of the skin. Dermatol. Res. Pract. **2012**(4951), 810749 (2012)

113. S. Seidenari, F. Arginelli, C. Dunsby, P. French, K. König, C. Magnoni, M. Manfredini, C. Talbot, G. Ponti, Multiphoton laser tomography and fluorescence lifetime imaging of basal cell carcinoma: morphologic features for non-invasive diagnostics. Exp. Dermatol. **21** (11), 831–836 (2012)

114. S. Seidenari, F. Arginelli, C. Dunsby, P.M.W. French, K. König, C. Magnoni, C. Talbot, G. Ponti, Multiphoton laser tomography and fluorescence lifetime imaging of melanoma: morphologic features and quantitative data for sensitive and specific non-invasive diagnostics. PLoS ONE **8**(7), e70682 (2013)

115. M.C. Skala, K.M. Riching, A. Gendron-Fitzpatrick, J. Eickhoff, K.W. Eliceiri, J.G. White, N. Ramanujam, In vivo multiphoton microscopy of NADH and FAD redox states, fluorescence lifetimes, and cellular morphology in precancerous epithelia. Proc. Natl. Acad. Sci. U. S. A. **104**(49), 19494–19499 (2007)

116. M.C. Skala, K.M. Riching, D.K. Bird, A. Gendron-Fitzpatrick, J. Eickhoff, K.W. Eliceiri, P.J. Keely, N. Ramanujam, In vivo multiphoton fluorescence lifetime imaging of protein-bound and free nicotinamide adenine dinucleotide in normal and precancerous epithelia. J. Biomed. Opt. **12**(2), 024014 (2007)

117. K. Sugata, S. Sakai, N. Noriaki, O. Osanai, T. Kitahara, Y. Takema, Imaging of melanin distribution using multiphoton autofluorescence decay curves. Skin Res. Technol. **16**(1), 55–59 (2010)

118. C. Suihko, L.D. Swindle, S.G. Thomas, J. Serup, Fluorescence fibre-optic confocal microscopy of skin in vivo: microscope and fluorophores. Skin Res. Technol. **11**, 254–267 (2005)

119. M. Szaszak, P. Steven, K. Shima, R. Orzekowsky-Schröder, G. Hüttmann, I.R. König, W. Solbach, J. Rupp, Fluorescence lifetime imaging unravels C. trachomatis metabolism and its crosstalk with the host cell. PLOS Pathogens **7**, e1002108-1–12 (2011)

120. C.A Thorling, X. Liu, F.J. Burczynski, L.M. Fletcher, M.S. Roberts, W.Y. Sanchez, Intravital multiphoton microscopy can model uptake and excretion of fluorescein in hepatic ischemia-reperfusion injury. J. Biomed. Opt. **18**(10), 101306 (2013)

121. C.A. Thorling, L. Jin, M. Weiss, D. Crawford, X. Liu, F.J. Burczynski, D. Liu, H. Wang, M.S. Roberts, Drug Metab. Dispos. 43(1), 154–62 (2015). doi:10.1124/dmd.114.060848

122. A. Uchugonova, M. Zhao, M. Weinigel, Y. Zhang, M. Bouvet, R.M. Hoffman, K. König, Multiphoton tomography visualizes collagen fibers in the tumor microenvironment that maintain cancer-cell anchorage and shape. J. Cell. Biochem. **114**(1), 99–102 (2013)

123. J.A. Veiro, P.G. Cummins, Imaging of skin epidermis from various origins using confocal laser scanning microscopy. Dermatology **189**(1), 16–22 (1994)

124. J. Vergen, C. Hecht, L.V. Zholudeva, M.M. Marquardt, R. Hallworth, M.G. Nichols, Metabolic imaging using two-photon excited NADH intensity and fluorescence lifetime imaging. Microsc. Microanal. **18**(4), 761–770 (2012)

125. M. Wakita, G. Nishimura, M. Tamura, Some characteristics of the fluorescence lifetime of reduced pyridine nucleotides in isolated mitochondria, isolated hepatocytes, and perfused rat liver in situ. J. Biochem. **118**(6), 1151–1160 (1995)

126. H.-W. Wang, V. Gukassyan, C. Chen, Y. Wei, H.-W. Guo, J.-S. Yu, F.-J. Kao, Differentiation of apoptosis from necrosis by dynamic changes of reduced nicotinamide adenine dinucleotide fluorescence lifetime in live cells. J. Biomed. Opt. **13**(5), 054011 (2008)

127. M. Weinigel, H.G. Breunig, J. Lademann, K. König. In vivo histology: optical biopsies with chemical contrast using multiphoton/CARS tomography. Laser Phys. Lett. 11055601 (2014)

128. J. Welzel, E. Lankenau, R. Birngruber, R. Engelhardt, Optical coherence tomography of the human skin. J. Am. Acad. Dermatol. **37**, 958–963 (1997)

129. D.H. Williamson, P. Lund, H.A. Krebs, The redox state of free nicotinamide-adenine dinucleotide in the cytoplasm and mitochondria of rat liver. Biochem. J. **103**(2), 514–527 (1967)

130. T. Wilson, C. Sheppard, Theory and Practice of Scanning Optical Microscopy (Academic Press, London, 1984)

131. X. Wu, J.A. Hammer, Melanosome transfer: it is best to give and receive. Curr. Opin. Cell Biol. **29**, 1–7 (2014)

132. W. Ying, NAD+/NADH and NADP+/NADPH in cellular functions and cell death: regulation and biological consequences. Antioxid. Redox Signal. **10**(2), 179–206 (2008)

133. J.-S. Yu, H.-W. Guo, C.-H. Wang, Y.-H. Wei, H.-W. Wang, Increase of reduced nicotinamide adenine dinucleotide fluorescence lifetime precedes mitochondrial dysfunction in staurosporine-induced apoptosis of HeLa cells. J. Biomed. Opt. **16**(3), 036008 (2011)

134. Y. Yu, A.M.D. Lee, H. Wang, S. Tang, J. Zhao, H. Lui, H. Zeng, Imaging-guided two-photon excitation-emission-matrix measurements of human skin tissues. J. Biomed. Opt. **17**(7), 077004 (2012)
135. G. Zhang, D.J. Moore, C.R. Flach, R. Mendelsohn, Vibrational microscopy and imaging of skin: from single cells to intact tissue. Anal. Bioanal. Chem. **387**, 1591–1599 (2007)
136. W.R. Zipfel, R.M. Williams, R. Christie, A.Y. Nikitin, B.T. Hyman, W.W. Webb, Live tissue intrinsic emission microscopy using multiphoton-excited native fluorescence and second harmonic generation. Proc. Natl. Acad. Sci. U. S. A. **100**(12), 7075–7080 (2003)
137. A.V. Zvyagin, X. Zhao, A. Gierden, W. Sanchez, J.A. Ross, M.S. Roberts, Imaging of zinc oxide nanoparticle penetration in human skin in vitro and in vivo. J. Biomed. Opt. **13**(6), 064031 (2008)

Chapter 16
Fluorescence Lifetime Imaging in Ophthalmology

Dietrich Schweitzer and Martin Hammer

Abstract Beginning disease in biological systems is often accompanied or preceded by changes in the metabolism of the tissue. Changes in the metabolism induce changes in the fluorescence decay functions of endogenous fluorophores. FLIM of the fundus of the eye is therefore a promising technique of detection early stages of eye diseases. Ophthalmic FLIM faces the problem that the excitation power is limited to very low levels, the transmission wavelength range of the ocular media is limited, the fluorescence intensities are low, and the decay functions of the fluorophores in the fundus are multi-exponential with extremely fast decay components. The task is further complicated by strong fluorescence of the lens of the eye and by the fact that the eye is constantly moving. We will show that these problems can be solved by a combination of multi-dimensional TCSPC with confocal scanning. This chapter gives a survey of the fluorophores found in the fundus, their fluorescence-decay properties, and the options to excite them at wavelengths that pass the ocular medium. It discusses the technical requirements to an ophthalmic FLIM system and describes the implementation of TCSPC FLIM in an ophthalmic scanning system for clinical use. The results obtained with the system allow fluorescence decay components to be associated to lateral and longitudinal anatomical structures of the eye, and to pathological alterations in the fundus. Statistical evaluation of the fluorescence decay parameters improves the identification of eye diseases and even detects phospho τ 181 protein in the eyes of Alzheimer patients.

D. Schweitzer (✉) · M. Hammer
Klinik für Augenheilkunde, Friedrich-Schiller-Universität Jena, 07743 Jena, Germany
e-mail: dietrichschweitzer@googlemail.com

M. Hammer
e-mail: Martin.Hammer@med.uni-jena.de

© Springer International Publishing Switzerland 2015
W. Becker (ed.), *Advanced Time-Correlated Single Photon Counting Applications*,
Springer Series in Chemical Physics 111, DOI 10.1007/978-3-319-14929-5_16

16.1 Introduction

Optical imaging of the fundus of the eye is the most frequently used diagnostic technique in ophthalmology. Images are obtained by fundus cameras, by ophthalmic scanners and by OCT imagers. The information recorded by these devices preferentially aims at the detection of morphological defects of the fundus tissue. However, when morphological changes have occurred it is often too late to reverse pathological processes. The need to detect pathological processes before they have resulted in severe tissue damage has led to fluorescence imaging. Fluorescence imaging may use exogenous fluorophores, such as fluorescein or ICG. In these cases, the fluorophore is mainly used to mark the vascular system, and to detect vessel occlusion or bleeding. More interestingly, fluorescence images can be obtained form the fluorescence on endogenous fluorophores. These may change their fluorescence behaviour depending on the local molecular environment of the fluorophore molecules and thus indicate early staged of changes in the metabolism. A typical example is the 'Redox Ratio', i.e. the intensity ratio of FAD-FADH2 (Flavin adenine dinucleotide) and NAD-NADH (Nicotinamide adenine dinucleotide) [7, 8, 41, 43, 44], see also Chaps. 13 and 14 of this book. Fluorophores may also exist in a protein-bound and in an unbound form. The balance between the forms can be a sensitive indicator of the metabolic state of the cells. This has been found especially for NADH and FAD [9, 25, 42]. Other fluorophores, such as Lipofuscin and AGE (advanced glycation end products) accumulate in the ocular fundus [26] in the process of age-related macular degeneration [10, 17] and diabetes mellitus. In these cases, the intensity ratio between the emission from different fluorophores may be used as a diagnostic parameter.

Moreover, the fluorescence quantum efficiency of most fluorophores depends on the molecular environment, especially the concentration of various ions, oxygen concentration, viscosity, and refractive index. Fluorophores may also exist in slightly different chemical constitutions, with slightly different fluorescence spectra and fluorescence quantum efficiencies. Changes in the fluorescence signals induced by these effects may be indicators of the tissue state as well.

The problem of simple autofluorescence intensity imaging is that most of these changes cannot be distinguished from changes in the concentration of the fluorophores. An improvement is obtained by spectral imaging. However, the fluorescence spectra of endogenous fluorophores are broad and poorly defined. Moreover, the apparent spectra can change due to variable absorption in the tissue.

The situation improves considerably if the *fluorescence lifetimes* or, more exactly, the *fluorescence decay functions* are used as imaging parameters. The fluorescence lifetime does not depend on the concentration, and it remains unchanged when a part of the fluorescence light is absorbed, see Chap. 3 of this book. Changes in the bound-unbound ratios, changes in the relative concentrations of the fluorophore fractions, or changes in the fluorescence quantum efficiencies of fluorophores therefore show up more clearly in fluorescence lifetime data than in intensity data or fluorescence spectra.

16.2 Boundary Conditions

Fluorescence imaging of the human ocular fundus is restricted by a number of boundary conditions. The most important one is, of course, a limitation in the excitation intensity. The eye as sensitive organ for electro-magnetic radiation in the visible range must not be damaged by the examination.

The effect of wavelength on optical damage at the molecular level can be estimated by comparing the radiation energy according $E = \hbar v$ with bond energy of atoms in molecules. Figure 16.1 shows the binding energy of a number of characteristic bonds (different colours), the photon energy (dark blue), and the transmission curve of the optical pathway of the eye (red) as a function of wavelength.

As can be seen from Fig. 16.1, the natural transmission of the ocular media, predominantly of the crystalline lens, protects the fundus against direct molecular damage. Only the C–S binding can be broken by radiation in the visible range. Nevertheless, electron transfer from excited states can cause radicals which, in turn, induce chemical reactions in the ocular tissue. The excitation power should therefore be kept as low as possible, see [1, 16].

Reflectance signals recorded from the fundus are on the order of 2 % of the incident light intensity. Autofluorescence signals from the fundus are at least three orders of magnitude weaker. Considering the fact that the excitation power is limited by eye safety considerations, a sufficiently large number of fluorescence photons can only be recorded by accumulating the photons for a longer period

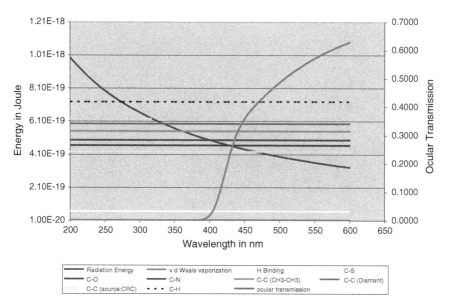

Fig. 16.1 Comparison between radiation energy of excitation and binding energy of atoms. If the energy of the exciting radiation is greater than the binding energy, the corresponding bond can be broken. The bond homolytic dissociation energy was calculated according to [20, 29]

oftime. Typical acquisition times are on the order of a few 10 s to a few minutes (see also Chap. 2, Sect. 2.2.4). However, long acquisition time causes a problem with eye motion: It is impossible for a patient to keep his eye fixed on a target for longer than a few seconds. Then the eye wanders off, and eventually jumps back to the target. To obtain clear images of the fundus the motion of the eye must be tracked during the image acquisition, and compensated in the recording process.

Another problem occurs due to the limited spectral transmission range of the ocular media, see Fig. 16.1, red curve. The excitation maxima of several endogenous fluorophores are in the spectral range below 400 nm. Light in this range is blocked by the absorption of the ocular media. A number of fluorophores can still be excited in their absorption tails above 400 nm [35]. However, no selective excitation via the excitation wavelength is possible. Thus, unavoidably fluorescence from several fluorophores contributes to the measured spectrally or temporally resolved data.

Another requirement is caused by the layered anatomy of the eye, see Fig. 16.2. The inset on the right shows a magnified drawing if the fundus. The beam path length from the anterior part (crystalline lens, cornea) to the posterior part (fundus) is about 22 mm whereas the fundus itself is only a few 300 μm thick. The lens and the cornea are fluorescent. Due to the large length of the beam path inside the lens and the cornea the relative amount of fluorescence from these structures is substantial. It must be suppressed to obtain clear fluorescence data from the fundus. The fundus on its part consists of several layers, having thicknesses down to some μm. The fundus layers are anatomically and functionally different. To select the desired layer for imaging the imaging process must have some kind of optically sectioning capability.

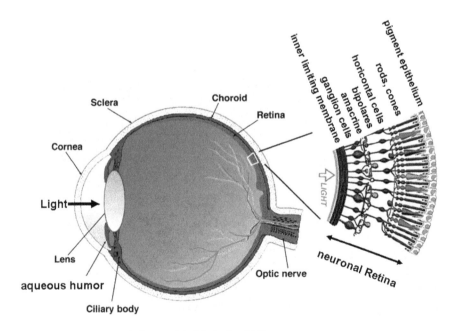

Fig. 16.2 Section through the eye. Modified, after [21]

Ophthalmic scanners therefore use a confocal principle: The fundus is scanned with a focused laser beam, and the fluorescence returned from the fundus is projected into a confocal pinhole. Only light from the focal plane passes the pinhole unimpeded. Light from other planes is not focused into the pinhole and thus substantially suppressed. However, the out-of focus suppression is not as good as in confocal microscopy: The numerical aperture of the lens of the eye is much lower than that of a microscope lens, and the optical quality of the eye is far from being perfect. The suppression, U, of the fluorescence of the lens depends on the diameter, d_A, of the aperture of the detection system (diameter of the pupil), the diameter, d_F, of the illumination/detection pinhole at the fundus field, as well as on the thickness of the lens, D_L, and of the layered fundus, D_F [33].

$$U = \frac{d_F^2 \cdot D_L}{d_A^2 \cdot D_F} \tag{16.1}$$

As long as the pinhole remains larger than the point spread function of the eye-scanner combination the suppression of lens fluorescence increases with decreasing pinhole diameter. However, also, the amount of photons detected from the fundus decreases. Therefore, a compromise between out-of-plane suppression and in-plane detection efficiency must be found.

16.2.1 Lifetime Imaging Technique

Fluorescence lifetime imaging techniques can be classified into time-domain techniques and frequency-domain techniques, wide-field techniques and point-scanning techniques, and analog and photon counting techniques. Time-domain techniques can be further classified into techniques that scan a time gate over the decay functions, record the data simultaneously into a few time gates, or record simultaneously into a large number of time channels. Frequency-domain techniques can work with sinusoidal modulation of the excitation light, or with short light pulses of high frequency. Almost all combinations are in use. The combinations vary considerably in photon efficiency, i.e. the number of photons needed to achieve a given lifetime accuracy, time resolution, compatibility with fast scanning, ability to resolve multi-exponential decay profiles into their components, and the range of light intensities they can be applied to [3, 6].

The requirements for ophthalmic FLIM are high photon efficiency (the excitation power is limited by eyes safety considerations), high time resolutions (autofluorescence decay components can be faster than 100 ps), capability to resolve multi-exponential decay functions into their components (autofluorescence decays are multi-exponential), compatibility with the fast scanning used in ophthalmoscopes (frames times are on the order of 100 ms), tolerance to intensity changes during the acquisition (the patient may change the focus of the out of the confocal plane) and a fast way to compensate eye movement. Moreover, Schweitzer et al. have shown that even in an ideal optical system no more than one photon can obtained from the fundus for one excitation pulse [32, 36]. Thus, both the requirements and the measurement conditions are perfectly in the reach of multi-dimensional TCSPC [2, 4], see Chap. 1 of this book. TCSPC FLIM is clearly the technique of choice for ophthalmic FLIM. The implementation into a fluorescence lifetime imaging ophthalmoscope (FLIO) will be described in Sect. 16.3.

16.2.2 Previous Investigations

To find out the excitation and emission range as well as the interval of decay time of endogenous fluorophores, spectral and time-resolved measurements were performed on purified substances. In addition, the excitation and emission spectra as well as the fluorescence decay times were determined for anatomical structures of porcine eyes. These data are given in [34, 35, 38]. As the apparent excitation spectra of the endogenous fluorophores are distorted by the ocular transmission [45], the effective maxima for excitation are all around 450 nm. That means, no wavelength selective excitation of the individual fluorophores is possible.

Fluorescence lifetimes of some endogenous fluorophores are given in Table 16.1. Data found in the literature vary considerably; the results of an extended recherché can be found in [27].

Investigating the fluorescence of porcine eyes, we found fluorescence from virtually all anatomical structures of the eye. For the experiments, we used picosecond pulse excitation by a diode laser at a wavelength of 446 nm, and recorded the fluorescence decay function by TCSPC in the spectral range from 490 to 700 nm. Table 16.2 shows the modal value of the distribution of mean fluorescence lifetimes of anatomical structures of porcine eyes as result of bi-exponential fit.

The results show that all anatomical ocular structures fluoresce, though with different intensities. Although an exact mechanical separation of retinal pigment epithelium (RPE) from Bruch's membrane and choroid is difficult, the shortest lifetime were found in this tissue. The lifetime of neuronal retina is also shorter than in the other ocular tissues. As expected, the fluorescence lifetimes of the cornea and the lens are very similar and correspond to the mean lifetime of collagen 2. The long lifetimes of choroid and sclera correspond well to the mean lifetime of collagen 1.

Due to the difference in the emission spectra, a separation of the signals from the expected fluorophores can be obtained if the lifetime is detected in a short-wavelength interval (490–560 nm) and in a long-wavelength interval (560–700 nm). Based on this knowledge, the fundus layers were investigated in to by two-photon excitation at 760 nm and the fluorescence lifetime was detected in two spectral intervals. The mean lifetimes of a double-exponential fit are given in Table 16.3.

These measurements show that the RPE has the shortest lifetime of all layers, and that the lifetime is virtually the same in both spectral channels. The lifetimes of

Table 16.1 Fluorescence lifetimes of endogenous substances, expected at the fundus

Substance	T_1 (ps)	A_1 (%)	T_2 (ps)	A_2 (%)	References
FAD free	330	18	2810	82	Schweitzer [35]
Free			2068		Skala [41]
FAD protein bound	130 ± 20	Mono.			Nakashima [28]
Protein bound	40 ± 10	Dimer			
Protein bound	106				Skala [41]
NADH free	387	73	3650	27	Schweitzer [35]
Free	325				Skala [41]
Free extended	155				Vishwasrao [46]
Free folded	600–700				Vishwasrao [46]
NADH bound	599				Vishwasrao [46]
Bound			2154		Vishwasrao [46]
Bound			2356		Skala [41]
Bound			6040		Vishwasrao [46]
Bound			>1 ns		Wu [47]
Collagen 1	670	68	4040	32	Schweitzer [35]
Collagen 2	470	64	3150	36	
Collagen 3	345	69	2800	31	
Collagen 4	740	70	3670	30	
Elastin	380	72	3590	28	
AGE	865	62	4170	28	
A2E	170	98	1120	2	
Lipofuscin	390	48	2240	52	
DOPA melanin	40		1200		Ehlers [15]
Pheomelanin	340		2300		
Eumelanin	30		900		
Melanocyten naevus	136 ± 33	94 ± 3	1061 ± 376	6 ± 3	Dimitrow [12]
Melanin powder	280	70	2400	30	Schweitzer [35]

Table 16.2 Mean (amplitude-weighted) fluorescence lifetimes of anatomical structures of porcine eyes

Tissue	RPE	Retina	Vitreous	Lens	Cornea	Choroid	Sclera	Aqueous humor
τ_{mean} in ps	260	460	960	1320	1400	1700	1780	2520

Excitation by picosecond pulses at 446 nm, emission range 490–700 nm

Table 16.3 Mean (amplitude-weighted) lifetime of layers of porcine eyes, detected in to [30]

(nm)	Layer	NFL	GCL	GCL-e	GCL-m	INL	ONL	PRIS	RPE
500–550	τ_{mean} in ps	929	548	732	1111	927	1096	882	79
550–700	τ_{mean} in ps	521	449	410	570	454	386	308	78

Excitation by femtosecond pulses at 760 nm and bi-exponential fit. *NFL* nerve fibre layer, *GCL* ganglion cell layer, *GCLe* cell body, *GCLm* mueller cell, *INL* inner nuclear layer, *ONL* outer nuclear layer, *PRIS* photoreceptor inner segments the other structures are much longer in the short-wavelength channel than in the long-wavelength one.

Although the anatomical structure is comparable between porcine and human eyes, these data are only qualitative relations. In contrast to human eyes, there is no accumulation of ageing pigment lipofuscin or of advanced glycation end products. There is also no cataract formation in porcine eyes of pigs younger than 8 months.

16.2.3 Required Number of Photons

For an estimation of the required number of photons for separating fluorophores by their lifetimes, the basic (16.2) for a sum of exponential decay components and for the relative intensity contributions of the components, Q_i, were used (16.3).

$$\frac{I_1(t)}{I_0} = \sum_i a_i \cdot e^{-\frac{t}{\tau_i}} + b \qquad (16.2)$$

with:

$I_1(t)$ fluorescence intensity at time t
I_0 fluorescence intensity at time t = 0
a_i amplitude of component i
τ_i lifetime of component i
b   background (offset)

$$Qi = \frac{\tau_i \cdot a_i}{\sum_j \tau_j \cdot a_j} \qquad (16.3)$$

In the program system FLIMX [23] a simulation procedure was performed at different number of photons by the following steps:

Step 1: Giving a sum of e-functions with determined lifetimes and amplitudes, corresponding to selected relative contributions.
Step 2: Modifying the sum decay by random noise
Step 3: Fit of the noisy decay by a sum of e-functions
Step 4: Comparison of given and calculated model parameters

The sum of relative errors, ES, between the correct, τ_{set} and the calculated lifetimes, τ_{actual} was determined according to (16.4):

$$ES = \sum \left| \frac{\tau_{set1} - \tau_{actual1}}{\tau_{set1}} \right| \cdot 100 + \left| \frac{\tau_{set2} - \tau_{actual2}}{\tau_{set2}} \right| \cdot 100 \qquad (16.4)$$

The relative difference of lifetimes, not their absolute difference is crucial for the determination of fluorophores. Generally, two fluorophores can best discriminated if their relative contribution Qi = 0.5. The dependence of ES on the relative contribution is demonstrated in Fig. 16.3, with the number of photons as a parameter. Here, the relative difference of lifetimes was $(\tau_1 - \tau_2)/\tau_1 = 0.75$.

Figure 16.3 shows that the sum of relative errors ES is about 20 %, when 10,000 photons are collected and the relative contribution is equal of both components and the relative lifetime difference is 0.75. For an ES of 10 % about 40,000 photons are necessary. This result corresponds well with the data given in [24]. To reduce ES down to about 5 %, about 80,000 photons are required. This small error sum was than calculated for an extended interval $0.25 < Q_1 < 0.75$.

Considering mixtures of three components with parameters of Table 16.4, determined in vivo on healthy subjects, the errors of lifetimes are comparable in both spectral channels (CH1 490–560 nm, CH2 560–700 nm).

Figure 16.4 shows the lifetime error for triple-exponential decay analysis for different numbers of photons. It can be seen that at least 50,000 photons are required for the determination of lifetimes with acceptable errors. In this case, the relative errors for τ_1, τ_2, and τ_3 are about 5, 10, and 30 %. To calculate lifetimes with a sum of relative errors ES < 10 % in a three-exponential fit, about 400,000 photons must be accumulated.

Based on these simulations, the measuring time can be calculated for accumulation of certain numbers of photons. The measuring time depends on the repetition rate of the pulse laser, on the acquisition time of an image, on the image resolution,

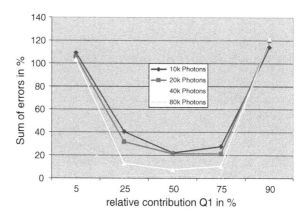

Fig. 16.3 Dependence of relative error sum ES on relative contribution Q_1 in a mixture of two fluorophores with relative difference of lifetimes 0.5

Table 16.4 Fit parameters of FLIO measurements on healthy subjects for simulation of lifetime errors depending on the collected photons

	CH1	CH2
τ_1 in ps	60	70
τ_2 in ps	400	380
τ_3 in ps	3150	2250
a_1 in %	86	78
a_2 in %	12	20
a_3 in %	2	2

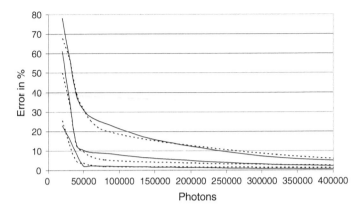

Fig. 16.4 Lifetime errors in triple-exponential fit results using data measured in vivo on healthy subjects. *Dotted Line* Channel 1, *dashed line* channel 2

on the number of spectrally different images, and on the assumed detection probability of fluorescence photons. In a raw estimation, about 2.9 min are necessary for collecting 400.000 photons in a limit case, when 80 MHz pulse frequency, 0.1 detection probability, a field of 150 × 150 pixel in two spectral different images and 25 binned pixels (binning factor B = 2 in SPCImage, BH software) are assumed. Under realistic conditions, the number of collected photons is at least 10 times lower during this measuring time [22].

16.3 Technical Implementation

The ophthalmic FLIM system uses an ophthalmic scanner of Heidelberg Engineering in combination with a Becker & Hickl picosecond diode laser and Becker and Hickl TCSPC FLIM modules. The system was developed in cooperation of the University of Jena, Heidelberg Engineering GmbH, Heidelberg, Germany, and Becker & Hickl GmbH, Berlin, Germany. Prototypes are currently used in clinical [11, 13] and pre-clinical [14] studies.

The system uses a confocal scanning technique with the pupil of the eye placed in the exit pupil of the scanning system [40]. Confocal detection suppresses light from outside the desired focal plane inside the eye, and significantly reduces the sensitivity to ambient light. The principle of the optical system is shown in Fig. 16.5.

A 473 nmpicosecond diode laser is used for fluorescence excitation. The laser is delivered into the scanner via a single-mode optical fibre. The light from the fibre is collimated by a lens, L1. The collimated laser beam passes a beam combiner, BC1, and a dichroic beam splitter, BS2. The beam is then deflected by a fast optical scanner which performs a raster scan in x andy. In the system made by Heidelberg Engineering, the beam is scanned by galvanometer mirrors. The mirror for the X scan works in resonance mode, with a speed of about 5000 lines per second.

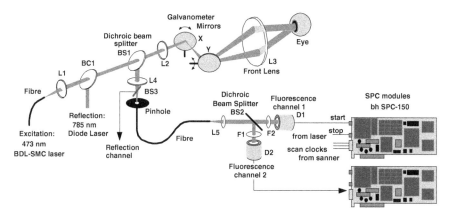

Fig. 16.5 Functional principle of the ophthalmic FLIM system. Simplified, details of real optics may differ from drawing

The beam is scanned over the aperture of the front lens of the scanner, L3. The front lens sends a parallel beam of light into the eye. The angle of this beam changes with the scan. The lens of the eye focuses the beam on the fundus, scanning a spot of laser light over the fundus.

Light emitted from the illuminated spot is collected by the front lens of the scanner, de-scanned by the galvanometer mirrors, separated from the excitation light by the dichroic beam splitter, BS1, and projected into the pinhole by a lens, L4. The light passing the pinhole is send to the detector assembly via an optical fibre. There it is further split into two detection wavelength intervals by another dichroic beam splitter, BS2, and two filters, F1 and F2. The wavelength channels are from 490 to 560 nm and from 560 to 700 nm, respectively [33]. The signals in the two spectral intervals are detected by fast TCSPC detectors, D1 and D2. Early systems used MCP PMTs, recent systems use hybrid detectors [5].

The single-photon pulses delivered by the detectors are recorded by the TCSPC modules. The early systems used a single bh SPC-830 TCSPC module [4] with a router. (Please see Chap. 1, Sect. 1.4.1 for principle of routing). However, the efficiency of recent systems is so good that the photons cannot be reasonable processed by a single TCSPC channel. The photons detected by the two detectors are therefore processed in two fully parallel SPC-150 TCSPC channels [4].

An important part of the instrument is the eye motion compensation system. Eye motion is detected by sending an NIR laser of 785 nm wavelength into the eye together with the excitation laser. The reflected light from this laser is separated from the fluorescence by a dichroic beam splitter, BS3. The light is used to record a reflection image of the fundus. The intensity of the reflection signal is sufficient for obtaining a reflection image for every frame of the scan. The reflection images are displayed on-line to allow the operator to focus on the desired fundus structures and to keep the system in focus during the FLIM acquisition. More importantly, subsequent reflection images are analysed to determine the displacement of the fundus image between the frames. A new displacement vector is determined for every single frame, and used to correct the x-y position in the TCSPC imaging process.

The principle of the signal processing in the TCSPC modules is shown in Fig. 16.6. It differs from the general TCSPC FLIM principle [2] (Chap. 1, Sect. 1.4.5), in two details.

Because the scan in the ophthalmic scanner is extremely fast it is no longer possible to transfer a pixel clock pulse for every single pixel. Instead, the pixel position inside the line is determined by measuring the times of the photons after the lineclock pulses. This 'internal pixel clock' mode is implemented in bh TCSPC FLIM modules [4] but not normally used for FLIM—it would be inconvenient to use when the scanner is operated at variable line frequency. For ophthalmic FLIM the internal pixel clock is no disadvantage: The X scan is performed by a resonance scanner the line frequency of which does not change.

The second difference is in the determination of the pixel locations for the photons inside the FLIM data array. As usual, the FLIM data are built up by software accumulation. However, the pixel positions, X, X, are not taken directly from the scanning interface of the TCSPC module. Instead, the destination of the photons is the sum of the xy coordinates determined by the TCSPC FLIM process and the displacement vector for the NIR imaging system of the ophthalmoscope.

FLIM data are usually recorded at a resolution of 512×512 pixels, and 1024 time channels per pixel. The large number of time channels is necessary because the decay functions contain components from about 5 ns down to less than 50 ps. On the one hand, long decay components should by recorded over a sufficiently long time interval, on the other hand, fast decay components should be sampled at a time-channel width no larger than 1/5 of the decay time constant. That means about 1000 time channels are needed for artefact-free recording of the decay functions.

A diagram of the signal flow in the ophthalmic FLIM system is shown in Fig. 16.7. Figure 16.8 shows a photo of the system and a typical fundus FLIM image recorded by the system.

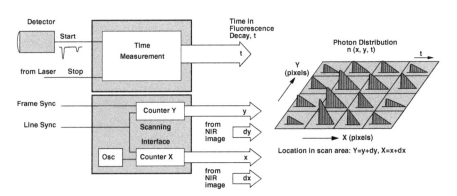

Fig. 16.6 TCSPC FLIM system for ophthalmic FLIM

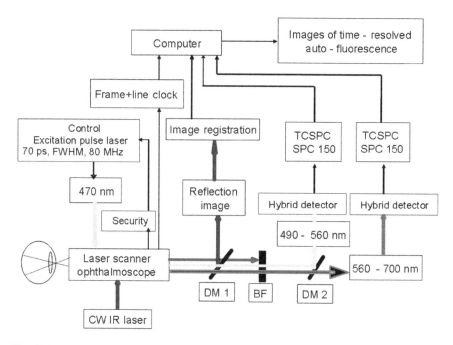

Fig. 16.7 Signal-flow scheme of the fluorescence lifetime imaging ophthalmoscope (FLIO) in the version of Heidelberg Engineering

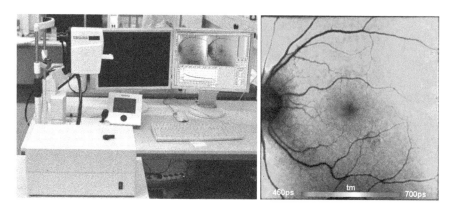

Fig. 16.8 *Left* Photo of the FLIO ophthalmic FLIM system of Heidelberg Engineering GmbH. *Right* FLIM image of a healthy eye recorded by the FLIO system. 512 × 512 pixels, 1024 time channels per pixel

16.4 Fluorescence Decay Functions in the Fundus

16.4.1 Lateral Distribution of Lifetimes

Typical decay signatures of the ocular fundus are shown in the FLIO images in Fig. 16.9. The shortest lifetimes are found in the macula, the longest lifetimes in the optic disc (papilla). The lifetimes in the range outside the macula and optic disc (PMB) range between the macula and papilla lifetimes. The lifetimes in the short-wavelength channel are somewhat longer than in the long-wavelength channel. The reason is the fluorescence spectrum of the crystalline lens, which is dominating in

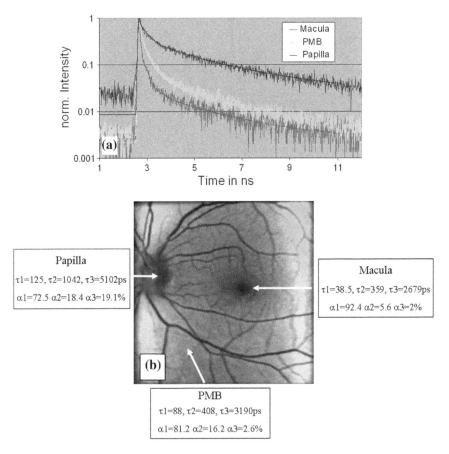

Fig. 16.9 Lifetimes and amplitudes of decay components of a healthy subject obtained by triple-exponential fit. Excitation at 448 nm, fluorescence detected from 490–560 nm. *Top* Fluorescence decay at selected fundus sites, b-fit results. *Bottom* Lifetime image of a human ocular fundus. Analysis by bh SPCImage, *colour* represents amplitude-weighted lifetime of triple-exponential decay

the short-wavelength channel. However, the long decay time of connective tissue is also detectable in the short-wavelength channel. The fluorescence intensity at the papilla is much weaker than at the other locations.

16.4.2 Lens Fluorescence

The influence of the fluorescence of the lens of the eye is demonstrated in Fig. 16.10. Here, the fluorescence decay in the macula of a subject wearing an artificial intra-ocular lens (IOL) is compared with that of a young subject with natural crystalline lens (CL) and a subject with cataract (CAT).

Amplitudes and lifetimes of these 3 cases are given in Table 16.5. The subjects were selected randomly.

It is somewhat surprising that the decay of the ocular lens has a considerable influence especially on τ_2 and τ_3, although the fluorescence was detected confocally. In reflection images, where the fundus light is detected through a small pinhole, the reflection at the surface of the crystalline lens is highly suppressed. The conditions for detection of fundus fluorescence in laser scanning ophthalmoscopy have been calculated in [33]. In addition to the ratio of solid angle for detection of light originating from the fundus and from the lens, the thickness of the fluorescent structures must be taken into account, see (16.1). As the crystalline lens is much thicker than the fundus, lens fluorescence is excited in a much larger volume than in the fundus. As a result, the signal from the lens is stronger than the signal from the fundus, and not entirely suppressed by the pinhole.

The influence of the lens fluorescence is also visible in the rising edge of recorded decay profiles: The lens fluorescence starts earlier than the fundus fluorescence. This fact should be taken into account when the left border of the fitting interval is selected.

As demonstrated in Fig. 16.11, different values are obtained for amplitudes and lifetimes when the left border of the fitting interval is set before the rise of the

Fig. 16.10 Influence of the ocular lens on fundus fluorescence decay. IOL-artificial intra ocular lens, CL-natural crystalline lens, CAT-cataract lens

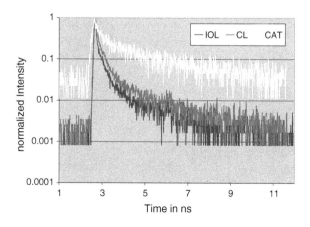

Table 16.5 Influence of lens fluorescence on the decay parameters of the macula

		α_1	τ_1	α_2	τ_2	α_3	τ_3
CH 1	IOL	81.1	42	10.7	231.4	5.1	764.6
	Cryst. lens	92.4	38	5.6	359	2	2679
	Cataract	72.8	84	20.6	1905.6	6.6	10,962
CH 2	IOL	94.4	27	4.8	283.9	0.8	1633.9
	Cryst. lens	57.7	68	37.2	262.5	5.1	1717
	Cataract	63.9	289	28.1	1877	7.8	5823.3

Fig. 16.11 Influence of the *left* border of the fitting interval in data containing lens fluorescence. *Top* Fit interval starts before the rise of the fluorescence. *Bottom* Fit interval starts at the maximum of the fluorescence

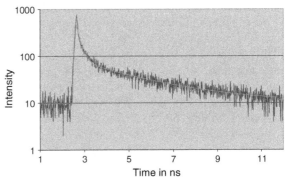

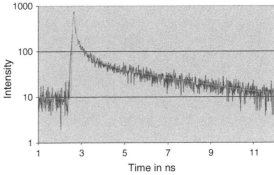

Table 16.6 Fit results, depending on the location of the left border of the fit interval

	α_1	τ_1	α_2	τ_2	α_3	τ_3	Red. chi square
Fit of complete decay	90.8	156	3.5	1877	5.7	3417	2.78
Fit starting at maximum	91.6	59	4.6	593	3.8	3929	1.16

fluorescence, or to the maximum of the fluorescence in the SPCImage data analysis software. In both cases the shift parameter was changed until the reduced chi square was minimal. The macular decay was fitted under both conditions. As shown in Table 16.6, the parameters amplitudes and lifetimes are considerably different, despite a minimum of the reduced chi^2 was reached in both cases. In case (a), the fit in the maximum is quite bad and the lifetimes τ_1 and especially τ_2 are quite long. The parameters of macular fit, reached in tail fit (b), the reduced chi^2 is much smaller. The differences are not surprising. They are simply a result of the fact that the model function (a sum of exponentials) does not describe the fluorescence decay accurately. To eliminate the influence of the lens fluorescence as far as possible the results presented in this chapter were obtained by setting the left boarder of the fit interval to the maximum of the fluorescence.

16.4.3 Correspondence of Decay Parameters with Vertical Anatomical Structures

Although probably more than three fluorophores are excited in the FLIO measurements, triple-exponential fit results of the fluorescence decay shows a certain correspondence of the decay components with the anatomical structures of the eye. The crystalline lens has mean lifetimes $2.5 < \tau_m < 3.5$ ns after bi-exponential fit. Such measurements are possible if the Laser Scanning Ophthalmoscope is focussed at the anterior part of the eye.

Besides previous measurements on isolated ocular structures and on 2-photon investigation of the layered fundus, intersection in images of amplitudes or lifetime confirm the relation between parameters of 3-exponential fit and fundus layers.

Figure 16.12 shows a cross section of the fundus of a human eye recorded by optical coherence tomography (OCT). As can be seen from the figure, the retinal pigment epithelium (RPE) is homogenously distributed across the fundus, but it is missing at the optic disc. The neuronal retina is thinned in the macula. That means, the relative amplitude of the strongly emitting RPE is high in the macula. On the other hand, the amplitude of the weakly emitting neuronal retina has a minimum in the macula. The anatomical structure of the optic disc contains predominantly connective tissue. As a consequence, the lifetimes τ_1, τ_2, and τ_3 are much longer than in the other fundus regions.

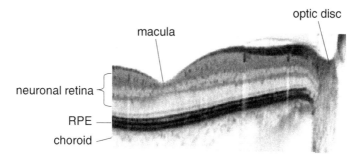

Fig. 16.12 Cross section through the macula and optic disc, recorded by OCT

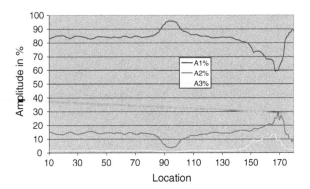

Fig. 16.13 Macular cross sections of amplitude of decay components of a subject with IOL

More can be concluded from FLIO images of a subject wearing an implanted artificial lens (IOL). An IOL is virtually free of fluorescence. Figures 16.13, 16.14, and 16.15 are cross sections through the macula and parts of the optic disc. Figure 16.13 shows a cross section of amplitudes, a_1, a_2, a_3, of the fast, medium, and slow fluorescence decay component.

The amplitude of the fast component, α_1, is dominating and reaches up to 95 % in the macula and decreases considerably in the optic disc. No melanin is present in the optic disc. The component α_1 corresponds with the highly fluorescent retinal pigment epithelium.

The amplitude of the medium component, α_2, is about 15 %, but decreases in the macula. The neuronal retina is very thin in the macula. Only receptor axons are present there. Thus, the component corresponds with the neuronal retina. It is interesting that a third component with an amplitude of some percent is required for an optimal fit.

Figure 16.14 shows the lifetimes of the fast, medium, and slow decay component in a macular cross section in logarithmic scale. All lifetimes come from mixtures of different fluorophores. The lifetime τ_1 is very short, about 60 ps, and further decreases in the macula. In the optic disc, the lifetime increases because the ana-tomical structure is completely different in comparison with the main part of the fundus. The lifetime τ_2 is about 350 ps at the whole fundus outside the optic disc. As there is no change of τ_2 in the macula, it confirms that the lifetime is independent of the concentration, here independent of thickness. The lifetime τ_2 relates to neuronal retina. The lifetime τ_3 is about 2.5 ns, but increases up to 7 ns at the optic disc. This lifetime predominantly comes from connective tissue (collagen, elastin).

The relative contribution Q_i indicates the relative intensity (or the relative number of photons) in the corresponding exponential component of the decay function. Figure 16.15 shows that the relative numbers of photons in the fast and medium decay component are nearly identical outside the macula and the optic disc. The fluorescence in the macula is dominated by the photons from the fast component; the fluorescence in the optic disc is dominated by photons from the slow component.

These examples demonstrate the existence of a third component at the fundus. If eyes with crystalline lenses were measured, the contribution of the third component

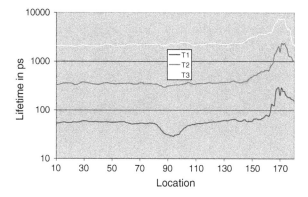

Fig. 16.14 Macular cross sections of FLIO lifetimes of a subject with IOL. *blue* T_1, lifetime of fast decay component, *pink* T_2, lifetime of medium decay component, *yellow* lifetime of slow decay component

Fig. 16.15 Macular cross sections of relative contributions, Q_1, Q_2, Q_3, of fast, medium and slow decay components for a subject with IOL

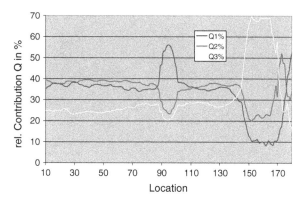

increases and the value of τ_3 is nearly the lifetime of the crystalline lens. In case of cataract, the third component dominates. This influence of the crystalline lens acts predominantly in the short-wavelength channel, where the fluorescence intensity of the lens is much higher than in the long-wavelength channel.

16.5 Presentation of FLIO Results

16.5.1 Presentation of Local Alterations in Images of Decay Parameters

The results of FLIO measurements are presented as images of fluorescence decay parameters or by statistical comparison between groups of patients and of healthy subjects. Local alterations in amplitudes or in lifetimes are visible in images of advanced stages of a disease. In early stages of diseases, differences with healthy subjects are demonstrable by statistical methods.

Figures 16.16 and 16.17 show images of the the mean lifetime, τ_m, and the fast decay component, α_1, of a healthy subject. The lifetimes and amplitudes were obtained by a triple-exponential tail fit of the decay functions in the short-wavelength channel.

In healthy subjects, the longest lifetime is generally detected in the optic disc and the shortest lifetimes, τ_m and τ_1, the macula. The relative amplitude α_1 of the

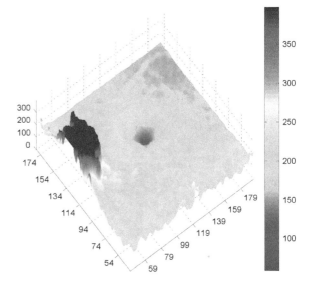

Fig. 16.16 Image of a healthy subject, mean (amplitude-weighted) lifetime, τ_m, in ps, in the short-wavelength channel (490–560 nm)

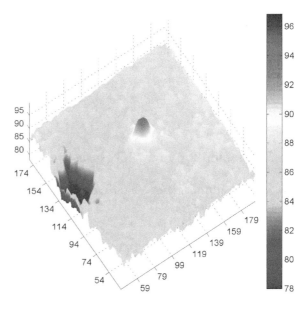

Fig. 16.17 Image of a healthy subject, amplitude of the fast decay component, α_1, in %

Fig. 16.18 Image of lifetime, τ_2, in channel 1, of a healthy subject

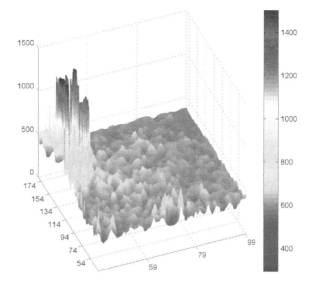

Fig. 16.19 Lifetime of fast decay component, τ_1, in channel 1 of a patient with dry AMD

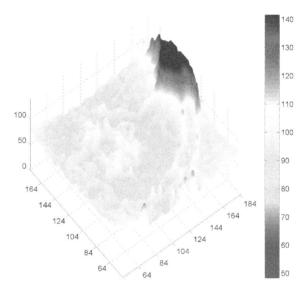

shortest component is maximal in the macula. The lifetime τ_2 is homogenous outside the optic disc as demonstrated in Fig. 16.18.

The following figures demonstrate images of FLIO parameters of fundus diseases. Figures 16.19, 16.20 and 16.21 are lifetime images recorded at patients with age-related macular degeneration (AMD). Weak signs of pathological changes are visible in the τ_1 image of the short-wavelength channel in dry stage of AMD (Fig. 16.19). The macula is surrounded by regions of increased values of τ_1. Increased values up to

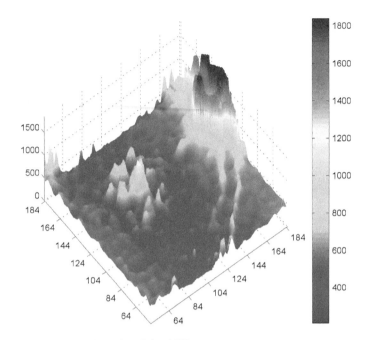

Fig. 16.20 Lifetime τ_2 in channel 1 of dry AMD

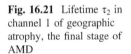

Fig. 16.21 Lifetime τ_2 in channel 1 of geographic atrophy, the final stage of AMD

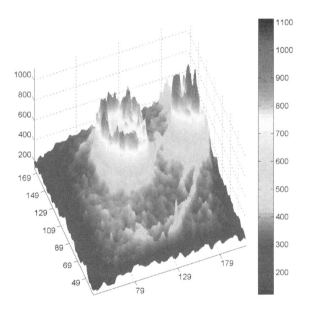

1 ns are more clearly visible in τ_2 images in he short-wavelength channel (Fig. 16.20). In geographic atrophy, the final stage of AMD, the lifetime τ_2 is increased up to the values of the optic disc in the whole macula (Fig. 16.21).

These images show local pathological alterations in severe states. Early metabolic changes occur weakly in the entire ocular tissue. It can be assumed that such alterations change the profile of histograms of FLIO parameters. That means, for example, a missing substance or a more frequently detectable fluorophore influences the histogram of lifetime τ_i, which is a sum of several fluorophores, at a certain range of time. Such comparison of differences in histograms can be statistically evaluated, see section below.

16.5.2 Statistical Evaluation of Decay Parameters

The program FLIMX [23] was developed for the specific evaluation of FLIO measurements. As demonstrated in Fig. 16.9 through Fig. 16.21, images of FLIO parameters and cross sections in these images as well as histograms of FLIO parameters, and statistical parameters modal value, median, mean, standard deviation as well as kurtosis and skew ness can be calculated for single measurements or for groups of subjects. Regions of the same position and size at the fundus should be selected for statistical comparison. Figure 16.22 demonstrates amplitude images of an AMD patient, which are different in the short-wavelength channel (left) and in the long-wavelength channel (right). Histograms of amplitudes are also given as well as parameters of descriptive statistics. Extended FLIO measurements were performed by Quick [31].

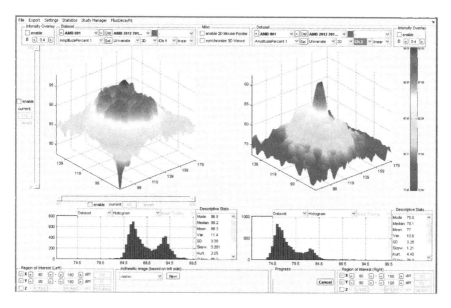

Fig. 16.22 Presentation of FLIO results by program FLIMVis

For statistical comparison between patients and healthy subjects data are required for the clinical characterization of each subject. Such data are e.g. age, gender, type of disease, classification and duration of disease, type of ocular lens, and parameters of internal medicine. These data are connected with the FLIO parameters. In this way, comparisons are possible between patients and healthy subjects e.g. of the same age or between different stages of a disease.

The Holm-Bonferroni method [18] is used for the statistical comparison of histograms of FLIO parameters between patients and healthy subjects. To apply this method, each range of FLIO parameter is divided in a number of intervals. The distribution of pathological and healthy values is compared by the Wilcoxon test in each interval. Significant differences exist, if the error probability α is smaller than the significance level p = 5 % divided by the number of intervals. Sensitivity and specificity are calculated for each statistical different parameter interval. The calculated cutting value allows a classification as healthy or diseased for each individual.

There is currently a clinical study [11] to detect early metabolic alterations in the fundus of diabetic patients having no signs of diabetic retinopathy yet. Histograms of amplitudes, lifetimes, and or relative contributions Q were calculated in equal regions. Such a histogram shows how many pixels of each value were detected in this region. The FLIO parameters of forty diabetic patients type II were compared with 30 healthy subjects. The most sensitive discrimination between both groups was found for the lifetime τ_2 in the short-wavelength channel. Figure 16.23 shows the FLIMVis window of this comparison. The red coloured intervals permit a

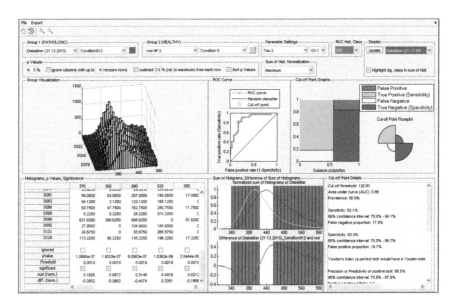

Fig. 16.23 Presentation of Holm-Bonferroni test between diabetic patients, having no signs of diabetic retinopathy, and healthy subject

significant discrimination between diabetic patients and healthy subjects. The highest value for sensitivity and of specificity of more than 80 % were reached in the interval $\tau_2 = 370$ ps.

An interpretation of substance-specific changes is possible based on the difference between histograms of diabetic patients having no signs of diabetic retinopathy with healthy subjects. This difference histogram means that pixels are less frequent at the fundus of diabetic patients having lifetime $\tau_2 = 370$ ps. On the other hand, more pixels are detectable in diabetic patients having lifetime $\tau_2 > 460$ ps. The lifetime of about 370 ps corresponds to free NADH. For free HADH the lifetime is in the range of 2–3 ns.

Free NADH acts as electron donor in metabolic basis processes, e.g. respiratory chain, whereas protein-bound NADH is not active in metabolism. That means, metabolic alterations appear predominantly in the neuronal retina of diabetic patients, already when no signs of diabetic retinopathy are detectable by other methods.

This interpretation was confirmed by FLIO measurements in branch arterial occlusion [39]. In this case, the metabolism is disturbed in the neuronal retina, which is supplied by retinal vessels. As a kind of internal normalization, none-supplied regions were compared with supplied regions at the same fundus. In the short-wavelength channel, the lifetime of the second decay component, τ_2, increased from about 450 ps in the supplied region to more than 1 ns in the none-supplied region.

A further promising application of FLIO was found in a pilot study comparing patients suffering from Alzheimer's disease with healthy subjects [19]. Although only a small group of Alzheimer's patients was investigated, correlations were found between FLIO parameters with subjective as well as objective diagnostic parameters. As demonstrated in Fig. 16.24, the MMSE score decreases with increasing relative

Fig. 16.24 Correlation between the relative contribution Q_2 in the long-wavelength channel with the MMSE score in Alzheimer patients

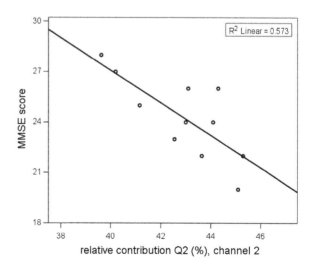

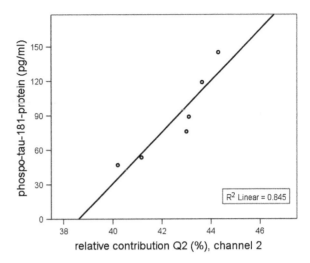

Fig. 16.25 Correlation between concentration of phosphor τ 181 protein in liquor and the relative contribution Q_2 in the long-wavelength channel

contribution Q_2 in the long-wavelength channel. This score characterizes the individual cognitive properties of a subject. A healthy person reaches a score value 30.

The concentration of phospho tau 181 protein in liquor is an objective mark for the degree of severity of Alzheimer's disease. Again the relative contribution Q_2 in the long-wavelength channel shows a positive correlation with the concentration of this protein, as demonstrated in Fig. 16.25. If these relations will be confirmed in further studies, FLIO measurements can be used as an objective non-invasive test for Alzheimer's disease. Following the assumption that the second component corresponds to the fluorescence in the neuronal retina in 3-exponential fit, Alzheimer's specific changes occur also in the neuronal retina.

16.5.3 Improved Decay Model for Ophthalmic FLIM Data

Good results are already attained by a tail fit of the fluorescence decay originating from the layered ocular structure. The long fluorescence lifetime of the crystalline lens and of connective tissue at the fundus are merged as lifetime τ_3 in the model function according to (16.2). The contribution of the fluorescence of the crystalline lens can be accounted for in different ways.

Lifetime measurements at the fundus are influenced by the fluorescence of the crystalline lens. There is a distance of 18.8 mm between the plane of the lens and the fundus. The distances and the thickness of the fundus layers are much smaller. The neuronal retina is about 300 μm thick and contains a number of different layers (e.g. receptors, bipolares, plexiform layers, ganglion cell layer). The retinal pigment epithelium is about 10 μm thick, some membranes only 1 μm. The fluorescence of the lens is therefore visible as a step in the rising edge of the fluorescence intensity (Fig. 16.26).

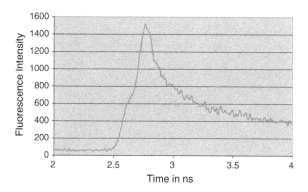

Fig. 16.26 Stepped rising edge of fluorescence intensity caused by the different transit time of fluorescence originating from the crystalline lens and from the fundus

The approximation of slope and fluorescence decay with the left border of the fitting interval at the beginning of the rising edge results in a poor fit.

The question is how the fluorescence of the lens can be taken into account. In common microscopic applications the left border of the fitting interval is set before or at the beginning of the rising edge of the fluorescence signal. The earlier detection of the fluorescence of the lens is not considered in a tail fit, where the left border of the fitting interval is set to the maximum of fluorescence signal. Using the model function in (16.5), both the fluorescence of the lens and the difference in detection time of fluorescence originating from lens and from fundus is taken into account [37]:

$$\frac{I(t)}{I_0} = \sum_{i,j=1}^{i=p,j=r} \alpha_{i,j} \cdot e^{\frac{t-tc_i}{\tau_{i,j}}} \tag{16.5}$$

In this equation means:
tc$_i$ distance between layers
i   layer i
j   component j in layer i
$\tau_{i,j}$ lifetime j in layer i
$\alpha_{i,j}$ amplitude j in layer i.

In Fig. 16.27 the blue light excites the yellow fluorescence of the crystalline lens first and a short time later the red fundus fluorescence.

The distance d between layers can be determined using the time Δtci in (16.6):

$$d = \frac{(tc_i - tc_j) \cdot c}{2 \cdot n} \tag{16.6}$$

Here the ratio c/n stands for the velocity of light in the medium. The (16.5) permits a combination of functionality (lifetime) with tomography (distances). To

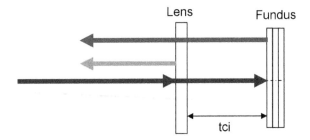

Fig. 16.27 Scheme of fluorescent layers in the eye for the application of the fit according to (16.5)

Fig. 16.28 Fit of stepped slope and fluorescence decay by (16.5)

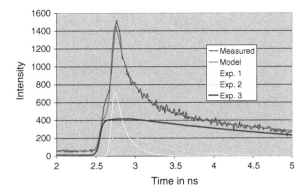

reach geometrical resolution comparable with OCT in ophthalmology, detection of differences of appearance times of fluorescence would be required at an accuracy level of better than 100 fs. The principle of the fit according to (16.5) is demonstrated in Fig. 16.28. The decay of exponent 3, which corresponds to the fluorescence in the lens, is shifted in relation to the decay of exponents 1 and 2.

The influence of the fluorescence of the crystalline lens can be investigated by separate measurement of the lens fluorescence. The fundus fluorescence is then biexponentially fitted. As shown in Fig. 16.29, the model curve fits the measured fluorescence quite well, both the stepped slope and the decay.

Although the 2-exponential fit of the fundus fluorescence with separately measured lens fluorescence is a very good approximation of the real decay profile a 3-exponential fit might be required. In case of artificial intra ocular lens, which are not fluorescent, a 3-exponential model fits the fundus fluorescence best.

To compare the fit results for different models and fit conditions, the fluorescence decay was fitted at the same pixel of the same fluorescence image. In Tables 16.7 and 16.8 the lifetimes and the amplitudes are given for the short-wavelength channel, calculated by left border at the beginning of the rising edge, the tail fit, a fit with the model described by (16.5) (tci-fit), lens fluorescence and 2-exponential model (lens + 2 exp.), and lens fluorescence and 3-exponential model (lens + 3-exp.).

Fig. 16.29 Fit of stepped slope and fluorescence decay using separate measurement of fluorescence of the crystalline lens

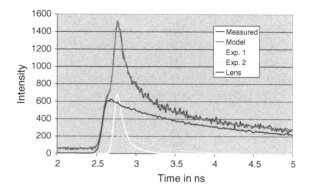

Table 16.7 Comparison of lifetimes obtained by different fit conditions and models

	Fit from start of rising edge	Fit from maximum	tc$_i$ fit	Lens + 2 exp.	Lens + 3 exp.
τ_1	372	101	100	60	80
τ_2	2935	453	429	407	465
τ_3	4036	2963	3412		9972

Table 16.8 Comparison of amplitudes obtained by different fit conditions and models

	Fit from start of rising edge	Fit from maximum	tc$_i$ fit	Lens + 2 exp.	Lens + 3 exp.
α_1	65.5	55.3	49.5	45.5	46.8
α_2	14.6	19.1	20.9	13.5	12.5
α_3	18.9	25.6	29.6		1.4

As expected, the fit from the rising edge approximation delivers poor results. The best correspondence, both in lifetimes and in amplitudes, exists between the results of a fit from the maximum and a tc$_i$ fit. The shortest lifetimes τ_1 were calculated by the "lens + 2 exp." and "lens + 3 exp." fits. The lifetime τ_2 was comparable for the fit from the maximum, the tc$_i$ fit and the "lens + 2 exp." and "lens + 3 exp." fits.

16.6 Summary

Fluorescence lifetime imaging ophthalmoscopy is a new technique which is on its way to become a diagnostic tool. In contrast to lifetime measurements on microscopic samples, eye movement must be compensated during measurements. Although the autofluorescence of ocular tissue is extremely weak, characteristic fluorescence lifetimes can be determined for some ocular layers, when the fluorescence decay is determined by time correlated single photon counting. The fluorescence decay is detected in two spectral channels for better separation of endogenous fluorophores. A change of lifetimes and of amplitudes in FLIO images is found for local pathological alterations in fundus layers. Early alterations in metabolism are detectable by comparing histograms of FLIO parameters of patients and healthy subjects via statistical methods. The crucial advantage of FLIO is the possibility to investigate cellular processes non-invasively at the living human eye.

References

1. ANSI American National Standard for the safe use of laser. ANSI Z 136.1-2000. Orlando, Suite 128, 13501 Ingenuite Drive, FL 32826, Laser Institute of America (2000)
2. W. Becker, *Advanced Time-Correlated Single-Photon Counting Techniques* (Springer, Berlin, Heidelberg, New York, 2005)
3. W. Becker, A. Bergmann, In *Lifetime-Resolved Imaging in Nonlinear Microscopy* ed. by B.R. Masters, P.T.C. So. Handbook of Biomedical Nonlinear Optical Microscopy (Oxford University Press, Oxford, 2008)
4. W. Becker, The bh TCSPC handbook. 5th edn. Becker & Hickl GmbH (2012), available on www.becker-hickl.com, Printed copies available from Becker & Hickl GmbH
5. W. Becker, B. Su, K. Weisshart, O. Holub, FLIM and FCS detection in laser-scanning microscopes: increased efficiency by gaasp hybrid detectors. Micr. Res. Tech. **74**, 804–811 (2011)
6. W. Becker, Fluorescence lifetime imaging—techniques and applications. J. Microsc. **247**, 119–136 (2012)
7. B. Chance, Pyridine nucleotide as an indicator of the oxygen requirements for energy-linked functions of mitochondria. Circ. Res. **38**, I31–I38 (1976)
8. B. Chance, B. Schoener, R. Oshino, F. Itshak, Y. Nakase, Oxidation–reduction ratio studies of mitochondria in freeze-trapped samples. NADH and flavoprotein fluorescence signals. J. Biol. Chem. **254**, 4764–4771 (1979)
9. D. Chorvat Jr, A. Mateasik, Y.G. Cheng, N.Y. Poirier, J. Miro, N.S. Dahdah, A. Chorvatova, Rejection of transplanted hearts in patients evaluated by the component analysis of multi-wavelength NAD(P)H fluorescence lifetime spectroscopy. J. Biophotonics **3**, 646–652 (2010)
10. F.C. Delori, C.K. Dorey, G. Staurenghi, O. Arend, D.G. Goger, J. Weiter, In vivo fluorescence of the ocular fundus exhibits retinal pigment epithelium lipofuscin characteristics. Invest. Ophthalmol. Vis. Sci. 36718–36729 (1995)
11. L. Deutsch. Evaluierung des Fluorescence Lifetime Imaging vom Augenhintergrund bei Patienten mit Diabetes mellitus, Dissertation, Jena (2012)
12. E. Dimitrow, I. Riemann, A. Ehlers, M. Koehler, J. Norgauer, J. Elsner, P. König, K. Katz, Spectral fluorescence lifetime detection and selective melanin imaging by multiphoton laser tomography for melanoma diagnosis. Exp. Dermatol. **18**, 509–515 (2009)
13. C. Dysli, G. Quellec, M Abegg, M.N. Menke, U. Wolf-Schnurrbusch, J. Kowal, J. Blatz, O. La Schiazza, A.B. Leichtle, S. Wolf, M.S. Zinkernagel, Quantitative analysis of fluorescence lifetime measurements of the macula using the fluorescence lifetime imaging ophthalmoscope in healthy subjects. IOVS **55**, 2107–2113 (2014)
14. C. Dysli, M. Dysli, V. Enzmann, S. Wolf, M.S. Zinkernagel, Fluorescence lifetime Imaging of the ocular fundus in mice. IOVS **55**, 7206–7215 (2014)
15. A. Ehlers, I. Riemann, M. Stark, K. König, Multiphoton fluorescence lifetime imaging of human hair. Microsc. Res. Tech. **70**, 154–161 (2007)
16. EN ISO 15004-2 (2007)

17. L. Feeney-Burns, E.R. Berman, H. Rothman, Lipofuscin of human retinal pigment epithelium. Am. J. Ophthalmol. **90**, 783–791 (1980)
18. S. Holm, A simple sequentially rejective multiple test procedure. Scand. J. Stat. **6**(2), 65–70 (1979)
19. S. Jentsch, d. Schweitzer, K.U. Schmidtke, S. Peters, J. Dawczynski, K.J. Bär, M. Hammer, Retinal fluorescence lifetime imaging ophthalmoscopy measures depend on the severity of Alzheimer´s disease. Acta Ophthalmol. (2014). doi:10.1111/aos12609
20. http://en.wikipedia.org/wiki/Bond-dissociation_energy
21. http://webvision.med.utah.edu/book/part-i-foundations/simple-anatomy-of-the-retina/
22. M. Klemm, A. Dietzel, J. Haueisen, E. Nagel, M. Hammer, D. Schweitzer, Repeatability of autofluorescence lifetime imaging at the human fundus in healthy volunteers. Curr. Eye Res. **38**(7), 793–801 (2013)
23. M. Klemm, Theoretische und experimentelle Untersuchungen zur Erfassung von Parametern des zellulären Stoffwechsels vom Augenhintergrund auf der Grundlage der zeitaufgelösten Autofluoreszenz. Manuscript Dissertation. Technical University Ilmenau, Department of Biomedical Technique and Informatics (2012). FLIMX is available at http://www.FLIMX.de
24. M. Köllner, J. Wolfrum, How many photons are necessary for fluorescence-lifetime measurements? Chem. Phys. Lett. **200**(1, 2), 199–204 (1992)
25. J.R. Lakowicz, *Principles of fluorescence spectroscopy*, 3rd edn. (Kluwer Academic/Plenum, New York, 2007)
26. M. Lu, M. Kuroki, S. Amano, M. Tolentino, K. Keough, I. Kim, R. Bucala, A.P. Adamis, Advanced glycation end products increase retinal VEGF. J. Clin. Invest. **101**(6), 1219–1224 (1998). http://www.jci.org
27. J. Meyer, Modellierung der integralen zeitaufgelösten Autofluoreszenz des Augenhintergrundes durch die spektralen Eigenschaften und Parameter des Fluoreszenzabklingverhaltens einzelner endogener Fluorophore-Vergleich mit in vivo Messungen. Masters Arbeit Ernst-Abbe-Fachhochschule Jena, (2012)
28. N. Nakashima, K. Yoshihara, F. Tanaka, K. Yagi, Picosecond fluorescence lifetime of the coenzyme of D-amino acid oxidase. J. Biol. Chem. **255**(11), 5261–5263 (1980)
29. Neufingerl: Chemie 1—Allgemeine und anorganische Chemie, Jugend & Volk; Wien; ISBN 978-3-7100-1184-9. S.47, (2006)
30. S. Peters, M. Hammer, D. Schweitzer, Two-photon excited fluorescence microscopy application for ex vivo investigation of ocular fundus samples. SPIE ECBO Munich 2011, 8086-4-1–8086-4-10, (2011)
31. S. Quick, Untersuchungen zum klinischen Wert des Fluoreszenz Lifetime Imaging am menschlichen Augenhintergrund. Dissertation. University of Jena, Experimental Ophthalmology (2009)
32. D. Schweitzer, A. Kolb, M. Hammer et al., Tau-mapping of the autofluorescence of the human ocular fundus. SPIE **4164**, 79–89 (2000)
33. D. Schweitzer, M. Hammer, F. Schweitzer, Limits of confocal laser scanning technique in measurements of time-resolved autofluorescence of the ocular fundus. German. Biomed. Technik **50**(9), 263–267 (2005)
34. D. Schweitzer, S. Jentsch, S. Schenke, M. Hammer, C. Biskup, E. Gaillard, Spectral and time-resolved studies on ocular structures. SPIE-OSA **6628**, 662807-1–662807-12 (2007a)
35. D. Schweitzer, S. Schenke, M. Hammer, F. Schweitzer, S. Jentsch, E. Birckner, W. Becker, A. Bergmann, Towards metabolic mapping of the human retina. Microsc. Res. Tech. **70**, 410–419 (2007b)
36. D. Schweitzer, Quantifying fundus autofluorescence. Chapter 8 in *Fundus Autofluorescence*, ed. by N. Lois, J.F. Forster. Wolters Kluwer/Lippincott Williams and Wilkins (2009a)
37. D. Schweitzer, M. Klemm, M. Hammer, S. Jentsch, F.Schweitzer, Method for simultaneous detection of functionality and tomography. Proc. SPIE **7368**, 736804-1–736804-9 (2009b)
38. D. Schweitzer, Metabolic mapping. Chapter 10 in *Medical Retina Focus on Retinal Imaging*, ed. by F.G. Holz, R.F. Spaide. Essentials in Ophthalmology, Series Editors G.K. Kriegelstein, R.N. Weinreb. (Springer, Berlin Heidelberg, 2010a), pp. 107–123

39. D. Schweitzer, S. Quick, M. Klemm, M. Hammer, S. Jentsch, J. Dawczynski, Time-resolved autofluorescence in retinal vascular occlusions. Der. Ophthalmol. **107**(12), 1145–1152 (2010b)
40. P.F. Sharp, A. Manivannan, H. Xu, J.V. Forrester, The scanning laser ophthalmoscope—a review of its role in bioscience and medicine. Phys. Med. Biol. **49**, 1085 (2004)
41. M.C. Skala, K.M. Riching, A. Gendron-Fitzpatrick, J. Eickhoff, K.W. Eliceiri, J.G. White, N. Ramanujam, In vivo multiphoton microscopy of NADH and FAD redox states, fluorescence lifetimes, and cellular morphology in precancerous epithelia. Proc. Nat. Acad. Sci. **104**(49), 19494–19499 (2007)
42. M.C. Skala, K.M. Riching, D.K. Bird, A. Gendron-Fitzpatrick, J. Eickhoff, K.W. Eliceiri, P. J. Keely, N. Ramanujam, In vivo multiphoton fluorescence lifetime imaging of protein-bound and free nicotinamide adenine dinucleotide in normal and precancerous epithelia. J. Biomed. Opt. **12**(2), 024014–024014-10 (2007)
43. M.C. Skala, N. Ramanujam, Multiphoton redox ratio imaging for metabolic monitoring in vivo. Methods Mol. Biol. **594**, 155–162 (2010)
44. L. Stryer, *Biochemie* (Spectrum Akademischer Verlag GmbH, Heidelberg, Berlin, New York, 1991)
45. J. van de Kraats, D. van Norren, Optical density of the aging human ocular media in the visible and the UV. J. Opt. Soc. Am. A **24**, 1842–1857 (2007)
46. H.D. Vishwasrao, A.A. Heikal, K.A. Kasischke, W.W. Webb, Conformational dependence of intracellular NADH on metabolic state revealed by associated fluorescence anisotropy. J. Biol. Chem. **280**(26), 25119–25126 (2005)
47. Y. Wu, W. Zheng, J.Y. Qu, Sensing cell metabolism by time- resolved autofluorescence. Opt. Lett. **31**(21), 3122–3124 (2006)

Chapter 17
Dynamic Mapping of the Human Brain by Time-Resolved NIRS Techniques

Adam Liebert, Michal Kacprzak, Daniel Milej, Wolfgang Becker, Anna Gerega, Piotr Sawosz and Roman Maniewski

Abstract Dynamic mapping of the human brain by time-resolved near-infrared-spectroscopy (trNIRS), or functional NIRS (fNIRS), is based on the injection of picosecond or sub-nanosecond laser pulses into the head and the measurement of the pulse shape and the intensity after diffusion through the tissue. By analysing the pulse shape and the intensity of the signals at different detector and source positions and different wavelengths, changes in the oxy- and deoxy-haemoglobin concentration are obtained for extracerebral and intracerebral tissue layers and for different depth in the brain. The technique can by combined with the injection of a bolus of an exogenous absorber. By recording either absorption or fluorescence, the in- and outflow of the absorber in different brain compartments can be monitored. The in- and outflow dynamics reveal differences in the blood flow caused by impaired

A. Liebert (✉) · M. Kacprzak · D. Milej · P. Sawosz · R. Maniewski
Polish Academy of Sciences, Institute of Biocybernetics and Biomedical Engineering, 02-106 Warsaw, Poland
e-mail: adam.liebert@ibib.waw.pl

M. Kacprzak
e-mail: michal.kacprzak@ibib.waw.pl

D. Milej
e-mail: dmilej@ibib.waw.pl

P. Sawosz
e-mail: psawosz@ibib.waw.pl

R. Maniewski
e-mail: roman.maniewski@ibib.waw.pl

W. Becker
Becker & Hickl GmbH, Berlin, Germany
e-mail: becker@becker-hickl.de

A. Gerega
Polish Academy of Sciences, Institute of Biocybernetics and Biomedical Engineering, Warsaw, Poland
e-mail: anna.gerega@ibib.waw.pl

© Springer International Publishing Switzerland 2015
W. Becker (ed.), *Advanced Time-Correlated Single Photon Counting Applications*,
Springer Series in Chemical Physics 111, DOI 10.1007/978-3-319-14929-5_17

perfusion or stroke. In this chapter, we describe the technical principle of TCSPC-based fNIRS and the associated data processing techniques. Typical results are shown for the haemodynamic response of the brain on visual-cortex stimulation, and for brain perfusion measurement by ICG bolus injection.

17.1 Introduction

Biological applications of near-infrared spectroscopy, or NIRS, aim at the determination of the of spectroscopic parameters in deep layers of biological tissue and their spatial distribution [1, 2]. The technique is based on the illumination of the tissue by near-infrared (NIR) light, and the detection of diffusely transmitted or reflected light or fluorescence. Clinical applications of NIRS techniques are optical mammography, see Chap. 19, static brain imaging [3, 4] and 'dynamic' or 'functional' brain imaging (fNIRS). Dynamic brain imaging experiments are focusing either on the determination of the brain oxygenation and its change with brain activity, or on measurements of the brain perfusion. Brain oxygenation measurements use the fact that oxy- and deoxyhemoglobin have different absorption spectra in the near infrared range [5]. Changes in the oxygenation are derived from the relative absorption changes at two different wavelengths. Brain perfusion measurements are based on the recording of the change in the intensity of the transmitted or diffusely returned light or the fluorescence emitted by a near-infrared dye after injection into the circulatory system.

Classical NIRS measurement at the human brain is performed by sending CW near-infrared light through the head and analysing variations in the attenuation of the light for different wavelengths and source and detector positions. The light is delivered and detected through optical fibres or fibre bundles. The source-detector separation is typically in the range of 2–4 cm. Changes in the light attenuation are measured at several wavelengths and used to calculate changes in the relative concentration of oxy- and deoxyhemoglobin. A distinct drawback of this technique, called continuous wave NIRS (cw NIRS), is that the measured brain oxygenation signals are contaminated by changes in the oxygenation of the extracerebral tissue. Measurements at several source-detector separations are used for better differentiation of the signal originating from the brain [6, 7], but signals from superficial layers cannot entirely be suppressed.

Significant improvements are obtained by using time-resolved techniques. The time the photons need to reach the detector position (the 'time of flight') not only depends on the scattering and absorption parameters of the tissue layers they have passed but also on the path they have taken inside the tissue.

The resulting signals can be recorded by frequency-domain and by time-domain techniques. Frequency-domain techniques use light modulated at sub-GHz frequency and determine the phase shift and the amplitude of the detected signal. The phase shift depends on the path length of the photons between source and detector and on the absorption and scattering properties of the tissue [8–10] and thus delivers information on the depth where the observed absorption changes have occurred.

Time-domain techniques are based on the injection of short (typically picosecond or sub-nanosecond duration) laser pulses into the tissue and the measurement of the pulse shape (or distribution of times of flight, DTOF) after diffusion through the tissue [11, 12]. By analysing the pulse shape and the intensity of the signal, absorption changes can be attributed to different depths in the tissue. The technique is thus able to differentiate information originating from the brain and from extracerebral tissue [13, 14].

Dynamic changes in the time-of-flight distributions acquired at the surface of the head are caused by the heart beat, variable oxy- and deoxyhemoglobin concentration induced by brain activity, and effects of associated physiological regulation. The haemodynamic response to brain stimulation occurs on the time scale of a few seconds [15–18]. The changes in the average time of flight of the photons are on the order of a few picoseconds. Changes in the arrival times of the photons on the required accuracy level can be recorded by time-correlated single photon counting (TCSPC) [19–21]. Analysis of the distributions of time-of-flight (DTOFs) recorded by TCSPC delivers the desired discrimination of oxygenation changes in extra- and intracerebral compartments of the tissue [13, 14, 22, 23].

The measurement of dynamic changes in the brain oxygenation requires to record a sequence of DTOFs for different wavelengths and different source and detector positions over a longer period of time at a time resolution of 50–100 ms per DTOF. This is a task that is difficult to perform with classic TCSPC devices. Multidimensional TCSPC [19, 21], however, delivers the desired data by recording photon distributions over several parameters simultaneously, see Chap. 1. The general architecture of the instrument is shown in Sect. 17.3, Fig. 17.3. Results obtained with this and a number of similar instruments are described in [24–28]. Currently, different instruments are compared in terms of accuracy, systematic errors, signal-to-noise ratio, and other performance parameters [29, 30].

17.2 Depth-Resolved Data Analysis

The analytical form of the diffuse reflectance R in function of absorption coefficient μ_a and diffusion coefficient D in semi-infinite homogenous medium is given by [31]:

$$R(r, t) = \frac{z_0}{(4\pi Dv)^{3/2} t^{5/2}} \exp\left(-\frac{r^2}{4Dvt} - \mu_a vt\right) \tag{17.1}$$

where: $z_0 = /\mu'_s$, $D \approx 1/3\mu'_s$, r—source-detector separation, v—speed of light in the medium, t—time.

This formula is derived from the diffusion equation and allows to determine the absorption coefficient, μ_a, and the reduced scattering coefficient, μ'_s, by fitting the diffuse reflectance function to the measured distribution of times of flight of photons. In this procedure the instrumental response function (IRF) of the instrument must be taken into account and convolved with the theoretical distribution before fitting [32, 33].

Alternatively, the optical properties of the homogeneous medium can be estimated by analysis of the moments of the measured DTOFs as described in [32]. Moment analysis may deliver slightly lower absolute accuracy (about 5 % discrepancy with fitting of the diffusion model) but is very robust (no fitting is used) and easily takes into account the influence of the IRF (only the first and the second moment of the IRF are needed).

Changes in the absorption of the tissue can be determined as a function of depth by using statistical moments of the distributions of time of flight. This method is based on sensitivity factors of the moments. These describe the relation between the changes in the total number of photons, the mean time of flight of the photons, the variance of the time of flight, and the changes in the absorption coefficient occurring at a certain depth in the tissue [13].

Figure 17.1 shows the spatial sensitivity profiles to changes in the absorption coefficient, μ_a, for the number of photons, the mean time of flight, and the variance of the time of flight. The profiles were obtained from a perturbation approach based on the diffusion approximation for the propagation of photon density [34]. For this simulation the reduced scattering coefficient, absorption coefficient and the refractive index were taken as $\mu_s' = 10 \text{ cm}^{-1}$, $\mu_a = 0.1 \text{ cm}^{-1}$ and $n = 1.33$, respectively.

As can be seen from Fig. 17.1, the number of photons (i.e. the intensity) delivers tissue properties preferentially from the surface of the tissue, especially from the closer ranges around the source and detector position. The mean time of flight and, even more, the variance deliver information from large depth in the tissue. Importantly, the sensitivity to μ_a changes in superficial layers is much lower than for the photon number.

The sensitivity factors are the basis of a method provided by Steinbrink et al. in which a relation between the measured changes in the moments of the DTOF were used for estimation of depth-resolved changes in absorption [14], see Fig. 17.2.

The sensitivity factors of the three moments for each layer indexed by l are described as:

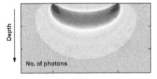  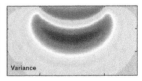

Fig. 17.1 Spatial sensitivity profiles for the number of photons (*left*), the mean time of flight (*middle*) and the variance of the time of flight (*right*). Images courtesy of Heidrun Wabnitz, Physikalische Bundesanstalt Berlin, Germany

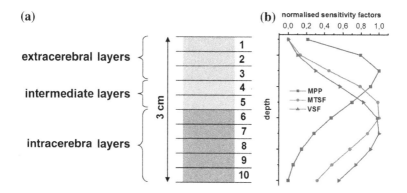

Fig. 17.2 Model of the human head (**a**) where the medium is represented by ten homogeneous layers of the same thickness (0.3 cm) divided into three compartments: extracerebral, intermediate, and intracerebral and, corresponding to this model, the courses of sensitivity factors of the moments of DTOFs as a function of depth (**b**). Mean partial path length (MPP), mean time of flight sensitivity factor (MTSF) and variance sensitivity factor (VSF) indicate sensitivity of each moment (integral, mean time of flight and variance) for absorption changes in the equivalent layer [77]

$$MPP_l = \frac{\Delta A_t}{\Delta \mu_{a,l}} \tag{17.2}$$

$$MTSF_l = \frac{\Delta \langle t_t \rangle}{\Delta \mu_{a,l}} \tag{17.3}$$

$$VSF_l = \frac{\Delta V_t}{\Delta \mu_{a,l}} \tag{17.4}$$

where MPP_l is the mean partial path length, $MTSF_l$ is the mean time of flight sensitivity factor, and VSF_l is the variance sensitivity factor and $\Delta \mu_a$ is the change of absorption coefficient in the layers indexed by l.

The changes of the sensitivity factors as a function of depth are presented in Fig. 17.2b. It can be noted that the variance is the most sensitive moment for the changes of absorption coefficient in the deeper layers and the statistical moment, which is most sensitive for superficial changes of absorption, is the integral of the DTOF. The profiles of the sensitivity factors as a function of depth may depend on the wavelengths utilized. However, it can be assumed that for both wavelengths used in the experiment, the same profiles can be adopted.

The changes in μ_a occurring in three layers of the medium can be estimated by solving the following system of equations:

$$\begin{bmatrix} MPP_1 & MPP_2 & MPP_3 \\ MTSF_1 & MTSF_2 & MTSF_3 \\ VSF_1 & VSF_2 & VSF_3 \end{bmatrix} \cdot \begin{bmatrix} \Delta\mu_{a,1} \\ \Delta\mu_{a,2} \\ \Delta\mu_{a,3} \end{bmatrix} = \begin{bmatrix} -\log(N_{totC}/N_{tot0}) \\ \langle t \rangle_C - \langle t \rangle_0 \\ V_C - V_0 \end{bmatrix} \tag{17.5}$$

where MPP_k, $MTSF_k$, and VSF_k correspond to the sensitivity factors of the moments summed for the extracerebral layers (k = 1, compartment 1), intermediate layers (k = 2, compartment 2) and intracerebral layers (k = 3, compartment 3); $\mu_{a,k}$ represents the unknown change of the absorption coefficient in each compartment; N_{tot}, $\langle t \rangle$, and V are the moments calculated from the measured DTOFs indexed 0 and C, respectively, for the measurement before and after the blood pressure changes.

Recently, the uncertainty in the estimation of $\Delta\mu_a$ for the method of moments was analyzed [35], and an improved algorithm for a two-layered model was proposed [36]. In several studies the authors reported that time-resolved measurement allows to obtain depth-resolved information on changes in absorption via analysis by Mellin-Laplace transform of the DTOF [37] or consideration of the time-windows [38].

17.3 Brain Oxygenation Studies

An instrument for dynamic brain oxygenation imaging based on 8 parallel TCSPC channels is described in [39]. The general architecture of the instrument is shown in Fig. 17.3.

The setup uses two Becker & Hickl SPC-134 four-channel TCSPC packages [21] in one industrial computer. Two semiconductor lasers are operating at wavelengths located symmetrically at both sides of the isosbestic point of the oxy- and deoxyhemoglobin absorption. The pulses of the two lasers are interleaved, and the DTOFs at the two wavelengths recorded as a single waveform in the same recording time interval of the TCSPC modules (see Chap. 1, Sect. 1.4.5.3).

The DTOFS at both sides of the head are recorded simultaneously. The position of tissue illumination is multiplexed by two 1–9 fibre switches. The current position of the fibre switches is transferred into the TCSPC modules through their routing input signals, and used to record the DTOFs into separate memory blocks. Please see Chap. 1 for details).

Subsequent DTOFs were recorded in the continuous-flow (memory swapping) mode of the SPC-134 TCSPC systems [21]. Please see Chap. 1, Sect. 1.4.4 for

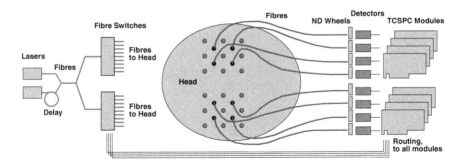

Fig. 17.3 Architecture of the 8-channel parallel NIRS system described in [39]

technical details. For every source position the DTOFs are acquired for 95 ms. Thus, one switching cycle through all 9 source positions is completed within 0.9 s. The setup was tested in phantom experiments and validated during in vivo tests on health volunteers. Changes in oxygenation of the brain cortex following motor stimulation were successfully imaged in the motor cortex area [39].

The measurement of the instrument response function (IRF) needs to take into account the dependence of the pulse dispersion in the detection fibres on the numerical aperture the fibres are illuminated with [40]. The illumination geometry for the IRF measurement must therefore be the same as for the experiment. This was achieved by positioning the source fibres in front of the detecting bundles covered by a sheet of white paper [40]. The instrumental response function (IRF) as measured for every source-detector pair at the both wavelengths was about 800 ps (FWHM). The IRF contains the laser pulse width, the temporal dispersion of the light propagation in the source fibres, the temporal dispersion in the detection fibres, and the temporal response of the detectors used. The largest contribution comes from the temporal dispersion in the detection fibres. These are used at high NA in order to obtain a high photon collection efficiency, which leads to large temporal dispersion [40]. It should be noted, that this relatively wide IRF has little influence on the accuracy on the measurement of absorption and scattering coefficients. More important than the absolute width of the IRF is its stability [20].

Figure 17.4 gives an impression of how the data of dynamic brain mapping experiments look like. Time-of flight curves of photons transmitted through a forehead were recorded in the continuous-flow mode of an SPC-134 system [21]. 20 subsequent time-of-flight curves from the continuous-flow sequence are shown in Fig. 17.4. The acquisition time was 100 ms per curve, the ADC resolution 1024 channels. The left sequence was detected at a source-detector distance of 5 cm, the right sequence at a distance of 8 cm. The count rates were 1.8×10^5 and $4.5 \times 10^6\,s^{-1}$, respectively.

It should be noted that a count rate of $4.5 \times 10^6\,s^{-1}$ per TCSPC channel is at the very limit of currently available TCSPC devices. Intensity measurements at rates this high require a correction for counting loss [19, 21]. The first and second moments of the time-of-flight distributions are not noticeably influenced by

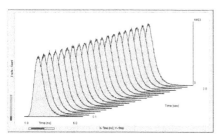
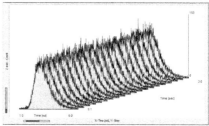

Fig. 17.4 20 steps of a TOF sequence recorded in the continuous flow mode of an SPC-134. Acquisition time 100 ms per curve, ADC resolution 1024 channels. *Left* Source-detector distance 5 cm, count rate $4.5 \times 10^6\,s^{-1}$. *Right* Source-detector distance 8 cm count rate $1.8 \times 10^5\,s^{-1}$. From [21]

counting loss but can be impaired by pile-up effects. It has been shown, however, that the influence of pile-up is smaller than commonly believed, and can be neglected up to count rates in the range of several MHz [19, 21]. IRF measurements over a wide range of count rates do not show noticeable changes in the IRF shapes [19–21].

The fluctuations in the moments of DTOFs caused by instability of light sources and detection electronics are relatively small [39]. In principle, brain activation response curves could be obtained by recording only a moderate number of stimulation events. However, in practice there is a strong variation in the data due to heart beat, respiration and vasomotion. The haemodynamic brain response can only be separated from these effects by recording sequences of DTOFs over a large number of stimulation events. All the experiments mentioned above used memory swapping (see Chap. 1, Sect. 1.4.4) in the 'continuous flow' mode of the TCSPC cards. Synchronization with the stimulation was obtained by triggering the memory bank swapping with the stimulation.

A number of instruments based on the same principle are in use. Quaresima et al. used a single SPC-630 TCSPC module and a multi-anode PMT to record sequences of time-of-flight curves in eight parallel channels [41]. The acquisition time per step of the sequence was 166 ms. The data of five steps were averaged. Values of μ_s' and μ_a were calculated from the averaged data by using a standard model of diffusion theory. An improved instrument of this type used an SPC-134 package with HRT-41 four-channel routers [21] connected to each TCSPC channel [42]. Thus, simultaneous detection in 16 detector channels was achieved. Two lasers were multiplexed by pulse interleaving; 16 source positions were multiplexed by fibre switches. Applications to functional brain imaging are described in [43–45].

An instrument that multiplexes lasers of different wavelength by fibre switches has been described by Re et al. [46]. The system avoids crosstalk of different laser wavelengths by reflections, mutual pile-up, counting loss or detector afterpulsing [19, 21]. An application to haemodynamics of the human brain is described by Aletti et al. [47]. The authors show how signal components caused by skin vasomotion and by cerebral perfusion can be separated.

Typical results of haemodynamic response measurements performed by Liebert et al. [22, 23] are shown in Fig. 17.5. Visual stimulation was used. An annular black and white checkerboard alternating at 8 Hz on a computer screen was shown to the patient. During the rest period a dark grey screen was presented while fixation was maintained. The stimulation periods lasting 30 s were repeated 20 times separated by 30 s of rest. Signals from 20 stimulation periods were averaged. Three multiplexed laser wavelengths were used. The photons were detected by four Hamamatsu R7400 PMTs at different source-detector positions. The DTOF curves were recorded by an SPC-134 system in the continuous-flow mode. In Fig. 17.5 the intensity change, the change in the mean time of flight, and the change in the variance of the time of flight over the stimulation period are shown. Depth-resolved intra- and extra-cerebral changes of the oxy- and deoxyhemoglobin concentrations calculated from the data of the four detectors at the three wavelengths are shown in Fig. 17.6.

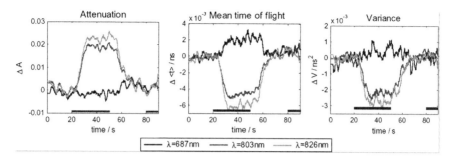

Fig. 17.5 *Left to right* Intensity change, change in the mean time of flight, and change in the variance of the time of flight over the stimulation period. The *horizontal bars* indicate the periods of stimulation. From [23]

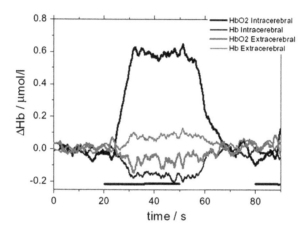

Fig. 17.6 Intra- and extra-cerebral changes of oxy- and deoxy-haemoglobin concentrations during visual stimulation obtained from DTOFs measured at 3 wavelengths and four source-detector separations. The *horizontal bars* indicate the stimulation period. From [23]

17.4 Brain Perfusion Assessment

Exogenous absorbers can be used by detecting either their absorption or their fluorescence. The only NIR dye currently approved for use at human patients is indocyanine green (ICG) [48, 49]. Absorption spectra of ICG in water are shown in Fig. 17.7, left.

ICG absorbs strongly between 650 and 820 nm. The absorption of ICG monomers peaks at about 780 nm. In water ICG tends to form aggregates which cause a second absorption band around 700 nm. In blood ICG binds to the serum albumin. Aggregation is then partially suppressed and the absorption around 780 nm dominates. Due to the binding to albumin ICG stays in the blood. It is therefore used as a contrast agent to mark blood vessels.

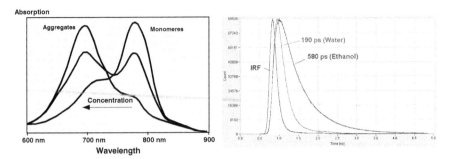

Fig. 17.7 *Left* Absorption spectra for ICG in water, after [48]. *Right* Fluorescence decay of ICG in water and ethanol, BHL-600-650 laser, PMC-100-20 detector and SPC-630 TCSPC module

Fluorescence decay curves of ICG are shown in Fig. 17.7, right. The fluorescence lifetime in Ethanol is about 580 ps. In water the lifetime is about 160–190 ps [50]. In both cases the decay functions are multi-exponential, probably due to the presence of monomeric and aggregated forms. The short lifetimes indicate that the fluorescence quantum efficiency is low. Values given in the literature are around 4 % [48].

17.4.1 Diffuse Reflectance

Monitoring of ICG inflow and washout after its intravenous injection can be used to detect blood-flow dynamics in the brain. This technique was successfully used for bed-side assessment of cerebral perfusion in stroke patients [51–53]. A bolus of ICG was injected and the subsequent absorption changes in the tissue monitored. The instrument was essentially the same as used for visual [23] and motor stimulation experiments, see Fig. 17.3. Two multiplexed lasers operating at the same wavelength (about 800 nm) were used for the two brain hemispheres. The diffusely reflected signals were detected by four detectors and recorded by the four parallel channels of an SPC-134 TCSPC system. By using the Continuous-Flow Mode of the SPC-134, time-of-flight distributions were recorded at a rate of 50 ms per curve. From the time-of-flight distributions the changes in the attenuation, in the mean time of flight, and in the variance of the time of flight were calculated. Figure 17.8 shows results for a healthy volunteer and a stroke patient [51]. The curves are sliding averages over 20 subsequent 50 ms recordings. The moments were scaled to a range from 0 to 1 by subtracting the minimum value and normalising the result to its maximum value.

For the healthy subjects the changes in the moments appear virtually simultaneously in both brain hemispheres. For the stroke patients the changes in the moments at the location of the stroke are delayed. The delay is most pronounced in the variance, ΔV_{norm}. A thrombolytic therapy of the stroke patient resulted in normal perfusion after 30 h.

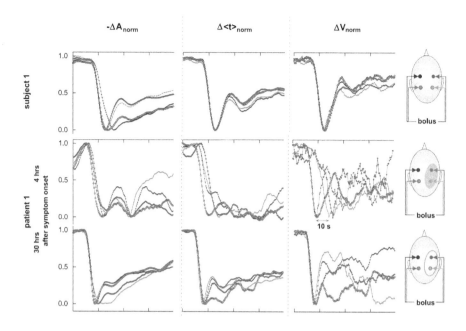

Fig. 17.8 Changes of the normalised attenuation, $-\Delta A_{norm}$, the mean time of flight, $\Delta\langle t\rangle_{norm}$, and the variance of the time of flight, ΔV_{norm} for a healthy subject and stroke patient. The curves were scaled to a range from 0 to 1 by subtracting the minimum value and normalising the result to its maximum. From [51]

Absorption changes are not only caused by inflow dynamics of the ICG bolus but also by physiological effects like heart beat, breathing, and blood pressure variations. These effects can be seen as quasi-periodic oscillations in Fig. 17.8. The mechanisms of these effects and their impact on NIRS tissue spectroscopy are discussed in [54].

Diop et al. [55–57] used the ICG bolus technique to determine cerebral blood flow in piglets under normocapnia, hypercapnia, and carotid occlusion. They compared time-resolved (TCSPC) and CW data recorded under similar conditions. Although the general correlation of the TR and CW data was good the authors also found systematic differences. After discussing possible instrumental effects the authors conclude that the techniques deliver information from different depth in the brain. This is supported by the fact that in the time-resolved data the amplitude of the attenuation change during the bolus is larger for late-arriving photons.

Recently, an instrument that records ICG concentration changes at extremely large source-detector distance was reported [58]. By using high incident power from a Ti:Sa laser, and placing a detector directly at the surface of the tissue the authors were able to record at 9 cm source-detector distance. The instrument was used both on phantoms and on human volunteers. The phantom consists of a fish tank filled with a mixture of milk, water and an absorber. The tank contained a transparent flow-tube through which the same mixture was pumped. The ICG bolus was simulated by injecting ICG in the flow system. The authors showed that the ICG bolus was detectable down to a depth of 5 cm. Measurements at the human head confirmed this result in that they showed physiological effects expected in deep brain layers.

17.4.2 Fluorescence

For a given number of recorded photons, a fluorescence measurement in general yields a better intrinsic SNR for the dye concentration than an absorption measurement. The reason is that the dye concentration measured via fluorescence is directly related to the number of detected photons, whereas the concentration derived from absorption is the (usually small) difference of two large photon numbers. However, compared to the diffusely transmitted or reflected intensity the fluorescence intensity is lower. Therefore, the SNR actually obtained depends on the efficiency of the optics and the detection system, the tissue thickness, the fluorophore concentration and quantum yield, and the acceptable acquisition time.

Milej et al. compared fluorescence and diffuse reflection measurements in phantom experiments [59]. The instrument was similar to the one shown in Fig. 17.3 but with a laser of 780 nm and filters to suppress the laser wavelength. They used a phantom with tubes inserted at different depths for modelling of the dynamic inflow of indocyanine green (ICG). They showed that the use of fluorescence signals delivered better sensitivity and signal-to-noise ratio compared to the use of diffuse reflectance data. A similar study for a fluorescent inclusion within non-fluorescent and weakly fluorescent phantoms came to the same result [60].

Liebert et al. and Milej et al. have shown that the fluorescence of ICG can well be detected from inside the human brain [61–63]. They used R7400 PMTs connected to the channels of an SPC-134 package. One group of channels detected the diffusely reflected light, the other the fluorescence. Measurements were performed at a rate of 10 s^{-1}. A result is shown in Fig. 17.9.

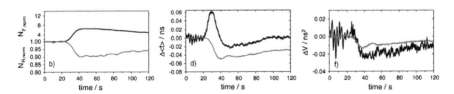

Fig. 17.9 Normalised number of diffusely reflected photons N_{Rnorm} (*grey lines*) and fluorescence photons (*black line*) N_{Fnorm}, changes in mean time of flight, $\Delta\langle t \rangle$, and changes in variance of distribution, ΔV, after injection of an ICG bolus. Both hemispheres of brain. *Gray curve* Diffuse reflection. *Black curve* Fluorescence. From [61], copyright Elsevier Ltd.

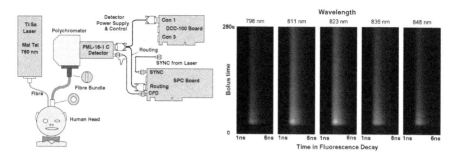

Fig. 17.10 *Left* Setup for time- and wavelength-resolved bolus detection used by [67]. *Right* Result of monitoring of inflow and washout of the ICG by means of fluorescence at 5 emission wavelengths. From [67]

The shapes of the curves of the mean time of flight and the variance are significantly different for the fluorescence and the diffuse reflectance. These differences can be attributed to the time differences of the inflow of the ICG in the intra- and extra-cerebral compartments [64, 65]. Intra- and extra-cerebral fluorescence components during an ICG bolus have been investigated in detail by Jelzow et al. [66]. The authors present an algorithm to separate these components based on the bolus function and on changes in the DTOFs over the bolus time.

Gerega et al. [67] have demonstrated multi-wavelength fluorescence detection of ICG boli. The authors used a Becker & Hickl PML-SPEC (MW FLIM) 16-channel detector [21] connected to an SPC-830 TCSPC module. The light from the tissue was collected by a fibre bundle and transferred into the input slit of the PML-SPEC polychromator, see Fig. 17.10, left. A time-series of multi-wavelength recordings was performed in the Scan Sync Out mode of the SPC-830 module. A result is shown in Fig. 17.10, right. The authors discuss the effects of absorption, re-absorption, ICG concentration, and depth in the tissue on the recorded data.

As mentioned above, it can be assumed that changes in the first and second moment of the DTOFs are related mainly to changes the intracerebral, changes in the intensity to changes in the extracerebral blood flow. This observation combined with the analysis of the pattern of blood flow in the extra- and intra-cerebral tissues in healthy subjects and patients with brain perfusion impairment lead to a data analysis algorithm which is able to detect disorders in the brain perfusion. The principle of this algorithm is shown schematically in Fig. 17.11.

In earlier publications it has been shown that for evaluation of differences in cerebral blood flow between hemispheres with impaired and unimpaired perfusion, the difference in the time of the inflow of an optical contrast medium between both hemispheres ΔTTP (time-to-peak) can be used [51, 52]. In the proposed algorithm the delay, ΔT_{MAX}, between the inflow of ICG into extra- and intracerebral tissues (represented by the signals of number of photons and variance of the DTOF) is analyzed. It was observed that in healthy subjects a delay in ΔT_{MAX} between 1 and 5 s can be observed whereas in most of the patients with brain perfusion disturbance

Fig. 17.11 Principle of the algorithm for time-resolved data analysis. In healthy subjects the inflow to the brain occurs earlier than to the extracerebral tissue. In the patient with disturbed brain perfusion the inflow to the brain is delayed which leads to a reduced delay between the signals from extra- and intracerebral tissue compartments

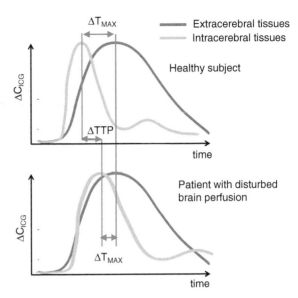

this delay is reduced [68]. It should be emphasized that the size of the delay depends on a distance between the point of emission and detection of light pulses [58]. It was shown in a study on patients with posttraumatic brain injury that the proposed method of signal processing can be used both for the analysis of diffuse reflectance and fluorescence signals [68].

17.5 Summary

The advances in TCSPC technology together with the development of small-size pulsed laser systems allowed us to build up compact time-resolved NIRS systems for brain monitoring. The systems can be used for in vivo experiments in neurophysiology as well as for clinical tasks. Several other technical solutions were proposed for brain studies in the last years. Systems based on time-gated image intensifiers use the CCD camera as a multichannel detector [69, 70] or directly take time-gated image the light re-emitted from the tissue [71–74]. Recently, Mazurenka et al. reported that a scanning system with supercontinuum laser for imaging of changes in the absorption in the brain [75, 76], see Chap. 18. This system uses TCSPC FLIM (see Chap. 1, Sect. 1.4.5) and a supercontimuum laser with fast multiplexing of the laser wavelength via an acousto-optical tuneable filter (AOTF). Both approaches solve the problem of spatial under-sampling. However, direct imaging at the surface of the head can be applied only in areas without hair. Another problem is the influence of the topography of the head, and motion artefacts. Both problems can possibly be solved by recording a reference signal directly from the illuminated spot. Moreover, the numerical aperture of the detection in an imaging system is necessarily small. It is therefore difficult to detect a sufficient number of photons within a given acquisition time interval.

References

1. F.F. Jobsis, Noninvasive, infrared monitoring of cerebral and myocardial oxygen sufficiency and circulatory parameters. Science **198**, 1264–1267 (1977)
2. A. Villringer, B. Chance, Non-invasive optical spectroscopy and imaging of human brain function. Trends Neurosci. **20**, 435–442 (1997)
3. J.C. Hebden, A. Gibson, R.M. Yusof, N. Everdell, E.M.C. Hillman, D.T. Delpy, S.R. Arridge, T. Austin, J.H. Meek, J.S. Wyatt, Three-dimensional optical tomography of the premature infant brain. Phys. Med. Biol. **47**, 4155–4166 (2002)
4. F.E.W. Schmidt, M.E. Fry, E.M.C. Hillman, J.C. Hebden, D.T. Delpy, A 32-channel time-resolved instrument for medical optical tomography. Rev. Sci. Instrum. **71**, 256–265 (2000)
5. S. Wray, M. Cope, D.T. Delpy, J.S. Wyatt, E.O. Reynolds, Characterization of the near infrared absorption spectra of cytochrome aa3 and haemoglobin for the non-invasive monitoring of cerebral oxygenation. Biochim. Biophys. Acta **933**, 184–192 (1988)
6. P.G. al-Rawi, P. Smielewski, P.J. Kirkpatrick, Preliminary evaluation of a prototype spatially resolved spectrometer. Acta Neurochir. Suppl. **71**, 255–257 (1998)
7. D.M. Hueber, M.A. Franceschini, H.Y. Ma, Q. Zhang, J.R. Ballesteros, S. Fantini, D. Wallace, V. Ntziachristos, B. Chance, Non-invasive and quantitative near-infrared haemoglobin spectrometry in the piglet brain during hypoxic stress, using a frequency-domain multidistance instrument. Phys. Med. Biol. **46**, 41–62 (2001)
8. B. Chance, M. Cope, E. Gratton, N. Ramanujam, B. Tromberg, Phase measurement of light absorption and scatter in human tissue. Rev. Sci. Instrum. **69**, 3457–3481 (1998)
9. B.W. Pogue, M.S. Patterson, Frequency-domain optical absorption spectroscopy of finite tissue volumes using diffusion theory. Phys. Med. Biol. **39**, 1157–1180 (1994)
10. T. Tu, Y. Chen, J. Zhang, X. Intes, B. Chance, Analysis on performance and optimization of frequency-domain near-infrared instruments. J. Biomed. Opt. **7**, 643–649 (2002)
11. R. Cubeddu, A. Pifferi, P. Taroni, A. Torricelli, G. Valentini, Experimental test of theoretical models for time-resolved reflectance. Med. Phys. **23**, 1625–1633 (1996)
12. J.C. Hebden, E.W. Schmidt, M.E. Fry, M. Schweiger, E.M.C. Hillman, D.T. Delpy, Simultaneous reconstruction of absorption and scattering images by multichannel measurement of purely temporal data. Opt. Lett. **24**, 534–536 (1999)
13. A. Liebert, H. Wabnitz, J. Steinbrink, H. Obrig, M. Moller, R. Macdonald, A. Villringer, H. Rinneberg, Time-resolved multidistance near-infrared spectroscopy of the adult head: intracerebral and extracerebral absorption changes from moments of distribution of times of flight of photons. Appl. Opt. **43**, 3037–3047 (2004)
14. J. Steinbrink, H. Wabnitz, H. Obrig, A. Villringer, H. Rinneberg, Determining changes in NIR absorption using a layered model of the human head. Phys. Med. Biol. **46**, 879–896 (2001)
15. T. Durduran, G. Yu, M.G. Burnett, J.A. Detre, J.H. Greenberg, J. Wang, C. Zhou, A.G. Yodh, Diffuse optical measurement of blood flow, blood oxygenation, and metabolism in a human brain during sensorimotor cortex activation. Opt. Lett. **29**, 1766–1768 (2004)
16. M.A. Franceschini, S. Fantini, J.H. Thompson, J.P. Culver, D.A. Boas, Hemodynamic evoked response of the sensorimotor cortex measured noninvasively with near-infrared optical imaging. Psychophysiology **40**, 548–560 (2003)
17. M.A. Franceschini, V. Toronov, M. Filiaci, E. Gratton, S. Fantini, On-line optical imaging of the human brain with 160-ms temporal resolution. Opt. Express **6**, 49–57 (2000)

18. V. Toronov, A. Webb, J.H. Choi, M. Wolf, L. Safonova, U. Wolf, E. Gratton, Study of local cerebral hemodynamics by frequency-domain near-infrared spectroscopy and correlation with simultaneously acquired functional magnetic resonance imaging. Opt. Express **9**, 417–427 (2001)
19. W. Becker, *Advanced Time-Correlated Single Photon Counting Techniques* (Springer, Berlin, 2010)
20. W. Becker, A. Bergmann, Timing stability of TCSPC experiments. Proc. SPIE **6372**, 637209-1–637209-7 (2006)
21. W. Becker, *The bh TCSPC Handbook*, 5th edn. (Becker & Hickl GmbH, Berlin, 2012), electronic version available on www.becker-hickl.com
22. A. Liebert, H. Wabnitz, M. Möller, R. Macdonald, H. Rinneberg, H. Obrig, J. Steinbrink, A. Villringer, Assessment of cerebral perfusion by time-resolved diffuse NIR-reflectance following injection of a dye-bolus, in Optical Society of America Biomedical Optics Topical Meeting (Miami Beach, Florida, USA, 2004), pp. CD-ROM: SESSION FA, FB 5
23. A. Liebert, H. Wabnitz, J. Steinbrink, H. Obrig, M. Möller, R. Macdonald, H. Rinneberg, Intra- and extracerebral changes of hemoglobin concentrations by analysis of moments of distributions of times of flight of photons. Proc. SPIE **5138**, 126–130 (2003)
24. B.M. Mackert, S. Leistner, T. Sander, A. Liebert, H. Wabnitz, M. Burghoff, L. Trahms, R. Macdonald, G. Curio, Dynamics of cortical neurovascular coupling analyzed by simultaneous DC-magnetoencephalography and time-resolved near-infrared spectroscopy. Neuroimage **39**, 979–986 (2008)
25. T.H. Sander, A. Liebert, M. Burghoff, H. Wabnitz, R. Macdonald, L. Trahms, Cross-correlation Analysis of the Correspondence between Magnetoencephalographic and Near-infrared Cortical Signals. Methods Inf. Med. **46**, 164–168 (2007)
26. T.H. Sander, A. Liebert, B.M. Mackert, H. Wabnitz, S. Leistner, G. Curio, M. Burghoff, R. Macdonald, L. Trahms, DC-magnetoencephalography and time-resolved near-infrared spectroscopy combined to study neuronal and vascular brain responses. Physiol. Meas. **28**, 651–664 (2007)
27. J. Steinbrink, A. Liebert, H. Wabnitz, R. Macdonald, H. Obrig, A. Wunder, R. Bourayou, T. Betz, J. Klohs, U. Lindauer, U. Dirnagl, A. Villringer, Towards noninvasive molecular fluorescence imaging of the human brain. Neurodegener Dis **5**, 296–303 (2008)
28. H. Wabnitz, M. Moeller, A. Liebert, H. Obrig, J. Steinbrink, R. Macdonald, Time-resolved near-infrared spectroscopy and imaging of the adult human brain. Adv. Exp. Med. Biol. **662**, 143–148 (2010)
29. H. Wabnitz, D.R. Taubert, M. Mazurenka, O. Steinkellner, A. Jelzow, R. Macdonald, D. Milej, P. Sa-wosz, M. Kacprzak, A. Liebert, R. Cooper, J. Hebden, A. Pifferi, A. Farina, I. Bargigia, D. Contini, M. Caffini, L. Zucchelli, L. Spinelli, R. Cubeddu, A. Torricelli., Performance assessment of time-domain optical brain imagers, part 1: basic instrumental performance protocol. J. Biomed. Opt. **19**(8), 086010 (2014)
30. H. Wabnitz, A. Jelzow, M. Mazurenka, O. Steinkellner, R. Macdonald, D. Milej, N. Zolek, M. Kacprzak, P. Sawosz, R. Maniewski, A. Liebert, S. Magazov, J. Hebden, F. Martelli, P. Di Ninni, G. Zaccanti, A. Torricelli, D. Contini, R. Re, L. Zucchelli, L. Spinelli, R. Cubeddu, A. Pifferi, Performance assessment of time-domain optical brain imagers, part 2: nEUROPt protocol. J. Biomed. Opt. **19**(8), 086012 (2014)
31. M.S. Patterson, B. Chance, B.C. Wilson, Time resolved reflectance and transmittance for the non-invasive measurement of tissue optical properties. Appl. Opt. **28**, 2331–2336 (1989)
32. A. Liebert, H. Wabnitz, D. Grosenick, M. Moller, R. Macdonald, H. Rinneberg, Evaluation of optical properties of highly scattering media by moments of distributions of times of flight of photons. Appl. Opt. **42**, 5785–5792 (2003)
33. V. Ntziachristos, B. Chance, Accuracy limits in the determination of absolute optical properties using time-resolved NIR spectroscopy. Med. Phys. **28**, 1115–1124 (2001)

34. S. Carraresi, T.S. Shatir, F. Martelli, G. Zaccanti, Accuracy of a perturbation model to predict the effect of scattering and absorbing inhomogeneities on photon migration. Appl. Opt. **40**(25), 4622–4632 (2001)
35. A. Liebert, H. Wabnitz, C. Elster, Determination of absorption changes from moments of distributions of times of flight of photons: optimization of measurement conditions for a two-layered tissue model. J. Biomed. Opt. **17**, 057005 (2012)
36. A. Jelzow, H. Wabnitz, I. Tachtsidis, E. Kirilina, R. Bruhl, R. Macdonald, Separation of superficial and cerebral hemodynamics using a single distance time-domain NIRS measurement. Biomed. Opt. Express **5**, 1465–1482 (2014)
37. A. Puszka, L. Herve, A. Planat-Chretien, A. Koenig, J. Derouard, J.M. Dinten, Time-domain reflectance diffuse optical tomography with Mellin-Laplace transform for experimental detection and depth localization of a single absorbing inclusion. Biomed. Opt. Express **4**, 569–583 (2013)
38. L. Zucchelli, D. Contini, R. Re, A. Torricelli, L. Spinelli, Method for the discrimination of superficial and deep absorption variations by time domain fNIRS. Biomed. Opt. Express **4**, 2893–2910 (2013)
39. M. Kacprzak, A. Liebert, P. Sawosz, N. Zolek, R. Maniewski, Time-resolved optical imager for assessment of cerebral oxygenation. J. Biomed. Opt. **12**, 034019 (2007)
40. A. Liebert, H. Wabnitz, D. Grosenick, R. Macdonald, Fiber dispersion in time domain measurements compromising the accuracy of determination of optical properties of strongly scattering media. J. Biomed. Opt. **8**, 512–516 (2003)
41. V. Quaresima, M. Ferrari, A. Torricelli, L. Spinelli, A. Pifferi, R. Cubeddu, Bilateral prefrontal cortex oxygenation responses to a verbal fluency task: a multichannel time-resolved near-infrared topography study. J. Biomed. Opt. **10**, 11012 (2005)
42. D. Contini, A. Torricelli, A. Pifferi, L. Spinelli, F. Paglia, R. Cubeddu, Multi-channel time-resolved system for functional near infrared spectroscopy. Opt. Express **14**, 5418–5432 (2006)
43. M. Butti, D. Contini, E. Molteni, M. Caffini, L. Spinelli, G. Baselli, A.M. Bianchi, S. Cerutti, R. Cubeddu, A. Torricelli, Effect of prolonged stimulation on cerebral hemodynamic: a time-resolved fNIRS study. Med. Phys. **36**, 4103–4114 (2009)
44. E. Molteni, D. Contini, M. Caffini, G. Baselli, L. Spinelli, R. Cubeddu, S. Cerutti, A.M. Bianchi, A. Torricelli, Load-dependent brain activation assessed by time-domain functional near-infrared spectroscopy during a working memory task with graded levels of difficulty. J. Biomed. Opt. **17**, 056005 (2012)
45. A. Torricelli, D. Contini, A. Pifferi, L. Spinelli, R. Cubeddu, Functional brain imaging by multi-wavelength time-resolved near infrared spectroscopy. Opto-Electron. Rev. **16**, 131–135 (2008)
46. R. Re, D. Contini, M. Caffini, R. Cubeddu, L. Spinelli, A. Torricelli, A compact time-resolved system for near infrared spectroscopy based on wavelength space multiplexing. Rev. Sci. Instrum. **81**, 113101 (2010)
47. F. Aletti, R. Re, V. Pace, D. Contini, E. Molteni, S. Cerutti, A. Maria Bianchi, A. Torricelli, L. Spinelli, R. Cubeddu, G. Baselli, Deep and surface hemodynamic signal from functional time resolved transcranial near infrared spectroscopy compared to skin flowmotion. Comput. Biol. Med. **42**, 282–289 (2012)
48. T. Desmettre, J.M. Devoisselle, S. Mordon, Fluorescence properties and metabolic features of indocyanine green (ICG) as related to angiography. Surv. Ophthalmol. **45**, 15–27 (2000)
49. S. Mordon, J.M. Devoisselle, S. Soulie-Begu, T. Desmettre, Indocyanine green: physicochemical factors affecting its fluorescence in vivo. Microvasc. Res. **55**, 146–152 (1998)
50. A. Gerega, N. Zolek, T. Soltysinski, D. Milej, P. Sawosz, B. Toczylowska, A. Liebert, Wavelength-resolved measurements of fluorescence lifetime of indocyanine green. J. Biomed. Opt. **16**, 067010 (2011)

51. A. Liebert, H. Wabnitz, J. Steinbrink, M. Moller, R. Macdonald, H. Rinneberg, A. Villringer, H. Obrig, Bed-side assessment of cerebral perfusion in stroke patients based on optical monitoring of a dye bolus by time-resolved diffuse reflectance. Neuroimage **24**, 426–435 (2005)
52. O. Steinkellner, C. Gruber, H. Wabnitz, A. Jelzow, J. Steinbrink, J.B. Fiebach, R. Macdonald, H. Obrig, Optical bedside monitoring of cerebral perfusion: technological and methodological advances applied in a study on acute ischemic stroke. J. Biomed. Opt. **15**, 061708 (2010)
53. O. Steinkellner, H. Wabnitz, A. Jelzow, R. Macdonald, C. Gruber, J. Steinbrink, H. Obrig, Cerebral Perfusion in Acute Stroke Monitored by Time-domain Near-infrared Reflectometry. Biocybern. Biomed. Eng **32**, 3–16 (2012)
54. E. Kirilina, A. Jelzow, A. Heine, M. Niessing, H. Wabnitz, R. Bruhl, B. Ittermann, A.M. Jacobs, I. Tachtsidis, The physiological origin of task-evoked systemic artefacts in functional near infrared spectroscopy. NeuroImage **61**, 70–81 (2012)
55. M. Diop, K.M. Tichauer, J.T. Elliott, M. Migueis, T.Y. Lee, K. St Lawrence, Comparison of time-resolved and continuous-wave near-infrared techniques for measuring cerebral blood flow in piglets. J. Biomed. Opt. **15**, 057004 (2010)
56. J. Elliott, D. Milej, A. Gerega, W. Weigl, M. Diop, L. Morrison, T.-Y. Lee, A. Liebert, K. St. Lawrence, Variance of time-of-flight distribution is sensitive to cerebral blood flow as demonstrated by ICG bolus-tracking measurements in adult pigs. Biomed. Opt. Express, accepted for publication (2012)
57. J.T. Elliott, M. Diop, K.M. Tichauer, T.Y. Lee, K. St Lawrence, Quantitative measurement of cerebral blood flow in a juvenile porcine model by depth-resolved near-infrared spectroscopy. J. Biomed. Opt. **15**, 037014 (2010)
58. A. Liebert, P. Sawosz, D. Milej, M. Kacprzak, W. Weigl, M. Botwicz, J. Maczewska, K. Fronczewska, E. Mayzner-Zawadzka, L. Krolicki, R. Maniewski, Assessment of inflow and washout of indocyanine green in the adult human brain by monitoring of diffuse reflectance at large source-detector separation. J. Biomed. Opt. **16**, 046011 (2011)
59. D. Milej, M. Kacprzak, N. Zolek, P. Sawosz, A. Gerega, R. Maniewski, A. Liebert, Advantages of fluorescence over diffuse reflectance measurements tested in phantom experiments with dynamic inflow of ICG. Opto-Electron. Rev. **18**, 208–213 (2010)
60. M. Kacprzak, A. Liebert, P. Sawosz, N. Zolek, D. Milej, R. Maniewski, Time-resolved imaging of fluorescent inclusions in optically turbid medium - phantom study. Opto-Electron. Rev. **18**, 37–47 (2010)
61. A. Liebert, H. Wabnitz, H. Obrig, R. Erdmann, M. Moller, R. Macdonald, H. Rinneberg, A. Villringer, J. Steinbrink, Non-invasive detection of fluorescence from exogenous chromophores in the adult human brain. Neuroimage **31**, 600–608 (2006)
62. D. Milej, A. Gerega, M. Kacprzak, W. Weigl, R. Maniewski, A. Liebert, Time-resolved multi-channel optical system for assessment of brain oxygenation and perfusion in the human brain. Opto-Electron. Rev. **22**(1), 55–67 (2013)
63. D. Milej, A. Gerega, N. Zolek, W. Weigl, M. Kacprzak, P. Sawosz, J. Maczewska, K. Fronczewska, E. Mayzner-Zawadzka, L. Krolicki, R. Maniewski, A. Liebert, Time-resolved detection of fluorescent light during inflow of ICG to the brain-a methodological study. Phys. Med. Biol. **57**, 6725–6742 (2012)
64. A. Liebert, H. Wabnitz, N. Zolek, R. Macdonald, Monte Carlo algorithm for efficient simulation of time-resolved fluorescence in layered turbid media. Opt. Express **16**, 13188–13202 (2008)
65. D. Milej, A. Gerega, H. Wabnitz, A. Liebert, A Monte Carlo study of fluorescence generation probability in a two-layered tissue model. Phys. Med. Biol. **59**, 1407–1424 (2014)
66. A. Jelzow, H. Wabnitz, H. Obrig, R. Macdonald, J. Steinbrink, Separation of indocyanine green boluses in the human brain and scalp based on time-resolved in-vivo fluorescence measurements. J. Biomed. Opt. **17**, 057003 (2012)

67. A. Gerega, D. Milej, W. Weigl, M. Botwicz, N. Zolek, M. Kacprzak, W. Wierzejski, B. Toczylowska, E. Mayzner-Zawadzka, R. Maniewski, A. Liebert, Multi-wavelength time-resolved detection of fluorescence during the inflow of indocyanine green into the adult's brain. J. Biomed. Opt. **17**, 087001 (2012)
68. W. Weigl, D. Milej, A. Gerega, B. Toczylowska, M. Kacprzak, P. Sawosz, M. Botwicz, R. Maniewski, E. Mayzner-Zawadzka, A. Liebert, Assessment of cerebral perfusion in post-traumatic brain injury patients with the use of ICG-bolus tracking method. Neuroimage **85,** 555–565 (2014)
69. J. Selb, D.K. Joseph, D.A. Boas, Time-gated optical system for depth-resolved functional brain imaging. J. Biomed. Opt. **11**, 044008 (2006)
70. J. Selb, J.J. Stott, M.A. Franceschini, A.G. Sorensen, D.A. Boas, Improved sensitivity to cerebral hemodynamics during brain activation with a time-gated optical system: analytical model and experimental validation. J. Biomed. Opt. **10**, 11013 (2005)
71. C. D'Andrea, D. Comelli, A. Pifferi, A. Torricelli, G. Valentini, R. Cubeddu, Time-resolved optical imaging through turbid media using a fast data acquisition system based on a gated CCD camera. J. Phys. D Appl. Phys. **36**, 1675–1681 (2003)
72. I. Sase, A. Takatsuki, J. Seki, T. Yanagida, A. Seiyama, Noncontact backscatter-mode near-infrared time-resolved imaging system: Preliminary study for functional brain mapping. J. Biomed. Opt. **11**, 054006 (2006)
73. P. Sawosz, M. Kacprzak, N. Zolek, W. Weigl, S. Wojtkiewicz, R. Maniewski, A. Liebert, Optical system based on time-gated, intensified charge-coupled device camera for brain imaging studies. J. Biomed. Opt. **15**, 066025 (2010)
74. P. Sawosz, N. Zolek, M. Kacprzak, R. Maniewski, A. Liebert, Application of time-gated CCD camera with image intensifier in contactless detection of absorbing inclusions buried in optically turbid medium which mimic local changes in oxygenation of the brain tissue. Opto-Electron. Rev. **20**, 309–314 (2012)
75. M. Mazurenka, L. Di Sieno, G. Boso, D. Contini, A. Pifferi, A.D. Mora, A. Tosi, H. Wabnitz, R. Macdonald, Non-contact in vivo diffuse optical imaging using a time-gated scanning system. Biomed. Opt. Express **4**, 2257–2268 (2013)
76. M. Mazurenka, A. Jelzow, H. Wabnitz, D. Contini, L. Spinelli, A. Pifferi, R. Cubeddu, A.D. Mora, A. Tosi, F. Zappa, R. Macdonald, Non-contact time-resolved diffuse reflectance imaging at null source-detector separation. Opt. Express **20**, 283–290 (2012)
77. M. Kacprzak, A. Liebert, W. Staszkiewicz, A. Gabrusiewicz, P. Sawosz, G. Madycki, R. Maniewski, Application of a time-resolved optical brain imager for monitoring cerebral oxygenation during carotid surgery. J. Biomed. Opt. **17**, 016002 (2012)

Chapter 18
Time-Domain Diffuse Optical Imaging of Tissue by Non-contact Scanning

Heidrun Wabnitz, Mikhail Mazurenka, Laura Di Sieno,
Gianluca Boso, Wolfgang Becker, Katja Fuchs, Davide Contini,
Alberto Dalla Mora, Alberto Tosi, Rainer Macdonald
and Antonio Pifferi

Abstract We present the concept, design and first in vivo tests of a novel non-contact scanning imaging system for time-domain near-infrared spectroscopy of tissues. Employing a supercontinuum laser in combination with an acousto-optic

H. Wabnitz (✉) · M. Mazurenka · K. Fuchs · R. Macdonald
Physikalisch-Technische Bundesanstalt (PTB), Abbestr. 2-12, 10587 Berlin, Germany
e-mail: heidrun.wabnitz@ptb.de

M. Mazurenka
e-mail: mmazurenka@gmail.com

K. Fuchs
e-mail: katja_fuchs1@freenet.de

R. Macdonald
e-mail: rainer.macdonald@ptb.de

L. Di Sieno · D. Contini · A. Dalla Mora · A. Pifferi
Dipartimento di Fisica, Politecnico di Milano, Piazza Leonardo da Vinci 32, 20133 Milan,
Italy
e-mail: laura.disieno@polimi.it

D. Contini
e-mail: davide.contini@polimi.it

A. Dalla Mora
e-mail: alberto.dallamora@polimi.it

A. Pifferi
e-mail: antonio.pifferi@polimi.it

W. Becker
Becker & Hickl GmbH, Nahmitzer Damm 30, 12277 Berlin, Germany
e-mail: becker@becker-hickl.de

G. Boso · A. Tosi
Dipartimento di Elettronica, Informazione e Bioingegneria, Politecnico di Milano, Piazza
Leonardo da Vinci 32, 20133 Milan, Italy
e-mail: gianluca.boso@polimi.it

A. Tosi
e-mail: alberto.tosi@polimi.it

© Springer International Publishing Switzerland 2015
W. Becker (ed.), *Advanced Time-Correlated Single Photon Counting Applications*,
Springer Series in Chemical Physics 111, DOI 10.1007/978-3-319-14929-5_18

561

tunable filter as light source, the tissue was scanned by a galvanometer scanner from a distance of more than 10 cm. The distance between the illumination spot (source) and the detection spot from which the diffusely remitted photons were collected was small (few mm) and kept fixed during the scan. A fast-gated single-photon avalanche diode was employed to eliminate the intense early part of the diffusely remitted signal and to detect late photons only. Polarization-selective detection was additionally applied to suppress specular reflections from the object. An array of gated time-of-flight distributions of photons was recorded by imaging TCSPC synchronized with the movement of the galvanometer scanner. A tissue area of several cm^2 was scanned with 32×32 pixels within a frame rate of 1 s^{-1}. The wavelength was switched line by line between two bands centred at 760 and 860 nm. Concentration changes of oxy- and deoxy-haemoglobin were derived from changes in photon counts in a selected time window of the gated distributions at the two wavelengths. First in vivo tests included the recording of haemodynamics during arm occlusion as well as brain activation tasks. These tests demonstrated the successful non-contact imaging of haemoglobin concentration changes in deeper tissues. Additional applications seem feasible by increasing the spectral information content of the non-contact scanning approach. To this end we implemented and tested the non-contact scanning in combination with eight-wavelength multiplexing.

18.1 Introduction

Near-infrared light has been employed to image heterogeneities or monitor dynamic processes in biological tissues of dimensions in the centimetre range. Diffuse optical imaging and spectroscopy has become an important and well-studied field in biomedical optics, with a growing number of clinical applications, as in optical mammography [25] and optical brain imaging and monitoring. In this context, optical methods provide a portable and less expensive alternative to established neuroimaging methods. During 20 years of development of functional near-infrared spectroscopy (fNIRS) of the brain [4], numerous approaches and technical solutions have been reported, among them continuous wave (CW) optical topography and tomography employing multiple sources and detectors [37] or time-domain fNIRS imaging [43]. The fNIRS methodology has found a wide range of applications [13]. Moreover, NIRS cerebral and muscle oximetry [14] are on their way to become tools of clinical relevance. Other applications of diffuse optical techniques target diagnostic information related to bone, joints and skin.

While optical tomography, e.g. of the neonatal brain [19] or of the breast [16] as well as fNIRS and tissue oximetry originally rely, in general, on direct contact between fibre-based optodes and the tissue surface, there is a trend towards non-contact approaches. They allow problems that arise from pressure and unstable coupling between the optodes and the skin to be avoided. They are particularly advantageous where physical contact causes additional pain to the patient, as in the case of the diagnosis of burn wound severity [23]. Another advantage of non-contact approaches is the opportunity to obtain images that contain a larger number and more densely spaced pixels compared to arrays of multiple optodes, by flying spot scanning or 2D recording by cameras. In this way, the lateral spatial under-sampling intrinsic to optode arrays can be overcome. Non-contact techniques have been already applied, e.g. to study oxygen saturation in skin or muscle tissue [32, 42]. In fluorescence diffuse optical imaging, non-contact schemes are rather common. Examples are fluorescence imaging of indocyanine green (ICG) tracer kinetics for detection of rheumatoid arthritis in finger joints [15] and small animal imaging with fluorescent contrast agents [10].

The CW scanning technique "laminar optical tomography" [20] that is based on diffuse scattering enables depth resolution of up to a few mm beneath the tissue surface, by varying the distance between the spots of light incidence and detection. In this way, 3D images, e.g., of the rat cortex [20] or of skin lesions [31] were obtained. This CW technique, however, is not suitable to reach the adult human cortex ~ 1 to 2 cm beneath the surface. For that purpose larger source-detector separations between ~ 2 and 4 cm are necessary. In this case the detection of light having travelled inside the tissue is hampered by the crosstalk from the comparably large intensity at the source-tissue interface into the sensitive detection channel [17].

Picosecond time resolution provides additional information compared to CW recording by measuring the flight time of each photon detected. This is relevant in diffuse optical imaging and spectroscopy of deep tissues where typical lengths of trajectories of photons in tissue are on the order of a few tens of centimetres. The additional information is useful to better localize and characterize inhomogeneities as, e.g. tumours in the female breast [18], see Chap. 19 of this book. In time-domain optical brain imaging [43] the time-of-flight information is employed to distinguish between absorption changes in the cerebral cortex and extracranial tissues [27, 40], in particular in fNIRS [21, 22, 24] and in perfusion monitoring by ICG bolus tracking [26, 41].

The idea to remotely record activation in the human brain has been discussed many years ago. The group of Britton Chance suggested a system for recording prefrontal brain activity from a distance of 1 m, using time-resolved detection based on TCSPC [48]. Several groups employed time-gated intensified CCD cameras to perform non-contact fNIRS imaging, in particular, for late photons [35, 36]. A hybrid scheme was proposed where the injection was provided by a single fibre in contact with the head and shielded by a 1 cm diameter patch to avoid directly reflected light, while collection was obtained via a time-gated CCD camera [47].

A specific development for time-resolved diffuse reflectance measurements for the characterization of deep tissue compartments is the null source-detector separation (NSDS) approach [34, 44] in conjunction with a single-photon avalanche diode (SPAD) operated in fast-gated mode [9, 45]. By gating off early photons, late photons that visited deep tissue regions can be recorded even for short source-detector separation. This zero or small-distance concept was first applied with fibre-based technology. On this basis, a method for interstitial time-of-flight spectroscopy with a single fibre was developed and demonstrated on phantoms [1]. Moreover, successful in vivo measurements to detect brain activation were performed with short interfibre distance [12].

The novel non-contact scanning approach described in this chapter combines the following features introduced above, (i) non-contact imaging, (ii) a dense and flexible grid of measurement positions, (iii) depth-selective detection (up to a few cm depth) by time-resolved recording. This combination was feasible by using fast spatial scanning, the imaging capability of multi-dimensional TCSPC [2, 3] and the NSDS approach. Proof-of-principle tests of our non-contact method on phantoms were reported in [28]. The implementation of the scanning setup and first in vivo tests were described in [29].

In this chapter we introduce the concept of the non-contact deep-tissue scanning in detail and describe its experimental implementation as well as examples of in vivo applications to record haemodynamic changes, in particular in the brain of adult subjects. In addition, we present a demonstration of non-contact scanning in combination with eight-wavelength multiplexing.

18.2 Technical Concept

18.2.1 Scanning and TCSPC Recording

Images are obtained by scanning the object under investigation with a pulsed laser beam. The laser beam is deflected by a fast galvanometer scanner. The light remitted from the tissue goes back through the scan mirrors and is guided to the detector. The spot from which the light is detected can be offset from the laser spot by an adjustable distance (source-detector separation). The scanner runs a raster scan at a resolution of up to 256×256 pixels at a frame rate of typically 1 frame per second. This is fast enough to record haemodynamic changes in the tissue.

The distributions of times of flight of photons (DTOFs) in the particular pixels are recorded by multi-dimensional TCSPC. The principle is identical with the one used for TCSPC FLIM: the TCSPC module builds up a photon distribution over the coordinates of the scan and the arrival times of the photons in the laser pulse period. The result is an array of pixels, each containing a time-of-flight distribution in a large number of time channels [2, 3]. Data of subsequent frames can be saved consecutively and analyzed as time series or can be accumulated over a longer time into the same array of DTOFs. In the first case, the data can be used to record haemodynamic changes. In the second case, time-of-flight distributions with high signal-to-noise ratio are generated, e.g., for model-based fitting to determine tissue optical properties.

18.2.2   Wavelength Multiplexing

Measurements at a minimum of two wavelengths are necessary to determine the concentration changes in oxy- and deoxy-haemoglobin. Time-domain systems in diffuse optics often use wavelength multiplexing by pulse-interleaved illumination (see Chap. 19). This technique cannot be used here for two reasons. The first one is that the system must be able to use gated detection (see below), but present time-gated detector technology [5] can only select one of the pulses in the interleaved pulse train. Unless several gated detectors are used, gating is thus not compatible with pulse-interleaved illumination. The second reason is that the optical setup uses the same beam path for excitation and detection. This causes reflection and scattering at optical surfaces. Since the gated detection of late photons is performed with largely increased sensitivity [9], even second or third order reflections from the large amount of early photons present in the remitted light can spoil the true late-photon signal. Thus a single wavelength at a time is easier to handle than multiple interleaved wavelengths.

Therefore, wavelength multiplexing was performed in larger groups of pulses (sometimes called burst mode). The principle is shown in Chap. 1, Fig. 1.16. It has the advantage that there is no crosstalk by overlapping of the signals of different wavelengths by optical reflections, pile-up, counting loss, or afterpulsing. The multiplexing rate is compatible with the switching speed of acousto-optical tunable filters (AOTF). Wavelength multiplexing can thus be performed by an AOTF in the beam path of a super-continuum laser. To avoid aliasing with the scanning the multiplexing is synchronised with the frames, lines or pixels of the scan.

18.2.3   Reduction of Source-Detector Separation and Gated Detection

Increasing the number of spatial sampling points has maximum effect on the data quality if the photons detected have travelled though a reasonably confined volume within the turbid medium under investigation. From this point of view, recording should be performed at short source-detector separation.

In functional optical imaging of the brain, the source-detector separation r_{SD} is typically between 2 and 4 cm. Such large r_{SD} is necessary in CW measurements to obtain a reasonable signal contribution from deep absorption changes, e.g. from the cerebral cortex in adults at a depth of about 15 mm. With decreasing r_{SD} the maximum of the overall (time-integrated) sensitivity moves closer to the surface since the probability of short photon trajectories from source to detector increases. For $r_{SD} = 0$, the overall fraction of photons that visit a region deeply beneath the tissue surface is extremely small. Therefore in such case an absorption change in the cortex could not be detected by a CW method.

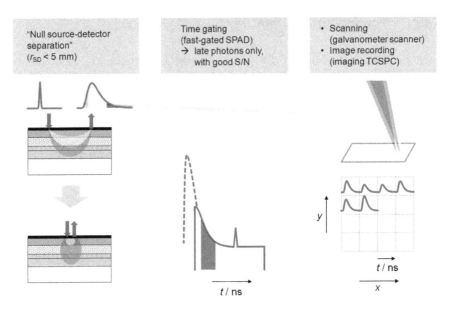

Fig. 18.1 Overview of the concept of time-domain non-contact scanning imaging

Time-resolved detection allows absorption changes in deep and superficial compartments to be separated due to the different regions preferentially visited by early and late photons (see Fig. 18.1, upper left, and Chap. 17). This concept was first introduced for time-domain fNIRS at large source-detector separation. It remains valid for short r_{SD}. It was found [44] that the contrast for a deep, localized absorption change at a fixed late time does not decrease with decreasing r_{SD}, but is even largest for $r_{SD} = 0$ at any time. In addition, the absolute number of (late) photons increases with decreasing r_{SD}. Simultaneously, the fraction of early photons increases dramatically with decreasing r_{SD} [44]. Thus it is necessary to eliminate the early part of the DTOF when measuring deep absorption changes at short r_{SD}, as illustrated in Fig. 18.1, centre. This requirement results from the limited dynamic range of detection.

The lateral spatial resolution depends on time since the size of the volume visited by the photons that reach the detector grows with time. However, comparing the cases with r_{SD} of 3 cm and 0, the zero distance case has always the most confined sensitivity volume at a given time [39]. Moreover, for $r_{SD} = 0$ this volume is exactly rotationally symmetric. Its asymmetry remains small for r_{SD} of a few mm.

The consequences of decreasing r_{SD} are illustrated in Fig. 18.2 for the contrast due to deep absorption changes in a two-layered model. Del Bianco et al. [11] showed that the time-dependent mean partial path length and thus the contrast caused by an absorption change in a layer is independent of r_{SD}. This is shown in Fig. 18.2 for two different source-detector separations by Monte-Carlo simulations based on the code originally described in [40]. The optical properties in the initial (baseline) state were assumed as follows: reduced scattering coefficient

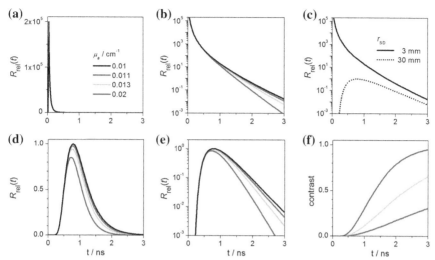

Fig. 18.2 Time-resolved reflectance $R(t)$ and contrast for various absorption values (see legend) in the lower layer of a two-layered model, obtained from Monte-Carlo simulations for a point-like source and effective source-detector separation of 3 mm (**a**, **b**) and 30 mm (**d**, **e**). **c** DTOFs for both r_{SD} values, for the baseline state $\mu_a = 0.1$ cm^{-1}, plotted on the same scale. **f** Time-dependent contrast, the curves for both r_{SD} values are indistinguishable

$\mu_s' = 10$ cm^{-1}, absorption coefficient $\mu_a = 0.1$ cm^{-1} and refractive index $n = 1.4$. The absorption coefficient in the lower layer was changed to $\mu_a/$cm$^{-1} = 0.11; 0.13; 0.2$. The thickness of the upper layer was 10 mm. Figure 18.2a, b depict the time-resolved reflectance $R(t)$ for $r_{SD} \approx 3$ mm in linear and logarithmic scale. To this end, the photons were collected within a ring-shaped detector area with inner and outer radius 2 and 4 mm, respectively. Figure 18.2d, e depict $R(t)$ for a ring between 28 and 32 mm radius ($r_{SD} \approx 30$ mm). All sets of $R(t)$ curves were divided by the maximum of the DTOF in the baseline state in case of the source-detector separation $r_{SD} \approx 30$ mm. The width of the DTOFs changes substantially when increasing the source-detector separation. However, at late times, the slopes in both cases become similar, dominated by the absorption in the lower layer. Figure 18.2c directly compares the curves for the baseline state and both source-detector separations. The maximum reflectance at short r_{SD} is by orders of magnitude higher than at large r_{SD}. Nevertheless, the influence of absorption changes in the lower layer is the same, leading to the same time-dependent contrast function where contrast was defined as $1 - R(t)/R_0(t)$ (Fig. 18.2f). It should be noted that although the contrast is the same for each r_{SD}, its detectability (contrast-to-noise ratio) depends on the number of photons available in the particular time window.

A fast-gated SPAD detector [5] is able to cut off the early part of the DTOF measured at short r_{SD} without running the risk of damage by the large amount of early photons that hit the detector without being detected. It should be noted, however, that there is still an influence of the carriers that are generated by these early photons during the gate-off state preceding the gate-on window under consideration. Indeed, they lead to an afterpulse-like effect known as "memory effect" [8], inherent to thin-junction silicon SPADs, thus increasing the background level. Therefore it is preferable to work at small rather than zero r_{SD} to reduce the amount of early photons arriving at the detector to some extent. A source-detector separation of up to 5 mm can be tolerated without significant losses in terms of contrast and spatial resolution compared to $r_{SD} = 0$ [39, 44].

Having eliminated the early part of the DTOF, the signal-to-noise ratio (SNR) of the late part can be improved by increasing the laser power impinging on the tissue. This is illustrated in Fig. 18.3 for a phantom experiment employing a solid phantom with $\mu_s' = 13.4$ cm^{-1}, $\mu_a = 0.086$ cm^{-1}, $n = 1.55$. At $r_{SD} = 0$ the laser power had to be attenuated dramatically. While moving the gate to later times in steps of 100 ps, the filter attenuation was gradually decreased to keep a nearly constant count rate ($\sim 4 \times 10^5$ s^{-1}, collection time 10 s). Figure 18.3a shows a selection of the recorded DTOFs. It is obvious that the SNR for late photons is enormously increased. At the same time, residual reflections (around ~ 2.5 and 4.3 ns) in the optical system outgrow the noise. From all curves, the dark background (taken from the measurement of the non-gated DTOF) was subtracted. For the highest power steps, an additional constant background component is visible that is due the "memory effect" [8] and finally limits the dynamic range. Figure 18.3b illustrates the complete DTOF (black curve) that is reconstructed by making use of the rescaled gated DTOFs [9]. It exhibits a tremendously increased dynamic range compared to the non-gated DTOF (grey curve in Fig. 18.3a, note the different scales).

The output of the fast-gated SPAD could, in principle, be processed with time-integrated recording. However, a combination with subsequent time-resolved detection by TCSPC (see Fig. 18.1, centre) is advantageous for several reasons. Having recorded the full temporal profile of the gated DTOF allows one to arbitrarily select specific time windows within this profile at the analysis stage, thus

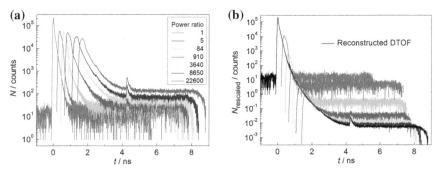

Fig. 18.3 DTOFs recorded at increasing gate delays on a solid phantom (details see text), with null source-detector separation ($r_{SD} = 0$). **a** Original DTOFs after subtraction of dark background; *grey curve* non-gated DTOF, *coloured curves* gated DTOFs with early part increasingly cut off by the gate, at increased power (ratio see legend); **b** DTOFs rescaled according to the power ratio; *black curve* reconstructed DTOF over the full dynamic range

improving depth selectivity. This may also be important to avoid the influence of reflections in the optical components. In addition, the signal time course within the gate interval, i.e. for late photons, can explicitly be analyzed.

18.3 Dual-Wavelength Implementation

18.3.1 Experimental Setup

The optical setup and the signal flow of the non-contact scanning system [29] are shown in Fig. 18.4.

18.3.1.1 Switchable Light Source

A supercontinuum (SC) laser (SC500-6, Fianium Ltd, UK) equipped with an 8-channel AOTF for the NIR spectral range provided picosecond pulses (<100 ps) at a repetition rate of 40.5 MHz. The wavelength of each AOTF channel can be tuned independently. Switching between wavelengths can be performed in two ways,

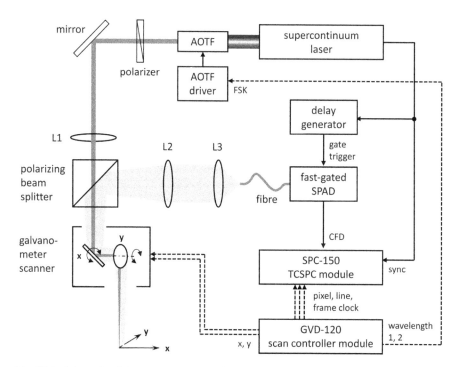

Fig. 18.4 Schematic of the non-contact scanning system

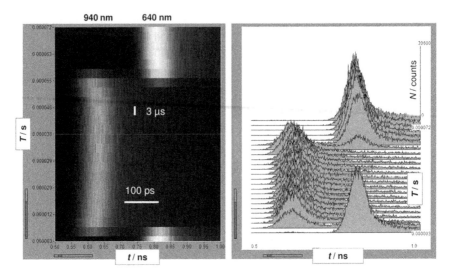

Fig. 18.5 Determination of the switching time of the AOTF output

(1) by using the "FSK" (Frequency Shift Keying) mode that enables for each channel fast switching between two preset wavelengths, triggered by an external digital signal, (2) by turning individual channels on and off by a modulation voltage. In the dual-wavelength implementation the first option was used. For an application of the second option see Sect. 18.4.

To estimate the switching time in the FSK mode, the following experiment was performed (see Fig. 18.5). A 10 kHz trigger pulse train from a pulse generator was applied to the FSK input. The AOTF output was recorded in the triggered accumulation mode of a TCSPC board (SPC 130) [3] with the same 10 kHz trigger, and counts from 20,000 trigger periods were accumulated. A pigtailed silicon SPAD (PDM series, MPD, Italy) was used as a detector. FSK switching was performed between wavelengths of 640 and 940 nm. The temporal position of the corresponding signals differed by ~160 ps due to different delays with respect to the sync pulse train obtained from the oscillator of the SC laser. This wavelength-dependent time shift is mainly caused by the SC laser. The switching time was found to be of the order of 3 μs, see Fig. 18.5.

In addition to the switching time, it is important to know the possible delay between the digital pulse applied for switching and the response of the optical output. A delay of about 10 μs was found, presumably caused by the propagation times of the acoustic waves in the AOTF crystal. This delay is insignificant as long as line-by-line or frame-by-frame multiplexing is employed.

To maximize the laser power transmitted by the AOTF, all its eight channels were stacked together, with a wavelength spacing (~4 nm) also optimized for maximum output. The maximum available power was ~32 mW at the sample surface. Both wavelength bands toggled by the FSK trigger are shown in Fig. 18.6.

Fig. 18.6 Spectra of
oxy- and deoxy-haemoglobin
[7] (*red* and *blue curves*,
respectively); alternating
AOTF output spectra with
central wavelengths 760 nm
(*cyan*) and 860 nm (*magenta*)

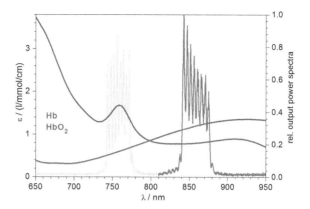

The wavelengths around 760 and 860 nm were chosen as a compromise between
the achievable output power and the differences between the absorption spectra of
oxy- and deoxy-haemoglobin.

18.3.1.2 Optical System

The output beam of the AOTF was slightly focused by lens L1 and directed to the
tissue by the mirrors of the galvanometer scanner, a 2-axis laser beam deflection
unit (Superscan-7, Raylase, Germany, aperture 7 mm). Light diffusely reemitted
from the tissue passed the scanner again and was imaged by the lenses L2
(f = 300 mm) and L3 (f = 35 mm) onto a multimode fibre (diameter 200 μm,
NA = 0.22, length 2 m) that guided it to the detector. The magnification factor
determined the size of the spot from which light was collected. The focal ratio of
lens L2 and the aperture of the galvanometer scanner defined the acceptance angle.
The distance between scan head and tissue was 13 cm.

The position of the detection spot relative to the incident light spot (r_{SD}) was
adjusted by the position of the entrance fibre tip. A separation of r_{SD} = 4 mm was
chosen to reduce the early-photon contribution, as explained above. The output face
of the detection fibre was imaged onto the active area (100 μm diameter) of the
SPAD detector by a pair of aspheric lenses.

The optical system was designed such that reflections from the tissue surface as
well as from optical components were suppressed as far as possible. This is particularly important for the sensitive measurement at late times in the presence of a
large amount of early photons. Unlike with fluorescence detection, reflected light
cannot be rejected by spectral filtering. A substantial suppression of direct reflections as well as weakly scattered light from the sample was achieved by polarization-selective detection. The linearly polarized AOTF output was further cleaned
by a polarizer. Diffusely scattered light from the tissue is randomly polarized, and
half of it passes the polarization splitting cube and can be detected. The use of

separate lenses in both arms (L1, L2) was an additional measure to avoid the occurrence of parasitic laser light in the detection path.

18.3.1.3 Gated Detector

A second generation compact fast-gated SPAD module (Politecnico di Milano, Italy) with embedded gating and signal conditioning circuitry [5] was employed for selective detection of late photons. Its active area had a diameter of 100 μm. The module included a fast pulse generator which achieved gating transitions below 200 ps at repetition rates up to 50 MHz. The gate width was 6 ns. The gate delay was adjusted by a home-built transmission-line based delay generator with 25 ps steps. A delay of about 1.3 ns with respect to the early-photon peak was typically chosen in the in vivo experiments. This delay still allowed a maximum count rate of 3×10^6 s^{-1} to be achieved for late photons, to maintain a high SNR.

The fast-gated SPAD provided standard NIM pulses for single-photon timing. Both gate trigger and TCSPC sync input were derived from the pulse train monitor output of the supercontinuum laser.

18.3.1.4 Scanning and Laser Wavelength Multiplexing

A GVD-120 scan controller card (Becker & Hickl, Germany) [3] was employed to control the scan as well as the image acquisition. Scan parameters included the size and shape (square or rectangular) of the scan area and the acquisition time. Corresponding X and Y signals were applied to the galvanometer scanner. In parallel, the actual frame, line and pixel information was communicated to an imaging TCSPC module (SPC-150, Becker & Hickl, Germany) via the Pixel Clock, Line Clock and Frame Clock signals. The GVD-120 scan controller also provided the FSK signal for laser wavelength switching via the AOTF. For the results presented in this chapter we used line-by-line multiplexing. Simultaneously with the FSK signal a routing signal to the SPC-150 module was provided. The GVD and SPC modules were controlled by SPCM software (Becker & Hickl, Germany) [3].

18.3.1.5 Data Acquisition

For the in vivo results presented in Sect. 18.3.3, the parameters were adjusted to cover a scan area of 4×4 cm^2 on the tissue with a 32×32 pixel frame size in step scanning mode. The frame time was chosen to be 1 s, resulting in a pixel time of ∼1 ms. The Laser Routing signal was applied to the FSK input of the AOTF to switch between the two wavelength bands from line to line. Thus for each frame two 32×16 pixel images corresponding the two wavelengths were recorded quasi-simultaneously. The TCSPC data were recorded in the Scan Sync In (hardware accumulation) mode [3], see Chap. 1, Sect. 1.3.2.

18.3.2 Data Analysis

The data processing was based on a time-window (TW) analysis within the late-photon part of the DTOF that was selected by the electronic gate. The analysis was performed in MATLAB®. For each experiment a TW was chosen in which the photons detected were integrated for each DTOF. The limits of this time window were defined such to include a part of the DTOF with a good signal-to-noise ratio while avoiding the influence of residual reflections in the optical path due to early photons. The result were two time series (time T in steps of the frame time 1 s) of 32×16 pixel intensity images (photon counts within the TW) corresponding to the two wavelengths.

Since the time series contained the response to multiple repetitions of the same stimulation paradigm (see Sect. 18.3.3, In Vivo Results) of brain activation, the signals of all trials were summed up (block averaging) to improve the signal-to-noise ratio. This approach is common in fNIRS of the brain where the signal changes are small and vary from trial to trial.

The concentration changes of oxygenated and deoxygenated haemoglobin in each pixel were estimated on the basis of the time-resolved (or microscopic) Beer-Lambert law [33]

$$\frac{I_T(t)}{I_0(t)} = \exp(-\Delta\mu_a v\, t), \tag{18.1}$$

where I_T and I_0 are the intensities at (macro) time T during the activation paradigm and in the baseline state, respectively. I_T and I_0 were obtained as total photon count in the time window under consideration. The time t (on the picosecond time scale) was approximated by taking the time at the centre of the TW. The time-dependent path length is $L = v\, t$ where v is the speed of light in the medium (refractive index 1.4). The time origin $t = 0$ was determined as the maximum position of the complete DTOF in a reference measurement without gate delay.

From the absorption changes at the two wavelengths (1, 2), the changes in oxy- and deoxyhemoglobin concentrations (Δc_{HbO2}, Δc_{Hb}) were retrieved by solving the system of equations

$$\Delta\mu_{a,1,2} = \left(\varepsilon_{1,2}^{HbO2}\Delta c_{HbO2} + \varepsilon_{1,2}^{Hb}\Delta c_{Hb}\right) \ln(10) \tag{18.2}$$

where $\varepsilon_{1,2}^{HbO2}$ and $\varepsilon_{1,2}^{Hb}$ are the mean values over the respective wavelength intervals of the molar absorption coefficients for oxy- and deoxyhemoglobin [7], respectively (see Fig. 18.6).

This simplified approach provides quantitative concentration changes (in μM), however, it involves a number of approximations. Notably, the absorption change is assumed to be homogeneous within the whole tissue volume sampled. In particular, a separation between absorption changes in brain and superficial tissue cannot be achieved, and an absorption change in the brain is underestimated due to the partial path length effect.

The procedure described yielded 32×16 pixel images of block-averaged time traces (T) for oxy- and deoxy-haemoglobin. An optional 4×2 pixel binning was applied to obtain 8×8 pixel maps.

18.3.3 In Vivo Results: Forearm Occlusion and Brain Activation

18.3.3.1 Arterial Occlusion on the Forearm

Several stimulation paradigms were performed on the forearm and the head of healthy adult subjects. As first basic test, haemodynamic changes in the forearm were induced by cuff occlusion, similar to the procedure of a blood pressure measurement. In particular, arterial occlusion was induced by cuff pressure applied to the upper arm to block the arterial inflow into as well as the venous outflow from the forearm. The paradigm for arterial occlusion consisted of 128 s of baseline recording, 96 s of occlusion (pressure applied: 250 mmHg), and 128 s of recovery, without repetitions.

The results of the arterial occlusion experiment on a female subject (24 years) are shown in Fig. 18.7. The measurement area was located on the upper inner forearm. The three panels in the upper row of Fig. 18.7 display the time courses of the Hb and HbO_2 concentration changes for three regions of interest. For six selected time points before, during and after occlusion complete 32×16 pixel images of the whole area scanned are provided.

The arterial occlusion leads to large changes of the HbO_2 and Hb concentrations, compared to the brain activation that will be discussed below. The changes in the total photon count (not shown) typically dropped to 40 % of the initial level at 760 nm and 50 % at 860 nm. Such large changes enable the retrieval of a signal with reasonable signal-to-noise ratio on a single-trial level. During the build-up of the pressure applied by a hand pump the time traces between 130 and 140 s resemble the behaviour of venous occlusion, i.e. the concentrations of HbO_2 and Hb rise together due to blocked outflow of blood. When the pressure is large enough to occlude the arteries, HbO_2 starts decreasing (panels a and c) while the concentration of Hb further increases. This behaviour can be explained by the conversion of HbO_2 into Hb due to tissue metabolism. At the end of occlusion, the cuff is deflated again during several seconds. While Hb quickly returns to baseline, HbO_2 exhibits a marked overshoot before returning at a slower pace.

The 2D images in the lower part of Fig. 18.7 reveal a substantial heterogeneity in the haemodynamic response and differences for HbO_2 and Hb. Such local differences might be due to the presence of superficial, but also deeper and thus less resolved vessel structures. In particular, the Hb images display a slightly curved line

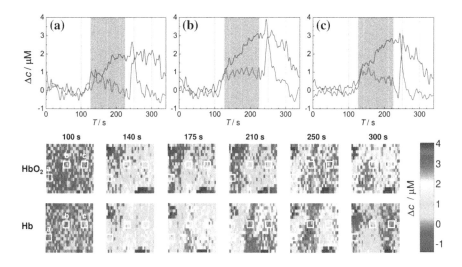

Fig. 18.7 Results of an arterial occlusion measurement. *Top row* time courses of HbO$_2$ (*red line*) and Hb (*blue line*) for three different regions of binned (4 × 2) pixels, marked by *white squares* on the images below. A sliding average of 5 s was applied. *Grey shaded* areas mark the time of occlusion. *Bottom row* 32 × 16 pixel images (pixel separation 1.25 mm in X and 2.5 mm in Y direction) of HbO$_2$ (*top*) and Hb (*bottom*) recorded at selected times (marked by *green lines* in the *upper panels*), averaged over 5 frames (5 s), without spatial smoothing. The dark spots on the *upper left* and *lower right* of the images are due to black markers fixed to the skin. Reprinted, with permission, from [29]

coming from the top of the image to the bottom, left of the midline of the images, showing elevated values compared to the surrounding tissue as well as different time traces (panel b). This structure can be attributed to a superficial vein that was visible beneath the skin.

This example demonstrates the advantages of an imaging approach with lateral spatial resolution of a few mm. The clearly heterogeneous behaviour of the haemoglobin changes would affect results obtained by NIRS techniques, based on a few single optodes or even optode arrays with separations in the centimeter range, in an unknown manner.

18.3.3.2 Brain Activation by a Cognitive Task

Activation in the left frontal lobe was induced by asking the subjects to solve a sequence of simple math problems. The paradigm included 32 s of baseline measurements, 32 s of activation, followed by 32 s of rest. The whole cycle was repeated 20 times. The scan area was centred about 5 cm to the left from the centre of the forehead. The subjects were resting in supine position, and their head was fixed by a vacuum cushion.

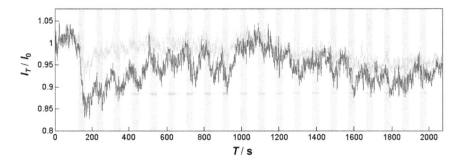

Fig. 18.8 Cognitive brain activation by solving simple math tasks. Time series of photon counts at 760 nm (*cyan*) and 860 nm (*magenta*), normalized to the signals during the initial baseline phase and averaged over the whole image, activation periods are marked by *grey-shaded* areas

Figures 18.8 and 18.9 present the results obtained for a single subject (female, 24 years). In Fig. 18.8 the time evolution of the signal during the whole experiment is illustrated as time series of total photon counts in the gated DTOF for both wavelengths, integrated over the whole image. At the longer wavelength, mainly influenced by HbO_2, clear responses can be observed for almost all trials. These changes correspond to an increase in HbO_2 that is visible in the 8×8 map of block-averaged data in Fig. 18.9. An activation pattern with increased HbO_2 and decreased Hb concentration due to the stimulation can be discerned throughout the whole area scanned, with a smaller magnitude of changes in the upper left part of the image. The most pronounced Hb response is found in the upper right part. The substantial noise in the lower row was due to the black eye shield touching the scan area.

By recording late photons only it cannot be excluded that the signals are also affected by a superficial response which is known to be particularly strong with cognitive paradigms and in the HbO_2 response [24, 30]. Selective sensitivity to changes in the brain could be achieved by combining the different information carried by late and early photons, thus eliminating superficial signals [6, 38].

18.3.3.3 Motor Activation

The left motor cortex was stimulated by a finger tapping task executed with the right hand. The motor paradigm consisted of 20 trials of 32 s of finger tapping followed by 32 s of rest. The subjects were resting in supine position, with the head tilted to centre the scan area at the motor cortex (C3 position according to the 10–20 system).

The results obtained for a subject with an almost bald head (male, 52 years) are presented in Fig. 18.10. In an area above the centre of the image (approximate C3 position) the time traces show the typical pattern of a cerebral activation, i.e. an

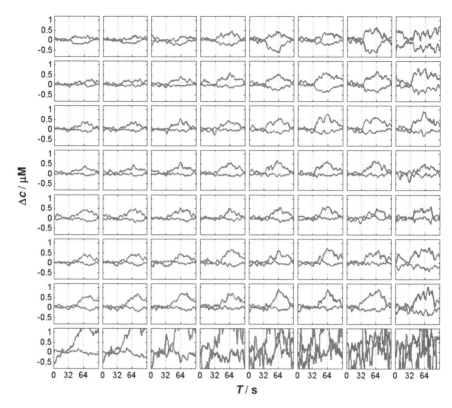

Fig. 18.9 Cognitive brain activation by solving simple math tasks. Map of time traces of changes of HbO$_2$ (*red*) and Hb (*blue*) on the *left forehead*, activation from $T = 32$ to 64 s. Each pixel of the 4×4 cm^2 image corresponds to an area of 5×5 mm^2. A sliding average of 5 s was applied to the block-averaged traces

increase in oxy- and a (smaller) decrease in deoxy-haemoglobin. No such response is visible in the upper left part of the image where the SNR is similar. The identification of a localized response is another indication that the signal is indeed of cerebral origin, while systemic changes would exhibit a more global behaviour. The lower right part of the image is impaired by the presence of noise. The count rate in this area was lower by a factor of four compared to the top area of the image, due to the presence of very short hair. The experience with motor activation measurements on other subjects showed that a useful signal is detectable only if there is absolutely no hair present in the area of detection. Even hair of a length of only a few mm absorbs and scatters too many photons and confounds sufficient signal levels.

This example is another illustration of the advantage of dense imaging. The position and extension of the activated area can be determined with much better

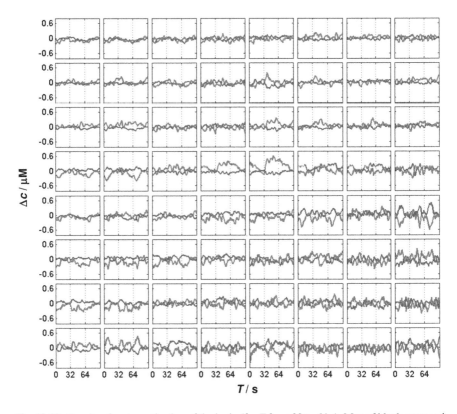

Fig. 18.10 Results of motor activation of the brain (for T from 32 to 64 s). Map of block-averaged time traces of changes of HbO$_2$ (*red*) and Hb (*blue*) within a 4×4 cm^2 scan area (each pixel corresponds to a an area of 5×5 mm^2), with approximate position of the *left* motor cortex (C3). In the *lower right* part of the area scanned very short hair was present

resolution than with a sparse array of source and detector fibres. A measurement with only a single optode pair relies on the a priori knowledge of the location of the motor cortex which may vary from subject to subject.

18.4 Eight-Wavelength Multiplexing

The dual-wavelength multiplexing as described in Sect. 18.3 is the simplest spectroscopic approach for fast dynamic recording of functional changes in oxy- and deoxy-haemoglobin concentrations. The use of multiple wavelengths is required if more than two chromophores are to be analyzed or if the wavelength dependence of optical properties of the tissue is to be studied. We demonstrated the feasibility of combining the non-contact scanning approach with eight-wavelength multiplexing.

18.4.1 Setup

The principles of a tissue scanning system for recording time-of-flight distributions with eight multiplexed laser wavelengths are shown in Fig. 18.11. The optical setup and the major components were the same as used above. The measurement object is scanned by a laser beam via a fast galvanometer scanner. The scanner is driven by a GVD-120 scan controller. The light returning from the sample is separated from the excitation beam by a beam splitter. The light passes a variable optical attenuator and enters an optical fibre that transfers the light to the detector. The entrance face of the fibre acts as a pinhole in the confocal arrangement. The spot on the tissue that is imaged onto the fibre can be overlapped with the laser spot or offset by a few millimetres.

The demonstration was performed with a non-gated detector. The large intensity ratio between early and late photons requires a detector with a high dynamic range. We used a hybrid detector (Becker & Hickl HPM-100-50) to record complete time-of-flight distributions with a superior dynamic range. It should be noted that a gated SPAD detector can be used with the eight-wavelength multiplexing option equally well. The single-photon pulses from the detector were recorded by an SPC-150 imaging TCSPC module [3]. The pixel, line, and frame clock pulses for synchronisation with the scan were provided by the GVD-120 scan controller [3].

The supercontinuum laser delivered the picosecond pulse train. Unlike in the setup used above, the eight AOTF channels were configured to provide individual wavelengths that could be freely selected. The 'MOD' inputs of the AOTF controller were employed to achieve multiplexing by activating the eight individual AOTF channels one after another. The switching signals were provided by a

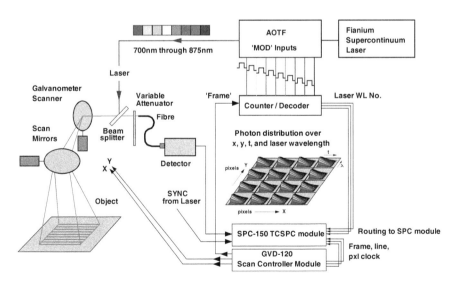

Fig. 18.11 Scanning system with 8 multiplexed wavelengths

counter that was driven by the frame clock pulses of the scan controller. As a result, the AOTF cycled through the 8 programmable wavelengths for subsequent frames of the scans. Simultaneously, the routing inputs of the SPC-150 module were driven by signals from the counter outputs. Thus, the SPC module built up separate images for the eight individual laser wavelengths, see Chap. 1, Fig. 1.16. It should be noted that, alternatively, line-by-line or pixel-by-pixel wavelength switching can be used.

18.4.2 In Vivo Demonstration

A result of an in vivo demonstration experiment is shown in Fig. 18.12. The hand of a volunteer was scanned at a resolution of 128 × 128 pixels and eight diffuse-reflection images were recorded for wavelengths from 700 to 875 nm. The source-detector separation was 4 mm.

For a simple check of the behaviour of the DTOFs in the images, SPCImage FLIM analysis software (Becker & Hickl, Germany) [3] was employed. Figure 18.13 displays two images obtained by the 8-wavelength setup. Both are taken from the 850-nm channel, and display the first moment, M1, of the time-of-flight distribution relative to the first moment of the IRF. The left image is related to a source-detector separation $r_{SD} = 0$, the right one to $r_{SD} = 4$ mm. Both images show differences in M1 between a pixel located on a superficial vein and one on the surrounding tissue. As expected, the difference depends on the source-detector separation: For $r_{SD} = 0$ it is only 3.5 ps, whereas it is almost 22 ps for $r_{SD} = 4$ mm. Although these differences in the mean time of flight are clearly detectable by TCSPC there is no doubt that they are caused by relatively superficial tissue

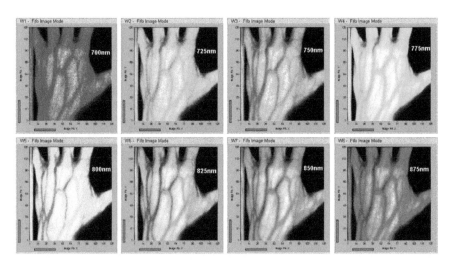

Fig. 18.12 Diffuse-reflection scan of a human hand. Source-detector separation 4 mm, wavelengths indicated in the images. False colour, selected to indicate different laser wavelengths

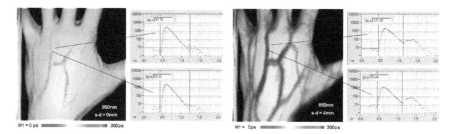

Fig. 18.13 M1 images of the data in the 850-nm channel. $r_{SD} = 0$ (*left*) and $r_{SD} = 4$ mm (*right*) obtained by SPCImage FLIM analysis software. The *green curves* are synthetic IRFs, the *red curves* are single-exponential fits which are not relevant here

structures. Deeper penetration can only be obtained by wider source-detector separation (which causes blurring) or by suppressing early photons by a fast-gated SPAD.

Figure 18.13 also illustrates two other problems of non-contact scanning. The first one is that the measurement object is not flat. Consequently, there are differences in the optical path length in air that show up in the M1 data. This problem could be solved by taking a reference scan at $r_{SD} = 0$ and using the timing of the leading edge of the (non-gated) DTOF as a time origin for scans at larger source-detector distances. Another problem comes from reflection and scattering at optical surfaces in the beam path, see the bumps at late times in the DTOFs shown in Fig. 18.13. These reflections can only be minimised by optimising the optical system.

18.5 Conclusion

This chapter presented the concept and the instrumental setup of our novel time-domain non-contact scanning system for diffuse reflectance imaging of deep tissues as well as first results of its successful in vivo testing. The scanning scheme with a frame rate of 1 s^{-1} over a 4×4 cm^2 area and the quasi-simultaneous acquisition at two wavelengths allowed dynamic changes in the oxy- and deoxy-haemoglobin concentrations to be recorded. This included fNIRS, i.e. the study of haemodynamic changes as a response to functional stimulation of the brain where the rise-time of the haemoglobin concentration changes is typically a few seconds.

A major advantage of the scanning system is its flexibility with respect to the number of pixels, the scan area on the tissue, and the frame rate. The inherently sequential measurement and the low duty cycle at each pixel necessitate a careful experiment design. The available photon count per pixel depends on the total count rate, the frame rate, the number of scan positions (pixels) and wavelengths. For the dynamic (~ 1 s^{-1}) measurements mentioned above, recording at high count rates up to 3×10^6 s^{-1} turned out to be essential to achieve a good signal-to-noise ratio.

Thanks to the fast-gated detector and the rather high laser power (several tens of mW) obtained by AOTF channel stacking, such high count rates could be achieved selectively for late parts of the DTOF. This enormously facilitated the detection of deep absorption changes. We have shown that both motor and cognitive activation of the brain were clearly detectable on a single-subject level, after block averaging of 20 repetitions of the stimulation paradigm. Such block averaging is a standard approach in fNIRS where the signal changes due to brain activation are small and variable. For the considerably larger Hb and HbO_2 changes due to arm occlusion, even single trial data were sufficient to record the dynamic changes with good signal to noise ratio.

If a frame rate on the order of 1 s^{-1} is too slow, e.g. if heart-beat related changes are to be studied, a reduction from a two-dimensional scan to a line scan (like in FLITS, see Chap. 1, Sect. 1.4.6) is an option. On the other hand, if the dynamic requirements are less challenging, e.g. for the determination of stationary tissue oxygenation or the content of various tissue constituents, the overall recording time can be increased. In such case a multi-spectral approach is an option, the feasibility of which has been demonstrated for eight-wavelength multiplexing.

An important issue with a non-contact method is the overall light collection efficiency. It depends on the area and the acceptance angle from which light from the tissue can be transferred to the detector. While the small size of the detector (100 µm diameter) remains a major limitation, there is definitely room for improvement of the transfer optics and the aperture of the scanner. The measurement of the responsivity of the detection system [46] is a useful tool to optimize the system.

The selective recording of late photons increases the sensitivity to absorption changes occurring deeply within the tissue. However, these late photons also propagate through superficial tissues which leads to a related signal contamination. We pursue an advancement of the scanning system by including a second, "early photon" detection channel with a non-gated detector. Thus signal changes in early diffusely scattered photons can be detected and utilized to probe superficial haemodynamics.

Non-contact scanning without mechanical fixation of the subject is inherently prone to movement artefacts. However, time-domain image recording potentially enables tracking of movements and related signal correction: lateral movements can be tracked by attaching markers to the skin. Information on the distance from the scan head can be derived from the arrival time of the early photons.

Although its applicability is restricted to hairless parts of the body, the non-contact scanning approach may be advantageous for a number of applications where high density functional optical mapping of deep tissues is required or helpful. Examples are the localization of functional activation in the prefrontal cortex, the study of peripheral vascular pathologies or intraoperative monitoring.

References

1. E. Alerstam, T. Svensson, S. Andersson-Engels, L. Spinelli, D. Contini, A. Dalla Mora, A. Tosi, F. Zappa, A. Pifferi, Single-fiber diffuse optical time-of-flight spectroscopy. Opt. Lett. **37**, 2877–2879 (2012)
2. W. Becker, *Advanced Time-Correlated Single-Photon Counting Techniques* (Springer, Berlin, 2005)
3. W. Becker, *The bh TCSPC Handbook*, 5th edn (Becker & Hickl GmbH, 2012). www.becker-hickl.com. Printed copies available from Becker & Hickl GmbH
4. D.A. Boas, C.E. Elwell, M. Ferrari, G. Taga, Twenty years of functional near-infrared spectroscopy: introduction for the special issue. NeuroImage **85**(Pt 1), 1–5 (2014)
5. G. Boso, A. Dalla Mora, A. Della Frera, A. Tosi, Fast-gating of single-photon avalanche diodes with 200 ps transitions and 30 ps timing jitter. Sens. Actuators, A **191**, 61–67 (2013)
6. D. Contini, L. Spinelli, A. Torricelli, A. Pifferi, R. Cubeddu, Novel method for depth-resolved brain functional imaging by time-domain NIRS. Proc. SPIE **6629**, 662908 (2007)
7. M. Cope, The development of a near infrared spectroscopy system and its application for non invasive monitoring of cerebral blood and tissue oxygenation in the newborn infant, PhD Thesis, University College London (1991)
8. A. Dalla Mora, D. Contini, A. Pifferi, R. Cubeddu, A. Tosi, F. Zappa, Afterpulse-like noise limits dynamic range in time-gated applications of thin-junction silicon single-photon avalanche diode. Appl. Phys. Lett. **100**, 241111 (2012)
9. A. Dalla Mora, A. Tosi, F. Zappa, S. Cova, D. Contini, A. Pifferi, L. Spinelli, A. Torricelli, R. Cubeddu, Fast-gated single-photon avalanche diode for wide dynamic range near infrared spectroscopy. J. Select. Topics Quantum Electron. **16**, 1023–1030 (2010)
10. C. Darne, Y. Lu, E.M. Sevick-Muraca, Small animal fluorescence and bioluminescence tomography: a review of approaches, algorithms and technology update. Phys. Med. Biol. **59**(1), R1–R64 (2014)
11. S. Del Bianco, F. Martelli, G. Zaccanti, Penetration depth of light re-emitted by a diffusive medium: theoretical and experimental investigation. Phys. Med. Biol. **47**(23), 4131–4144 (2002)
12. L. Di Sieno, D. Contini, A. Dalla Mora, A. Torricelli, L. Spinelli, R. Cubeddu, A. Tosi, G. Boso, A. Pifferi, Functional near-infrared spectroscopy at small source-detector distance by means of high dynamic-range fast-gated SPAD acquisitions: first in-vivo measurements. Proc. SPIE **8804**, 880402 (2013)
13. M. Ferrari, V. Quaresima, A brief review on the history of human functional near-infrared spectroscopy (fNIRS) development and fields of application. NeuroImage **63**(2), 921–935 (2012)
14. M. Ferrari, V. Quaresima, Review: near infrared brain and muscle oximetry: from the discovery to current applications. J. Infrared Spectrosc. **20**(1), 1 (2012)
15. T. Fischer, B. Ebert, J. Voigt, R. Macdonald, U. Schneider, A. Thomas, B. Hamm, K.-G.A. Hermann, Detection of rheumatoid arthritis using non-specific contrast enhanced fluorescence imaging. Acad. Radiol. **17**(3), 375–381 (2010)
16. M.L. Flexman, M.A. Khalil, R. Al Abdi, H.K. Kim, C.J. Fong, E. Desperito, D.L. Hershman, R.L. Barbour, A.H. Hielscher, Digital optical tomography system for dynamic breast imaging. J. Biomed. Opt. 16(7), 076014 (2011)
17. T. Funane, H. Atsumori, A. Suzuki, M. Kiguchi, Noncontact brain activity measurement system based on near-infrared spectroscopy. Appl. Phys. Lett. **96**, 123701 (2010)
18. D. Grosenick, H. Wabnitz, K.T. Moesta, J. Mucke, P.M. Schlag, H. Rinneberg, Time-domain scanning optical mammography: II. Optical properties and tissue parameters of 87 carcinomas. Phys. Med. Biol. **50**(11), 2451–2468 (2005)
19. J.C. Hebden, T. Austin, Optical tomography of the neonatal brain. Eur. Radiol. **17**(11), 2926–2933 (2007)

20. E.M.C. Hillman, D.A. Boas, A.M. Dale, A.K. Dunn, Laminar optical tomography: demonstration of millimeter-scale depth-resolved imaging in turbid media. Opt. Lett. **29** (14), 1650–1652 (2004)

21. A. Jelzow, H. Wabnitz, I. Tachtsidis, E. Kirilina, R. Brühl, R. Macdonald, Separation of superficial and cerebral hemodynamics using a single distance time-domain NIRS measurement. Biomed. Opt. Express **5**(5), 1465–1482 (2014)

22. M. Kacprzak, A. Liebert, W. Staszkiewicz, A. Gabrusiewicz, P. Sawosz, G. Madycki, R. Maniewski, Application of a time-resolved optical brain imager for monitoring cerebral oxygenation during carotid surgery. J. Biomed. Opt. **17**(1), 016002 (2012)

23. M. Kaiser, A. Yafi, M. Cinat, B. Choi, A.J. Durkin, Noninvasive assessment of burn wound severity using optical technology: a review of current and future modalities. Burns **37**, 377–386 (2011)

24. E. Kirilina, A. Jelzow, A. Heine, M. Niessing, H. Wabnitz, R. Brühl, B. Ittermann, A.M. Jacobs, I. Tachtsidis, The physiological origin of task-evoked systemic artefacts in functional near infrared spectroscopy. NeuroImage **61**, 70–81 (2012)

25. D.R. Leff, O.J. Warren, L.C. Enfield, A. Gibson, T. Athanasiou, D.K. Patten, J. Hebden, G.Z. Yang, A. Darzi, Diffuse optical imaging of the healthy and diseased breast: a systematic review. Breast Cancer Res. Treat. **108**(1), 9–22 (2008)

26. A. Liebert, H. Wabnitz, J. Steinbrink, M. Möller, R. Macdonald, H. Rinneberg, A. Villringer, H. Obrig, Bed-side assessment of cerebral perfusion in stroke patients based on optical monitoring of a dye bolus by time-resolved diffuse reflectance. NeuroImage **24**(2), 426–435 (2005)

27. A. Liebert, H. Wabnitz, J. Steinbrink, H. Obrig, M. Möller, R. Macdonald, A. Villringer, H. Rinneberg, Time-resolved multidistance near-infrared spectroscopy of the adult head: intracerebral and extracerebral absorption changes from moments of distribution of times of flight of photons. Appl. Opt. **43**(15), 3037–3047 (2004)

28. M. Mazurenka, A. Jelzow, H. Wabnitz, D. Contini, L. Spinelli, A. Pifferi, R. Cubeddu, A. Dalla Mora, A. Tosi, F. Zappa, R. Macdonald, Non-contact time-resolved diffuse reflectance imaging at null source-detector separation. Opt. Express **20**(1), 283–290 (2012)

29. M. Mazurenka, L. Di Sieno, G. Boso, D. Contini, A. Pifferi, A. Dalla Mora, A. Tosi, H. Wabnitz, R. Macdonald, Non-contact in vivo diffuse optical imaging using a time-gated scanning system. Biomed. Opt. Express **4**(10), 2257–2268 (2013)

30. E. Molteni, D. Contini, M. Caffini, G. Baselli, L. Spinelli, R. Cubeddu, S. Cerutti, A.M. Bianchi, A. Torricelli, Load-dependent brain activation assessed by time-domain functional near-infrared spectroscopy during a working memory task with graded levels of difficulty. J. Biomed. Opt. **17**, 056005 (2012)

31. T.J. Muldoon, S.A. Burgess, B.R. Chen, D. Ratner, E.M.C. Hillman, Analysis of skin lesions using laminar optical tomography. Biomed. Opt. Express **3**(7), 1701–1712 (2012)

32. M. Niwayama, H. Murata, S. Shinohara, Noncontact tissue oxygenation measurement using near-infrared spectroscopy. Rev. Sci. Instrum. **77**, 073102 (2006)

33. Y. Nomura, O. Hazeki, M. Tamura, Relationship between time-resolved and non-time-resolved Beer–Lambert law in turbid media. Phys. Med. Biol. **42**, 1009 (1997)

34. A. Pifferi, A. Torricelli, L. Spinelli, D. Contini, R. Cubeddu, F. Martelli, G. Zaccanti, A. Tosi, A. Dalla Mora, F. Zappa, S. Cova, Time-resolved diffuse reflectance using small source-detector separation and fast single-photon gating. Phys. Rev. Lett. **100**, 138101 (2008)

35. I. Sase, A. Takatsuki, J. Seki, T. Yanagida, A. Seiyama, Noncontact backscatter-mode near-infrared time-resolved imaging system: preliminary study for functional brain mapping. J. Biomed. Opt. **11**(5), 054006 (2006)

36. P. Sawosz, M. Kacprzak, N. Zolek, W. Weigl, S. Wojtkiewicz, R. Maniewski, A. Liebert, Optical system based on time-gated, intensified charge-coupled device camera for brain imaging studies. J. Biomed. Opt. **15**(6), 066025 (2010)

37. F. Scholkmann, S. Kleiser, A. J. Metz, R. Zimmermann, J. Mata Pavia, U. Wolf, M. Wolf, A review on continuous wave functional near-infrared spectroscopy and imaging instrumentation and methodology. NeuroImage **85 Pt 1**, 6–27 (2014)

38. J. Selb, J.J. Stott, M.A. Franceschini, A.G. Sorensen, D.A. Boas, Improved sensitivity to cerebral hemodynamics during brain activation with a time-gated optical system: analytical model and experimental validation. J. Biomed. Opt. **10**, 011013 (2005)
39. L. Spinelli, F. Martelli, S. Del Bianco, A. Pifferi, A. Torricelli, R. Cubeddu, G. Zaccanti, Absorption and scattering perturbations in homogeneous and layered diffusive media probed by time-resolved reflectance at null source-detector separation. Phys. Rev. E Stat. Nonlin. Soft Matter Phys **74**, 021919 (2006)
40. J. Steinbrink, H. Wabnitz, H. Obrig, A. Villringer, H. Rinneberg, Determining changes in NIR absorption using a layered model of the human head. Phys. Med. Biol. **46**(3), 879–896 (2001)
41. O. Steinkellner, C. Gruber, H. Wabnitz, A. Jelzow, J. Steinbrink, J.B. Fiebach, R. Macdonald, H. Obrig, Optical bedside monitoring of cerebral perfusion: technological and methodological advances applied in a study on acute ischemic stroke. J. Biomed. Opt. **15**(6), 061708 (2010)
42. A.A. Stratonnikov, N.V. Ermishova, V.B. Loschenov, Influence of red laser irradiation on hemoglobin oxygen saturation and blood volume in human skin in vivo. Proc. SPIE **4257**, 57–64 (2001)
43. A. Torricelli, D. Contini, A. Pifferi, M. Caffini, R. Re, L. Zucchelli, L. Spinelli, Time domain functional NIRS imaging for human brain mapping. NeuroImage **85**(Pt 1), 28–50 (2014)
44. A. Torricelli, A. Pifferi, L. Spinelli, R. Cubeddu, F. Martelli, S. Del Bianco, G. Zaccanti, Time-resolved reflectance at null source-detector separation: improving contrast and resolution in diffuse optical imaging. Phys. Rev. Lett. **95**(7), 078101 (2005)
45. A. Tosi, A. Dalla Mora, F. Zappa, A. Gulinatti, D. Contini, A. Pifferi, L. Spinelli, A. Torricelli, R. Cubeddu, Fast-gated single-photon counting technique widens dynamic range and speeds up acquisition time in time-resolved measurements. Opt. Express **19**, 10735–10746 (2011)
46. H. Wabnitz, D.R. Taubert, M. Mazurenka, O. Steinkellner, A. Jelzow, R. Macdonald, D. Milej, P. Sawosz, M. Kacprzak, A. Liebert, R. Cooper, J. Hebden, A. Pifferi, A. Farina, I. Bargigia, D. Contini, M. Caffini, L. Zucchelli, L. Spinelli, R. Cubeddu, A. Torricelli, Performance assessment of time domain optical brain imagers, part 1: basic instrumental performance protocol. J. Biomed. Opt. **19**(8), 086010 (2014)
47. Q. Zhao, L. Spinelli, A. Bassi, G. Valentini, D. Contini, A. Torricelli, R. Cubeddu, G. Zaccanti, F. Martelli, A. Pifferi, Functional tomography using a time-gated ICCD camera. Biomed. Opt. Express **2**(3), 705–716 (2011)
48. Z. Zhao, X.C. Wang, B. Chance, Remote sensing of prefrontal cortex function with diffusive light. Proc SPIE **5616**, 103–111 (2004)

Chapter 19
Breast Monitoring by Time-Resolved Diffuse Optical Imaging

Giovanna Quarto, Alessandro Torricelli, Lorenzo Spinelli,
Antonio Pifferi, Rinaldo Cubeddu and Paola Taroni

Abstract The aim of this chapter is to provide an overview on the applications of time-domain diffuse optics to the assessment of breast physiology and pathology. A number of different implementations to optical mammography have been evaluated ranging from lesion detection and characterisation using endogenous or exogenous contrast, to breast density assessment as strong cancer risk factor, and to monitoring during neoadjuvant chemotherapy. The time-domain approach is the common factor of all these applications, which permits to uncouple absorption from scattering contributions and to derive tissue optical properties in vivo. The Time-Correlated Single-Photon Counting (TCSPC) is the measurement technique of choice to acquire fast and weak optical signals at the picosecond time scale. For what concerns the time-resolved optical mammography with endogenous contrast, results obtained within the European project Optimamm are reported. The aim of the clinical trial was to understand the detection breast tumor capability of the developed multi-wavelength time-domain scanning systems operating in transmittance geometry on compressed breast. Results of the Berlin and Milan groups, have shown a similar detection rates for malignant breast lesions, with a high contrast at short wavelengths due to the presence of high blood volume. Regarding the optical mammography with exogenous contrast agent, a time domain optical mammography with Indocyanine

G. Quarto · A. Torricelli · L. Spinelli · A. Pifferi · R. Cubeddu · P. Taroni (✉)
Dipartimento di Fisica, Politecnico di Milano, 20133 Milan, Italy
e-mail: paola.taroni@polimi.it

G. Quarto
e-mail: giovanna.quarto@polimi.it

A. Torricelli
e-mail: alessandro.torricelli@polimi.it

L. Spinelli
e-mail: lorenzo.spinelli@fisi.polimi.it

A. Pifferi
e-mail: antonio.pifferi@polimi.it

R. Cubeddu
e-mail: rinaldo.cubeddu@polimi.it

© Springer International Publishing Switzerland 2015
W. Becker (ed.), *Advanced Time-Correlated Single Photon Counting Applications*,
Springer Series in Chemical Physics 111, DOI 10.1007/978-3-319-14929-5_19

Green (ICG) including also fluorescence measurements is presented. Results have shown a good sensitivity of the fluorescent optical mammography in detecting tumor. A good discrimination between healthy and tumor tissue was achieved. Concerning the optical mammography for therapy monitoring, a time-domain optical tomography system and very preliminary results were presented, showing good results on monitoring tumor response to chemotherapy using this approach. The other application presented in this chapter is based on the use of a multi-wavelength time-domain optical mammography for the assessment of breast density since it is a recognized independent risk factor for the development of breast cancer. A good correlation between optical data and mammographic breast density (provided by X-ray mammograms) was obtained, showing an increase of water and collagen in subjects with high breast density as expected from the physiological point of view and demonstrating the capability of the time-domain system to identify subjects with a high risk to develop breast cancer.

19.1 Introduction

Breast cancer is one of the most common tumours and one of the leading causes of death in women [9]. According to estimates of lifetime risk by the U.S. National Cancer Institute [21], in the U.S about 13 % of women, which means 1 in 8 women, will develop breast cancer in their lifetime. Many countries (e.g. UK, Italy, Australia) offer screening programs to women for prevention, typically between 50 and 70 years of age, since early diagnosis and therapy can have huge impact in terms of death rate and quality of life [36].

Breast screening essentially relies on X-ray mammography, which is the first line of defence in the early diagnosis of the breast cancer. However, mammography is less accurate in patients with dense glandular breasts [26], including young women, with reported sensitivity as low as 48 % [24].

Magnetic resonance imaging (MRI) is the most sensitive technique to diagnose invasive breast cancer, with good performance even in dense breast tissue, but because of the increased sensitivity it can also lead to false positive results [25]. Moreover, MRI is characterized by high costs and long examination times, which prevent its routine use for screening purposes.

Another important technique, non-invasive and with an excellent spatial resolution which is used in the clinical practice, is breast ultrasonography (US). It is often applied as complementary breast imaging technique in women with mammographically dense breast tissue, permitting detection of small occult breast cancers [30]. The technique is not particularly expensive, but US data interpretation is strongly operator's dependent [5].

Positron Emission Tomography (PET) is also applied to breast imaging. It is important in order to obtain information on the nature of the detected lesion. PET is also effectively applied in cancer staging. However, it is an expensive technique and involves the administration of a radioactive tracer [44].

It is worth mentioning that breast imaging aims not only at diagnosing diseases, but also at monitoring therapy. For the latter use MRI is typically applied, but mammography is still the gold standard among imaging techniques that aim at diagnosing breast pathologies. Therapy monitoring involves several imaging sessions concentrated in a short time period. Since the use of ionizing radiation can have adverse effects that should be avoided, it is evident that new breast imaging methods are needed.

Optical mammography is an emerging diagnostic tool which can provide information on breast tissue composition and related physiological or pathological parameters as well as tissue structure. The technique is absolutely non-invasive thanks to the use of visible and near-infrared (NIR) light, which is a non-ionizing radiation, able to cross some centimetres of biological tissue, offering a real opportunity to investigate the whole breast volume in vivo. Moreover, other advantages of optical mammography are its relative low instrumentation costs and capability to investigate dense breasts, typical of young women, covering X-ray mammography limits. Indeed, optical mammography could be effectively applied as a complementary diagnostic technique. Further, optical mammography can be performed at multiple wavelengths so as to combine imaging and spectroscopic information for lesion detection and characterization at the same time.

Besides to all these advantages, optical mammography has also some limitations. The technique is able to provide breast images with low spatial resolution, making the detection of tumours smaller than 1 cm difficult. This is an intrinsic limit of the technique, depending on NIR light diffusion in biological tissue. To overcome this problem, multimodality imaging approaches are being developed, in which optical mammography is combined with the conventional ones (X-ray, MRI, US and PET) [2, 14, 35, 46], providing complementary information. Another limit affecting optical mammography is linked to the signal acquired in breasts of big size: it has often low information content, resulting useless for diagnostic purposes. This is a contingent limit which could be overcome by the technological advances.

The physical basis of optical mammography, operating in the visible and near infrared light spectrum (600–1100 nm), is the difference in propagation of light through healthy tissue and tumours. Light is attenuated by two main phenomena: absorption, which destroys the incoming photons, reducing the intensity of the transmitted light, and scattering, which changes the direction of propagating photons. Light absorption is caused by chromophores, such as haemoglobin, water, lipid and collagen, which are present in the breast tissue; light scattering originates, instead, from changes of the refractive index occurring between connective tissues and cell constituents. Both absorption and scattering phenomena are strongly wavelength-dependent: at short visible wavelengths, they occur with comparable probability, while at longer wavelengths (red and NIR), the scattering events become dominant, but both processes are much less probable, allowing a significantly increased light penetration. As a consequence, to get access to deeper tissue structures for diagnostic purposes, wavelengths in the range between 600 and 1100 nm, named 'therapeutic window', are usually chosen.

Light transport through highly scattering media can in principle be described using the electromagnetic theory directly based on Maxwell's equations and taking into account the wave nature of light. However, due to the complex structure of tissue, the application of the electromagnetic theory is extremely difficult and not of practical use. Alternatively, the radiative transport theory can be considered. Relying on the particle nature of light, it describes the photon transport in tissue as the propagation of a particle flow. Further approximations of the transport theory lead to the 'Diffusion theory', which is mathematically less rigorous than the electromagnetic theory, yet providing an analytical solution of the problem. The scattering events are supposed to be isotropic and strongly dominant over the absorption events, and several scattering events need to occur into the medium before the description provided becomes accurate enough. Actually, exact solutions of the radiative transport equation have recently been introduced for specific geometries [27] and other theoretical models are available [15]. However, the diffusion approximation is still the most widely used theoretical approach in the biomedical field, because it provides a simple analytical solution to the problem of photon propagation in highly scattering media, such as biological tissues [28].

Different substances typically absorb at different wavelengths, so the measurement of the absorption spectrum of a medium allows in principle for the identification of its constituents. Then, to obtain information on tissue composition, measurements must be performed at several wavelengths. In fact, the absorption coefficient μ_a of tissue at any wavelength λ is due to the superposition of the contributions from the different constituents, given by the product of the intrinsic absorption ε, known in literature, and the concentration C of each constituent, as in the following formula (Beer's Law):

$$\mu_a(\lambda) = \sum_{i=1}^{n} \varepsilon_i(\lambda) C_i \qquad (19.1)$$

Concerning breast tissue, there are five main constituents to consider (oxy- and deoxygenated haemoglobin, water, lipid and collagen, see Fig. 19.1, so to estimate their concentration, measurements at a minimum of five wavelengths have to be performed.

As shown in Fig. 19.1, blood strongly absorbs in the red spectral range, while lipids, water and collagen are characterized by maximum absorption at wavelengths longer than 900 nm. The need to operate at several wavelengths led to a careful choice of their value, which had to take into account also their commercial availability. Generally, the choice fell on red wavelengths (e.g. 635–685 nm) because they are most sensitive to the blood components, in particular to the deoxy-haemoglobin. Higher wavelengths below 800 nm are usually used for the oxy-haemoglobin. The wavelengths in the near-infrared range, up to 900 nm are used for the other two important components of biological tissues, such as lipids and

Fig. 19.1 Normalized extinction spectra of the main constituents of the breast tissue

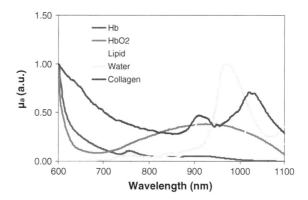

water. In particular the wavelengths around 930 and 1040 nm, where dominant peaks are present, are used for lipids; the wavelengths around 975 nm are used for the discrimination of water, because here it reaches a higher absorption value than the other chromophores. If, on one hand, the blood investigation is useful for the breast cancer detection due to its higher vascularization than healthy tissue, another important component to be investigated for diagnostic purposes is the collagen, whose absorption is dominant around 1100 nm. In fact, it is one of the main components of soft and hard tissues, and in particular of the breast one. Its contribution to tissue absorption has to be considered for a correct quantification of the tissue composition. Furthermore, collagen seems to be involved in the development of breast cancer and thus sensitivity to collagen could be relevant for cancer detection.

As mentioned above, scattering is the other light attenuation phenomenon due to the presence at microscopic level of refractive index discontinuities in biological tissues. In particular both the size and density of scattering centres affect the scattering properties, so their assessment can provide information on the structure of tissues. The reduced scattering coefficient μ_s' decreases monotonically upon increasing wavelength, without characteristic peaks. It is related to the scattering amplitude a, which provides information on the density of scattering centres, and to the scattering power b, which gives information on their size, according to the Mie theory:

$$\mu_s'(\lambda) = a(\lambda)^{-b} \tag{19.2}$$

High values of a correspond to denser tissues, whereas smaller scattering centres, so high values of b, lead to steeper slope. An example of reduced scattering spectra related to two different breast patterns is reported in Fig. 19.2. The steepest spectrum (blue) refers to fibrous breast which is characterized by smaller scattering centres with respect to the adipose breast, which has instead a flat spectrum (pink).

Since both absorption and scattering contribute to light attenuation and have independent origins, it is important to use a technique capable to discriminate between these two optical phenomena.

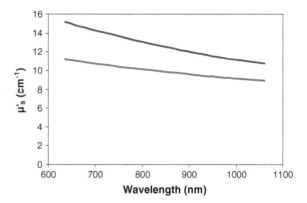

Fig. 19.2 Reduced scattering spectra of two different breast patterns. *Blue line* refers to fibrous breast and *pink line* to adipose one

A number of different implementations to optical mammography have been evaluated. The first procedure of transillumination of breast was diaphanography, where light from a lamp was diffused throughout the tissue and randomly scattered [10]. Opaque lesions formed shadows on the surface of the breast which acted as a screen. The problem was that the deeper the lesion, the greater the distance from the 'screen', and the less the contrast. It had sensitivity only at the detector side. Improved contrast and resolution was achieved by employing a narrow-beam light source and by scanning it in tandem with a localized optical detector. This approach enhances detectability of deep optical inhomogeneities and provides high sensitivity to superficial structures near both the source and the detector side. More recently, the approach based on the use of steady state light sources (i.e. continuous wave) and detectors allowed for dynamic measurements and spectral information over a broad continuous spectrum. A drawback is its inability to discriminate between absorption and scattering properties unless a complex tomographic scheme is adopted.

To overcome these limitations, richer information content can be achieved by performing spectroscopy or imaging in either the frequency domain, where the light source intensity is sinusoidal modulated, or in the time domain, where the light source is pulsed. These two implementations differ in the instrumentation used.

The frequency domain approach is based on modulating the intensity of the light source at a certain frequency (usually of the order of 100 MHz) and detecting the changes in amplitude and phase of the optical signal emerging from the tissue. These parameters can provide information to characterize both absorption and scattering properties of the tissue. Accurate estimates of the absorption and scattering properties generally require measurements to be performed on a wide range of the modulation frequency.

The time domain approach consists of injecting a short light pulse (typically with a duration of the order of 100 ps) into the tissue and measuring the distribution of the photon time-of-flight emerging from at a certain distance from the injection point. Scattering and absorption events, occurring during light propagation through the tissue, cause attenuation, delay and broadening of the injected pulse. The scattering essentially delays the collected pulse, as each scattering event changes the direction of photon propagation and, thus, photons move along trajectories that are much longer than the distance between the injection and the detection points: the higher the scattering, the longer the delay. The absorption influences the steepness of the temporal tail of the detected photon time-of-flight. The effects of absorption can be mostly seen at long times, that is on the pulse tail, because the longer the photons stay in the medium, the higher the probability of an absorption event. Thus, strong absorption means that many photons are removed from the temporal tail of the pulse and its slope becomes steeper.

The aim of this chapter is to provide an overview on the applications of time-domain diffuse optics to the assessment of breast physiology and pathology. As we will see different applications were proposed, ranging from breast lesions detection and characterization, using endogenous (Sect. 19.2) or exogenous (Sect. 19.3) contrast, to monitoring neoadjuvant chemotherapy (Sect. 19.4) and to the assessment of breast density as strong cancer risk factor (Sect. 19.5). As mentioned, the time-domain approach permits to uncouple absorption from scattering contributions and to derive tissue optical properties in vivo. Furthermore, Time-Correlated Single-Photon Counting (TCSPC) is the measurement technique often chosen to acquire time-resolved fast and weak optical signals: it is well suited for biological tissue spectroscopy because the overall attenuation of biological tissues is in the order of 8–10 OD and the temporal dynamics of photon migration in diffusive media last for 1–5 ns [3].

19.2 Time-Resolved Optical Mammography with Endogenous Contrast

Diffuse optical imaging techniques use measurements of transmitted light to produce spatially resolved images. Images of the absorption or scattering properties of the tissue, or other physiological parameters (such as haemoglobin, lipid, water, collagen) may be generated. As reported in the introduction, a wide spectral range in the NIR region is necessary in order to get information on the main constituents of the breast tissue.

Over the past, the use of wavelengths longer than 850 nm was often prevented by limitations on available commercial detectors, in fact most initial breast imaging studies were performed at 2–4 wavelengths within the range of 650–850 nm. More recently, longer wavelengths have became available, allowing operating over an extended spectral range.

An example of set-up of a time domain scanning optical mammographs, working at 7 wavelengths in the extended spectral range 635–1060 nm, is reported in Fig. 19.3.

The set-up was developed by the research group at Politecnico di Milano, Italy, and it is currently applied in a clinical trial [37]. Briefly, seven pulsed diode lasers (PDL Heads, PicoQuant, Germany) are presently used as light sources emitting at 635, 680, and 785 nm (VIS), and at 905, 930, 975 and 1060 nm (NIR), with average

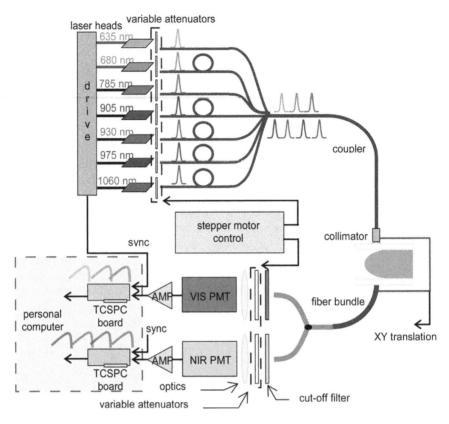

Fig. 19.3 Instrument set-up. *NIR* near-infrared; *PMT* photomultiplier tube; *TCSPC* time-correlated single photon counting

output power of ∼ 1–4 mW, temporal width of ∼ 150–400 ps (full width at half maximum, FWHM), and repetition rate of 20 MHz. A photomultiplier tube (PMT) for the detection of VIS wavelengths (sensitive up to 850 nm, R5900U-01-L16, Hamamatsu Photonics KK, Japan) and a PMT for NIR wavelengths (sensitive up to 1100 nm, H7422P-60, Hamamatsu Photonics KK, Japan) are used for the detection of the transmitted light.

The simultaneous collection of time-resolved curves at several wavelengths has been accomplished by means of a pulse-interleaved laser operation and the use of a bifurcated fibre bundle coupled to the two different PMTs and two independent TCSPC acquisition boards (mod. SPC130, Becker & Hickl GmbG, Germany [4]). Proper delays (realized by electronic cables from the master oscillator to each laser head driver) were inserted to allow firing of the laser pulses with nanosecond delays (see Fig. 19.4). More details on the instrument are reported in [38].

A recent alternative to this approach (probably more efficient in terms of photon counting capability) consists in the wavelength multiplexing with a supercontinuum laser (see Chap. 18).

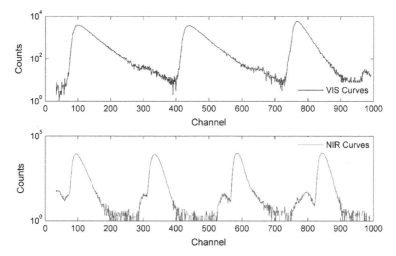

Fig. 19.4 Typical waveforms at the seven wavelength acquired by the two TCSPC boards by the instrument

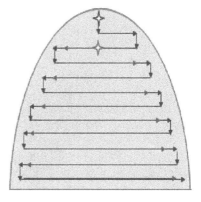

Fig. 19.5 Scan procedure during measurement. The *blue* (*red*) path mimics the scan performed with the VIS (NIR) wavelengths

The breast is softly compressed between two parallel plates. The illumination fibre and collecting bundle are scanned in tandem. The measurement starts searching the top of the breast, corresponding to the nipple area. A feedback on the total number of counts per point, ruled by an adjustable threshold, restricts the scan to the breast area. Independent thresholds are used for the two PMTs. The VIS PMT controls the scan, but, if the NIR threshold is reached, light to the NIR PMT can be shut off even during scanning to prevent damaging due to high signal levels. Once the tops for VIS and NIR are defined, the acquisition measurement starts. The NIR attenuator will close before and open after than VIS one, performing the different scan area, as reported in Fig. 19.5. The choice of the multianode PMT for detecting the VIS time resolved curves is dictated by its better time transit spread (i.e. jitter of the time of arrival of photoelectrons) as compared to the single anode solution (approximately 180 ps for the multianode as compared to 250 ps).

In order to reduce dead time, continuous acquisition is performed and data are stored every millimeter of path, i.e. every 25 ms with the current settings of the translation stages. For each point both VIS and NIR curves are recorded on the two TCSPC boards respectively, as reported in Fig. 19.4. A complete scan with a count rate of about 106 counts/s typically requires 5 min (depending on the size of breast).

Important results on the potential of time resolved transmittance imaging for optical mammography were obtained within the European project 'Optimamm: Imaging and characterization of breast lesions by pulsed near-infrared laser light', performed in 2000–2004. In particular, within this project, extensive clinical data were collected by partners in Germany [17] and Italy [39]. Time domain scanning optical mammographs operating in transmittance geometry on the compressed breast were developed and tested by both research groups. Images were acquired from both breasts in two views (cranio-caudal and medio-lateral, or cranio-caudal and oblique), in agreement with routine clinical practice for X-ray mammography for easier comparison between optical images and standard X-ray mammograms for the retrospective assessment of the optical technique.

In this work, the attention is focused on the results of the most extensive clinical trials performed in Berlin and Milan. In particular, the group in Berlin [17] developed a triple wavelength (670, 785, 843/884 nm) instrument and applied it on 154 patients, suspected of having breast cancer. Off-axis mammograms, with 2 cm offset between the transmitting and detecting fibre bundle, were also recorded upgrading the instrument. Off-axis scans were limited to a region of interest, presumably containing a tumour, to reduce the overall recording time. This approach was performed in order to infer the location of the tumour along the compression direction. Time-window analysis of distributions of times of flight of photons recorded at a large number of scan positions was used to obtain the optical mammograms. Additionally, absorption and reduced scattering coefficients were used to generate mammograms. These coefficients were derived by the time-resolved transmittance curves measured at each scan position and calculated within the diffusion approximation for a homogeneous tissue slab. They were then used to derive information on the constituent concentration. Setting the percentage content of water and lipids to fixed values, total haemoglobin concentration and blood oxygen saturation were estimated from 2 to 3-wavelength data. Seventy-two out of 102 histological confirmed tumours (71 %) were detected retrospectively in both optical projection mammograms, and 20 more lesions (20 %) were detected in one projection only.

The group in Milan [39] initially developed an instrument operating at 4 wavelengths (683, 785, 913 and 975 nm). In particular, 2 wavelengths were longer than 900 nm (i.e. 915 and 975/985 nm) to increase the sensitivity to lipids and water, which show major absorption peaks in that wavelength range (Fig. 19.1). To investigate the diagnostic potential of shorter wavelengths, the instrument was modified adding other wavelengths shorter than 700 nm (637 and 656 nm). The several upgrades of the instrument led to collect mammograms at a different number of wavelengths, from 4 to 7, even if the final version operated at four wavelengths (637, 785, 905 and 975 nm).

Table 19.1 General information of the 2 main clinical trials within the European project Optimamm

Group	N° patients	N° wavelengths	Spectral range (nm)	N° lesions	Detected lesions in 2 views	Detected lesions in 1 view
PTB	154	3	670–884	102[a]	72	20
Polimi	194	4	637–985	225	130	21

[a] Only lesions validated by histology

The instrument, in the several configurations, was tested on a total of 194 patients with 225 malignant and benign lesions. Forty-one out of 52 cancers (79 %) were detected in both views and 9 more (17 %) in just one view. Concerning benign lesions, a significant number of cysts was analysed ($n = 82$), achieving a detection rate of 83 % in both views, which came to 90 % if detection in a single view was accepted.

A synthetic scheme of the main information of the two clinical trial is reported in Table 19.1.

Images at all wavelengths were constructed by plotting the number of photons collected within a selected time window as well as the reduced scattering coefficient. Also in this case, data analysis was performed using the time gated approach. As first step, a reference position was selected far from boundaries and possible inhomogeneities. This was performed to avoid the dependence of gated intensity images on the temporal position and width of single transmittance curves. The time distribution of data acquired in that position was divided into 10 time windows, each one collecting 1/10 of the total number of counts. The same time gates were then used in any other position of the scanned area to build gated intensity images. The 8th gate, on the tail of the pulse, was routinely applied for breast imaging. The estimated absorption and reduced scattering coefficients (μ_a and μ'_s, respectively) were average values along the line of sight, as obtained from the best fit of the experimental data with the analytical solution of the diffusion equation, with extrapolated boundary condition, for a homogeneous slab [20, 31].

For each breast projection and wavelength, the estimate of bulk optical properties was limited to a reference area excluding the boundaries and marked inhomogeneities, but still including most of the breast. To select that area, the time-of-flight (i.e. the first moment of the time-resolved transmittance curve) was calculated for each image pixel, and only pixels with time-of-flight greater than or equal to the median of the distribution were included in the reference area. The absorption and reduced scattering values of bulk tissue were then obtained by averaging the absorption and reduced scattering coefficients over the reference area.

An example of gated intensity images acquired at six wavelengths from a patient with a lobular invasive carcinoma in her left breast is reported in Fig. 19.6. The tumour is visible at all wavelengths between 637 and 785 nm, but the contrast reduces upon increasing wavelength [40]. Adipose or fibrous structure can be easily observed at longer wavelengths, 916 and 975 nm respectively. For example, the mammary gland, together with water-rich structures such as liquid cysts, can generally be

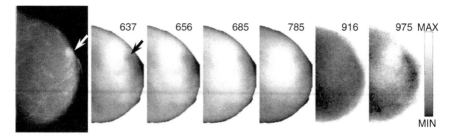

Fig. 19.6 X-ray mammogram and late gated intensity images (637, 656, 685, 785, 916, and 975 nm) of the *left* breast (*cranio-caudal view*) of patient #146, bearing a lobular invasive carcinoma (max. diameter = 15 mm)

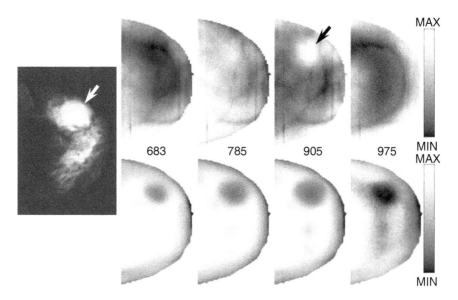

Fig. 19.7 X-ray mammogram, (*top*) late gated intensity and (*bottom*) scattering images (683, 785, 905, and 975 nm) of the *left* breast (*cranio-caudal view*) of patient #185, bearing multiple cysts (max. diameter of the cyst indicated by the arrow = 38 mm)

identified at 975 nm. In the case of the adipose breast reported in Fig. 19.6, a stronger absorption at 975 nm is observed in the lower and anterior part of the breast. A same high absorption area is more evident in the denser breast reported in Fig. 19.7, top. In fact, the mammary gland, which is detected as an opaque area in the X-ray mammogram, is well observed as a strongly absorbing region of similar shape in the gated intensity image at 975 nm. In that image, the mammary gland overlaps a cyst, masking the contribution of the lesion to the overall light attenuation. However, the presence of the cyst is revealed by the weak absorption at 905 nm, in agreement with a lower lipid content than in the surrounding tissue. Moreover, cyst with a liquid nature, are usually characterized by low scattering. In fact, it is evident from all scattering images reported in Fig. 19.7, bottom.

Figure 19.8 reports images from a patient with a fibroadenoma in her right breast. The detection of fibroadenomas often proved to be problematic. When identified, they are generally characterized by higher absorption at 975 nm and sometimes even in the red (637–683 nm), as observed in Fig. 19.8. The marked absorption at 975 nm is likely related to the high water content, but it might also be at least in part due to collagen, with absorption increasing both at short (<700 nm) and long (>900 nm) wavelengths [32]. From Fig. 19.8, we can see how absorption at 975 nm follows, even though with clearly lower spatial resolution, the patterns identified in the X-ray mammography, with a region of strong absorption starting at the fibroadenoma location and a second area of marked absorption in retroareolar position. In contrast, data acquired on the lipid peak (905–916 nm) typically show opposite behaviour, adipose tissue being transparent to X-rays. However, in the latter case the correspondence is often not complete, possibly due to some contribution from haemoglobin around 900 nm. In Fig. 19.8, dominant absorption is observed in median position, both at 683 and 975 nm. The area corresponds to an X-ray dense region, and is possibly highly vascularized fibro-glandular tissue.

In conclusion, both studies had similar detection rates for malignant breast lesions. On average, late gated intensity images at short wavelengths (i.e. <700 nm) provided the highest tumour-to-healthy contrast and also the highest contrast-to-noise ratio. At short wavelengths, blood absorption is dominant, so the contrast was attributed to high blood volume in the lesion. For the estimate of tissue composition and specifically to quantify constituents in the detected lesions, simple perturbative approaches were used, considering a single spherical lesion in an otherwise homogeneous background [16, 34]. Both studies confirmed significantly higher haemoglobin content in the lesion, as compared to the surrounding healthy tissue, while blood oxygen saturation proved to be a poor discriminator.

In parallel, within the same project, also the potential of time-resolved optical tomography of the uncompressed breast as a diagnostic tool was tested [45]. This limited study was performed on 24 subjects, including 19 cases with a specific lesion. In this approach, the patient lies prone with her breast pendent through a

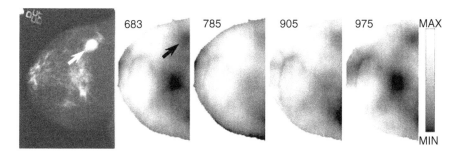

Fig. 19.8   X-ray mammogram and late gated intensity images (683, 785, 905, and 975 nm) of the *right* breast (cranio-caudal view) of patient #180, bearing a fibroadenoma (max. diameter = 20 mm)

hole, while sources and detectors (16 in total) are distributed along one or more rings around the breast. For each source, the time-of-flight distributions of the photons were recorded by all detectors simultaneously. The system operated at 2 wavelengths (780 and 815 nm). Thus, for lesion detection and identification they relied on absorption and scattering images. Seventeen out of 19 cases of lesions of different nature were detected by optical images.

In 2005, another time-domain optical mammograph (softscan, Art, Canada) operating at four wavelengths (760, 780, 830, and 850 nm) was developed and applied on 49 women of different age [22]. For all patients absolute bulk and local values of breast constituent concentrations were retrieved. In the 23 cases imaged with suspicious masses, the optical images were consistent with the mammographic findings. Also consistent differences between malign and benign cases were found. In particular, the discrimination based on deoxy-haemoglobin content was statistically significant between malignant and benign cases.

A multi-channel time-resolved optical mammography prototype for breast cancer screening, operating at three wavelengths (765, 800 and 835 nm), was developed by Ueda et al. [43]. A tomographic approach was used to collect data by using a hemispherical cup and a liquid-coupled interface since the breast was uncompressed and a non-contact optical fibre approach was used. Image reconstruction was performed by using the time-resolved photon path distribution (PPD). Early arrival photons were used to improve the spatial resolution. Few cases of tumour detection were evaluated, showing that the lesion could not be detected using the peak photon region, whereas high concentration of haemoglobin could be derived using the early arrival photons.

19.3 Optical Mammography with Exogenous Contrast Agent

Together with the development of experimental methods for NIR imaging, the use of contrast agents for diffuse optical imaging was considered, aiming at improving the detection of carcinoma. Ideally, the contrast agent should accumulate selectively in the tumour, but not in healthy tissue, in order to increase the contrast between the two areas. Fluorescent contrast agents with excitation and emission in the NIR spectral range are usually considered for optical application, since they allow for deep penetration into breast tissue and are coupled to a limited tissue autofluorescence. Thus, the fluorescence signal coming from the contrast agent localized in the lesion would be dominant with respect to the one coming from the background, leading to an increased contrast.

Nowadays, no contrast agents have these specific features. The only contrast agent that is approved for clinical use and is suitable for breast imaging is Indocyanine Green (ICG). While it is characterized by strong absorption and fluorescence in the NIR wavelength range, it does not provide selectivity for tumour tissue. It operates as blood pool contrast agent: it stays in the vascular system for a long time period and the tumour-healthy contrast is achieved by exploiting its slower kinetics in the tumour, due to an increased resistance of the tumour vasculature to blood flow.

An exploratory study of time domain optical mammography with ICG was performed by the Berlin research group [19]. In particular, for this purpose the previously developed multichannel laser pulse mammograph, (briefly described in Sect. 19.2), was upgraded for fluorescence measurements both in transmission and reflection [18]. The same planar geometry, with the breast slightly compressed between two parallel glass plates, was used to detect fluorescence of a contrast agent in the breast at high sensitivity. A picture of the final device is shown in Fig. 19.9.

The instrument was upgraded also in terms of spectral range: a new picosecond diode laser with emission wavelength at 1066 nm was added for a better reconstruction of the constituent concentration, whereas the other 3 wavelengths were modified, emitting in this case at 660, 797 and 934 nm. The most important upgrade was the introduction of the excitation laser emitting at 780 nm for fluorescent images. Details on the instrumentation are reported in [18]. Preliminary measurements were performed administrating the contrast agent intravenously by a bolus followed by an infusion and data were recorded 30 min after the end of the infusion of the contrast agent, corresponding to the extravascular phase. Raw absorption and fluorescence mammograms were dominated by the absorption of the excitation and fluorescence radiation by blood. The carcinoma, cannot be detected in the fluorescence mammogram because of cancellation effects. The higher fluorescence intensity of the carcinoma is in fact compensated for by the higher absorption of laser and fluorescence radiation due to its increased blood content. In order to eliminate the inhomogeneous background absorption of the breast tissue and

Fig. 19.9 Fluorescence
mammograph. Reprinted,
with permission, from [18]

recognize fluorescent objects with a higher contrast, ratio images obtained dividing the normalized total fluorescence counts by the normalized total laser photon counts were calculated.

Preliminary results in which the system was tested on a patient bearing a tumour demonstrated the good sensitivity of the fluorescent optical mammography. The small amount of the extravasated ICG in the extracellular space of the carcinoma, which generates the fluorescent signal, could be detected, allowing one to distinguish between malignant and benign lesions.

Fluorescence ratio mammograms in 68 years old woman with 22 mm invasive ductal carcinoma are reported in Fig. 19.10.

Figure 19.10 shows that in ratio fluorescence image the detection of the tumour is possible with a better contrast. Moreover other structures such as blood vessels that are visible in absorption and fluorescence images are largely eliminated, enabling a clear demarcation of tumour. Of course, a bigger population was needed to test the capability of the fluorescence optical mammograph in discriminating breast tumours.

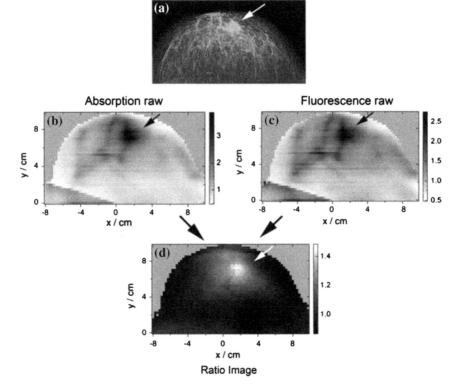

Fig. 19.10 (**a**) Conventional X-ray mammogram (**b**) Absorption and (**c**) Fluorescence raw mammograms (**d**) Ratio mammograms obtained dividing normalized fluorescence image data by normalized absorption image data. Lesion is indicated by the *arrow*. Reprinted, with permission, from [33]

A further study was then performed by the same research group with a similar instrument set-up [33]. The aim of latter study was to perform fluorescence NIR imaging with early and late enhancement of ICG, corresponding to the vascular and extravascular phases of contrast agent enhancement, respectively, for the detection of breast cancer and the discrimination between benign and malignant breast lesions. Twenty women having 21 suspicious breast lesions were investigated, including 13 malignant and 8 benign lesions. The breast was imaged by on-axis movement of the source and detection fibre bundle at a step size of 2.5 mm, and optical mammograms were recorded in 5–10 min. For the measurements, the contrast agent was administered intravenously by a bolus followed by an infusion and data were recorded before, during and after the end of the infusion of the contrast agent. The images acquired before the injection of the contrast agent were used to measure the intrinsic absorption and scattering properties of the breast and the corresponding autofluorescence. To get information on the early and late fluorescence, absorption and fluorescence data were recorded during and after the infusion of ICG, and then the ratio of the images was calculated. For data analysis the total photon counts of the transmittance curve, instead of the photon counts in a selected time window, were used to improve the signal-to-noise ratio.

Only for two lesions (one benign and one malignant) increased autoflorescence was observed with respect to the background tissue, whereas no significant variations were observed in the remaining 19 cases. Malignant lesions were correctly defined in 11 or 12 (depending on the reader) of 13 cases, and benign lesions were correctly defined in 6 or 5 (depending on the reader) of 8 cases with late-fluorescence imaging. Significant differences in the discrimination between malignant and benign lesions were observed using the early-fluorescence ratio mammograms with respect to the late one, indicating that late fluorescence ratio mammograms highlight malignant tumour well with a good contrast against the healthy tissue.

The here presented study on optical mammography with ICG led to good results for what concerns the discrimination between healthy and tumour tissue. Of course, the effectiveness of the system in detecting lesions and the method of imaging reconstruction need to be further tested on a wider population.

19.4 Optical Mammography for Therapy Monitoring

Chemotherapy is widely used in the treatment of locally advanced breast cancer before surgery to reduce the tumour size. Monitoring the response to therapy can improve survival and reduce morbidity. Moreover, the immediate knowledge of the individual response to the treatment reduces useless exposure to ineffective drugs with heavy side effects.

Most of the work performed in the field of therapy monitoring concerned chemotherapy was performed using continuous-wave and/or frequency domain set-ups [7, 23].

However, recently time domain optical imaging of the response to hormone therapy has also been developed. In particular, the UCL three-dimensional time-resolved optical imaging system (briefly described in Sect. 19.2) was used to monitor the response to neoadjuvant chemotherapy in 3 patients bearing advanced breast cancer. The system is characterized by 2 pulsed lasers emitting at 780 and 815 nm, which are transmitted to the breast through 31 optical fibres. Details on the instrument set-up can be found in [12]. The patient lies prone with her breast pendant in the liquid-coupled patient interface, as shown in Fig. 19.11.

Three optical imaging sessions were performed. The first scan was performed before the start of the treatment; a second scan in the middle of the treatment, and the third one before surgery, when the treatment was completed. Both breasts were scanned, taking 3 min for each scan.

Three-dimensional (3D) images were generated from the data using the time-resolved optical absorption and scatter tomography (TOAST) reconstruction package developed at UCL [1].

After the reconstruction of the 3-D images of absorption, scattering and blood parameters, a two-dimensional (2D) coronal slice was extracted from each one in correspondence of the maximum contrast point in the lesion. Then, two circular areas of that image were selected: one containing the tumour and another one related to the background medium (out of the tumour area and the coupling fluid). Similarly, equivalent areas from the contralateral healthy breast were also analysed. In this way, changes in physiological and optical properties within the tumour and in the rest of the breast were evaluated during the course of therapy [13].

All three patients showed changes in both optical and physiological properties in the region of the tumour compared to the rest of the breast. In two patients, a good response to the chemotherapy treatment led to a reduction of the tumour. A decrease of the tumour total haemoglobin was observed in the optical images performed within the treatment. In the third patient a lack of response to therapy was observed and this type of response was also seen in the optical images scan within the treatment.

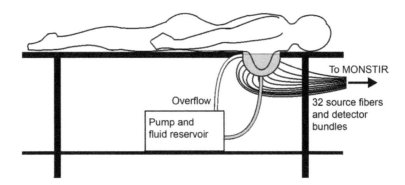

Fig. 19.11 Fluid-coupled patient interface for the 3D optical breast imaging. The patient lies with her breast pendant in a hemispherical cup filled with scattering fluid. Reprinted with permission, from [13]

The present study showed good preliminary results on monitoring tumour response to chemotherapy using time domain optical tomography. Of course, a large number of cases has to be evaluated to establish the capability of the technique and confirm the strong correlation with treatment outcome obtained with very few study cases.

19.5 Optical Mammography with Breast Density Assessment

Breast density is a recognized independent risk factor for developing breast cancer [6, 29]. At present, breast density assessment is based on the radiological appearance of breast tissue, requiring the use of ionizing radiation. Its non-invasive estimation would certainly be of diagnostic interest. Optical techniques could contribute to assess breast density in a non-invasive way. A study on the direct correlation of breast density with optically derived main breast constituents (water, lipids, collagen, and haemoglobin) could lead to a better understanding of the role of mammographic density in breast cancer risk assessment. In particular, collagen is an important constituent of breast tissue that seems to be involved in the onset and progression of breast cancer. Furthermore, collagen, as a major constituent of stroma, is expected to be related to breast density [8, 29, 41]. Thus, sensitivity to collagen content could be relevant for breast cancer development risk and its quantification by optical means could provide useful diagnostic information.

In this line, an on-going clinical study performed with a 7-wavelength (635–1060 nm) time domain instrument (shown in Fig. 19.3) developed by the research group at Politecnico di Milano has given promising preliminary results [37, 38].

Data were collected from 45 patients (age range 31–78). According to the Breast Imaging and Reporting Data System (BI-RADS) mammographic density categories, reported in Table 19.2, breast types were classified by an expert radiologist.

For each patient, optical images in cranio-caudal and oblique views were recorded for both breasts. Tissue composition and scattering parameters were estimated from time-resolved transmittance data by using a global spectral fit procedure [11] and averaged over each image in order to investigate their dependence on mammographic density.

As shown in Fig. 19.12, several parameters obtained from optical data proved to correlate with mammographic breast density: water, lipid and collagen content, as well as the scattering power b [37].

Table 19.2 BI-RADS mammographic density categories and their description

BI-RADS	Description
1	Almost entirely fat
2	Scattered fibroglandular densities
3	Heterogeneously dense
4	Extremely dense

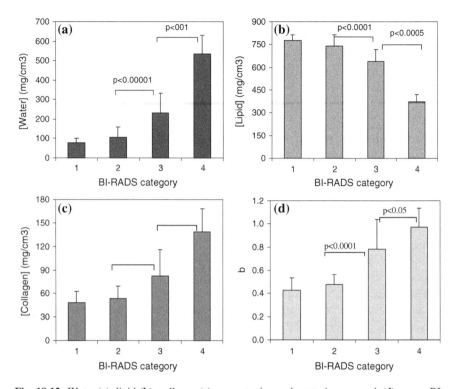

Fig. 19.12 Water (**a**), lipid (**b**), collagen (**c**) concentration and scattering power *b* (**d**) versus BI-RADS categories. Results are expressed as mean ± SD and refer to the intersubject variability. Mann-Whitney test was used to estimate statistical significance of the difference between the different categories

Water content increases progressively and significantly with breast density. Correspondingly, a gradual decrease is observed in lipid content. The amount of collagen also increases for increasing breast density. For all the three main constituents, the difference between categories is always highly significant (Mann-Whitney test), except in some cases between category 1 and 2. Since the scattering power *b* provides information on the microscopic tissue structure that is at the origin of breast density, also this parameter was considered and a progressive increase is observed for increasing density.

Since their contributions may not be fully independent, their combination could result in improved correlation with mammographic density. All that parameters that showed significant dependence on mammographic breast density were combined in a single optical index *OI*, defined as follows:

$$OI = \frac{[Water]\,[Collagen]\,b}{[Lipid]} \tag{19.3}$$

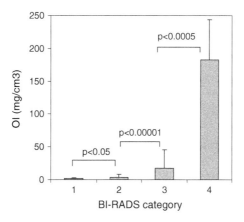

Fig. 19.13 OI versus BI-RADS categories. Results are expressed as mean ± SD and refer to the intersubject variability. Mann-Whitney test was used to estimate statistical significance of the difference between the different categories

where parameters that are expected to increase with breast density (i.e., the concentrations of water and collagen, and the scattering slope b) are all multiplied, and divided by lipid concentration, which conversely is expected to decrease upon increasing breast density. A good correlation was achieved between breast density estimated using conventional mammography and the optical index, as shown in Fig. 19.13.

In particular, Fig. 19.13 shows that the optical index can significantly separate category 4 (highest cancer risk) from category 3, with a very good significance ($p < 0.0001$) and category 4 from all the others with even higher confidence.

For the first time, the direct correlation between mammographic density and tissue composition, in terms of the main constituents estimated from optical data, was performed. The clinical study is still in progress, so data from a wider population are becoming available for deeper investigations. Moreover, more sophisticated statistical methods have been used for the correlation between optical and radiological data, showing very promising results for the identification of subjects at high risk to develop breast cancer [42].

19.6 Conclusions and Future Perspectives

In vivo time-resolved diffuse optical imaging and spectroscopy of breast tissue has been widely investigated in the last few decades. Technological advances in both pulsed light sources, fast detector and time-resolved acquisition electronics allowed the fabrication of compact and accurate devices suitable for the clinical environment.

From what concerns light injection, the time domain approach can benefit from pulsed laser sources with tens of picosecond duration, tens of MHz repetition rate, and average power in the order of few mW. On the detection side, it can benefit from state-of-the-art TCSPC electronics with picosecond resolution and acquisition frequency in the order of few MHz. Moreover, this approach has the advantage to uncouple absorption from scattering contributions, leading to reconstruct the constituent concentrations of the investigated tissue and thus providing a wide range of information useful for diagnostic purposes and for the study of physiological processes.

Different research directions of the time-resolved technique were considered, such as lesion detection by time-gated analysis [16, 39], tumour detection with contrast agents [18], monitoring of neoadjuvant chemotherapy [13], and assessment of cancer risk [37].

Most of the presented imaging studies were based on the use of laser sources in the spectral range below 800 nm, where absorption of blood is dominant. In fact, currently the detection of lesions is essentially based on the quantification of haemoglobin since tumours are characterized by an increased vascularisation with respect to the healthy tissue. It is also emerging that an increase of the connective component (water and collagen related) could be associated to the presence of a tumour. Furthermore, for a correct quantification of the breast tissue composition, it is important that all chromophores which contribute to the absorption of the breast tissue are considered. Therefore, the extension of the spectral range to the near infrared region for the quantification also of lipid, water and collagen is necessary. The Milan group is the only one operating in the time domain that focused the attention on the wide spectral range and extended it over the years up to covering the whole therapeutic window.

Even though for most of the presented research directions the technique needs to be validated on a wider population, as outlined in this chapter, time-resolved optical imaging techniques using near infrared light may have the potential to assist the identification and characterization of breast lesions as well as monitor the response to therapy and estimate cancer risk.

In the future, the spectral range extension could be one of the fundamental keys for the investigation of lesion components to distinguish them between benign and malignant, thus providing a specific diagnostic tool. This aspect of course involves the use of several or tuneable laser sources. Moreover, appropriate efficient detectors need to be identified to effectively cover the spectral range of interest.

References

1. S.R. Arridge, J.C. Hebden, M. Schweiger, F.E.W. Schmidt, M.E. Fry, E.M.C. Hillman, H. Dehghani, D.T. Delpy, A method for 3D time-resolved optical tomography. Int. J. Imaging Syst. Technol. **11**, 2–11 (2000)
2. F.S. Azar, K. Lee, A. Khamene, R. Choe, A. Corlu, S.D. Konecky, F. Sauer, A.G. Yodh, Standardized platform for coregistration of nonconcurrent diffuse optical and magnetic resonance breast images obtained in different geometries. J. Biomed. Opt. **12**,051902 (2007)
3. W. Becker, *Advanced Time-Correlated Single Photon Counting Techniques* (Springer, Heidelberg, 2005)

4. W. Becker, *The bh TCSPC handbook*, 5th edn. Becker & Hickl GmbH (2012)
5. W.A. Berg, Rationale for a trial of screening breast ultrasound: American college of radiology imaging network (ACRIN). Am. J. Roentgenol. **180**, 1225–1228 (2003)
6. N.F. Boyd, H. Guo, L.J. Martin, L. Sun, J. Stone, E. Fishell, R.A. Jong, G. Hislop, A. Chiarelli, S. Minkin, M.J. Yaffe, Mammographic density and the risk and detection of breast cancer. N. Engl. J. Med. **356**, 227–236 (2007)
7. D.R. Busch, R. Choe, M.A. Rosen, W. Guo, T. Durduran,.M.D. Feldman, C. Mies, B.J. Czerniecki, J. Tchou, A. De Michele, M.D. Schnall, A.G. Yodh, Optical malignancy parameters for monitoring progression of breast cancer neoadjuvant chemotherapy. Biomed. Opt. Express, **4**, 105–121 (2012)
8. C. Byrne, Studying mammographic density: implications for understanding breast cancer. J. Natl Cancer Inst. **89**, 531–533 (1997)
9. F. Bray, P. McCarron, D.M. Parkin, The changing global patterns of female breast cancer incidence and mortality. Breast Cancer Res. **6**, 229–239 (2004)
10. M. Cutler, Transillumination of the breast. Surg. Gynecol. Obstet. **48**, 721–727 (1929)
11. C. D'Andrea, L. Spinelli, A. Bassi, A. Giusto, D. Contini, J. Swartling, A. Torricelli, R. Cubeddu, Time-resolved spectrally constrained method for the quantification of chromophore concentrations and scattering parameters in diffusing media. Opt. Express **14**, 1888–1898 (2006)
12. L.C. Enfield, A.P. Gibson, N.L. Everdell, D.T. Delpy, M. Schweiger, S.R. Arridge, C. Richardson, M. Keshtgar, M. Douek, J.C. Hebden, Three-dimensional time-resolved optical mammography of the uncompressed breast. Appl. Opt. **46**, 3628–3638 (2007)
13. L. Enfield, G. Cantanhede, M. Douek, V. Ramalingam, A. Purushotham, J. Hebden, A. Gibsona, Monitoring the response to neoadjuvant hormone therapy for locally advanced breast cancer using three-dimensional time-resolved optical mammography. J. Biomed. Opt. **18**:056012(1–6) (2013)
14. Q. Fang, J. Selb, S.A. Carp, G. Boverman, E.L. Miller, D.H. Brooks, R.H. Moore, D.B. Kopans, D.A. Boas, Combined optical and X-ray tomosynthesis breast imaging. Radiology **258**, 89–97 (2011)
15. J.B. Fishkin, E. Gratton, Propagation of photon-density waves in strongly scattering media containing an absorbing semi-infinite plane bounded by a straight edge. J. Opt. Soc. Am. A **10**, 127–140 (1993)
16. D. Grosenick, H. Wabnitz, K.T. Moesta, J. Mucke, P.M. Schlag, H. Rinneberg, Time-domain scanning optical mammography: II. optical properties and tissue parameters of 87 carcinomas. Phys. Med. Biol. **50**, 2451–2468 (2005)
17. D. Grosenick, K.T. Moesta, M. Möller, J. Mucke, H. Wabnitz, B. Gebauer, C. Stroszczynski, B. Wassermann, P.M. Schlag, H. Rinneberg, Time-domain scanning optical mammography: I. Recording and assessment of mammograms of 154 patients. Phys. Med. Biol. **50**, 2429–2449 (2005)
18. D. Grosenick, A. Hagen, O. Steinkellner, A. Poellinger, S. Burock, P.M. Schlag, H. Rinneberg, R. Macdonald, A multichannel time-domain scanning fluorescence mammograph: performance assessment and first in vivo results. Sci. Instrum. **82**, 024302 (2011)
19. A. Hagen, D. Grosenick, R. Macdonald, H. Rinneberg, S. Burock, P. Warnick, A. Poellinger, P.M. Schlag, Late-fluorescence mammography assesses tumor capillary permeability and differentiates malignant from benign lesions. Opt. Express **17**, 17016–17033 (2009)
20. R.C. Haskell, L.O. Svaasand, T.T. Tsay, T.C. Feng, M.S. McAdams, B.J. Tromberg, Boundary conditions for the diffusion equation in radiative transfer, J. Opt. Soc. Am., A, **11**, 2727–41 (1994)
21. http://www.cancer.gov/
22. X. Intes, Time-domain optical mammography SoftScan: initial results. Acad. Radiol. **12**, 934–947 (2005)
23. D.B. Jakubowski, A.E. Cerussi, F. Bevilacqua, N. Shah, D. Hsiang, J. Butler, B.J. Tromberg, Monitoring neoadjuvant chemotherapy in breast cancer using quantitative diffuse optical spectroscopy: a case study. J. Biomed. Opt. **9**, 230–238 (2004)

24. T.M. Kolb, J. Lichy, J.H. Newhouse, Comparison of the performance of screening mammography, physical examination and breast US and evaluation of factors that influence them: an analysis of 27, 825 patient evaluations. Radiology **225**, 165–175 (2002)

25. C.K. Kuhl, S. Schrading, C.C. Leutner, N. Morakkabati-Spitz, E. Wardelmann, R. Fimmers, W. Kuhn, H.H. Schild, Mammography, breast ultrasound, and magnetic resonance imaging for surveillance of women at high familial risk for breast cancer. J. Clin. Oncol. **23**, 8469–8476 (2005)

26. E. Marshall, Brawling over mammography. Science **327**, 936–938 (2010)

27. F. Martelli, S. Del Bianco, A. Pifferi, L. Spinelli, A. Torricelli, G. Zaccanti, Hybrid heuristic time dependent solution of the radiative transfer equation for the slab. Proc. SPIE **7369**, 73691C (2009)

28. F. Martelli, *Light Propagation through Biological Tissue and Other Diffusive Media* (SPIE Press, Washington, 2010)

29. V.A. McCormack, I.S. dos Silva, Breast density and parenchymal patterns as markers of breast cancer risk: a meta-analysis. Cancer Epidemiol. Biomark. Prev. **15**, 1159–1169 (2006)

30. M. Nothacker, V. Duda, M. Hahn, M. Warm, F. Degenhardt, H. Madjar, S. Weinbrenner, U.-S. Albert, Early detection of breast cancer: benefits and risks of supplemental breast ultrasound in asymptomatic women with mammographically dense breast tissue: a systematic review. BMC Cancer **9**, 335 (2009)

31. M.S. Patterson, B. Chance, B.C. Wilson, Time-resolved reflectance and transmittance for the noninvasive measurement of tissue optical properties. Appl. Opt. **28**, 2331–2336 (1989)

32. A. Pifferi, A. Torricelli, P. Taroni, D. Comelli, R. Cubeddu, Optical characterization of primary tissue components by time-resolved reflectance and transmittance spectroscopy. *Int. Symp. on Biomedical Optics*, 20–26 Jan (San Jose, CA, 2001)

33. A. Poellinger, S. Burock, D. Grosenick, A. Hagen, L. Lüdemann, F. Diekmann, F. Engelken, R. Macdonald, H. Rinneberg, P.-M. Schlag, Breast cancer: early- and late-fluorescence near-infrared imaging with indocyanine green—a preliminary study. Radiology **258**, 409–416 (2011)

34. L. Spinelli, A. Torricelli, A. Pifferi, P. Taroni, G. Danesini, R. Cubeddu, Characterisation of female breast lesions from multi-wavelength time-resolved optical mammography. Phys. Med. Biol. **50**, 2489–2502 (2005)

35. S. Srinivasan, C.M. Carpenter, H.R. Ghadyani, S.J. Taka, P.A. Kaufman, R.M. Diflorio-Alexander, W.A. Wells, B.W. Pogue, K.D. Paulsen, Image guided near-infrared spectroscopy of breast tissue in vivo using boundary element method. J. Biomed. Opt. **15**, 061703 (2010)

36. L. Tabar, M.-F. Yen, B. Vitak, H.-H.T. Chen, R.A. Smith, S.W. Duffy, Mammography service screening and mortality in breast cancer patients: 20-year follow-up before and after introduction of screening. Lancet **361**, 1405–1410 (2003)

37. P. Taroni, A. Pifferi, G. Quarto, L. Spinelli, A. Torricelli, F. Abbate, A. Villa, N. Balestreri, S. Menna, E. Cassano, R. Cubeddu, Non-invasive assessment of breast cancer risk using time-resolved diffuse optical spectroscopy. J. Biomed. Opt. **15**, 060501 (2010)

38. P. Taroni, A. Pifferi, E. Salvagnini, L. Spinelli, A. Torricelli, R. Cubeddu, Seven-wavelength time-resolved optical mammography extending beyond 1000 nm for breast collagen quantification. Opt. Express **17**, 15932–15946 (2009)

39. P. Taroni, A. Torricelli, L. Spinelli, A. Pifferi, F. Arpaia, G. Danesini, R. Cubeddu, Time-resolved optical mammography between 637 and 985 nm: clinical study on the detection and identification of breast lesions. Phys. Med. Biol. **50**, 2469–2488 (2005)

40. P. Taroni, A. Pifferi, A. Torricelli, L. Spinelli, G.M. Danesini, R. Cubeddu, Do shorter wavelengths improve contrast in optical mammography? Phys. Med. Biol. **49**, 1203–1215 (2004)

41. P. Taroni, D. Comelli, A. Pifferi, A. Torricelli, R. Cubeddu, Absorption of collagen: effects on the estimate of breast composition and related diagnostic implications. J. Biomed. Opt. **12**, 014021 (2007)

42. P. Taroni, G. Quarto, A. Pifferi, F. Ieva, A.M. Paganoni, F. Abbate, N. Balestreri, S. Menna, E. Cassano, R. Cubeddu, Optical identification of subjects at high risk for developing breast cancer. J. Biomed. Opt. **18**, 060507-3 (2013)
43. Y. Ueda, K. Yoshimoto, E. Ohmae, T. Suzuki, T. Yamanaka, D. Yamashita, H. Ogura, C. Teruya, H. Nasu, E. Ima, H. Sakahara, M. Oda, Y. Yamashita, Time-resolved optical mammography and its preliminary clinical results. Technol. Cancer Res. Treat. **10**, 393–401 (2011)
44. D. Wu, S.S. Gambhir, Positron emission tomography in diagnosis and management of invasive breast cancer: current status and future perspectives. Clin. Breast Cancer **4**, S55–S63 (2003)
45. T. Yates, J.C. Hebden, A. Gibson, N. Everdell, S.R. Arridge, M. Douek, Optical tomography of the breast using a multi-channel time-resolved imager. Phys. Med. Biol. **50**, 2503–2517 (2005)
46. Q. Zhu, P.U. Hedge, A. Ricci, M. Kane, E.B. Cronin, Y. Ardeshirpour, C. Xu, A. Aguirre, S.H. Kurtzman, P.J. Deckers, S. Tannenbaum, Early-stage invasive breast cancers: potential role of optical tomography with US localization in assisting diagnosis. Radiology **256**, 367–378 (2010)

Errata to: Advanced Time-Correlated Single Photon Counting Applications

Wolfgang Becker

Errata to:
W. Becker (ed.), *Advanced Time-Correlated*
Single Photon Counting Applications, **Springer Series**
in Chemical Physics 111, DOI 10.1007/978-3-319-14929-5

The affiliation of the authors Dmitri A. Rusakov and Kaiyu Zheng was incorrect in the Contributors section under Front matter. The correct information is given below:

Institute of Neurology, University College London, London, UK.

The name "Richard Dimble" should not be listed among the authors in the Contributors section under Front matter and Chapter 3.

The online version of the original book can be found under
DOI 10.1007/978-3-319-14929-5

W. Becker (✉)
Becker & Hickl GmbH, Berlin, Germany
e-mail: becker@becker-hickl.com

© Springer International Publishing Switzerland 2015
W. Becker (ed.), *Advanced Time-Correlated Single Photon Counting Applications*,
Springer Series in Chemical Physics 111, DOI 10.1007/978-3-319-14929-5_20

Erratum to: Fluorescence Lifetime Imaging (FLIM): Basic Concepts and Recent Applications

Klaus Suhling, Liisa M. Hirvonen, James A. Levitt, Pei-Hua Chung, Carolyn Tregido, Alix le Marois, Dmitri A. Rusakov, Kaiyu Zheng, Simon Ameer-Beg, Simon Poland and Simao Coelho

Erratum to:
Chapter 3 in: W. Becker (ed.), *Advanced Time-Correlated Single Photon Counting Applications*, **Springer Series in Chemical Physics 111, DOI 10.1007/978-3-319-14929-5_3**

The spelling of the author "Simon Coelho" in Chapter 3 was incorrect. The name should read as "Simao Coelho".

The online version of the original chapter can be found under
DOI 10.1007/978-3-319-14929-5_3

K. Suhling (✉)
Department of Physics, King's College London, London WC2R 2LS, UK
e-mail: klaus.suhling@kcl.ac.uk

L.M. Hirvonen · J.A. Levitt · Pei-Hua Chung · C. Tregido · A. le Marois
Department of Physics, King's College London, London WC2R 2LS, UK
e-mail: liisa.2.hirvonen@kcl.ac.uk

J.A. Levitt
e-mail: james.levitt@kcl.ac.uk

Pei-Hua Chung
e-mail: m9314001@gmail.com

C. Tregido
e-mail: carolyn@tregidgo.com

A. le Marois
e-mail: alix.le_marois@kcl.ac.uk

The affiliations were incorrect for the last three authors—Simon Ameer-Beg, Simon Poland and Simao Coelho of Chapter 3 was incorrect. The correct information is given below:

Randall Division of Cell and Molecular, Biophysics, King's College London, London SE1 1UL, UK

and

Department of Cancer, Research, Division of Cancer Studies, New Hunt's House, Guy's Campus, King's College London, London SE1 1UL, UK.

D.A. Rusakov · K. Zheng
Institute of Neurology, University College London, London WC1N 3BG, UK
e-mail: d.rusakov@ucl.ac.uk

K. Zheng
e-mail: k.zheng@ucl.ac.uk

S. Ameer-Beg · S. Poland · S. Coelho
Randall Division of Cell and Molecular, Biophysics, King's College London, London SE1 1UL, UK

S. Ameer-Beg · S. Poland · S. Coelho
Department of Cancer, Research, Division of Cancer Studies, New Hunt's House, Guy's Campus, King's College London, London SE1 1UL, UK

Index

A

A2E, 539
  fluorescence lifetime, 514, 515
Absorption coefficient, 544, 573, 590
Absorption dipole, 387
AcAc, 414
Acceptor, for FRET, 251, 282, 319
Accumulation
  of continuous-flow mode data, 21
  of FITS data, 217
  of FLIM data, 215
  of FLIM data, by hardware, 22, 572
  of FLIM data, by software, 23, 75, 520
  of FLIM data, ophthalmic FLIM, 520
  of FLITS data, 33
  of mosaic FLIM data, 32, 96
  of TCSPC time series, 19, 570
Accuracy of FLIM, 22, 77
Acquisition time
  of FLIM, 22, 31, 33, 77, 101, 102, 108
  of PLIM, 237
  of TCSPC, 6, 19, 22, 32, 33, 468
  ophthalmic FLIM, 512
Acriflavine, 486
Advanced glycation end products, 510
AFM, 108
AGC, 510
AGE
  fluorescence lifetime, 514, 515
Alzheimer disease, 155, 533
AMD, 529
Amplitude weighted lifetime, 261, 263
Anisotropy, 158
  anisotropy decay, 345, 349
  decay time, 158, 396
  depolarisation factors, 399
  fundamental anisotropy, 399
  G-factor, 345, 349, 396

  limiting anosotropy, 157, 390, 395, 396
Anti-bunching, 8
AOTF, 565, 569
Apoptosis, 440
Arterial occlusion, 574
ATP, 408, 418
ATP synthase
  conformational changes, 54, 318
  c-ring rotation, 54, 320, 323
  single-molecule FRET, 52, 54, 315, 318, 327
Autocorrelation function
  brightness analysis, 47, 286, 294, 313
  calculation from TCSPC data, 42
  linear-tau algorithm, 43
  multi-tau algorithm, 43
  online calculation, 41, 43
Autofluorescence, 152
  A2E, 514
  AGE, 514
  carcinoma, 446
  cardiac myocytes, 411
  collagen, 514
  elastin, 514
  endogenous fluorophores, 461
  FAD, 476
  FLIM, 478
  keratin, 478
  lifetimes, 442, 446, 461, 514
  lipofuscin, 514
  melanin, 514
  NADH, 410, 436, 466, 514
  ocular fundus, 514
  redox ratio, 466
  sample deterioration, 484
  sample storage conditions, 483
  separation of fluorescence components, 417
  skin, 461, 476

© Springer International Publishing Switzerland 2015
W. Becker (ed.), *Advanced Time-Correlated Single Photon Counting Applications*,
Springer Series in Chemical Physics 111, DOI 10.1007/978-3-319-14929-5

CPSIA information can be obtained
at www.ICGtesting.com
Printed in the USA
LVHW081927310319
612462LV00005B/25/P